Foodservice Organizations

A Managerial and Systems Approach

THIRD EDITION

Marian C. Spears

Professor Emeritus, Kansas State University

Merrill,
an imprint of Prentice Hall
Upper Saddle River, New Jersey   Columbus, Ohio

Library of Congress Cataloging-in-Publication Data

Spears, Marian C.
    Foodservice organizations : a managerial and systems approach / Marian C. Spears.—
3rd ed.
      p. cm.
    Includes bibliographical references and indexes.
    ISBN 0-02-414282-4
    1. Food service management. I. Title.
TX911.3.M27S69   1995
647.95'068—dc20   94-1320
                                         CIP

Editor: Kevin Davis
Production Editor: Jonathan Lawrence
Text Designer: Mia Saunders
Cover Designer: Patricia Cohan
Production Manager: Pamela D. Bennett
Electronic Text Management: Marilyn Wilson Phelps, Matthew Williams, Jane Lopez,
   Karen L. Bretz
Illustrations: Academy ArtWorks, Inc.

This book was set in Garamond by Prentice Hall and was printed and bound by R. R. Donnelley
& Sons Company. The cover was printed by Phoenix Color Corp.

© 1995 by Prentice-Hall, Inc.
A Simon & Schuster Company
Upper Saddle River, NJ 07458

Earlier editions © 1985, 1991 by Macmillan Publishing Company.

Printed in the United States of America
10 9 8
ISBN 0-02-414282-4

Prentice-Hall International (UK) Limited, *London*
Prentice-Hall of Australia Pty. Limited, *Sydney*
Prentice-Hall Canada Inc., *Toronto*
Prentice-Hall Hispanoamericana, S. A., *Mexico*
Prentice-Hall of India Private Limited, *New Delhi*
Prentice-Hall of Japan, Inc., *Tokyo*
Simon & Schuster Asia Pte. Ltd., *Singapore*
Editora Prentice-Hall do Brasil, Ltda., *Rio de Janeiro*

Preface

The third edition of *Foodservice Organizations: A Managerial and Systems Approach* is a complete rewrite, not merely a revision of the second edition. This edition reflects the many changes that have taken place in business in recent years. The foodservice organization goes through the same cycles as other business organizations. When economic conditions are good, restaurant business booms, but when the economy declines, restaurants have problems.

Management philosophies are changing for businesses throughout the world. Most organizations, including those in foodservice, practice either total quality management (TQM) or total quality service (TQS). Customer satisfaction is more strongly emphasized than it was 5 years ago. The concept of customers has expanded from those who are external to a business to include employees, who are identified as internal customers. Employees, as team members, are being empowered to make decisions that increase customer satisfaction, thus flattening the organization by eliminating middle managers.

The hospitality industry has stabilized after a few years of fast growth. The numbers of students in hospitality programs have decreased slightly since the early 1990s, but opportunities for graduates have expanded from work in restaurants to work in retirement communities, schools, and hospital foodservice operations. The numbers of students in dietetics programs, however, have increased slightly, primarily because of society's emphasis on a healthy life-style. Dietitians currently are being hired for corporate positions in chain restaurants.

An alliance has been formed between the National Restaurant Association (NRA) and the American Dietetic Association (ADA), which is a breakthrough in the foodservice industry. The NRA has a dietitian on staff, and Ferdinand Metz, the president of the Culinary Institute of America (CIA), was awarded an honorary membership in the ADA in 1993 for his commitment to incorporating nutrition principles and applications throughout the CIA curriculum. Another recent change in the foodservice industry is that it is no longer divided into commercial and noncommercial segments, but it has become one industry with the goal of satisfying customers while making a profit. Even in nonprofit organizations such as hospitals and schools, profit from the foodservice can be recycled into buying new equipment or establishing nutrition programs for customers.

This text was written primarily for junior- and senior-level students and as a resource for graduate students and instructors. Because of budgetary problems in universities, instructors often have such heavy assignments that they do not have time to research each topic; perhaps this text will be of help to them. The text also is being used by foodservice managers both in the United States and in other countries.

Every effort was made by the author to keep the text short by providing a quick review of information given in basic courses and then concentrating on the application of theory to the foodservice organization. To reinforce this application, case studies have been added to this edition. Important concepts, such as TQM and Hazard Analysis Critical Control Points (HACCP), have been woven through all pertinent chapters. Chapter 22, "Computer Assisted Management," which appeared in the second edition, was eliminated because most students today are computer literate. Information on computer applications, however, has been added to those chapters in which the technology assists managers in making decisions. A case study on a Marriott Lifecare Retirement Community follows each part to help students apply theoretical concepts to actual problems in a well-organized, profit-making foodservice operation. The Marriott Corporation was chosen because it hires both hotel and restaurant management and dietetics program graduates for its retirement communities.

Organization of Text

Management of a foodservice system is the base of the text. The foodservice system is a conceptual framework that shows the pathway resources follow to meet the goals of the organization.

The text is divided into six parts. Part 1 is the **Introduction**, in which the concept of the foodservice system is explained in depth, as is the management and marketing of foodservice operations. In the second part, **Designing the Foodservice System**, the flow of food through the various types of foodservice operations to produce and deliver menu items to the customer is described. **Procurement**, the third part, consists of purchasing, receiving, storage, and inventory control. **Production** is the fourth part and includes production planning, ingredient control, quantity food production and quality control, food safety, labor control, and energy control. The fifth part, **Distribution, Service, Sanitation, and Maintenance**, covers those subsystems of the foodservice system. Part 6, **Management of Foodservice Organizations**, is the capstone of the book and should be the last course before entering practice. It emphasizes management of the organization, employees, and finances.

An updated bibliography, which includes many 1993 publications, appears at the end of each chapter. Tables and figures have been revised for clarity, and extensive portions of text have been deleted, rearranged, and added. Most concepts are illustrated with examples from actual situations to help students relate the theory to practice. A glossary of approximately 500 key terms from **ABC inventory method** to **yield** has been added to the text. The text is designed to meet the needs of all courses in a foodservice curriculum and can serve as the only text with appropriate supplemental readings. Following is a suggestion of how the text can be used in a college or university foodservice program.

Course	**Chapters**
Quantity Food Production	1–3, 6, 10–12, 13 (productivity), 16, 18 (communications)
Foodservice Systems	1–9, 12, 14–16
Organization and Management	2–4, 13, 17–21

Chapters are integrated and interrelated, but they can be used as independent units and arranged in various sequences to meet the needs of individual courses.

Specific chapters can become the outline for graduate-level courses. For example, a course in resource procurement could be developed from chapters 2, 7, 8, 11; and a course in food production management could be drawn from chapters 2, 5, 9–12, 14. In addition, graduate students should be expected to abstract and discuss the literature.

Chapter Revisions

The following is a description of the intent and scope of the revisions, chapter by chapter.

Chapter 1—The Foodservice Industry
- Deletion of segmentation of foodservice organizations into commercial and noncommercial operations
- Discussion of trends in the menu, staff, customer, and kitchen

Chapter 2—Systems Approach to Foodservice Organizations
- Addition of a section on TQM and its impact on management functions

Chapter 3—Managing Foodservice Systems
- Incorporation of TQM philosophy into management functions

Chapter 4—Marketing Foodservice
- Addition of atmospherics to tangible attributes in service marketing
- Emphasis on marketing principles pertinent to foodservice operations

Chapter 5—Food Product Flow
- Change of title from "Types of Foodservice Systems" to avoid confusion of the term *systems*
- Emphasis on flow of food through four types of foodservice operations
- Introduction of the HACCP and the NRA's SERVSAFE® model, which are discussed in subsequent chapters on subsystems

Chapter 6—The Menu, the Primary Control of the Foodservice System
- Expansion of nutrition component of the menu to include the USDA Food Guide Pyramid
- Update of examples to show current menu trends
- Discussion of computer applications in menu pricing

Chapter 7—Purchasing
- Identification of the role of purchasing in TQM
- Revision of marketing channel model to reflect current terminology, such as *producer* for *supplier* and *customer* for *consumer*

- Addition of irradiation, genetically engineered foods, and nutrition labeling in market regulations
- Change of terminology from *vendor* to *supplier* and from *purchaser* to *buyer,* according to the National Association of Purchasing Management
- Suggestions for computer applications for bid analysis, items needed, and placing orders

Chapter 8—Receiving, Storage, and Inventory Control

- Discussion of critical control points for safe receiving and storage of food products
- Addition of electronic receiving method
- Importance of the just-in-time (JIT) philosophy and strategy on purchasing and inventory control

Chapter 9—Production Planning

- Addition of batch cooking to production scheduling
- Revision of forecasting section to include new examples

Chapter 10—Ingredient Control

- Discussion of computer programs that support the ingredient room
- Expansion of section on recipe standardization

Chapter 11—Quantity Food Production and Quality Control

- Discussion of critical control points for safe cooking of food products
- Introduction of induction as a method of heat transfer
- Information on the combination convection oven/steamer, referred to as the "combo" or "combi," the impinger, and the clamshell
- Update of Figure 11.23 to reflect current terminology for textural attributes of food products

Chapter 12—Food Safety

- Addition of new tables and text on pH of common foods, foodborne bacterial diseases, and emerging pathogens causing foodborne illness
- Discussion of the role of various organizations in establishing food safety standards for foodservice operations
- Updated information on changes in state and local health inspections, including internal and external audits

Chapter 13—Labor Control

- Addition of federal restrictions for hiring teenagers
- Current information on temporary employees and leasing staff
- Treatment of employees as internal customers by management

Chapter 14—Energy Control

- Endorsement by the NRA of an Environmental Protection Agency program, Green Lights
- Reduction of energy costs by using insulated equipment and conserving water

Chapter 15—Distribution and Service

- Discussion of critical control points for safe holding and service of menu items
- Expansion of information on vending machines
- Suggestion by Albrecht that emphasis be changed from management (TQM) to service (TQS), causing a paradigm shift from quality to total customer value by cross-training employees

- Management of the cultural diversity of the current work force
- Changing demographics of customers

Chapter 16—Sanitation and Maintenance of Equipment and Facilities

- Responsibility of foodservice managers to society for environmental issues, such as solid waste
- Control of contamination of food products by developing a HACCP program

Chapter 17—The Organization Structure

- Discussion of the traditional and the new organization
- Emphasis on quality of work life and corporate culture in the new organization
- Change to a flattened organization by focusing on the customer rather than the process

Chapter 18—Linking Processes

- Update of cost-effectiveness model and example
- Addition of focus groups to group decision-making methods

Chapter 19—Leadership and Organizational Change

- Expansion of title to include organizational change
- Addition of trends in organizational leadership
- New section on organizational change including mentoring, women in leadership positions, and work/home transitions

Chapter 20—Human Resources Management

- Change of title from "Management of Personnel" to reflect new policies in TQM organizations with a defined corporate culture
- Examination of legal environment including sexual harassment, Immigration Reform and Control Act, and Americans with Disabilities Act
- Discussion of the new model of unionism

Chapter 21—Management of Financial Resources

- Deletion of superfluous information
- Emphasis on accounting principles pertinent to foodservice operations

Acknowledgments

Sincere appreciation is expressed to Sharon Morcos, a faculty member in the Department of Foods and Nutrition, Kansas State University, for the many hours she spent in helping with the third edition of the text. She has taught the junior-level foodservice systems course many times for both hotel and restaurant management and dietetics students; thus she was especially helpful in identifying passages needing clarification to improve student understanding. Her knowledge of current resources is encyclopedic. She spent many hours in library research to develop chapter bibliographies. Her organizational skills are excellent, and preparing the final manuscript for the publisher could not have been accomplished without her help. Sharon Morcos deserves much credit for her aid in preparing the third edition of the book.

Special thanks go to Mary B. Gregoire, PhD, RD, associate director of the Division of Applied Research at the University of Southern Mississippi, National Food Service Management Institute, for revising chapter 21. She also was responsible for adding computer applications to pertinent chapters throughout the book. Thanks also go to the faculty of Kansas State University, who supported the book. Carole Setser, PhD, in the Department of Foods and Nutrition, revised the text for the food product evaluation section in chapter 11. Deborah Canter, PhD, RD, Department of Hotel, Restaurant, and Institution Management and Dietetics, provided resources for part 6 and also reviewed and made suggestions for the case studies. Michael Petrillose, instructor in the Hotel and Restaurant Management Program, had a special knack for tracking down current information from various agencies.

Ruby Puckett, MS, RD, director, Food and Nutrition Services, Shands Hospital at the University of Florida, provided menus for chapter 6, as did Catherine Powers, MS, RD, team leader for curriculum development at the Culinary Institute of America. Linda Lafferty, PhD, RD, director of Food and Nutrition Services, Rush-Presbyterian-St. Luke's Medical Center, provided organization charts depicting the transition of the Department of Food and Nutrition Services from a traditional to a new foodservice organization, which are discussed in chapter 17.

W. Milt Santee and Associates, foodservice equipment and layout consultants, in Kansas City, Missouri, assisted by identifying appropriate equipment and suppliers for illustrations and examples in several chapters. Data and typical examples from practice were provided by professional colleagues in the Kansas State University Housing and Dining Services and the University of Kansas Medical Center, College of Health Sciences and Hospital.

A special thank-you to Denise Banach, RD, consulting/administrative dietitian, and Dick Macedonia, senior vice president of operations for the Marriott Senior Living Services communities, for their cooperation in making possible the inclusion of actual case studies at the end of each part of the book. They not only provided information but also set up telephone interviews with Will Brucker, general manager, and John Brady, director of the Food and Beverage Department, in The Fairfax Marriott Senior Living Services community, Fort Belvoir, Virginia. John Brady, who now directs food and beverage operations for the 18 Marriott Senior Living Services communities in the United States, has been especially helpful in providing information and Marriott's printed materials for the cases.

I especially appreciate the suggestions made by the reviewers of the second edition—Claire Dobson Schmelzer, University of Kentucky; Jerrold K. Leong, Oklahoma State University; Thomas W. Long, Virginia Tech; Michael J. McCorkle, University of Kentucky; and Ken Smith, Colorado State University. As I revised each chapter, I reviewed their comments and incorporated most of them into the text. The comments were excellent, and certainly helped me clarify the concepts their students had difficulty in understanding.

Finally, I want to thank Kevin Davis, the Prentice Hall editor for the third edition, for his help in preparing the text for publication. Thanks also go to his assistant, Susan Wesner, who always is cheerful and willing to help a frustrated author. Jonathan Lawrence, production editor, and Janet Willen, copy editor, deserve much praise for their thorough review of the original text and preparation of it for publication.

Marian C. Spears

About the Author

Marian C. Spears, PhD, RD, Professor Emeritus, Kansas State University, formerly head of the Department of Hotel, Restaurant, Institution Management and Dietetics, is a native of Ohio. She holds Bachelor's and Master's degrees from Case Western Reserve University, followed some years later by a PhD from the University of Missouri–Columbia. Seventeen years of professional practice before entering academe included manager of a commercial cafeteria, chief dietitian of a nationally known children's home, and chief dietitian of a private hospital, all in Cleveland, Ohio. She later was associate director of dietetics in Barnes Hospital, St. Louis, Missouri. Academic experience began as assistant professor of home economics at the University of Arkansas, Fayetteville, in 1959. During the years of residence in Arkansas, she and her husband maintained an extensive consulting practice in the design and operation of hospital foodservice facilities. In 1971 she became associate professor and director of education, Food Systems Management Coordinated Program in Dietetics at the University of Missouri–Columbia.

Dr. Spears's professional memberships include The American Dietetic Association, American School Food Service Association, Council of Hotel, Restaurant, Institutional Education, Foodservice Systems Management Education Council, National Restaurant Association, and Academy of Management. She has authored and coauthored numerous publications in refereed journals. Honors include Sigma Xi, Phi Kappa Phi, Gamma Sigma Delta, and Omicron Nu. Dr. Spears received The American Dietetic Association Marjorie Hulsizer Copher award in 1989. This is the highest honor conferred upon one of 58,000 members. She also is listed in *Who's Who in America* and *Who's Who in American Men and Women of Science*.

Contents

2. *Designing the Foodservice System*   *123*

3. *Procurement*   199

4. *Production*   301

1

Introduction

The four chapters in part one provide an overview of the foodservice organization. The industry is no longer divided into commercial and noncommercial operations because each is using ideas from the other and the two groups are working together with the goal of meeting customers' expectations.

- **Chapter 1, The Foodservice Industry.**   The scope and status of the foodservice industry are defined, and the most recent information on the 18 segments is presented. Trends in the industry are categorized according to the menu, staff, customer, and kitchen.

- **Chapter 2, Systems Approach to Foodservice Organizations.**   The conceptual framework, identified as the system, is presented and discussed in depth. Each of the remaining chapters is based on a component of the system. The total quality management (TQM) philosophy is introduced in this chapter and is applied to major concepts throughout the text.

- **Chapter 3, Managing Foodservice Systems.**   Because the term *managerial approach* is in the title of the text, the management process is discussed early. One of the author's goals is to prepare students for careers in managing foodservice operations.

- **Chapter 4, Marketing Foodservice.**   The marketing concept is a management philosophy that affects all activities of an organization and emphasizes satisfying customers' needs. Marketing both food products and services will increase sales and profits and is necessary to stay in business.

Overview of Marriott Senior Living Services

A case study on one foodservice organization in a Marriott continuing care retirement community (CCRC) will be presented at the end of each of the six parts of the text. The Marriott Corporation was chosen because during the last decade it has built a position in the hospitality industry that is unparalleled in scope. It is the world's largest lodging operator and the leading provider of contract services in North America. The corporation is also positioned well for growth in its other diverse operations: foodservice and facilities management; food, beverage, and merchandise concessions for airports and toll roads; and services for seniors. For its retirement communities, Marriott hires both hotel and restaurant management and dietetics program graduates, which is another reason for selecting it for these case studies.

Marriott employees serve more than 4 million meals and make an estimated 10 million customer contacts each day. The corporation purchases in excess of $2.5 billion of food products annually for its own operations, franchises, and external customers.

One of America's greatest challenges is to improve the quality of life of our oldest citizens (King, Schechter, & Friedland, 1993). Providers of eldercare services are looking for new ways to enrich the seniors' golden years, whether in their own homes, nursing homes, or in lifecare communities. Paralleling the population growth is the increase in the number of facilities and programs being developed to meet seniors' needs. According to the American Association of Homes for the Aging (AAHA), which represents nonprofit nursing homes and CCRCs, approximately 19,250 nursing homes and more than 800 CCRCs are now operating in this country. The trend in CCRCs is toward guaranteeing continuous care including independent, assisted living and skilled nursing to residents through one-price agreements instead of added fee-for-service provisions.

Now that the baby boomers are starting to think about retirement, the possibility of expansion of the foodservice industry into eldercare is imminent. Employment opportunities in designing facilities, managing foodservices, and purchasing supplies, therefore, are becoming available for graduates of hotel and restaurant management programs (Samenfink & Partain, 1993).

Graduates of dietetics programs also are finding career opportunities in eldercare. In addition to being able to manage a foodservice operation, dietitians understand the nutritional problems that stem from the biological effects of aging, such as the decreases in the senses of taste and smell, which often affect the appetite. Marriott currently has 18 Senior Living Services communities in 9 states (Arizona, California, Florida, Illinois, Maryland, New Jersey, Pennsylvania,

Texas, and Virginia) and additional ones are being planned. The senior vice-president for operations and administrative dietitian of the Marriott Senior Living Services communities chose The Fairfax and Belvoir Woods Health Care Center as the property for the case studies.

The Fairfax is a lifecare retirement community for retired officers and their spouses, widows, or widowers of the United States Army, Air Force, Navy, Marine Corps, and Coast Guard. The community was opened in 1989 on property adjacent to Fort Belvoir, Virginia, near the Washington, D.C., metropolitan area (Figure 1). It provides a full range of amenities including fine dining, healthcare, concierge service, scheduled transportation, housekeeping, maintenance, and security in spacious surroundings bordering a lake.

Figure 1. Location of The Fairfax community in relation to Washington, D.C.

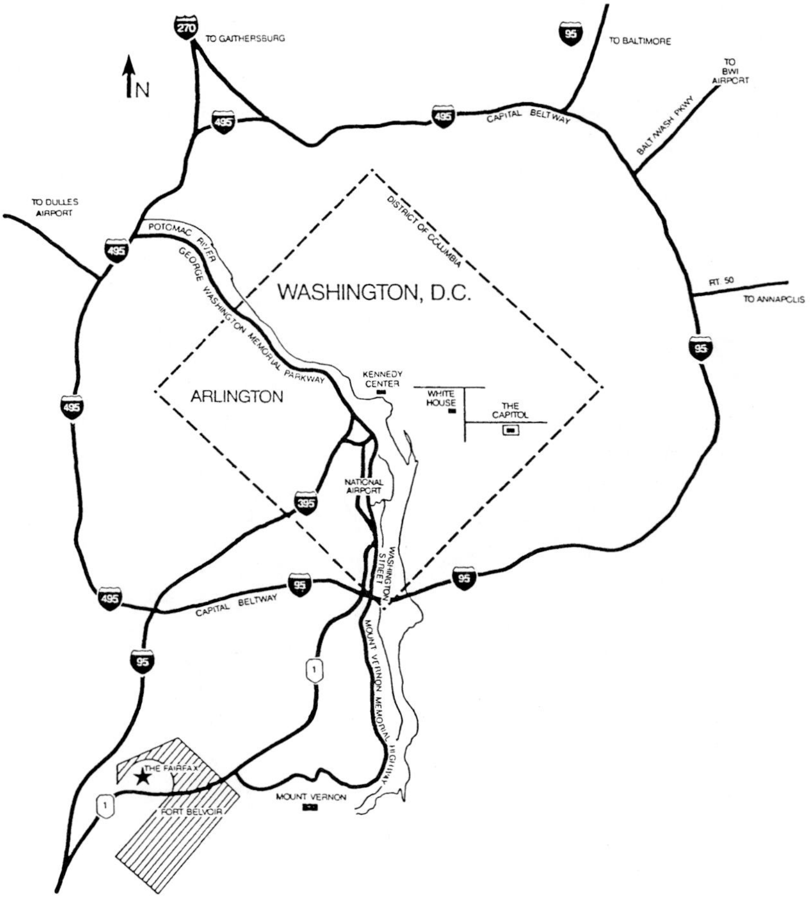

Source: Courtesy of Marriott Corporation.

The Fairfax Community

The Fairfax was developed and is operated by Marriott Senior Living Services at the request of the Army Retirement Residence Foundation–Potomac (ARRF-P). It was organized by a group of Army officers, led by Lieutenant General Frank A. Camm, USA (Retired), to meet the perceived need for a military retirement community in the Potomac area. ARRF-P was incorporated in 1983.

As of fall 1993, The Fairfax was 97% occupied, with over 500 residents from all services. The majority (about 53%) are retired U.S. Army officers, of which 116 are West Point alumni (including 47 widows). About 11% are Air Force, with lesser numbers from the Navy, Marine Corps, and Coast Guard. Ranks range from captain to general with about 56% colonels. Ages of the residents range from 65 to 99 years, with an average of 79 years.

The Fairfax community, as shown in Figure 2, consists of five low-rise apartment buildings (The Jefferson, The Madison, The Adams, The Washington West, and The Washington East) totaling 347 apartments and 35 cottages with an occupancy

Figure 2. Site plan of The Fairfax including the Community Center and the Belvoir Woods Health Care Center.

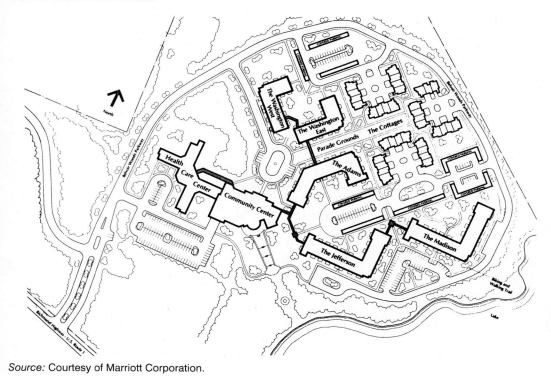

Source: Courtesy of Marriott Corporation.

4

Figure 3. Upper level of the Community Center.

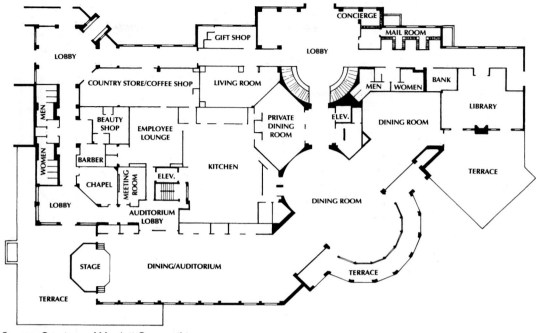

Source: Courtesy of Marriott Corporation.

rate of approximately 97%. Currently, 147 couples and 215 single residents occupy these independent living residences. The ratio of men to women residents is approximately 1 to 3. A community center for activities including dining in restaurants is located within walking distance of the apartments and cottages. Enclosed walkways connect apartments to the community and healthcare centers. Also, the Belvoir Woods Health Care Center, with 45 assisted living units and a 60-bed skilled nursing center, is available for residents needing additional services.

The Community Center

The Community Center has two levels. The lobby with a concierge station is on the upper level as are the kitchen and main dining room (Figure 3). Other rooms on that level are the auditorium, chapel, barber and beauty shops, country store/coffee shop/bistro, employee lounge and cafeteria, living room, private dining room, gift shop, mail room, bank, and library. The lower level has employee locker rooms, housekeeping, kitchen storage, administrative offices, and a heated indoor swimming pool, therapeutic whirlpool, and exercise room as well as game and arts and crafts rooms (Figure 4).

Figure 4. Lower level of the Community Center.

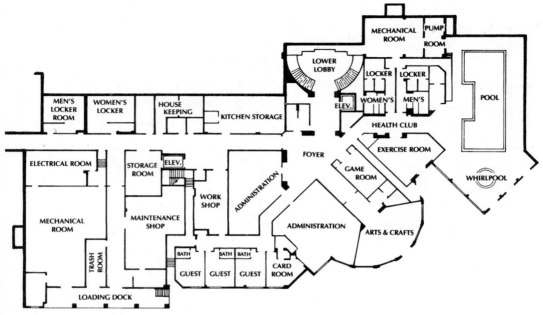

Source: Courtesy of Marriott Corporation.

Entry plan payments, which cover the long-term care cost protection program, provide for 95% repayment of the residents' entry fee upon their departure or to their estate. Costs for units range from $50,000 to $373,000, plus monthly fees starting at approximately $1,350.

Management

The Fairfax has a general manager and nine managers or directors of the following services: recreational, auxiliary, food and beverage, healthcare, sales and marketing, engineering, business, human resources, and resident services. Job descriptions developed by Marriott Senior Living Services are used in all properties. All managers are required to ensure high-quality resident services and care, within budgetary guidelines, while meeting, or exceeding, Marriott hospitality and service standards. Responsibilities and educational and experience qualifications for selected managers follow.

Job Title: General Manager, Independent Full Service
- *Responsibility.* Overall management and performance of a Marriott Senior Living Services full-service community that includes independent full-service units, assisted living units, and a healthcare center.

- *Job qualifications.* Related bachelor's degree and a minimum of 5 years' hospitality, senior housing, or healthcare management and administrative experience.

Job Title: Director, Food and Beverage
- *Responsibility.* Directs the overall operation of the food and beverage department in the community. Ensures compliance with all federal, state, and local regulatory bodies.
- *Job qualifications.* Bachelor's degree in foodservice management or 3 years' experience in foodservice production.

Job Title: Healthcare Administrator
- *Responsibility.* Manages an independent full-service health center that includes assisted living and nursing care.
- *Job qualifications.* Related bachelor's degree and valid administrator license within the respective state.

Job Title: Director of Engineering
- *Responsibility.* Directs the overall operation of the engineering department including security and loss prevention functions in the community.
- *Job qualifications.* High school diploma and a minimum of 6 years' experience in maintenance supervision.

Job Title: Director of Resident Services
- *Responsibility.* Develops and implements resident services programs for the independent living resident.
- *Job qualifications.* Related bachelor's degree and a minimum of 2 years' experience working with the elderly in resident services.

General managers of the retirement communities make decisions and establish programs that better serve the customer while contributing to the profit of the operation. The Marriott Corporation establishes guidelines through the budget and Standards Manual; the general manager of each property is responsible for meeting budgetary goals and maintaining or surpassing the standards. The Food and Beverage Department controls 34% of the total operating budget for The Fairfax and its healthcare center.

Continuing Education

The Marriott Corporation conducts many continuing education conferences for general managers and directors of services. A recent 4-day management conference, *Continuous Improvement—Continuous Learning and Change,* for Marriott Senior Living Services was conducted at Marriott's Grand Hotel in Point Clear, Alabama. Those in attendance included the Senior Living Services president and vice presidents; national directors of such areas as food and beverage, engineering, and human resources; and general managers from all 18 properties. The agenda included the following topics and speakers:

- 1993 Results and 1994 Business Plan: Senior Vice Presidents for Finance, Operations, and Human Resources
- 1994 Key Business Objectives: President of Marriott Senior Living Services
- Total Quality Management (4-hour session): President of Senior Living Services and Assistants
- Value-Driven Business (Re)Engineering (7-hour session): Kenneth Mifflin, Partner of Anderson Consulting
- Health & Wellness Seminar (4-hour session): Leonard A. Wisneski, M.D.
- The 7 Habits of Highly Effective People and Principle-Centered Leadership: Dr. Blaine N. Lee, Vice President of Covey Leadership Center
- Dinner & Presentation of Awards: J. W. Marriott, Jr., Keynote Speaker

Bibliography

King, P., Schechter, M., & Friedland, A. (1993). *Food Management*, *28*(12), 58–61, 64, 65, 68, 69.

Samenfink, W. H., & Partain, D. (1993). *Hospitality & Tourism Educator*, *5*(4), 49–51.

1

The Foodservice Industry

The foodservice industry is a virile force in world society, and all who aspire to managerial status should acquire a sound knowledge of the design, operation, and management of a foodservice system early in their education. This industry is customer driven, as manifested in the production of quality food and service for customer satisfaction with the ever-present financial concerns.

Jane Wallace (1993) wisely tells foodservice managers not to let anyone tell them foodservice is no longer a dynamic industry. All they need to do is find the right recipe to seize their own piece of a very big pie.

SCOPE AND STATUS OF THE INDUSTRY

The foodservice industry is exciting because it is in a constant state of change. Futurists, already predicting what will happen in the year 2000, are uncertain about what will happen tomorrow. The 90s are bound to have a lot of changes, especially with the president being a Democrat, for the first time in 12 years, and a baby boomer, for the first time ever. After years of wanting "our turn," the **baby boomers**—those people born between 1946 and 1964—have to prove themselves. Business leaders, including those in foodservice, will have to make a major adjustment to the political, legislative, and philosophical style of the Democratic party (Bartlett & Bertagnoli, 1993).

President Clinton has indicated that he wants to help small businesses, particularly restaurants, and he supports legislation to improve workers' lives. The family-leave bill, requiring employers to provide up to 12 weeks of unpaid leave during any 12-month period, already has been signed. A minimum wage to keep pace with inflation is being studied, and a healthcare plan for employees is being negotiated. Some of these laws have the potential to cut into foodservice profits. At the same time, government must reduce the deficit and create new revenue sources to improve the economy. After that happens, environmental legislation will become a priority. Stricter water controls are on the horizon; such controls will affect customers, foodservice operators, and food manufacturers (Bartlett & Bertagnoli, 1993).

Total foodservice industry real growth fell from 3.2% in 1988 to 0.9% in 1991. Preliminary 1992 data indicate 1.4% real growth, with 1.7% forecast for 1993. The average real growth for the 1990s is expected to be 2% (Bartlett & Bertagnoli, 1993).

The Bureau of Labor Statistics predicts that the real **gross national product (GNP)** will increase at an average annual rate of 2.3% per year between 1990 and 2005, down from a 2.9% average between 1975 and 1990. Changes in restaurant sales are closely related to changes in personal income. According to the Blue Chip Eco-

Table 1.1.   1993 Annual Forecast

Segments	1993 Real Growth[a]	1993 Sales ($ billions)[a]	1993 Market Share[a]	1992 Units[b]
Fast food[c]	3.4%	$86.121	31.9%	180,125
Full service	0.8	84.006	31.1	196,250
Employee feeding	0.6	17.913	6.6	16,225
Schools	1.7	15.242	5.6	89,000
Hospitals	0.2	11.872	4.4	6,730
Lodging	0.1	9.713	3.6	28,000
Colleges and universities	0.3	8.025	3.0	3,460
Military	−0.9	6.236	2.3	7,100
Recreation	1.8	4.475	1.7	31,075
Nursing homes	1.7	4.406	1.6	18,000
Supermarkets	3.8	4.054	1.5	23,400
Transportation	1.1	3.508	1.3	300
Social caterers	4.2	3.173	1.2	5,500
Convenience stores	1.3	2.889	1.1	66,925
Child care	1.5	2.807	1.0	29,000
Retail	0.4	2.149	0.8	16,100
Life care and elder care	2.5	1.112	0.4	10,000
All other foodservice[d]	2.2	2.528	0.9	9,975

[a] Forecast.
[b] Most available data.
[c] Excludes separate drinking places.
[d] Consists primarily of fraternal clubs, prisons, and recreational camps.
Source: Cahners Bureau of Foodservice Research. Used by permission.

nomic Indicators, real disposable income is expected to increase at a 2.5% average annual rate between 1994 and 2003, compared to 2.7% between 1975 and 1990. Despite slowing economic growth, consumer spending is expected to be healthy into the 21st century (Brault, Iwamuro, Mills, Riehle, & Welland, 1992). Foodservice sales, which were $42.8 billion in 1970, are expected to reach $267.2 billion in 1993 and $400 billion by the turn of the century (National Restaurant Association [NRA], 1992; Bartlett & Bertagnoli, 1993).

Americans 8 years of age or older eat out an average of 3.8 times a week, or 198 times a year. On a typical day almost 50% of adults are foodservice patrons, with 52% of men eating out compared to 47% of women. The per person check averaged $3.94 in 1991 (NRA, 1992). Table 1.1 shows projected foodservice industry growth for 1993 and number of units in 1992 (cited in Bartlett & Bertagnoli, 1993).

Without fanfare, the foodservice industry has become the number one retail employer in the nation; it employs more than 9 million persons, and 12.4 million employees are forecast by the year 2005. Almost 6 out of 10 employees currently in foodservice jobs are women, 12% are African American, and 12% are of Hispanic origin (NRA, 1992). Labor costs, which include salaries and benefits, probably will increase 4.8% in 1993 (Brault et al., 1992).

INDUSTRY SEGMENTATION

The National Restaurant Association divides the foodservice industry into two categories: commercial and noncommercial. In its 1993 annual forecast, however, *Restaurants and Institutions* no longer divided the industry into these categories. Actually, differences in menu item and facility ambience choices between categories are almost nonexistent. For example, people who have been used to fine dining can still enjoy it if, perchance, someday they live in a retirement community with a healthcare component. The terms *commercial* and *noncommercial* are still used, however, to indicate the degree of choice a customer has in selecting where to eat. Customers in hospitals and schools generally have little choice, and the foodservice in these industries is referred to as noncommercial, whereas restaurant customers have many choices. Foodservice operations currently are being segmented into the following types of industries.

Fast Food (Limited-Menu)

Fast-food restaurants are classified as refreshment places under the **Standard Industrial Classification (SIC),** Eating and Drinking places (*Standard Industrial Classification Manual,* 1987). SIC is used to promote the comparability of establishment data describing various facets of the U.S. economy. *Refreshment places* are defined as establishments primarily selling limited lines of refreshments and prepared food. Included in this group are establishments that prepare items such as chicken and hamburgers for consumption either on or near the premises or for consumption at home. The NRA is now identifying fast-food operations as **limited-menu restaurants,** which is the term being used in this text.

Limited-menu restaurants will be one of the brighter spots in the foodservice industry in the 1990s (Brault et al., 1992). As competition increases in the United

States, foodservice operators are going beyond the country's borders, for example, to Russia, France, China, and Japan. The United States has the potential to become the global leader in the foodservice industry. According to 1992 sales, McDonald's tops the list of limited-menu restaurants, followed by KFC, Burger King, Pizza Hut, Wendy's, Hardee's, and Taco Bell ("R&I 400: Ranking," 1993). Top trends include value-priced menus, discount wars, healthful menus, global expansion, and gargantuan pizzas (Bernstein, 1993a).

Limited-menu chains will decrease the size of their kitchens and purchase many products from a commissary, thus reducing labor costs. For example, Pizza Hut, Taco Bell, and KFC purchase all their products and supplies from Pepsico Food Systems. Introducing menu items for a short period of time keeps customer interest high. Combo meals, such as a hamburger on a bun, french fries, and a beverage, are popular sellers. Concentration on increasing dinner sales also brings in additional revenue; Burger King has created a dinner menu and table service (Chaudhry, 1993b).

Mandatory health legislation, menu labeling, and tighter environmental regulations are threats to chain profitability. Most chains are emphasizing good service, and they are training employees in the art of suggestive selling, which costs nothing. Store managers, because they know their market and need flexibility, are being empowered to do what is right for the customer. Limited-menu restaurant chains are participating in the so-called **branding evolution** by contracting with colleges, hospitals, food courts in malls, military bases, and even prisons to set up their own store, cart, or kiosk in the facility. A well-known chain can increase sales in a hospital or college by 40% if the location has good traffic and changing customers (Lorenzini & McCarthy, 1992).

Full Service

Full service includes casual, theme, family dining, and fine dining restaurants. Casual and theme restaurants have the highest sales records, with family restaurants close behind. Top trends in theme restaurants are fajitas in many forms and the return of red meat (Weinstein, 1993d). Top trends in family dining include home-style value meals, combo breakfasts, anything in a skillet, pancakes, and Southwestern seasoning (Straus, 1993d). High check, or upscale, white tablecloth establishments are not faring as well as the other full-service establishments, but they are increasing sales by adding lower-priced items to the menu.

Exceptional service is necessary to make a full-service restaurant better than all the others. Waitstaff must be trained and function well, or customers will go elsewhere. Menus for children or with suggestions for children as well as healthful alternatives should be available (Weinstein, 1993a). Managers should be in the dining room exploring possibilities for increased sales and adding a personal touch to the customers' dining experience. Because customers like to support one of their own, managers should participate in community events, such as feeding the homeless, and practice a little "neighborliness" (Riell, 1993).

What the combo meal has done for limited-menu restaurants, **prix fixe** (fixed price) can do for full service. An example is the participation by New York City restaurateurs during the national Democratic convention in the offering of complete meals for $19.92. The promotion was so popular with local people that many of the restaurateurs continued it through the end of the year (Weinstein, 1993a).

Employee Feeding

Employee feeding is undergoing great changes because of the rising cost of labor and the decrease in corporate subsidies. Managers are realizing that these operations must be self-supporting. Decreases in the labor force have hurt sales and profits, but employee feeding will follow the economy in 1993, improving slightly (Chaudhry, 1993a). The competition from contract companies has caused many organizations' foodservice managers to seek employment elsewhere.

Employee-feeding contractors are expected to participate in the quality management program of the organization that commissions them by satisfying the customer and empowering the foodservice employees. Contractors are strengthening relationships with brand name, or branded, concepts by becoming franchisees or administering franchise agreements for their principals (Brault et al., 1992). Marriott Management Services, which led the branding evolution, has signed an agreement with the Subway sandwich chain to develop both full-menu units and limited-menu carts and kiosks in its employee-feeding accounts. Sales will increase for "grab and go," such as coffee and donuts and ready-made sandwiches, and for hot take-out items. More employee feeders will become involved in catering to generate revenues and decrease company subsidies.

Business and industry leaders will expand wellness programs and encourage healthful food in employee dining rooms. The objective of an employee-feeding program is to give employees food that exceeds the quality in a restaurant so they will not leave the building (Chaudhry, 1993a).

Schools

Containing costs is the main focus of school foodservice managers (Stephenson, 1993d). Menu items and purchasing specifications need to be changed to eliminate unnecessary labor. Because employee benefits are expensive, more part-time employees who do not receive benefits will be used. Another way to contain costs will be to compare costs of holding a large inventory with paying for more deliveries. Many managers have large quantities of food left at the end of the school year.

School boards are wrestling with conflicting demands of more services without new taxes. School boards typically look favorably on school foodservice contractors that are willing to provide capital improvements, like new equipment, in exchange for long-term commitments. Contractors are aggressively pursuing school foodservice accounts; much potential exists to increase volume beyond the 30 million meals served daily nationwide through developing breakfast, à la carte sales, catering, and after-school foodservice programs. Also, schools usually have established programs that require little, if any, capital investment or start-up time for the contractor (Schuster, 1993).

The number of healthful menu items in a meal will be increased. Students like to have brand-name products on the menu and want to have meal deals like those offered by many limited-menu restaurants. Teenagers are especially brand conscious. A student soon learns that a government-reimbursable meal consisting of a slice of pizza, salad, fruit, and milk can be purchased for the same price as just the pizza and milk sold à la carte. The reimbursable meal becomes one more meal deal to the student and a source of revenue for the school lunch program (Opitz, 1993).

Hospitals

Hospital foodservice directors and dietitians, gearing up for the most comprehensive healthcare reform in U.S. history, hope to remain competitive by offering new patient-centered care services and developing cost-reduction programs (Schechter, 1993). They are learning how to manage change in order to respond to the changing goals and objectives of their hospitals ("Hospital Foodservice," 1992). Patient census counts are declining because the rise in hospital costs has made care affordable only for those with health insurance. Reducing labor costs by streamlining menus, staffing only one or two shifts, and relying on more convenience foods are some of the ways managers are meeting budget cuts (Cheney, 1993a). Emphasis will be on interdepartmental cross-training of employees; for example, an employee assigned to a particular floor may serve meals, do light cleaning, and perform nontechnical patient-care tasks (Schechter, 1993; Stephenson, 1993g).

Training multicultural employees will need to have top priority. According to Bureau of Labor Statistics projections for the year 2005, the most significant changes in the labor force, including those working or looking for work, will be a 26% increase from 1990 in women, 32% in African Americans, 75% in Hispanics, and 74% in Asians and others (Sabatino, 1993).

Many hospitals are using foodservice management contract companies, such as Marriott's Health Care Food and Nutrition Services, one of the largest in the healthcare segment (Carlino, 1991). Of the 1,185 respondents to the *1993 Hospital Contract Services Survey,* 12% said they use contract services for nutrition care and 10% for foodservice (Taylor, 1993).

Hospital cafeterias are becoming more competitive with restaurants. Catering services ranging from trays for staff meetings to full buffets for special events and take-out have become profit generators (Lang, 1993; Walkup, 1991b). A number of hospitals are hiring trained chefs, such as graduates from the Culinary Institute of America, for both patient service and extra events (Martin, 1991). Branding concepts like McDonald's, Pizza Hut, and Dunkin' Donuts are giving relief to the continuing budget squeeze in the healthcare industry. Chains are benefiting, too, with long-term leases and very attractive square footage rates compared to the cost of a free-standing restaurant site (Hayes, 1991a).

Lodging

Hotel and motel food and beverage operators are seeking ways to provide a valued service without sacrificing profit. Managers of food and beverage departments are trying to temper the dramatic decline in traffic in the early 90s, when hotel rooms were a glut on the market, by cutting costs. Menus have been rewritten to incorporate less expensive foods, purchasing strategies have been revised to permit more bargain hunting, and payrolls have been scrutinized to eliminate unnecessary positions (Romeo, 1991b).

The old stigma that hotel restaurants are inferior to free-standing ones is no longer true (Walkup, 1992). In the past, lodging restaurants were hidden in remote parts of the building, gearing them to captive guests who did not want to leave the property. Today, however, they have names unrelated to the hotel or motel and are given the most desirable locations in the building, for example, the street level with walls of windows and separate entrances.

Restaurants are becoming more casual and informal. Italian and Asian restaurants, which have low food costs, are replacing many of the fine dining establishments. Breakfast and lunch buffets with low prices are popular. **Sous vide** foods, specialties that have been cooked in a vacuum-sealed bag and then chilled or frozen, are being tested as a way to reduce labor costs. An employee produces a sous vide meal simply by reheating the contents of the pouch in boiling water (Romeo, 1991b). Banquets and catering account for approximately 35% of all food and beverage revenue industrywide (Chaudhry, 1993c).

The bed-and-breakfast (B&B) segment of the hospitality industry, long popular in European private homes and country inns, has expanded in the United States. An increasing number of B&Bs are being opened as primary businesses; a major change is the addition of restaurants offering service at periods other than breakfast and to customers other than overnight guests (Lanier & Berman, 1993).

Colleges

Enrollments have been declining and will continue to do so for the next few years. Campus foodservice directors are looking for ways to cut expenses and add sales (Straus, 1993b). They must make all-out efforts to control purchasing and other costs while making campus foodservice so attractive that students will not eat at off-campus facilities (Walkup, 1991a). The bottom line is customer satisfaction that translates into greater meal plan participation, more revenue, and better cost control (Hayes, 1991b). Marketing to off-campus customers is increasing through limited board-plan options and cash operations (Stephenson, 1993f). Customers come back to a place because of the way it feels; the feeling comes from the dining environment. All colleges and universities sponsor special dining events for students and their guests; these have been very successful.

Today's students are demanding and receiving fresh, healthful foods and lots of choices. A higher percentage of students are interested in good nutrition than in years past, and many have enrolled in nutrition courses. Surprisingly, students are seeking alternatives to hamburgers and fries. They even like vegetables. Most menus are planned to include a minimum of one vegetarian entrée for each meal. Ethnic foods, such as Italian, Mexican, and Asian, and grilled chicken are perceived by customers to be fresh and healthful (Straus, 1993b). An increasing number of students also are eating breakfast.

Foodservice directors are adjusting staffing formulas, reducing hours of operation, using fewer full-time and more student employees, reducing management staff, and empowering employees to be better decision makers. To compensate for fewer employees, managers must be well trained in food production techniques such as batch cooking, cook-chill, and vending. Food courts, remote kiosks, take-out and delivery services, and free-standing chain units reach more customers. Foodservice directors continue to supply catering services to the campus and community to increase revenue. Retail bakeries that satisfy collegiate-size appetites also are considered revenue generators (Straus, 1993b).

Military

Base closings present the biggest current challenge for military foodservice. Transferred troops mean more customers at those bases that do remain open (Lorenzini, 1993b). Better management skills are being emphasized, thus reducing the number of managers in each military operation. Managers are accountable for staying within the budget, which many do by using competitive bidding and controlling inventory levels, for example (Stephenson, 1993b).

Over the last decade, military foodservice managers have responded to the same customer preferences and trends that managers in other market segments have experienced. Soldiers, sailors, marines, and air force personnel are being offered a wide variety of menu choices and styles of service. Dining areas are decorated and colorful and have tables for four persons and booths instead of the long mess hall tables. Enlistees are bringing with them expectations formed by their exposure to commercial foodservice (King, 1991a). Branding, the natural accompaniment to the privatization of military foodservice, is being incorporated into many of the operations; McDonald's and Sbarro are on some bases, for example (Stephenson, 1993b).

Nutrition is a big trend in cafeterias and cash operations. The Department of Defense Armed Forces Recipe Service is modifying recipes to reduce fat, salt, and cholesterol (Lorenzeni, 1993b). The demand for more fresh fruits and vegetables, healthful entrées, and nutrition education and classes has increased on bases in all four service branches (King, 1991a).

Recreation

Americans are spending more of their leisure time at home visiting theme parks, baseball parks, and national parks, although foreign travel is increasing because of the favorable exchange rate caused by the weak U.S. dollar (Weinstein, 1993c). Very few foodservices are operated by recreation companies. Orlando-based Disney World and Universal Studios are examples of theme parks that operate their own foodservices (Romeo, 1991a; Peterson, 1992a, 1992b). Many concepts are used, such as the Yacht Club Galley and Cape May Cafe in Disney World and Studio Stars and Mel's Drive-in at Universal Studios Florida. The type of service varies from full service, to cafeteria, to counter, to kiosk. Serving healthful food is very important in both of these theme parks.

Contractors have launched furious bidding wars for lucrative ballpark concessions. Hot dogs, popcorn, peanuts in the shell, ice cream, soda, and beer have long been the items sold most often at ballparks. Today, however, fans can leave work and go straight to the ballpark to eat dinner that could include deli sandwiches, pizza, fajitas, frozen yogurt, and cheesecake (King, 1991b). Contractors, also, have accounts in zoos, aquariums, and at Olympic Games. At the summer Olympic Games in Barcelona, ARA Services served close to 3 million meals (Straus, 1992b).

Managing the culinary agenda for Madison Square Garden requires a manager who not only has many years of restaurant experience but also is a computer wiz, marketing maverick, and sharp business person (Howard, 1993). Approximately $20 million in food and beverage sales are generated there annually. Seats may be easy to fill when the Garden hosts an event, but the 165 "dark days" when events are not scheduled decreases revenue. Play-off games cause havoc with the food budget; if the team wins, large quantities of food are needed for the next game, but if it loses, the foodservice manager has to be creative with suppliers who sometimes allow partial orders to be returned.

The number of passengers taking cruises continues to grow because of increased berth capacity and promotional pricing (Brault et al., 1992). Carnival Cruise Lines, based in Miami, purchases more than $50 million in food for one year, plans to add a new ship annually, and has placed an order for the largest passenger vessel in history; the 300-yard-long ship is scheduled for completion in 1996 (Bernstein, 1993b; Hirschfeld, 1990). The trick to successful cruise line foodservice today is precision-engineered menu planning and reliance on a steady flow of the same high-volume weekly purchases to hold costs down. Demographics and mix of customers tend to be almost identical, which allows using the same menu for each cruise.

Nursing Homes

Nursing home occupancy will continue to increase as baby boomers age and people live longer (Stephenson, 1993c). Nursing home operators are asking if they can maintain the current level of service with the rising labor and food costs, the lack of federal money available for Medicaid reimbursements, and difficulties in hiring and retaining qualified staff (King, 1991c). These problems are being compounded by federal nursing home regulations and the **Omnibus Budget Reconciliation Act (OBRA) of 1987,** which are tied to Medicaid reimbursement rates and are designed to improve the quality of life and care for residents in nursing homes (King, 1991c).

Concerned foodservice directors foster the joys of eating and social interaction by providing favorite foods in an attractive dining room with small tables that encourage conversation (Keegan, 1992). Theme meals, outdoor barbecues, and a small dining room for entertaining families are some of the special events planned for residents. Contract companies often manage nursing homes.

Supermarkets

Supermarkets and restaurants are exchanging merchandising techniques (Stephenson, 1993e). Boundaries separating supermarkets from restaurants are blurring (Van Warner, 1992). Not only are supermarkets selling branded restaurant items, like Stouffer's products, but also restaurants are increasingly promoting well-known branded retail products on their menus, such as Grey Poupon Mustard. Several restaurant operators, recognizing the potential for selling their signature products in supermarkets, are entering the arena. ARA Services and Morrison's Hospitality Group also have gone into the supermarket business (Lorenzini, 1993a). Of the approximately 36,000 supermarkets nationwide profiled in the 1992 *Directory of Supermarket, Grocery and Convenience Store Chains*, 17% have in-store restaurants and more than 75% have deli departments (cited in Hayes, 1992).

Most supermarkets serve à la carte food from scramble-type areas, which have bakeries, delis, and juice bars, or entire meals from a single branded area, either part of a food court or a free-standing kiosk (Stephenson, 1993e). Many chefs and other foodservice employees are leaving restaurants for jobs in supermarkets (Hayes, 1992). Salaries are equal to or better than those in a restaurant, and hours are better. Customers are happy, too, because they like fresh food that has been recently prepared and also appreciate one-stop shopping convenience (Stephenson, 1993e).

Transportation

For the ninth consecutive year, transportation foodservice sales are forecast to register impressive growth (Brault et al., 1992). A slightly improved economy, a strong international tourism market, and the lowering of ticket prices by airlines should increase airline travel. Price wars between airlines have decreased, and in order to compete for passengers, food quality and service are being targeted.

Airlines have been revamping their menus to include fresh, colorful, nutritious food by consulting with some of the top chefs in the country ("No More Nuts," 1993). Most are decreasing the number of menu items and increasing the use of first-class ingredients and creative seasonings. Many of the entrées are cold, such as mesquite-grilled chicken with Southwestern pasta salad or roasted chicken with Caesar salad. Coach-class passengers are being offered more cost-effective, identifiable, and easier-to-serve fare: cold sandwiches made with special breads, finger foods, and branded snack products (Weinstein, 1993f). Fresh fruit and a cookie often are served for lunch. Some airlines also offer brand name pizzas and frozen yogurt. Amtrak also is treating passengers to upgraded foodservice. A club-service menu featuring new salads, deluxe sandwiches, and hot entrées has been introduced (Weinstein, 1993f).

Social Caterers

According to the *Standard Industrial Classification Manual* (1987), **social caterers** are establishments primarily engaged in serving prepared food and beverages for weddings and banquets at a hall or similar facility rather than at a fixed business location, such as a restaurant. Caterers' stock-in-trade is still the once-in-a-lifetime event, such as a bar mitzvah or big wedding with all the trimmings (Straus, 1992a). Because of stock-in-trade events, catering does well even in recessionary times when people will live on strict budgets in order to have an expensive catered event, such as a daughter's wedding. Business catering was downscaled a notch or two from 1990 to 1992, but it is back up again at an even higher level (Straus, 1992a).

Catering is hard work and requires excellent management skills. As one caterer said between smiles, "Catering is not a career for those who suffer from high blood pressure" (Cheney, 1992). The caterer has to visit the site of the event to find out if the kitchen can be used or if kitchen units have to be set up in the garage. If buffet service is requested, stations may need to be set up for liquor, meat, seafood, pasta, or dessert, according to the menu. Often guests place an order, and the menu items are prepared in front of them. Finally, delivering good service means hiring a good staff.

Convenience Stores

Convenience stores (C-stores), especially those that sell gasoline, are showing strong growth in the 90s (Brault et al., 1992). For many of these operations, a priority has been to improve their image, with women in particular, by having brighter stores, better food, and well-trained employees (McCarthy, 1993b). Convenience for the customer has to be given primary importance. Having coffee, newspapers, and bakery products available at the entrance makes breakfast on the go easy for business people. Checkout counters are uncluttered to make payment easier. At 7-Eleven stores in Austin, Texas, cash, credit cards, checks, food stamps, and coupons are accepted. At least two clerks are on duty at all times, one of whom is stationed at the checkout counter. Rigorous training programs for employees and managers are being conducted in some stores.

Many C-stores have deli areas that serve sandwiches and hot foods, such as entrées, soup, and chili. Branded foods are a good part of sales. Many of the stores are beginning to stock health and beauty-care products, thus appealing to women. New markets for C-stores are being explored. A C-store catering to students that stocks 2,500 to 3,000 items has been opened at Kansas State University. The store is divided into the following departments: homemade cookies and Rice Krispies squares, sandwiches purchased from a supplier, fountain drinks, beverages, groceries, ready-to-eat food (Deli Express, Tony's Pizza, chips, hot dog on a bun, nachos), bread and dairy, chewing tobacco, candy, automotive supplies (oil, wax), health and beauty products, and videotapes. A copy machine also is available.

Child Care

Foodservice must keep up with the shifting child-care market (McCarthy, 1993a). The birthrate is climbing, the number of single parents is increasing, and in more two-parent families both spouses work (Gall, 1991). The result is that America can no longer consider child care a luxury. In comparison to other less affluent nations, America is far behind in taking care of our children. Policymakers differ in their opinions about who should be responsible for child care: the employer or the county, state, or federal government.

A number of hospitals have child-care centers that operate on a 6 A.M. to midnight schedule, which makes it easier for nurses with young children to work (Schuster, 1990). Colleges and universities as well as corporations also have child-care programs available for students, faculty, and employees. Food sometimes is prepared on-site by cooks, or it is brought in by food contractors. ARA Services and Marriott Food and Services Management have entered into the child-care market for hospitals, colleges, corporations, and other clients that are already using their dining services. Frequently, dietitians help in developing menus that will meet federal and state nutritional requirements for licensed child-care facilities.

Federal, state, and county governments are developing sound child-care policies and helping to start responsible programs (Gall, 1991). Most school foodservice operations are involved in child care as providers to licensed Head Start, community recreation or child-care programs, church child-care centers, individual child-care home providers, or schools offering after-school care, and they are identified as sponsors under the Child and Adult Care Food Program. This program helps to provide foodservices for children and adults in nonresidential child-care institutions. Schools can qualify as a site for the Summer Food Service Program (SFSP) if 50% of the total school enrollment is eligible for free and reduced-price meals. Meals are based on the same pattern used for the National School Lunch Program except SFSP cannot provide offer versus serve, in which five menu items are offered but students only have to choose three.

Retail

Discount and upscale department store operators are adding foodservice to their operations to create a point of difference in the minds of customers. Ready-to-eat and ready-to-heat packaged foods are being marketed (Weinstein, 1993e). Retail foodservice depends upon retail sales. When sales increase, the number of buyers of products increases, and the sales of food and drink do likewise (Brault et al., 1992). Some retailers have in-store restaurants. For example, K mart has replaced cafeterias with its Eatery Express concept in stores that do not have a Little Caesar Pizza operation. Other retailers are adding brand-name carts and kiosks wherever they would be appropriate in the store (Straus, 1993c). Vending also is a large part of retail foodservice. Vending machines can be found at any place that has customers.

Retailers in shopping centers benefit from food courts and restaurants on the premises (Michalski, 1991). Competition among retailers to lure shoppers is increasing with the growth of the number of shopping centers in the United States. As a result, many retailers in the center are placing an emphasis on foodservice as a marketing tool. Not only do restaurants in shopping centers mean that customers can satisfy their hunger without leaving a center, but also restaurants are lucrative tenants.

Elder Care

Many older people do not want to lose their independence by going to a nursing home, but they need assistance in preparing meals (Schechter, 1990). The purpose of the Nutrition Services Program for Older Americans, as authorized by Title III of the Older American Act, is to provide nutritious, low-cost meals to homebound persons and congregate meals in senior centers. Many organizations also sponsor home-delivered meals including the Visiting Nurse Service and the National Association of Meal Programs, which is subsidized partially by the U.S. Department of Agriculture and the United Way. Typically, meals are prepared and packaged by outside contractors, hospitals, schools, or senior centers, and volunteers deliver them to people's homes at lunchtime five days a week. Recipients generally are charged what they can afford for the meal.

More and more empty nesters, people whose children have left home, are giving up their homes for the comforts of a retirement community (Stephenson, 1993a). Major hospitality corporations have identified **Continuing Care Retirement Communities (CCRCs)** as the growth area of the 1990s. Retired people, often affluent professionals, want quality residential services with healthcare available. CCRCs are on the cutting edge of two major trends in service delivery: the aging population and the emphasis on services.

Traditionally, such communities were sponsored by not-for-profit religious groups; however, corporate giants, such as Marriott and Hyatt, are entering the industry. They are hoping their reputations as dependable, service-oriented leaders in the hospitality industry will give them a competitive edge. Residents choose their meals from restaurant-style menus that have a wide variety of choices and meet their nutritional needs. As one foodservice manager said, "Our residents' attitudes range from highly nutrition conscious to 'I'm 85 years old and I'll eat what I want'" ("Life Care," 1990). Many foodservice staffs include professional chefs and dietitians who provide, in addition to daily meals, elaborate catering, room service, and elegant theme dinners for residents.

All Other Foodservice

Fraternal clubs, prisons, and recreational camps are included in this category. Of these three, prisons feed the greatest number of people. Foodservice directors have been asked to reduce the money spent per inmate despite a total prison population that has quadrupled in the last 20 years (King, 1993). Consolidation of inmate and employee feeding and centralized production are two ways directors are lowering food and labor costs (King, 1993; Stephenson, 1993h).

Large contract companies are entering the prison foodservice market, more often at the local than at the state or federal level. Contractors can reduce foodservice costs because of their purchasing power (King, 1993). ARA Services, Canteen, and Service America have a number of accounts.

Planning meals can be difficult for a predominantly male population with varying needs; younger inmates have greater protein and calcium requirements, and some inmates have specific medical or religious dietary needs ("RDs Working," 1992). Menu items must be familiar and acceptable to most of the inmates to prevent problems because most riots start in the dining room. Foodservice employees need to be trained not only in foodservice but also in proper security procedures and prison etiquette. State and federal guidelines require the expertise of a Registered Dietitian on staff to sign off on all menus. A current trend is to reduce fat in the preparation of menu items. Nutrition education also is being increased by some correctional facilities; the Federal Bureau of Prisons in Washington, D.C., lists the portion size, fat, cholesterol, and sodium content, and calories of items being served on both staff and inmate serving lines (Stephenson, 1993h).

TRENDS IN THE INDUSTRY

The 1990s is a trendy decade according to the trade and research journals in the foodservice industry. Surveys are being conducted, forecasts are being made, and new ideas for satisfying the customer are rampant. All of this is with the basic objective of staying within a budget or making a profit in order to stay in business. Trying to classify all the trends is almost impossible, but after the editorial staff of *Restaurants and Institutions* (R&I) talked to its readers, held focus groups, and conducted surveys, they settled on the Core Four as a format for the R&I publication (Bartlett, 1992). Readers' interests revolved around four areas: Food/Menus, Service/Management, Marketing/Merchandising, and Equipment/Design. Very quickly, the editors determined that the following areas are really the four basics, or cores, of foodservice: The Menu, The Staff, The Customer, and The Kitchen. Thus, trends in this text will be discussed under each of the Core Four areas.

The Menu

The menu includes both food and menus. Food in the 90s is not dull, and menus are exciting. The old-fashioned cook, who was usually a woman with a big family, is being replaced by a chef who has been educated and trained in a culinary school. Predictions made by R&I were divided into three trend categories: "Sure Bets," on a roll; "Worth Watching," coming on strong; and "Wait and See," out on a limb (Ryan, 1993a).

Sure Bets

The most exciting trend takes America out of the melting pot and into the melding pot (Ryan, 1993a). For some time, restaurants like Carlos O'Kelly's and the Chinese Jewish Delicatessen have exposed customers to a mixture of cultures. Pizzas are not only Italian, but also Mexican or Hawaiian. Stir-fry has gone from Chinese to Italian in such menu items as spicy stir-fried fettucini, and grains have gone from Italian to Chinese in Chinese polenta.

Coffee, especially premium, has gained great popularity. Coffeehouses are popping up around universities and theaters as places to meet friends, and espresso drive-thru stands are common in the Pacific Northwest. The emerging trend is for "tan" coffee, like cappuccino and café au lait.

The most basic trend is bread. The American breadbasket containing fresh, top quality, regional, international, and signature breads can enhance the image of a foodservice in the eyes of a customer (Ryan, 1993a). Many foodservice operations are going back to in-house bakeries. Once a bread is introduced to a customer, it remains popular. For example, bagels, croissants, muffins, pita, sourdough, scones, and cinnamon rolls have had staying power. The prediction is that whole wheat focaccio, old-fashioned oatmeal, Irish soda, cornmeal yeast-dough, and Italian, not too sweet, chocolate breads will be appearing on menus.

American beers will have their day. The brewing industry has a strong foothold in America, and **microbreweries** are springing up from coast to coast (Ryan, 1993a). Signature and seasonal beers are being developed, and beer-and-food-matched menus will appear in restaurants.

Customer-friendly menus already are being used in foodservice operations. Food quality has improved and prices are very competitive, making service the difference among operations. The menu is identified as a service vehicle. Customers with special dietary needs are encouraged to request that their food be prepared accordingly. More nutritional information is included in menu item descriptions and unfamiliar names and ingredients are explained.

Worth Watching

The controversy over organic and bioengineered foods will continue. Chefs are taking sides, either pro or con. *Organically grown foods* presumably are free of pesticides, fertilizers, and hormones and are produced "naturally." *Bioengineered foods* are produced by inserting genes from animals or vegetables into another food, thus altering the genetic code.

The challenge of the 90s is to retrain foodservice managers and chefs to meet the nutritional menu options demanded by customers. This is both a science and an art. Measuring, calorie and fat counting, and portion control are a science; creating good-tasting food without fat and salt by grilling, steaming, poaching, or roasting is an art.

Fruit is being used in sauces, marinades, and salsas for entrées of meat, poultry, and fish because it is not cloyingly sweet but complementary to the food and contributes a whole new range of flavors. Examples are walnuts and sea scallops in a grapefruit juice reduction with tomatillo sauce (Straus, 1993a) and Chinese mu shu pork with plum sauce (Ryan, 1993a).

Wait and See

Emerging trends of the 90s are on the horizon. Vegetables are gaining prominence on menus and will continue to be highlighted. Simple foods, such as macaroni and cheese or vegetarian sandwiches, have been on most restaurant, college and university, and delicatessen menus for a long time to satisfy those customers who do not eat meat. Today outstanding chefs not only include vegetarian items on menus but also are offering a prix fixe vegetarian menu daily, often using only organically grown produce (Liddle, 1993). Many chefs also offer what are known as **Vegan-approved menu items** for those customers who are strict vegetarians and will not eat animal meat or by-products, including dairy products and eggs.

Ethnic cuisine will continue to grow with many new regional dishes from Italy, Mexico, and Asia. Italian cuisine ranks number one both in terms of availability and in the quantity of items offered (Brault et al., 1992). The Olive Garden restaurant chain has been instrumental in bringing the Italian style of eating to the masses, with its signature "Hospitaliano" phrase reflecting its corporate culture of customer-driven service (Keegan, 1993). Certain Olive Garden restaurants are gravitating toward more sophisticated Italian food, especially in large cities and in the Southwest where highly spiced food is the norm. More vegetarian food generally is offered in Denver and more seafood in the Northwest.

The growth of the Mexican segment, the fastest among all limited-menu restaurants, suggests that America's appetite for low-priced Mexican fare is far from satisfied. Creative chefs are seeking out authentic regional dishes from Mexico, Asia, Italy, and other Mediterranean countries. The Japanese breakfast is considered a rising star because of the number of Japanese tourists visiting the United States (Parseghian, 1993).

Steak is back! Even though red-meat consumption per capita has decreased, sales in steak houses have increased. Customers treat themselves by eating a good steak occasionally, and a really good steak can be found only at a good steak house (Ryan, 1993b). Meats that are lower in fat and cholesterol are replacing the traditional red meats in recipes. Lean-bred beef, venison, rabbit, and farmed buffalo are beginning to appear on menus.

Spices and herbs are no longer limited to cinnamon, oregano, and chili powder but also include cumin, fresh cilantro, fresh basil, fresh red and green chilies, dried chilies, and whole black peppercorns (Straus, 1993e). Garlic is the herb most commonly identified in entrée descriptions (Brault et al., 1992). These flavor enhancers even are being found in hospital menu items because of the ethnic patient mix. As they begin to feel better, patients do not want bland food but want food they eat at home.

The prediction is that curry, a mixture of spices, will be the next hot spice on menus (Ryan, 1993b). Curry dishes have wide ethnic variety including Thai, Indian, Malaysian, and Cambodian. Chefs will have a field day creating a variety of chutneys to enhance the curried menu items.

Cookies baked while the customer watches have been top sellers in food courts, convenience stores, and college and university foodservices. Some airlines are even baking chocolate chip cookies on board. Restaurants have used cookies to supplement desserts such as ice cream, but cookies are beginning to appear on their own as desserts. Platters with signature cookies currently are being offered in upscale restaurants.

Beer tastings are becoming popular in midscale restaurants and brew pubs. They are less formal and more fun than wine tastings in upscale restaurants. Tastings educate customers and help them decide which brews they prefer. **Grappa,** a pungent brandy distilled from the pulp that remains after the grapes have been pressed and the juice run off, is predicted to become a popular after-dinner drink for customers who are not afraid of beverages that require an acquired taste (Ryan, 1993a, 1993b).

The Staff

Foodservice is the top retail employer. It employs more than 9 million people, and the number is expected to increase to 12.4 million by the year 2005 (NRA, 1992).

Foodservice is a ladder to management opportunity; the number of foodservice and lodging managers will increase by 33 percent from 1990 to 2005. The Educational Foundation of the National Restaurant Association conducted a futuristic study of the foodservice manager in the year 2000 (Riehle, 1992). A panel of experts used the **Delphi technique** to identify issues that managers might face and then assessed them for importance. According to the panel, managers in the year 2000 will:

- Need greater computer proficiency
- Supervise a more culturally diverse staff
- Find that service will become a more competitive point of difference
- Need better teaching and training skills
- Possess greater people-management skills

Management style is changing rapidly. The customer is king. Managers can no longer sit in their offices and make decisions, but they need to create a fun environment where managers, employees, and customers all enjoy themselves. The foodservice business is a people business, and managers should mingle with customers and employees. Managing a multiethnic staff will be quite a challenge in foodservice operations (Stern, 1991). A global work force is inevitable (Johnston, 1991). Many countries will make immigration easier, and, therefore, many workers will travel the globe. As a result, personnel policies and practices will be standardized.

Empowerment is the leading management buzzword in the 90s (Weinstein, 1993b). Employees will have to make more decisions for meeting customer expectations because economic pressures have forced cutbacks in layers of managers. Employers finally are realizing that employees serving customers can fix a problem immediately without going through the chain of command. Labor trends that go along with empowerment are training and incentives to control rising costs. Training is the key to success for instilling in employees the confidence to act independently. Managers also need to reinforce their own confidence in employees by discussing the outcome of decisions with them.

Once the food reaches the quality of competitors, service becomes the dominant concern of customers. Astute managers know that service training is a necessary cost that can be compensated for by a reduction in advertising costs. Daily meetings include not only a review of the menu items but also a discussion on how to treat customers. Cross-training of front-of-the-house and back-of-the-house employees is being emphasized in many foodservice operations. Waitstaff needs to spend time in the kitchen to be able to answer customers' nutrition questions about ingredients and preparation methods. Also, chefs and cooks should know directly from customers if products coming out of the kitchen meet their expectations. Chefs, many of whom have been trained at top culinary schools, roam the dining room to determine immediately which menu items are successful and which need more work (Herlong, 1993). School foodservice has a unique situation because most of the cooks not only prepare menu items but also serve them to the children and know their reactions instantly.

Foodservice operators are looking for ways to cut costs and reward employees who have ideas for saving money on food and labor (Weinstein, 1993b). Hourly employees are compensated for not using all of their sick leave days or for coming to work on time for a specific number of days. Employers are trying to make employment opportunities as attractive as possible in the competitive labor market. Health-care insurance is still very expensive, but as an alternative many employers are trying to improve the life-style of employees by limiting the work week to five days and by supporting in-house wellness programs.

Foodservice operators also will have to contend with issues they have little control over, such as government mandates, IRS crackdowns, and lawsuits from new regulations, such as the Americans with Disabilities Act. The family-leave bill already has been passed, but child labor regulations and a higher minimum wage could be passed in the future. Stiff penalties for underreporting tip income also are being suggested by the IRS.

The Customer

"Knowing the customer" should have top priority for foodservice managers. Who is the customer? Is the customer the buyer who purchases a meal in a restaurant or the personnel department employees who review requests for an additional staff member? Many organizations today are committed to the total quality management philosophy, in which the focus is on the external as well as the internal customer. Regardless of an employee's job in the organization, that employee is a link in a chain of internal customers and suppliers that leads, eventually, to the company's external customer (Lee, 1991). A foodservice manager cannot meet the expectations of the external customer unless the internal processes are in alignment. Managers are committed to customer satisfaction, but often they do not apply that idea to the way they treat employees in the next department.

External Customer

The external customer of the 90s is quite different from the one in the 80s. The baby-boom generation has caused the major demographic changes that have occurred in the last half of the 20th century (Brault et al., 1992). The number of young adults in the United States exploded in the 1980s as the baby boomers came of age, making up 3 out of every 10 residents. As they begin to reach retirement age, the number of those aged 65 or over is expected to nearly double from the present level of 13% to 22% by the year 2030. Baby boomers will inherit the wealth of the World War II generation in addition to being part of a dual-income family in their peak earning years, ages 35 to 55. The average income of husband-wife families is approximately 25% higher than the average for all households; baby boomers should contribute to affluent households.

The baby-bust generation is composed of persons ages 18 to 29, who were born after the baby-boom years. **Baby busters** account for 18.1% of the total population as compared to baby boomers, who account for 33%, according to the Census Bureau. Baby busters are demanding customers, perhaps due to the independence they had to experience at an early age. They are the first generation of latchkey children, the product of dual-career households. Busters have relatively high disposable incomes even though their incomes are low, because they are postponing marriage and childbearing and many are opting to remain in the family nest rather than starting out on their own. They have credit, which they are willing to use, and they spend more than half of their food budget on food away from home. Restaurant operators would be wise to woo this group of customers now to increase patronage in the future (Iwamuro, 1993).

Restaurateurs and managers of retirement communities and nursing homes need to understand the demands of this group, which responds to specific values, including personal growth and revitalization, autonomy and self-sufficiency, and altruism and social connectedness, according to studies by the *American Demographics* magazine (cited in Cheney, 1993b). Other important trends are the higher levels of education of the baby-bust generation and the rising number of the elderly living alone. More adults are attending college. As the number of older students increases, college menus will be more sophisticated and consideration should be given to supermarket-type take-home dinners for students' families.

As more mothers work outside the home, children are making more decisions regarding household expenditures (Cheney, 1993b). Children control many millions of dollars of the family budget. Single-parent families eat out more often than the traditional family, and children often are the ones who select the restaurant. As a result, children are being targeted in advertising campaigns. Children from 6 to 12 years of age will be the key target group in the echo-boom generation through the year 2000. **Echo-boom children** of the baby boomers range from newborn to their 20s. Limited-menu restaurants have been running campaigns for children in the target group, and **full-service restaurants** are starting to do the same. Child-care centers and school foodservices will have to accommodate even more children as the youngest echo-boom children enter first grade in 1995.

The number of single adults increased 12% in 1990, and 61% of those who live alone are women (Cheney, 1993b). Divorce and aging baby boomers have increased the number of middle-age singles today. Restaurants will profit from this change in demographics, because 42% of the singles under 45 years of age are dining out more frequently in the 1990s compared to 28% of married persons in the same age bracket and 35% of older singles.

According to the Census Bureau, the U.S. population is projected to include 9 million immigrants who entered the country after 1986 and their descendants (Brault et al., 1992.) Their numbers are expected to more than triple to reach 32 million, or 12% of the population, by the year 2030. A percentage comparison of the origin of immigrants in 1990 with 2040 projections is as follows.

Origin	1990	2040
European	75.0%	less than 60.0%
African-American	12.1%	12.4%
Asian	3.0%	7.0%
Hispanic	6.2%	18.0%

The West is the most racially diverse region of the United States, and the Midwest is the least. One example of how immigration is affecting the foodservice industry is the recent gain in the number of ethnic restaurants. Ethnic foods can be found on almost all menus including those of nonethnic restaurants, colleges and universities, school foodservices, hospitals, nursing homes, supermarkets, and recreational facilities. Perhaps by the year 2040, ethnic menu items will be considered American food.

Internal Customer

Two major trends in the past decade are in quality and customer service. These have led to a third trend, which has many labels including total quality service, customer-centered quality, and the customer-driven organization. The focus, however, remains constant: The customer defines quality (Lee, 1991). This concept prescribes an approach to total quality management (TQM) that begins by looking at the expectations of the external customer and then works backwards through internal customer–supplier links. A chain is only as strong as its weakest link. Products and services must go through many customer–supplier relationships before reaching the external customer. Opportunities for mistakes or inefficiencies can occur in each relationship, and by examining the weak link, quality improvement should occur. For example, in a foodservice operation, the purchaser supplies food to the internal customer, who is a chef, and then the chef supplies menu items to the internal customer, the waitperson. Finally, the waitperson serves the external customer. If the TQM philosophy has been adopted and the menu item does not meet the standard, the waitperson should discuss the product with the chef rather than serving the product and making the external customer unhappy.

Thinking of employees as internal customers and suppliers is a very important concept to understand in quality improvement. Corporate culture plays a role in improving quality. Two rather simple statements sum up the service culture:

"Do unto your employees as you expect them to do unto your customers."
"If you are not serving the customer, you'd better be serving someone who is."

The message is clear; foodservice employees will not try to meet or exceed external customers' expectations if the manager does not extend the same consideration to them. Once employees become sensitive to the needs of their internal customers and suppliers, they want to know about the needs of external customers, whom they seldom meet (Lee, 1991).

The Kitchen

Customers in the 90s want to have fun when they go out to eat (Weinstein & McCarthy, 1993). Concept creators, so dubbed by *Restaurants & Institutions*, are visionaries who know what customers want. This group of 60 or more individuals and partnerships takes an idea and transforms it into a restaurant that has great appeal to customers. An example is the Wolfgang Puck Restaurants group, which has developed eight units and has others on the drawing board. Customers are always wondering what goes on in the kitchen beyond the swinging doors and are excited if they have an opportunity to watch the chef perform. As a result, more and more kitchens will go "on display" during this and the next decade (Stephenson & Chaudhry, 1993). In addition, foodservice managers are examining their kitchens to see if they can cut costs or increase revenues.

Sure Bets

Many foodservice operations use pizza ovens and rotisseries as display equipment (Stephenson & Chaudhry, 1993). With displayed equipment, customers know the ingredients are fresh and the food is being prepared especially for them. Chefs mixing ingredients and cooking menu items in front of the customer add to the dining experience. Employee and guest cafeterias in many hospitals and foodservice operations in colleges and universities offer customers the choice of raw ingredients, displayed on a refrigerated counter similar to a salad bar, that they place in a bowl and hand to a chef to stir-fry.

Kitchens are shrinking in size. Foodservice operators are realizing that making kitchens smaller provides more space for additional seating and eliminates some of the labor needed to keep a large kitchen in operation. Purchasing prepped foods from a supplier or having it done in a commissary, especially for chain operations, saves revenue-producing space. Many operations are going back to baking cookies and breads on the premises using frozen dough. Customers find it hard to resist temptation as they smell these products baking. As a result of the use of convenience products, fewer highly skilled employees are needed in the kitchen. The trend also is toward less storage space and more just-in-time deliveries. Food is fresher, and inventory costs are reduced. Walk-in coolers and large storage areas could become obsolete in many foodservice operations.

Carts and kiosks are being used in high-traffic areas because they are low cost and mobile. Express units in airports, including international ones, are doing big business; enormous real estate and occupancy costs are eliminated.

Worth Watching

More automation will appear in kitchens. Precise and greater communication will occur between customers and waitstaff, waitstaff and cooks or chefs, and cooks and chefs and suppliers (Stephenson & Chaudhry, 1993). Less waste should be the result. Automation also should decrease costs of labor and employee benefits. Limited-menu operations, which depend on low prices and consistency, are committed to automation.

Kitchen equipment, like computers, will be more user-friendly. A different set of controls currently is found on each piece of equipment. Manufacturers are beginning to think about standardizing controls, especially because the foodservice labor pool includes many non-English-speaking and -reading persons. User-friendly equipment also will be designed for disabled workers, making compliance with the Americans with Disabilities Act easier for foodservice managers.

Manufacturers will have to be more competitive with the durability of equipment and service contracts. Capital purchasing budgets need to be contained, which means that equipment will be used longer. As a result, foodservice operators will be more discriminating when they make equipment purchases.

As health and ethnic food items are added to menus, more specialty equipment will be necessary; juicers, grills, rotisseries, steamers, poachers, woks, and perhaps automated tortilla makers will likely be needed. Scales also will be required in the cooking area because nutritious food often requires exact measurements of ingredients. More nutrition analysis software is being produced to help foodservice managers comply with possible Food and Drug Administration labeling laws (Stephenson & Chaudhry, 1993).

Wait and See

Many noncommercial foodservice operations close down between mealtimes to decrease labor costs. A number of hospitals have no foodservice available after 7:00 P.M. until the next morning. Vending machines, however, are available for customers who want food. Vending will move from a static, change-resistant business into one that will provide food wherever customers may be, whether in trains and airports, offices, or beaches. New vending machines are being built that grind coffee beans to brew, accept debit cards, and offer frozen foods that can be heated in a nearby microwave. The type of food in the machines will change, too. Frozen, fresh pasta and high-grade frozen yogurt are examples of the quality of food that can be expected in the future.

A greater emphasis on security in response to increases in crime will be required. More sophisticated systems will be used such as video cameras and card-identification for entry into different areas of the kitchen. For large operations in high-crime areas, entry identification screens that read fingerprints will be installed (Stephenson & Chaudhry, 1993).

SUMMARY

The foodservice industry is customer driven, as recognized in the production of quality food and service for customer satisfaction with the ever-present financial concerns. The industry has become the number one retail employer in the nation.

Until recently, the industry had been divided into two categories, commercial and noncommercial. *Restaurants & Institutions*, a trade magazine, departed from this tradition in its 1993 annual forecast because the differences between the categories are almost nonexistent. The industry is classified, however, into the following segments: fast food (limited-menu), full service, employee feeding, schools, hospitals, lodging, colleges, military, recreation, nursing homes, supermarkets, transportation, social caterers, convenience stores, child care, retail, elder care, and all other foodservices.

Trends are based on the four cores of the industry: the menu, the staff, the customer, and the kitchen. Menus will reflect a mixture of cultures. Coffee is gaining popularity, as is the American breadbasket and American beers. Controversy between organic and bioengineered foods will continue, and chefs will have to be retrained to meet the nutritional menu options demanded by customers.

Foodservice is the number one retail employer. The staff will be empowered to make decisions for meeting customer expectations and will have to be trained to act independently. This gives the manager more time to mingle with customers rather than spending time in the kitchen or office.

Many organizations are committed to the total quality management philosophy, in which the focus is on the external as well as the internal customer. The external customer is the one who purchases a meal, and the internal customer is an employee. Each employee is a link in a chain of internal customers that leads to the external customer.

The kitchen will be on display in this and the next decade. Customers like to know that food is fresh and is being prepared specially for them. Kitchens are smaller, thus providing extra space for seating customers. More automation will appear in kitchens and equipment will be user-friendly.

BIBLIOGRAPHY

Aaker, D. A. (1991). *Managing brand equity: Capitalizing on the value of a brand name.* New York: Macmillan.

Bartlett, M. (1992). R & I's core four: Foodservice in a nutshell. *Restaurants & Institutions, 102*(1), 9.

Bartlett, M., & Bertagnoli, L. (1993). '93 R & I forecast: Operators work to keep business costs in line as the economy slowly perks up. *Restaurants & Institutions, 103*(1), 14–15, 18, 22, 26.

Bernstein, C. (1993a). R & I 400: Burgers: V-menu strategy: Value, veggies, variety. *Restaurants & Institutions, 103*(17), 54–55, 60.

Bernstein, C. (1993b). R & I 400: Recreation: Recreation rides to the top. *Restaurants & Institutions, 103*(17), 197.

Brault, D., Iwamuro, R., Mills, S., Riehle, H., & Welland, D. (1992). National Restaurant Association 1993 foodservice industry forecast. *Restaurants USA, 12*(11), 13–36.

Carlino, B. (1991). Marriott proves a point: Hospital food *can* taste good. *Nation's Restaurant News, 25*(28), 27–29.

Cetron, M., & Davies, O. (1989). *American renaissance: Our life at the turn of the 21st century.* New York: St. Martin's Press.

Chaudhry, R. (1991). Bursting prisons strain feeding programs. *Restaurants & Institutions, 101*(22), 22.

Chaudhry, R. (1993a). '93 R & I forecast: Employee feeding. *Restaurants & Institutions, 103*(1), 47.

Chaudhry, R. (1993b). '93 R & I forecast: Fast food. *Restaurants & Institutions, 103*(1), 36–37.

Chaudhry, R. (1993c). '93 R & I forecast: Lodging. *Restaurants & Institutions, 103*(1), 54, 59.

Cheney, K. (1992). Service tips for successful catering. *Restaurants & Institutions, 102*(27), 136, 138.

Cheney, K. (1993a). '93 R & I forecast: Hospitals. *Restaurants & Institutions, 103*(1), 52, 54.

Cheney, K. (1993b). '93 R & I forecast: The new U.S. household. *Restaurants & Institutions, 103*(1), 106–108.

Gáll, A. L. (1991). Child care: The new frontier. *School Food Service Journal, 45*(9), 34–36.

Hayes, J. (1991a). Branded fast-feeders: The Rx for ailing health-care budgets? *Nation's Restaurant News, 25*(40), 37–38.

Hayes, J. (1991b). Campus operators put on their creative thinking caps to woo student diners. *Nation's Restaurant News, 25*(26), 87, 90.

Hayes, J. (1992). Supermarkets lure more workers into fold. *Nation's Restaurant News*, *26*(24), 33–36.

Herlong, J. E. (1993). Research chefs: Out of the kitchen and into the lab. *Restaurants USA*, *13*(4), 10–12.

Hirschfield, J. (1990). How cruiseline f/s is performing. *FoodService Director*, *3*(4), 43, 46.

Hollingsworth, P. (1993). Foodservice trends: Convenient, casual, and comfy. *Food Technology*, *46*(4), 32–33.

Hospital foodservice: Under examination. (1992). *Food Management*, *27*(6), 78–79.

Howard, T. (1993). Under the toque. Paul R. Cyr: Feeding the fans. *Nation's Restaurant News*, *27*(28), 39, 43.

Iwamuro, R. (1993). Here come the "baby busters." *Restaurants USA*, *13*(2), 40–43.

Johnston, W. B. (1991). Global work force 2000: The new world labor market. *Harvard Business Review*, *69*(2), 115–127.

Keegan, P. O. (1992). Elmhurst's innovative ideas help residents feel young again. *Nation's Restaurant News*, *26*(10), 48.

Keegan, P. O. (1993). Italian competitors pour into dinner-house arena. *Nation's Restaurant News*, *27*(7), 39, 42.

King, P. (1991a). A salute to military foodservice. *Food Management*, *26*(12), 78–81, 84, 86, 88.

King, P. (1991b). Chicago: A city at play. *Food Management*, *26*(8), 120–122, 124–125, 130–131, 134.

King, P. (1991c). Coming to terms with OBRA. *Food Management*, *26*(3), 112–115, 118, 120.

King, P. (1993). Adjusting to the inmate explosion. *Food Management*, *28*(8), 92–93, 96, 98, 100.

Lang, J. (1993). Hospital launches 'Perfectly . . . Practical': Combining take-out and catering. *FoodService Director*, *6*(2), 118.

Lanier, P., & Berman, J. (1993). Bed-and-breakfast inns come of age. *Cornell Hotel and Restaurant Administration Quarterly*, *34*(2), 14–23.

Lee, C. (1991). The customer within. *Training*, *28*(7), 21–26.

Liddle, A. (1993). Vegetarian cuisine gets 'dressed' for prime time. *Nation's Restaurant News*, *27*(15), 39–41.

Life care. (1990). *Food Management*, *25*(3), 102–105, 108, 110, 114–116.

Lorenzini, B. (1993a). R & I 400: Contractors: Gaining a stronger retail edge. *Restaurants & Institutions*, *103*(17), 65, 68, 72.

Lorenzini, B. (1993b). R & I 400: Military: Armed for healthful dining. *Restaurants & Institutions*, *103*(17), 155–157.

Lorenzini, B., & McCarthy, B. (1992). The branding evolution. *Restaurants & Institutions*, *102*(21), 87, 90, 95, 98, 102, 106.

Martin, R. (1991). Health-care facilities get a dose of foodservice technology. *Nation's Restaurant News*, *25*(5), 29, 34.

McCarthy, B. (1993a). '93 R & I forecast: Child care. *Restaurants & Institutions*, *103*(1), 76, 80.

McCarthy, B. (1993b). '93 R & I forecast: Convenience stores. *Restaurants & Institutions*, *103*(1), 76.

Michalski, N. (1991). Update on shopping-center restaurants. *Restaurants USA*, *11*(4), 39–41.

Naisbitt, J., & Aburdene, P. (1990). *Megatrends 2000: Ten new directions for the 1990's*. New York: William Morrow.

National Restaurant Association. (1992). 1992–1993 National Restaurant Association foodservice industry pocket factbook. *Restaurants USA*, *12*(11), attachment.

"No more nuts." (1993, March 10). *Manhattan Mercury*, p. C1.

Opitz, A. (1993). Overcoming the "evils" of a la carte. *School Food Service Journal*, *47*(6), 20, 22–23.

Parseghian, P. (1993). Japanese breakfast: Rising star. *Nation's Restaurant News*, *27*(3), 29, 32.

Peterson, L. C. (1992a). Recreational foodservice: Staging a taste of Hollywood in Orlando. *Food Management*, *27*(6), 116–118, 122, 124.

Peterson, L. C. (1992b). Recreational foodservice: Universal appeal. *Food Management*, *27*(6), 112–114.

R & I 400: Ranking. *Restaurants & Institutions*, *103*(17), 25, 30, 34, 38, 44, 46, 48, 50.

RDs working in correctional facilities: Dietitians face the challenges of an inmate population. *ADA Courier*, *31*(2), 3–4.

Riehle, H. (1992). The foodservice manager in the year 2000. *Restaurants USA*, *12*(2), 36–37.

Riell, H. (1993). National chains build local loyalty. *Restaurants USA*, *13*(4), 1–15.

Romeo, P. (1991a). Disney newcomers offer a world of value for their guests. *Nation's Restaurant News*, *25*(9), 33–34.

Romeo, P. (1991b). Hotel chains mount cost-cutting assault against traffic slump. *Nation's Restaurant News*, *25*(30), 1, 116–120.

Ryan, N. R. (1993a). '93 fare turns on flair: Basics are served with verve; old ethnics add up to new. *Restaurants & Institutions*, *103*(1), 90–91, 94, 96.

Ryan, N. R. (1993b). Under the silver dome: Menu trends. *Restaurants & Institutions*, *103*(16), 45, 48, 52.

Sabatino, F. (1993). Culture shock: Are U.S. hospitals ready? *Hospitals*, *67*(10), 23–25, 28.

Schechter, M. (1990). Home care. *Food Management*, *25*(3), 134–138, 140, 142, 144.

Schechter, M. (1993). Healthcare foodservice report 1993: Getting ready for reform. *Food Management*, *28*(6), 70–79.

Schuster, K. (1990). Child care. *Food Management*, *25*(3), 120–123, 126, 130–132.

Schuster, K. (1993). Independents & contractors: The competition heats up. *Food Management*, *28*(8), 74–75, 78, 80, 82, 84, 86, 88.

Standard industrial classification manual. (1987). Washington, DC: U.S. Department of Commerce.

Stephenson, S. (1993a). '93 R & I forecast. Elder care. *Restaurants & Institutions*, *103*(1), 64, 68.

Stephenson, S. (1993b). '93 R & I forecast. Military. *Restaurants & Institutions*, *103*(1), 59, 60.

Stephenson, S. (1993c). '93 R & I forecast. Nursing homes. *Restaurants & Institutions*, *103*(1), 64.

Stephenson, S. (1993d). '93 R & I forecast. Schools. *Restaurants & Institutions*, *103*(1), 47, 52.

Stephenson, S. (1993e). '93 R & I forecast. Supermarkets. *Restaurants & Institutions*, *103*(1), 64, 68.

Stephenson, S. (1993f). R & I 400: Education: School operators optimistic. *Restaurants & Institutions*, *103*(17), 201, 204.

Stephenson, S. (1993g). R & I 400: Healthcare: Health care awaits reform. *Restaurants & Institutions*, *103*(17), 191, 194.

Stephenson, S. (1993h). R & I 400: Institutions: Prison dollars down, counts up. *Restaurants & Institutions*, *103*(17), 187.

Stephenson, S., & Chaudhry, R. (1993). Small kitchens, big ideas: '93 R & I forecast. Display equipment, kiosks and homey touches help cut costs. *Restaurants & Institutions*, *103*(1), 115–116, 120.

Stern, G. M. (1991). Managing a multiethnic staff. *Restaurants USA*, *11*(5), 32–34.

Straus, K. (1992a). Catering to every whim. *Restaurants & Institutions*, *102*(27), 36–37, 42, 44.

Straus, K. (1992b). Olympian feed. *Restaurants & Institutions*, *102*(29), 124–125, 128, 132.

Straus, K. (1993a). A taste of 1993: Restaurants share trendsetting recipes and idea: '93 R & I forecast. *Restaurants & Institutions, 103*(1), 122–123, 126, 128.

Straus, K. (1993b). '93 R & I forecast: Colleges. *Restaurants & Institutions, 103*(1), 59.

Straus, K. (1993c). '93 R & I forecast. Retail. *Restaurants & Institutions, 103*(1), 80, 88.

Straus, K. (1993d). R & I 400: Family dining: Family-dining chains scramble. *Restaurants & Institutions, 103*(17), 90–91, 94.

Straus, K. (1993e). The flavor explosion. *Restaurants & Institutions, 103*(13), 16–17, 24, 26–28.

Taylor, K. S. (1993). Contracting gains ground. *Hospitals, 67*(10), 32, 34, 36, 38, 40, 42.

Van Warner, R. (1992). Time to take a fresh look at supermarket foodservice. *Nation's Restaurant News, 26*(12), 25.

Walkup, C. (1991a). Dorm food: You've come a long way, baby! *Nation's Restaurant News, 25*(37), 25, 28.

Walkup, C. (1991b). Four-star hospital food?!! Health care gives restaurants a run for their money. *Nation's Restaurant News, 25*(16), 25, 28.

Walkup, C. (1992). Hotels go head to head with Chicago's best eateries. *Nation's Restaurant News, 26*(18), 16, 39.

Wallace, J. (1993). $400 billion by 2000. *Restaurants & Institutions, 103*(1), 11.

Weinstein, J. (1993a). '93 R & I forecast. Full service. *Restaurants & Institutions, 103*(1), 42.

Weinstein, J. (1993b). '93 R & I forecast: Power to the employees. *Restaurants & Institutions, 103*(1), 99–100, 102.

Weinstein, J. (1993c). '93 R & I forecast. Recreation. *Restaurants & Institutions, 103*(1), 60.

Weinstein, J. (1993d). R & I 400: Dinner houses. Dinner houses beef up, trim down. *Restaurants & Institutions, 103*(17), 116–117, 120.

Weinstein, J. (1993e). R & I 400: Retail. Retailers reinvest, reposition. *Restaurants & Institutions, 103*(17), 207, 120.

Weinstein, J. (1993f). R & I 400: Transportation: Carriers lighten their loads. *Restaurants & Institutions, 103*(17), 173, 175.

Weinstein, J., & McCarthy, B. (1993). Concept creators. *Restaurants & Institutions, 103*(15), 34–35, 38–39, 42, 46, 50, 54, 56–60.

2

Systems Approach to Foodservice Organizations

The systems approach to the management of organizations provides the basic conceptual framework for this book. The basic theory of a **conceptual framework** is defined as a loosely organized set of ideas, some simple and some complex, that provides the fundamental structure of an organization. In this chapter, systems concepts are applied to foodservice organizations, and assurance of quality and total quality management are discussed.

The term *system* has been in vogue for some time to describe everything from families to philosophy to education to engineering. It has been applied to a broad spectrum of our physical, biological, and social world (Kast & Rosenzweig, 1985). For example, in the universe, there are galactic, geophysical, and molecular systems. In biology, an organism is described as a system of mutually dependent parts, each of which includes many subsystems. The most complex organism, the human body, includes a skeletal, digestive, circulatory, and nervous system. Each day we come into contact with transportation, economic, computer application, and communication systems. Similarly, viewing organizations and management as systems simplifies their analysis in today's complex environment.

THE SYSTEMS CONCEPT

Systems theory provides a basis for understanding and integrating knowledge from a variety of widely specialized fields, as well as a macro view from which we may look at all kinds of systems. Systems may be viewed as *closed* or *open*. Physical and mechanical systems are closed in relation to their environment; biological and social systems, which are in constant interaction with their environment, are considered open.

Foodservice managers are seeking means to deal with both the complexities of organizations and the continually changing technological, economic, political, and sociological demands of today's world. According to Senge (1990), humans now have the capacity to create more information than can be absorbed, to foster greater interdependency than can be managed, and to accelerate change faster than anyone can keep pace. Drucker (1985) described the current economic and social scene as turbulent and contended that an organization has to be managed well both to withstand sudden changes and to profit from sudden, unexpected opportunities.

The application of systems concepts has been used to facilitate problem solving and decision making for managers. The systems approach focuses on the totality of the organization, rather than its processes or parts. It considers the impact of both the internal and external environment on the organization and on the managing process.

With the advent of the computer age in the late 1950s and early 1960s, the usefulness of the systems approach to managing organizations became evident. It was first applied to the foodservice industry in the mid-1960s when authorities in the field began discussing foodservice operations as systems.

A number of models of foodservice systems have been published in trade and professional literature. A **model** is a conceptual simplification of a real situation in which extraneous information is excluded and analysis is simplified. The systems approach has been used to design foodservice systems models for foodservices and hospital dietetics departments. Subsequently, a descriptive model of a foodservice system was developed that enabled managers, suppliers, and others to evaluate current practices and the impact of proposed changes on the foodservice operation (Freshwater, 1969; Gue, 1969; Konnersman, 1969; Livingston, 1968). David (1972), in describing a systems model of a foodservice organization developed at the University of Wisconsin–Madison, emphasized that the systems concept is an organized, practical application of commonsense techniques to the design and analysis of an operation.

Before 1960, analytical fact-finding approaches were used to examine organizations; however, the need for synthesis and integration became obvious. The systems era has been described as an age of synthesis, which is the act of combining separate parts into a conceptual whole. Managers must be capable of coordinating complex organizations by focusing on interactions and interrelationships of components and subsystems of the organization.

A **system** is defined as a collection of interrelated parts or subsystems unified by design to obtain one or more objectives. Luchsinger and Dock (1976) listed fundamental implications of the term *system*.

- A system is designed to accomplish an objective.
- Subsystems of a system have an established arrangement.
- Interrelationships exist among the elements.
- Flow of resources through a system is more important than basic elements.
- Organization objectives are more important than those of the subsystems.

The **systems approach** to management is simply keeping the organization's objectives in mind throughout the performance of all activities. It requires a communication network and coordination among all parts of the organization. Essentially, then, management is a process whereby unrelated resources are integrated into a total system for objective accomplishment.

According to Johnson, Kast, and Rosenzweig (1973), the **systems approach** encompasses three fundamental concepts: systems philosophy, systems analysis, and systems management. **Systems philosophy** is a way of thinking about an event in terms of parts or subsystems, with emphasis on their interrelationships. **Systems analysis** is a decision-making process aiding the manager in making the best choice among several alternatives. **Systems management** involves the application of systems theory to managing organizations. Thus, the systems approach is a way of thinking, a method of analysis, and a managerial style.

THE ORGANIZATION AS A SYSTEM

As indicated previously, organizations are described as **open systems** because they are in continual interaction with the environment. They can also be described as employee-equipment systems, that is, systems in which employees interact with equipment to achieve desired objectives. Equipment ordinarily is used to replace activities performed by employees or to assist employees in performing a task more effectively. Employees are relieved from routine or heavy physical tasks, thus permitting them to perform more creative or innovative kinds of work.

Davis (1980) discussed new designs for organizations and emphasized the importance of optimizing technical and social systems in the organization. By integrating the productive capacity of technical systems with the problem-solving capabilities of the social system, the organization should function at its optimum.

The **basic systems model of an organization** is shown in Figure 2.1. A system has a specific goal, major parts, and unique characteristics. A **goal** can be described as the aim or purpose of the system. It determines the major parts of a system: input, transformation, and output, as shown in the model.

The **input** of a system may be defined as any human, physical, or operational resource required to accomplish objectives of the system. **Human resources** are the skills, knowledge, and energies of people required for the system to function.

Figure 2.1. Basic systems model of an organization.

Physical resources include materials and facilities. **Operational resources** refer to such items as money, time, utilities, and information.

Transformation involves any action or activity utilized in changing input into output, such as activities involved in procurement and production of food. The **output** is the result from transforming the input, and it represents achievement of the system's goal. For example, the primary output in a foodservice system is the production of the desired quantity and quality of food to meet customers' needs. Other outputs of a foodservice system include customers' satisfaction with the food and employees' satisfaction with the job. Also, from a managerial perspective, achieving financial accountability is a necessary outcome.

The expanded systems model of an organization includes three additional parts: control, memory, and feedback (see Figure 2.2).

Control, both internal and external, provides guidance for the system. *Internal control* consists of plans including the goals and objectives of the organization, standards, and policies and procedures. *External control* consists of local, state, and federal regulations and contracts with outside companies.

The control element performs three functions in a system. It ensures that resources are used effectively and efficiently in accomplishing organizational objectives; it ensures that the organization is functioning within legal and regulatory constraints; and, finally, it provides standards to be used in evaluation of operations.

Memory includes all stored information and provides historical records of the system's operations. Analysis of past records can assist the manager in making plans and avoiding repetition of past mistakes. Rapid advances in computer technology are revolutionizing the memory capability of all types of systems. Rather than rely on filing cabinets for storage of information, managers increasingly are relying on computer disks for rapid access to records.

Feedback includes those processes by which a system continually receives information from its internal and external environment. If utilized, feedback assists the system in adjusting to needed changes. For instance, feedback from customers' comments could be valuable information to the manager. Organizations without effective feedback mechanisms become relatively closed systems and may go out of business.

Figure 2.2. Expanded systems model of an organization.

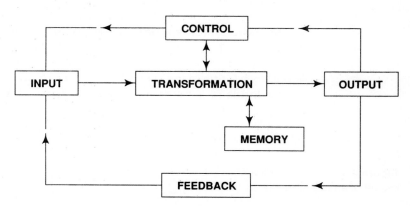

CHARACTERISTICS OF OPEN SYSTEMS

An open system has a number of unique characteristics:

- Interdependency of parts, leading to integration and synergy
- Dynamic equilibrium
- Equifinality
- Permeable boundaries
- Interface of systems and subsystems
- Hierarchy of the system

These characteristics are coordinated by linking processes in the transformation of inputs to outputs.

Interdependency is the reciprocal relationship of the parts of a system; each part mutually affects the performance of the others. This characteristic emphasizes the importance of viewing the organization as a whole rather than the parts in isolation. For example, in a foodservice system a decision to purchase a new piece of automated equipment may affect the menu, type of food purchased, and employee schedules.

Interaction among units of an organization is implied by interdependency. Units do not operate in a vacuum but continually relate with others. For example, for the organization to function as an effective system, the purchasing department must interact with the production unit and advertising with the sales department. The result of effective interaction is **integration,** in which the parts of the system share objectives of the entire organization. Integration leads to **synergy,** meaning that the units or parts of an organization acting in concert may have greater impact than if you were to combine the impact of each of them operating separately.

Dynamic equilibrium, or steady state, is the continuous response and adaptation of a system to its internal and external **environment,** which includes all the conditions, circumstances, and influences affecting the system. To remain viable, an organization must be responsive to social, political, and economic pressures. A foodservice director continually must evaluate cost and availability of food, labor, and supplies as well as advances in new technology. Change then is required to adapt to these new conditions and maintain viability. Feedback processes discussed in the previous section of this chapter are important in maintaining dynamic equilibrium.

The term **equifinality** is applied to the organization as a system. It means that a same or similar output could be achieved by using different inputs or by varying the transformation processes. In other words, various alternatives may be used to attain similar results. In a foodservice organization, a decision to change from conventional to convenience foods will affect inputs and the transformation processes; however, a similar output, meals for a given clientele, will be achieved from these different inputs and processes.

Permeability of boundaries is the characteristic of an open system that allows the system to be penetrated or affected by the changing external environment. **Boundaries** define the limits of a system, and permeability allows the system to interact with the environment. For example, a hospital constantly interrelates with

Figure 2.3.   Levels of the organization.

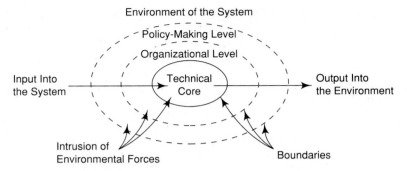

Source: Adapted from "A Behavioral Theory of Management" by T. A. Petit, 1967, *Academy of Management Journal, 10*(4), p. 346. Reprinted by permission.

the community, other healthcare institutions, and government agencies, all of which are part of the external environment. An organization can be described as having three different levels. The model shown in Figure 2.3 was adapted from that proposed initially by Parsons (1960) and modified by Petit (1967) and Kast and Rosenzweig (1985) as a way of conceptualizing the organization as a system.

The internal level, also called the *technical core* or production level, is where goods and services of the organization are produced. The *organizational level,* in which coordination and services for the technical operations are provided, is responsible for relating the technical and policy-making levels. *Policy-making,* the third level, is primarily responsible for interaction with the environment and long-range planning. An example is the corporate headquarters for a chain of restaurants. Although all three levels have permeable boundaries and environmental interaction, the degree of permeability increases from the technical to the policy-making level.

The concept of *boundaries* among levels of a system, between subsystems, or between systems is a rather nebulous one. Previously, boundaries were defined as the limits of a system. Boundaries also are described as demarcation lines or regions of system activity; they set the domain of organizational activity. According to Kanter (1991), however, walls between subsystems are crumbling. Rigid boundaries are fading because of change. For example, activities of the food production and service units provide boundaries for each subsystem. Despite separate realms of activity, the boundaries between the two subsystems are disappearing because the goal of the foodservice system is to satisfy the customer. The production and service subsystems, therefore, must be interdependent to meet this goal.

The area of interdependency between two subsystems or two systems is often referred to as the *interface.* The example just cited illustrates the interface between the production and service subsystems of a foodservice organization. The overall organizational system has many interfaces with such other systems as suppliers, government agencies, community organizations, and unions.

A point of friction often occurs when two moving parts come together. Similarly, the interface between two subsystems within an organization is likely to be characterized by tension. Whyte (1948), in a landmark study of the restaurant industry, and later Slater (1989) identified the area between the front and back of the house as a point of maximum tension between waitstaff and cooks. In hospitals, a classic example of interface is patient tray service. Dietetic and nursing services are in direct contact with each other, and conflict frequently occurs. An age-old argument in many hospitals has been, "Who is responsible for clearing the patient's bedside table at mealtime?" These interface areas often require special attention by managers.

Another characteristic of a system is **hierarchy.** A system is composed of subsystems of lower order; the system is also part of a larger suprasystem. In fact, the ultimate system is the universe. For purposes of analysis, however, the largest unit with which one works generally is defined as the system, and the units thereof become subsystems. A **subsystem,** a complete system in itself but not independent, is an interdependent part of the whole system.

For example, a hospital may be viewed as a system; dietetic services, nursing, radiology, and other departments are considered subsystems. By the same token, a college or university may be viewed as a system, and academic units and student services as two of the subsystems. The foodservice department would then be viewed as a component of the student services division. One may, however, wish to analyze foodservice departments in a hospital and college in more detail and thus view them as systems; the units within the foodservice would then become the subsystems.

Linking processes are needed to coordinate the characteristics of the system in the transformation from resources to goals. These processes are decision making, communication, and balance. **Decision making** is defined as the selection by management of a course of action from a variety of alternatives. **Communication,** which is the vehicle whereby decisions and other information are transmitted, includes all the oral, written, and computer forms used throughout the system. **Balance** refers to management's ability to maintain organizational stability under shifting technological, economic, political, and social conditions.

A FOODSERVICE SYSTEMS MODEL

A foodservice systems model (see Figure 2.4) was developed to illustrate applications of systems theory to a foodservice organization. An examination of the model reveals that it is based on the basic systems model of an organization (Figure 2.1), which includes input, transformation, and output. The additional components of control, memory, and feedback, which are from the expanded systems model of an organization (Figure 2.2), are integral parts of the foodservice systems model.

Arrows in the model represent the flow of materials, energy, and information throughout the foodservice system. Gaps in the arrows from output to input on the periphery of the model represent the permeability of the boundaries of the foodservice system and reflect the environmental interaction inherent in the effectiveness of the system. The bidirectional arrows represent environmental interactions, both internal and external to the system.

The input of the foodservice system is the human and physical resources that are transformed to produce the output. Traditionally, these resources have been referred

Figure 2.4.   A foodservice systems model.

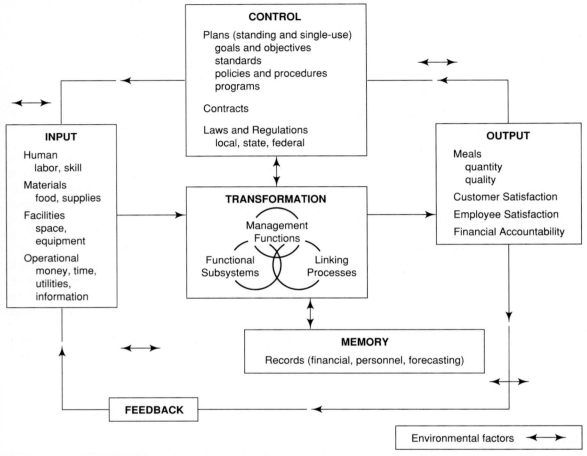

Source: Adapted from *A Model for Evaluating the Foodservice System* by A. G. Vaden, 1980. Manhattan, KS: Kansas State University. ©1980. Used by permission.

to as men, materials, money, and minutes. This traditional definition has been expanded by defining the following four types of **resources:**

- Human: labor and skill
- Materials: food and supplies
- Facilities: space and equipment
- Operational: money, time, energy, and information

Input requirements are dependent upon and specified by the objectives and plans of the organization. For example, the decision to open a full-service restaurant serving fine cuisine rather than a limited-menu operation with carryout service would have a major impact on type and skill of staff, food and supplies for production of menu items, capital investment, and type of foodservice facility and layout.

In the foodservice systems model (Figure 2.4), **transformation** includes the functional subsystems of the foodservice operation, managerial functions, and linking processes. These are all interdependent parts of transformation that function in a synergistic way to produce the output of the system.

The subsystems of a foodservice system (see Figure 2.5) are classified according to their purpose and may include procurement, production, distribution and service, and sanitation and maintenance. Depending on the type of foodservice system, the subsystems within the system may vary.

The type of system determines the characteristics and activities of the subsystems. In the example given above, the full-service restaurant serving fine cuisine would have a more sophisticated and elaborate production unit than that of the limited-menu restaurant. Distribution and service in hospitals represent a very complex and difficult subsystem to control; the appropriate food at the correct temperature and quality must be delivered to patients in many locations. Food contractors providing meals for several airlines face the complexities of different menus and schedules, varying numbers of passengers, and problems such as delayed and canceled flights. Designing subsystems to meet the unique characteristics of these various foodservice organizations requires a systems approach in which the overall objectives of the organization are considered as well as interrelationships among parts of the system.

Management functions, an integral component of the transformation element, are performed by managers to coordinate the subsystems in accomplishing the system's objectives. In this text, management functions are defined as planning, organizing, staffing, leading, and controlling. For example, the foodservice manager plans a sanitation program, organizes the activities, staffs the program with employees,

Figure 2.5. Functional subsystems of a foodservice system.

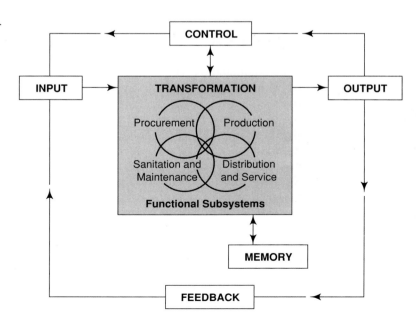

leads employees to meet goals, and finally controls the program by conducting a sanitation inspection.

Linking processes, previously described, are shown in the transformation portion of the model. Decision making, communication, and balance are critical to the effective coordination of activities within the system.

The output is the products and services that result from transforming the input of the system and expresses how objectives are achieved. The primary output in the foodservice system is meals in proper quantity and quality. In addition, customer and employee satisfaction and financial accountability are desired outcomes.

A meal consists of many menu items; the number of meals dictates the quantity of each item. Overproduction of menu items results in loss of income, and underproduction, in loss of customer goodwill.

Traditionally, textbooks in this field have stated that the objective of a foodservice is to produce the highest possible quality food. In this text, however, the objective of production of food is to satisfy the expectations, desires, and needs of customers, clients, or patients. A customer at an office snack bar, for example, might be content with a grilled cheese sandwich and tomato soup; that evening at an upscale restaurant, however, the customer will have quite different expectations of the cuisine.

Customer satisfaction is closely related to the types and quality of food and services provided and to customer expectations. For example, a college student, pleased with pizza on the luncheon menu of a college residence hall, would be unhappy if that same item were served at a special function of a social fraternity at a country club, even though in both instances the product may be of high quality. The student's expectations in these two situations are quite different.

Defining employee satisfaction as an output of the system might be viewed by some as surprising. Currently, management considers employees as customers. Chefs, for example, are beginning to understand that waitpersons are their customers (Wood, 1993). Managers also should be concerned about assisting employees in achieving and coordinating personal and organizational objectives. The effectiveness of any system, in large measure, is related to the quality of work done by the people staffing the organization.

Financial accountability is an output applicable to either a for-profit or not-for-profit foodservice organization. A foodservice manager must control costs in relation to revenues regardless of the type of operation. In the profit-making organization, a specific profit objective generally is defined as a percent of income. In a not-for-profit organization, the financial objective may be to break even; in other words, costs must equal revenues. In others, however, a certain percentage of revenues in excess of expenses must be generated to provide funds for renovations, replacement costs, or expansion of operations.

Control encompasses the goals and objectives, standards, policies and procedures, and programs of the foodservice organization. The **menu** is considered the most important internal control of a foodservice system. The menu controls food and labor costs, type of equipment needed, customer and employee satisfaction, and profit. All plans, however, are internal controls of the system and may be either standing or single use. Standing plans are those used repeatedly over a period of time and updated or reviewed periodically for changes. Single-use plans are those designed to be used only one time for a specific purpose or function.

A cycle menu is an example of a standing plan. For example, a hospital may have a 2-week menu cycle that is repeated throughout a 3-month seasonal period. Many restaurants use the same menu every day; some might add a "daily special." Various types of organizational policies also are examples of standing plans. The menu for a special catered function, however, is an example of a single-use plan. A particular single-use plan may provide the basis for a subsequent event of a similar type, but is not intended to be used in its exact form on a second occasion.

Contracts and various local, state, and federal laws and regulations are other components of control. Earlier in this chapter, examples of the effect of contractual and legal controls on the foodservice system were provided. Contracts are either internal or external controls. Internal controls may be for security, pest control, and laundry services; legal requirements are externally imposed controls on the foodservice system. The foodservice manager must fulfill various contractual and legal obligations in order to avoid litigation. For example, in constructing a foodservice facility, local, state, and federal building and fire codes must be followed in both design and construction. New federal regulations were imposed with the recent passage of the Americans with Disabilities Act, which requires that public accommodations and private businesses serving the public must remove barriers that interfere with access to the facilities and services provided (National Restaurant Association, 1992). Controls, then, are the standards for evaluating the system, and they provide the basis for the managerial process of controlling.

Memory stores and updates information for use in the foodservice system. Inventory, financial, forecasting, and personnel records, as well as copies of menus, are among the records that should be maintained by management. Review of past records provides information to management for analyzing trends and making adjustments in the system.

Feedback provides information essential to the continuing effectiveness of the system and for evaluation and control. As stated earlier, a system continually receives information from its internal and external environment that, if utilized, assists the system in adapting to changing conditions. Effective use of feedback is critical to maintaining viability of the system. A few examples of feedback that a foodservice manager must evaluate and utilize on a regular basis are comments from customers, plate waste, patronage, profit or loss, and employee performance and morale.

QUALITY IN THE FOODSERVICE SYSTEM

Quality has intensely personal connotations and thus is a very difficult word to define. The American Society for Quality Control defines **quality** as "the totality of features and characteristics of a product or service that bears on its ability to satisfy a customer's given needs" (Vaughn, 1990). Fortuna (1990) stated that quality can be summarized by one of the following statements:

- Conformance to specifications—quality is defined by the relative absence of defects.
- Meeting customer expectations—quality is measured by the degree of customer satisfaction with a product's characteristics and features.

Juran succinctly defined quality as "fitness for use" encompassing both freedom from defects and the many elements required to meet the needs of a customer (cited in Ciampa, 1992).

Concern about quality began in the 1940s, with the advent of World War II. American statisticians W. Edwards Deming, Joseph Juran, and W. A. Shewhart developed new methods for managing quality in wartime industries that had to produce high-quality armaments with a largely unskilled labor force. The thought was that if inventories are kept low, good relationships with suppliers are made, and jobs are performed more efficiently, then better quality products would be produced at lower cost (Beasley, 1991a).

After the war, the United States reverted to prewar manufacturing practices that depended on large inventories, labor-intensive production processes, and top-down management. The Japanese, however, were committed to rebuilding their country by following the teachings of Deming, Juran, and Shewhart. Their priority was to change customers' associations of "made in Japan" with poor quality. The Japanese learned to produce high-quality products that were cost-effective (Beasley, 1991a).

An abundance of literature is available on dealing with quality issues in organizations. The terms *quality assurance, total quality management, continuous quality improvement,* and *quality improvement process* have been used interchangeably, but often incorrectly, to describe management styles in which processes are examined and refined with the goal of improving performance. Definitions for each term are given below.

- **Quality assurance (QA).** A procedure that defines and ensures maintenance of standards within prescribed tolerances for a product or service (Thorner & Manning, 1983).
- **Total quality management (TQM).** A management philosophy directed at improving customer satisfaction while promoting positive change and an effective cultural environment for continuous improvement of all organizational aspects (Gift, 1992).
- **Continuous quality improvement (CQI).** A focused management philosophy for providing leadership, structure, training, and an environment to continuously improve all organizational processes (Shands, 1992).
- **Quality improvement process (QIP).** A structured problem-solving approach that focuses on operating processes and involves staff in analyzing current situations and developing recommendations for problem resolution (Gift, 1992).

QA has been used in healthcare for many years. Although the concepts were developed during World War II, TQM and its variations currently are being emphasized in the United States.

Quality Assurance

Figure 2.6 presents the foodservice system model with emphasis on quality assurance as a component of the control element. Organizational goals and objectives provide the beginning point for a quality assurance program. Quality customer service is the goal of both profit-oriented and not-for-profit organizations.

Goals and objectives also provide the basis for defining quality standards, which in turn are used for developing policies and procedures for quality control. The key to a QA program is continuous monitoring and evaluation to determine if quality standards and control are being maintained in all aspects of operations. Feedback mechanisms are critical to providing information on the quality of both processes and products. QA requires feedback to determine the need for changes in inputs and transformation processes to meet goals and verify that necessary changes have been made.

Among the first to give impetus to the establishment of formalized QA programs was the healthcare segment of the foodservice industry. In 1979, the Joint Commission on Accreditation of Hospitals (JCAH) published the original QA standards for hospitals. Since 1987, the scope of JCAH has been expanded, coincident with the name change to Joint Commission on Accreditation of Healthcare Organizations

Figure 2.6. Quality assurance in the foodservice system.

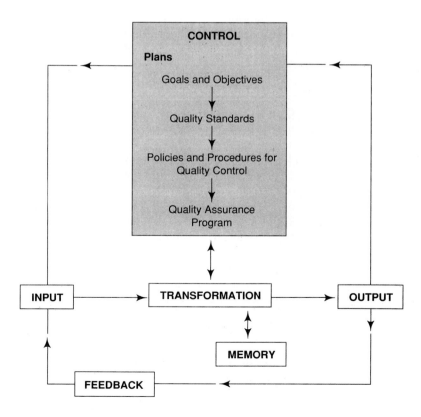

(JCAHO, 1992). Interestingly, JCAHO is phasing out QA terminology in favor of quality assessment and improvement concepts (Learning the Language, 1993).

JCAHO serves as a regulatory agency for the voluntary QA program by sending a team of healthcare professionals to determine the degree to which an organization complies with established control standards. Twelve standards for handling and preparation of food are included in the foodservice sanitation category under dietetic services; standard 3.3.2, for example, is "Foods are stored at proper temperatures, utilizing appropriate thermometers and maintaining temperature records." Each category of standards is scored from 1, meaning the department is in substantial compliance with all 12 standards, to 5, the department is in substantial compliance with fewer than three standards (JCAHO, 1992). An organization is expected to correct problems in order to become accredited.

Controlling quality in a hospital foodservice operation is a major component in assuring quality of patient care. This concept also can be applied to a restaurant. Quality control of menu items prepared in the kitchen is very important to the manager who is concerned with assuring customer satisfaction. Increased emphasis on quality control is evident in other segments of the foodservice industry as well. The National Association of College and University Foodservices published a professional standards manual that defines standards and criteria to be used for self-monitoring or for a voluntary peer review program (National Association of College and University Foodservices [NACUFS], 1991).

Throughout this text, controls in all aspects of the system are emphasized. The critical importance of defining quality standards and monitoring operations to ensure adherence to these standards is stressed repeatedly. Achieving the goal of producing an acceptable quality product that is served in an acceptable manner to the customer requires continuous diligence and surveillance. The foodservice manager is primarily responsible for ensuring the effectiveness of the QA program.

Quality control is one aspect of the management function of controlling, which is a continuous process of checking to determine if standards are being followed and of taking corrective action if they are not. Quality assurance is output oriented; it includes the process of defining measurable quality standards. Quality controls thus are essential at every step of the operation, from the development of procurement specifications through distribution and service of food to customers (Pickworth, 1987). Effective quality controls can mean the difference between substantial financial losses and profits. Customer satisfaction, employee productivity, and financial accountability all depend on quality control throughout the foodservice system.

The theme of all industries in the 1990s is assuring and improving quality. Concern about the declining quality of goods and services in the United States has led to a revival of managerial emphasis on quality standards and quality control in all organizations. Kenneth Heymann, vice president of The Hospitality Group, a consulting company in Denver, stated the following:

> The key difference between quality control and quality management is that quality control is a reactive process and quality management is a proactive process. Quality control is predicated on follow-up and inspection and finding error after the fact. Quality management is devoted to organizing and delivering service in a way that allows you to ensure that you don't make an error. (Wood, 1993)

Controversy has arisen over whether QA, which is based on QC, is reactive or proactive. QA, as originally defined, is a reactive process. Once an organization accepts the idea that customers define quality, however, the responsibility for quality reaches beyond that of QA/QC. It begins with a thorough understanding of customers' needs and carries through the controlled production of the product or service (Huge, 1990). QA flows into TQM, and the controversy over reactive and proactive occurs in the gap between the two.

Total Quality Management

TQM and related concepts, such as CQI and QIP, seem to be the latest and most promising methods for focusing attention on a commitment to quality in both products and services. TQM really is a management philosophy in which processes are refined with the goal of improving the performance of an organization. It builds upon other management philosophies such as management by objectives, quality circles, and strategic planning (Schmidt & Finnigan, 1992). The strength of TQM is that it consists of specific steps that can be followed to identify and remedy problems before they happen (Zabel & Avery, 1992). QA is in the control element of the foodservice systems model, and TQM permeates management functions in the transformation element. Planning, organizing, staffing, leading, and controlling are essential to make TQM succeed in an organization.

A Management Philosophy

TQM is a management philosophy and has six components, as shown in Figure 2.7 (Zabel & Avery, 1992). TQM originated in manufacturing industries but currently is being applied to public institutions such as colleges and universities, governmental agencies, and other nonprofit organizations.

Often, the way work is accomplished creates problems. TQM is built on the assumption that people want to do a good job and are concerned about quality but that little thought is given to the process of doing work or the reasons for performing particular tasks. Tools used for producing a better product or service are benchmarking and Pareto analysis. **Benchmarking** is a label for the concept of setting goals based on knowing what has been achieved by others (Juran, 1992). It involves rating a company's products and services against those of the front-runners in the industry. For example, an independent limited-menu restaurant manager might improve service by analyzing the methods used by McDonald's or Wendy's. **Pareto analysis** is often called the 80-20 rule because marketers have found that approximately 80% of the total sales volume comes from approximately 20% of the customers. These vital few customers are often called key customers, because they are responsible for 80% of the income (Juran, 1992).

Customers are people who use products or receive services, and TQM helps the organization focus on them by identifying and satisfying their needs and expectations. A **customer** is anyone who is affected by a product or service and may be

Figure 2.7. Components of total quality management.

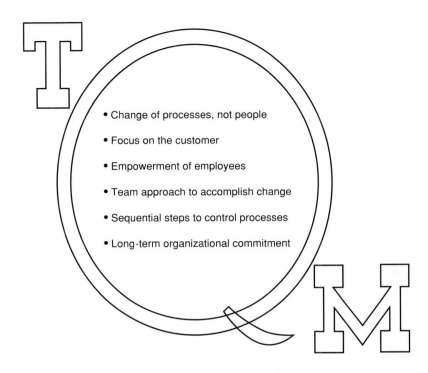

- Change of processes, not people

- Focus on the customer

- Empowerment of employees

- Team approach to accomplish change

- Sequential steps to control processes

- Long-term organizational commitment

external or internal (Juran, 1992). **External customers**, like those who buy a hamburger or a pizza from a restaurant, are affected by the product but do not belong to the organization that produces it. **Internal customers** are both affected by the product and belong to the organization that produces it. Employees are internal customers, as are suppliers of food products. If a new supplier is added to the list, sales volume probably will decrease for the regular suppliers.

TQM encourages employees' participation in identifying problems and finding solutions for the improvement of the organization's overall performance. Management is responsible for facilitating and supporting each employee's growth, thus empowering them to use their best abilities and improve their weaknesses (Ebel, 1991). **Empowerment** is the level or degree to which managers allow their employees to act within their job descriptions. It occurs when employees are permitted to make decisions in the areas where they work. Everyone in the organization requires training, retraining, and the opportunity to acquire and develop skills necessary to do their jobs better. Waitstaff in a restaurant often are not permitted to take care of customer complaints immediately without first checking with the manager. For example, suppose the last customer in a restaurant wants a refill of decaffeinated coffee, but the pot is empty. The manager tells the waitperson to use regular coffee. The waitperson is concerned about potential customer dissatisfaction, which could be easily prevented if she were empowered to do her job. She then could make a new pot of decaffeinated coffee, ask the customer if regular coffee is acceptable, or not charge for the first cup. Regardless of which choice she makes, customers would be pleased to know that waitpersons care enough to include them in the decision (Kapoor, 1991).

TQM requires teams consisting of a leader, who is part of management, employees with experience in the process, and a facilitator for accomplishing change. Quality improvement teams are assigned specific projects aimed at improving product or service quality while reducing costs. Facilitators assist the teams and usually are supervisors or specialists with special training for the role. Being a facilitator is a part-time assignment included in one's regular job. The facilitator helps trainees learn how to be team members by teaching them to communicate with one another, contribute information, challenge decisions, and share experiences. The facilitator also can serve as a consultant to the team chairperson (Juran, 1992).

The number of steps organizations use for implementing TQM processes varies depending upon the setting and the problem. TQM usually requires interviewing customers, identifying a problem, determining causes, developing measurable improvement, selecting and implementing the best solution, collecting data to measure results statistically, refining the solution, and repeating the cycle.

TQM requires a long-term commitment from the organization, as well as a more complete understanding of its operation. A systems approach should be used because TQM transcends the entire organization, and implementation requires vision and commitment. The unifying goal of TQM is to provide customer satisfaction by increasing the quality of products and services. The output of the process goes to either internal or external customer satisfaction. As a result, the benefits of TQM are increased productivity, worker satisfaction, and decreased cost. TQM cannot be implemented overnight; results take time to appear. Without upper-management commitment and participation, TQM initiatives are destined for failure. Free and open lines of communication among all departments must be established. Poor quality often can be traced to voids or overlaps of responsibility (Pfau, 1989).

TQM in Foodservice

TQM programs are in the planning or implementation stages in many organizations, such as hospitals, restaurants, and colleges and universities. Foodservice managers already are very much involved in the process. If, however, the foodservice operation is independent, as at many fine restaurants, foodservice managers are responsible for planning and implementing their own TQM program.

Quality currently is one of the major goals of planning and action in the healthcare industry. Interest in it is the result of the unprecedented rise in healthcare expenditures that instigated a customer movement for demanding high-quality healthcare for less cost. This quality movement has served as a catalyst for what the JCAHO calls its "Agenda for Change," which has an outcome focus, and the organization's overwhelming interest in industrial quality management programs. Since 1992, the Joint Commission changed from a quality assurance program to a quality assessment and improvement program by revising standards to help hospitals use their current commitment, resources, and approaches to improving the quality of patient care more effectively and efficiently (Russo, 1993a). Organization leaders including the governing body, chief executive officer, administrators, and department heads are expected to prepare a written plan for "assessing and improving quality that describes the objectives, organization, scope, and mechanisms for overseeing the effectiveness of monitoring, evaluation, and improvement activities" (JCAHO, 1992).

Veterans Administration (VA) hospitals that have been chosen as TQM pilot sites and private-sector hospitals that have begun TQM programs can be used as benchmarks by other healthcare organizations (Russo, 1993b). Marjorie Beasley, food and nutrition services director at Bloomington Hospital in Indiana, where a continuous quality improvement management program was initiated in 1989, stated that managers are also facilitators and coaches. Managers, as coaches, often have to ask questions instead of just stating what they think is right (Boss, 1992).

Restaurant managers are very much aware of the importance of both product and service quality. To be successful they realize that they must focus on the customer. They ask customers what they want and then find a way to meet their expectations. Many operators use comment cards that they pass out personally while telling the customer they really want to know what they think about the food and service. Customers who are willing to provide their names and addresses on the cards receive a letter telling them the improvements made because of their comments (Wood, 1993).

Taco Bell did a quality-led turnaround, which was used as a case study in the Harvard Business School. After a top-management change in the early 1980s, consumer studies were conducted in 1987 and 1989 to find out what customers really want. Before that, upper-level management thought that sales would increase if new items were added to the menu frequently. Surprisingly, customers were more interested in certain fundamentals than in menu changes. By analyzing and classifying customers' responses, management came up with the acronym FACT: *F*ast service and *A*ccurate orders with food served in a *C*lean restaurant and at the right *T*emperature. The company improved performance by redesigning the layout of its stores, preparing more menu items off site to decrease preparation time, and scheduling employees more efficiently (Wood, 1993). According to sales, Taco Bell currently is in the top category for limited-menu operations.

John Engstrom, director of Culinary Services at Virginia Polytechnic Institute and State University in Blacksburg, says that the hardest decision he has ever made is to use participative management. Resistance occurs because the process blows up the management pyramid; change is pain for all involved, according to Engstrom (King, 1992). Engstrom introduced TQM by leading a process he calls "Triad of Excellence" (TOE) for improving the department's quality, service, and cooperation.

The second phase of the process at Virginia Polytechnic is a statistical analysis, called an "environmental scan." Management and employees look at the department internally and externally, their customers, and the total market to evaluate their accomplishments. Since Engstrom initiated TOE in 1992, the foodservice operation has gone from a deficit of $1.5 million to a $1 million profit center (King, 1992).

The Malcolm Baldrige National Quality Award

In 1987, the U.S. government established the **Malcolm Baldrige National Quality Award,** named after the former U.S. Secretary of Commerce. This award has risen from a little-known measure of quality achievement to the most prestigious award in American business. The criteria for the Baldrige award and the associated weights are shown in the chart (Juran, 1992).

Category	Weight
Customer satisfaction	30
Quality results	15
Human resource utilization	15
Quality assurance of products and services	15
Leadership	12
Planning for quality	8
Information and analysis	5
Total	100

Many leading businesses have used the rigorous criteria for the award as the measure of quality improvement (How Significant Is the Baldrige?, 1991). The only service industries to receive the award as of 1993 were the Federal Express Corporation and the Ritz-Carlton Hotel Corporation. A basic belief throughout the Ritz-Carlton corporation is that "customer satisfaction begins with employee satisfaction. Employees who are well treated and empowered will treat customers in a courteous manner that will result in job security and various kinds of rewards and opportunities for advancement" (Schmidt & Finnigan, 1992).

The International Organization for Standardization (ISO) in Geneva, Switzerland, has developed ISO 9000, a set of five universal quality standards. More than 35 countries have adopted the standards, which include generic requirements for 20 basic elements that affect quality. The European Community has adopted ISO 9000 to ensure cross-border quality among its members (Hockman, 1992).

Achieving integration of quality, strategy, and financial management is critical to the future of TQM. To be successful, quality management must be integrated with the strategic management process and blended into financial planning. One of the best tools to assist in strategic planning is the environmental scan including governmental changes caused by elections, demographic changes brought on by fluctuating population characteristics, and social changes in taste and morals (Odiorne, 1987). Many companies try to change everything at once and then expect immediate results, thus causing the failure of TQM.

Companies need to focus resources on projects with high potential for success that build on a foundation of measurable resources (Davis, 1992). One common mistake in implementing TQM is a failure to link quality to the bottom line (Port, Carey, Kelly, & Forest, 1992). Davis (1992) stated that perhaps the best lesson learned from Baldrige Award recipients is the vital importance of human resource practices that support quality work and encourage people at the middle and lower levels to create ideas for improvement.

SUMMARY

The systems concept has practical applications for foodservice organizations meeting the challenges and demands of today's world and facilitates problem solving and decision making. A system is defined as a collection of interdependent parts or subsystems unified by design to obtain one or more objectives. The major parts of a system are input, transformation, and output. The system, a variant of a conceptual framework defined as a loosely organized set of ideas that provides the fundamental structure of an organization, constitutes a model for this text.

Inputs are the human and physical resources that are transformed to produce the outputs of the system. Inputs are human, materials, facilities, and operational resources referring to money, time, and information. Transformation includes the functional subsystems (procurement, production, distribution and service, sanitation and maintenance), the managerial functions (planning, organizing, staffing, leading, and controlling), and the linking processes (decision making, communication, and balance). These are all interdependent parts of transformation and function synergistically to produce the outputs, the primary one of which is meals in the correct quantity and of the right quality to satisfy the customer. In addition, personnel satisfaction and financial accountability are desired outcomes. Other elements in the foodservice model are control, memory, and feedback. Control encompasses the plans for goals and objectives, standards, policies, procedures, and programs. The menu, the primary plan of the foodservice system, functions as an internal control of the system. Memory stores and updates information by reviewing past records for analyzing trends and making adjustments in the system. Feedback provides information essential to the continuing effectiveness of the system. The foodservice system is an open one and, therefore, has both internal and external feedback that is critical to maintaining viability.

Concern about the declining quality of products and services has led to a revival of managerial emphasis on quality standards and control as well as quality management in all organizations. Quality assurance (QA) is outcome oriented and includes the process of defining measurable quality standards. Quality control is a reactive process predicated on follow-up and inspection and finding errors that have occurred. Quality management is a proactive process devoted to organizing and delivering service in a way that ensures an error is not made. Total quality management (TQM) is a management philosophy in which processes are refined with the goal of improving the performance of an organization. For TQM to be successful in foodservice, management must focus on the customer, empower employees, use a team approach for problem solving, and ensure a long-term commitment from the organization. Effective controls can assure profits and prevent financial losses. In the foodservice systems model, QA is a component of the control element, and TQM permeates management functions in the transformation element.

BIBLIOGRAPHY

Albrecht, K. (1990). *Service within: Solving the middle management leadership crisis*. Homewood, IL: Dow Jones-Irwin.

Albrecht, K. (1992). *The only thing that matters*. New York: HarperBusiness.

Barker, J. A. (1992). *Future edge: Discovering the new paradigms of success*. New York: William Morrow.

Beasley, M. A. (1991a). The story behind quality improvement. *Food Management, 26*(5), 52, 56–57, 60.

Beasley, M. A. (1991b). Quality improvement process—Part II. *Food Management, 26*(6), 21–26.

Boss, D. (1992). Phantoms of the workplace. *Food Management, 27*(4), 26.

Brown, W. B. (1966). Systems, boundaries and information flow. *Academy of Management Journal, 9*, 318–327.

Carter, J. R., & Narasimhan, R. (1993). *Purchasing and materials management's side in total quality management and customer satisfaction.* Tempe, AZ: Center for Advanced Purchasing Studies.

Ciampa, D. (1992). *Total quality: A user's guide for implementation.* Reading, MA: Addison-Wesley.

Cox, D. R. (1990). Quality and reliability: Some recent developments and a historical perspective. *Journal of the Operational Research Society, 41*(2), 95–101.

David, B. D. (1972). A model for decision making. *Hospitals, 46*(15), 50–55.

David, B. D. (1979). Quality and standards—The dietitian's heritage. *Journal of The American Dietetic Association, 75*, 408–412.

Davis, L. E. (1980). Individuals in the organization. *California Management Review, 22*(3), 5–14.

Davis, T. (1992). Conference report: Baldrige winners link quality, strategy, and financial management. *Planning Review, 20*(6), 36–40.

Drucker, P. F. (1985). *Managing in turbulent times.* New York: HarperCollins.

Duncan, W. J., & Van Matre, J. G. (1990). The gospel according to Deming: Is it really new? *Business Horizons, 33*(4), 3–9.

Eacho, B. (1992). Quality service through strategic foodservice partnerships: A new trend. *Hosteur, 2*(2), 22–23, 28.

Ebel, K. E. (1991). *Achieving excellence in business: A practical guide to the total quality transformation process.* Milwaukee: American Society for Quality Control, and New York: Marcel Dekker.

Eubanks, P. (1992). The CEO experience: TQM/CQI. *Hospitals, 66*(11), 24–36.

Fortuna, R. M. (1990). The quality imperative. In E. C. Huge (Ed.), *Total quality: An executive's guide for the 1990s* (pp. 3–25). Homewood, IL: Dow Jones-Irwin.

Freshwater, J. F. (1969). Future of food service systems. *Cornell Hotel and Restaurant Administration Quarterly, 10*(3), 28–31.

Garwin, D. A. (1991). How the Baldrige Award really works. *Harvard Business Review, 69*(6), 80–93.

Gift, B. (1992). On the road to TQM. *Food Management, 27*(4), 88–89.

Glover, W. G. (1988). Managing quality in the hospitality industry. *FIU Hospitality Review, 6*(1), 1–14.

Gue, R. L. (1969). An introduction to the systems approach in the dietary department. *Hospitals, 43*(17), 100–102.

Hart, W. L., & Bogan, C. E. (1992). *The Baldrige, what it is, how it's won, how to use it to improve quality in your organization.* New York: McGraw-Hill.

Heymann, K. (1992). Quality management: A ten-point model. *Cornell Hotel and Restaurant Administration Quarterly, 33*(5), 51–60.

Hockman, K. K. (1992). Taking the mystery out of quality. *Training & Development, 46*(7), 35–39.

How significant is the Baldrige? (1991). *Training, 28*(Suppl. 3), 22.

Huge, E. C. (Ed.). (1990). *Total quality: An executive's guide for the 1990s.* Homewood, IL: Dow Jones-Irwin.

Jackson, R. (1992). Continuous quality improvement for nutrition care. Amelia Island, FL: American Nutri-Tech.

Johnson, J. A., Jones, W. J., & Schilling, L. M. (1992). Commentary: Perspectives on quality service management in health care. *Journal of Allied Health*, (Winter), 23–30.

Johnson, R. A., Kast, F. E., & Rosenzweig, J. E. (1973). *The theory and management of systems* (3rd ed.). New York: McGraw-Hill.

Joint Commission on Accreditation of Healthcare Organizations. (1992). *Accreditation manual for hospitals. Volume II: Scoring guidelines*. Chicago: Author.

Juran, J. M. (1992). *Juran on quality by design: The new steps for planning quality into goods and services.* New York: Free Press.

Kanter, R. M. (1991). Transcending business boundaries: 12,000 world managers view change. *Harvard Business Review, 69*(3), 151–164.

Kapoor, T. (1991). A new look at ethics and its relationship to empowerment. *Hospitality & Tourism Educator, 4*(1), 21–24.

Kast, F. E., & Rosenzweig, J. E. (1985). *Organization and management: A systems and contingency approach* (4th ed.). New York: McGraw-Hill.

King, P. (1992). A total quality makeover. *Food Management, 27*(4), 96–98, 102, 107.

Konnersman, P. M. (1969). The dietary department as a logistics system. *Hospitals, 43*(17), 102–105.

Learning the language of quality care. (1993). *Journal of The American Dietetic Association, 93*, 531–532.

Livingston, G. E. (1968). Design of a food service system. *Food Technology, 22*(1), 35–39.

Luchsinger, V. P., & Dock, V. T. (1976). *The systems approach: A primer*. Dubuque, IA: Kendall/Hunt.

National Association of College and University Food Services. (1991). *National Association of College and University Food Services professional standards manual* (2nd ed.). East Lansing: Michigan State University.

National Restaurant Association. (1992). *The Americans with Disabilities Act: Answers for foodservice operators*. Washington, DC: National Restaurant Association and National Center for Access Unlimited.

Niven, D. (1993). When times get tough, what happens to TQM? *Harvard Business Review, 71*(3), 20–22, 24–26, 28–29, 32–34.

Odiorne, G. S. (1987). The art of crafting strategic plans. *Training, 24*(10), 94–97.

Parsons, T. (1960). *Structure and process in modern societies*. New York: Free Press.

Petit, T. A. (1967). A behavioral theory of management. *Academy of Management Journal, 10*, 341–350.

Pfau, L. D. (1989). Total quality management gives companies a way to enhance position in global marketplace. *Industrial Engineering, 21*(4), 17–18, 20–21.

Pickworth, J. R. (1987). Minding the Ps and Qs: Linking quality and productivity. *Cornell Hotel and Restaurant Administration Quarterly, 28*(1), 40–47.

Port, O., Carey, J., Kelly, K., & Forest, S. A. (1992). Quality: Small and midscale companies seize the challenge—not a moment too soon. *Business Week*, #3295, pp. 66–72.

Powers, T. F. (1978). A systems perspective for hospitality management. *Cornell Hotel and Restaurant Administration Quarterly, 19*(1), 70–76.

Puckett, R. P. (1991). JCAHO's agenda for change. *Journal of The American Dietetic Association, 91*, 1225–1226.

Puckett, R. P. (1992). Continuous quality improvement: Where are we going in health care? *Topics in Clinical Nutrition, 7*(4), 60–68.

Puckett, R. P., & Miller, B. B. (1988). *Food service manual for health care institutions*. Chicago: American Hospital Publishing.

Russo, P. M. (1993a). The health care quality imperative. *Market•Link*, *12*(11), 1–4.

Russo, P. M. (1993b). The health care quality imperative. *Market•Link*, *12*(2), 5–6.

Schmidt, W. H., & Finnigan, J. P. (1992). *The race without a finish line: America's quest for total quality*. San Francisco: Jossey-Bass.

Senge, P. (1990). *The fifth discipline: The art and practice of the learning organization*. New York: Doubleday.

Shands Hospital at the University of Florida. (1992). *Management initiative for continuous quality improvement*. Gainesville, FL: Author.

Slater, D. (1989). Coming to a truce: Ways to bridge the gap between the front and the back of the house. *Restaurants USA*, *9*(10), 26, 28–29.

Thorner, M. E., & Manning, P. B. (1983). *Quality control in foodservice* (rev. ed.). Westport, CT: AVI.

Vaughn, R. C. (1990). *Quality assurance*. Ames, IA: Iowa State University Press.

Walton, M. (1990). *Deming management at work*. New York: Putnam Sons.

Whitley, R. (1991). *The customer driven company: Moving from talk to action*. Reading, MA: Addison-Wesley.

Whyte, W. F. (1948). *Human relations in the restaurant industry*. New York: McGraw-Hill.

Williams, A. G., & Marquis, E. (1993). The "how to's" of quality assurance. *Hosteur*, *2*(2), 20–21.

Wood, T. (1993). Total quality management: Learning from other industries. *Restaurants USA*, *13*(2), 16–19.

Zabel, D., & Avery, C. (1992). Total quality management: A primer. *RQ*, *32*(2), 206–216.

3

Managing Foodservice Systems

In this chapter, management concepts are reviewed and applied to the foodservice operation. The importance of management functions in the transformation element of the foodservice system is emphasized.

THE MANAGEMENT PROCESS

In the previous chapter, **management** was defined as a process whereby unrelated resources are integrated into a total system for accomplishment of objectives. Management, involving the basic functions of planning, organizing, staffing, leading, and controlling, also was described as the primary force that coordinates the activities of subsystems within organizations. Management was explained by Hersey and Blanchard (1993) as the act of working with and through individuals and groups to accomplish organizational goals. All these definitions underscore how important it is that managerial activity be directed toward achieving the goals and objectives of the organization.

Managing Organizations

Although this book focuses on managing foodservice organizations, management concepts have broad applications because much of an individual's activity takes place within an organizational context. We are all members of a family, which is the basic unit of our society and the first organization in which most of us interact as members. We have spent a great deal of time in educational institutions as well as informal groups. Such groups develop spontaneously when several people with mutual interests pursue a common objective, which could be a shopping expedition, a fishing trip, or a picnic. Work accounts for a large part of our time; professional organizations, social clubs, and churches provide activities that encompass much of our leisure time.

The tendency to develop cooperative and interdependent relationships is a basic human characteristic. All organizations, ranging on a continuum from informal, ad hoc groups to formal, highly structured organizations, require managing. An **organization** is defined as a group of people working together in a structured and coordinated way to achieve goals (Griffin, 1993). Resources come together in an organization; the manager is responsible for coordinating them in a sensible way by acquiring, organizing, and combining resources to accomplish goals. **Management** is a set of activities (planning, organizing, staffing, leading, and controlling) that is directed at an organization's resources (human, materials, facilities, and operational) for achieving goals effectively and efficiently (Griffin, 1993). *Efficiently* means using resources wisely without unnecessary waste; *effectively* means achieving goals successfully.

Managerial Efficiency and Effectiveness

Management requires coordination of human and material resources while maintaining concern for morals, ethics, and ideals. Goals are determined by values and preferences, but the method for reaching them must be socially and morally acceptable. A manager's job is unpredictable and full of challenges, but it also is filled with opportunities to make a difference.

Many of the situations that contribute to managers' uncertainty about how to act stem from the environment in which organizations function. Of particular interest are those specific groups that are likely to affect the organization, such as owners, competitors, customers, suppliers, and regulators (Griffin, 1993). Stockholders of major limited-menu restaurant corporations are beginning to take more active roles in influencing upper-level management. Burger King, Hardee's, McDonald's, and Wendy's are competitors for customer dollars. Suppliers provide material resources to the organization and often can update managers on trends in the foodservice industry. Regulators, such as the Food and Drug Administration, have the potential to control, regulate, and influence an organization's practices.

Authority, responsibility, and accountability are concepts important to the process of management. **Authority** is delegated from the top level to lower levels of management and is the right of a manager to direct others and take actions because of his or her position in the organization. **Responsibility** is the obligation to perform an assigned activity or see that someone else performs it. Because responsibility is an obligation a person accepts, it cannot be delegated or passed to another; essentially, the obligation remains with the person who accepted the responsibility. **Accountability** is the state of being responsible to one's self, to some organization, or even to the public. In the systems context, management was described as a process for accomplishment of objectives, implying, therefore, that accountability is an integral aspect of the managerial role.

Managers must show results in an era when scarce resources are an increasing concern. Efficient and effective use of these resources to produce desired results is a requisite for a viable organization. In contemporary jargon, **efficiency** is described as "doing things right," and **effectiveness** as "doing the right things."

According to Kast and Rosenzweig (1985), managerial efficiency, the ability to get things done correctly, is an "input-output" concept. An efficient manager achieves outputs by using minimum inputs or resources. Foodservice managers who are able to reduce the cost of food products to attain goals are acting efficiently. Effectiveness, in contrast, is the ability to choose appropriate objectives; an effective foodservice manager selects the right things to accomplish certain ends, such as interviewing customers to determine if quality expectations have been met.

The foodservice manager who plans a menu featuring grilled orange roughy when the customer would prefer fish and chips may be performing efficiently but not effectively. No amount of efficiency can compensate for lack of effectiveness. Drucker (1964), one of the first management authorities to discuss efficiency and effectiveness in relation to managerial performance, stated that the question is not how to do things right but how to find the right things to do. Thus, effectiveness is at the heart of accountability.

TYPES OF MANAGERS

The term *manager* has been used up to this point to refer to anyone who is responsible for people and other organizational resources. An organization, however, has different types of managers with diverse tasks and responsibilities. Managers can be classified by the level of their jobs in the organization and also by the nature of their organizational responsibilities.

Managerial Levels

Most organizations have first-line, middle, and top managerial levels (Figure 3.1). *First-line,* or first-level, managers generally are responsible for supervising employees. In the foodservice organization, these managers usually are referred to as *foodservice supervisors,* whereas in manufacturing organizations the term frequently used is *foreman.* Functional responsibilities may be indicated as part of their title. For example, in a college residence hall foodservice, first-line supervisors may be assigned to production, service, or sanitation.

In chapter 2, a model (see Figure 2.3) was presented that showed three levels of an organization: technical, organizational, and policy-making. Figure 3.1 shows how the management levels work within the organizational levels. First-line managers function at the technical core and are responsible for day-to-day operational activities.

Middle management may refer to more than one level in an organization, depending on the complexity of the organization. The primary responsibility of middle managers is to coordinate activities that implement policies of the organization and to facilitate activities at the technical level. Middle managers direct the activities of other managers and sometimes those of functional employees. This level of management also is responsible for facilitating communication between the lower and upper levels of the organization, and it functions at the organizational level.

In a college residence hall foodservice, one or more middle managers may be responsible for coordinating each of the various units that make up the foodservice operation. Multiple foodservice centers, each with a unit manager and often an assistant manager, are essential because of the size of the campus and the convenience of the customer. Within each of the units, the first-line managers report to the unit managers, who in turn report to the campus foodservice director.

Figure 3.1. Coordination of managerial and organizational levels.

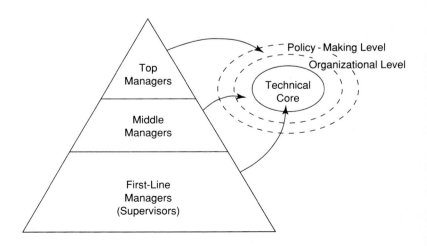

Top managers make up the relatively small group of executives that control the organization. They develop the vision for the organization's future, are responsible for its overall management, establish operating policies, and guide organizational interaction with the environment. These managers operate at the policy-making level of the organization. Multiunit managers in restaurant chains are responsible for policy implementation, sales promotions, facility appearance and maintenance, financial control, and personnel management (Umbreit, 1989).

Educational Preparation of Managers

The lower-level managers in foodservice organizations are often employees who work up through the ranks of the organization and may not have formal management training. In the noncommercial segment of the industry, however, many first-line supervisors have completed a 1-year training program for dietary managers at vocational technical schools, junior or community colleges, or by correspondence. Increasingly, in healthcare organizations, dietetic technicians are being employed as first-line supervisors because of the responsibility of patient nutritional care. Generally, a dietetic technician has completed a 2-year associate degree program.

Historically, middle- and upper-level managers in noncommercial foodservices have been the most likely to be professionally educated. The current trend, however, is to recruit managers with college degrees, especially in corporate offices of multiunit restaurant operations. Top managers of these operations or university foodservices generally are identified as chief executive officers (CEOs). Although the term *administrator* is frequently used, the titles of president or CEO are becoming more popular.

TQM Managerial Levels

Changes in the foodservice organization will be imperative if managers are planning to adopt a TQM philosophy. The traditional organization model is a pyramid with first-line managers as a base (see Figure 3.1). Inverting the pyramid provides a model for an organization committed to TQM implementation, as shown in Figure 3.2 (Gufreda, Maynard, & Lytle, 1990). Note that employees have been added to this new management model. The first-line managers and employees become the most important workers in the organization because they are producing menu items and serving customers. Top and middle managers in their new roles as planners, coaches, and facilitators should support and guide the supervisors and employees.

Top managers must focus on creating a vision for the future of the foodservice organization by developing a change strategy (Huge & Vasily, 1990). The goal of the organization will be to satisfy customers, exciting them about the food and service. This probably will not happen if the employees who prepare and serve the food are not excited, too.

Responsibilities of managers and employees will change as the structure of organizations becomes leaner and flatter, possibly causing managers to feel unrest in the organization (Kanter, 1989). According to Johnson and Frohman (1989), frustration often is evident among middle managers because of a gap in the organization struc-

Figure 3.2.   Model for an organization committed to TQM implementation.

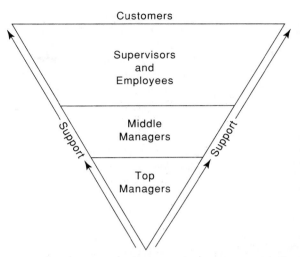

Source: Adapted from "Employee Involvement in the Quality Process" by J. J. Gufreda, L. A. Maynard, and L. N. Lytle, 1990. In *Total Quality: An Executive's Guide for the 1990s* (p. 168), edited by E. C. Huge for Ernst & Young. Homewood: IL: Dow Jones-Irwin. Reprinted by permission of the publisher.

ture. Cetron and Davies (1989) predicted that middle management will be phased out of organizations in favor of highly trained personnel.

In some cases, however, traditional top-management activities probably will need to be shared and performed by middle managers. For example, at the beginning of the quality process, top managers should appoint a steering committee to oversee the various teams that will be necessary for TQM. Team leaders will need to report their team's activities to the committee.

Middle managers should translate top management's vision and provide feedback on the status of activities being performed by first-line managers and employees. This might threaten middle managers who are used to making decisions and telling employees what to do. These managers are now asked to create an environment that encourages all employees to solve problems and make improvements and that empowers them to implement solutions (Kanter, 1989). Managers should share business or competitive information with all employees because they need to understand where their organization stands in terms of profit and loss and market share. Only then can employees make good decisions that fit into the mission of the organization (Gufreda, Maynard, & Lytle, 1990).

General and Functional Managers

Earlier in this chapter, managers were classified by their level in the organization; managers also can be classified according to the range of organizational activities for which they are responsible. In this second classification, managers can be considered either general or functional managers.

A **general manager** is responsible for all the activities of a unit. In a restaurant, everything that happens on a specific shift is the responsibility of the general manager on duty. A **functional manager** is responsible for only one area of organizational activity, such as the bar. If the bartender is absent, the bar manager must make arrangements for coverage. If, however, the bar manager is absent, the general manager would be responsible for covering the position or appointing someone to take the manager's place.

Although a small organization may have only one general manager, a larger, more complex organization may have several. A college foodservice director and unit managers and assistant unit managers at university foodservice centers typically are all considered general managers. Depending on the size of the units, two or more functional managers may be responsible for various areas of activity within each of the units.

ROLES OF MANAGERS

Mintzberg (1980) described the manager's job in terms of various roles, which he referred to as organized sets of behaviors identified with a position. He depicted the manager's position as being composed of 10 different but closely related roles, shown in Figure 3.3. The formal authority of a manager gives rise to interpersonal, informational, and decisional roles.

Interpersonal Roles

Interpersonal roles of figurehead, leader, and liaison focus on relationships. The *figurehead* role has been described by some management experts as the representational responsibility of management. By virtue of a manager's role as head of

Figure 3.3.   Managerial roles.

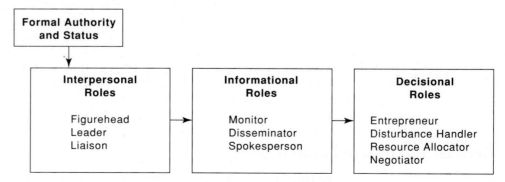

Source: Adapted from "The Manager's Job: Folklore and Fact" by H. Mintzberg, 1975, *Harvard Business Review, 53*(4), p. 55. Used by permission.

an organization or unit, ceremonial duties must be performed and may involve a written proclamation or an appearance at an important function. For example, a manager's ceremonial tasks may include greeting a group of touring dignitaries or signing certificates for a group of employees who have completed a training program.

The manager in charge of an organization or unit also is responsible for the work of the staff; this constitutes the *leader* role. Functions of this role range from hiring and training employees to creating an environment that will motivate the staff. Mintzberg (1975) contended that the influence of the manager is most clearly seen in the role of leader. Although formal authority vests the manager with great potential power, leadership determines, in large measure, how much is realized. A manager must encourage employees and assist them in reconciling personal needs with organizational goals.

The manager must also assume the interpersonal role of *liaison* by dealing with people both inside and outside the organization. Managers must relate effectively to peers in other departments of the organization and to suppliers and clients. Depending on a manager's level in the organization, responsibility for liaison relationships will vary. In Mintzberg's (1975) research, 44% of the time that company chief executives spent with people was spent with people outside their organizations. The liaison role is important in building a manager's information system.

Informational Roles

Mintzberg (1975) suggested that communication may be the most important aspect of a manager's job. A manager needs information to make sound decisions, and others in a manager's unit or organization depend on information they receive from and transmit through the manager. According to Mintzberg (1975), the **informational roles** of a manager are those of monitor, disseminator, and spokesman.

As *monitor,* the manager constantly searches for information to use to become more effective. The manager queries liaison contacts and subordinates and must be alert to unsolicited information that may result from the network of contacts previously developed. The manager collects this information in many forms and must discern implications of its use for the organization.

In the *disseminator* role, the manager transmits information to subordinates who otherwise would probably have no access to this information. An important aspect of this role is to make decisions concerning the information needs of staff members. The manager must assume responsibility to disseminate information that helps staff members become well informed and more effective.

The *spokesman* role of the manager is closely akin to the figurehead role. In the spokesman role, the manager transmits information to people inside and outside the organization or unit. For example, the director of dietetics in a hospital should keep the administrator up to date about problems in the department, and the food and beverage manager in a hotel should relay information to the general manager. The spokesman role may also include providing information to legislators, suppliers, and community groups.

Decisional Roles

The manager occupies the major role in decision making within the organization. Because of vested formal authority, a manager may commit the unit to new courses of action and determine unit strategy. As Mintzberg (1975) indicated, informational roles provide a manager with basic inputs for decision making. The **decisional roles** include those of entrepreneur, disturbance handler, resource allocator, and negotiator.

As *entrepreneur*, the manager is the voluntary initiator of change. In the informational role, the manager serves as a monitor, continually searching for new ideas that may have implications for the overall organization. In the decisional role, the manager networks by forming informal ties with other managers (Charan, 1991). The entrepreneur role may involve a decision to change the menu after networking with other restaurateurs or customers.

In the role of *disturbance handler,* the manager responds to situations that are beyond his or her control. In this role, the manager must act because the pressures of the situation are too severe to be ignored; for example, a strike looms, or a supplier fails to provide goods or services. Although a good manager attempts to avoid crisis situations, no organization is so well run or systematized that every contingency in the uncertain environment can be avoided. Disturbances may arise because poor managers ignore situations until a crisis arises; good managers also must deal with occasional crises.

As *resource allocator,* the manager decides how and to whom the resources of the organization will be distributed. In authorizing important decisions, the manager must be mindful of the needs of the unit while considering priorities of the overall operation. Such decisions often will require compromise.

In the *negotiator role,* the manager participates in a process of give and take until a satisfactory compromise is reached. Managers have this responsibility because only they have the requisite information and authority to develop complex contracts with suppliers or less formal negotiations within the organization. For example, the unit manager of a limited-menu restaurant chain might negotiate with the parent company about local advertising.

MANAGEMENT SKILLS

Katz (1974) identified three basic types of skills—technical, human, and conceptual—which he said are needed by all managers. The relative importance of these three skills varies, however, with the level of managerial responsibility. Katz defined a **skill** as an ability that can be developed and that is manifested in performance. He described the manager as one who directs the activities of others and undertakes the responsibility for achieving certain objectives through these efforts. Technical, human, and conceptual skills are interrelated, but they are examined separately in the following paragraphs.

Technical Skill

A **technical skill** involves an understanding of, and proficiency in, a specific kind of activity, particularly one involving methods or techniques. Such skill requires specialized knowledge, analytical ability, and expertise in the use of tools and procedures. Managers need sufficient technical skill to understand and supervise activities in their areas of responsibility. For example, the foodservice manager must understand quantity food production and operation of equipment. Managers must have technical expertise to develop the right questions to ask subordinates as well as the abilities to evaluate operations, train employees, and respond in crisis situations.

Human Skill

Human, or **interpersonal**, **skill** concerns working with people and understanding their behavior. Human skill, which requires effective communication, is vital to all the manager's activities and must be consistently demonstrated in actions. As Katz (1974) indicated, human skill cannot be a "sometime thing." Such skillfulness must be a natural, continuous activity that involves being sensitive to the needs and motivations of others in the organization.

Katz (1974) described two aspects of human skill: leadership within the manager's own unit and skill in intergroup relationships. This description of human skill is similar to Mintzberg's (1975) interpersonal roles of leader and liaison. Both authors emphasized the importance of a manager working effectively with staff within the organizational unit and with people outside the unit. The campus foodservice director described previously must work effectively with both the unit managers within the department and the housing director, head of maintenance, and campus purchasing director.

Conceptual Skill

Conceptual skill is the ability to view the organization as a whole, recognizing how various parts depend on one another and how changes in one part affect other parts. Conceptual skill also involves the ability to understand the organization within the environmental context; a good example is the relationship of the organization to other similar organizations and to suppliers within the community. It also includes understanding the impact of political, social, and economic forces on the organization. From this description, conceptual skill is obviously a systems approach to management. A manager needs conceptual skill to recognize how the various forces in a given situation are interrelated to ensure that decisions are made in the best interest of the overall organization.

In summary, Katz (1974) stated that effective management depends on three basic skills: technical, human, and conceptual. Adequate technical skill is needed to accomplish the mechanics of the job, sufficient human skill is necessary in working with others to enable development of a cooperative effort, and conceptual skill is required to recognize interrelationships of factors involved in the job.

Figure 3.4. Management skills necessary at various organizational levels.

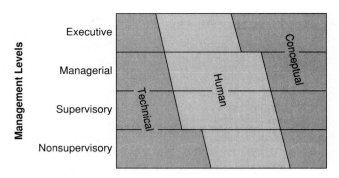

Source: Adapted from *Management of Organizational Behavior* (p. 8) by P. Hersey and K. Blanchard, 1993, Englewood Cliffs, NJ: Prentice-Hall. Copyright 1993. Used by permission.

Managerial Levels and Skills

Although all three skills are important at every managerial level, the technical, human, and conceptual skills used by managers vary at different levels of responsibility (see Figure 3.4). Technical skill is most important at the lower levels of management, identified as nonsupervisory and supervisory by Hersey and Blanchard (1993), and it becomes less important in the higher levels. The nonsupervisory level includes employees who participate in on-the-job training of other employees. The foodservice production supervisor at the supervisory level, for example, is called upon to use technical skills frequently in supervising employees in daily operations. These technical skills are important in evaluating products, in training employees, and in problem solving.

The middle manager at the managerial level uses technical skills in performing the tasks of evaluating operations and selecting employees who have appropriate skills to perform various jobs. Also, in crisis situations, the middle manager's technical skills may be called into action. Top-level managers at the executive level, although generally not involved in daily operations, need understanding of technical operations to enable effective planning.

Human skill, the ability to work effectively with others, is essential at every level of management, as reflected in Figure 3.4. The first-line manager, who is responsible for daily supervision of operating employees, must be effective in guiding and leading these individuals to accomplish the activities for which they are responsible. These employees must be motivated to produce quality products, to serve customers cheerfully, and to wash dishes properly. Morale and satisfaction are important to each employee's effective performance.

Middle managers, because of their pivotal role in the organization, must be especially accomplished in human skills. These managers must effectively lead their own groups and appropriately relate to other parts of the organization. At the top level, the manager must be equally effective in dealing with people outside the organization.

The importance of conceptual skill increases with movement up the ranks of the organization. The higher a manager is in the hierarchy, the greater the manager's involvement in broad, long-range decisions affecting large parts of the organization. At this level, conceptual skill becomes the most important one for successful performance.

MANAGEMENT FUNCTIONS

The five management functions are planning, organizing, staffing, leading, and controlling. Managers perform these functions in the process of coordinating activities of the subsystems of the organization. Now that insights have been provided into managers' many roles and responsibilities, the activities or functions of foodservice managers are examined in greater detail.

The interrelationship of management functions and the integral nature of linking processes in performing these functions are illustrated in Figure 3.5. Although in practice managers perform several management functions simultaneously, these functions are described separately. The two-way arrows in the diagram are used to illustrate the interdependency of the functions. For example, managerial plans provide the basis for control, and staff selected in the staffing function become the employees who must be led in achieving organizational objectives.

Figure 3.5. Management functions.

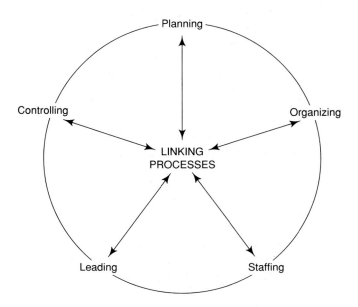

Figure 3.6. Hierarchy of plans.

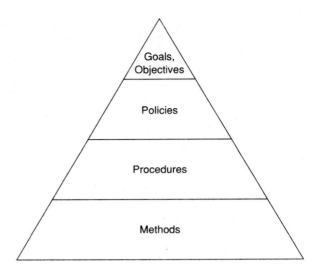

Planning

Plans, which are the result of the managerial process of planning, establish organizational objectives and set up procedures for reaching them. Plans provide for acquiring and committing resources to attain objectives and for assigning members their activities. Plans also provide standards for monitoring performance of the organization and taking corrective action when necessary.

Definition of Planning

Planning is defined as determining in advance what should happen. Planning is essential as a manager organizes, staffs, leads, and controls. For example, a supervisor or manager prepares the menu, which is a basic plan that indicates the organization of the food preparation unit (organizing), the number (staffing) and assignments (leading) of employees, and the quality and cost of the product (controlling).

A hierarchy of plans is shown in Figure 3.6. The initial plans are the goals and objectives of the organization, thus providing the basis for objectives of the various subsystems. Goals represent the desired future conditions that individuals, groups, or organizations strive to achieve (Kast & Rosenzweig, 1985). Objectives are merely goals, or end points, and set the direction for all managerial planning (Fulmer, 1988). Once objectives are determined, specific plans such as policies, procedures, and methods can be established for achieving them in a more systematic manner. Policies are the guidelines for action in an organization, and procedures and methods define steps for implementation.

An organization will have only a few broad plans but many specific plans, as depicted in Figure 3.7. District school foodservices, for example, may have only two broad goals, one concerned with provision of nutritious meals within federal and state guidelines and budgetary constraints, and another with nutrition education.

Many policies would be needed, however, to achieve these goals and assure uniformity of operations throughout the various schools in the district. An even greater number of procedures would be needed to give school foodservice employees specific instructions on implementation of policies.

Dimensions of Planning

Kast and Rosenzweig (1985) identified four dimensions of planning: repetitiveness, time span, level of management, and flexibility. In chapter 2, plans were described as internal controls of the system, and the two types of plans were defined as standing and single use. This distinction of standing versus single-use plans illustrates the planning dimension of repetitiveness. Planning can be thought of as a continuum of repetitiveness—planning for novel, one-time projects in comparison to development of policies and procedures for activities occurring repeatedly in organizations. Another dimension of planning is time span. Planning should be considered in relation to daily activities or long-range goals and objectives from the aspects of managerial level and flexibility. Plans made at the higher managerial level have a wider influence than the more specific ones made at the lower level. Some plans may be highly fixed, and others may be more flexible and capable of adaptation to a variety of changing conditions.

Repetitiveness.   Figure 3.8 shows the relationship of standing and single-use plans to organizational objectives. *Standing plans,* or plans for repetitive action, are used over and over again; *single-use plans,* also called *single purpose,* are not repeated but remain as part of historical records of the organization. Standing plans result in policies, operating procedures, methods, and rules, all of which are important to any organization. These plans, which develop into habit patterns, are a primary cohesive force connecting the various subsystems of an organization.

Figure 3.7.   Relationship of number and specificity of plans.

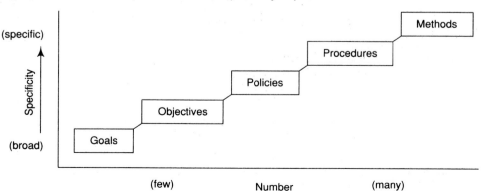

Figure 3.8. Relationship of standing and single-use plans to objectives.

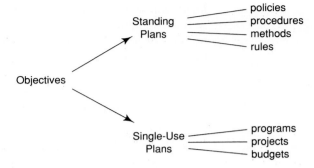

A **policy,** the broadest of the standing plans, is a general guide to organizational behavior developed by the governing body or top-level management. Organizations should have a wide variety of policies covering the most important functions. Frequently these policies are formalized and available in policy and procedure manuals. Characteristics of policies and procedures are listed in Table 3.1.

Procedures and methods establish more definite steps for the performance of certain activities and are developed especially for use at the technical level of the system. A **procedure** shows a chronological sequence of activities; a **method** is even more detailed, relating to only one step of a procedure.

Rules specify action by stating what must or must not be done, whenever or wherever they are in effect. Some examples of rules are prohibitions against smoking and the requirements to wear a specific uniform and hair restraint in the production area of a foodservice operation. A policy and its procedures for a hospital department of dietetics are shown in Figure 3.9.

An advantage of the standing plan is that it ensures uniformity of operations throughout the system. Once established, understood, and accepted, the standing plan provides similarity of action in meeting certain situations; on the negative side, however, standing plans may create resistance to change.

Management by exception is an important concept in relation to standing plans. Although standing plans serve as guidelines for decision making, upper levels of management must become involved whenever the application of policy is questioned.

Table 3.1. Characteristics of policies and procedures

Policies	Procedures
Guide decision making throughout the organization	Specify guides to action
Delimit an area within which a decision can be made	Delineate steps in descending order for completion of a task
Activate goals and objectives of the organization	Order sequential actions for performance of workers
Give direction for action	

Figure 3.9.   An example of a policy and its procedures for a small hospital.

POLICY: SCHEDULING TIME OFF

Policy Statement: Prior supervisory approval is required when scheduling time off in order for an employee to obtain a satisfactory attendance record. Work is scheduled to satisfy department workload and to accommodate employee's need for time off. This policy supplements the hospital-wide absenteeism policy.

Procedure:

1. Time off should be requested from the supervisor at least two weeks prior to the posting of the work schedule in which the absence will occur.

    *   Time off will be granted whenever possible, depending on the needs of the department and the availability of adequately trained substitute workers.
    *   In general, no more than one employee from any area will be granted time off at any one time. Requests usually will be considered on a first come, first serve basis.

2. Occasionally, time off will be requested after the work schedule has been posted. In such cases, supervisory approval will be granted for the requested time off only if ALL the following conditions are met:

    The substitute employee

    *   is asked to work by the person initiating the request;
    *   has worked the position previously;
    *   does not accrue overtime except with supervisor's permission; and
    *   is not allowed to work more than six (6) days in a row, or three (3) weekends in a row, because of the change.

Single-use, or single-purpose, plans are designed to attain specific objectives, usually within a relatively short period of time. A single-use plan in a foodservice organization might be a major program for the design, development, and construction of a central food-processing facility for a restaurant chain, a plan for a "monotony breaker" in a college residence hall foodservice, or a New Year's Eve celebration at a country club.

Time Span.   The time span for planning refers to short-range versus long-range planning. Short-range, or operational, planning covers a period of 1 year or less, and often the plan is the operating budget. Long-range planning in most organizations encompasses a 5-year cycle; however, a longer time span may be essential for some aspects of planning, such as a major building program. Long-range planning begins with an assessment of the current conditions and projections about changes. Managers must be able to see the connections between actions in one place and consequences in another (Kanter, 1992). Effective long-range planning requires a mission statement of the long-range vision of the organization.

The model for long-range planning shown in Figure 3.10 indicates the progression from premising to planning to implementing and reviewing the resulting plan.

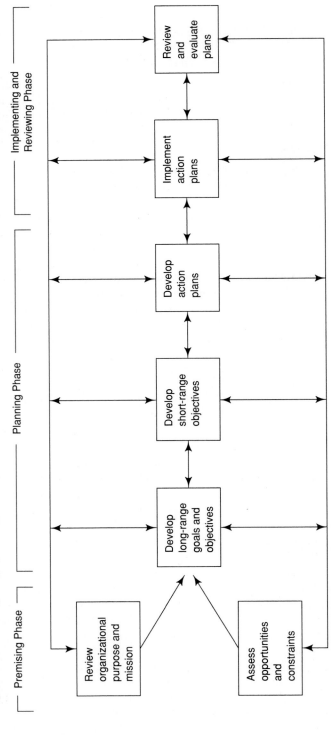

Figure 3.10. A long-range planning model.

In the premising phase, the basis for the plan is considered in terms of the mission and opportunities of the organization. The planning phase consists of developing long-range goals and objectives, short-range objectives, and action plans logically leading to implementation of the long-range plan. At this time, a final review and evaluation are necessary and may result in revision.

According to Kast and Rosenzweig (1985), plans also may be strategic. Both long-range and strategic plans are an integral part of the total planning process and establish the basic framework on which more detailed programming and operational planning take place. A **strategic plan** concentrates on decisions, not on documented plans, analyses, forecasts, and goals. Strategic planning is part of the management process and deals with decisions regarding the broad technological and competitive aspects of the organization, the allocation of resources over an extended period, and the eventual integration of the organization within the environment. According to Nanus and Lundberg (1988), a workable strategic plan is based on the evaluation and interpretation of the future in a systematic manner. Strategic planning is essential to survival in foodservice today (Feltenstein, 1992). Foodservice managers must understand the forces that are shaping the future and consider alternative strategies resulting from anticipated changes in the organization.

As used today, strategic planning has a strong connotation of overcoming obstacles, as can be seen in the derivation of the word "strategy" from the Greek *stratego*, meaning to plan the defeat of an enemy through effective use of resources. In modern business terminology, an organization must develop a competitive edge over its rivals by planning the effective use of personnel, materials, facilities, and operational resources. The outcome of the strategic planning process is a brief working document that unifies action of participants toward achievement.

An example of strategic planning is the introduction of prepackaged salads by McDonald's (Alva, 1988). Long-range planning was predicated upon the desirability of serving salads; strategic planning was involved with overcoming the flaws of salad bars, namely, the difficulty of quality control and the impossibility of service from a drive-thru window. During the strategic planning stage, McDonald's tested various ways to present salads for 10 years before launching its prepackaged products. The goal was quality control and satisfactory service of salads for its drive-thrus, which account for more than 50% of total sales. McDonald's prepackaging consists of a single or half-size serving in a transparent plastic dish with a snap-on cover and a separate foil pouch of dressing. After a massive media campaign, prepackaged salads rapidly became popular and now account for an estimated 8% of McDonald's sales. The effect of McDonald's advertising campaign was a public response that induced other limited-menu restaurant operators to follow suit. Burger King, Hardee's, Arby's, KFC, and Wendy's International currently offer prepackaged salads in response to competitive pressures.

Level of Management.   A relationship exists between the hierarchy of plans and the level of management involved in the planning effort (Figure 3.11). Generally, top managers, who function at the policy-making level of the organization, as shown in Figure 3.1, are responsible for broad, comprehensive planning involving goals and objectives. Middle managers, at the organizational or coordinative level, are responsi-

Figure 3.11.   Planning responsibilities by level of management.

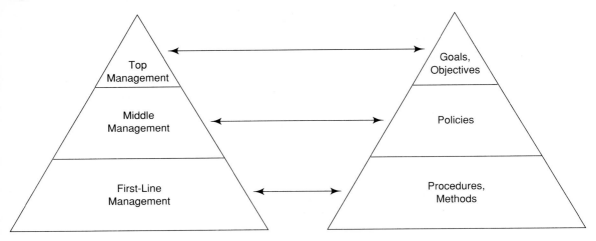

ble for developing policies; first-line managers at the technical or operational level are responsible for developing procedures and methods.

The differences in planning responsibility associated with managerial levels help explain the required skills distribution needed by managers. In the discussion of the three managerial skills—technical, human, and conceptual—the importance of conceptual skills at top levels of the organization and, conversely, of technical skills at the lower level were emphasized. The responsibility of these managers for broad versus specific operational plans should make these concepts clear. Because long-range and strategic plans are broad and time oriented, upper levels of managers must have conceptual ability, enabling them to view the overall organization in relation to its environment. Similarly, the first-line manager must be well versed on technical operations because of the responsibility for planning daily production and service activities.

To apply these concepts to a foodservice organization, the top management of a large national limited-menu restaurant company is concerned with such issues as identifying sites for new locations, assessing the impact of adding new menu items on costs, revenues, and profits, and projecting capital required for expansion. The manager of one of the units, however, is concerned with scheduling employees, predicting the impact of bad weather on customer traffic, and ordering an adequate amount of frozen yogurt for the next day. As managers move up in the organization, they must develop skills in long-range planning.

Flexibility.   One of the major considerations in planning is the permissible degree of flexibility. Long-range planning involves decision making that commits resources over an extended period of time. Rapidly changing technology, competitive and market situations, and political pressures make forecasting extremely difficult. Rigid planning at early stages involves the risk of inability to cope with changes. Organizations may have to compromise on rigidity versus flexibility by developing relatively fixed short-range operation plans and more flexible long-range strategic plans.

Organizing

After developing objectives and plans to achieve them, managers must design and develop an organization capable of activating the plan. Thus the purpose of the organizing function is to create a structure of task and authority relationships (Huber & McDaniel, 1986). **Organizing** is the process of grouping activities, delegating authority to accomplish activities, and providing for coordination of relationships, both horizontally and vertically. An overview of the organizing function is included in this chapter; organization charts, departmentalization, job descriptions, and related concepts are discussed in depth in chapter 17.

Formal Organization

The outcome of organizing is the development of the formal organization, which is usually depicted in the form of a chart. An example of an organizational chart for a university residence hall foodservice department is shown in Figure 3.12. Once managers have established objectives and developed plans to reach them, they must design an organization to activate these plans.

Different objectives will require different kinds of organizations. For example, an organization for a limited-menu restaurant operation will be far different from one for an upscale gourmet restaurant. Similarly, the organization of a 50-bed nursing home foodservice department will differ markedly from that of a 500-bed teaching hospital.

Figure 3.12.   Organization chart for a university residence hall foodservice department.

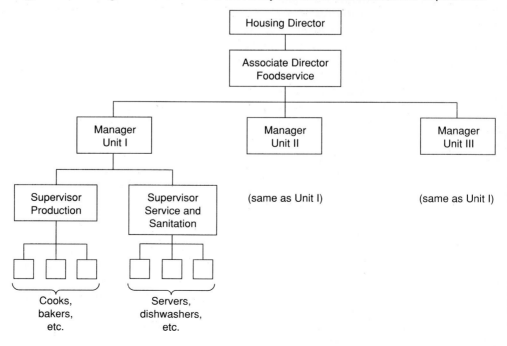

The TQM philosophy, however, forces review of organizational structure. According to Hamilton (1993), a more flexible organizational structure that removes layers of management, empowers employees, tears down communication barriers, and fosters work-force creativity is needed. Managers must have the ability to accomplish a given set of objectives. The process of organizing is determining the way in which work is to be arranged and allocated among organization members for effective attainment of goals.

Concepts of Organizing

Organizing is a dynamic process based on two concepts, span of management and authority. Span of management, often referred to as span of control, is concerned with the number of people any one person can supervise effectively. Because that number is limited, organizations must be departmentalized by areas of activity, with someone in charge of each area. Authority is the basis through which work is directed.

Span of Management.   Managers cannot supervise effectively beyond a limited number of persons, and determination of the appropriate number is difficult. Several factors are involved in determining the proper number, which is referred to as the span of management. The factors are the following:

- **Organizational policies.** Clearly defined policies can reduce the time managers spend making decisions; the more comprehensive the policies, the greater the span of management.
- **Availability of staff experts.** Managers can have increased span if staff experts are available to provide advice and services.
- **Competence of staff.** Well-trained workers can perform their jobs without close supervision, thus freeing competent managers to increase their span of management.
- **Objective standards.** In organizations with objective standards and standardized procedures, workers have a basis by which to gauge their own progress, thus allowing managers to concentrate on exceptions. As a result, larger spans are possible.
- **Nature of the work**. Less complicated work tends to require less supervision than more complicated work. Generally, the simpler and more uniform the work, the larger the possible span.
- **Distribution of work force**. The number of areas where supervised workers are on duty may inhibit severely a manager's ability to visit all work sites. The greater the dispersion of workers, the shorter is the span.

Authority.   Authority is defined as the right of a manager to direct others and to take action because of the position held in the organization. This authority is delegated down the hierarchy of the organization as designated by upper management. A sound organizing effort, therefore, includes defining job activities and scopes of authority for each position in the organization.

As indicated previously, responsibility is a concept closely related to authority and refers to an obligation for performing an assigned activity. In accepting a job, a person agrees to discharge the duties or be responsible for their accomplishment by others. Because responsibility is an obligation a person accepts, it cannot be delegated or passed to a subordinate; that is, the obligation remains with the person who accepted the job. Authority, however, must be delegated to enable individuals to carry out their responsibilities or obligations. Without proper authority, first-line and middle managers may find completion of delegated job activities difficult.

Delegation is the process of assigning job activities to specific individuals within the organization. Through this process, authority and responsibility are transferred to lower-level personnel within an organization. In a sense, delegation is the essence of management because management has been defined as getting work done with and through other people.

Failure to delegate is a weakness common to many managers. They often do not delegate enough because of the time and effort required to communicate to others. All too frequently, managers believe the saying "If you want a job done right, do it yourself."

Effective delegation, however, has many advantages. It is one of the most important means managers have of developing the potential of their subordinates; also, as subordinates accept additional responsibility, managers are freed for planning and other tasks that require conceptual skill. For effective delegation to occur, the following three elements are important:

- Specific tasks must be assigned clearly. A manager must communicate the nature of a task to ensure that a subordinate clearly understands it.
- Sufficient authority must be granted. The subordinate must be given the power to accomplish the assigned duties. This authority must be understood by the subordinate and also by others whose cooperation is required to complete the task.
- Responsibility must be created. A sense of accountability must be engendered in the subordinate to ensure responsible completion of the assigned task. The subordinate must be empowered to make decisions on quality issues to meet customer expectations.

Organizing is the division of labor. Within organizations, labor can be divided both horizontally through departmentalization and vertically through the delegation of authority. In designing organizations, **line and staff** authority relationships are created. Generally, line personnel are in a linear responsibility relationship; a superior has supervision over a subordinate. Staff personnel serve in an advisory capacity to line managers.

With the growing complexity and size of organizations, the staff role has become especially important. Staff personnel may function to extend the effectiveness of line personnel, as the administrative assistant does for an organization president. Other staff personnel provide advice and service throughout an organization. As an example, the personnel manager relieves the line manager of such functions as recruiting and screening new employees. Staff may also provide advice or services to only a segment of an organization. A dietetic consultant in a nursing home provides advice and counsel primarily to the dietetic services department.

The distinction between line and staff is not as clear as it might appear in the foregoing paragraphs. In many organizations, managers have both line and staff responsibilities. In a small foodservice operation, separate staff probably would not be available to perform such functions as personnel and quality control; thus the manager would perform these functions in addition to being responsible for production of goods and services. In larger organizations, middle managers may be advisers to various other units or departments in the organization. For example, a food and beverage manager may serve on a TQM team for an entire hotel.

Mintzberg (1989) agreed that the distinction between line and staff is becoming blurred. He stated that an innovative organization has an administrative component and an operating core. The administrative component consists of line managers and staff experts and the operating core of employees. In the past few years, administrative positions have been eliminated and the responsibility for decision making has been shifted to the operating core. CEOs, who support a TQM program, have given administrative responsibility in the form of empowerment to operation employees. Their ultimate goal is to satisfy customers while reducing costs.

Staffing

The most valuable resources of an organization are its human resources—the people who provide the organization with their work, talent, drive, and commitment. Among the most critical tasks of a manager is **staffing**: the recruitment, selection, training, and development of people who will be most effective in helping the organization meet its goals. Competent people at all levels are required to ensure that appropriate goals are pursued and that activities proceed in such a way that these goals are achieved.

In the organizing process, various jobs in the organization are defined. The staffing process then involves a series of steps designed to supply the right people to the right positions at the right time. This process is performed on a continuing basis because organizational personnel change over time due to resignation, retirement, and other reasons.

In many organizations, staffing is carried out primarily by a personnel department. The responsibility for staffing, however, lies with line managers. Every line manager, even one not involved in recruiting and selecting personnel, is responsible for training and development and other aspects of staffing.

Staffing consists of several steps (Figure 3.13): human resource planning; recruitment and selection; orientation, training, and development; performance appraisal; and compensation. Closely linked to these steps are a variety of staffing functions concerned with maintenance of the work force. These functions include promotions, demotions, transfers, layoffs, and dismissals. **Human resources planning** is designed to ensure that the organization's labor requirements are continuously met. It is a process involving both forecasting of staffing needs and analysis of labor market conditions.

Figure 3.13. Steps in the staffing process.

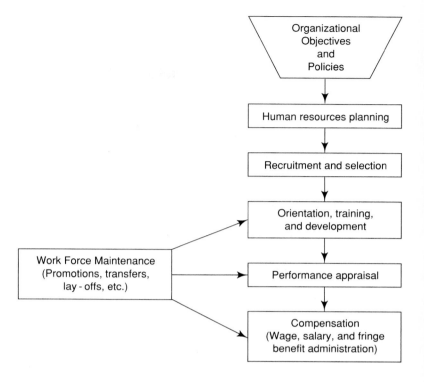

Recruitment and **selection** are concerned with developing a pool of job applicants and evaluating and choosing among them. These processes have become increasingly complex because of legislative employment mandates. They are discussed in part 6.

Orientation, training, and **development** are processes designed first to acquaint newcomers with the organization and its goals and policies and to inform them of their responsibilities. Later, training is designed to improve job skills, and development programs are utilized to prepare employees for increased responsibilities.

Performance appraisal is concerned with comparison of an individual's performance with established standards for the job. It also involves determination of rewards for high performance and corrective action to bring low performance in line with standards. Rewards may include bonuses, pay increments, or more challenging work assignments. Additional training is often necessary for low performers.

Compensation encompasses all activities concerned with administration of the wage and salary program. Fringe benefits are an important part of the program and include insurance programs, leave time, and retirement programs.

Leading

The leading function of management involves directing and channeling human effort for the accomplishment of objectives. **Leading** is the human resources function particularly concerned with individual and group behavior. All managing involves interaction with people and thus demands an understanding of how we affect and are affected by others. When managers lead, they use that understanding to accomplish tasks through the work of other members of the organization.

Management and *leadership* often are thought of as synonymous terms. Effective managers generally will be effective leaders. Leading is certainly one of the most critical functions of management, and success in management is closely related to success in leading, which also has been described as the interpersonal aspect of managing.

Leading is primarily concerned with creating an environment in which members of the organization are motivated to contribute to achieving goals. It has many dimensions, including morale, employee satisfaction, productivity, and communication. As stated in chapter 2 in the discussion of personnel satisfaction as a system output, managers must be concerned with assisting employees in achieving personal objectives at work and with coordinating personal and organizational objectives.

The traditional view of the organization is centered on chain of command, negative sanctions, and economic incentives to motivate workers. In the late 1920s, the famous Hawthorne studies conducted by Western Electric and Harvard University researchers revealed that the factors that influenced worker performance included such things as social and psychological conditions, informal group pressure participation in decision making, and recognition (Roethlisberger & Dickson, 1956). Since that time, the behavioral sciences have added new dimensions to the understanding of motivation and behavior in the workplace. Today, leading is viewed as being concerned with interpersonal and intergroup relationships. The role of the manager includes influencing these relationships to create cooperation and enlist commitment to organizational goals. The evolution of a better-educated work force today has significantly increased the use of participative management in organizations. In chapter 19, these and other aspects of leadership, including factors affecting leadership style, are described in more detail.

Controlling

Controlling is the process of ensuring that plans are being followed. It involves comparing what should be done with what was done and then taking corrective action, if necessary. Controlling must be a continuous process that affects and is affected by each of the other managerial functions. For example, the goals, objectives, and policies established in the planning process become control standards. When comparing performance to these standards, the need for new goals, objectives, and policies may become obvious. Effective organizing, staffing, and leading result in more effective control. Likewise, more effective control also leads to better

Figure 3.14.   The planning-controlling cycle.

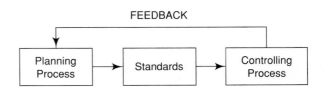

Figure 3.15. The controlling function of management.

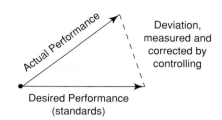

Actual Performance

Deviation, measured and corrected by controlling

Desired Performance
(standards)

organizing, staffing, and leading. Within this interrelatedness and interdependency of managerial functions, controlling relates most closely to planning.

Standards created in the planning process define the dimensions of what is expected to happen. These expected performance standards are the criteria that managers use to control performance; in turn, feedback from the controlling process is the information managers use to evaluate and adjust plans (Figure 3.14). The controlling function of management involves the following three steps, as depicted in Figure 3.15:

- Measuring actual performance and comparing it with desired performance or standards
- Analyzing deviations between actual and desired performance and determining whether or not deviations are within acceptable limits
- Taking action to correct unacceptable deviations

Taking corrective action is a process that cuts across both the leading and controlling functions because many deviations from expected standards are related to performance of personnel. For example, fewer portions than expected from a particular recipe might be caused by a foodservice worker using an inappropriate portioning tool. Appropriate controls within each of the functional subsystems of a foodservice organization are discussed in chapters 7 through 16.

SUMMARY

Management is the process by which unrelated resources are integrated into a system for the accomplishment of an organization's objectives. It is also the primary force within organizations that coordinates the activities of subsystems, involving the functions of planning, organizing, staffing, leading, and controlling. Managers serve as change agents and continuously evaluate how changes in social, economic, political, and technological conditions will affect the organization.

In most organizations, first-line, middle, and top levels of management exist. If a total quality management philosophy is adopted, the goal of the foodservice organization will be to provide quality in products and services that meet the expectations of customers, and employees will be empowered to function as first-line managers. Managers can also be classified as general, with responsibility for all activities of a unit, or functional for only one area of activity. The manager's job can be described in terms of various roles or organized sets of behaviors identified with a position. Mintzberg's 10 different but closely related roles are categorized into interpersonal, informational, and decisional roles. All managers need three basic skills—technical, human, and conceptual.

Planning, the first management function, is defined as determining in advance what should happen. Short-range plans cover 1 year or less, and long-range plans generally encompass a 5-year cycle. Strategic planning is always long range and concentrates on decisions and not on documented plans, analyses, forecasts, and goals. Organizing is based on span of management, which is concerned with the number of people any one person can supervise effectively, and authority is defined as the right of a manager to direct others and take action.

Staffing includes the recruitment, selection, training, and development of people who will be most effective in helping the organization meet its goals. The staffing function is performed by the personnel department, but the final responsibility for training and development of workers lies with first-line managers especially. The leading function involves directing and channeling human effort for the accomplishment of objectives. Leading is concerned primarily with creating an environment in which members of the organization are motivated to contribute to achievement of goals. Controlling, the process of ensuring that plans are being followed, involves comparing what should be done with what was done and then taking corrective action. It is a continuous process that affects and is affected by each of the other managerial functions. These five management functions performed by managers aid in coordinating the subsystems and facilitating transformation of resources into desired outputs.

BIBLIOGRAPHY

Alva, M. (1988). The death of the salad bar. *Nation's Restaurant News, 22*(17), F7–F8.

American Institute of Management. (1959). *What is management?* New York: American Management Association.

Boettinger, H. M. (1975). Is management really an art? *Harvard Business Review, 53*(1), 54–55, 57–59, 61, 63–64.

Cetron, M., & Davies, O. (1989). *American renaissance: Our life at the turn of the 21st century.* New York: St. Martin's Press.

Charan, R. (1991). How networks reshape organizations—for results. *Harvard Business Review, 69*(5), 104–115.

Donnelly, J. H., Gibson, J. L., & Ivancevich, J. M. (1990). *Fundamentals of management* (7th ed.). Homewood, IL: BPI/Irwin.

Drucker, P. F. (1964). *Managing for results.* New York: Harper & Row.

Feltenstein, T. (1992). Strategic planning for the 1990s: "Exploring the inevitable." *Cornell Hotel and Restaurant Administration Quarterly, 33*(3), 50–54.

Ferguson, D. H., & Berger, F. (1984). Restaurant managers: What do they *really* do? *Cornell Hotel and Restaurant Administration Quarterly, 25*(1), 27–38.

Fulmer, R. M. (1988). *The new management* (4th ed.). New York: Macmillan.

Griffin, R. W. (1993). *Management* (4th ed.). Boston: Houghton Mifflin.

Gufreda, J. J., Maynard, L. A., & Lytle, L. N. (1990). Employee involvement in the quality process. In E. C. Huge (Ed.), *Total quality: An executive's guide for the 1990s* (pp. 162–176). Homewood, IL: Dow Jones-Irwin.

Hamilton, J. (1993). Toppling the power of the pyramid. *Hospitals, 67*(1), 38, 40–41.

Hart, C. W. L., Spizizen, G. S., & Muller, C. C. (1988). Management development in the foodservice industry. *Hospitality Education and Research Journal, 12*(1), 1–20.

Hersey, P., & Blanchard, K. H. (1993). *Management of organizational behavior: Utilizing human resources* (6th ed.). Englewood Cliffs, NJ: Prentice-Hall.

Hoover, L. W. (1983). Enhancing managerial effectiveness in dietetics. *Journal of The American Dietetic Association, 82*,58–61.

Huber, G. P., & McDaniel, R. R. (1986). The decision-making paradigm of organizational design. *Management Science, 32*(5), 572–589.

Huge, E. C., & Vasily, G. (1990). Leading cultural change: Developing vision and change strategy. In E. C. Huge (Ed.), *Total quality: An executive's guide for the 1990s* (pp. 54–69). Homewood, IL: Dow Jones-Irwin.

Hunt, V. D. (1992). *Quality in America: How to implement a competitive quality program.* Homewood, IL: Business One Irwin.

Johnson, L. W., & Frohman, A. L. (1989). Identifying and closing the gap in the middle of organizations. *Academy of Management Executive, 3*(2), 107–114.

Jones, P. (1993). Foodservice operations management. In M. Khan, M. Olsen, & T. Var (Eds.), *VNR's encyclopedia of hospitality and tourism* (pp. 27–36). New York: Van Nostrand Reinhold.

Kanter, R. M. (1989). The new managerial work. *Harvard Business Review, 67*(6), 85–92.

Kanter, R. M. (1992). The long view. *Harvard Business Review, 70*(5), 9–11.

Kast, F. E., & Rosenzweig, J. E. (1985). *Organization and management: A systems and contingency approach* (4th ed.). New York: McGraw-Hill.

Katz, R. L. (1974). Skills of an effective administrator. *Harvard Business Review, 52*(5), 90–102.

Lefever, M. M. (1989). Multi-unit management: Working your way up the corporate ladder. *Cornell Hotel and Restaurant Administration Quarterly, 30*(1), 61–67.

Logan, H. H. (1966). Line and staff: An obsolete concept? *Personnel, 43*(1), 26–33.

Mintzberg, H. (1975). The manager's job: Folklore and fact. *Harvard Business Review, 53*(4), 49–61.

Mintzberg, H. (1980). *The nature of managerial work.* Englewood Cliffs, NJ: Prentice-Hall.

Mintzberg, H. (1987). Crafting strategy. *Harvard Business Review, 65*(4), 66–75.

Mintzberg, H. (1989). *Mintzberg on management: Inside our strange world of organizations.* New York: Free Press.

Nanus, B., & Lundberg, C. (1988). Strategic planning. *Cornell Hotel and Restaurant Administration Quarterly, 29*(2), 18, 20–23.

Odiorne, G. S. (1987). The art of crafting strategic plans. *Training, 24*(10), 94–97.

Palacio, J. P., Spears, M. C., Vaden, A. G., & Dayton, A. D. (1985). The effect of organizational level and practice area on managerial work in hospital dietetic services. *Journal of The American Dietetic Association, 85,* 799–806.

Peters, T. (1992). *Liberation management: Necessary disorganization for the nanosecond nineties.* New York: Knopf.

Primozic, K. I., Primozic, E. A., & Leben, J. (1991). *Strategic choices: Supremacy, survival, or sayonara.* New York: McGraw-Hill.

Roethlisberger, F. J., & Dickson, W. J. (1956). *Management and the worker.* Cambridge, MA: Harvard University Press.

Semler, R. (1989). Managing without managers. *Harvard Business Review, 67*(5), 76–84.

Sultemeier, P. M., Gregoire, M. B., & Spears, M. C. (1989). Managerial functions of college and university foodservice managers. *Journal of The American Dietetic Association, 89*(7), 924–928.

Tse, E. C., & Olsen, M. D. (1988). The impact of strategy and structure on the organizational performance of restaurant firms. *Hospitality Education and Research Journal, 12*(2), 265–276.

Ulrich, D., & Wiersema, M. E. (1989). Gaining strategic and organizational capability in a turbulent business environment. *Academy of Management Executive, 3*(2), 115–122.

Umbreit, W. T. (1989). Multi-unit management: Managing at a distance. *Cornell Hotel and Restaurant Administration Quarterly, 30*(1), 53–59.

Uyterhoeven, H. E. R. (1992). General managers in the middle. *Harvard Business Review, 70*(2), 75–85.

West, J. J. (1990). Strategy, environmental scanning and firm performance: An integration of content and process in the foodservice industry. *Hospitality Research Journal, 14*(1), 87–100.

West, J. J., & Olsen, M. D. (1988). Environmental scanning and its effect upon firm performance: An exploratory study of the foodservice industry. *Hospitality Education and Research Journal, 12*(2), 127–136.

4

Marketing Foodservice

Marketing has become a philosophy and a way of doing business in many industries, including foodservice. It has a much wider scope than advertising and selling, which often have been erroneously equated with marketing. These activities are only subsets of marketing and may not be necessary, as shown by the many successful industries that never advertise or practice direct selling. For example, only in recent years has the Hershey Corporation advertised its chocolate products. Once advertising was initiated, Hersey utilized every medium, from magazines, trade journals, and newspapers to radio and television to create customers.

Managers in both the hospitality and healthcare industries currently recognize marketing as a component of management. Marketing, long defined in terms of the product, is currently being expressed in terms of the customer. As an integral part of planning, organizing, staffing, leading, and controlling, marketing assists in the transformation of resources for providing quality meals for customer satisfaction.

DEFINITION OF MARKETING

Marketing has been defined in many ways, as have most developing disciplines. The definition given by Pride and Ferrell (1993) has greater significance for foodservice operations than most other definitions because it is not limited to business organizations. **Marketing** consists of individual and organizational activities that facilitate and expedite satisfying exchange relationships in a dynamic environment through the creation, distribution, promotion, and pricing of goods, services, and ideas.

Marketing Products

Many activities are needed to market products, or goods. Producers, sellers, and buyers of products are all involved in marketing. Because of escalating costs in products and labor in the past decade, noncommercial foodservice managers have become cognizant of the value of using marketing principles. Competition for survival has become a priority for healthcare organizations, universities, and other institutions. Commercial foodservice managers early learn to be competitive because of the high failure rate of restaurants.

For an *exchange* to occur between two or more individuals or organizations, each must be willing to give up "something of value" for "something of value" (Pride & Ferrell, 1993). Both the buyer and seller have to communicate with each other to make their "something of value" available, as shown in Figure 4.1. In most situations, the seller has products, and the buyer has financial resources such as money or credit. In an exchange, products are traded for other products or money. The exchange must be satisfying to both the buyer and the seller. The buyer must be pleased with the product received from the seller and the seller with the reimbursement.

Figure 4.1.   Exchange between buyer and seller.

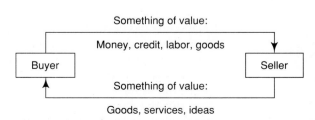

Source: Pride, William, and O. C. Ferrell, *Marketing: Concepts and Strategies*, Eighth Edition. Copyright © 1993 by Houghton Mifflin Company. Used by permission.

Marketing Environment

The **marketing environment** consists of all the forces outside the organization that influence marketing activities and exchanges. Uncontrollable forces include consumers, competition, government, the economy, technology, and media (Evans & Berman, 1992). An organization may have to change a marketing plan to comply with the uncontrollable environment. According to Pride and Ferrell (1993), marketing focuses both on making the product available at the right place at the right time at a price that is acceptable to customers and in assisting customers in determining if the product meets their needs. In the definition of marketing, a *product* is a good, service, or an idea, each of which will be discussed later in this chapter.

MARKETING CONCEPT

The **marketing concept** is a management philosophy that affects all activities of an organization and emphasizes satisfying customers' needs. First, a producer must determine what will satisfy a customer and then create products and sell them to that customer. This process must continue to keep pace with customers' changing desires and preferences. Satisfying the customer, however, cannot be the only objective. A producer must achieve overall goals of increasing profits, marketing shares, or sales to stay in business. Implementing the marketing concept should benefit the organization as well as its customers.

Evolution of Marketing

Satisfying customers has not always been the philosophy of business. This concept was preceded by the production and sales eras. During the late 1800s, the Industrial Revolution marked the beginning of the modern concept of marketing. With mass production, better transportation, and more efficient technology, products could be manufactured in greater quantities and sold at lower prices. In the initial stages of the Industrial Revolution, output was limited, and marketing was devoted to the physical distribution of products. Because demand was high and competition low, businesses did not have to conduct consumer research, modify products, or otherwise adapt to consumer needs. The production era of marketing had the goal of increasing production to keep up with demand (Evans & Berman, 1992).

Once a company was able to maximize its production capabilities, it hired a sales force to sell its inventory. Beginning in the 1920s, the sales era appeared and customers' desires were altered to accept the products being produced. Business executives decided that advertising and sales were the major means of increasing profits.

By the early 1950s, business executives recognized that efficient production and promotion of products did not guarantee that customers would buy them. The marketing era began with the creation of the marketing department to conduct consumer research and advise management in how to design, price, distribute, and promote products. During the past 20 years, marketing managers have been represented on organizations' decision-making teams because of their ability to conduct consumer research.

Competition is intense, and companies must draw sophisticated customers to their products and retain them (Evans & Berman, 1992). According to Pine (1993), today's competitive edge is found in the concept of mass customization, which is the development, production, marketing, and delivery of affordable goods and services with enough variety that most people will find exactly what they want. For example, Burger King's advertising campaigns have included anti-mass-production slogans such as "Have It Your Way!" and "Sometimes You've Gotta' Break the Rules." The marketing concept, with its emphasis on satisfying the customer, thus forms the basis of the marketing era.

Implementation of Marketing

Once the management of an organization has adopted a marketing philosophy, the development and implementation of that philosophy is based on the marketing concept. According to Lewis and Chambers (1989), the marketing concept derives from the premise that the customer is king, has a choice, and does not have to buy the product. Thus the best way to earn a profit is to serve the customer better.

The marketing concept affects all types of business activities and should be adopted entirely by top-level management. These executives must incorporate the marketing concept into their personal philosophies of business management so completely that customers become the most important concern in the organization. Support of managers and employees at all levels of the organization is required for implementation of the marketing concept (Pride & Ferrell, 1993).

First, management must establish an information system to determine the customers' real needs and use the information to develop products that satisfy them. This is expensive and requires money and time to make the organization customer oriented. Second, the organization must be restructured to coordinate all activities. The head of the marketing department should be a member of the top-level management team in the organization (Pride & Ferrell, 1993).

Problems can occur with this new marketing approach. Most operations cannot make products specific to the needs of each customer in our mass-production economy. Regardless of the great amount of time and money spent for research, products still are produced that do not sell. Occasionally, satisfying one segment of the population makes another dissatisfied. Limited-menu restaurants early realized that one menu would not appeal to all members of a family. Applebee's Neighborhood Grill and Bar and Denny's have capitalized on this fact by recognizing that young families will become repeat customers if children are occupied and prices are low (Chaudhry, 1993). Applebee's has developed a larger, more colorful menu, new games, and greater flexibility in menu choices that let restaurant managers choose food items that please their customers. Denny's has lowered food prices by 25% for children's entrees.

Employee morale might decrease during the organizational restructuring required for coordinating activities in all departments. If the foodservice manager has adopted the TQM philosophy, however, employees empowered to make customer satisfaction decisions could be very helpful in the transition.

MARKETING MANAGEMENT

Marketing management is a process of planning, organizing, implementing, and controlling marketing activities to facilitate and expedite exchanges effectively and efficiently (Pride & Ferrell, 1993). The managerial functions in the transformation element of the foodservice system have an important role in marketing management. *Effectiveness* refers to the degree to which an exchange helps to achieve an organization's objectives; the quality of the exchanges may range from highly desirable to highly undesirable. *Efficiency* refers to the minimization of resources that an organization must spend to achieve a specific level of desired exchanges. Pride and Ferrell (1993) summarized these definitions by stating that the overall goal of marketing management is to facilitate highly desirable exchanges and to minimize, as much as possible, the costs of doing so.

To achieve the goal of facilitating and expediting desirable exchanges, marketing management is responsible for developing and managing marketing strategies. Strategy encompasses key decisions for reaching an objective. A **marketing strategy** pertains to the selection and analysis of a group of people, identified as a target market, which the organization wants to reach, and includes the creation and maintenance of an appropriate **marketing mix** that will satisfy those people.

Marketing Mix

To manage marketing activities, managers must deal with variables relating to the marketing mix and the marketing environment. The marketing mix is defined as the specific combination of marketing elements used to achieve an organization's objectives and satisfy the target market (Evans & Berman, 1992). The **target market** is a group of persons for whom an organization creates a marketing mix that specifically meets the needs of that group (Pride & Ferrell, 1993). An organization has control over the marketing mix decision variables, which are constructed around the buyer, but it has limited control over the marketing environment, as shown in Figure 4.2. A target market must be identified before a marketing mix can be developed. Resources must be scrutinized carefully and organizational goals analyzed to verify that the marketing mix consists of four elements: product, distribution, promotion, and price. Many decisions have to be made concerning the activities required for each element included in the mix.

Product

As stated in the definition of marketing, a *product* can be a good, a service, or an idea. Even though the manufacturing of products is not a marketing activity, research

Figure 4.2.   Components of the marketing mix and marketing environment.

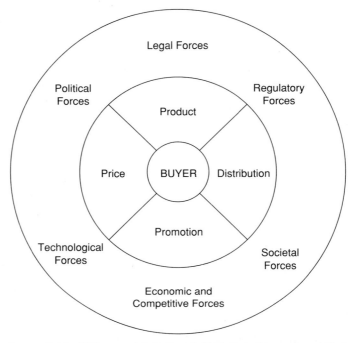

Source: Pride, William, and O. C. Ferrell, *Marketing: Concepts and Strategies*, Eighth Edition. Copyright © 1993 by Houghton Mifflin Company. Used by permission.

on customer needs and product designs is. The marketing manager must be sensitive to these needs and must be able to design new products, modify old ones, or eliminate those that are no longer desired by the customer or are not making a profit.

Distribution

Products must be *distributed* properly for availability at the right time in accessible locations. A marketing manager also has to make products available to customers in the quantities needed and, at the same time, keep inventory, transportation, and storage costs as low as possible.

Promotion

This element is used to facilitate exchanges by informing prospective customers about an organization and its products. *Promotion* is used to increase public awareness about a new product or brand; also, it is used to renew interest in a product that is waning in popularity.

Price

Marketing managers usually are involved in establishing pricing policies for various products because consumers are concerned about the value obtained in the exchange. *Price* is a critical component of the marketing mix and is often used as a competitive tool. Price also helps establish a product's image.

Dodd (1992), by substituting the word *place* for distribution, suggested that a fifth *p, pizzazz,* should be added to the marketing mix. She stated that consumers of the 90s are demanding and getting choice. An individual's perception of product worth may be the deciding factor in choosing a product. The energy, creativity, style, or flair that a product or service represents to customers affects perceived value. A favorable first impression thus may lead to customer satisfaction and repeat business.

Environmental Forces

As shown in Figure 4.2, the marketing environment surrounds the buyer and the marketing mix (Pride & Ferrell, 1993). Political, legal, regulatory, societal, economic and competitive, and technological forces in the environment affect the marketing manager's ability to facilitate and expedite change. The marketing environment influences customers' preferences and needs for products. These forces also directly influence how a marketing manager should perform certain marketing activities. Finally, a manager's decisions may be affected by environmental forces that influence customers' reactions to the organization's marketing mix.

Political forces influence the economic and political stability in the country as well as decision making, which in turn affects domestic matters, negotiation of trade agreements, and determination of foreign policy. Political trends can have tremendous impact on the hospitality and healthcare industries. Organizations such as the National Restaurant Association, American Hotel and Motel Association, American Dietetic Association, and School Food Service Association maintain lobbyists in Washington, D.C.

Legal forces are responsible for legislation and interpretation of laws. Marketing is controlled by numerous laws designed to preserve competition and protect the consumer; interpretation of laws by the marketers and courts has a great effect on marketing mix components.

Local, state, and federal regulatory forces develop and enforce regulations that can affect marketing decisions. Quite often, regulatory agencies, especially at the federal level, encourage industries to develop guidelines to stop questionable practices. Industry leaders usually cooperate in order to avoid government regulations. Individual industries and trade organizations also put regulatory pressures on themselves and their members.

Societal forces cause marketers to be responsible for decisions. Thousands of consumer groups have been formed to discuss such issues as environmental pollution and the use of pesticides on fruit and vegetable crops. Consumer groups also have been active in discussions on food labeling.

Economic forces have a major influence on competition, which is affected by the number of industries controlling the supply of a product, the ease by which a new operation can enter the industry, and the demand for the product relative to the supply. Demand is determined by buyers' abilities and willingness to purchase.

Technological forces have an impact on everyday living, influencing consumers' desires for products and desires for products and the stability of the marketing mix. The technologies of communication, transportation, computers, and packaging influence the types of products being produced.

Managers must be able to adjust marketing strategies to major changes in the environment. If they want to develop effective strategies, managers must recognize the dynamic environmental forces that cause marketing problems and opportunities.

Market Segmentation

Swinyard and Struman (1986) stated that separating customers into "natural" market groups provides the basis for successful strategy development in marketing a restaurant. According to NPD Crest, the Illinois-based research firm, the key market segments of the future will be the middle-aged and older customers; these segments currently account for 26% of all restaurant business (Allen, 1992). A market is extremely difficult to keep once it has been captured. Capturing a market occurs when foodservice managers determine that what they have to sell is what their customers want (Vance, 1992). According to Sullivan (1991), the most important cost in a restaurant or bar is an empty chair.

Market segmentation is the process of dividing a total market into groups of people with similar product needs (Pride & Ferrell, 1993). A market, in the context used here, is not a place but rather a group of people; as individuals or organizations, the group needs products and possesses the ability, willingness, and authority to purchase them. A market segment is a mixture of individuals, groups, or organizations that shares one or more characteristics, which causes them to have similar product needs.

In a homogeneous market, which consists of individuals with similar product needs, a marketing mix is easier to design than one for a heterogeneous group with dissimilar needs. Choosing the correct variable for segmenting a market is important in developing a successful strategy. Variables have been grouped into four categories for the segmentation process: demographic, geographic, psychographic, and product related.

Demographic Variables

Demographic variables consist of population characteristics that might influence product selection, for example, age, sex, race, ethnicity, income, education, occupation, family size, family life cycle, religion, social class, and price sensitivity. Only those demographic variables pertinent to the population segment under consideration must be ascertained.

A useful classification system for segmenting customers in terms of a broad range of demographic and life-style factors is the **Values and Life-Styles (VALS) research program**, which is sponsored by SRI International in Menlo Park, California. The current version of the program tries to predict customer behavior by defining segments, based on self-orientation and resources, according to product use. Customers are asked what they do with their time and money. *Self-orientation* includes the attitudes and activities that help people define their social self-image. *Principle-oriented* customers make choices by beliefs and principles; *status-oriented* customers are influenced by actions, approval, and opinions of others; and *action-oriented* customers desire social or physical activity, variety, and risk taking (Evans & Berman, 1992).

Geographic Variables

Geographic variables include climate, terrain, natural resources, population density, and subcultural values that influence customers' product needs. In addition, the size of the region, city, county, or state and whether the area is urban, suburban, or rural have an effect on the market. Population in these areas, customer preferences, and spending patterns also need to be considered in marketing decisions. As an example of geographical variables, Cajun food, which is part of the culture in Louisiana, has been accepted in New York City and in California. The astute foodservice manager knows when and where to introduce such food items on the menu by doing a market analysis before making a decision.

Psychographic Variables

Psychographic variables include many factors that can be used for segmenting the market, but the most common are motives and life-styles. When a market is segmented according to a motive, it is done in recognition of the reason a customer makes a purchase. A current example is the emphasis on eating healthful foods and the inclusion on menus of foods that are high in fiber and low in fat, sodium, and calories. Life-style segmentation categorizes people according to what is important to them, for example, entertaining at home instead of going out for dinner. In recognition of this trend, many foodservice managers have added a catering component, which has become a profit generator in their operations. The use of psychographic segmentation has been limited because measurement of success is difficult.

Behavioristic Variables

Behavioristic variables are the basis of some feature of consumer behavior toward a product. To satisfy a specific group of customers, a special product might have to be produced, for example caffeine-free diet cola. How the customer uses a product also may determine segmentation. Frozen menu items are being packaged in single servings to meet the needs of people living alone.

SERVICE MARKETING

In the definition of marketing used in this chapter, a product refers to a good, a service, an idea, or any combination of the three. A good is a tangible product that a customer can physically touch; a service is the application of human or mechanical efforts to people or objects. The most satisfactory definition of service is that given by Kotler (1991): A service is any act or performance that one party can offer to another that is essentially intangible and does not result in the ownership of anything.

Manufacturing firms still exist, but services have replaced goods in the U.S. economy, accounting for 53% of the **Gross Domestic Product (GDP)** and 56% of consumer expenditures ("National Income," 1992). Several reasons for this growth in the services sector are apparent. The United States now is an information rather than an industrial society; services are the primary material of an information society. Services are no longer by-products in a manufacturing or production process, but often are the products. A third reason is the prosperity in the United States that has led to growth in financial, travel, entertainment, and personal-care services. Life-style changes, such as more women in the work force, have created increased demand for child-care, domestic, and other time-saver services including meal preparation. The elderly require increased healthcare services, and at the same time the younger population demands increased fitness and health maintenance services. The technology explosion has created more complex goods that require servicing and repair; the business environment is more specialized, justifying more business and industrial services.

Lewis and Chambers (1989) emphatically stated that marketing and management in a service business, such as the hospitality industry, are one and the same. Many marketing experts argue that service marketing is different from goods marketing and requires different strategies and tactics and that a pure good without some elements of service attached to it is impossible.

Characteristics of Services

Zeithaml, Parasuraman, and Berry (1985) stated that the problems of service marketing are not the same as those of goods marketing. A look at the four basic characteristics of service marketing—intangibility, inseparability of production and consumption, perishability, and heterogeneity—explains why. Services generally are sold before they are produced, and goods generally are produced before they are sold (Berry & Parasuraman, 1991). Moreover, services marketing has a more limited influence on customers before purchase than goods marketing.

Intangibility

Generally, **intangibility of services** is defined in terms of what services are not; they cannot be seen, touched, tasted, smelled, or possessed (Pride & Ferrell, 1993). Services are performances, and therefore, intangible; products are tangible. Services, however, have a few tangible attributes. **Atmospherics** is an example of a tangible attribute and has been used as a marketing tool for many years.

Atmospherics describes the physical elements in an operation's design that appeal to customers' emotions and encourages them to buy (Pride & Ferrell, 1993). Berry and Parasuraman (1991) stated that the principal responsibility for the service marketer is to manage tangibles to convey the proper signals about service.

The atmosphere of the exterior and interior of the operation may be friendly, exciting, quiet, or elegant. Exterior atmospherics is important to new customers who often judge a restaurant by its outside appearance. If windows are foggy or the grounds unkempt, customers might decide that service would be unacceptable, too. Interior atmospherics includes lighting, wall and floor coverings, furniture, and rest rooms. A pleasing and clean interior probably would indicate to the customer that the service would be impeccable. Sensory elements also contribute to atmosphere. Color can attract customers. Many limited-menu restaurants, for example, use bright colors, such as red and yellow, because they have been shown to make customers feel hungrier and eat faster, thus increasing turnover. Sound is important; a very noisy restaurant probably would not be chosen by customers who are celebrating a wedding anniversary. Odor also might be relevant; the scent of freshly baked bread makes customers feel cared for and wanted. Again, these tangible sensory elements might reflect on the customer's perception of the intangible service (Pride & Ferrell, 1993).

Levitt (1981) stated that the most important thing to know about intangible service is that "customers don't know what they're getting until they don't get it." He continued his explanation by saying that a customer is more precious than the tangible assets shown on a balance sheet. Companies that understand this concept continually tailor products and services to their customers. They look beyond the value of a single transaction to the customer's lifetime value to the company (Treacy & Wiersema, 1993).

The intangibility of a service creates a dilemma for one who wishes to market it. Often service providers can induce an aura of tangibility by using illustrations to improve a prospective customer's confidence (Kotler, 1991). Television spots can be used to introduce a new menu item. As an example, Dave Thomas, the founder of Wendy's, makes a chicken cordon bleu signature sandwich (that is a sandwich for which the company is known) for his daughter, Wendy, but she doesn't answer his call to try it out. Dave, with a big smile on his face, eats the sandwich. Jim Near, president of Wendy's International and *Restaurants & Institutions'* 1992 CEO of the Year, said, "Dave's so honest, so benevolent. He really is. What you see is what you get. He's the best promoter, flag bearer, standard bearer that you could have" (Chaudhry, 1992).

Inseparability

According to Pride and Ferrell (1993), **inseparability** of production and consumption is related to intangibility. Services are normally produced at the same time they are consumed. In a commercial foodservice operation, the waitstaff, bartender, and **maître d'hôtel** are producing services at the same time the customer is consuming them. The knowledge and efficiency of the waitstaff in taking the order and serving the meal, the desire of the bartender to mix a drink exactly the way the customer wants it, and the concern of the maître d' that the customer is satisfied are examples of inseparability of production and consumption. In hospitals, foodservice personnel deliver trays to patients who are either satisfied or dissatisfied immediately with the attitude or concern of the delivery person. Likewise, the warmth of a smile from a cook serving a child a school lunch is strongly associated with the acceptance of the meal.

Perishability

The **perishability of services** means that those services cannot be stored for future sale (Evans & Berman, 1992). Unused capacity cannot be shifted from one time to another. Because service is produced and consumed simultaneously, it is perishable. The service supplier must try to regulate customer usage to develop consistent demand throughout various periods. The service operation must have the capacity and capability to produce when demand occurs; if demand does not occur, however, that capacity and capability are lost and wasted, resulting in losses in the bottom line (Lewis & Chambers, 1989). For example, if overstaffing occurs, the labor cost is too high; if understaffing occurs and demand increases, service becomes too slow.

One alternative is to charge prices high enough to permit overstaffing. Restaurant managers of operations known for a high level of service often use this scheme. Most foodservices cannot afford such a solution to the problem, and the result is irate customers. Reduction of staff is both a marketing and management decision, and the impact on the customer must be the first consideration. If service is being marketed, it becomes an expectation of the customer, and management must accept the risk of overstaffing and the higher cost to the customer. A customer, however, often makes a sacrifice beyond cost, which is time. Waiting for room service, for lunch in a restaurant, for a bottle of wine to be served, or for the check to come causes the customer to become irritated. An alternative is the limited-menu restaurant, which has been very successful because it capitalizes on the time saved. If marketing creates expectations, makes promises, offers value, and reduces risk, management needs to understand that the cost of keeping a customer is far less than that of creating a new one.

Heterogeneity

Heterogeneity of service is concerned with the variation and lack of uniformity in the performance of people. This is different from the poor service caused by an insufficient number of staff; rather, it is fluctuations in service caused by unskilled employees, customer perceptions, and the customers themselves. Variations might occur between services within the same organization or in the service provided by one employee from day to day or from customer to customer. Most services are labor intensive, and the performance of each employee is different. Managers have difficulty in predicting how employees with different backgrounds and personalities will react in various circumstances. Use of the TQM philosophy and empowerment of employees to make decisions allows many of these problems to be solved at team meetings.

Marketers of services who make promises to the customer never know how employees will handle a situation or how the customer will perceive a service. Lewis and Chambers (1989) stated that the consequence is that good service may equal bad service. One customer may be pleased that the waitstaff never permits the coffee cup to become empty, but another customer may be annoyed because the cup never becomes empty and refilled with fresh hot coffee. Sometimes, less service is more service. An example is the popular salad bar. Many people like the idea that they can select what they want, although this is less service because they have to get their own food. The marketer, who has many problems trying to cope with heterogeneity of services, must know both the market and the customers to be able to customize services. Without question, the emphasis should be on the customer, not the service.

Components of Service Products

Lewis and Chambers (1989) stated that customers are concerned with the components of goods, services, and environment when purchasing the hospitality product. Goods are mostly physical factors over which management has direct, or almost direct, control and are usually tangible. The manager's expertise determines the quality level of goods, as illustrated by the employment of a competent chef. Quality of goods also could depend on the willingness of managers to support financially the menu items desired by the target market they are trying to reach. Lewis and Chambers (1989) define price as being tangible, although it is a cost of services as well as goods. To the customer, however, price is tangible in any purchase decision.

Service includes nonphysical, intangible attributes that management should control. Personal elements provided by employees such as friendliness, speed, attitude, and responsiveness are very important components of service. Additional factors might depend on the specific foodservice operation, for example, handling of reservations. In the environment category, items over which management may have some control may be included. These items may or may not be tangible but are something the customer feels. That feeling is what the manager is marketing. Decor, atmosphere, comfort, ambience, and architecture are attributes included in this category.

Lewis and Chambers (1989) used hotel room service to illustrate the interrelationship of these three components of the hospitality product. Management first must decide if offering room service will meet customers' needs, based on the particular property and target market. If the answer is yes, then demand, cost, resources, and facilities should be analyzed, as well as potential customer dissatisfaction if the service is not offered. Expectations of the customer should be studied in such matters as the attitude of employees who answer the telephone and deliver the tray, the length of time for delivery, amount of service in the room, and the length of time the finished tray will remain in the hall before pickup. The goods also must meet customer expectations of, for example, fresh orange juice, hot coffee, clean silverware, unchipped china, properly cooked food, and fair price. The environment must be conducive to room service and include a table to put the food on without rearranging the room, correct height of chairs for the table, a flower on the tray, and correct presentation. Finally, if the customer is satisfied, will the customer praise the service? Probably not, but if it is not up to expectations, management and probably future customers will be told.

Service Marketing Mix

Managers desiring to market service must provide benefits that satisfy the needs of the customer. Target markets should be defined and the marketing mix identified before a marketing strategy can be finalized. The four elements required for a marketing mix for goods are applicable to service: product, distribution, promotion, and price.

Services are intangible products and thus difficult for customers to evaluate. If a limited-menu restaurant chain can standardize a service and market it more effectively than other chains can market the service, the chain generally will gain a greater share of the market. An example has been the serving of certain food items, such as baked-to-order pizza within 15 minutes of placing the order.

Distribution in the service context refers to making services available to prospective users. Instead of taking the goods to the customer, customers must come to the service. Pizza, hamburger, chicken, fish, and other specialty limited-menu restaurants distribute products that are the same in many locations.

Lewis and Chambers (1989) define *promotion* as marketing communication that serves specifically as an incentive to stimulate sales on a short-term basis. Promotions are frequently used to stimulate business in off periods when normal business flow has decreased.

Establishing a price for service can be difficult because of its intangibility. The more standardized service becomes, the easier pricing is for the manager. Pricing service in a limited-menu restaurant with well-defined procedures for employees to follow is much easier than pricing in an upscale restaurant in which employees are encouraged to satisfy individual customer's desires.

STRATEGIC MARKETING

As stated previously, a marketing strategy pertains to the selection and analysis of a target market and the creation and maintenance of an appropriate marketing mix. Any strategic planning process begins with the organization's **mission statement** and objectives and ends with a marketing plan.

Strategic Planning Process

Evans and Berman (1992) outlined seven interrelated steps in the strategic planning process, as shown in Figure 4.3. *Organizational mission* refers to a commitment to a type of organization and a place in the market. In addition, the organization has competitive, consumer, and individual objectives. The mission statement is a summation of the organization's purpose, competition, target market, product, and service as well as of the recipients of the service, including consumers, employees, owners, and the community. For example, the mission of a hospital could be to provide both inpatient and outpatient medical service to the people in the community within budgetary limitations. Each department would have a mission compatible with that of the organization. The mission of the foodservice department thus would be to provide food within the departmental budget to the patients, employees, and visitors.

After defining a mission, an organization should establish **strategic business units (SBUs)**. Each SBU is a separate component of the organization and has a specific market focus and a manager with responsibility for placing all functions into a strategy. SBUs are the basis for a strategic marketing plan. In business terminology, the hospital is the corporate level, and the foodservice department is the strategic business unit. Every SBU has a clearly defined market segment with a strategy consistent with that of the corporation, its own mission, and its own competitors.

A **marketing objective** is a statement of what is to be accomplished through marketing activities. Each SBU in an organization needs to set its own objectives in clear, simple terms for marketing performance. Objectives generally are described in both quantitative terms (dollar sales, percentage profit growth, market share) and qualitative terms (image, uniqueness, customer service). Many foodservice organizations, such as restaurants, combine quantitative and qualitative goals, for example, dollar sales and uniqueness based on a new theme.

Situation analysis is the identification of marketing opportunities and potential problems confronting an organization. The manager needs to know the current condition of the organization and the direction in which it is going. The environment, opportunities, and strengths and weaknesses of competitors should be studied for the analysis.

Figure 4.3.   Strategic planning process.

Source: Reprinted with the permission of Macmillan College Publishing Company from *Marketing*, Fifth Edition, by Joel R. Evans and Barry Berman. Copyright © 1992 by Macmillan College Publishing Company, Inc.

A **marketing strategy** encompasses selecting and analyzing a target market and creating and maintaining an appropriate marketing mix that will satisfy that market (Pride & Ferrell, 1993). A strategy should be as specific as possible. For example, to increase dessert sales, a poor strategy might be something imprecise like, "The addition of low-fat frozen yogurt to the menu will be advertised." A better strategy would provide more guidance: "Dessert sales will be increased by 10% within 3 months by adding low-fat frozen yogurt to the menu and increasing advertising to health-conscious consumers."

The marketing strategy is implemented through a series of *tactics*, which are specific actions. According to Lewis and Chambers (1989), strategy is the way to gain and keep customers; tactics are the step-by-step procedures on how to do it. They gave a good example citing a hotel, but the illustration could well be a restaurant. The objective could be "To be perceived as the restaurant of choice" and the strategy, "To give customers better value." Some of the tactics could include having a table ready for customers who have made reservations, calling customers by name, having the print on the menu large enough to read, and offering a selection for customers with special dietary needs. Tactics flow from strategy, which means the appropriate strategy must be developed first.

Monitoring results involves the comparison of performance standards against actual performance over a definite time. Budgets, timetables, sales, and cost analyses may be used to analyze results. If actual performance does not meet the standards, corrective action should be taken in problem areas. Many organizations have contingency plans if performance standards are not met.

Marketing Research

The foundation of a successful marketing plan is research (Yesawich, 1987). Only through research can proper judgments be made about the best combination of product, distribution, promotion, and price. Market research can help an establishment succeed, rather than merely survive, by attracting new customers, keeping up with trends, and tailoring menus to meet customer needs (Stern, 1990). Intuition and past experience, rather than scientific decision making, often govern marketing decisions, however.

Marketing research has been defined by the American Marketing Association as the systematic gathering, recording, and analyzing of data about problems relating to the marketing of goods and services (Alexander, 1960). To be effective, marketing research must be systematic and not haphazard or disjointed. Marketing research involves a series of steps including data collection, recording, and analysis (Evans & Berman, 1992). Data may be available from different sources: the organization itself, an impartial marketing research company, or a research specialist working for the organization.

The scientific method based on objectivity, accuracy, and thoroughness should be followed in conducting research (Boyd, Westfall, & Stach, 1989). *Objectivity* means that research is conducted in an unbiased, open-minded manner and conclusions are not reached until all data have been analyzed. *Accuracy* refers to the use of research tools that are carefully constructed and utilized. The sample should repre-

Figure 4.4. Marketing research process.

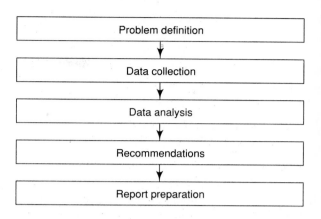

sent the population; a questionnaire, if used, should be pretested; and the analysis of data should be statistically correct. Conclusions may not be correct if the research does not probe deeply or widely enough. Despite the value of formal research, marketing decisions often are made without it.

The marketing research process, as shown in Figure 4.4, consists of five steps for logically solving a problem: problem definition, data collection, data analysis, recommendations, and preparation of the report. Foodservice managers conducting research should think about each of these steps and tailor them to fit the problem.

The Marketing Plan

Pride and Ferrell (1993) defined **marketing plan** as a written document or blueprint governing an organization's marketing activities, including the implementation and control of those activities. *Marketing planning* is a systematic process involving the assessment of marketing opportunities and resources, the determination of marketing objectives, the development of a marketing strategy, and planning for implementation and control. A marketing plan needs to be integrated and evaluated.

Development of Plan

Strategic planning should be done before a marketing plan is developed. Lewis and Chambers (1989), in discussing marketing in hospitality industries, stated that in strategic marketing, market segments are identified as those the organization can best serve to competitive advantage. In marketing management that leads to a marketing plan in which the segments are defined in terms of marketing mix variables. Strategic marketing is an overall view of marketing in the organization in which resources are allocated and objectives set after defining the market; marketing management develops the marketing mix to service-designated markets in accordance with those resources and objectives. Strategic marketing deals with the long-term view of the market for a specific operation, and marketing management emphasizes implementing that operation within the constraints established by the strategy.

A marketing plan should be real and workable and should be easy to execute (Lewis & Chambers, 1989). It also should be flexible but have a certain amount of stability. Specific responsibilities with times and dates for accomplishment should be designated in the plan. Finally, a marketing plan needs to be constantly reviewed and evaluated to keep it current.

Types of Plans

Marketing plans can be categorized according to duration, scope, and method of development (Evans & Berman, 1992). Marketing plans typically are developed for 1 year and are considered short range; medium-range plans from 2 to 5 years are sometimes used; and those over 5 years, long-range marketing plans, are seldom developed. Short- and medium-range plans are more detailed and more geared to the operation than long-range plans. A 1-year plan, for example, may detail the marketing of a new menu item, but a 10-year plan will be concerned with changes in the environment.

The scope of marketing plans varies tremendously. Separate marketing plans may be developed for individual menu items and special services and often are used in limited-menu restaurants. Hamburger, fried chicken, and pizza operations generally have plans for marketing their special items. Also, the limited-menu concept probably requires a separate marketing plan. A single, integrated marketing plan is occasionally used for all menu items and services and is most often used by service operations such as hotels.

Finally, the method of development of plans may be bottom-up, top-down, or a combination of the two. In the bottom-up approach, which uses the TQM philosophy, information from employees is used to establish objectives, budgets, forecasts, timetables, and marketing mixes. Bottom-up plans are realistic and good for morale. Coordination of each bottom-up plan in one integrated plan may be difficult to achieve because of conflicts, for example, in estimates of the impact of marketing a new menu item. In the top-down approach, top management directs and controls planning activities. Top-level managers understand the competition and environment and provide direction for marketing efforts. If input from lower level managers is not sought, however, morale may diminish. A combination of these two approaches could be the best solution; top management could set the overall objectives and policy, and lower-level managers could establish the plans for implementing the policy (Evans & Berman, 1992).

Integration of Plans

Integration of marketing is necessary if the product, distribution, promotion, and price elements of the marketing mix are to be synchronized (Evans & Berman, 1992). An *integrated marketing plan* is one in which all the various components are unified, consistent, and coordinated; it consists of seven elements, as shown in Figure 4.5. A clear organizational mission defines an organization's type of business and place in the market. The mission is involved each time products or services are

Figure 4.5. Elements leading to a well-integrated marketing plan.

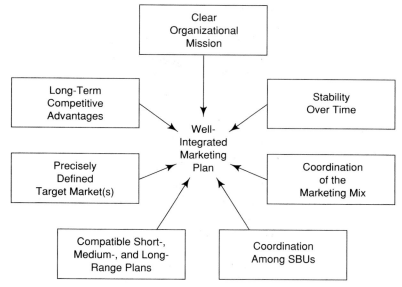

Source: Reprinted with the permission of Macmillan College Publishing Company from *Marketing*, Fifth Edition, by Joel R. Evans and Barry Berman. Copyright © 1992 by Macmillan College Publishing Company, Inc.

added or deleted or new target markets are sought or abandoned. A marketing plan must show stability over time for it to be implemented and evaluated correctly.

The product, distribution, promotion, and price components of the marketing mix need to be coordinated within each department (Evans & Berman, 1992). Coordination among individual departments or SBUs is increased when the strategies and resources allocated to each are described in short-, medium-, and long-range plans. Compatible plans for each department or SBU become the broad marketing plan for the organization. The target market needs to be identified in a marketing plan to guide marketing efforts and future direction. If two or more distinct target markets are present, each should be clearly defined. For example, a foodservice organization catering to children and parents probably will have separate strategies for each of these segments.

Customer expectations and quality of the product and service—the TQM philosophy—need to be emphasized in developing long-term plans for successful competition in the market. Internal customers, like the maître d', chef, waitperson, and cashier, all need to work together while keeping the external customer in focus.

Control and Evaluation

Control.   Control is as necessary in marketing as in managing all facets of the foodservice organization. The manager should establish performance standards for marketing activities based on goals of the organization. Internal standards generally are expressed as profits, sales, or costs. Most organizations use external individuals or organizations, such as consultants or marketing research firms, for marketing assistance.

When foodservice managers attempt to control marketing activities, they frequently have problems because information is not always available. Even though controls should be flexible enough to allow for environmental changes, the frequency and intensity of changes may curtail effective control. Because marketing overlaps other activities, the precise costs of marketing are difficult to define.

Evaluation.   Sales analysis can be used for evaluating the actual performance of marketing strategies (Pride & Ferrell, 1993). A **sales analysis** is the detailed study of sales data for the purpose of evaluating the appropriateness of a marketing strategy (Evans & Berman, 1992). It is only one method for measuring the effectiveness of a marketing strategy and can be extremely costly. Performance standards, usually profits, sales, or costs, should be stated in the marketing plan. Sales analysis requires sales figures, either volume or market share, to evaluate an organization's current performance. Dollar volume sales are frequently used because the dollar is the common denominator of profits, sales, and costs (Pride & Ferrell, 1993). Price increases and decreases, however, affect total sales figures. For example, if a restaurant increases prices by 10% this year and its sales volume is 10% greater than last year's, it has not had any increase in unit sales. A restaurant marketing manager should factor out the effects of price changes.

Market share is stated as the percentage of industry sales for a product. The rationale for using market share is to estimate if sales changes occurred because of the organization's marketing strategy or from uncontrollable environmental factors. The assumption is that industry sales decrease when restaurant sales decrease and market share remains constant. If a restaurant suffers a decrease in both sales and market share, however, the marketing strategy is not effective. Market share analysis should be interpreted with caution because it is based on uncontrollable factors, for example, differing objectives among companies.

Marketing cost analysis classifies costs to determine which are associated with specific marketing activities. A foodservice organization should know the marketing costs of a strategy that is being used to achieve a certain sales level. Costs need to be broken down and classified to determine those that are associated with specific marketing activities. The marketing manager evaluates the effectiveness of the current marketing strategy by comparing sales and costs. Determining marketing costs can be difficult, and ascertaining the cost of marketing a product is seldom adequate. Usually the costs must be broken down into specific geographic areas, market segments, or even specific customers. Accounting records need to be examined to conduct a marketing cost analysis adequately.

SUMMARY

Marketing is an integral part of managing and operating a business. Long defined in terms of the product, it is currently being expressed in terms of the customer. An exchange occurs between a buyer and seller in a marketing environment that consists of all forces outside an organization.

The marketing concept is a management philosophy that affects all activities of an organization and emphasizes satisfying the needs of customers. Implementing the marketing concept should benefit the business as well as the customer. The evolution of marketing went through three eras: production, sales, and marketing.

Marketing management is a process of planning, organizing, implementing, and controlling management activities to facilitate exchanges and minimize costs effectively and efficiently. To do this, management is responsible for developing and managing marketing strategies that pertain to creation of an appropriate market mix and the selection and analysis of a target market.

An organization has control over the marketing mix variables of product, distribution, promotion, and price, but it has only limited control over the environment. A product can be a good, a service, or an idea, and it must be distributed for availability at the right time in accessible locations. Promotion is used to facilitate exchanges by informing prospective customers about the product. Price is a critical component in the marketing mix and is often used as a competitive tool. The marketing environment surrounds the buyer and the marketing mix, and it consists of political, legal, regulatory, societal, economic, and technological forces.

Market segmentation is the process of dividing a total market into groups of people sharing characteristics causing them to have similar product needs. Variables for segmenting the market are demographic, geographic, psychographic, and behavioristic. Demographic variables consist of population characteristics such as age, sex, race, and education. Geographic variables, including the size of the region, city, county, or state and whether the area is urban, suburban, or rural, have an effect on the market. The most common psychographic variables are personality characteristics, motives, and life-styles. A market can also be segmented on behavioristic variables, which are the basis of some features of customer behavior toward a product.

A service is any activity or benefit that one person can offer another. Service marketing has four basic characteristics: intangibility, inseparability of production and consumption, perishability, and heterogeneity. Intangibility means services cannot be seen, touched, tasted, or smelled. Inseparability of production and consumption indicates that services are produced at the same time they are consumed. Perishability means services cannot be stored for future sale, and heterogeneity of service is concerned with the variation and lack of uniformity in the performance of people.

The four elements required for a marketing mix for goods are applicable to service, but with a different emphasis. A service is intangible and thus difficult for customers to evaluate. Distribution refers to making services available to prospective users. Promotions frequently are used to stimulate sales on a short-term basis, and pricing a service can be difficult because of its intangibility.

A marketing strategy pertains to selection and analysis of a target market and creation of an appropriate mix, and it needs to be done before a marketing plan is developed. Research is the foundation of a successful marketing plan. Marketing research is the systematic gathering, recording, and analyzing of data about problems relating to the marketing of goods and services. The marketing plan is the written document for implementing and controlling marketing activities during the forthcoming year. A method for assessing and improving the effectiveness of a marketing strategy should be stated in the plan, as well as performance standards generally expressed as profits, sales, or costs. Actual performance needs to be measured in terms of these standards, making a comparison possible.

BIBLIOGRAPHY

Albrecht, K. (1988). *At America's service*. Homewood, IL: Dow Jones-Irwin.

Albrecht, K., & Zemke, R. (1985). *Service America! Doing business in the new economy*. Homewood, IL: Dow Jones-Irwin.

Alexander, R. S. (1960). *Marketing definitions: A glossary of marketing terms*. Chicago: American Marketing Association.

Allen, R. L. (1992). The world of target marketing. *Nation's Restaurant News, 26*(11), 25, 62.

Berry, L. L., & Parasuraman, A. (1991). *Marketing services: competing through quality*. New York: Free Press.

Bonoma, T. V. (1989). Marketing performance: What do you expect? *Harvard Business Review, 67*(5), 44–46, 48.

Boyd, H. W., Westfall, R., & Stach, S. F. (1989). *Marketing research: Text and cases* (7th ed.). Homewood, IL: Richard D. Irwin.

Chaudhry, R. (1992). James Near cleans up Wendy's. *Restaurants & Institutions, 102*(17), 73, 78, 82.

Chaudhry, R. (1993). Food for tot. *Restaurants & Institutions, 103*(7), 131, 134.

Congram, C. A., & Friedman, M. L. (Eds.). (1991). *The AMA handbook of marketing for the service industries*. New York: AMACOM.

Davidow, W. H., & Uttal, B. (1989). Service companies: Focus or falter. *Harvard Business Review, 67*(4), 77–85.

Dodd, J. (1992). President's page: The fifth P. *Journal of The American Dietetic Association, 92*, 616–617.

Evans, J. R., & Berman, B. (1992). *Marketing* (5th ed.). New York: Macmillan.

Gronroos, C. (1990). *Service management and marketing: Managing the moments of truth in service competition*. Lexington, MA: Lexington Books.

Hart, C. W., Casserly, G., & Lawless, M. J. (1984). The product life cycle: How useful? *Cornell Hotel and Restaurant Administration Quarterly, 25*(3), 54–63.

Heskett, J. L. (1986). *Managing in the service economy*. Boston: Harvard Business School Press.

Houston, F. S. (1986). The marketing concept: What it is and what it is not. *Journal of Marketing, 50*(2), 81–87.

Jacobs, P. (1993). Staying focused: Strategic plans need constant care. *Nation's Restaurant News, 27*(20), 22.

Kanter, R. M. (1992). Think like the customer: The global business logic. *Harvard Business Review, 70*(4), 9–10.

Kashani, K. (1989). Beware the pitfalls of global marketing. *Harvard Business Review, 67*(5), 91–98.

Kotler, P. (1977). From sales obsession to marketing effectiveness. *Harvard Business Review, 55*(6), 67–75.

Kotler, P. (1991). *Marketing management: Analysis, planning and control* (7th ed.). Englewood Cliffs, NJ: Prentice-Hall.

Levitt, T. (1981). Marketing intangible products and product intangibles. *Harvard Business Review, 59*(3), 94–102.

Lewis, R. C. (1984). Theoretical and practical considerations in research design. *Cornell Hotel and Restaurant Administration Quarterly, 24*(4), 25–35.

Lewis, R. C. (1989). Hospitality marketing: The internal approach. *Cornell Hotel and Restaurant Administration Quarterly, 30*(3), 41–45.

Lewis, R. C., & Chambers, R. E. (1989). *Marketing leadership in hospitality: Foundations and practices.* New York: Van Nostrand Reinhold.

Lovelock, C. H. (1980). Why marketing management needs to be different for services. In J. H. Donnelly & W. R. George (Eds.), *Marketing of services* (pp. 708–719). Chicago: American Marketing Association.

Lovelock, C. H. (1983). Classifying services to gain strategic marketing insights. *Journal of Marketing, 47*(3), 9–20.

Lovelock, C. H. (1991). *Services marketing.* Englewood Cliffs, NJ: Prentice-Hall.

McKenna, R. (1988). Marketing in an age of diversity. *Harvard Business Review, 66*(5), 88–95.

McKenna, R. (1991). Marketing is everything. *Harvard Business Review, 69*(1), 65–79.

National income and product accounts. (1992). *Survey of Current Business, 72*(9), 5–47.

Pine, B. J., II. (1993). *Mass customization: The new frontier in business competition.* Boston: Harvard Business School Press.

Pride, W. M., & Ferrell, O. C. (1993). *Marketing: Concepts and strategies* (8th ed.). Boston: Houghton Mifflin.

Solomon, J. (1993). Homemade marketing strategies. *Restaurants USA, 13*(1), 17–19.

Star, S. H. (1989). Marketing and its discontents. *Harvard Business Review, 67*(6), 148–154.

Stern, G. M. (1990). The case for marketing research. *Restaurants USA, 10*(7), 26–29.

Sullivan, J. (1991). Market your restaurant as you work the floor. *Nation's Restaurant News, 25*(16), 22.

Swinyard, W. R., & Struman, K. D. (1986). Market segmentation: Finding the heart of your restaurant's market. *Cornell Hotel and Restaurant Administration Quarterly, 27*(1), 89–96.

Treacy, M., & Wiersema, F. (1993). Customer intimacy and other value disciplines. *Harvard Business Review, 71*(1), 84–93.

Uhl, K. P., & Upah, G. D. (1986). The marketing of services. In J. N. Sheth & D. E. Garrett (Eds.), *Marketing management: A comprehensive reader* (pp. 999–1026). Cincinnati: South-Western Publishing.

Vance, D. E. (1992). Capture your market—then work to keep it. *Nation's Restaurant News, 26*(6), 52.

West, J. J., & Olsen, M. D. (1989). Competitive tactics in foodservice: Are high performers different? *Cornell Hotel and Restaurant Administration Quarterly, 30*(1), 68–71.

Yesawich, P. C. (1987). Hospitality marketing for the '90s: Effective marketing research. *Cornell Hotel and Restaurant Administration Quarterly, 28*(1), 49–57.

Yesawich, P. C. (1988). Planning: The second step in market development. *Cornell Hotel and Restaurant Administration Quarterly, 28*(4), 71–81.

Yesawich, P. C. (1989). The final steps in market development: Execution and measurement of programs. *Cornell Hotel and Restaurant Administration Quarterly, 29*(4), 83–91.

Zeithaml, V. A., Parasuraman, A., & Berry, L. L. (1985). Problems and strategies in services marketing. *Journal of Marketing, 49*(2), 33–46.

Zeithaml, V. A., Parasuraman, A., & Berry, L. L. (1990). *Delivering quality service: Balancing customer perceptions and expectations.* New York: Free Press.

The Fairfax Food and Beverage Department

This case study covers the first 4 chapters of this book, which include an overview of the foodservice industry, the foodservice systems model with an emphasis on total quality management (TQM), and the managing and marketing of the foodservice system. All of these chapters should be used in solving the case problem.

Background Information

The mission statement for the food and beverage service is "to make available to the residents and guests of The Fairfax and the Belvoir Woods Health Care Center a varied, well-balanced and nutritious menu, served in a prompt, courteous, and sanitary manner." Figure 1 is the organization chart for the Food and Beverage Department located in the Community Center; four managers are shown: director of the Food and Beverage Department (F&B), dining room manager, executive chef, and the healthcare foodservice manager.

The F&B director is responsible for the overall operation of all foodservice in the retirement community, as shown in the job description (Figure 2). The executive chef manages and directs all phases of food production and is responsible for keeping within budgetary guidelines and for ensuring that all meals meet Marriott's standards. The chef also ensures compliance with all federal, state, and local regulatory bodies. The dining room manager is responsible for the service and delivery of all food and beverages in the Community Center, including the main and private dining rooms, auditorium, coffee shop/country store/bistro as well as the assisted-living and skilled-nursing dining rooms in the healthcare center. The certified dietary manager, identified as the healthcare foodservice manager on the organization chart, manages the foodservice operation and dietary functions in the Belvoir Woods Health Care Center and serves as liaison between the consulting dietitian, the F&B department, and nursing services.

The consulting dietitian spends one day each week with the certified dietary manager to assure the F&B director that the foodservice is in compliance with regulatory issues, as shown on the job description (Figure 3). The consulting dietitian also consults at The Jefferson Senior Living Services community in Arlington, Virginia, and Bedford Court in Silver Spring, Maryland, 1 day each week and serves 2 days a week as the administrative dietitian under the direction of the national director of food and beverage in the Washington, D.C., corporate office.

Figure 1. Organization chart for the Food and Beverage Department in The Fairfax community.

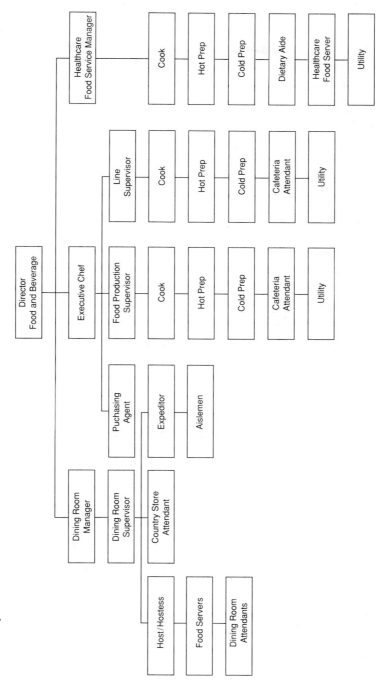

Source: Courtesy of Marriott Corporation.

Figure 2. Job description for the director of the Food and Beverage Department.

JOB TITLE: Director, Food and Beverage
OCCUPATION CODE:
DEPARTMENT NAME: Food and Beverage
DEPARTMENT NUMBER:

Overall Responsibilities

Directs the overall operation of the food and beverage department in the community. Operates food and beverage department profitably, within budgetary guidelines, while maintaining, or exceeding, Marriott hospitality and service standards for all aspects of foodservice. Ensures compliance with all federal, state, and local regulatory bodies. Responsible for the human resources management of the department.

Working Relationships

Reports to: General Manager
Supervises: Food and Beverage Department managers
Interfaces with: Community department heads, residents, regulatory agencies, Sr. Director, Food and Beverage, National Director, Food and Beverage

Primary Job Duties

Managing the Work

- Maintain efficiency of food preparation while meeting, or exceeding, Marriott hospitality and service standards.
- Establish effective relationships with other department heads and communicate information regarding new developments and trends in foodservice.
- Direct menu planning using Marriott hospitality and service guidelines to meet the needs and desires of the community's resident population.
- Direct department production control system and procedures within Division policy and procedures and within budgetary guidelines.
- Ensure all special dietary needs are met as they pertain to resident needs and healthcare guidelines.
- Assist dietitian in establishing and revising dietary policies and procedures.
- Manage the department's purchasing program in an efficient and effective manner; authorize suppliers; establish procedures for receiving food, supplies and proper storage.
- Regularly evaluate foodservice in all service delivery areas—dining rooms (IL, AL, Health Care) as well as tray delivery; periodically visit residents to evaluate foodservice—quality, quantity, temperature, and appearance; analyze feedback and take corrective action when necessary to promote high-quality foodservice.
- Conduct internal quality assurance audits on all aspects of foodservice on an ongoing basis as directed by Division standards.
- Direct department safety and loss prevention program; monitor adherence to safety rules and regulations and take remedial action when necessary.

Figure 2. *continued*

Managing Costs

- Plan and develop department's operating budget working within Division, community, and F&B guidelines.
- Plan and execute department operating plans according to budgetary guidelines.
- Regularly monitor department performance; prepare weekly productivity reports; analyze the department's weekly distribution summaries and period Profit & Loss Reports in a timely manner and take corrective action when necessary.
- Utilize the Division's F&B reporting forms and the Management Control Record (M.C.R.) system; establish a system to track and record department costs.
- Train department managers in pertinent cost control techniques and methods.
- Responsible for inventory control and establishing efficient and effective inventory control systems.

Managing Human Resources

- Responsible for the recruitment and selection of all employees in the food and beverage department.
- Ensure adequate and effective orientation and training of all F&B employees in their job-specific duties, in Marriott's hospitality and service standards, and in understanding the community resident population.
- Supervise and evaluate all F&B managers by providing ongoing, timely feedback.
- Ensure that all F&B employees meet or exceed Marriott's standards of appearance; monitor sanitation of department, hygiene and health standards of personnel.
- Establish positive employees relations programs and practices; implement Marriott's Guarantee of Fair Treatment Policy.
- Responsible for managing F&B employee compensation; recommend to General Manager wage increases and adjustments for the department.
- Fairly and consistently apply Division Human Resources Policy and Procedures, especially in the areas of scheduling, training, promotions, transfers, etc.
- Attend inservices as required.
- Maintain and protect the confidentiality of resident information at all times.

Job Qualifications

Required

- Bachelor Degree in foodservice management or three years experience in foodservice production
- Ability to provide food and beverage service to the elderly which meets or exceeds Marriott hospitality and service standards
- Ability to effectively supervise a diverse employee work group
- Desire to work with the elderly

Preferred

- Experience in healthcare foodservice operations.

Figure 2. *continued*

OSHA Occupational Exposure Category

After careful analysis, it has been determined that this position falls into OSHA Occupational Exposure Category _____ and requires the following protective equipment be worn by anyone filling this position:

Training will be provided in how to properly and effectively use the equipment listed above, in addition to education regarding precautionary measures, epidemiology, modes of transmission and prevention of HIV/HBV (Human Immunodeficiency Virus/Hepatitis B Virus).

I have read and agree that the contents of this job description accurately reflect what is expected of me in my current position.

_____   _____
Employee's Signature   Date

_____
Employee's Printed Name

_____   _____
Immediate Supervisor's Signature   Date

Source: Courtesy of Marriott Corporation.

Figure 3. Job description for the administrative/consulting dietitian.

JOB TITLE: Administrative/Consulting Dietitian
OCCUPATION CODE:
DEPARTMENT NAME: Food and Beverage
DEPARTMENT NUMBER: 07

Overall Responsibilities

Assist National Director of Food and Beverage with the development, implementation and monitoring of all the division's clinical nutrition programs. Act as a resource for the consulting dietitians. Act as consulting dietitian for The Jefferson, The Fairfax and Bedford Court with the remainder of time working on national responsibilities.

Working Relationships

Reports to: As Administrative Dietitian—National Director of Food and Beverage
Supervises: Consulting Dietitians and Certified Dietary Managers
Interfaces with: National Director of Food and Beverage, National Director of Health and Wellness, General Managers, Food and Beverage Directors, Consulting Dietitians and State Regulatory Surveyors

Primary Job Duties

As Administrative Dietitian

- Keep all the required clinical resources updated and current with the latest nutritional requirements for the Older Adult, i.e., Diet Manual, Nutrition's Analysis Manual, etc.
- Develop and standardize the clinical forms, recording and reporting procedures.
- Provide the leadership needed to assure the National Food and Beverage Director that the dietary clinical program of each community is current and in compliance with state regulatory agencies.
- Develop and assist with the implementation of Nutritional Health and Wellness programs that promote the division policy on wellness.
- Work with the National Director of Food and Beverage to assure that the dietary clinical programs are properly integrated into our Health Care Centers.
- Act as a resource for all the Regional Nurses and consulting Dietitians, and Certified Dietary Managers.
- Determine division clinical standards by incorporating the division's hospitality philosophy with state regulatory guidelines.

As Consulting Dietitian

- Work with the community's Certified Dietary Manager to assure the F&B Director that the department is in compliance with regulatory issues.
- Complete charting requirements of community dietitian.
- Provide the community's General Manager, Health Care Administrator, and F&B Director with a visitation report outlining your work, observations and recommendations upon completing your visit.

Figure 3.   *continued*

Job Qualifications

Required

- B.S. degree in Dietetics
- Currently a Registered Dietitian
- Three years of clinical experience
- Desire to work with the elderly
- Able to work independently
- Good communication skills

Preferred

- Understanding of Marriott's hospitality and service standards for Senior Living Services.

OSHA Occupational Exposure Category

After careful analysis, it has been determined that this position falls into OSHA Occupational Hazard Category _____ and requires the following protective equipment be worn by anyone filling this position:

Training will be provided in how to properly and effectively use the equipment listed above, in addition to education regarding precautionary measures, epidemiology, modes of transmission and prevention of HIV/HBV.

I have read and agree that the contents of this job description accurately reflect my understanding of what is expected of me in this position.

_____   _____
Employee's Signature   Date

_____
Employee's Printed Name

_____   _____
Immediate Supervisor's Signature   Date

Source: Courtesy of Marriott Corporation.

Dining Facilities

Two dining rooms are available for the independent living residents in apartments or cottages. The main dining room features a select menu and full service with a maître d', food servers, and dining room attendants. A different menu originally was used for lunch and dinner until the residents asked to have the dinner menu served both at noon and in the evening. Many of the residents prefer to have a full-course meal at noon and a light evening meal in their apartment or cottage. Also, older friends, whom the residents often invite to lunch instead of dinner, prefer to drive in the daylight and avoid the heavy traffic in Washington, D.C.

The coffee shop and bistro, which is an extension of the coffee shop and located in the alcove between the country store/coffee shop and lobby, are less formal than the main dining room. Breakfast is served every day but Sunday, because most residents attend the brunch in the main dining room, and dinner is served every day of the week. Two breakfast menus, specials and à la carte, are the same every morning (Figure 4) and the dinner menu is the same select menu as the one in the main dining room. Waitstaff serve and clear tables.

The coffee shop also is a morning meeting place for residents and sells coffee and doughnuts. Coffee can be ordered by residents at any time of the day or evening in the public rooms, such as the living room and library. Residents eat in the more casual coffee shop when they do not feel like dressing up to go to the main dining rom. Originally, the coffee shop served food all day until labor costs had to be decreased to lower expenses.

In the assisted-living dining room located on the second floor of the healthcare center, the main dining room select menu is available for residents without dietary restrictions as well as waited service. In the skilled-nursing unit on the third floor, residents with minimal dietary restrictions order from the Belvoir Woods Health Care Center's select menu. Nursing assistants are available for those who require help. In the restorative dining room, also on the third floor, nursing assistants actually feed or assist in the feeding of those residents requiring this service. Tray service is offered to those residents who cannot come to the dining room.

Dining Hours and Services

Hours open for serving in the main dining room and coffee shop are as follows:

Main Dining Room

Sunday brunch	11:00 A.M. – 3:00 P.M.
Monday through Saturday lunch	12:00 A.M. – 1:30 P.M.
Monday through Saturday dinner	5:00 P.M. – 7:30 P.M.

Coffee Shop

Monday through Saturday breakfast	8:00 A.M. – 10:00 A.M.
Daily dinner	5:00 P.M. – 7:00 P.M.

Room Service

Sunday (guest rooms and apartments)	8:00 A.M. – 10:00 A.M.

Figure 4. The coffee shop breakfast specials and à la carte menus.

Country Store Breakfast Specials

One or Two Eggs,
Any Style
Includes Breakfast
Potatoes and Buttered
Toast
One Egg $2.25
Two Eggs $2.50

Continental Breakfast
Serve yourself from our
Continental Cart $3.25

**The following Specials
include Juice, Fresh Fruit
and Coffee $4.00**

The B-52 Bomber
Three Pancakes with
Bacon or Sausage.

The All American
Two Eggs, any style with
Bacon or Sausage,
Breakfast Potatoes and
Toast or Muffins

The Battleship
Two pieces of French Toast
with Bacon or Sausage

The Gun Swinger
Two Egg Western
Omelette with Breakfast
Potatoes and Toast

S.O.S
Creamy Sausage Gravy
over Toast or Biscuits

À la Carte Breakfast

Juices
Orange, Grapefruit, Apple,
Prune, Tomato, V-8,
Pineapple or Grape $.75

Fruits
Grapefruit Half, Pineapple
Wedge, Honeydew,
Cantelope, Strawberries or
Prunes $1.00

The Bakery
Buttered Toast (Your
Choice), Croissant,
Danish, Donuts,
Buttermilk Biscuits,
English or Bran Muffins
$.50
Bagel with Cheese $1.00

Beverages
Coffee, Tea, Decaf, Hot
Cocoa or Assorted Milks
$.50

Cereals
Your Favorite Cold Cereal,
Granola or Oatmeal (We
will add Bananas or Berries,
if you wish) $.75

Breakfast Meats
Bacon Strips, Sausage
Links or Ham Steak $1.25

Yogurt
Plain or Fruited
$1.00

Source: Courtesy of Marriott Corporation.

Saturday night dinner is a theme buffet in the main dining room and coffee shop. On Sunday, a brunch is the only meal served in the main dining room. Breakfast room service, however, is available for residents in apartments and persons in guest rooms. The coffee shop serves dinner on Sunday evening for those residents desiring a meal at that time. In the Belvoir Woods Health Care Center, three meals are served 7 days a week.

A private dining room is available to provide residents of The Fairfax a place for special or formal entertaining. A special menu can be requested with at least 7 days' notice. Arrangements for catering or special functions, such as menus, beverage selections, room set-ups, decorations, servers, and clean-up, can be made with the dining room manager. The cost of the function is subject to Virginia sales taxes. Room service is available for all meals and has a $3 delivery charge. Family and friends are welcome to dine with residents either in the main or private dining room or the coffee shop.

Beer, wine, and liquor may be purchased in the Community Center at all meals. Bartenders can be hired for special functions. Prices for beverages and hot or cold hors d'oeuvres are listed in the dining section of the Residents' Handbook. Liquor laws in the Commonwealth of Virginia are very strict and any infractions could result in the loss of the liquor license, fines, imprisonment, or any combination of the three.

Take-out orders are allowed and may be picked up in the main dining room for lunch or the coffee shop for dinner. Doggie bags are allowed for plated items that have not been consumed.

The Fairfax and Belvoir Woods Health Care Center have two different meal plans. The Fairfax, which is independent living, is on a meal-a-day plan or 28 meals in 28 days. In Marriott properties, 4 weeks are in one period and the year consists of 13 periods. A daily attendance record (Figure 5) has the names listed alphabetically of every resident in the independent living facilities and their apartment or unit numbers. It has three columns to be checked by the hostess indicating if the resident is having brunch/lunch or dinner, and if he or she has guests. This list also is used in the coffee shop. The next morning, a clerk enters by computer these figures into a weekly meal attendance record (Figure 5). At the end of 28 days, the figures are added and if they are over 28 meals, the resident is billed for the extra. In the healthcare center, residents receive three meals a day and snacks and nourishments as desired or prescribed.

Total Quality Management

The Marriott Corporation has made total quality management (TQM) a top priority in its 5-year plan. Top management believes the TQM concept pulls together, in an organized way, most of the key management principles that have been followed for 30 years. Each division is expected to improve levels of product and service quality to distinguish it from competitors', while providing the best value for the customer or resident and reducing costs. By reducing layers of management,

Figure 5. The Fairfax daily and weekly meal attendance records for the main dining room.

Resident	Apt/Unit	Brunch	Dinner	Guests
Abbott, Mrs. B.	J209			
Abel, Mrs. J.	M206			
Abernathy, Col. J.	WW401			
Abernathy, Mrs. D.	WW401			
Ackerman, LtGen. R.	M114			
Ackerman, Mrs. A.	M114			
Adams, Mrs. H.	J218			
Andrews, Mrs. C.	A103			
Antonelli, Capt. J.	WE302			
Antonelli, Mrs. M.	WE302			
Armbruster, Cdr. E.	C9018			
Armbruster, Mrs. D.	C9018			
Arnold, Col. J.	WE114			
Arthur, Brig Gen. T.	J316			
Arthur, Mrs. N.	J316			
Atkinson, Mrs. E.	WE301			
Avery, Col. M.	A110			
Avery, Mrs. O.	A110			

Daily meal attendance record

Resident	Apt/Unit	11/6	11/7	11/8	11/9	10	11	12	Total
Abbott, Mrs. B.	J209								
Abel, Mrs. J.	M206								
Abernathy, Col. J.	WW401								
Abernathy, Mrs. D.	WW401								
Ackerman, LtGen. R.	M114								
Ackerman, Mrs. A.	M114								
Adams, Mrs. H.	J218								
Andrews, Mrs. C.	A103								
Antonelli, Capt. J.	WE302								
Antonelli, Mrs. M.	WE302								
Armbruster, Cdr. E.	C9018								
Armbruster, Mrs. D.	C9018								
Arnold, Col. J.	WE114								
Arthur, BrigGen. T.	J316								
Arthur, Mrs. N.	J316								
Atkinson, Mrs. E.	WE301								
Avery, Col. M.	A110								
Avery, Mrs. O.	A110								

Weekly meal attendance record

activities that do not contribute to taking care of external customers will be eliminated. The corporation uses the term *associate* to describe all its employees. The associate is considered an internal customer in organizations using the TQM philosophy. Marriott has been experimenting with hotel restaurants run by teams of associates, who do labor scheduling, purchase food, and resolve customer problems on the spot. Marriott also has formed teams with suppliers to identify areas for improving quality and developing new delivery programs.

The Problem

A number of residents, especially the men, do not like to dress up for lunch and would be happy to have the coffee shop open all day. Some would like to have it open until midnight so they can visit and have a snack before going to bed. Then others, who are early risers, would like to have breakfast at 6:00 in the morning. They asked the F&B director to meet with them to discuss extending the coffee shop's hours and increasing its services. Prepare a proposal on the pros and cons of extending the hours the coffee shop is open that you can present to the F&B director.

Points for Discussion

- How would each component of the foodservice systems model be affected by your proposal?
- What do you see as the pros and cons of your proposal?
- If The Fairfax general manager adopts the TQM philosophy in the operation, how would this affect your proposal?
- If the plan is not feasible, how might this affect residents' satisfaction? What compromises might be made to alleviate their disappointment?

2

Designing the Foodservice System

Foodservice managers must have an understanding of how food flows through the operation. Both managers and employees have the major responsibility of ensuring the safety of food from the time it is purchased until it is served to the customers. Also, the importance of a well-planned menu cannot be emphasized enough because it controls everything that happens in the foodservice organization.

- **Chapter 5, Food Product Flow.** The Hazard Analysis Critical Control Point (HACCP) model has been given high priority. Flow of food through the following four types of foodservice operations is discussed: conventional, commissary, ready prepared, and assembly/serve. Safety of food products at each step of the food flow is emphasized by using the HACCP model.

- **Chapter 6, The Menu, the Primary Control of the Foodservice System.** The menu is the primary control of the system and is very complex. The foodservice manager has many decisions to make including availability of foods in the market, food cost, production capability, and type of service when planning a menu. Menu pricing also is a consideration by the manager.

5

Food Product Flow

Food is the primary resource in the foodservice system. In the early 1970s, developments in food technology influenced changes in food product flow through the subsystems—procurement, production, and distribution and service. **Food product flow** refers to the alternative paths within foodservice operations that food and menu items may follow, initiating with receiving and ending with service to the customer (Unklesbay, Maxcy, Knickrehm, Stevenson, Cremer, & Matthews, 1977). Alternative paths have been aimed primarily at increasing productivity, decreasing cost, or strengthening control of operations. Physical, chemical, and microbiological changes occurring in food throughout all stages of procurement, production, and service must be controlled to ensure the quality and safety of the finished products.

HAZARD ANALYSIS CRITICAL CONTROL POINT

Hazard Analysis Critical Control Point (HACCP) is the systematic analysis of all process steps in the foodservice subsystems starting with food products from suppliers to consumption of menu items by customers. Analysis is applied throughout food product flow to establish critical controls and eliminate hazardous conditions and procedures (Snyder, 1993). HACCP was developed in the 1970s by major industrial food processors to monitor and control contamination risks in their facilities.

Hazard analysis identifies which specific foods are at risk and the locations in the food product flow where mishandling is likely to occur. **Critical control points** are established to ensure that the hazards have been corrected.

The original HACCP model was modified by Bobeng and David (1978) to include not only microbiological but also nutritive and sensory qualities. They applied the model to quality control of entrée production in various types of hospital foodservices.

Nervous restaurant operators have turned to The Educational Foundation of the National Restaurant Association for help since front page news has been made about illnesses and deaths of customers served undercooked hamburgers in limited-menu restaurants (Mermelstein, 1993). In addition, they are concerned about customers who drink and drive and about the violent murders of employees and customers during robberies (Walkup, 1994). The foundation developed the Servsafe Risk Management Series program, which covers three areas—serving safe food, responsible alcohol service, and employee and customer safety. Of these, safe food handling continues to be the one that generates the most interest. More than 300,000 employees take the food safety course and receive certification each year. The current program is based on implementing the HACCP model to guard against foodborne illness. The issue is so critical that the foundation recently formed the Industry Council on Food Safety, a coalition of industry leaders representing major foodservice companies, to promote food safety and training. The HACCP/Servsafe method identifies points in an operation where contamination or growth of microorganisms can occur and then implements a control procedure based on that hazard (Applied Foodservice Sanitation, 1992).

Managers in school foodservice also are beginning to use the HACCP model to provide menu items free of contaminants responsible for foodborne illnesses (Eck & Ponce, 1993). Considerable managerial competence is required to design controls to monitor the quality and safety of foods and the time and temperature relationships of menu items along the path from procurement to consumption (Klein, Matthews, & Setser, 1984).

A HACCP flowchart for beef stew appears in *Applied Foodservice Sanitation* (1992), the textbook for the NRA National Food Safety Certification Program (see Table 5.1). The critical control points for the beef and vegetables include receiving, storage, preparation, cooking, holding and service, cooling, and reheating. These points will be discussed in depth throughout this text.

To implement the program, NRA suggests that the foodservice manager select two high-risk menu items such as beef stew and chicken salad and follow them from delivery by the supplier to service to the customer (NRA, 1991). In following through each step, the manager should identify critical control points where the food might become contaminated, bacteria might survive cooking temperatures, and bacteria might grow if food is held at incorrect temperatures. The manager then has to develop corrective measures and train employees in how to monitor the critical control points for continuous high quality and safety. The purchase of some thermometers is about the only cost involved. The cost of this project is minimal compared to a lawsuit involving foodborne illness. HACCP should be used in all types of foodservices.

TYPES OF FOODSERVICES

Faced with both labor costs and a shortage of highly skilled employees, foodservice managers have been receptive to using new forms of food with built-in convenience or labor-saving features. New food products, available in various forms and stages of preparation, have appeared on the market in increasing numbers each year. Many require specialized equipment for final production, delivery, and service. Microwave ovens have had a revolutionary impact on the use of frozen food products.

Table 5.1.   A HACCP flowchart for beef stew

Critical Control	Hazard	Standards	Corrective Action If Standard Not Met
		Receiving	
Receiving beef	Contamination and spoilage	Accept beef at 45°F (7.2°C) or lower; verify with thermometer	Reject delivery
		Packaging intact	Reject delivery
		No off odor or stickiness, etc.	Reject delivery
Receiving vegetables	Contamination and spoilage	Packaging intact	Reject delivery
		No cross-contamination from other foods on the truck	Reject delivery
		No signs of insect or rodent activity	Reject delivery
		Storage	
Storing raw beef	Cross-contamination of other foods	Store on lower shelf	Move to lower shelf away from other foods
		Label, date, and use FIFO rotation	Use first; discard if maximum time is exceeded or suspected
	Bacterial growth and spoilage	Beef temperature must remain below 45°F (7.2°C)	Discard if time and temperature abused
Storing vegetables	Cross-contamination from raw potentially hazardous foods	Label, date, and use FIFO rotation	Discard product held past rotation date
		Keep above raw potentially hazardous foods	Discard contaminated, damaged, or spoiled products

Table 5.1. *continued*

Critical Control	Hazard	Standards	Corrective Action If Standard Not Met
		Preparation	
Trimming and cubing beef	Contamination, cross-contamination, and bacteria increase	Wash hands	Wash hands
		Clean and sanitize utensils	Wash, rinse, and sanitize utensils and cutting board
		Pull and cube one roast at a time, then refrigerate	Return excess amount to refrigerator
Washing and cutting vegetables	Contamination and cross-contamination	Wash hands	Wash hands
		Use clean and sanitized cutting board, knives, utensils	Wash, rinse, and sanitize utensils and cutting board
		Wash vegetables in clean and sanitized vegetable sink	Clean and sanitize vegetable sink before washing vegetables
		Cooking	
Cooking stew	Bacterial survival	Cook **all** ingredients to minimum internal temperature of 165°F (73.9°C)	Continue cooking to 165°F (73.9°C)
		Verify final temperature with a thermometer	Continue cooking to 165°F (73.9°C)
	Physical contamination during cooking	Keep covered, stir often	Cover
	Contamination by herbs and spices	Add spices early in cooking procedure	Continue cooking at least 1/2 hour after spices are added
		Measure all spices, flavor enhancers and additives, and read labels carefully	
	Contamination of utensils	Use clean and sanitized utensils	Wash, rinse, and sanitize all utensils before use
	Contamination from cook's hands or mouth	Use proper tasting procedures	Discard product
		Holding and Service	
Hot holding and serving	Contamination, bacterial growth	Use clean and sanitary equipment to transfer and hold product	Wash, rinse, and sanitize equipment before transferring food product to it
		Hold stew above 140°F (60°C) in preheated holding unit, stir to maintain even temperature	Return to stove and reheat to 165°F (73.9°C)
		Keep covered	Cover
		Clean and sanitize serving equipment and utensils	Wash, rinse, and sanitize serving utensils and equipment

Table 5.1. *continued*

Critical Control	Hazard	Standards	Corrective Action If Standard Not Met
Cooling			
Cooling for storage	Bacterial survival and growth	Cool rapidly in ice water bath and/or shallow pans (<4" deep)	Move to shallow pans
		Cool rapidly from 140°F (60°C) to 45°F (7.2°C) in 4 hours or less	Discard, or reheat to 165°F (73.9°C) and re-cool one time only
		Verify final temperature with a thermometer; record temperatures and times before product reaches 45°F (7.2°C) or less	If temperature is not reached in less than 4 hours, discard; or reheat product to 165°F (73.9°C) and re-cool one time only
	Cross-contamination	Place on top shelf	Move to top shelf
		Cover immediately after cooling	Cover
		Use clean and sanitized pans	Wash, rinse, and sanitize pans before filling them with product
		Do not stack pans	Separate pans by shelves
	Basterial growth in time or after prolonged storage time	Label with date and time	Label with date and time or discard
Reheating			
Reheat for service	Survival of bacterial contaminants	Heat rapidly on stove top or in oven to 165°F (73.9°C)	Reheat to 165°F (73.9°C) within 2 hours
		Maintain temperature at 140°F (60°C) or above; verify temperature with a thermometer	Transfer to preheated hot holding unit to maintain 140°F (60°C) or above
		Do not mix new product into old product	Discard product
		Do not reheat or serve leftovers more than once	Discard product if any remains after being reheated

Source: Reprinted with permission from *Applied Foodservice Sanitation,* 4th edition (pp. 82–84). Copyright © 1992 by The Educational Foundation of the National Restaurant Association, Chicago, Illinois.

Four types of foodservice operations were identified by Unklesbay et al. (1977):

- Conventional or traditional
- Commissary
- Ready prepared
- Assembly/serve

These types of foodservice operations were identified as foodservice systems in the original research. Because the conceptual framework of this text is the managerial foodservice system, the term *food product flow* is being used to avoid confusion. Conceptual diagrams, developed by Unklesbay et al. (1977), illustrate the four paths for food product flow within various foodservice operations and will be discussed.

Coincidental with the evolution of food product flow, the interdependence between food processing and foodservice industries has become more evident, requiring coordination of functions. Food processing is a commercial industry in which food is processed, prepared, packaged, or distributed for consumption in the foodservice operation. The foodservice industry includes restaurants, hospitals, schools, and other specialized operations.

Figure 5.1 is the basis for the separate food product flow diagrams. The distinguishing feature of this diagram is the food processing continuum and its interface with a particular food product flow path and the consumer. At the left side of the continuum, the point of origin of the elongated black triangle indicates no processing of the food entering the food product flow. The increasing altitude of the triangle represents the amount of food processing, reaching a maximum at the far right, required for complete processing. For example, fresh apples, sugar, flour, and shortening used in baking an apple pie are foods with almost no initial processing, and a frozen baked apple pie is purchased completely processed. The continuum is related to food procurement alternatives, depending upon the needs of a particular food product path.

The central block of the diagram, labeled Foodservice Type, covers the production phase of a food product flow path. The requisite interface with the food processing continuum is shown by lines from the appropriate point on the continuum to the

Figure 5.1. Food product flow in a foodservice operation.

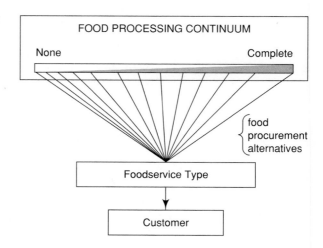

foodservice type block. The distribution of completed menu items is shown by the detached connected block representing customers.

Conventional Foodservice

Conventional foodservice traditionally has been used in foodservice operations. Foods are purchased in various stages of preparation for an individual operation, and production, distribution, and service are completed on the same premises. Following production, foods are held hot or refrigerated, as appropriate for the menu item, and served as soon as possible. The critical control points in Table 5.1 are receiving, storage, preparation, cooking, and holding and service.

In the past, conventional foodservices often included a butcher shop, bake shop, and vegetable preparation unit. Currently, many conventional foodservice operations purchase preportioned meats, baked goods, and fresh, canned, frozen, or pre-processed fruits and vegetables instead of completely processing raw foods on premises.

Although alternative food product flow paths have evolved, the conventional foodservice remains the dominant one in the United States. Managers of conventional foodservices have frequently made changes in purchased ingredients and menu items in order to reduce labor costs in production. Types of food procured for conventional operations vary from no to little or complete processing (Matthews, 1982).

Figure 5.2 illustrates the food product flow for a conventional foodservice. Foods with varying degrees of processing, as shown by the interface lines, are procured and prepared for service in the food production subsystem. As the arrows show, some foods are merely purchased and held chilled before service, such as milk or butter pats, and other menu items are produced from raw foods and held either heated or chilled until time of service. Items such as cereal, bread, and condiments do not go through this cycle but are placed in dry storage for withdrawal as needed.

Following receipt and appropriate storage of food items and ingredients, menu items are prepared as near to service time as possible in order to assure quality. Because food subjected to hot-holding conditions is affected by temperature, humidity, and length of holding time, however, nutritional and sensory quality can be affected adversely.

Foods prepared in the conventional foodservice may be distributed for service directly to an adjacent or nearby serving area, such as a cafeteria or dining room. In hospitals and other healthcare facilities, food may be served on trays, using centralized or decentralized service. In centralized service, individual patient trays are assembled in or close to the production area. Trays are then distributed by carts or conveyors for delivery to patients' rooms. In decentralized service, food is distributed in bulk quantities for tray assembly in an area close to patients' rooms, such as a galley located in a hospital wing. In some facilities, a combination of these two approaches is used.

The systems model in chapter 2 is directly applicable to the conventional foodservice. Foods are brought directly into the operation, menu items are produced using conventional methods, and the meals are then served without extensive holding. In discussions of the other three types of foodservice operations, modifications of the systems model are indicated.

Figure 5.2.   Food product flow in a conventional foodservice.

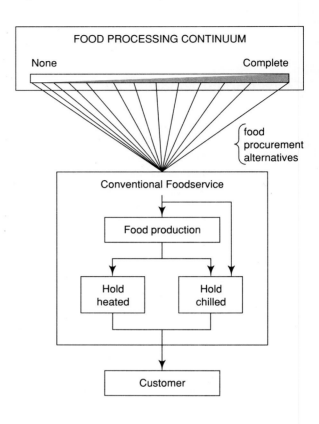

Traditionally, a skilled labor force for food production has been utilized in conventional foodservices for long periods each day. According to Unklesbay et al. (1977), with constantly rising labor costs within the foodservice industry, the current trend in conventional foodservices is to procure more extensively processed foods.

Commissary Foodservice

Technological innovations and the design of sophisticated foodservice equipment have led to the evolution and development of commissary foodservices. Unklesbay et al. (1977) described **commissary foodservices** as centralized procurement and production facilities with distribution of prepared menu items to several remote areas for final preparation and service. Centralized production facilities often are referred to as central commissaries, commissariats, or food factories, and the service units are known as satellite service centers. The menu items usually are delivered hot or cold and served immediately, although some commissaries chill or freeze the items and then reheat them at the remote site. The potential for economies from large-scale purchasing and production in a central facility is a common justification for design and construction of these facilities. In addition, expensive automated equipment for production of foods from unprocessed states is required.

In commissary foodservices, foods purchased have received little or no processing, as indicated in the continuum at the top of the diagram in Figure 5.3. These foods generally are purchased in large quantities and held after delivery under appropriate environmental conditions in dry, refrigerated, or frozen storage. Among the advantages of large-scale purchasing are increased supplier competition and cooperation and volume discounts. Operational advantages include centralized receiving, storage, and inventory control.

Most menu items in commissary foodservices are processed completely in the central facility. Because of the large quantities produced, the equipment for preprocessing and production is often different from that used in conventional foodservices. These large central production units may be designed using equipment frequently seen in food industry operations, such as canneries or frozen food processing plants. Large-scale production quantities also require major modifications of recipes and food production techniques. For example, cooking is actually done in two stages: first to a satisfactory and safe condition for transport and second for complete doneness and acceptable temperature.

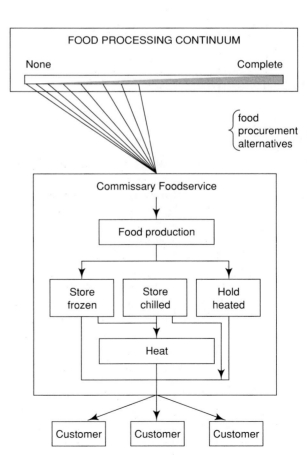

Figure 5.3.   Food product flow in a commissary foodservice.

As Figure 5.3 also shows, foods are held after production either frozen, chilled, or heated for distribution to the satellite service centers. These menu items may be stored in bulk or in individual portions. The type of storage may depend on the time necessary between production and service. In many instances, however, the type of storage for prepared menu items guides the design of the foodservice. For example, a decision to use frozen storage for menu items would be made before proceeding with the design. Many menu items that are held chilled or frozen require additional heating to reach desirable service temperatures. Highly specialized distribution equipment may be needed, depending on the type and location of satellite service centers. Chilling and freezing of food are discussed in detail under ready prepared foodservices.

In Figure 5.4, a modification of the foodservice systems model in chapter 2 illustrates the uniqueness of commissary foodservices. As indicated, the major changes affecting inputs to the system are the type of food and facilities used. In the transformation element, the nature of the functional subsystems differs greatly from the conventional foodservice, primarily due to larger production capacity, storage for prepared menu items, and distribution capabilities for transporting prepared menu items to many satellite centers.

Figure 5.4.   Uniqueness of a commissary foodservice.

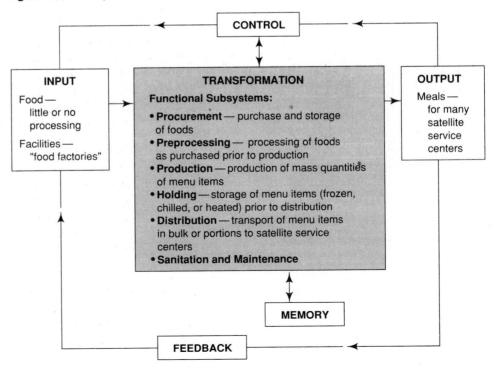

The packaging and storage of prepared menu items present challenges for control in commissary foodservices. Various packaging materials and techniques are now being used, ranging from individual pouches or serving dishes designed for chilled or frozen holding to disposable or reusable metal pans for different types of distribution and transportation equipment. In addition, specialized equipment is required for packaging, storing, and distributing products. Preserving the microbiological, nutritional, and sensory qualities of foods during holding and heating at point of service can present problems. The importance of establishing critical control points for hot holding and service or for cooling and reheating need to be emphasized. In large operations, a food technologist or microbiologist is frequently on staff and is responsible for quality control.

Commissary foodservice principles have been adopted in operations in which service centers are remote from, yet accessible to, the production unit. Reducing the duplication of production, labor, and equipment that occurs if production is done at each service center has been the objective. Space requirements at the service centers can also be minimized because of limited production equipment. The high construction costs of commissaries and transportation equipment are current concerns in evaluating the cost-effectiveness of these foodservices.

Commissary foodservices are especially suitable for service in unique situations. For example, Sky Chefs, Inc., the in-flight caterer, recently built a 122,500 square-foot flight kitchen at Miami International Airport to service American Airlines' more than 100 daily flights out of Miami (King, 1992d). The company's first kitchen, built in 1977, is located in the same airport and services more than 100 daily flights for four airlines. The new kitchen has state-of-the-art equipment and computer capability, which generates accurate requisitions and improves inventory control.

Commissary foodservice has long been used in schools, although many districts have combined it with conventional foodservice. In recent years, centralized production facilities frequently have been constructed in urban districts with a large number of schools. More often, however, large secondary school kitchens serve as commissaries, producing meals for transport either in bulk or individual portions to elementary schools. In this case, the operation is a combination of commissary and conventional foodservices because secondary students also are served in an adjacent cafeteria.

One of the most appealing advantages of a commissary is that the equipment and personnel operate at a high efficiency rate during the day with no idle periods. Also, a commissary often operates 8 hours a day for 5 days a week. With a greater number of dependent service centers, the operation may be extended to multiple shifts, 24 hours a day, 7 days a week. The commissary cannot be operated efficiently unless highly skilled personnel are employed.

The use of a commissary is not restricted to a metropolitan area. Several of the large national restaurant chains operate a mass production commissary at one location and transport their products all over the United States. An example is Pepsico Inc., the parent company for approximately 6,000 Pizza Hut, 6,000 KFC, and 4,000 Taco Bell stores. Pepsico Food Systems (PFS), a subsidiary of Pepsico, service all the 16,000 stores through approximately 23 nationwide commissaries. The mission of PFS is to supply each Pizza Hut, KFC, and Taco Bell unit with all products, equipment, and supplies to do business. PFS provides one-stop shopping for managers, thus giving them more time for the customer. Pepsico is a proponent of TQM and also encourages managers to empower employees.

Ready Prepared Foodservice

Ready prepared foodservices have evolved because of increased labor costs and a critical shortage of skilled food production personnel. In **ready prepared foodservices**, menu items are produced and held chilled or frozen until heated for service later. A significant difference between ready prepared and conventional foodservices is that menu items are not produced for immediate service but for inventory and subsequent withdrawal. Food items, upon receipt, are stored and recorded in a storage inventory. When needed for production, the food item is withdrawn. After production, menu items are stored in refrigerators or freezers and entered in the distribution inventory.

Many of the production, packaging, and storage techniques are similar to those used in commissary foodservices. As first used, the ready prepared foodservice was designed for a single operation. Currently a number of ready prepared foodservices are functioning as commissaries for other facilities within a limited area. Instead of being merely a cost saver, a ready prepared foodservice can be a revenue generator by delivering products to other facilities. As an example, the University of Kansas Medical Center in Kansas City, a 483-bed hospital, operates a ready prepared cook-freeze foodservice in an adjoining building, serving all floors by elevator. In addition to providing meals for patients, the foodservice operation distributes food to the hospital employee and guest cafeterias and the catering department. Meals also are transported to a mental-health and treatment center a few miles away.

Figure 5.5 shows the food product flow in ready prepared foodservices. As indicated in the diagram, foods from the entire spectrum of the food processing continuum are used. Completely processed foods brought into the operation are stored either chilled or frozen, as appropriate to the food item. Items such as cereal and bread do not go through production but are put into dry storage for withdrawal as needed. Foods procured with little or no processing are used to produce menu items that are stored either chilled or frozen. A distinct feature of these foodservices is that prepared menu items are readily available at any time for final assembly and heating for service.

Center-of-the-plate menu items and hot vegetables require two phases of heat processing in ready prepared foodservice. The first occurs during the production of menu items, and the second occurs after storage, when items are brought to the appropriate temperature for service to the consumer.

Cook-chill and cook-freeze are two methods used in ready prepared foodservices. **Cook-chill** technology is moving into all noncommercial market sectors. Although industrywide statistics regarding its overall popularity are lacking, sales of equipment to support the new technology are growing rapidly (Schuster, 1993). In cook-chill operations, most menu items are maintained in the chilled state for as short a time as possible. Most cook-chill foodservice operations are located in hospitals (King, 1991; King, 1992c; King, 1993; Matthews, 1991; Seelye, 1991), but school foodservice

Figure 5.5.   Food product flow in a ready prepared foodservice.

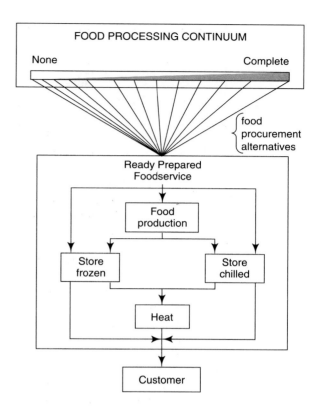

operations also are beginning to use cook-chill technology (Friedland, 1992; King, 1992a). In the **cook-freeze** method, menu items are stored in the frozen state for periods generally ranging from 2 weeks to 3 months.

Sous vide, from the French term for *under vacuum,* is classified as ready serve because it involves chilling and sometimes freezing menu items. Sous vide entails sealing raw, fresh food in impermeable plastic pouches with special equipment. Air is forced out of the pouches by sealing them, and a vacuum is created. The foods are then partially or completely cooked slowly in low-temperature circulating water, rapidly cooled, and stored in temperature controlled refrigerators (32° to 38°F), extending shelf life to about 21 days (Educational Foundation of the NRA, 1992; Jones, 1993; Snyder, 1993). Menu items then are heated for service. The Food and Drug Administration (FDA) states that only licensed food processors can perform the sous vide process. Foodservice operations are allowed to purchase sous vide products only from a reliable supplier that has a HACCP program, as improper handling of the products can cause microbiological health hazards (Birmingham, 1992; Educational Foundation of the NRA, 1992).

As indicated in the diagram (Figure 5.5), procured foods are placed in appropriate storage units and, after production, are held in one of the two forms previously discussed. Accurate forecasting and careful production scheduling are needed to maintain quality of foods and to avoid prolonged holding, especially in the chilled state.

In the cook-chill method, the cooked food is chilled in bulk and then portioned for service several hours or as much as a day in advance of the serving period. In the cook-freeze method, however, the food is portioned immediately after cooking and held in the frozen state until ready for service. In both cook-chill and cook-freeze operations, final heating occurs just before service, often in units near the serving areas. For example, some hospitals have galleys on patients' floors in which specialized heating equipment, generally microwave ovens, is used to finish the preparation of menu items and bring them to the correct serving temperature. Minor preparation, such as toasting bread or making coffee, also may occur in these galleys. Menu items may be slightly undercooked during initial preparation to avoid overcooking and loss of sensory quality in the final heating for service.

In both methods, special recipe formulations are needed for many menu items because of the changes that occur, especially in freezing. Development of off-flavors may be a problem with some food items. Some of these changes may be controlled by substituting more stable ingredients, by exercising greater control of storage time, temperature, and packaging, or by adding stabilizers. To thicken sauces, for example, waxy maize starch, which has much greater thickening ability than flour, may be used to prevent product breakdown during thawing and heating (McWilliams, 1993).

Challenges in ready prepared foodservices are retention of microbiological, nutritional, and sensory qualities of food; the critical control points for cooling and reheating are extremely important in this type of foodservice (Table 5.1). Note that if the time and temperature standards are not adhered to in the ready prepared system, the corrective action is to discard the product. If thawing of frozen products is required prior to heating, it must be done under appropriate conditions in tempering boxes held at a temperature of 40°F and not scheduled too long before service. Prolonged holding is avoided, and careful control in the final heating stage before service is critical. In addition to microwave ovens for this final heating process, immersion techniques and convection ovens have been proven effective. Immersion techniques involve heating pouches of food in boiling water or in steamers and are used for moist food items, such as entrées in sauces. **Convection ovens**, in which air is circulated during the heating process, are effective for heating foods held in bulk because appropriate temperatures are reached more rapidly than in conventional ovens. Further discussion of heating equipment may be found in chapter 15.

The foodservice systems model in chapter 2 has direct application to the ready prepared foodservice. Procurement of food items in various stages of processing greatly affects the activities of the subsystems in the transformation process. Food with little processing passes through production before being placed in either chilled or frozen storage; processed food items bypass production and go directly to storage. The ready prepared foodservice is unique in that it requires separate inventories for purchased and for ready-to-serve food. In the foodservice systems model, the subsystems of distribution and service are emphasized for the ready prepared foodservice.

Ready prepared foodservice, adopted in many operations to reduce labor expenditures and utilize staff more effectively, are expected to increase ("World of Noncommercial," 1992). Peak demands for labor are removed because production is designed to meet future rather than immediate needs. Production personnel can be scheduled for regular working hours rather than during early morning, late evening, and weekend shifts required in conventional foodservices. The heating and service of menu items do not require highly skilled employees, and thus reductions in labor costs are often possible. Food procurement in volume also may decrease food costs in these foodservices. Integration of computer controls with large-scale production allows management to react quickly and accurately to changes in the production environment and service needs of units (Alexander, 1991).

Any food production operation using cook-chill or cook-freeze methods requires special equipment to hold food after the initial production phase. After production, food is portioned into either bulk or individual containers, wrapped, and chilled or frozen. Covered or wrapped pans of bulk food or plates of portioned food items are first placed in a blast chiller or blast freezer, depending on the type of ready pre-pared foodservice. In both kinds of blast equipment, very rapid circulation of cold air reduces the temperature quickly, thus preserving the microbiological, nutrient, and sensory qualities of the food. Bulk food items should be stored no more than 2 inches deep in pans for cooling. In 90 minutes, most foods can be cooled in a blast chiller to temperatures ranging from 33° to 37° F; they should then be transferred to a cold storage unit. According to Longree and Armbruster (1987), refrigerated stor-age for 1 day is ideal. Longer storage times should be used only in rigidly controlled cook-chill foodservices. Once food is removed from refrigeration, it should not be held more than 30 minutes before heating.

Food items prepared for the cook-freeze method are completely frozen in a blast freezer before being moved to a freezer for holding at 0° F or below. Foods frozen in bulk need to be tempered in a rapid thawing refrigerator, which uses forced air or a slightly higher than normal temperature to defrost food safely and quickly. Plated food begins to defrost quickly after removal from the freezer and generally can be held in a regular refrigerator.

Machines may be used for portioning some food items, such as mashed potatoes with melted butter on a plate, and may be part of a sequential service on a conveyor terminating at a shrink-wrap machine. The plate with the portion of food is covered with a special plastic wrap and passed through a shrink tunnel that, by removing air, shrinks the wrap to the shape of the food. The wrapped portions are then put through the blast freezer and finally stored in the inventory freezer.

Assembly/Serve Foodservice

The development of **assembly/serve** foodservices—called *convenience-food food-services* or the *minimal cooking concept*—occurred primarily because of the mar-ket availability of foods that are ready-to-serve or require minimum cooking. Another factor has been the chronic shortage of skilled personnel in food production and the increasing cost of labor.

Food product flow in assembly/serve foodservices is illustrated in Figure 5.6. As the diagram shows, food products are brought into the operation with a maximum degree of processing. Only storage, assembly, heating, and service functions are commonly performed in these foodservices, thus reducing labor and equipment costs. Little if any preprocessing is done on-site, and production is very limited.

Fresh, frozen, and dried items like shredded lettuce, sliced carrots, beef stew, pre-portioned roast beef, and lasagna, for example, are purchased in an assembly/serve foodservice.

The three market forms of foods used predominantly in these foodservice operations are bulk, preportioned, and preplated. The bulk form requires portioning before or after heating within the foodservice operation, whereas the preportioned market form requires only assembly and heating. The preplated products require only heating for distribution and service, and thus are the most easily handled of the three forms.

In many assembly/serve operations, a combination of foods is used, some requiring a limited degree of processing in the foodservice operation and others requiring none. Often, partially prepared foods are purchased to be combined with other ingredients before heating or chilling. In many operations, completely processed foods may be enhanced in assembly/serve foodservices as a way of individualizing menu items; for example, a sauce may be added to an entrée.

Following procurement, food items are held in dry, refrigerated, or frozen storage. When menu items are heated in either bulk or preportioned form, quality control is a definite concern. Thawing of frozen food must be done with rigorous control of time and temperature. The same technology that made ready prepared foodservices feasible also is utilized in assembly/serve operations.

Figure 5.6. Food product flow in an assembly/service foodservice.

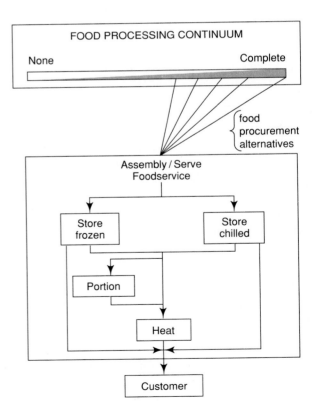

The assembly/serve foodservice has gained some degree of acceptance because it appears to offer an easy solution to labor and production problems. Nevertheless, a readily available supply of highly processed, high-quality food products is a prerequisite for a successful assembly/serve operation. Availability of such items is often a problem, especially for special diets. Although modified dietary products have been developed in recent years, these items are not always readily available, particularly in rural and small communities. Therefore, if an assembly/serve foodservice is used for hospital patients or nursing home residents, menu items may be needed to be prepared for those on modified diets.

Another common complaint about assembly/serve operations is the lack of individuality of food products. These complaints duplicate the comments frequently heard about the "sameness" of convenience foods available on the retail market. As discussed in the section on conventional foodservices, a trend toward the use of foods with some degree of processing is evident. This trend appears to be more predominant than total adoption of the assembly/serve foodservice. In some instances, however, an assembly/serve foodservice meets the needs of particular operations in which space is very limited for production facilities or competent labor is not available.

Introducing the assembly/serve foodservice into a conventional facility is not economically feasible because of unused equipment and excess personnel. The combination of the higher cost of the menu items and the excess labor cost in a facility not designed for a convenience foodservice can make it very expensive. For realization of full economic benefits, fewer pieces of equipment and a greatly reduced number of personnel are needed.

The ultimate in a hospital convenience operation is the "kitchenless" patient foodservice program developed by Jerome Berkman, registered dietitian, at Cedars-Sinai Medical Center in Los Angeles (Berkman & Schechter, 1991). By using primarily frozen retail convenience products for patients' menus, 5,700 square feet of medical center space were released for revenue-generating medical services, and the kitchen labor force was reduced (Boss, 1991). With these new products and the increased contact of the foodservice staff with patient customers, satisfaction scores reached levels never achieved before. Berkman made the statement that hospital foodservice has entered a time when activities must emphasize foodservice, not food production. Perhaps hospital foodservice is leaving the era of high-tech, central production, and entering the age of high-touch, enhanced customer service (Berkman & Schechter, 1991).

SUMMARY

Changes in food technology have influenced the food product flow in the foodservice subsystems. To ensure the quality and safety of menu items, physical, chemical, and microbiological changes have to be monitored by using the Hazard Analysis Critical Control Point model.

Spiraling labor costs and technological innovation in both food products and equipment have led to new types of foodservice operations. Production and service have been separated, permitting production to occur independently of service time, whether on or off site. Four major types of foodservices are identified: conventional or traditional and the more recent commissary, ready prepared, and assembly/serve. The advent of food processing as a commercial industry has had a great deal of influence on quantity food production. The customer is the basis for a model showing the interface of food processing at various stages.

The conventional foodservice traditionally has been used in most operations. Foods are purchased in various stages of processing, and production, distribution, and service are completed on the same premises.

In commissary foodservices, foods are purchased with little or no processing but are processed completely in the central facility. Menu items generally are transported in bulk to satellite service centers in the local area.

The ready prepared foodservice is similar to the commissary and was originally designed for a single operation. Foods with varying degrees of processing are purchased, and the amount of production depends upon the state of the purchased food. Menu items are stored either chilled or frozen and are readily available at any time for final assembly and heating for service. Challenges in ready prepared systems are retention of microbiological safety, nutrient content, and sensory qualities of food requiring monitoring of critical control points.

Completely processed food items are purchased in the assembly/serve foodservice, and only storage, assembly, heating, and service functions are performed in the operation. High labor cost and a shortage of competent employees have led to the use of these foodservice operations.

BIBLIOGRAPHY

Alexander, J. (1991). Cook-chill automation. *Hospital Food Service, 24*(3), 7–8.

Berkman, J., & Schechter, M. (1991). Today, I closed my kitchen. *Food Management, 26*(11), 110–114, 118, 122.

Birmingham, J. (1992). Whatever happened to sous vide? *Restaurant Business, 91*(6), 64–65, 68, 72.

Bobeng, B. J., & David, B. D. (1978). HACCP models for quality control of entreé production in hospital foodservice systems. I. Development of Hazard Analysis Critical Control Point models. II. Quality assessment of beef loaves utilizing HACCP models. *Journal of The American Dietetic Association, 73*, 524–535.

Boss, D. (1991). A kitchenless future? *Food Management, 26*(11), 16.

Eck, L. S., & Ponce, H. (1993). HACCP: A food safety model. *School Food Service Journal, 47*(2), 50–52.

Educational Foundation of the National Restaurant Association. (1992). *Applied foodservice sanitation* (4th ed.). Chicago: Author.

Escueta, E., Fiedler, K., & Reisman, A. (1986). A new hospital foodservice classification system. *Journal of Foodservice Systems, 4*, 107–116.

Fox, M. (1993). Quality assurance in foodservice. In M. Kahn, M. Olsen, & T. Var (Eds.), *VNR's encyclopedia of hospitality and tourism* (pp. 148–155). New York: Van Nostrand Reinhold.

Frakes, E. M., Arjmandi, B. H., & Halling, J. F. (1986). Plate waste in a hospital cook-freeze production system. *Journal of The American Dietetic Association*, 86, 941–942.

Franzese, R. (1981a). Food service systems of 79 hospitals studied. *Hospitals*, 55(3), 64–66.

Franzese, R. (1981b). Survey examines hospitals' use of convenience foods. *Hospitals*, 55(2), 109–110, 112.

Franzese, R. (1984). Food services survey shows delivery shift. *Hospitals*, 58(16), 61, 64.

Friedland, A. (1993). 3 districts, 1 kitchen. *Food Management*, 47(5), 42–43.

Glew, G., & Armstrong, J. (1981). Cost optimization through cook-freeze systems. *Journal of Foodservice Systems*, 1(3), 235–254.

Greathouse, K. R., & Gregoire, M. B. (1988). Variables related to selection of conventional, cook-chill, and cook-freeze systems. *Journal of The American Dietetic Association*, 88, 476–478.

Greathouse, K. R., Gregoire, M. B., Spears, M. C., Richards, V., & Nassar, R. F. (1989). Comparison of conventional, cook-chill, and cook-freeze foodservice systems. *Journal of The American Dietetic Association*, 89, 1606–1611.

Halling, J. F., & Frakes, E. M. (1981). Case history: Product oriented production in a cook freeze system. *Journal of Foodservice Systems*, 1(4), 355–361.

Hospital patient feeding systems. (1982). Washington, DC: National Academy Press.

Hsu, D. M. (1977). The minimal cooking concept: A system in review. *Journal of The American Dietetic Association*, 70, 510–513.

Jones, P. (1993). Foodservice operations management. In M. Khan, M. Olsen, & T. Var (Eds.), *VNR's encyclopedia of hospitality and tourism* (pp. 27–36). New York: Van Nostrand Reinhold.

Jones, P. & Heulin, A. (1990). Foodservice systems—generic types, alternative technologies, and infinite variation. *Journal of Foodservice Systems*, 5, 299–311.

King, P. (1991). Merging for efficiency. *Food Management*, 26(6), 52, 54.

King, P. (1992a). Central production scores in America's heartland. *Food Management*, 27(9), 58, 63.

King, P. (1992b). Implementing a HACCP program. *Food Management*, 27(12), 54, 56, 58.

King, P. (1992c). Massachusetts General's 'patient-flexible' system. *Food Management*, 27(5), 70, 72, 74.

King, P. (1992d). Sky Chefs enters new era with central kitchen. *Food Management*, 27(1), 48, 50.

King, P. (1993). A shift in philosophy. *Food Management*, 28(8), 40–41.

King, P., & Boss, D. (1991). The politics of a renovation. *Food Management*, 26(7), 104–117.

Klein, B. P., Matthews, M. E., & Setser, C. S. (1984). *Foodservice systems: Time and temperature effects on food quality*. North Central Regional Research Bulletin No. 293, Urbana-Champaign: University of Illinois Agricultural Experiment Station.

Kooser, R., & Forgac, J. (1985). Commissary contemplation. *Restaurant Hospitality*, 69(11), 117–118, 120.

Kujava, G. (1991). Cook-chill technology: Serving savings on every tray. *The Consultant*, 24(2), 22–24.

Livingston, G. E. (1968). Design of foodservice system. *Food Technology*, 22(1), 35–39.

Livingston, G. E. (1990). Foodservice: Older than Methuselah. *Food Technology*, 44(7), 54, 56, 58–59.

Longree, K., & Armbruster, G. (1987). *Quantity food sanitation* (4th ed.). New York: Wiley.

Lundberg, D. E. (1984). A look at restaurant commissaries. *FIU Hospitality Review*, 2(1), 7–13.

Matthews, L. (1991). Gearing up for cook-chill. *Food Management*, 26(3), 39.

Matthews, M. E. (1977). Quality of food in cook/chill foodservice systems: A review. *School Food Service Research Review, 1*(1), 15–19.

Matthews, M. E. (1982). Foodservice in healthcare facilities. *Food Technology, 36*(7), 53–55, 58–60, 62–64, 71.

McWilliams, M. (1993). *Foods: Experimental perspectives* (2nd ed.). New York: Macmillan.

Mermelstein, N. H. (1993). Controlling *E. coli* 0157:H7 in meat. *Food Technology, 47*(4), 90–91.

National Restaurant Association. (1991). *Make a S.A.F.E. choice: Sanitary assessment of food environment.* Washington, DC: Author.

Niepold, C. (1986). Commissaries: Large opportunities for small operators. *Restaurants USA, 6*(9), 25–26.

Norton, C. (1991). What is cook-chill? *Hospital Food Service, 24*(2), 5–7.

Palmer, J. (1984). The logic and logistics of commissaries. *Cornell Hotel and Restaurant Administration Quarterly, 25*(1), 104–109.

Rethermalization—A hot topic! (1991). *Hospital Food Service, 24*(4), 6–7.

Rhodehamel, E. J. (1992). FDA's concerns with sous vide processing. *Food Technology, 46*(12), 73–76.

Rinke, W. J. (1976). Three major systems reviewed and evaluated. *Hospitals, 50*(4), 73–78.

Schuster, K. (1993). Is your future in cook-chill? *Food Management, 28*(7), 90–91, 94, 96, 98, 123–124.

Seelye, K. H. (1991). The downsizing of cook-chill. *Food Management, 26*(11), 50.

Snyder, O. P. (1993). Hazard analysis and critical control point in foodservice. In M. Khan, M. Olsen, & T. Var (Eds.), *VNR's encyclopedia of hospitality and tourism* (pp. 185–219). New York: Van Nostrand Reinhold.

Snyder, O. P., Gold, J. I., & Olson, K. A. (1987). Quantifying design parameters for foodservice systems in American hospitals. *Journal of Foodservice Systems, 4*(3), 171–186.

Stephenson, S. (1992). New cook-chill system earns high grades. *Restaurants & Institutions, 102*(7), 112, 114.

Unklesbay, N. (1977). Monitoring for quality control in alternate foodservice systems. *Journal of The American Dietetic Association, 71*, 423–428.

Unklesbay, N., Maxcy, R. B., Knickrehm, M., Stevenson, K., Cremer, M., & Matthews, M. E. (1977). *Foodservice systems: Product flow and microbial quality and safety of foods.* North Central Regional Research Public. No. 245. Columbia: University of Missouri-Columbia Agricultural Experiment Station.

Walkup, C. (1994). NRA Educational Foundation: Industry's classroom. *Nation's Restaurant News.* [The National Restaurant Association 1919–1994. Special Commemorative issue.] 84, 89, 96.

Woodburn, J. M. (1986). Is a commissary right for your operation? *Restaurants USA, 6*(9), 27–29.

World of noncommercial foodservice according to . . . (1992). *Food Management, 27*(1), 121–123, 128.

The Menu: The Primary Control of the Foodservice System

The **menu,** a list of food items, serves as the primary control of the foodservice operation and is the core common to all functions of the system (Figure 6.1). The menu controls each subsystem and is the major determinant for the budget. In a new operation, the menu governs the layout and equipment to produce the specified items. To the production employee, the menu indicates work to be done; to the waitstaff,

Figure 6.1. The menu: the primary control of the foodservice system.

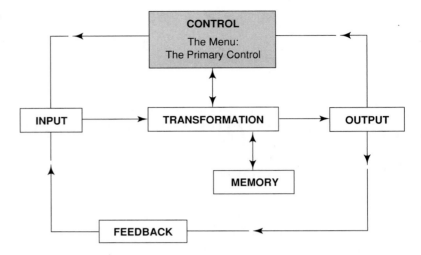

the foods to be served; and to the dishroom staff, the number and types of dishes, glasses, and flatware requiring washing and sanitizing. According to Kotschevar (1987), a good menu should lead customers to food selections that satisfy their dining preferences as well as the merchandising priorities of management. A hospital, school foodservice, and limited-menu or gourmet restaurant have different menus because their objectives and customers are different. For example, most multiunits use a limited menu to aid the customer in making quick selections. Such a menu promotes high turnover to compensate for lower check averages. The menu, therefore, expresses the character of a foodservice operation and is largely responsible for its reputation, good or bad.

MENU PRESENTATION

The menu provides a list of food items available for selection by a customer, thus becoming a sales tool for foodservice operations. An additional function of the menu is customer education. For example, the menu is a valuable tool for teaching children about good nutrition through the school lunch program and for instructing hospital patients on modified diets. Menus may be presented in many different forms ranging from the visual, like the typical printed copies, to the spoken, in which a waitperson recites it. McDonald's introduced braille menus in 1979 and picture menus in 1988 for customers with vision, hearing, or speech impairments. These menus were updated in 1992 for its 8,700 domestic restaurants ("McD Rolls Out," 1992). Red Lobster has had braille menus since the mid-70s and recently developed large-type menus for the growing population of people over age 55, which is the largest consumer group with visual impairments (American Express, 1992). To help businesses comply with the Americans with Disabilities Act, The Lighthouse, a New

York organization that helps people overcome visual impairments, will translate menus into braille for a small charge (Lorenzini, 1992b).

The use of a menu board is excellent for restaurants in which menus are planned daily according to product availability. Many restaurants use this technique to advertise daily specials. Menus can be handwritten on chalkboards or on fluorescent illuminated blackboards. Letters and numbers are commercially available in a wide selection of colors for mounting on grooved felt boards.

The video menu board, a recent innovation in menu merchandising, has the advantage of being readily changed. Current menus are displayed on video monitors that are placed in locations easily seen by customers. The video menu system has been used in plant and corporate dining facilities and has application for hospitals to inform patients of daily menu offerings. Video menu software is available and may be obtained in a range from black and white to complex color graphics. Ordering by computer currently is being developed. Some corporate cafeterias have made their menus available to business computer monitors, with workers able to order a meal for delivery to their offices. The major disadvantage of the video menu system is the high cost of software and equipment, which may be reduced as more systems are developed.

Facsimile machines electronically transmit printed or photographed material over telephone lines for reception and reproduction by the receiving machine. Many restaurants are providing a fax number to customers to place orders for take-out, delivery, or catered meals ("Tableservice Restaurants," 1993).

Often, when table d'hôte meals are served, a verbal menu recited by the waitstaff is used. *Table d'hôte*, a French term for the host's table, is a complete meal consisting of several courses at a fixed price. A very popular five-course prix fixe dinner is served once a month at the Goose Island Brewing Co., an independent brewery restaurant in Chicago (Straus, 1992a). Hearty German food, such as oxtail soup, sausages, veal shanks, and potato pancakes, is served with ample amounts of microbrewery and imported beers for $35. Jazz or polka music also is played on prix fixe nights.

MENU PATTERN

A *menu pattern* is an outline of the menu item categories for each meal, for example appetizers, entrées, and desserts on the dinner menu. One of three basic types of menus generally is used, and the number of menu item choices in each of these can vary according to the goals of the foodservice operation.

Types of Menus

The three types of menus are static, cycle, and single use. Deciding which type to use depends upon expectations of the customer when eating away from home. For example, college students away from home eat many of their meals on campus and, therefore, like variety; when they go off campus, however, they often choose a limited-menu restaurant because they want a cheeseburger and french fries.

Static

A **static menu** is one in which the same menu items are offered every day. Traditionally, the static menu has been characteristic of many restaurants, but today many hospitals are using **restaurant-type menus**, as shown in Figure 6.2.

The entire concept for a restaurant frequently is built around the menu, including the decor, advertising campaign, and market segment identified as the target audience. An example is the Broker Restaurant that has been a Denver, Colorado, tradition in romantic dining since 1972. The restaurant, nestled in a turn-of-the-century bank vault, is located in the old Denver National Bank, built around 1903. Beyond the 23-ton round-shaped vault door, customers dine in cherry wood booths once used by bank customers to inspect safety deposit boxes. The European antiques and the vault, coupled with the flavor of Wall Street, continue to make The Broker one of Denver's most original restaurants. The dinner menu is an 8½-by-14-inch four-page booklet printed in brown ink on tan parchmentlike paper (Figure 6.3). The page labeled "Preferred Stock" consists of **center-of-the-plate** menu choices with a suggested wine for each. Accompaniments for each preferred-stock menu item are on the "Dividends" page, and they include the Famous Broker Shrimp Bowl, which consists of a pound and a half of Gulf shrimp in the shell on a bed of ice; choice of soup du jour or Broker Caesar salad; oven-baked breads; fresh vegetables; and dessert. Prices vary according to the center-of-the-plate choice; alcoholic beverages are priced separately.

On a larger scale, Applebee's Neighborhood Grill & Bar has introduced a new children's menu that lists five menu items (Figure 6.4). On the back of the menu are different games, like tic-tac-toe, word search, connect-the-dots, and mazes. Young customers receive crayons and stickers with the menu. The menu was pilot tested by children, and their ideas were incorporated into the menu (Allen, 1993).

Cycle

A **cycle menu** is a series of menus offering different items each day on a weekly, biweekly, or some other basis, after which the cycle is repeated. In many noncommercial operations, seasonal cycle menus are common; for example, a 3-week menu for winter, spring, summer, and fall may be repeated during each season. Cycle menus typically are used in healthcare institutions and schools, offering variety with some degree of control over purchasing, production, and cost. Figure 6.5 shows a typical day from a 9-day cycle menu developed by Food and Nutrition Services in Shands Hospital at the University of Florida, Gainesville. Concern for the environment is emphasized by featuring the manatee, an endangered species, on the recycled paper menu and place mat. A cook-chill foodservice is used to feed 564 patients. In addition, the department is responsible for five satellite cafeterias and 10 off-site clinics plus vending and catering services.

Figure 6.2. Example of a restaurant-style hospital menu.

NOON MEAL CHOICES
Choose one Entree from either the Chilled or Hot Entrees

ᔆGREEN PEPPER stuffed with Ground Beef and Onion in a Tomato Sauce on a bed of Rice.
ᔆHAMBURGER on a Bun with Tomato & Lettuce.
ITALIAN SPAGHETTI with Meat Sauce accompanied by crisp Broccoli Spears.
ᔆROAST BEEF RIBEYE accompanied by Peas with Pearl Onions and Carrots.
SWISS STEAK braised in a zesty Tomato Sauce accompanied by a Baked Potato and Leaf Spinach.
FILET OF FISH baked in a Dill Sauce and served with Broccoli Spears.
SALMON LOAF served with a Vegetable Medley.
BARBEQUED HAM on a Bun with Tomato and Lettuce.
ᔆBAKED BREAST OF CHICKEN with Scalloped Potatoes and Mixed Greens.
ROAST BREAST OF TURKEY topped with a mild Cheese and Mushroom Sauce and served with Corn and Broccoli Spears.
VEGETABLE QUICHE with a Hot Spiced Peach.
HOT DOG on a Bun served with Tomato & Lettuce.
MACARONI WITH CHEDDAR CHEESE accompanied by Seasoned Green Beans.
HOT HAM & TURKEY ON RAISIN BREAD SANDWICH slices of Ham and Turkey on Raisin Bread covered with a light Cheese Sauce and accompanied by Hot Spiced Peach.
FRIED CHICKEN THIGHS served with Peas.
PIZZA Thick Crust Individual with Pepperoni and Cheese.

ᔆROAST BEEF GROUND served with Mashed Potatoes and choice of Carrots or Peas.
ᔆROAST TURKEY GROUND served with Mashed Potatoes and choice of Carrots or Peas.
PUREED FOOD COMBINATION including Meat, Vegetable and Fruit.
ᔆMASHED POTATOES Brown Gravy or Cream Gravy.

GARDEN SPOT SALADS
Fancy Mixed Fruit Salad
Assorted Crisp Garden Relishes
Fresh Spinach Salad & Dressing of your choice
Tossed Green Salad and Dressing of your choice
Gelatin Salad
(See Noon and Evening Choices below)
Cottage Cheese
Hard Cooked Egg

SOUPS
Beef Broth
Chicken Broth
Chicken Noodle
Cream of Tomato
Cream of Mushroom

DESSERTS
Cherry Pie
Carrot Cake with Cream Cheese Icing
Peanut Butter Cookies

EVENING CHOICES
Choose one Entree from either the Chilled or Hot Entrees.

ᔆHAMBURGER on Bun with Tomato & Lettuce.
ROAST RIBEYE OF BEEF au jus served with Mashed Potatoes and seasoned Green Beans.
SAVORY BEEF STEAK smothered in a Mushroom Sauce plus Au Gratin Potatoes and Vegetable Blend.
TACOS all the "fixins" for two Mexican Tacos-Meat Sauce, a salad of Tomatoes, Lettuce and Cheese, two Tortilla Shells and Taco Sauce.
BREADED FILET OF FISH baked, served with Peas and Mushrooms on Rice.
ᔆTUNA NOODLE CASSEROLE with Egg Noodles and Mushrooms accompanied by Carrots.
RED BEANS AND RICE flavored with mild spices and Ham accompanied by Steamed Rice
ROAST TENDERLOIN OF PORK on Noodles with Sour Cream, Onion and Mushroom sauce with Carrots on the side.
ᔆGINGER CHICKEN with Broccoli, Carrots, Water Chestnuts in a light Oriental Sauce accompanied by Steamed Rice.
FRIED BREAST OF CHICKEN served with Mashed Potatoes and a special Broccoli Casserole.
ᔆROAST TURKEY BREAST served with Cracked Wheat Dressing and Green Beans.
SCALLOPED VEGETABLE CREPES with Sour Cream Celery Sauce.
MEAT LOAF served with Parsley seasoned Potatoes.
HOT DOG on a Bun served with Tomato and Kidney Beans.
CHILI mildly spiced with Ground Beef and Kidney Beans.

ᔆROAST BEEF GROUND served with Mashed Potatoes and choice of Green Beans or Mashed Winter Squash.
ᔆROAST TURKEY GROUND served with Mashed Potatoes and choice of Green Beans or Mashed Winter Squash.
PUREED FOOD COMBINATION including Meat, Vegetable and Fruit.
ᔆMASHED POTATOES Brown Gravy or Cream Gravy.

GARDEN SPOT SALADS
Combination Fruit Cup
Creamy Coleslaw
Tossed Green Salad and Dressing of your choice
Gelatin Salad
(See Noon and Evening Choices below)
Cottage Cheese
Hard Cooked Egg

SOUPS
Beef Broth
Chicken Broth
Vegetable Beef
Cream of Potato
Chicken Noodle

DESSERTS
Apple Pie
Lemon Layer Cake
Oatmeal Raisin Cookies

─── NOON OR EVENING CHOICES ───

LIGHTER MEALS

DELUXE CLUB SANDWICH Turkey, Ham and Cheese on Whole Wheat and White Bread with Potato Chips and Pickles.
CHEF'S SALAD PLATE crisp Greens topped with Turkey, Ham and Cheese strips, Tomato and Hard-Cooked Egg Half with your choice of Dressing.
COTTAGE CHEESE FRUIT PLATE with a slice of Pineapple in its own juice and an unsweetened Peach half
ᔆTUNA SALAD PLATE Tuna Salad with a Whole Wheat Bun and Pineapple Rings in their own juice.
PASTA CHICKEN SALAD PLATE White Meat Chicken, Spiral Pasta, and Vegetables in a Ranch Dressing served with a Whole Wheat Honey Bagel.

CREATE YOUR OWN SANDWICHES
or
Order Individually

SANDWICH FIXINGS
MEATS & CHEESES
1oz. Served on Lettuce Leaf
Bologna Slice
ᔆTurkey Slice
Ham Slice
ᴸRoast Beef Slice
American Cheese

Peanut Butter

SPREADS
Polyunsaturated Margarine
Butter
Honey
Jelly
Apple Spread

BREADS
Banana Nut Bread
Cinnamon Roll
Dinner Roll
White Bread
Whole Wheat Bread
Saltine Crackers

THE LITTLE EXTRAS
Catsup
Mustard
Tartar Sauce
Pickle Relish
Salad Dressing
Light Mayonnaise

GELATIN SALADS

Monday	Strawberry
Tuesday	Carrot Pineapple
Wednesday	Pineapple Coconut
Thursday	Cranberry
Friday	Applesauce & Cream Cheese
Saturday	Bing Cherry
Sunday	Fruit Cocktail

SALAD DRESSINGS AND SEASONINGS
French Dressing
Low Calorie French Dressing
Thousand Island Dressing
Italian Dressing
Low Calorie Italian Dressing
Creamy Horseradish Dressing
Lemon Juice
Vinegar
Salt
Salt Substitute
ᔆSaltless Seasoning
*Pepper

BEVERAGES
Apple Juice
Cranberry Juice
Grape Juice
Grapefruit Juice
Orange Juice
Pineapple Juice
Prune Juice
V-8 Juice
Loᵥy-8 Juice
Skim Milk
2% Milk
Whole Milk
Buttermilk
*Chocolate Milk
*Hot Chocolate
Vanilla Milkshake
*Coffee
*Decaffeinated Coffee
*Hot Tea
*Decaffeinated Hot Tea
*Iced Tea
Lemon Wedge
Coffee Creamer
Non Dairy Creamer
Sugar
Sugar Substitute

CHILLED FRUITS AND DESSERTS
Apple
Orange
Fruit in Season
Apricot Halves
Applesauce
Peach Halves
Pear Halves
Pureed Fruit
Graham Crackers
Flavored Gelatin
Baked Custard
Vanilla Pudding
*Chocolate Pudding
Orange Sherbet
Pink Lemonade Sorbet
Frozen Yogurt
Vanilla Ice Cream
*Chocolate Ice Cream

Source: From the University of Kansas Medical Center, College of Health Sciences and Hospital, Kansas City, KS. Used by permission.

149

Figure 6.3.   Example of a static menu from a full-service restaurant.

Source: From The Broker, Denver, CO. Used by permission.

Single Use

The last of the three basic menu types, the **single-use menu,** is planned for service on a particular day and is not used in the exact form a second time. This type of menu is used most frequently in noncommercial foodservice in which the customer does not vary from day to day. Because of the stability of a student population, many college and university foodservices utilize a single-use menu as a "monotony breaker." An example of a unique single-use menu is shown in Figure 6.6. Kansas State University parodied the television show *Saturday Night Live* for the school's annual special dinner in three residence hall dining rooms. The theme was "LIVE from K-State . . . It's Thursday Night! a dining experience celebrating 18 years of Saturday Night Live." Dining tables were covered with recyclable paper to be used by creative customers. Centerpieces were small easels with bathtub drawings; play-dough balls (made in the kitchens) and wax bathtubs with bubbles (whipped paraffin wax) held crayons. Some employees were dressed as Toonces the Cat, Samurai Warrior, the Blues Brothers, Mr. Bill, and Killer Bee, characters from the program for whom menu items were named.

Figure 6.4. Example of a static children's menu.

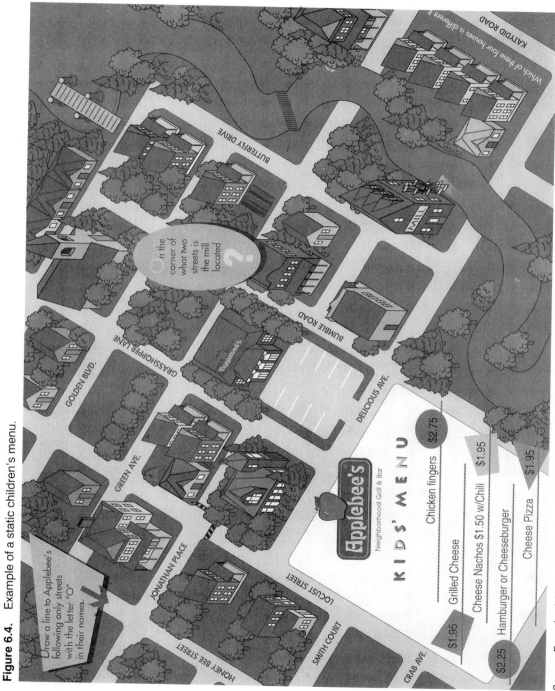

Source: From Applebee's Neighborhood Grill & Bar, Overland Park, KS. Used by permission.

151

Figure 6.5. Example of a general menu from a 9-day cycle.

Breakfast

Please Circle Your Selections

Eye Openers

•Orange Juice
Apple Juice
Cranberry Juice
Prune Juice
Banana
Fresh Fruit in Season

Cereals "n" Starches

•Southern Style Grits
Hot Oatmeal
Cornflakes
Raisin Bran
All Bran
Rice Krispies

Breakfast Entrees

•Scrambled Eggs
Scrambled Egg Substitute
Pancakes with Syrup
Bacon
Sausage Patty
Fruited Yogurt

Breads "n" Spreads

•Hot Biscuit
Blueberry Muffin
Donut
Bran Muffin
Whole Wheat Toast
White Toast

•Margarine   •Jelly   Honey

Beverages

•Coffee
Caffeine Free Coffee
Hot Tea
Herbal Tea
Caffeine Free Tea
Milk
Skim Milk
Buttermilk
Hot Chocolate
Lemon
•Non-Dairy Creamer
Artificial Sweetener

REGULAR _____   Day 1 _____

Name _____   Room _____

Lunch

Please Circle Your Selections

Savory Beginnings

Cream of Broccoli Soup
•Tossed Garden Salad with Dressing   Diet Italian
Melon Ring with Fresh Fruit

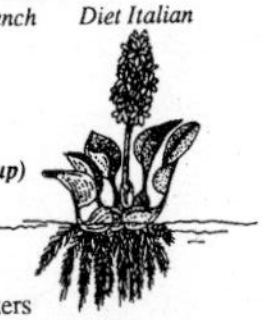

Ranch   1000 Island   Italian   French   Diet French

Main Course

(Limit selection to one in this group)

•Fried Chicken
Beef Stew
L •Herb Broiled Fish
Chef Salad with Assorted Crackers   1000 Island   Italian

Specialty Dressings: Ranch   1000 Island

Hot Vegetables of the Day

Whipped Potatoes with Gravy
•Macaroni and Cheese
•Green Peas w/ Garlic & Onions
Stewed Apples
Seasoned Cabbage

Breads "n" Spreads

•Hot Biscuit
Dinner Roll
White Bread
Whole Wheat Bread
Crackers
Honey

•Margarine   •Jelly   Honey

Sweet Endings

Lemon Coconut Cake
•Cherry Pie
Sliced Peaches
Vanilla Ice Cream

Beverages

•Iced Tea
Hot Tea
Herbal Tea
Caffeine Free Tea
Coffee
Caffeine Free Coffee
Milk
Skim Milk
Buttermilk
Chocolate Milk
Hot Chocolate
Lemon
Non-Dairy Creamer
Artificial Sweetener

L Indicates entrees that are lower in fat, calories and cholesterol

REGULAR _____   Day 1 _____

Name _____   Room _____

Dinner

Please Circle Your Selections

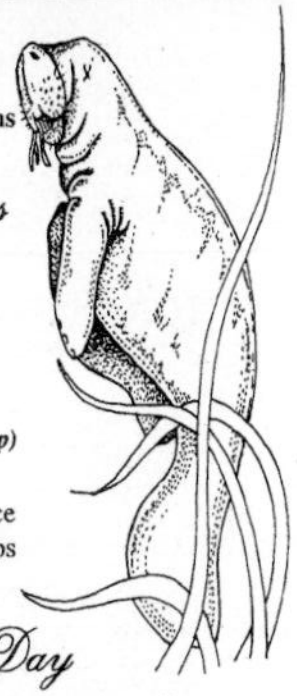

Savory Beginnings

Herbed Tomato Soup
Marinated Cucumber Salad
•Arabian Peach Gelatin Salad

Main Course

(Limit selection to one in this group)

L •Roast Beef
L Sweet and Sour Pork with Rice
Club Sandwich with Potato Chips

Hot Vegetables of the Day

Steamed Rice with Gravy
•Scalloped Potatoes
•Green Beans Almondine
Sauteed Peppers, Onions
& Mushrooms
Spinach Souffle

Breads "n" Spreads

•Dinner Roll
Whole Wheat Bread
White Bread
Crackers
Potato Chips

•Margarine   •Jelly   Honey

Sweet Endings

•Banana Pudding
Peanut Butter Cookie
Sherbet
Fresh Fruit in Season

Beverages

•Iced Tea
Hot Tea
Herbal Tea
Caffeine Free Tea
Coffee
Caffeine Free Coffee
Milk
Skim Milk
Buttermilk
Chocolate Milk
Hot Chocolate
Lemon
Non-Dairy Creamer
Artificial Sweetener

L Indicates entrees that are lower in fat, calories and cholesterol

REGULAR _____   Day 1 _____

Name _____   Room _____

Source: From Shands Hospital at the University of Florida, Gainesville, FL. Used by permission.

152

Figure 6.6. Example of a single-use menu.

LIVE from K-State...
It's Thursday Night!

MID-WEEK UPDATE

This just in ...April 15, 1993 Kansas State University Housing and Dining Services will present - without dress rehearsal - a live Dining experience celebrating 18 years of Saturday Night Live. Go ahead, "Draw On" your personal memories of SNL, and doodle bathtub drawings at your table! For your enjoyment there will be Cheeseburger... Cheeseburger... No Coke, Pepsi - NOT!

"Church Lady" Punch...This recipe was deemed heavenly by the "Church Lady." It's devilishly sour with Cranberry Juice, and innocently sweet with Sherbet islands. Now, isn't that special?

ENTREES AND ACCOMPANIMENTS

"Not Ready for Prime Time" Prime Rib...but it will be ready for you no matter how much "moo" you want left in your beef.

"Toonces" Favorite" Chicken Breast...Go on, drive on the wild side - go over the edge with our Creamy Garlic Basil Sauce drizzled on top.

"Samurai Warrior" Swine with Sweet Sour Mustard or Barbecue Sauce... Step right up and enjoy this Whole Roast Pig, but look out - the Samurai Warrior may be ready to carve more than sow.

"Blues Brothers'" New Potatoes with Lemon Butter Sauce...You won't be singing the blues after a taste of these soulful potatoes.

"Father Guido Sarducci" Spinach Fettuccine...This fettuccine was rated as the best by the good Father. Definite strong ties to Rome.

"Two Wild & Crazy Rices"...The Czech Brothers discovered these rices in America, and cooked them every time foxes came to dinner.

"Mr. Bill's Creamed (OHH NOOO) Corn"...Sluggo visited Mr. Bill just as he was sitting down to his favorite meal of Whole Kernel Corn - OHH NOOO!

"Jane, You Ignorant... Stir Fry"...Actually Jane knew that this Stir Fry of Sugar Snap Peas and fresh Tomatoes was a perfect accompaniment to any meal - sorry Dan - you'll have to eat it this time!

BREADS

"Mrs. Loopner's" Secret Crescent Rolls...These rolls were a staple in Lisa's nerdy household, but were never served to Todd - Todd preferred day-old store-bought buns.

"Roseanne Roseannadanna" Banana Muffins...It just goes ta show you, it's always somethin' - Roseanne Roseannadanna's Dad used to say this after her Grandmother baked all day to prepare these beauties.

"Killer Bee" Honey...Or for a less risky condiment - try Whipped Butter, Strawberry Preserves, or Cream Cheese.

SALADS

"Sluggo Salads"...Sluggo got his strength to "play" with Mr. Bill from eating this array of greens and toppings. Maybe it'll work for you too!

"Mr. Subliminal's" Fresh (last year) Oriental (American) Asparagus (OH WOW) Salad (try it)!

"Hans & Franz Pump You Up" Pickled Eggs...A daily dining ritual of Hans & Franz - these eggs are guaranteed to PUMP YOU UP!

"Baba Wawa Stwawbewwy Banana Jehwo Mode"...With fresh strawberries and bananas, Baba Wawa reported this salad to be "extwaordinawy."

"Bass O Matic" Salmon Cheese Ball...After an extended search, we found an original "Rovco Bass O Matic '76" machine in mint condition to create this salmon ball. We've sliced, we've diced, and you'll love it.

DESSERTS

"Opera Man's" Fresh Fruit and Cheese Platters...Revealed on these platters will be fresh fruits not yet sung about this season. As for the cheese use the Camp Granada tune..."Is it Cheddar or is it Jack - is it Cheddar or is it Jack - you decide, but if you can't - we know you'll enjoy Baby Swiss!"

"Mr. Robinson's Neighborhood" Ice Cream Parlor...Like everything Mr. Robinson did, this is a trifle "questionable"...ok, it's ten Call Hall flavors, not 31 - but "can you say DELICIOUS, boys and girls?"

"Conehead" Sugar Cones...And you thought Beldar, Prymaat, and Connie only fried chicken embryos and served massive quantities in six packs! MIPS, MIPS...

The "Widettes'" Homemade Brownies...The "Liar" gave these recipes to the Widettes. Yeh, there's no calories . . . Yeh, these brownies will make you smart. . .Yeh, that's the ticket!

"Emily Litella's" Oreo Cheesecake..."Gross, why would anyone want to eat Greasecake, it gets in your teeth, it oozes out,...what? Oh, CHEESECAKE - never mind!"

And now . . .

LIVE FROM K-STATE . . . It's Thursday Night!

Source: From Kansas State University Housing and Dining Services, Manhattan, KS. Used by permission.

Degree of Choice

Menus also may be categorized by the *degree of choice*. The no-choice menu is seldom used today, especially if the TQM philosophy has been adopted by the foodservice organization and customer's expectations have priority. Many nursing homes had no-choice menus until the Health Care Financing Administration in accordance with the Department of Health and Human Services developed a series of regulations recently implemented by the Omnibus Budget Reconciliation Act (OBRA). Residents now have a voice in what they eat, when they eat, and how they eat (Keegan, 1992). The new OBRA makes foodservice more customer- and service-oriented and empowers the residents to state their opinions.

The number of choices in each menu item category is determined by the goals of each foodservice operation. Choices may vary from many to few in static, cycle, and single-use menus. A static menu might have few choices in a limited-menu restaurant. A cycle or single-use menu may provide selections for some menu items but not for others. The menu of a catered business lunch, for example, may include no choice for the center-of-the-plate item and salad but offer the customer a choice of bread, beverage, and dessert.

Customers used to go to a restaurant for a big meal consisting of soup or salad, an entrée, and dessert, but today they are eating differently. They often order only a soup and salad or an appetizer and pasta (Lorenzini, 1992c). As a result, foodservice managers are using menu subcategories, such as starters, pasta, small salads, entrée salads, snacks, small and late-night meals, and family platters. Subcategories accent variety for the customer and enable managers to sell more foods that are low in cost for the operation as well, as in the case of pasta and salads.

MENU STRUCTURE

The menu always has been the heart of every foodservice operation, but today it is more crucial to the operation than ever (Wallace, 1991). The menu is the sales tool that leads customers to the operation. Bartlett (1991) stated that a well-executed menu is like a television series. It has captivating lead characters, a strong supporting cast, good production values, and plot twists that keep viewers coming back for more. As a bonus, it produces profitable spin-offs. Foodservice managers must check their menus often to be sure they satisfy cost-conscious but value-greedy customers.

"Making a menu is a balancing act," says Twyla Fultz, executive chef at the Quality Hotel Capitol Hill, in Washington, D.C. (Ryan, 1993). The menu can be approached from many angles: location, the name of the restaurant and what it says to potential customers, primary target market, menu priorities, balancing traditional and hearty with light and healthful menu items, and the capabilities of the staff.

Managers of various types of foodservice operations spend many hours in planning menus that will meet customer expectations. Multiunit restaurant chains need menus that have broad appeal but are short because customers want fast service and do not want to waste time making decisions. Menus must have regional appeal. The

Olive Garden chain, for example, has 92 different menus for 350 units in 42 states. Hospital menus not only serve patients with special nutritional needs, but also healthy, active customers, such as patients' families, doctors and nurses, and employees. Because most foodservice operations serve a minimum of two, and often three, meals a day, menus have to be geared to breakfast, lunch, and dinner as well as to style of service and menu price range. Balancing labor and food costs also is a challenge. Some other concerns of the menu planner include adding variety to seasonal menus and keeping cycle menus exciting, including new items along with old favorites, and offsetting high-priced items with those that are low priced (Ryan, 1993).

Typical American meals used to be a quick breakfast, a light lunch, and a heavy dinner in the evening, particularly for those living alone or in dual-career families. Three meals a day, popular in the 1940s and 1950s, is atypical for many people today. "Grazing," eating small amounts of food throughout the day, has become habitual for many people. Adams (1987) succinctly described grazing in mathematical terms as

$$grazing = flexibility + frequency + food.$$

Customer emphasis on quality, convenience, nutrition, and value with changing life-styles affects meal patterns. Whatever the meal of the day may be, it is the primary factor influencing menu planning.

Breakfast and Brunch

The good news is that breakfast traffic increased by 19% in the last decade and grew 2% from 1990 to 1991 (Ryan, 1992a). At least once a week, 20% of adults, especially senior citizens, eat breakfast away from home; customers come from high-income households or from households in which the head is a woman (Ryan & Stephenson, 1991). Although still the smallest piece of the commercial foodservice pie, breakfast posted the most gains in consumer traffic in 1992, 6% (Iwamuro, 1993).

Breakfast items range from those that are considered light to traditional items. Bob Evans Farm Restaurants, which emphasize breakfast menu items all day, has an oatmeal breakfast on its light menu consisting of a bowl of oatmeal, a blueberry muffin, and orange juice; the traditional favorite is sausage gravy and biscuits with home fries (McCarthy, 1993). To be responsive to customers, the restaurant chain has added new items, such as the Border Skillet, which draws on the popularity of Southwestern cuisine.

"Grab and go" appears to be the trend in weekday breakfasts, leaving the weekends for a more leisurely breakfast (Chaudhry, 1992a). The director of dining services at Wichita State University in Kansas is adding a breakfast bar for students who think 5 minutes is too long to walk around cafeteria stations picking up fruit, juice, pastry, and coffee. A limited-menu concept will be used for the bar; menu items, most of which travel well, include fresh fruit, juice, coffee, milk, granola, cereal, yogurt, home-made muffins, doughnuts and Danish pastries, one hot cereal, perhaps scrambled eggs, and one breakfast meat. The goal is that the healthier, fast items at Wichita State's breakfast bar will draw students away from off-campus limited-menu restaurant chains where they've been having breakfast (Lorenzini, 1992a).

Both commercial and noncommercial foodservice operations usually offer traditional breakfast items as well as light and healthful options. Menus reflect customers' changing food preferences and life-styles: greater meal flexibility, a back-to-basics trend, interest in spicy and ethnic foods, and a more relaxed approach to nutrition

Figure 6.7.   Example of a Sunday brunch menu.

THE RITZ-CARLTON
KANSAS CITY

COLD DISPLAY

Smoked Salmon
Cured Meats
Pates and Terrines
Fresh Fruit and Berries
Imported and Domestic Cheeses
4 Assorted Salads
Ice Carving
Shrimp
Belugu, Salmon and Whitefish Caviar

HOT DISPLAY

Carved Meat Station
Belgian Waffle Station, Assorted Toppings
Sausage and Bacon
Scrambled Eggs
Poached Eggs
Vegetable, Rice, Potato and Pasta
Fresh Seafood
Chicken or Beef

PASTRIES

Pastries, Bread Pudding, Bagels
Breads, Danish, Croissants

Source: From The Ritz-Carlton Rooftop Restaurant, Kansas City, MO. Used by permission.

that recognizes all food is healthful ("Best of Breakfast," 1992). Hearty breakfast menu items—eggs, sausage, bacon, hash browns, pancakes, waffles, and breakfast breads and pastries—are available for customers who want them. Health and fitness awareness and the high profile of carbohydrates and fiber have increased consumption of cold and hot cereals. McDonald's is given credit for creating the Egg McMuffin, which is considered the most popular breakfast sandwich at limited-menu restaurants. Other foodservice operations quickly followed and today customers have their choice of biscuit, croissant, bagel, and tortilla sandwiches, often with traditional breakfast foods as fillings. Coffee, both regular and decaffeinated, is the most popular breakfast beverage, but soft drinks are edging into the breakfast market. Fresh fruit and fruit juices also are popular.

Brunches, low-cost meals with a high perceived value, traditionally were held in hotels because of the built-in advantage of large staffs and facilities. An example of a brunch menu from The Ritz-Carlton in Kansas City®, Missouri, is shown in Figure 6.7. Seldom are printed menus used because chefs are empowered to choose the menu items based on what is fresh in the market.

In many types of foodservice operations, brunch is often served on weekends or for catered events. Because brunch is a hybrid of breakfast and lunch, menus are varied and may include both breakfast and lunch items. Many restaurants have turned Saturday and Sunday prelunch hours into a profitable brunch business by creating signature brunches, like the New Orleans jazz brunch by Ella Brennan at the Commander's Palace (Ryan & Stephenson, 1991). Noncommercial operations, particularly retirement centers and colleges and universities, may serve brunch in lieu of both breakfast and lunch on weekends and for special occasions, such as a birthday or commencement.

The basic brunch menu pattern generally begins with fruit and juice; when alcoholic beverages are desired, the Bloody Mary and Screwdriver are popular. The champagne brunch, often associated with wedding parties, usually is held in elegant restaurants or hotels or in country clubs. Menu selections will vary depending on type of service; the menu for a buffet generally will include a greater variety of selections than that planned for a sit-down brunch. Entrée brunch offerings usually include egg dishes and breakfast meats, in addition to such typical lunch and dinner entrées as steamship round of beef, chicken breast in mushroom wine sauce, or lobster Newburg. A variety of hot breads is included as well as an assortment of cheeses, fruits, and salads.

Lunch

The restaurant lunch menu is hard to balance between trends and tradition, healthy and hearty foods, variety, and the just-right menu mix. Lunch is a difficult meal to deliver to the customer because menu items are more complicated than those served at breakfast and have to be produced faster than dinner items (Ryan, 1993). Lunch is the meal most often eaten away from home; approximately 6 in 10 individuals consume a commercially prepared lunch at least once a week (Gordon, 1992). In a survey conducted by the National Restaurant Association (NRA), french fries topped the list for the restaurant menu item consumers order most frequently during lunch, at 3 orders in 10. Following are the next most popular menu items by the percentage of lunch orders including the item: hamburgers/cheeseburgers (25%), side-dish salads (11%), chicken entrées (10%), Mexican food (8%), and pizza (7%) (Gordon, 1992).

Lunch-to-go is a quickly growing trend for workers who run errands at noon or who want to relax on a park bench while enjoying a salad and a change of scenery (McCarthy, 1992). These customers want a good meal that travels well but do not want to go into debt paying for it, and they also want it fast. Sandwiches are the quintessential fare for people on the go (Solomon, 1993). The secret to a successful lunch-to-go program is the packaging required to make eating the lunch easy for customers; packaging, however, adds to the cost of the lunch.

Catering lunches remains one of the big profit makers in the foodservice industry. In no other operation is time more important than in catering. The customer waiting for a catered brown bag or boxed lunch watches the clock because timing is part of the contract. Catering is booming in the Au Bon Pain sandwich/bakery chain, which franchises many units, including 20 in Chicago and Minneapolis. A 9,000-square-foot catering commissary was built recently in Chicago's Loop to free unit managers from customer calls at lunchtime. This commissary has a custom voice-mail system, which identifies calls by unit and takes orders. Completed orders are delivered by truck to the unit for delivery to the customer (Chaudhry, 1992b).

In noncommercial foodservices, lunch menu items are becoming more like those served in commercial operations, especially limited-menu restaurants. The idea that people should eat whatever is put before them is gone. Managers of hospital, nursing home, college and university, and school foodservices are as eager to satisfy customers as are restaurateurs. Thus pizza, fajitas, chili, sandwiches, soup, chicken nuggets, and pasta appear regularly on menus.

The menu pattern for the USDA National School Lunch Program is defined in federal regulations. The USDA recommends, but does not require, that portions be adjusted by age/grade group to better meet the food and nutritional needs of children according to their ages, as shown in Figure 6.8. Schools must follow the defined menu pattern and offer the complete menu to all children in order to receive federal reimbursement. Current regulations, however, specify that secondary school students may decline two of the five food components, defined as offer-versus-serve, and the meal still will qualify for reimbursement. **Offer-versus-serve** is permitted in elementary schools at the discretion of the school administration.

An innovative approach to increase student participation occurred when Spring Valley Senior High School in New York was chosen for the initial project of transforming the cafeteria into a seven-station food court from which students select their own meals for $1.50 per lunch ("Lunch at Court," 1993). ARA Services, the contract company that provides the meal service, adapted the food court concept to the school lunch menu pattern. Each of the station names, such as Itza Pizza, Esta Fiesta, and Gretel's Bake Shop, is trademarked by ARA. The school experienced a 90% increase in the number of reimbursable meals served over the previous year and a 20% increase in à la carte sales.

Figure 6.8.   National School Lunch Program menu patterns.

CHART 1

SCHOOL LUNCH PATTERNS FOR VARIOUS AGE/GRADE GROUPS

U.S. Department of Agriculture, National School Lunch Program

USDA recommends, but does not require, that you adjust portions by age/grade group to better meet the food and nutritional needs of children according to their ages. If you adjust portions, Groups I–IV are minimum requirements for the age/grade groups specified. If you do not adjust portions, the Group IV portions are the portions to serve all children.

COMPONENTS	MINIMUM QUANTITIES				RECOMMENDED QUANTITIES[1]	SPECIFIC REQUIREMENTS
	Preschool ages 1-2 (Group I)	ages 3-4 (Group II)	Grades K-3 ages 5-8 (Group III)	Grades 4-12[1] age 9 & over (Group IV)	Grades 7-12 age 12 & over (Group V)	
MEAT OR MEAT ALTERNATE A serving of one of the following or a combination to give an equivalent quantity:						• Must be served in the main dish or the main dish and one other menu item. • **Vegetable protein products, cheese alternate products, and enriched macaroni with fortified protein** may be used to meet part of the meat/meat alternate requirement. Fact sheets on each of these alternate foods give detailed instructions for use.
Lean meat, poultry, or fish (edible portion as served)	1 oz	1½ oz	1½ oz	2 oz	3 oz	
Cheese	1 oz	1½ oz	1½ oz	2 oz	3 oz	
Large egg(s)	½	¾	¾	1	1½	
Cooked dry beans or peas	¼ cup	3/8 cup	3/8 cup	½ cup	¾ cup	
Peanut butter	2 Tbsp	3 Tbsp	3 Tbsp	4 Tbsp	6 Tbsp	
VEGETABLE AND/OR FRUIT Two or more servings of vegetable or fruit or both to total	½ cup	½ cup	½ cup	¾ cup	¾ cup	• No more than one-half of the total requirement may be met with full-strength fruit or vegetable juice. • Cooked dry beans or peas may be used as a meat alternate or as a vegetable but not as both in the same meal.
BREAD OR BREAD ALTERNATE Servings of bread or bread alternate A serving is: • 1 slice of whole-grain or enriched bread • A whole-grain or enriched biscuit, roll, muffin, etc. • ½ cup of cooked whole-grain or enriched rice, macaroni, noodles, whole-grain or enriched pasta products, or other cereal grains such as bulgur or corn grits • A combination of any of the above	5 per week	8 per week	8 per week	8 per week	10 per week	• At least ½ serving of bread or an equivalent quantity of bread alternate for Group I, and 1 serving for Groups II-V, must be served daily. • Enriched macaroni with fortified protein may be used as a meat alternate or as a bread alternate but not as both in the same meal. NOTE: *Food Buying Guide for Child Nutrition Programs, PA-1331* (1983) provides the information for the minimum weight of a serving.
MILK A serving of fluid milk	¾ cup (6 fl oz)	¾ cup (6 fl oz)	½ pint (8 fl oz)	½ pint (8 fl oz)	½ pint (8 fl oz)	At least one of the following forms of milk must be offered: • Unflavored lowfat milk • Unflavored skim milk • Unflavored buttermilk NOTE: This requirement does not prohibit offering other milks, such as whole milk or flavored milk, along with one or more of the above.

[1] Group IV is highlighted because it is the one meal pattern which will satisfy all requirements if no portion size adjustments are made.

[1] Group V specifies recommended, not required, quantities for students 12 years and older. These students may request smaller portions, but not smaller than those specified in Group IV.

Dinner

Dinner does not follow the distinctive pattern of food choices that characterizes breakfast and lunch menus. The traditional dinner menu includes an entrée of meat, fish, or poultry, potato or substitute, vegetable, and salad. For lighter or late evening meals, often referred to as *supper,* menus may be similar to those served at breakfast, brunch, or lunch.

In many restaurants, menus are getting shorter, but the appetizer section is getting longer. On many menus, appetizers equal or outnumber entrées. Customers are sharing appetizers or ordering two appetizers instead of an entrée. A classic example is the Chinese tradition of an all-appetizer feast, the dim sum, which is shared by many people (Ryan, 1992d). Tappas, Spanish appetizers, and mezzes, little dishes originating in the Middle East, also are found on many ethnic menus today. The appetizer in a hospital or nursing home usually is soup, seafood cocktail, fruit cup, or fruit or tomato juice.

Foodservice operators are trying to solve a difficult puzzle: how to bring in a newer, younger customer and still retain the regulars. In the publication, *Sysco's Menus Today* ("What's to be in '93," 1993), prime-time foods that satisfy the younger customer are listed (Figure 6.9). These menu items were popular in the early 90s and are expected to sell well in the future. The Italian and Asian influence and the move toward spicy and strong flavors are evident. Most of these items have a low

Figure 6.9.   Prime-time foods: Some of the most popular menu items for the 90s.

- Quesadillas stuffed with everything from bacon to smoked duck breast
- Wontons stuffed with crab, crawfish, and spinach, with or without cheese
- Pizzas topped with grilled vegetables
- Deep-fried plantain chips
- Focaccia
- Poached and grilled Pacific Rim fish, like tilapia
- Oriental chicken salad
- Antipasto plates
- Buffalo and venison
- Chicken and dumplings made with skinless chicken in broth instead of gravy
- Pasta platters to share around the table family-style
- Roast chicken with garlic mashed potatoes
- Stir-fry display cooking
- Prime aged steaks
- Chicken-fried steak
- Buy-by-the-ounce self-serve items
- Crusty peasant breads to dip in olive oil
- Beans and legumes
- Scones and muffins
- Crisps and cobblers

Source: From "What's to Be in '93," January 1993, *Sysco's Menus Today,* p. 22. Used by permission.

food cost, which is appreciated by foodservice operators. The center-of-the-plate menu items, usually for the regular and older customers, are the traditional favorites, such as steak, prime rib of beef, fried chicken, barbecued ribs, cheeseburgers, baked potatoes, pasta, tossed salad, apple pie, cheese cake, and ice cream. The wise food-service operator will include menu items to satisfy both newer and younger as well as regular and older customers.

Desserts commonly are included as a menu component in noncommercial operations. Enticing customers with rich desserts is easy, but the challenge is to create a light dessert that still dazzles. In a recent study by the NRA, 40% of the respondents said they were likely to choose fresh fruit desserts (Backas & Lorenzini, 1991). Examples of elegant fruit desserts are fresh pears poached in raspberry sauce that turn a brilliant pink color after being refrigerated overnight or baked apples with raisins and a little brown sugar baked in phyllo dough.

FACTORS AFFECTING MENU PLANNING

The crux of menu planning is that the menu is customer driven. The overriding concern in all facets of planning should be the satisfaction of customer desires. The concept of value cannot be ignored in menu planning; value prompts the clientele to select a particular item from the menu. Although satisfying customers is a primary concern for all foodservice managers, producing menu items at an acceptable cost takes priority. Both customer satisfaction and management decisions must be involved in menu planning.

Customer Satisfaction

Sociocultural factors should be considered in planning menus to satisfy and give value to the customer. Nutritional needs provide a framework for the menu and add to customer satisfaction. Probably the most important aspects for satisfying customers are the aesthetic factors of taste and appearance of the menu items. Will the customer be satisfied with the meal and want to return?

Sociocultural Factors

Sociocultural factors include the customs, mores, values, and demographic characteristics of the society in which the organization functions (Griffin, 1993). Sociocultural processes are important because they determine the products and services that people desire. Customers have food preferences that influence the popularity of menu items.

Food Habits and Preferences

Consideration of food habits and preferences should be a priority in planning menus for a particular population. Cultural food patterns, regional food preferences, and age are related considerations. Too often, menu planners are influenced by their own likes and dislikes of foods and food combinations rather than those of the customer.

Food habits are the practices and associated attitudes that predetermine what, when, why, and how a person will eat. Regional and cultural food habits still exist in the United States, but the mobility of the population and the sophistication of food marketing and distribution have lessened these distinctions. *Food preferences* express the degree of liking for a food item and are especially related to food habits (Peryam and Pilgrim, 1957).

Analysis of food habits and preferences should be conducted to provide data for menu planning. Formal and informal methods may be used to examine customer reactions to various menu items. Highly sophisticated market research studies are conducted by large national multiunit foodservice corporations before a major menu change or even the introduction of a new item.

Small-scale surveys, formal and informal interviews with customers, observations of plate waste, customer comment cards, and tallying of menu selections are methods used to collect data on food preferences and menu item popularity. Too often, menu planners offer variety for variety's sake.

The number one trend for the 1990s is captivating combinations of food on one plate, like the mixed grill entrée with variations of meat, seafood, and poultry, instead of ordering each item separately (Ryan, 1992b). For example, a jumbo grilled butterflied shrimp with a pork or veal medallion and grilled marinated chicken breast is arranged artistically on one plate.

Food preference surveys usually employ a hedonic scale in which foods are rated by an individual on a continuum from "like extremely" to "dislike extremely" (Peryam & Pilgrim, 1957). For measurement of food preferences among children, facial hedonic scales have been widely used, sometimes called the "smiley face" rating scale (Comstock, St. Pierre, & Mackiernan, 1981; Lachance, 1976). An example of a facial hedonic scale is shown in Figure 6.10. Wells (1965) found the facial method is easier to use with children than words or numbers because it allows good communi-

Figure 6.10.   Facial hedonic scale used for measuring children's food preferences.

Did you like what you ate?

Check (✓) the face that shows how you *felt* about the
food served today in the lunchroom

Figure 6.11.   Example of a sensory evaluation score card for menu items.

Circle **the one word that** *best* **describes how you feel about the food today.**

Food

1. Spaghetti with meat sauce	TASTES	Great	Good	So-So	Awful
	LOOKS	Great	Good	So-So	Awful
	TEMPERATURE	Just Right	OK	Too Cool	
	AMOUNT	Too Much	Right Amount	Not Enough	
2. Broccoli	TASTES	Great	Good	So-So	Awful
	LOOKS	Great	Good	So-So	Awful
	TEMPERATURE	Just Right	OK	Too Cool	
	AMOUNT	Too Much	Right Amount	Not Enough	

cation and understanding regardless of age, intelligence, education, or even the ability to speak English.

Frequency of acceptance of foods is another method used for studying food habits and preferences. Knickrehm, Cotner, and Kendrick (1969) studied frequency of acceptance of menu items by asking college students to indicate how often they would be willing to eat the food. Vegetables were found to be accepted less frequently than other menu items.

Sensory evaluation has been used to measure reactions to food by asking individuals to rate menu items on various dimensions such as flavor, appearance, temperature, and portion size. Figure 6.11 shows a sensory evaluation score card for evaluation of elementary students' reactions to menu items.

Plate waste, or the amount of food left on a plate, is a method used as a measure of food acceptability. Plate waste often is weighed to provide numerical results that can be used in many studies, particularly in the school lunch program. Plate waste also is being used in environmental control studies. It can be used to weigh uneaten menu items on an individual or group basis or the total waste for a meal.

Observation is a method that requires trained observers to estimate visually the amount of plate waste. Results of several studies indicate that the visual estimation of plate waste is a sufficiently accurate and simple method for assessing food acceptability. Kirks and Wolff (1985) said visual studies should be limited to informal studies in which broad generalizations will not be made regarding the effectiveness of a nutrition program. **Self-reported consumption** is another technique for measuring plate waste in which individuals are asked to estimate their plate waste using a scale similar to one used by trained observers. An example of this type of scale is shown in Figure 6.12.

Figure 6.12.   Scale for measuring self-reported consumption.

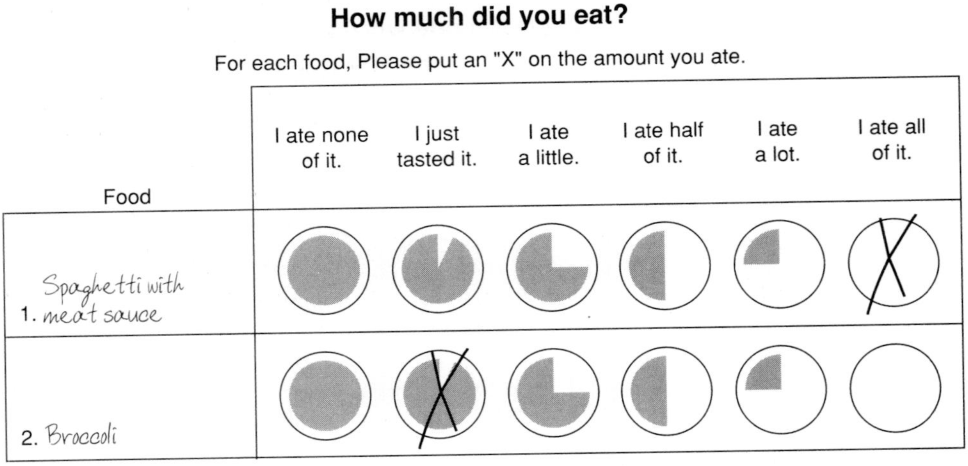

How much did you eat?

For each food, Please put an "X" on the amount you ate.

Menu planners must be cognizant of the food habits and preferences of the target population in order to plan menus that are acceptable and will generate sales and overall customer satisfaction. Age, cultural, and regional food patterns are important to consider, as well as changing food patterns over time.

Nutritional Influence

Nutritional needs of the customer should be a primary concern for planning menus for all foodservice operations, but they are a special concern when living conditions constrain persons to eat most of their meals in one place. In healthcare facilities, colleges and universities, and schools, for example, most of the nutritional needs of the customer are provided by the foodservice.

Increasing public awareness of the importance of nutrition to health and wellness also has motivated commercial foodservice operators to consider the nutritional quality of menu selections. Interest in nutrition is a trend that has been further prompted by the release of two reports: *The Surgeon General's Report on Nutrition and Health* (1988) and the report of the National Research Council (1989a), *Diet and Health: Implications for Reducing Chronic Disease Risk*. These publications increased customer awareness of the relationship of diet to such chronic diseases as heart disease, stroke, cancer, and diabetes (Dougherty, 1989). As a result, foodservice managers cannot afford to ignore customer demand for nutritionally adequate menu offerings. With all the emphasis on nutrition, however, steak houses are still very popular, and customers who otherwise want low-calorie meals choose high-calorie desserts.

Many people believe that healthful food is not delicious food. In a recent Gallup survey commissioned by The American Dietetic Association, approximately half of the respondents believed foods they like are not good for them; and 36% said they feel guilty when they eat what they like. The American Institute of Wine and Food, with Julia Child as a catalyst, conducted a conference, called Resetting the American Table: Creating a New Alliance of Taste & Health, to bridge the gap between health professionals and culinary leaders (see Figure 6.13). Selected chefs, food journalists, physicians, registered dietitians, and public health professionals attended. Participants defined the relationship between food and diet quality and provided the link between taste and health that supports the enjoyment of food ("Creating an Alliance," 1991). Members of the conference developed the following four core values when establishing standards for food and diet quality:

- Taste is a major determinant of consumer food choices in America.
- Dietary recommendations should respect culinary traditions that reflect and support our cultural and ethnic heritages.
- Nutrition and health begin around the dinner table.
- There are no good or bad foods—it's the overall diet that counts.

The Restaurants & Institutions National Consumer Survey of Eating Out Patterns confirmed that half the households indicated that low-fat, low-cholesterol foods would be ordered more frequently if offered; 60% indicated that concern with nutrition had increased during the past few years. Consumers today tend to focus principally on calories, fat, cholesterol, and sodium (Regan, 1987).

A market survey of current or potential customers is recommended to determine their needs, desires, and attitudes (Carlson, 1987). A sample survey questionnaire for a restaurant, shown in Figure 6.14, can be adapted for noncommercial operations. An informal method of determining the desires of customers is to review questions asked of the waitstaff concerning nutritional values and preparation of menu items.

Figure 6.13.   Logo for a historic coalition-building project on health and food.

Source: Organized by the American Institute of Wine & Food, San Francisco, CA, 1990. Used by permission.

Figure 6.14. Example of survey questionnaire.

Questions about Specific Menu Items

1. We are considering offering healthier menu selections. How likely would you be to order the following?

	Very likely	Likely	Neither likely nor unlikely	Very unlikely
a. Raw vegetable appetizers	_____	_____	_____	_____
b. Baked, broiled, or roasted rather than fried foods	_____	_____	_____	_____
c. Calorie-controlled entrees	_____	_____	_____	_____
d. Margarine rather than butter	_____	_____	_____	_____
e. Entrees cooked without salt	_____	_____	_____	_____
f. Whole-wheat breads, rolls and crackers	_____	_____	_____	_____
g. Vegetables seasoned with herbs and lemon juice rather than butter and sauces	_____	_____	_____	_____
h. Low-fat, low-calorie fruit-based desserts	_____	_____	_____	_____
i. Reduced-calorie salad dressing	_____	_____	_____	_____
j. Lean meats and fish	_____	_____	_____	_____
k. Poultry without skin	_____	_____	_____	_____
l. Dried or cooked cereal with skim milk	_____	_____	_____	_____
m. Whole-grain muffins	_____	_____	_____	_____
n. Whole-grain pancakes or waffles with fruit-yogurt topping	_____	_____	_____	_____
o. Fresh fruit	_____	_____	_____	_____
p. Low-cholesterol egg substitutes	_____	_____	_____	_____
q. Low-fat cream cheese	_____	_____	_____	_____
r. Bagel with low-fat cottage cheese	_____	_____	_____	_____

Demographic Information

2. How old are you?
   - _____ 18–24
   - _____ 25–34
   - _____ 35–44
   - _____ 45–54
   - _____ 55–64
   - _____ 65 or older

3. Are you:
   - _____ Male
   - _____ Female

4. Are you employed outside your home?
   - _____ Yes
   - _____ No

5. Approximately how much is your total annual household income?
   - _____ Under $10,000
   - _____ $10,000–$14,999
   - _____ $15,000–$19,999
   - _____ $20,000–$24,999
   - _____ $25,000–$29,999
   - _____ $30,000 and over

Source: From *A Nutrition Guide for the Restaurateur* (p. 64) by the National Restaurant Association, 1986, Washington, DC. Used by permission.

Figure 6.15. Dietary guidelines for Americans.

7 recommendations for good health.

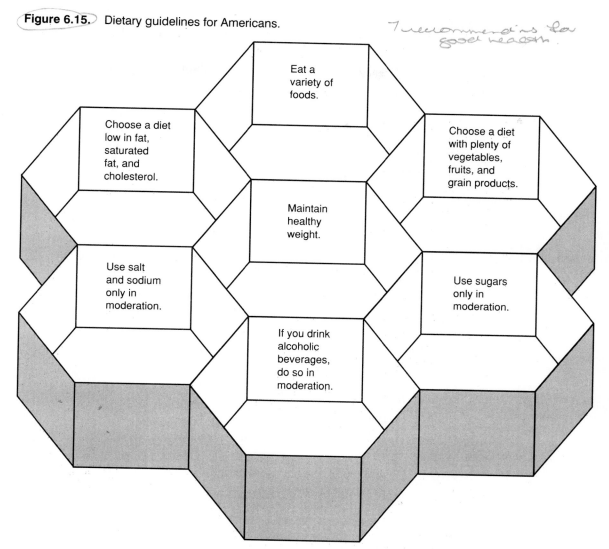

Eat a variety of foods.

Choose a diet low in fat, saturated fat, and cholesterol.

Choose a diet with plenty of vegetables, fruits, and grain products.

Maintain healthy weight.

Use salt and sodium only in moderation.

Use sugars only in moderation.

If you drink alcoholic beverages, do so in moderation.

Source: From *Nutrition and Your Health: Dietary Guidelines for Americans* (pp. 14–15), 1990, Washington, DC: U.S. Departments of Agriculture and Health and Human Services.

The third edition of *Nutrition and Your Health: Dietary Guidelines for Americans,* issued in 1990 by the United States Department of Agriculture (USDA) and the U.S. Department of Health and Human Services, made the seven recommendations for good health shown in Figure 6.15. In foodservice operations providing three meals a day, menus should satisfy the recommended dietary allowances (RDAs) defined by the National Research Council (1989b), which specify nutrient needs for various age groups by sex. The National Restaurant Association (1986) developed and published

A Nutrition Guide for the Restaurateur specifically for restaurant managers who have minimal knowledge of nutrition.

Because the **dietary guidelines for Americans** (Figure 6.15) are not easily understood by the public, the USDA launched the **Food Guide Pyramid** (Figure 6.16) in the spring of 1992. The pyramid was the first official attempt to illustrate the guidelines in a meaningful way. The Food Guide Pyramid is a complex illustration with many different food and nutrition messages. After much testing, the final version provides the numbers of servings in boldface type for each of the six food groups and also better illustrations of the groups. The research also revealed that consumers do not misinterpret the pyramid to mean that some foods are less impor-

Figure 6.16.   Food Guide Pyramid.

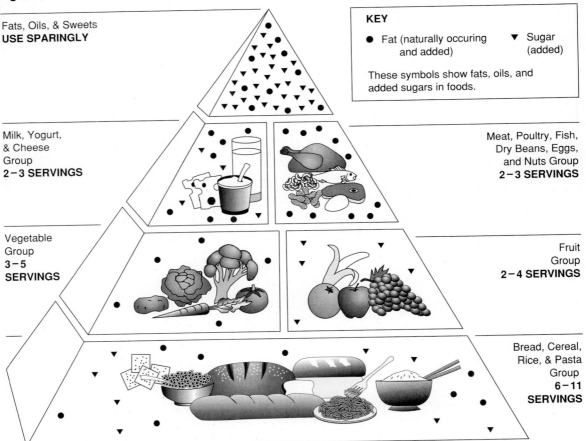

Source: From *The Food Guide Pyramid* by U.S. Departments of Agriculture and Health and Human Services prepared by Human Nutrition Information Service, Home and Garden Bulletin Number 252, August 1992, pp. 2–3.

tant in a healthful diet than others ("USDA Pyramid," 1992). As a result of the pyramid, the Produce for Better Health Foundation, a nonprofit organization incorporated in 1991, and the National Cancer Institute sponsored a **5 A Day program** to increase consumption of fruits and vegetables to at least five servings a day ("5 A Day," 1993). The program is being promoted on television and radio and in nationally syndicated newspaper columns to improve the partnership among the health community, government agencies, and the fruit and vegetable industry.

For some feeding programs, specific nutritional guidelines have been made mandatory. For example, the nutritional goal of the National School Lunch Program is to provide one third of the RDAs. According to the Menu Planning Guide for School Food Service (U.S. Department of Agriculture, 1983), each lunch is not expected to provide one third of the RDAs for all nutrients, but the average over a period of time should meet the goal. The menu pattern, specified by the USDA, was shown earlier in this chapter (Figure 6.8).

Most noncommercial foodservice operations have either a registered dietitian on staff or a consultant on nutritional aspects of menu planning. Commercial foodservice operators were slow to adopt nutritional emphasis in menus, but now are positioning and marketing nutrition to respond to consumer demand and to gain a competitive edge (Welland, 1991).

Chefs traditionally have been unconcerned about nutrition, and their training supported that tradition. The Culinary Institute of America (CIA) now has broken new ground. The General Foods Nutrition Center, the first of its kind in the country, was recently built on the CIA campus, and it offers valuable training in nutritional concepts and principles to students and foodservice professionals.

The center houses a state-of-the-art kitchen, two classrooms, a nutrition resources center, and the St. Andrew's Cafe. A registered dietitian is on the CIA staff and has the responsibility for teaching the nutrition course, planning the menu, and preparing a nutritional analysis for all menu items served in the St. Andrew's Cafe, one of four public restaurants on the campus. The 1993 CIA book, *Techniques of Healthy Cooking*, has been nominated for a James Beard Award. It includes approximately 300 recipes created by chef-instructors at St. Andrew's Cafe as well as some basic nutrition facts.

St. Andrew's changes and prints its menus approximately every 3 days to permit the chef to take advantage of fresh foods available on the market. At the bottom of the menu, the cafe prints its goal, which follows: "Our aim is that each dish be moderate in terms of the amount of calories, fat, cholesterol, and sodium it contains but also be prepared according to the culinary principles of taste and presentation." Table 6.1 is a computerized nutritional analysis of the dinner menu shown in Figure 6.17.

Upon reaching the decision to offer more nutritious fare, restaurateurs have two choices for menu planning: to either do it themselves or retain a consultant (Regan, 1987). A number of restaurateurs choose to work with a registered dietitian (Sneed & Burkhalter, 1991). Consultants can translate medical terminology related to specific diets, can assist in calculating nutrient content of menu items, and can determine the potential for promoting particular foods as nutritious, light, or healthful.

Table 6.1 Example of computerized nutritional analysis for items on the dinner menu shown in Figure 6.17

Menu Item	Calories (kcal)	Protein (gm)	Fat (gm)	Carbo. (gm)	Sodium (mg)	Chol. (mg)	Sat. Fat (gm)	Mono Fat. (gm)	Poly. Fat (gm)	Fiber (gm)
Appetizers										
Wood Roasted Salmon Cakes	196	113.0	6.4	21.3	22.6	3.6	1.5	1.3	3.3	4.4
Antipasti	198	8.8	6.3	28.8	319	3	1.6	2.1	2.0	6.7
McKenzie Jerk Chicken	179	13.1	8.8	15.3	108	2.6	1.2	1.9	4.8	2.5
Cod and Potato Ravioli	198	12.1	2.7	32.4	34.4	22.8	8.5	1.7	0.5	
Soups										
Wild Mushroom Chowder	74	3.0	0.9	13.3	159	3	0.5	0.2	0.1	1.2
Tortilla Soup	120	10.8	5.0	9.4	61	18	1.7	2.2	2.2	0.8
Chicken and Clam Consommé	65	9.2	2.2	2.1	46	22	0.8	1.0	0.4	0.2
Salads										
Hearty Greens w/Warm Vinaigrette	83	4.6	5.2	7.0	219	9	1.2	2.4	1.6	1.1
Mixed Greens/Herb Vinaigrette	37	1.2	2.7	2.5	33	tr.	0.4	1.1	1.2	1.2
Mixed Greens/Creamy Citrus Dressing	45	0.7	2.4	1.9	9	1.0	0.5	1.6	0.2	0.1
Pizzas										
Appetizer	239	14.8	5.7	32.1	387	51	2.8	1.8	0.5	2.9
Entrée	463	28.7	11.3	61.5	714	101	5.5	3.6	1.0	5.6
Entrees										
Stir-Fry of Duck and Shrimp	319	37.3	6.8	38.4	187	103	1.5	2.1	2.4	3.1
Chicken w/Mushrooms, Asparagus	398	33.5	11.6	42.7	118	69	3.9	5.4	1.4	5.5
Pan-Seared Salmon, Chardonnay Sc.	336	24.8	12.7	31.5	161	71	5.9	4.8	0.8	2.7
Beef w/BBQue Beans, Slaw	396	35.6	10.8	42.0	170	76.0	3.6	3.5	0.7	9.3
"Buttons and Bows"	365	25.7	6.9	51.8	48.9	52.7	3.2	1.7	1.2	8.3

Desserts

Angel Food Cake w/Passionfruit Sauce	200	4.6	1.9	42.7	60	3	0.8	0.4	0.1	6.8
Chocolate Ricotta Bavarian	195	5.5	6.0	31.1	61	23	3.1	2.0	0.4	2.2
Chocolate-Cherry Glacé	197	7.7	4.3	32.5	91	12	2.4	1.5	0.1	0.5
Chef Souffle w/Sauce	122	4.2	3.0	16.0	65	10	1.8	0.7	0.1	0.8

Accompaniments

Whole Wheat Roll	88	3.3	2.0	15.5	217	3	0.6	0.7	0.5	3.9
Butter	36	0	4.1	0	34	11	2.6	1.2	0.2	0
Margarine	36	0	4.1	0	48	0	0.8	1.8	1.3	0

Beverages

Dry White Wine	98	0.1	0	0.9	6	0	0	0	0	0
Red Wine	105	0.2	0	2.6	7	0	0	0	0	0
Rose Wine	104	0.2	0	2.1	6	0	0	0	0	0
Coffee	2	0.1	0	1.1	5	0	0	0	0	0
Tea	2	0	0	0.5	7	0	0	0	0	0
White Sugar 1 Packet	25	0	0	6	0	0	0	0	0	0
Beer	146	0.9	0	13.2	19	0	0	0	0	0.7
Light Beer	100	0.7	0	4.6	10	0	0	0	0	0.3
Fresh Orange Juice	111	1.7	0.5	25.8	2	0	0.3	0	0	1.0
Madras	78	0.6	0.2	18.7	2	0	0.1	0.1	0.1	0.6
Sea Breeze	72	0.5	0.1	17.5	2	0	0.6	0	0	0.6
Dessert Wine	136	0.4	0	10.4	8	0	0.6	0	0	0
Sparkling Wine	113	0.2	0	3.1	9	0	0	0	0	0

Source: From St. Andrew's Cafe, The General Foods Nutrition Center, The Culinary Institute of America, Hyde Park, NY. Used by permission.

Figure 6.17. Example of a dinner menu.

St. Andrew's Cafe
Dinner

Appetizers, Soups and Salads

Cod and Potato Ravioli with Warm Tomato Vinaigrette and Olives 4.00

Antipasti of Grilled Vegetables, Jaxberry Greens and Sonoma Dry Jack Cheese 4.25

McKenzie Jerk Chicken with Chive Roti and Cooling Island Salad 4.25

Wood Roasted Salmon Cakes with Wild Mushrooms and Sherry Vinaigrette 4.50

Chicken and Clam Consommé 3.50

Traditional Arizona Tortilla Soup 3.25

Wild Mushroom Chowder 3.50

Hearty Greens with Grilled Onions, Warm Vinaigrette, Maytag Blue Cheese and Toasted Nuts 3.50

Mixed Salad of Local Organic Lettuces 2.25
Choice of Herb Vinaigrette or Creamy Citrus

Wood-Fired Pizzas

White Beans, Shrimp, Roasted Onions, Mushrooms and Provolone Cheese on Whole Wheat Crust Appetizer 4.25 or Dinner 5.75

Entrees

Stir-Fry of Moulard Duck Breast and Gulf Shrimp with Asian Vegetables 11.95

Pan-Seared Salmon with Herbed Potato Puree and Citrus Chardonnay Sauce 12.25

Mesquite Grilled Tenderloin of Double J Limousin Beef with Barbequed Prairie Beans and Creamy Western Slaw 14.25

Grilled Breast of Chicken with Orechiette Pasta, Asparagus and Mushrooms 10.75

"Buttons and Bows," Breaded Baked Scallops and Spinach Bow Tie Pasta with Tomatoes, Pearl Onions and Yellow Squash 12.25

Desserts

Dark Chocolate Bavarian with Peanut Butter Hippen Cup and Raspberry Sauce 3.75

Grilled Angel Food Cake with Ripe Tropical Fruit and Passion Fruit Sauce 3.50

Chef's Hot Soufflé 3.75

Dried Cherry Chocolate Glace with Lace Cookie 3.25

A 12 percent service charge is included in your bill enabling your patronage to benefit the Student Council Scholarship Fund. Students receive 100 percent of these funds and additional tipping is not expected or required.

NOT-FOR-PROFIT STATEMENT
The Culinary Institute of America is an independent, not-for-profit educational organization pursuing its mission of providing the highest quality culinary education. This not-for-profit status enables us to focus on the quality of education rather than on satisfying the investment expectations of any shareholders.

At St. Andrew's Cafe, we believe that eating wisely should be enjoyable. Menus are planned to follow guidelines of balance, nutrition, and taste. Our aim is that each dish be moderate in terms of the amount of calories, fat, cholesterol, and sodium it contains but also be prepared according to the good culinary principles of taste and presentation. ✿ The Culinary Institute of America has served as a leader for many advances in the foodservice and hospitality field, including the incorporation of nutrition into fine cuisine. The General Foods Nutrition Center and St. Andrew's Cafe are examples of our commitment to the constant refinement of our educational mission. ✿ St. Andrew's Cafe is one of four restaurants at The Culinary Institute of America. All are staffed by students in the final semester of the school's associate's degree programs. We hope you enjoy your meal and encourage you to dine with us often.

For restaurant reservations and information, call (914) 471-6608

THE CULINARY INSTITUTE OF AMERICA 1946
AMERICA'S CENTER FOR CULINARY EDUCATION SINCE 1946

651 South Albany Post Road, Hyde Park, New York 12538-1499

Source: From St. Andrew's Cafe, The General Foods Nutrition Center, The Culinary Institute of America, Hyde Park, NY. Used by permission.

172

They also can make suggestions to chefs concerning the reduction of calories, fat, cholesterol, and sodium without sacrificing a good product. Computer software packages are available in which the nutritional content of a recipe is calculated and shown on a printout. The National Restaurant Association has recognized that the popular emphasis on health and good eating is not a fad but a solid trend by retaining a registered dietitian as research manager and nutritionist in its Washington, D.C., office.

Aesthetic Factors

Flavor, texture, color, shape, and method of preparation are other factors to consider in planning menus. A balance should be maintained among *flavors*, such as tart and sweet, mild and highly seasoned, light and heavy. Certain combinations have become traditional, such as turkey and cranberry sauce and roast beef and horseradish sauce. The flavors are complementary, and customers tend to expect these combinations to be served together. Foods of the same or similar flavors generally should not be repeated in a meal. A variety of flavors within a meal is more enjoyable than duplications, although there are exceptions to this rule. For instance, tomato in a tossed salad is an acceptable accompaniment to spaghetti and meatballs in a tomato sauce.

Texture refers to the structure of foods and is detected by the feel of foods in the mouth. Crisp, soft, grainy, smooth, hard, and chewy are among the descriptors of food texture, which should be varied in a meal. For example, a crisp salad served with soup on a luncheon menu is a more pleasing textural combination than guacamole with soup; stir-fried vegetables add crispness to any meal.

Consistency of foods is the degree of firmness, density, or viscosity. Runny, gelatinous, and firm describe the characteristics of consistency, as do thin, medium, and thick when referring to sauces.

Color on the plate, tray, or cafeteria counter has eye appeal and helps to merchandise the food. The combination of colors of foods always should be considered in selecting menu items. The orange-red of tomatoes and the purple-red of beets, for example, is an unappealing combination, and a menu with several white foods is unimaginative. A grilled steak with a baked potato and sauteed mushrooms is a wonderful meal but lacks eye appeal. The presentation could be improved by the addition of a broiled tomato half or small wedge of watermelon with the green rind.

The *shape* of food also can be used to create interest in a menu through the variety of forms in which foods can be presented. Food processors and mixers with attachments can produce a wide variety of shapes. As an example, the ever-popular french fried potato can be served as regular, curly, or steak fries.

Combinations of foods using different methods of preparation can add variety to the menu. Foods prepared in the same manner generally should not be served in the same meal, barring some common exceptions such as fish and chips, both of which are fried. Southern fried chicken with cream gravy and steamed asparagus would be appealing, but if the asparagus were dressed with hollandaise sauce, the appeal would be lost. Other ways to introduce variety and texture are to serve both cold and hot foods or both raw and cooked foods together.

Management Decisions

The menu should be viewed as a managerial tool for controlling cost and production. A number of management-related factors must be considered in its design: food cost, production capability, type of service, and availability of foods.

Food Cost

Food cost is the cost of food as purchased. Foodservice managers, in a competitive situation, must be cost conscious in all areas of operations. Because the menu is a major determinant of pricing for food items, the manager must be particularly aware of both raw and prepared food costs for each menu item.

A very important part of most food-cost accounting systems is the determination of the food-cost percentage figure, the ratio of the cost of food sold over the dollars received from selling the food (Keiser & DeMicco, 1993). For example, a 40% raw-food cost in relation to sales revenues has been a rule of thumb for many commercial operations. This objective does not necessarily apply to each menu item, but to the overall sales. Some menu items, such as beverages, may have a much lower food cost in relation to sales, and other items, especially entrées, may be higher. What the customer is willing to pay also must be considered in menu planning.

In most hospitals and college and university foodservices, a daily food cost per customer is provided in the budget. In school foodservice, the amount of federal and state reimbursement is an important factor in determining the budget available for food. Food cost control is discussed further in chapter 21.

Production Capability

To produce a given menu, several resources must be considered, of which a primary one is labor. The number of labor hours as well as the number and skill of personnel at a given time determine the complexity of menu items. Some menu items may be produced or their preparation completed during slack periods to ease the production load during peak service times; however, the effect on food quality may limit the amount of production in advance of service that could be completed. Employees' days off may need to be considered in menu planning because relief personnel may not have equal skill or efficiency. In limited-menu operations, however, variations in employees' skill are less important because of product standardization. Planning less complicated menu items or using convenience items may be alternatives.

Production capability also is affected by the layout of the food production facility and the availability of large and small equipment. The menu should be planned to balance the use and capacity of ovens, steamers, fryers, grills, and other equipment. Refrigeration and freezer capacity must also be considered. Many novice menu planners have planned menus that overtax the oven capacity in a foodservice facility. In a small hospital foodservice operation, for example, a menu including meat loaf, baked potato, rolls, and brownies may present a production problem because all the items require the oven.

Type of Service

Type of service is a major influence on the food items that can be included on a menu. A restaurant with table service will have a different menu from that of a school cafeteria.

Food items with longer holding capability should be selected for menus in establishments where last-minute preparation may present a problem, like healthcare facilities with patient tray service. As discussed in chapter 5, in some hospitals and other facilities galley kitchens may be available for last-minute preparation of menu items that do not hold well, such as toast and coffee.

Equipment for holding and serving will affect the menu selections that can be offered. Hot foods may be held in either stationary or portable heated equipment. Required temperatures are relatively low, and humidification is necessary for some foods. Cold foods may be held in refrigerated units or iced counters. Central commissaries and satellite units require insulated transport units for both hot and cold foods. The availability of sufficient china, flatware, and glassware is another problem. Certain menu items may require special serving equipment, such as chafing pans for flamed desserts or small forks and iced bowls for seafood cocktail appetizers. According to Bobeng (1982), this impact of menu items and combinations on dishroom capability is often overlooked.

The temperature, color, or texture of some menu items deteriorate during the time between food production and service. Items that present a particular problem should be eliminated or modified. For example, schoolchildren may like grilled cheese sandwiches, but that sandwich does not lend itself to a menu for a satellite service center unless a grill and enough employees are available for last-minute preparation.

Availability of Foods

Improvement in transporting food both nationally and internationally and in food preservation makes many foods that were once considered seasonal available during most of the year. Strawberries are imported into the United States from Mexico in January, and frozen foods are available all year. During the growing season, however, local food products often are of better quality and less expensive than those shipped from distant markets, and they should be added to the menu at that time. Foodservice managers in small communities also may need to consider frequency of delivery from various food distributors in planning menus.

MENU PLANNING

Thus far, this chapter has been devoted to the menu planner. In many foodservice operations, however, menu planning is often the responsibility of a team rather than an individual. This is especially true in large organizations in which the viewpoints of managerial personnel in both production and service are important in menu planning. In addition, personnel responsible for procurement have valuable input on availability of food, comparative cost, and new products in the market. In a healthcare organization, the clinical dietitian should be included on the menu planning team to ensure that patients' needs and food preferences are given appropriate consideration.

General Considerations

Computer-assisted menu planning is not widely used today because of the difficulty in quantifying the many variables involved in menu planning, such as flavor, color, and texture. Instead, the computer is more often used for analyzing menus for cost and nutrient composition.

Since the pioneer work on computer applications for foodservice, many developments have occurred. The most exciting is creating menus with desktop publishing. A desktop publishing system consists of computer hardware, software, and a laser printer, which can be either purchased or rented. Foodservice managers are encouraged to hire a menu designer to develop an attractive finished product (Kass, 1989). According to Miller (1988a), with computers menu items can be changed quickly in response to customer demands or increases in raw food costs. Greater opportunity exists for experimentation by adding new menu items or repositioning existing ones. Also, specials and other price changes can be incorporated into regular menus easily. Special events promotions can be advertised earlier with the potential of bringing in more customers. Finally, the overall cost for menu production with more flexibility in design is lowered.

Menu planning should proceed from the premise that the primary purpose of any foodservice is to prepare and serve acceptable food at a cost consistent with the objectives of the operation. Certain decisions must be made in advance and policies and procedures established for planning menus in a systematic manner. In addition to decisions on the design and format of the menu, other considerations are the number of choices to be offered, type of menu, and frequency of revision.

This planning must be far enough in advance of actual production to allow delivery of food and supplies and permit labor to be scheduled. Many operations also need time for printing or other reproduction of menus.

The initial decision is the design of the menu pattern or the outline of menu items to be served each meal. As stated when discussing menu structure, the meal of the day is the key influence on the menu pattern, although it may vary from commercial to noncommercial operations. Menus for either type of operation, however, should be designed to inform the customer of what items are available and, in many instances, their cost. Simple descriptions of menu items should be used, and confusing or overstated terms should be avoided. A pitfall for many menu planners is the use of interesting names to enliven a menu when these flowery terms often only confuse the patron. Any special names or menu items should be readily understood in the region where they are served. For example, Hopping John, a traditional southern New Year's Day dish consisting of black-eyed peas, rice, salt pork, and onions, would be puzzling to most New Yorkers.

Legislation in a number of states requiring truth in menus has had an impact on names of menu offerings. The California Restaurant Association, for example, has distributed to its membership a comprehensive special report explaining the labeling and advertising requirements of the California Business and Professional Code for food items served in restaurants. The National Restaurant Association and committees of restaurant operators defined menu terms and have distributed them in printed form, called *Accuracy in Menus*. The following are areas of potential misrepresentation with a short explanation of each (Miller, 1987):

Quantity. Proper procedures should preclude any concerns with misinformation of quantities (e.g., extra-large salad).

Quality. Federal and state standards of quality grades exist for many products including meat, poultry, eggs, dairy products, fruits, and vegetables (e.g., choice sirloin of beef).

Price. Extra charges for service or special requests for food items should be brought to the customer's attention.

Brand Names. Any product brand that is advertised must be the one served.

Product Identification. Substitutions for products must be on the menu, such as blue cheese for Roquefort cheese.

Points of Origin. Claims of origin should be documented (e.g., Maine lobster). Geographic names used in a generic sense, such as New England Clam Chowder, are permitted.

Merchandising Terms. Terms for specific products need to be qualified (e.g., flown in daily).

Means of Preservation. To preserve food, it may be canned, chilled, bottled, frozen, or dehydrated. If a method is identified on the menu, it should be correct (e.g., frozen orange juice is not fresh).

Food Preparation. Absolute accuracy is a must (e.g., charcoal broiled).

Verbal and Visual Presentation. If a picture of a meal is shown, the actual meal must be identical (e.g., if six shrimp are shown, six—not five—must be served).

Dietary or Nutritional Claims. Misrepresentation of nutritional content of food is not permitted (e.g., "low calorie" must be supportable by specific data).

The National Academy of Sciences has recommended that restaurants be required to label menus to identify ingredient and nutritional elements in menu items. Customers who are allergic to foods such as eggs, wheat, or shellfish want ingredients to be identified. The National Restaurant Association maintains that menu labeling is simply not a practical or applicable solution to this matter (Howat, 1991). In regulations implementing the Nutrition and Education Act of 1990, the FDA required that only nonmenu items that make nutrient-content or health claims, such as signs, placards, and posters, meet FDA criteria for such claims. Menus, which are subject to frequent, even daily change, were to be exempted whether or not they made any claims (Mermelstein, 1993). Under pressure from two consumer groups that filed a lawsuit challenging the menu claims exemption, the FDA announced its intent to expand the regulations to cover menus for health claims and nutrient-content claims in 1994 (NRA, 1993).

Keeping various types of records can assist not only in menu planning but also in the purchase and production of foods. Data on past acceptance of items, weather, day of the week, season, and special events that may have influenced patronage are essential for menu planning. Other resources should include files of standardized recipes with portion size and cost, market quotations, suggestions from clientele, lists of food items classified by category (vegetables, entrées, desserts, etc.), trade publications, and cookbooks.

Planning Process

The general principles of menu planning are applicable to both noncommercial and commercial foodservice operations. Diners in noncommercial operations eat most of their meals on the premises, unlike those who may choose any restaurant in commercial operations. Even though planning principles are similar in both operations, planners should remember that more variety needs to be incorporated into menus in the noncommercial foodservice than in the commercial foodservice. The person eating out often goes to a restaurant for particular menu items and would be disappointed if the menu completely changed every day or week.

Noncommercial Operations

Most noncommercial menus, with the exception of school foodservice, are designed on a three-meal-a-day plan. Some foodservice operations use a four- or five-meal plan built around brunch and an early dinner with some light, nutritious snack meals at other times of the day.

Cycle menus are used widely in noncommercial foodservice, with the length of the cycle varying from 1 to 3 weeks or longer. Also, cycles may change according to the season of the year to take advantage of plentiful foods on the market and to satisfy clientele expectations. The average length of stay is an important consideration in determining the length of the menu cycle in healthcare institutions. Cycles of 1 or 2 weeks have been used successfully in hospitals with a 4- or 5-day patient stay. Many larger hospitals, in recognition of the relatively short-term patient stay, use a restaurant-type static menu. In long-term care facilities, a 3- or 4-week cycle menu may be used. The general or regular menu provides the basis for planning menus for the various modified diets in hospitals.

The recording of the menus on a form designed for that purpose and suited to the needs of a particular foodservice is recommended. Figure 6.18 is the form for planning school foodservice lunches; the format can be modified for other noncommercial operations.

A step-by-step procedure for institutional menu planners is outlined on page 180 for a three-meal-a-day pattern. Note that the entrée is the main item around which the meal is planned and must therefore be selected before any complementary foods.

Figure 6.18.   Menu-planning worksheet.

Lunch Pattern	MONDAY	PORTION SIZE		TUESDAY	PORTION SIZE	
		Group	Group		Group	Group
Meat and Meat Alternate						
Vegetable and Fruit						
Bread and Bread Alternate						
Milk						
Other foods						
	WEDNESDAY			THURSDAY		
Meat and Meat Alternate						
Vegetable and Fruit						
Bread and Bread Alternate						
Milk						
Other foods						
	FRIDAY					
Meat and Meat Alternate						
Vegetable and Fruit						
Bread and Bread Alternate						
Milk						
Other Foods						

Source: From *Menu Planning Guide for School Food Service* (p. 80) by U.S. Department of Agriculture, Food and Nutrition Service, 1983, Washington, DC: Government Printing Office.

1. *Plan the dinner meats or other entrées for the entire cycle.* If a single-use menu is used, the entrées for at least 1 week should be planned. Because entrées are the most expensive foods on the menu, cost can be controlled to a great extent through careful planning at this stage. A balance between high- and low-priced items will average out the cost over the week or period covered by the cycle.

If choices of entrée are offered, the alternatives should include meat, chicken or other poultry, fish, a vegetarian entrée, and a meat extender, such as meat loaf or stew. Choices should be available for persons who have religious or medical dietary restrictions. Menus for preceding and subsequent days should be considered to preclude repetition.

2. *Select the luncheon entrées or main dishes, avoiding those used on the dinner menu.* Provide variety in method of preparation. A desired meal cost per day can be attained by serving a less expensive item at one meal of the day when a more expensive food has been planned for the other meal. Soups, sandwiches, main-dish salads, and casseroles are commonly served as luncheon entrées.

3. *Decide on the starch item appropriate to serve with the entrée.* Usually, if the meat is served with gravy, a mashed, steamed, or baked potato should be on the menu. Scalloped, creamed, or au gratin potatoes are most appropriate with meats having no gravy. Rice or pasta are common substitutes for potatoes. Variations in nonstarchy vegetables are obtained by serving them raw, cooked, peeled, or unpeeled; cutting vegetables in different sizes and shapes is another alternative. Methods of preparation can add variety to a vegetable, as can seasonings and sauces.

4. *Select salads, accompaniments, and appetizers next.* Work back and forth between the lunch and dinner meals to avoid repetition, introduce texture and color contrast into the meal, and provide interesting flavor combinations.

5. *Plan desserts for both lunch and dinner.* They may be selected from the following main groups: fruit, pudding, ice cream or other frozen desserts, gelatin, cake, pie, and cookies.

6. *After the luncheon and dinner meals have been planned, add breakfast and any others.*

7. *Review the entire day as a unit and evaluate if clientele and managerial considerations have been met.* Check the menu for duplication and repetition from day to day. The use of a checklist aids in making certain that all factors of good menu planning have been met. An example of a menu evaluation checklist developed by the USDA for use in the school lunch program is shown in Figure 6.19.

Commercial Operations

Merchandising is the primary consideration in planning menus for commercial operations. Because of the varied types of operations, the menu takes many forms. The static-choice menu is the predominant type used in commercial foodservices, including upscale restaurants, limited-menu operations, and coffeeshops. Menus are revised infrequently. Either all meals are included on one menu or separate printed menus are available for each meal with clip-ons for daily specials. Usually, entrées

Figure 6.19.   Menu evaluation checklist for school lunch menus.

Menu Evaluation

After you have planned the menu items and serving sizes for the various age/grade groups, use the checklist below.

Requirements   *Yes   No*

• Have you included all components of the meal?   __ __

• Have you planned serving sizes sufficient to provide all students the required quantity of:

Meat or meat alternate?   __ __

Two or more vegetables and/or fruits?   __ __

Whole-grain or enriched bread or bread alternate?   __ __

Fluid milk?   __ __

Recommendations   *Yes   No*

• Have you included an unflavored form of fluid lowfat milk, skim milk, or buttermilk?   __ __

• Have you included a vitamin A vegetable or fruit at least twice a week?   __ __

• Have you included a vitamin C vegetable or fruit at least 2 or 3 times a week?   __ __

• Have you included several foods for iron each day?   __ __

• Have you kept concentrated sweets and sugars to a minimum?   __ __

• Have you kept calories from fat to a moderate level?   __ __

• Have you kept foods high in salt to a minimum?   __ __

• If you have not planned choices, have you avoided serving any one meat alternate or form of meat more than 3 times per week?   __ __

Good Menu Planning Practices   *Yes   No*

• Do your lunches include a good balance of:

Color—in the foods themselves and in garnishes?   __ __

Texture—soft and crisp or firm textured foods?   __ __

—starchy and other type foods?   __ __

Shape—different sized pieces and shapes of foods?   __ __

Flavor—bland and tart or mild and strong flavored foods?   __ __

Temperature—hot and cold foods?   __ __

• Have you included whole-grain bread and cereal products?   __ __

• Have you included fresh fruits and vegetables?   __ __

• Are most of the foods and food combinations ones your students have learned to eat?   __ __

• Have you considered students' cultural, ethnic, or religious food practices?   __ __

• Have you included a popular food in a lunch which includes a "new" or less popular food?   __ __

• Do you have a plan to introduce new foods?   __ __

• Have you planned festive foods for holidays, birthdays, and school activities?   __ __

• Have you included different kinds or forms of foods (fresh, canned, frozen, dried)?   __ __

• Have you included seasonal foods?   __ __

• Have you included less familiar foods or new methods of preparation occasionally?   __ __

Good Management Practices   *Yes   No*

• Have you planned lunches so that some preparation can be done ahead?   __ __

• Have you balanced the workload among employees from day to day?   __ __

• Is oven, surface-cooking, or steam-cooking space adequate for items planned for each lunch?   __ __

• Are proper-sized cooking and serving utensils available?   __ __

• Can you easily serve foods planned for each meal?   __ __

• Will foods "fit" on dishes or compartment trays?   __ __

• Have you taken advantage of USDA-donated foods?   __ __

• Have you used foods in inventory to the extent possible?   __ __

• Do high and low-cost foods and meals balance?   __ __

Source: From *Menu Planning Guide for School Food Service* (pp. 36–37) by U.S. Department of Agriculture, Food and Nutrition Service, 1983, Washington, DC: Government Printing Office.

and main dishes are planned first, as in noncommercial operations. Many restaurants are renowned for signature items, such as special foods, preparation methods, or even type of service or presentation (DiDomenico, 1992). Flamed menu items often are highlighted as signature items on the menu. The tableside service and flamboyant presentation qualify such items as Steak Diane or Bananas Foster as signature items.

A restaurant's menu is a powerful merchandising and marketing tool, and a systematic approach for planning must be used. Gray (1986) developed the following procedures for menu planning:

- Conduct a market study.
- Perform a competitive analysis.
- Interview restaurant critics/reviewers.
- Attend food shows.
- Develop a unified theme.
- Include current trends.
- Analyze nutritional content.
- Ensure variety and balance of menu items.
- Price the menu accurately.
- Check on availability of food products.
- Match the menu with the skill level of kitchen personnel and balance the production stations.
- Control labor costs.
- Increase sales with menu merchandising of appetizers and desserts.
- Test recipes and make adjustments.
- Standardize recipes.
- Conduct a taste testing.
- Establish garnish, plating, and portion standards.

A new menu requires coordination of industry resources with established controls. The outcome of this planning process, according to Gray (1986), should be a menu that is efficiently and consistently produced in the kitchen and is pleasing to guests.

MENU PRICING

Pricing menu items follows planning the menu and can be one of the most difficult decisions management makes. **Menu pricing** should cover the cost of food and labor and additional operating costs, including rent, energy, and promotional advertising. Other important factors to be considered when menu prices are set are perception of value and competition (Keiser & DeMicco, 1993). *Perception of value* is what a customer believes the menu is worth. The foodservice manager also needs to be aware of what the competition is doing, for example, McDonald's often sets local standards for hamburger prices.

Setting menu prices is a constant problem for foodservice managers (Keiser & DeMicco, 1993). A restaurant manager wants to make the highest possible profit and retain repeat customers. Noncommercial foodservice managers want to serve the best food possible within the budget. Unless they are subsidized, noncommercial foodservice managers may need to develop a pricing strategy that covers all costs.

Menu pricing today generally is computerized; most food management software programs include a menu-pricing component. Menu items and portion count can be entered into the program, and it will calculate portion sizes, selling prices, item costs, and the raw food and markup percentages. Sales can be calculated for any period of time, as can total or per customer costs and profit margins (Keiser & DeMicco, 1993).

Pricing Methods

Various methods are used to price menus; the one most often used is based on establishing a percentage of the selling price for food and labor. Miller (1987) named seven methods for pricing menus; the three most often used in foodservice operations are discussed here: factor, prime cost, and actual cost methods.

Factor

The **factor pricing method** is also known as the *markup method* (Keiser & DeMicco, 1993). *Markup,* the difference between cost and selling price, varies among types of foodservice operations. First, the desired percentage of food cost must be selected and divided into 100 to give a pricing factor. By multiplying the raw food cost by this factor, a menu sales price will result.

Raw Food Cost × Pricing Factor = Menu Sales Price

If the operator chooses a 40% food cost, the pricing factor would be 100 ÷ 40, or 2.5. For a raw food cost of $2, the selling price would be $2 × 2.5 = $5.00.

The factor method often is used by foodservice managers because only very simple mathematics is involved. The principal disadvantage is that costs other than food are not known until the end of the month, and it disregards perception of value and the fact that customers will not pay a uniform markup on all menu items.

Prime Cost

Prime cost consists of raw food cost and direct labor cost of those employees involved in preparation of a food item but not service, sanitation, or administrative costs. An accurate determination of prime cost for each menu item would require calculating the raw food cost and direct labor cost for pricing. In addition to cost records on raw food purchased for each menu item, time studies of the amount of direct labor would be required. This would be a gigantic task for an entire menu and difficult to justify because it is so labor intensive. The total process seems thoroughly impractical, especially in commercial operations, because pricing would have to be done almost daily and menu prices changed accordingly.

To make this cost method practical, some assumptions must be made on the percentage of prime cost attributable to raw food, direct labor, and operating margin. Each restaurateur would need to decide what percentage of the selling price would be assigned to the raw food and direct labor costs to give a prime cost total. As an example, in examining financial records, the foodservice manager finds that for every $2 spent for raw food, $.095 is spent for direct labor, making the prime cost total $2.095. The manager decides that 40% of the selling price of a food item would be for raw cost and 8% for direct labor, leaving an operating margin of 52%. The relationship of selling price with a margin of 52% is 100 divided by 52, or 1.923, as a multiplier for the prime cost to yield the selling price. In the example, the prime cost of $2.095 multiplied by 1.923 gives a selling price of $4.028. In actuality, this would probably be changed to $4.25. The multiplier, 1.923, would be used for pricing most menu items.

Actual Cost

Actual cost is used in operations that keep accurate cost records. The initial step, as usual, is to establish the food cost from standardized recipes and labor costs, which are the principal variable costs and are actual. Other variable costs, fixed costs, and profit can be obtained as a percentage of sales from the profit-and-loss statement explained in chapter 21. The menu price consists of the actual food cost + actual labor cost + other variable costs + fixed cost + profit. The actual-cost method has the advantage of including all costs and the desired profit in the selling price of the menu item.

Pricing Psychology

Foodservice operators consider **pricing psychology** in determining what and how to charge. Psychological aspects of pricing affect customer perceptions, which then influence the purchase decision (Pavesic, 1988). Many schemes have been devised to entice the customer to buy; these schemes are apparent in all merchandising. Menu items are priced by the same general considerations that are apparent in other items for sale. Some of the schemes in use by foodservice operators are odd-cents pricing, cost by the ounce, two-tier foodservice, à la carte, and table d'hôte.

Odd-cents pricing follows the basic philosophy of creating an illusion of a bargain. According to Miller (1987), it is an attempt to maximize profit and psychologically affect the customer. The so-called magic numbers supposedly stimulate the consumer to buy. Kreul and Stock (1982) identified three methods practiced in odd-cents pricing:

- Price ends in an odd number (e.g., $4.75).
- Price ends in a number other than zero (e.g., $4.77).
- Price is just below a zero (e.g., $4.99).

Miller (1987) stated that odd-cents pricing is for people who eat out, not people who dine out.

Pricing by the ounce, according to Rose (1988b), is becoming common among noncommercial foodservice operators, especially for salad and sandwich bars. This concept also is popular for salad bars in supermarkets. Increasing menu prices causes dissatisfied customers, but when customers can weigh their own portions, the sense of control reduces complaints. The initial step in using this pricing system requires estimation of the total amount of raw food needed to serve an anticipated number of customers in order to determine the raw food cost per ounce. A markup factor per ounce of raw food to cover labor, other expenses, and profit needs to be established. The selling price per ounce is computed by multiplying the cost per ounce by a markup factor. The cost-per-ounce pricing scheme has satisfied many customers because they like paying only for what they eat.

Two-tier foodservice is being used in a number of healthcare centers around the country as hospitals compete for patients. Upscale amenities and menus are being prepared for wealthy patients who are willing to pay for special food items and elegant service. Special kitchens, private dining rooms, and suites have been added to many hospitals, with chefs preparing fine restaurant menu items to respond to individual desires.

The à la carte menu is actually not a separate type of menu but a method of pricing a menu. Menu items are priced, offered, and selected separately by the customer. The à la carte menu is typical in many commercial foodservices, ranging from upscale gourmet to limited-menu restaurant operations. Cafeterias, both noncommercial and commercial, generally offer à la carte pricing of menu items.

Table d'hôte menus group several food items together and offer them at a fixed price. The menu is a complete meal with several courses, and the only choices might be soup, salad, or dessert.

MENU EVALUATION

The menu has been identified as the primary control of the foodservice system, and because of its significance, it should be evaluated. **Menu evaluation** is a continuing process that is best conducted during production and service of each meal and after major menu-planning sessions. Menus should be evaluated not only for customer acceptability and food cost but also for profitability.

Food acceptability can be assessed using various methods, such as plate waste studies, data on menu selections, formal and informal customer surveys, and observations in the service areas. In noncommercial foodservices, nutritional evaluation of menus is especially important. A method as simple as checking for the presence of various food groups on the menu may be used. Today, however, computer-assisted nutrient analysis often is used to determine nutritional adequacy of menus. Aesthetic factors may be evaluated using a form similar to that in Figure 6.20.

Food cost should be estimated to determine if cost constraints were adequately considered in planning menus. Adjustments may be necessary if costs exceed the per person allowance. In a commercial operation, such estimation is important for adjusting menu selling prices at periodic intervals.

Figure 6.20.   Form for evaluating aesthetic appeal of menus.

	Mon.	Tues.	Wed.	Sun.

Color
    Contrast
        attractive combinations
        garnishes used				

Flavor
    Contrast something:
        bland, tart, sweet
        combinations acceptable				

Texture
    Contrast something:
        crisp, firm, soft				

Form, shape, size
    Vary something:
        flat, round, long, chopped				

Preparation type
    Not too many:
        starchy foods, sauces,
            mixtures, crunchy, chewy,
            some type fruit, veg.				

Repetition
    Do not repeat:
        same food in same meal
        food on same day of week				

Temperature Both hot and cold foods				

Another aspect in the preliminary assessment of menus is a review of the production demand on foodservice equipment and the estimate of labor requirements for producing menu items. Production and service supervisors should make notations on a regular basis of problems that occur in producing and serving food items. Also, difficulties in procurement of various food items should be noted by the person responsible for purchasing.

Menu engineering, introduced by Kasavana and Smith in 1982, is a computer-supported management information tool that focuses on the contribution to profit of a menu item rather than considering profit as a percentage of cost. Managers think they know which menu items make money and which do not, and quite often they are right. A computerized menu analysis might reveal a few surprises, however. The lunch special being promoted might turn out to be less profitable than an item hidden away on the regular menu (Lydecker, 1991).

The assumption is that popularity and profitability will determine the desirability of a menu item. Menu mix (MM) is concerned with the popularity of the menu item, as measured by customer demand, and the contribution margin (CM), which is the menu price of an item minus its cost, is concerned with the profit of each item. This information is used to create a matrix, as shown in Figure 6.21.

An essential procedure is calculation of the CM and MM percentages for each item on the menu and then assigning them to one of the four categories. A menu item classified as a star should be given top priority on the menu and perhaps made a signature item. A plow horse with a high customer demand could easily become a star if the price were increased (perhaps add more garnish), or a lower cost for the item is possible by serving smaller portions. A puzzle might become a star with better marketing or by reducing the selling price and creating additional sales. The item, for example a vegetarian entrée, in the dog category might be kept on the menu

Figure 6.21.   Menu engineering matrix.

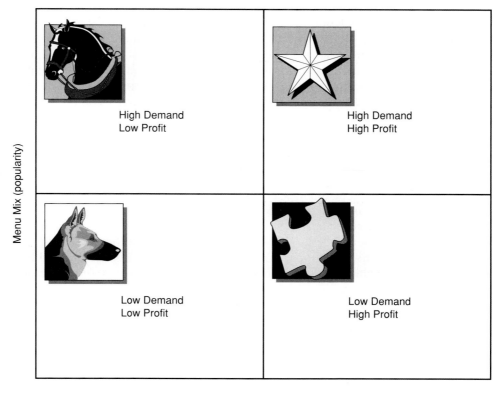

because it could be a popular item for frequent customers who often bring guests choosing stars.

The complexity of deciding which items to include on a menu has led to further refinement of the menu engineering process. Bayou and Bennett (1992) recommended that profitability analysis be determined by calculating contribution margins for not only individual menu items, but also various menu categories, meal periods, and the operation as a whole.

SUMMARY

The menu, defined as a list of food items, is the primary control of the foodservice operation and the basis for all the functional subsystems. Menus may be presented in many different forms ranging from printed to video to oral. A menu pattern consisting of categories of items needs to be established for each meal. The three basic types of menus are static, cycle, and single use, and the number of choices in each depends upon the expectations of the customer. Most menus have been based on three meals a day, but today breakfast and lunch often are eaten in a hurry, and only dinner might be a leisurely occasion.

The crux of menu planning is that the menu is consumer driven. Customer satisfaction and managerial decisions are elements involved in menu planning for both noncommercial and commercial foodservice managers. Sociocultural factors including food preferences and menu item popularity must be considered in menu planning. Satisfying nutritional needs of the customer should be a primary concern in planning; also, the aesthetic factors of taste and appearance of food items cannot be ignored. Management has many decisions it must make in menu planning depending on the type of foodservice system. Other factors influencing menu decisions are food and labor costs, production capability of labor and equipment, type of service, and availability of food items.

Menu planning is a complex process and may be done by an individual or a team. Truth-in-menu legislation has created problems for restaurateurs and has made them more conscious of their responsibilities. An innovative concept for the relief of menu planners is desktop publishing, which provides the opportunity for rapid changes in menus. Traditional principles of menu planning are applicable to all foodservice operations. A systematic approach, however, is required for restaurants because the menu is a powerful merchandising and marketing tool.

Pricing items after planning the menu can present some exasperating managerial decisions. Some of the methods available for pricing menus are factor, prime cost, and actual cost. Psychological aspects of pricing influence the purchase decision of customers, and many schemes have been devised to encourage the customer to buy. Menu evaluation is the final step in menu planning and pricing. Menus should be evaluated not only for customer acceptability and food cost but also for profitability.

BIBLIOGRAPHY

Adams, A. J. (1987). Consumers, what are they telling us? In *Agricultural research for a better tomorrow: Commemorating the Hatch Act centennial* (pp. 1887–1987). Washington, DC: U.S. Department of Agriculture.

Allen, R. L. (1993). Think big! Kids' menus help create future customers. *Nation's Restaurant News, 27*(3), 12.

American Express. (1992). Red Lobster menus combine braille/large type-face. *Restaurant Briefing, 17*(6), 3.

American Heart Association. *Dining out: A guide to restaurant dining.* (1984). Dallas: Author.

Backas, N., & Lorenzini, B. (1991). R & I's menus for the '90s: Menu ideas. *Restaurants & Institutions, 101*(8), 158, 164, 166, 169, 172, 177–178, 182, 186, 192–195, 198–201.

Bartlett, M. (1991). R & I's menus for the '90s. *Restaurants & Institutions, 101*(8), 8–9.

Bayou, M. E., & Bennett, L. B. (1992). Profitability analysis for table-service restaurants. *Cornell Hotel and Restaurant Administration Quarterly, 33*(2), 49–55.

Best of breakfast. (1992). *Restaurants & Institutions,* 102 (Suppl. to Issue #22).

Bobeng, B. (1982). A food service manager's perceptions of meal quality. In *Hospital patient feeding systems.* Washington, DC: National Academy Press.

Callaway, C. W. (1992). The marriage of taste and health: A union whose time has come. *Nutrition Today, 27*(3), 37–42.

Carlson, B. L. (1987). Promoting nutrition on your menu: Three myths, eight tarnished rules, and five hot tips. *Cornell Hotel & Restaurant Administration Quarterly, 27*(4), 18–21.

Chaudhry, R. (1992a). Best bets for breakfast. *Restaurants & Institutions,* 102 (Suppl. to Issue #22).

Chaudhry, R. (1992b). Catering come-ons. *Restaurants & Institutions, 102*(27), 141–142.

Comstock, E. M., St. Pierre, R. G., & Mackiernan, Y. D. (1981). Measuring individual plate waste in school lunches. *Journal of The American Dietetic Association, 79,* 290–296.

Creating an alliance between taste and health. (1991). *ADA Courier, 30*(12), 1–2.

Cummings, L. E., & Kotschevar, L. H. (1989). *Nutrition management for foodservices.* New York: Delmar Publishers.

DiDomenico, P. (1992). Adding a profit center that promotes: Selling signature items. *Restaurants USA, 12*(10), 18–21.

Dietary guidelines revisited. (1990). *The Nutrition Advisor, 3*(2), 1–2.

Donovan, M. D. (Ed.). (1993). *The professional chef's® techniques of healthy cooking.* New York: Van Nostrand Reinhold.

Dougherty, D. A. (1989). President's page: New emphasis on food. *Journal of The American Dietetic Association, 89*(6), 842–843.

Finn, S. (1992). *The real life nutrition book.* New York: Penguin.

Fintel, J. (1989). Just the fax. *Restaurants USA, 9*(6), 30–31.

5 a day—word is getting out! (1993). *5 A Day News, 2*(2), 1–5.

Ganem, B. C. (1990). *Nutritional menu concepts for the hospitality industry.* New York: Van Nostrand Reinhold.

Good food/good health: Do we have to choose? (1991). *The American Institute of Wine & Food,* (August), 1–8.

Gordon, E. (1992). Taking a look at lunch. *Restaurants USA, 12*(5), 36–38.

Gray, N. J. (1986). 17 steps to developing a winning menu. *NRA News, 6*(2), 16–20.

Griffin, R. W. (1993). *Management* (4th ed.). Boston: Houghton Mifflin.

Howat, M. (1991). Menu labeling: the pros and cons. *Nation's Restaurant News, 25*(4), 42.

Iwamuro, R. (1993). Looking back at 1992: The annual CREST report. *Restaurants USA, 13*(5), 36–37.

Kasavana, M. L., & Smith, D. l. (1982). *Menu engineering: A practical guide to menu analysis.* Lansing, MI: Hospitality Publications.

Kass, M. Computer age menus. (1989). *Restaurants & Institutions, 99*(16), 64–65, 68, 72–73, 76–77, 80.

Kass, M. (1990). Build a better menu. *Restaurants & Institutions, 100*(17), 36–39, 44, 48, 56, 58.

Keegan, P. O. (1992). OBRA gives nursing-home residents 'right to vote' on quality of care. *Nation's Restaurant News, 26*(10), 37, 40, 42.

Keiser, J., & DeMicco, F. J. (1993). *Controlling and analyzing costs in food service operations* (3rd ed.). New York: Macmillan.

Kirks, B. A., & Wolff, H. K. (1985). A comparison of methods for plate waste determinations. *Journal of The American Dietetic Association, 85,* 328–331.

Knickrehm, M. E., Cotner, C. G., & Kendrick, J. G. (1969). Acceptance of menu items by college students. *Journal of The American Dietetic Association, 55,* 117–120.

Kotschevar, L. H. (1987). *Management by menu* (2nd ed.). Dubuque, IA: Wm. C. Brown in cooperation with The Educational Foundation of the National Restaurant Association.

Kreul, L. M., & Scott, A. M. (1982). Magic numbers: Psychological aspects of menu pricing. *Cornell Hotel and Restaurant Administration Quarterly, 23*(2), 70–75.

Kroll, B. J. (1990). Evaluating rating scales for sensory testing with children. *Food Technology, 44*(11), 78–86.

Lachance, P. A. (1976). Simple research techniques for school foodservice. Part I: Acceptance testing. *School Food Service Journal, 30*(8), 54–56, 58, 61.

Lorenzini, B. (1992a). Breakfast at hand. *Restaurants & Institutions, 102*(15), 12–13.

Lorenzini, B. (1992b). Lighthouse offers braille menus. *Restaurants & Institutions, 102*(22), 22.

Lorenzini, B. (1992c). Menus that sell by design. *Restaurants & Institutions, 102*(6), 106, 108, 110, 116.

Lunch at court. (1993). *School Food Service Journal, 47*(1), 34–37.

Lydecker, T. (1987). Is your menu fit for customers? *Restaurants & Institutions, 97*(20), 34–37, 48–49, 54, 57, 62, 66.

Lydecker, T. (1991). Money-making menus. *Restaurants & Institutions, 101*(8), 96–97, 100, 102, 104, 109–110, 112.

Marshall, A. G., & Bellucci, E. C. (1984). Larceny by menu. *FIU Hospitality Review, 2*(1), 30–40.

McCarthy, B. (1992). Lunch to go. *Restaurants & Institutions, 102*(20), 61, 64, 66, 67.

McCarthy, B. (1993). Breakfast brings home the bacon. *Restaurants & Institutions, 103*(6), 26–27, 34, 39, 42, 46.

McCarthy, B., & Straus, K. (1992). Tastes of America 1992. *Restaurants & Institutions, 102*(29), 24–25, 28–29, 34, 36, 38, 44.

McD rolls out new braille, picture menus at U.S. units. (1992). *Nation's Restaurant News, 26*(12), 2.

Mermelstein, N. H. (1993). Nutrition labeling in foodservice. *Food Technology, 47*(4), 65–68.

Miller, J. E. (1987). *Menu pricing and strategy* (2nd ed.). New York: Van Nostrand Reinhold.

Miller, S. G. (1988a). Creating menus with desktop publishing. *Cornell Hotel and Restaurant Administration Quarterly, 28*(4), 32–35.

Miller, S. G. (1988b). Fine-tuning your menu with frequency distributions. *Cornell Hotel and Restaurant Administration Quarterly, 29*(3), 86–92.

Miller, S. G. (1992). The simplified menu-cost spreadsheet. *Cornell Hotel and Restaurant Administration Quarterly, 33*(3), 85–88.

National Research Council. (1989a). *Diet and health: Implications for reducing chronic disease risk. Executive summary*. Washington, DC: National Academy Press.

National Research Council. (1989b). *Recommended dietary allowances* (10th ed.). Washington, DC: National Academy Press.

National Restaurant Association. *A nutrition guide for the restaurateur*. (1986). Washington, DC: Author.

National Restaurant Association. (1987). *Guidelines for providing facts to foodservice patrons: Ingredient and nutrient information*. Washington, DC: Author.

National Restaurant Association. (January 1990). *Nutrition awareness and the foodservice industry. Current issues report*. Washington, DC: Author.

National Restaurant Association. (1993). FDA releases new proposal on menu claims, asks for public comment. *Washington Weekly, 13*(25), 1, 4.

OBRA '87 update. (1991). *Journal of The American Dietetic Association, 91*, 1381.

Pavesic, D. (1988). Taking the anxiety out of menu pricing. *Restaurant Management, 2*(2), 56–57.

Peryam, D. R., Palemis, B. W., Kamen, J. M., Eindhoven, J., & Pilgrim, E. J. (1960). *Food preferences of men in the U.S. Armed Forces*. Chicago: Institute for the Armed Forces.

Peryam, D. R., & Pilgrim, E. J. (1957). Hedonic scale method of measuring food preference. *Food Technology, 11*, 9.

Peterkin, B. B. (1990). Dietary guidelines for Americans, 1990 edition. *Journal of The American Dietetic Association, 90*, 172–177.

Pettus, E. M. (1989). Menu engineering. *Club Management, 68*(5), 24–25.

Position of The American Dietetic Association: Nutrition in foodservice establishments. (1991). *Journal of The American Dietetic Association, 91*, 480–482.

Powers, T. E. (1978). Menus: Poetry, the law, and the consumer. *Journal of Hospitality Education, 3*(1), 31–49.

Radice, J. (1989). The menu as star. *Restaurants USA, 9*(6), 32–34.

Regan, C. (1987). Adapting to the interest in nutrition. *Restaurants USA, 7*(2), 30–34.

Roberts, C. R., & Regan, C. (1991). Position of The American Dietetic Association: Nutrition in foodservice establishments. *Journal of The American Dietetic Association, 91*, 480–482.

Rose, J. C. (1988a). Pricing I: Three menu pricing systems. *Food Management, 23*(1), 40.

Rose, J. C. (1988b). Pricing II: By the ounce. *Food Management, 23*(2), 44.

Ryan, N. R. (1992a). Bowl them over with seasonal soups. *Restaurants & Institutions, 102*(19), 166–170.

Ryan, N. R. (1992b). Build a better breakfast. *Restaurants & Institutions, 102*(15), 36.

Ryan, N. R. (1992c). Menus of choice captivate diners. *Restaurants & Institutions, 102*(1), 66, 70, 74.

Ryan, N. R. (1992d). Reinventing the meal: À la carte leads to carte blanche. *Restaurants & Institutions, 102*(6), 14–15, 18, 22.

Ryan, N. R. (1993). Good 'mixes' break the lunch routine. *Restaurants & Institutions, 103*(6), 50–51, 56, 62, 64, 68.

Ryan, N. R., & Stephenson, S. (1991). Menu opportunities. *Restaurants & Institutions, 101*(8), 36–37, 40, 40–41, 44, 49, 54, 58, 64, 71, 74, 80–81, 85, 91–92.

Schultz, H. G. (1965). A good action rating scale for measuring food acceptance. *Journal of Food Science, 30*(2), 365–374.

Sensory evaluation and the consumer—points of interaction. (1990). *Food Technology, 44*(11), 153–172.

Sneed, J., & Burkhalter, J. P. (1991). Marketing nutrition in restaurants: A survey of current practices and attitudes. *Journal of The American Dietetic Association, 91*, 459–462.

Solomon, J. (1992). A guide to good menu writing. *Restaurants USA, 12*(5), 27–30.

Solomon, J. (1993). Savvy sandwiches. *Restaurants USA, 13*(4), 19–21.

Straus, K. (1992a). Dinner is served: Fixed price meals are hot and here's why. *Restaurants & Institutions, 102*(25), 36–37, 40, 44, 46.

Straus, K. (1992b). New-size entrees multiply sales. *Restaurants & Institutions, 102*(6), 62, 66, 68.

Tableservice restaurants offer fax service and other conveniences. (1993). *Restaurants USA, 13*(3), 36.

Thomas, P. R. (Ed.). (1991). *Improving America's diet and health: From recommendations to action.* Washington, DC: National Academy Press.

The Surgeon General's report on nutrition and health. (1988). DHHS (PHS) Publication Number 88-50210, Washington, DC: Government Printing Office.

USDA pyramid stacks up nutritional advice. (July/August 1992). *Food Insight,* 8.

U.S. Department of Agriculture, Food and Nutrition Service. (1983). *Menu planning guide for school food service.* Washington, DC: Author.

U.S. Department of Agriculture, Human Nutrition Information Service. (1989). *Eating better when eating out: Using the Dietary Guidelines.* House and Garden Bulletin No. 232-11. Washington, DC: Author.

U.S. Departments of Agriculture and Health and Human Services. (1990). *Nutrition and your health: Dietary guidelines for Americans* (3rd ed.). Washington, DC: Author.

Wallace, J. (1991). The menu challenge. *Restaurants & Institutions, 101*(8), 5.

Warshaw, H. S. (1993). America eats out: Nutrition in the chain and family restaurant industry. *Journal of The American Dietetic Association, 93,* 17, 19–20.

Welland, D. (1991). Are your recipes healthy? Making the most of nutrient analysis. *Restaurants USA, 11*(10), 33–34.

Wells, W. D. (1965). Communicating with children. *Journal of Advertising Research, 5*(2), 2–14.

Welsh, S., Davis, C., & Shaw, A. (1992). A brief history of food guides in the United States. *Nutrition Today, 27*(6), 6–11.

Welsh, S., Davis, C., & Shaw, A. (1992). Development of the food guide pyramid. *Nutrition Today, 27*(6), 12–23.

What's to be in '93. (1993). *Sysco's Menus Today,* (January), 20–24, 26, 28, 30, 32.

The Menu—Control of The Fairfax Foodservice System

This case study is based on chapters 5 and 6, which describe the type of foodservice operation and the menu. Safety of food from the time it is purchased until it is served to the resident is important in all types of foodservices and is of great concern to the director of the Food and Beverage Department. The menu needs careful planning because it is the control of the foodservice system. The information included in these chapters is needed to provide a solution to the problem.

Background Information

The Fairfax is primarily a conventional foodservice operation, but some menu items, such as lasagna and beef stew, are cooked and chilled to be served a day or two later. The kitchen also serves as a commissary because menu items are transported in bulk to the Belvoir Woods Health Care Center, where they are portioned and served to the residents living there.

Three types of foodservice—conventional, ready prepared, and commissary—operate at the same time at The Fairfax, which is one reason why the operation instituted the SERVSAFE program. Developed by the National Restaurant Association, SERVSAFE is the basis of Marriott's food safety program. Marriott's Manager's Training Tool Kit both in English and Spanish is available to F&B managers to be used in eight associate training sessions. Videos, flip charts, crossword puzzles, case studies, and questions are included in the kit as well as suggestions for teaching the course. At the end of each session, the associate receives a certificate indicating mastery of material. When the eight-section program is completed, the associate receives a Marriott Certificate of Merit and a SERVSAFE pin.

Types of Menus

A restaurant-type menu, as shown in Figure 1, is used in the main dining room for the independent-living residents. The F&B director, with the assistance of the chef, plans the menu, and the consulting dietitian reviews it for nutritional adequacy on her weekly visit. The same menu is used for both lunch and dinner. The

Figure 1. The main dining room menu.

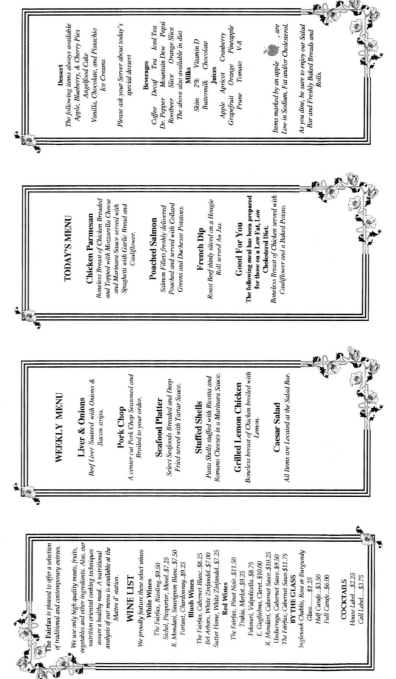

The Fairfax is pleased to offer a selection of traditional and contemporary entrees.

We use only high quality meats, fruits, vegetables and other ingredients. Also, our nutrition oriented cooking techniques assure a healthy meal. A nutritional analysis of our menu is available at the Maitre d' station.

WINE LIST

We proudly feature these select values

White Wines
The Fairfax, Riesling..$9.50
Sichel, Piesporter, Mosel..$7.25
R. Mondavi, Sauvignon Blanc..$7.50
Fortant, Chardonnay..$9.25

Blush Wines
The Fairfax, Cabernet Blanc..$8.25
Bel Arbors, White Zinfandel..$7.00
Sutter Home, White Zinfandel..$7.25

Red Wines
The Fairfax, Pinot Noir..$11.50
Trakia, Merlot..$9.25
Folonari, Valpolicella..$8.75
E. Guglielmo, Claret..$10.00
R. Mondavi, Cabernet Sauv..$10.25
Undurraga, Cabernet Sauv..$9.50
The Fairfax, Cabernet Sauv..$11.75

BY THE GLASS
Inglenook Chablis, Rose or Burgundy
Glass........$1.25
Half Carafe..$3.50
Full Carafe...$6.00

COCKTAILS
House Label...$2.25
Call Label....$3.75

WEEKLY MENU

Liver & Onions
Beef Liver Sauteed with Onions & Bacon strips.

Pork Chop
A center cut Pork Chop Seasoned and Broiled to your order.

Seafood Platter
Select Seafoods Breaded and Deep Fried served with Tartar Sauce.

Stuffed Shells
Pasta Shells stuffed with Ricotta and Romano Cheeses in a Marinara Sauce.

Grilled Lemon Chicken
Boneless breast of Chicken broiled with Lemon.

Caesar Salad
All Items are Located at the Salad Bar.

TODAY'S MENU

Chicken Parmesan
Boneless Breast of Chicken Breaded and Topped with Mozzarella Cheese and Marinara Sauce served with Spaghetti with Garlic Bread and Cauliflower.

Poached Salmon
Salmon Fillets freshly delivered Poached and served with Collard Greens and Duchesse Potatoes.

French Dip
Roast Beef thinly sliced on a Hoagie Roll served Au Jus.

Good For You
The following meal has been prepared for you on a Low Fat, Low Cholesterol Diet.
Boneless Breast of Chicken served with Cauliflower and a Baked Potato.

Dessert
The following items always available
Apple, Blueberry, & Cherry Pies
Angelfood Cake
Vanilla, Chocolate, and Pistachio Ice Creams

Please ask your Server about today's special dessert

Beverages
Coffee   Decaf   Tea   Iced Tea
Dr. Pepper   Mountain Dew   Pepsi
Rootbeer   Slice   Orange Slice
The above also available in diet

Milks
Skim   2%   Vitamin D
Buttermilk   Chocolate

Juices
Apple   Apricot   Cranberry
Grapefruit   Orange   Pineapple
Prune   Tomato   V-8

Items marked by an apple 🍎 , are Low in Sodium, Fat and/or Cholesterol.

As you dine, be sure to enjoy our Salad Bar and Freshly Baked Breads and Rolls.

Source: Courtesy of Marriott Corporation.

Figure 2. One week of a 5-week menu cycle at the Belvoir Woods Health Care Center.

WEEK I MENU

BELVOIR WOODS

SUNDAY	MONDAY	TUESDAY	WEDNESDAY	THURSDAY	FRIDAY	SATURDAY
CITRUS JUICE	CITRUS JUICE	CITRUS JUICE	CITRUS JUICE	CITRUS JUICE	CITRUS JUICE	CITRUS JUICE
PRUNES OR BANANAS	PRUNES OR CANTELOPE	PRUNES OR PEAR HALF	PRUNES OR HONEYDEW	PRUNES OR PEACHES	PRUNES OR PINEAPPLE	PRUNES/MIXED FRUIT CUP
COLD CEREAL/OATMEAL	COLD CEREAL/GRITS	CD CEREAL/CIN OATMEAL	COLD CEREAL/CRM WHEAT	COLD CEREAL/OATMEAL	COLD CEREAL OR GRITS	COLD CEREAL/CRM WHEAT
SCRAMBLED EGGS OR SUB	HAM&SCRAMBLED EGGS	SCRAMBLED EGGS OR SUB	SCRAMBLED EGGS OR SUB	SCRAMBLED EGGS OR SUB	CREAM EGG SCRAMBLE	SCRAMBLED EGGS OR SUB
HASHBROWNS	SALMON FRITATTA	FRENCH TOAST	APPLE PANCAKES	ITALIAN SAUSAGE SOUFFL	BLUEBERRY PANCAKES	FRENCH TOAST
BACON OR SAUSAGE	BACON/CANADIAN BACON	BACON OR HAM	BACON OR BEEF HASH	BACON OR SAUSAGE	BACON/CANADIAN BACON	BACON OR HAM
TOAST OR BRAN MUFFIN	TOAST/ENGLISH MUFFIN	TOAST /CINNAMON ROLL	TOAST/BLUEBERRY MUFF	TOAST/ENGLISH MUFFIN	TOAST/BANANA MUFFIN	TOAST/CHOCOLATE MUFF
COFFEE/MILK	COFFEE/MILK	COFFEE/MILK	COFFEE/MILK	COFFEE/MILK	COFFEE/MILK	COFFEE/MILK
CREAM OF MUSHRM SOUP	GARDEN VEG SOUP	CREAM OF TOMATO SOUP	CHICKEN RICE SOUP	SPLIT PEA SOUP	NEW ENGLAND CLAM	MINESTRONE SOUP
TROPICAL FRUIT SALAD	EGG SALAD SANDWICH	GRILLED CHEESE SANDWH	CUCUMBER&ONION SALAD	BAR-B-QUE PORK SANDWH	CHOWDER	SPAGETTI & MEATBALLS
PORK ROAST	PEA AND CHEESE SALAD	PEAR IN LIME GELATIN	IRISH LAMB STEW	GERMAN POTATO SALAD	COLE SLAW	ZUCCHINI SQUASH MEDLEY
SCALLOPED CORN	POTATO CHIPS		BRUSSEL SP. MORNAY	CORN	ORANGE ROUGHY	SUB:COTTAGE FRUIT
SWT &SOUR BEANS	SUB:MOROCCAN LEMON	SUB:BRAISED SHOR RIBS	SUB:CHICKEN SUPREME	SUB: BEER BATTER	FRENCH FRIES	PLATE
SUB:TURKEY	CHICKEN	DUCHESSE POTATOES	SANDWICH	SHRIMP	SUB:VEAL PICCATA	
W:IIPPED POTATO	RICE PILAF	GARDEN PEAS	ORIENTAL VEGETABLE	CARROTS	MEXICAN MEDLEY	
LIMA BEANS	ASPARAGUS		IRISH SODA BREAD OR			GARLIC BREAD OR
ASSORTED BREAD/MARG	ASSORTED BREAD/MARG	ASSORTED BREAD/MARG	ASSORTED BREAD/MARG	ASSORTED BREAD/MARG	ASSORTED BREAD/MARG	ASSORTED BREAD/MARG
BEVERAGE OF CHOICE	BEVERAGE OF CHOICE	BEVERAGE OF CHOICE	BEVERAGE OF CHOICE	BEVERAGE OF CHOICE	BEVERAGE OF CHOICE	BEVERAGE OF CHOICE
MACADAMIA NUT BAR	PINEAP UP/DOWN CAKE	BOSTON CREAM PIE	BAKED APPLES	BLACK FOREST TORTE	NEW YORK CHEESECAKE	CAFE TORTONI
FRENCH ONION SOUP	CURRIED CRM OF ZUCCHIN	LOUISIANA CORN CHOWDE	DUTCH GREEN BEAN SOUP	CHICKEN OKRA CREOLE	PEPPER POT SOUP	CHICKEN MULLIGAWTNY
MANDARIN CHICKEN	TOMATO MEAT LOAF	TOSSED SALAD	STUFFED CABBAGE ROLLS	CHICKEN"N"DUMPLINGS	ROAST PRIME RIB OF BEEF	LIGHT WALDORF SALAD
FRIED RICE	WHIPPED POTATO	CHICKEN POT PIE	BUTTERED LIMA BEANS	KALE	OVEN BROWNED POTATO	PARMESAN FRIED FISH
BROCCOLI CUTS	CARROTS&ONIONS	SLICED BEETS	FRENCH FRIES	SUB:SWISS STEAK	BAVARIAN BEANS	BAKED POTATO/SOUR CR
SUB:RED SNAPPER	SUB:PORK CHOP IN GRAVY	SUB:STUFFED PASTA	SUB:TURKEY	BAKED POTATO	SUB:DUCK ALA BIGARADE	PARSLEY CARROTS
STEWED TOMATO	GREEN BEANS	SHELLS	SWEET POTATO	PICKLED BEETS	BROCCOLI CUTS	SUB:CHICKEN BOLOGNAISE
		CAULIFLOWER	GREEN BEANS			PASTA AL PESTO
		GARLIC BREAD OR			YORSHIRE PUDDING	CAULIFLOWER FLORETS
ASSORTED BREAD/MARG	ASSORTED BREAD/MARG	ASSORTED BREAD/MARG	ASSORTED BREAD/MARG	ASSORTED BREAD/MARG	ASSORTED BREAD/MARG	ASSORTED BREAD/MARG
BEVERAGE/MILK	BEVERAGE/MILK	BEVERAGE/MILK	BEVERAGE/MILK	BEVERAGE/MILK	BEVERAGE/MILK	BEVERAGE/MILK
KEY LIME PIE/ICE CREAM	PIE/CAKE/ICE CREAM	PIE / CAKE/ ICE CREAM	PIE/CAKE/ICE CREAM	PIE/CAKE/ICE CREAM	PIE/CAKE/ICE CREAM	CENCI ALLA FIORENTINA

TOSS GREEN SALAD,FRESH FRUIT SALAD,ROAST TURKEY,BAKED CHICKEN AND FISH, MASHED POTATO, BAKED POTATO,SANDWICHES.   BEEF BARLEY SOUP,COTTAGE CHEESE, OFFERED DAILY

Source: Courtesy of Marriott Corporation.

195

four-part menu consists of a wine and cocktail list and weekly, daily, and dessert menus.

Each week has a new menu of items that are served each of the five days along with a menu that changes each day. The daily menu consists of a meat or poultry entrée, a catch of the day, and a sandwich of the day. Also, a Wellness and You (formerly Good for You) selection, in which cholesterol, fat, and sodium are controlled, is included in the daily menu. Both the weekly and daily menu changes allow the chef flexibility in purchasing foods that are being featured in the market.

A 5-week selective cycle menu is used in the Belvoir Woods Health Care Center, one week of which is shown in Figure 2. If nothing on the menu appeals to the resident, the chef will prepare a special menu item. Residents in the assisted-living unit on regular diets have the same menu as those in the independent-living unit; the cycle menu is used for those on modified diets.

Residents' dietary needs in the healthcare center are carefully monitored. The physician orders the type of diet required for each resident, and the certified dietary manager, who is supervised by the consulting dietitian, follows the order in planning the residents' menus. A special Dining with Dignity pureed food program (Figure 3) is used to make menu items more appetizing to those residents requiring this modification. For example, pureed carrots are put into a small, fluted mold, heated, and unmolded on the dinner plate.

The Fairfax enlivens dining with celebrations. For birthdays in the healthcare unit, once each month a party is held for all those who have a birthday that month. In the main dining room, on a resident's birthday a cake is brought to his

Figure 3.   Dining with Dignity pureed food program.

Dining with Dignity

Pureed Food Program

The main objective of the Dining with Dignity pureed food program is to provide the same variety and quality of foods in a thickened, pureed form that is visually appealing yet easily chewed and safe to swallow. This objective will accomplish the following goals:

- Eliminate the stigma associated with pureed foods
- Increase food consumption and at the same time decrease the need for nutritional supplements
- Market Marriott's efforts to provide quality dining to all residents regardless of diet modifications
- Contribute to resident's and family's satisfaction

The process includes combining the pureed food with a thickener (commercial or gelatin) and forming the food into its original form or an attractive shape by using molds, cookie cutters, or a pastry bag with various tips. As the number of pureed foods increases, pastry bags may become labor intensive. Using molds or cookie cutters is more efficient because shaped pureed food can be prepared a day in advance and frozen. When the staff becomes skilled at using a pastry bag, they can pipe the food on a cookie sheet in advance and then freeze and wrap it for future use.

Source: Courtesy of Marriott Corporation.

or her table. Each month a theme party with a "diet holiday," at which diets are not adhered to, is scheduled for residents in all levels of care.

The Problem

Most residents were accustomed to an expensive life-style, including fine dining, before moving to The Fairfax. Menu planning is a challenge for the F&B director, who has to stay within a budget while pleasing residents who have "champagne tastes." Food items such as truffles, caviar, and saffron can be found on purchase orders. Originally, the menu included filet mignon, prime rib of beef, or lobster tails twice a week paired with a less expensive menu item like broiled catfish or macaroni and cheese. Food costs were rapidly increasing because very few residents would order the less expensive menu item. The F&B director tried a few cost control ideas and finally found one that solved the budget problem and at the same time satisfied the residents. If you were the F&B director, how would you solve this problem?

Points for Discussion

- If the residents' satisfaction is the primary driving force in your foodservice operation, how can you as a manager respond to their demands, which are beyond your budget limitations?
- What can you do to make residents part of the solution rather than part of the problem?
- How does the fact that these residents are a quasi-captive audience affect your approach to this problem?
- If choice is important to residents, how can you effectively utilize this as a cost-control measure?
- How would you change the menu planning assumptions or guidelines?

3

Procurement

Procurement is the first subsystem in the foodservice system, and the first activity in it is purchasing. The quality of products brought into the foodservice operation is reflected in the quality of menu items served to the customer. Receiving and storage are two other activities that need to be monitored if serving quality food is the goal of the organization. Inventory is a tool for controlling both quality and cost of products.

- **Chapter 7, Purchasing.**   Good purchasing practices are critical to cost control and profit generation. Purchasing has been considered a cost center in many organizations, but today it is often considered a profit center because managers are making good decisions, such as selecting effective suppliers. Good purchasing managers guard against ethical problems that occur more often in purchasing than in any other activity in the foodservice organization.
- **Chapter 8, Receiving, Storage, and Inventory Control.**   Receiving, storage, and inventory control are the other activities included in the procurement subsystem. Good management also is needed in these areas because they are as critical to cost control and profit generating as is purchasing.

7

Purchasing

The goal of any foodservice system is to serve quality meals while maximizing value for both the operation and the customer. In simple terms, **value** is the perceived relationship between quality and price; as these relationships change, perceived value changes (Virts, 1987). An improvement in quality tends to increase the value, and an increase in price tends to lower the perceived value. Before the foodservice goal can be met, however, the necessary materials must be procured, preprocessed, and produced as menu items.

PROCUREMENT

The first step in this process, **procurement,** is the managerial function of acquiring material for production. These materials include nonfood items such as detergents and paper supplies, but the chief resource is food. According to the foodservice model presented in chapter 2, procurement is the first functional subsystem of the transformation element, as highlighted in Figure 7.1. Several important activities

Figure 7.1.   Foodservice systems model with procurement subsystem highlighted.

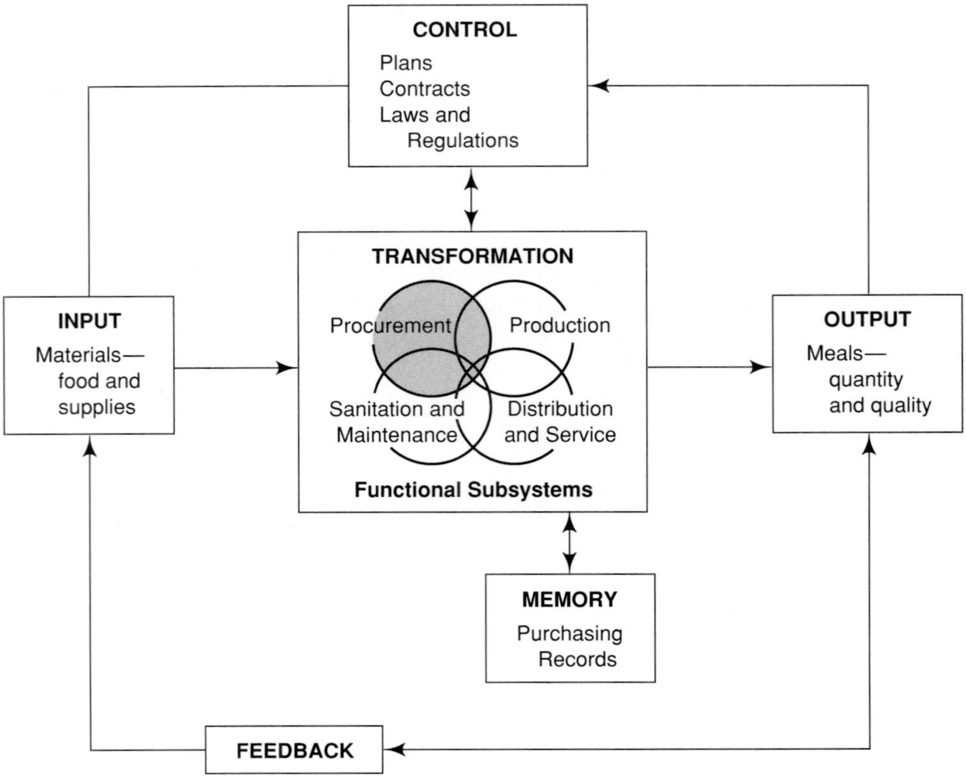

exist within this subsystem: purchasing is discussed in this chapter, and receiving, storage, and inventory control in chapter 8.

In common with other subsystems of the transformation element, the management of procurement involves planning, organizing, staffing, leading, and controlling. Because procurement is considered an important profit generator, those responsible for it should be members of the top management team and involved in high-level decision making.

ROLE OF PURCHASING MANAGERS

Purchasing managers should set priorities that are most important strategically. Reck and Long (1988) have placed the development of the purchasing activity on a continuum from no direction to being fully integrated into the strategic planning process. Purchasing executives across the United States were asked to rank skills of buyers that would be valuable in the year 2000 (Kolchin & Giunipero, 1993). The following 10 skills were rated in descending order of importance: interpersonal communication, customer focus, ability to make decisions, negotiation, analytical ability, managing change, conflict resolution, problem solving, influence and persuasion, and computer literacy.

According to the National Restaurant Association's Foodservice Purchasing Managers (FPM) executive study group, the role of purchasing managers is changing (Patterson, 1993). Three important topics were discussed during roundtable sessions at the recent FPM semiannual meeting in Chicago: managerial productivity, purchasing as a profit center, and the use of computers in purchasing.

Managerial productivity used to be based primarily on dollars saved, but today total quality management (TQM), partnering with suppliers, and ultimate customer satisfaction with products purchased also need to be measured (Patterson, 1993). Purchasing has a role in TQM and, therefore, a significant impact on quality, customer satisfaction, profitability, and market share (Carter & Narasimhan, 1993). In the Kolchin and Giunipero (1993) study, TQM topped the list for the common body of knowledge needed for a purchasing manager in the year 2000.

In many foodservice operations, purchasing is a profit center (Patterson, 1993). A large percentage of the sales revenue in a foodservice operation is spent for purchases. Better purchasing saves dollars paid to suppliers for needed products, supplies, and services. These savings go directly to the bottom line, before taxes, on the profit and loss statement (Leenders, Fearon, & England, 1989). Purchase dollars are high-powered dollars! Purchasing, therefore, could contribute to the profit of an operation. Until recently, purchasing in a large organization, for example a hospital, was considered a service to other departments, and top management thought of it as a cost center rather than a profit generator.

Reck and Long (1983) discussed the procedures for organizing the purchasing department as a profit center. A **profit center** is any department that is assigned both revenue and expense responsibilities. The purchasing manager is expected to manage expenses while creating profit for the organization. A **cost center,** however, is a department that is expected to manage expenses but not generate profits for the organization; it is expected to help other departments contribute to the creation of profit.

All purchasing managers in the Foodservice Purchasing Manager executive study group have personal computers, although not all were linked into a mainframe computer (Patterson, 1993). In general, most operations have used computers for purchasing for fewer than 3 years. A major problem is finding computer software programs useful for purchasing. Currently, computers are used by purchasing managers for pricing, inventory monitoring, product usage, tracking product movement, paying bills, and communicating with other departments and suppliers.

Expert systems, which are computer programs, already are being used by large companies, such as IBM, to improve the effectiveness and efficiency of purchasing management decision making (Cook, 1992). The use of expert systems in large limited-menu restaurant chains cannot be far away for helping purchasing managers make decisions in such areas as supplier selection.

PURCHASING AND THE MARKET

Purchasing is an activity concerned with the acquisition of products; it is often described as obtaining the right product, in the right amount, at the right time, and at the right price. To do this, food buyers must know the market and the products in addition to having general business acumen. They also rely on sales representatives to give advice on purchasing decisions and to relay valuable information about available food items and new products.

The **market** is the medium through which a change in ownership moves commodities from producer to consumer. Foodservice markets may be meat or fresh-produce establishments or locations in Chicago or Mexico.

Knowledge of the food market involves finding sources of supply and determining which food items can be obtained from which supplier. It also requires understanding the flow of supplies through the marketing channel and the effect of market regulations on the distribution process.

Marketing Channel

Exchange of ownership of a product occurs in the **marketing channel,** sometimes identified as the *channel of distribution,* as shown in Figure 7.2. Producers supply raw products to processors or manufacturers who finish and sell them to middlemen, who in turn sell them to the customer. For example, the rancher raises cattle for sale to a meat packing plant, which processes, portions, and sells to a wholesaler, who stores the product temporarily before selling to customers. Similar sequences are followed for metal, which is manufactured into equipment.

In today's economy, the cost of taking goods through the marketing channel often equals or even exceeds the initial cost of the product at the point of origin. Value is added to the product in the marketing channel. **Value added** is the increase in

Figure 7.2. The marketing channel.

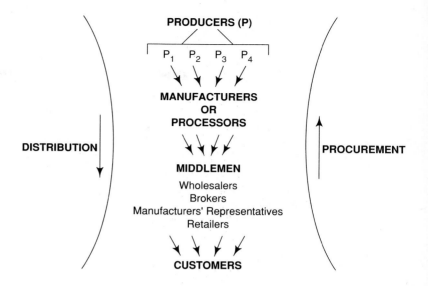

value caused by both processing or manufacturing and marketing, exclusive of the cost of products or overhead. Value-added agricultural products are raw commodities whose value has been increased through the addition of ingredients or processes that make them more attractive to the buyer or readily usable to the customer (Kansas State University, 1990). Simple examples would be producing a high-fiber breakfast cereal from wheat or adding a marinade to a cooked beef steak and selling it in a microwavable package. The process would create new jobs and keep more dollars in a community. In addition, the profit margin of a value-added product is generally higher than that of a raw commodity.

Occasionally, the wholesale price of a food item may be more than the "bargain special" at the local supermarket. For example, a large chain-affiliated supermarket may bypass many middlemen by purchasing a carload of coffee directly from the processor. The supermarket price then could be much less than the buyer would pay to the wholesaler. Such price differentials are part of buyers' concerns but are usually beyond their control.

Some small processing or manufacturing plants will sell products directly to the consumer, but the large size of most plants today precludes this type of selling. Instead, those who do not have their own sales forces must rely on middlemen to promote and distribute products. **Wholesalers** are middlemen who purchase from various plants, provide storage, and resell the products to retailers. They generally carry large lines of stock, permitting the buyer to purchase everything from frozen and canned products to kitchen equipment. Some wholesalers, however, deal with only one item category, such as meat, produce, dairy products, or detergents.

The latest entrant into the wholesale sector is the **superdistributor,** whose activities truly give meaning to "one-stop shopping." During the past decade, the rise of national restaurant chains generated the need for larger wholesale operations that could provide the coast-to-coast consistency those chains demand. The largest wholesalers, often identified as *distributors,* were eager to pursue the big dollars such multiunit accounts represent. But to service these customers adequately, the distributors needed greater geographic coverage and the inherent buying clout and cost efficiencies that accompany such growth. In response to these opportunities, these distributors have grown by normal expansion and corporate acquisitions to dimensions justifying the word *superdistributors* (Van Warner, 1988).

In the process of growth, particularly the acquisition of competitors, these superdistributors can offer an extremely wide variety of goods that foodservice operations need. Examples range from food products, service supplies and equipment, office supplies, to even furniture. Today, the total number of wholesale foodservice distributors is approximately 3,300 (National Food Service Management Institute, 1992). The SYSCO corporation was the largest distributor in 1991 and had sales of $8.3 billion. Kraft, which was second, had $3.4 billion, and Rykoff-Sexton, number three, had $1.5 billion in sales. Small independent foodservice distributors account for 62% of total sales and currently remain a major force in foodservice distribution.

Distribution is changing in favor of the foodservice operator. Services above and beyond delivery that help a manager do a better job are expected. Distributors providing such services at no additional cost differentiate themselves from their competition, giving foodservice managers more incentive to do business with them. These extra services have evolved from specialist help and merchandising aids for different menu items or products to seminars accredited toward professional certification, sophisticated order entry, and inventory control options. Services also could include nutritional analysis of food products, quality control testing, waitstaff training, and salad bar merchandising. Probably the most appreciated added service given to the buyer is product education. Not only are the operators introduced to new products, but they also are shown new and exciting ways to present them to the customer.

Other prime examples of middlemen in the food industry are brokers and manufacturers' representatives. **Brokers** are in business for themselves, but they do not take title to the goods being sold. They receive a commission for negotiating between the buyer and the supplier. A broker sales force offers to a producer or manufacturer a professional sales organization that is paid solely for what is sold. The focus of the broker is on marketing and selling the product. The broker organization employs the sales force and takes care of insurance costs, transportation, personnel management concerns, and whatever it takes to get the product sold. Broker organizations that are members of the National Food Brokers Association employ more than 40,000 sales experts. These organizations are local and attract sales persons because of limited travel. The producer or manufacturer, therefore, reaps the advantages of a stable sales force deeply rooted in the local community and familiar with needs of local buyers and customers (National Food Brokers Association, undated). **Manufacturers' representatives** do not take title, bill, or set prices, and they usually represent small manufacturing companies, often including foodservice equipment manufacturers. The companies pay a flat commission on sales volume. They represent fewer and more specialized lines and a minimum number of manufacturers; they have greater product expertise than brokers. **Retailers**, the final middlemen in the marketing channel, sell products to the ultimate buyer, the customer. The foodservice manager is considered a retailer who buys more often from the wholesaler than through a broker or manufacturer's representative because of the convenience of "one-stop shopping."

Market Regulation

The food industry is the most controlled industry in the United States today, covered as it is by comprehensive and complex federal regulations. The purpose of **market regulation** through federal legislation is to protect the consumer without stifling economic growth. Government, industry, and the consumer need to interact to accomplish this purpose; each is interdependent and mutually affects the performance of the other two. Government is responsible for enacting legislation that safeguards the consumer and at the same time promotes competition among industries. Industry, with the ultimate goal of satisfying the consumer while making a profit, is responsible for complying with this legislation. Consumers, who are becoming increasingly vocal about such issues as food safety, nutrition, and the environment, alert government about their concerns and expect industry to produce appropriate products. Although many federal agencies are responsible for regulations that directly affect the industry, the U.S. Department of Health and Human Services and the U.S. Department of Agriculture (USDA) are the most often involved. The U.S. Department of Commerce also is becoming involved because of the great concern for the microbiological safety of seafood for human consumption.

U.S. Department of Health and Human Services

Under the umbrella of the **U.S. Department of Health and Human Services** are the **Food and Drug Administration (FDA)** and the U.S. Public Health Services. The department cooperates closely with the USDA.

Food and Drug Administration.   The Food, Drug and Cosmetic Act, passed in 1938, is enforced by the FDA. The purposes of this act are to ensure that foods other than meat, poultry, and fish are pure and wholesome, safe to eat, and produced under sanitary conditions, and that packaging and labeling are in agreement with the contents.

The act provides three mandatory standards for products being shipped across state lines: identity, quality, and fill of container. *Standards of identity* establish what a given food product contains. Certain ingredients must be present in a specific percentage before the standard name may be used. For example, the consumer is assured that any product labeled mayonnaise, regardless of its manufacturer, contains 65% by weight of vegetable oil, as well as vinegar or lemon juice and egg yolk. Products labeled mayonnaise that do not contain egg yolk as a stabilizer, such as a low-cholesterol mayonnaise containing only egg white but with an added stabilizer, are in violation of FDA regulations. The FDA has initiated review of standards of identity for such products. The review will include public hearings, and the results will be either a change in regulations or prosecution of the current violators.

Standards of quality, or minimum regulatory standards for tenderness, color, and freedom from defects, have been set for a number of canned fruits and vegetables to supplement standards of identity. If a food does not meet the FDA quality standards, it must be labeled "Below Standard of Quality" and can bear an explanation such as "Good Food—Not High Graded," "Excessively Broken," or "Excessive Peel." *Standards of fill of container* tell the packer how full a container must be to avoid the charge of deception.

Two major amendments to the Federal Food, Drug and Cosmetic Act are the Food Additive and Color Additive Amendments of 1958 and 1960, respectively, both of which safeguard the consumer against adulteration and misbranding of foods. Food is considered *adulterated* if it contains substances injurious to health, was prepared or held under unsanitary conditions, or if any part is filthy, decomposed, or contains portions of diseased animals. Food is also considered to be adulterated if damage or inferiority is concealed, if its label or container is misleading, or if a valuable substance is omitted. Another addition was the Miller Pesticide Amendment of 1954, which provides procedures for establishing tolerances for residues of insecticides used with both domestic and imported agricultural products. Tolerances of permitted residues are established by the Environmental Protection Agency, but the FDA is responsible for their enforcement. The FDA samples dairy products, fish, fresh and processed fruits and vegetables, and animal foods for unsafe residue levels.

A food is considered *misbranded* if the label does not include adequate or mandatory information or gives misleading information. The objective of the Fair Packaging and Labeling Act of 1967 is to ensure that the consumer can obtain accurate quantity or content information from a food label, thus permitting value comparison. Specific kinds of information required on the label for each food are the following:

- Name and address of the manufacturer, distributor, or packer, including zip code
- Name of the food
- Net contents in terms of weight, measure or count
- A statement of ingredients listed by common or usual name in order of decreasing predominance by weight

The FDA also regulates food irradiation, genetically engineered foods, and nutrition labeling. Food irradiation is classified as a food additive and is regulated by the FDA. **Irradiation** refers to exposure of substances to gamma rays or radiant energy. The process for some foods has been approved since the 1960s, followed in 1985 by approval of irradiation to control trichinae in pork and, in 1992, to control foodborne pathogens and other bacteria in frozen poultry, including ground poultry products (Derr, 1993; Rubin, 1993). Ionizing radiation has energy high enough to change atoms by knocking an electron from them to form an ion, but not high enough to split atoms and cause foods to become radioactive. The amount of energy absorbed by a food is measured in units called *kilograys* (KGy). Less than 1 KGy inhibits the sprouting of tubers in potatoes, delays the ripening of some fruits and vegetables, controls insects in fruits and stored grains, and reduces parasites in foods of animal origin. One to 10 KGys control microbes responsible for foodborne illness and extend the shelf life of refrigerated foods. More than 10 KGys are used only on spices and dried vegetable seasonings.

In May 1992, the FDA proposed that the same policy for regulating the safety of all other foods be used for **genetically engineered foods.** The policy should be based on the characteristics of the foods and not the processes used to produce them ("FDA Issues," 1992). To bring these foods to the market more quickly, the FDA announced that genetically engineered fruits and vegetables produced through a process known as *recombinant DNA*, a new combination of genes, would need no special testing or labeling. By splicing in new genes or suppressing or eliminating existing genes, the genetic makeup of a food can be altered (Cheney, 1993). The gene that causes the softening process of tomatoes has been isolated, and scientists at Calgene Fresh, a biotechnology firm located in Evanston, Illinois, have created the "Flavr Savr" tomato that boasts a longer shelf life and, presumably, better taste (Ruggles, 1993). This should be a great improvement over the flavor, which usually is bland, of the luscious-looking tomato that is picked while green and later sprayed with ethylene gas to induce ripening.

Biotechnology soon will be among the most hotly debated issues in society (Cheney, 1993). The FDA is still seeking opinion on how genetically altered foods should be labeled and used (Ruggles, 1993). If food production is to keep pace with projected food needs over the next 40 years, technological innovation must continue (Etherton, 1993). Not surprisingly, it has won support from The American Dietetic Association, the Grocery Manufacturers of America, the International Food Information Council, the president of the American Culinary Federation, and the majority of the scientific community (Cheney, 1993). More than 1,000 chefs, however, have pledged to boycott genetically altered foods because they do not want to serve their customers foods that might cause dietary concerns. For example, would the addition of a flounder gene to a tomato make the item nonvegetarian? Also, the potential of breaking religious dietary restrictions or provoking food allergies has chefs worried.

New requirements for **nutrition labeling** are spelled out in regulations issued in late 1992 by the FDA and by the USDA's **Food Safety and Inspection Service (FSIS)** (Kurtzweil, 1993). FDA regulations meet the provisions of the Nutrition Labeling and Education Act of 1990, and the FSIS provisions regulate meat and poultry products not covered by the act. The FDA set May 8, 1994, for food processors to comply with the new regulations, and the FSIS required meat and poultry processors to relabel their products by July 6, 1993. The new label, identified as "Nutrition Facts," is shown in Figure 7.3. Consumers will see a new format on food products. Macronutrients (fat, cholesterol, sodium, carbohydrate, and protein) will be declared as a percent of the Daily Value. The amount, in grams or milligrams per serving, of these nutrients still must be listed to their immediate right, but for the first time a column headed "% of Daily Value" based on a 2,000-calorie diet will appear. Some nutrition labels will list daily values for a 2,000- and 2,500-calorie diet and the number of calories per gram of fat, carbohydrate, and protein. The content of micronutrients, vitamins and minerals, also will be expressed as a percentage.

According to food labeling regulations, specific definitions for all descriptors are now spelled out (National Center for Nutrition and Dietetics, 1993). A product described as "free" of fat, saturated fat, cholesterol, sodium, sugar, or calories must contain "no amount" or only a "trivial amount" of that component, according to the labeling regulations. For example, the regulations state that "calorie free = less than 5 calories per serving" and that "sugar or fat free = less than 1/2 gram per serving." The term *low* can be used with fat, saturated fat, cholesterol, sodium, and calories

Figure 7.3. Example of the new nutritional label.

Nutrition Facts

Serving Size ½ cup (114g)
Servings Per Container 4

Amount Per Serving

Calories 260   Calories from Fat 120

	% Daily Value*
Total Fat 13g	**20%**
Saturated Fat 5g	**25%**
Cholesterol 30mg	**10%**
Sodium 660mg	**28%**
Total Carbohydrate 31g	**10%**
Dietary Fiber 0g	**0%**
Sugars 5g	
Protein 5g	

Vitamin A 4%	•	Vitamin C 2%
Calcium 15%	•	Iron 4%

* Percent Daily Values are based on a 2,000 calorie diet. Your daily values may be higher or lower depending on your calorie needs:

	Calories:	2,000	2,500
Total Fat	Less than	65g	80g
Sat Fat	Less than	20g	25g
Cholesterol	Less than	300mg	300mg
Sodium	Less than	2,400mg	2,400mg
Total Carbohydrate		300g	375g
Dietary Fiber		25g	30g

Calories per gram:
Fat 9 • Carbohydrate 4 • Protein 4

only when the food could be eaten frequently without exceeding dietary guidelines for these nutrients, such as 3 grams or less of fat, less than 140 milligrams of sodium, or 40 or less calories per serving. The terms *lean* and *extra lean* can be used to describe the fat content of meat, poultry, and seafood. *Lean* is less than 10 grams of fat, 4 grams of saturated fat, and 95 milligrams of cholesterol per serving; *extra lean* is one-half the amount of fat and saturated fat and the same amount of cholesterol as in lean.

Restaurateurs might have to live up to their nutritional claims on menus and placards as they face mandatory compliance with the act (Keegan, 1993). A consumer group is concerned that exempting restaurant menus from the labeling act means customers will be unable to make informed nutrition health choices when they eat out. If the FDA does not counteract the suit, restaurants will be forced to use terminology defined by the FDA. Withdrawing health terms from menus could prove to be counterproductive. One of the directors of the NRA hopes to work with the FDA to develop a common goal to make the application easier for restaurant operators.

U.S. Public Health Service

Under the Public Health Service Act, the FDA advises state and local governments on sanitation standards for prevention of infectious diseases. The most widely adopted standards deal with production, processing, and distribution of Grade A milk. In contrast to USDA quality grade standards for food, the Public Health Service standard for Grade A milk is largely a standard of wholesomeness. The Grade A designation on fresh milk means that it has met state or local requirements that equal or exceed federal requirements.

U.S. Department of Agriculture

The **U.S. Department of Agriculture** has an important role in the food regulatory process. One of its most important functions, authorized by the Agricultural Marketing Act, is the grading, inspection, and certification of all agricultural products. U.S. grades are levels of quality, and U.S. grade standards, which are voluntary, define the requirements that must be met by a product to obtain a particular grade. The Food Safety and Inspection Service of the USDA is responsible for ensuring that meat and poultry products destined for interstate commerce and human consumption are wholesome, unadulterated, properly labeled, and do not pose any health hazards (NRA, 1989).

The USDA has established grade standards for several categories of food: fruits, vegetables, eggs, dairy products, poultry, and meat products. The varying terminology in standard grades has led to much consumer confusion; many believe a uniform grade terminology is needed. A summary of the grades in use today is shown in Table 7.1.

The USDA also has the responsibility for enforcing the Federal Meat, Poultry Products, and Egg Products Inspection Acts. These acts promulgate mandatory regula-

Table 7.1. Summary of major grades by food categories

Product	Products Graded[a]	Grading Criteria	Quality Grades Highest → Lowest					
			Highest					**Lowest**
Beef	54%	Eating quality Color of flesh Firmness and marbling	Prime, Choice, Select, Standard, Commercial, Utility, Cutter, Canner					
Veal	na[b]	Eating quality Flesh and bone color	Prime	Choice	Good	Standard	Utility	Cull
Lamb	na	Eating quality Bone-to-meat ratio Flesh color Firmness and marbling	Prime	Choice	Good		Utility	Cull
Pork	na	Primarily yield	No. 1	No. 2	No. 3		No. 4	Utility
Poultry	89% turkey 67% chicken and other	Confirmation Fleshing Fat covering	Grade A		Grade B			Grade C
Eggs	40%	Appearance of shell Size of air cell Condition of yolk and white	Grade AA		Grade A	Grade B		Grade C
Fish	na	Appearance Uniformity Absence of defects Texture Flavor and odor	Grade A			Grade B		Grade C

Product	Percent of products sold[a]	Quality factors used in grading	Grade (highest)			Grade (lowest)
Milk, fluid	na	Bacterial count Sanitary conditions	Grade A (only grade for human consumption)			
Milk, nonfat dry	na	Flavor and odor Bacterial count Scorched particle content Lumpiness Solubility	U.S. Extra			U.S. Standard
Milk, whole dry	na	Moisture content	U.S. Premium	U.S. Extra		U.S. Standard
Butter	63%	Flavor and odor Freshness Plasticity Texture	Grade AA	Grade A		Grade B
Cheese Cheddar Swiss Colby Monterey Jack	na	Flavor and odor Texture Body Appearance Finish Color	Grade AA	Grade A	Grade B	Grade C
Produce Frozen Fresh Canned	55% 45% 35%	Maturity Shape Color Size Uniformity Texture Presence of defects	U.S. Fancy (USDA reports 156 different grades for fruits, vegetables, and nuts)	U.S. No. 1	U.S. No. 2	U.S. No. 3

[a] Percent of products sold, Agricultural Marketing Service, USDA, 1987.

[b] na = data not available

tions to assure wholesomeness of the products. The major requirements of these three acts illustrate how this objective is met. The Meat Inspection Act, for example, provides for the destruction of diseased and unfit meat, regulates sanitation in meat plants, requires stamping of inspected meat, prevents addition of harmful substances in meat products, and eliminates false or deceptive labeling. The Meat Inspection Act was amended by the Wholesome Meat Act of 1967, which requires inspection of all meat if it is moved within or between states. In addition, the amendment provides for inspection of foreign plants exporting meats to the United States.

The Poultry Products Inspection Act was amended in 1968 and designated as the Wholesome Poultry Products Act, which requires inspectors to assess the cleanliness of plants and the maintenance of equipment. The inspection procedures for poultry are similar to those required for meat. Labels on poultry and poultry parts must also be approved. Under the Egg Products Inspection Act of 1978, plants that break and further process shell eggs also are inspected. In all three acts, monetary and technical assistance is provided to aid plants in meeting federal requirements.

The USDA Agricultural Marketing Service, in cooperation with state agencies, offers official grading or inspection for quality of manufactured dairy products, poultry and eggs, fresh and processed fruits and vegetables, and meat and meat products. Grading is based on U.S. grade standards developed by the USDA for these products. The Food Acceptance Service, developed by the USDA to simplify noncommercial food buying, is included in the grading and inspection programs. This service provides impartial evaluation and certification that food purchases meet contract specifications. Any healthcare organization, commercial foodservice, governmental agency, educational institution, or public or private groups buying food in large quantities may use the service on request. Suppliers often use the acceptance service to ensure that they meet contract specifications.

If purchases are to be certified by the USDA acceptance service, contracts with suppliers should include this provision. The supplier is then responsible for obtaining certification, which, like all grading services, is provided for a fee. Contract specifications for the products can either be based on USDA grade standards or tailored to meet the buyer's needs. They may include factors such as USDA quality grades, condition, type of refrigeration, cut, trim, size, packaging, weight, shape, and color.

To provide the acceptance service, an official grader employed by the Agricultural Marketing Service or a cooperating state agency examines the product at the manufacturing, processing, or packing plant or at the supplier's warehouse. If the product meets contract specifications, the grader stamps it with an official stamp. The grader also issues certificates indicating that the products comply with the contract specifications. A sample certificate is shown in Figure 7.4.

In 1986, Congress passed the Processed Products Inspection Improvement Act, which allows the FSIS to institute an improved processing inspection system utilizing a risk-based approach. The new system stresses industries' responsibility for the production of safe products and takes advantage of industrial quality assurance programs (NRA, 1989). Inspections changed from a continuous to a sampling approach that recognizes past plant performance in determining the frequency of inspection. These improved inspection procedures apply to meat and poultry processing plants only. Slaughter plants will still be under continuous inspection.

Figure 7.4.   Sample USDA Certificate of Quality and Condition.

UNITED STATES DEPARTMENT OF AGRICULTURE AGRICULTURAL MARKETING SERVICE CERTIFICATE OF QUALITY AND CONDITION (PROCESSED FOODS)	Please refer to this certificate by number and inspection office.
This certificate is receivable in all courts of the United States as prima facie evidence of the truth of the statements therein contained. It does not excuse failure to comply with any applicable Federal or State laws. **WARNING:** *Any person who knowingly falsely make, issue, alter, forge, or counterfeit this certificate, or participate in any such action, is subject to a fine of not more than $1,000 or imprisonment for not more than one year, or both (7U.S.C. 1622 (h)).* The conduct of all services and the licensing of all personnel under the regulations governing such services shall be accomplished without discrimination as to race, color, religion, sex, or national origin.	Z- 012345 DATE November 23, 1993

APPLICANT ABC Company	ADDRESS Peach Glenn, Ohio
RECEIVER OR BUYER Clark Company	ADDRESS Chicago, Illinois
SOURCE OF SAMPLES OFFICIALLY DRAWN	PRODUCT INSPECTED CANNED APPLES

CODE MARKS ON CONTAINERS
3320   3321
SAPP and SAPP....

PRINICIPAL LABEL MARKS   "BIG BAY SLICED APPLES, Net Wt. 6 lbs. 8 oz. (2.94 Kg), Distributed by ABC Company; Peach Glenn, Ohio 49533."

Net Weights:	MEETS label declaration.
Vacuum Readings:	5 to 15 inches.
Drained Weights:	MEETS requirements of 70 ounces.
Style:	Sliced.
Condition of Container:	MEETS applicable U.S. Standards for Condition of Food Containers.

OFFICIALLY SAMPLED
NOV 2 3 1993
U.S. DEPARTMENT OF AGRICULTURE
FV

GRADE:

U.S. GRADE A or U.S. FANCY.
Average Score - 92 points.

REMARKS:

This certificate covers 952 cases, 6/No.10 cans (Applicant's Count). Product packed in beaded cans with enamel-lined ends, cased in domestic corrugated cases. Lot located in applicant's warehouse, Peach Glenn, Ohio. Written statement from the packer indicates product is from the latest season's crop. All cases stamped with the USDA "OFFICIALLY SAMPLED" stamp as shown above.

Pursuant to the regulations issued by the Secretary of Agriculture under the Agricultural Marketing Act of 1946, as amended (7 U.S.C. 1621-1627), governing the inspection certification of the product designated herein, I certify that the quality and condition of the product as shown by samples inspected on the above date were as shown, subject to any restrictions specified above.

ADDRESS OF INSPECTION OFFICE 915 N. Trust Brigosh, IN 46509	SIGNATURE OF INSPECTOR Connie Dye

FORM FV-146CS (9-92)
☆U.S. GPO: 1992—333-847

U.S. Department of Commerce

Public attention has been focused on the fact that fish and seafood are not subject to mandatory continuous inspection as are meat and poultry. Fish are cold-blooded animals whose diseases are not threatening to humans, as are those of warm-blooded cattle and poultry that can transmit diseases such as salmonella. Fish can be exposed, however, to a variety of pathogens, toxins, and parasites that can cause human illness (NRA, 1989).

The **National Marine Fisheries Service (NMFS)**, under the U.S. Department of Commerce, offers a voluntary inspection program. The FDA is responsible for inspection and sanitation of U.S. fish and seafood plants; it also monitors imports and interstate shipments for compliance with the Federal Food, Drug, and Cosmetic Act (NRA, 1989). Because the FDA is responsible for so many foods, fish has received only sporadic oversight in the past. An Office of Seafood now has been established to give higher priority to seafood safety.

Seven senators have introduced a new bill calling for a national seafood safety program that would include mandatory inspection (Allen, 1992). The FDA would be charged with establishing safety and inspection guidelines; enforcement would be divided between the USDA and the U.S. Department of Commerce. The FDA would establish tolerances for seafood contaminants, standards for processors and importers based on the HACCP model discussed in chapter 5, and inspection agreements with other countries. The USDA would be responsible for monitoring processing plants outside the United States and imported seafood at the point of entry. The Department of Commerce would monitor all domestic processing plants as well as shellfish beds and fishing grounds in federal waters. Enforcing this national program could be difficult because seafood is harvested from many more sources than meat or poultry.

In the meantime, the FDA, the Department of Commerce, and the National Restaurant Association proposed a pilot inspection program to improve seafood safety at all points of the seafood chain, including processors, grocery stores, and restaurants ("Restaurants Find," 1993). Twelve restaurant owners volunteered to work with the FDA and the Department of Commerce to identify "critical control points" in seafood handling and come up with ways to avoid contamination at these points. The participants hope that the result will be a voluntary seafood inspection program that is expected to be available to restaurants nationwide by 1994.

Imported Food Regulations

The FSIS is responsible for the safety of meat and poultry imported into the United States. Foreign countries must impose inspection requirements at least equal to those in the United States in order to make products eligible for import. The point-of-entry inspection includes examination of net weight, condition of the container, incubation of bacteria in canned products, and the label. A laboratory analysis is performed, including testing for drug and chemical residues.

Nonmeat and poultry items in packages and requiring no further processing are under FDA jurisdiction. The FDA has limited authority to inspect foreign manufacturing facilities. The FDA has a low sampling rate of products but is improving its inspection of imported foods with a computer system called the Import Support and Information System (ISIS). ISIS will be able to profile firms, products, and countries of origin to identify problems and detect trends.

PRODUCT SELECTION

Purchasing for a foodservice operation is a highly specialized job function. Buyers must know not only the products to be procured but also the market, buying procedures, market trends, and how the materials are produced, processed, and moved to market. In addition, they must be able to forecast, plan, organize, control, and perform other management-level functions.

The primary function of the buyer is to procure the required products for the desired use at the minimum cost. The accomplishment of this function frequently involves some research by the buyer to aid in decision making. For this purpose, techniques have been adapted from industry, notably value analysis and make-or-buy decisions.

Value Analysis

In the most liberal sense, value analysis is virtually any organized technique applicable to cost reduction. More precisely, however, **value analysis** is the methodical investigation of all components of an existing product or service with the goal of discovering and eliminating unnecessary costs without interfering with the effectiveness of the product or service (Miles, 1972). This definition of value analysis is related primarily to industry, but it contains broad implications for foodservice as well. For example, a value analysis of a menu item may reveal that some quality features may be eliminated without detracting from the utility of the final product.

The essence of quality is suitability (Heinritz, Farrell, Giunipero, & Kolchin, 1991). The supplier gives the buyer the quality of the product identified in the specification. Part of the price may be for quality features that do not contribute substantially to the suitability of the product for the foodservice operation and make the expenditure wasteful. A foodservice, for example, may purchase an expensive, sophisticated, software package although the manager does not know what is needed and no employee has been trained in computer usage. Probably, many of the software features would never be used. A simple package would probably have been adequate in the beginning, and it certainly would have been less expensive.

Value analysis is an important element of scientific purchasing and has brought about the realization that purchasing is a profit-making activity. The concept of value analysis is important in enhancing efficiency and profitability (Williams, Lacy, &

Smith, 1992). Reductions in purchasing costs resulting from making the process more efficient have a strong impact on profitability, which is quite different from increasing sales by selling more products.

Value Analysis in Foodservice

Value is the result of the relationship between the price paid for a particular item and its utility in the function it fulfills. Value analysis permits the foodservice manager to look at a problem from a new perspective or a "fresh eye" approach (Williams et al., 1992). The tendency to become locked into a pattern can be reduced.

Although value analysis has often been used in the development of new products in industry, its effectiveness is not confined to such development; it is used much more frequently in evaluating existing product specifications and, thus, is readily applicable in foodservice. For example, a major university residence hall foodservice applied value analysis to a lasagna entrée after students complained that its flavor was not like that at a local Italian restaurant. An associated problem was the foodservice director's concern that the product was extremely labor-intensive. In conference with the director, the foodservice manager decided to compare lasagna made by the current, standardized recipe that called for American cheese with a revised recipe substituting mozzarella cheese for American. For the secondary problem of labor intensity, one well-known brand of frozen lasagna was evaluated.

Reporting on the Value Analysis

A brief value analysis report prepared for the foodservice director, based on the residence hall foodservice lasagna example, is shown in Figure 7.5. This example illustrates a primary requirement for reporting to management—brevity. Complete data on the methodology and the test procedures should be prepared as a reference and excerpted only for the report to administration. Value analysis results for quality and cost and recommendations are of major interest to the foodservice director. This portion of the report should be concise and definite.

This lasagna example illustrates a good approach to a secondary problem. Quite properly, the evaluator gave a recommendation on the primary problem, that of customer dissatisfaction, and suggested that a separate analysis should be made for several brands of frozen lasagna to examine the secondary problem, labor intensity of the on-site product.

Make-or-Buy Decisions

The procedure of deciding whether to purchase from oneself (make) or purchase from suppliers (buy) is continuous, and reviews of previous **make-or-buy decisions** should be conducted periodically. A foodservice manager has three basic choices for production of a menu item:

- Produce the item completely, starting with basic raw ingredients
- Purchase some of the ingredients and assemble them
- Purchase the item completely from a wholesaler

Figure 7.5.   Example of a value analysis report prepared for the Kansas State University Dining Services.

VALUE ANALYSIS
Report to Administration

Product Evaluated: Lasagna   **Evaluator:** Mary Smith
                                                                 Date: November 2, 1994

Statement of Problem
- Students criticize flavor of lasagna prepared in Residence Hall foodservice.
- Management is concerned about labor intensity of the product.

Products Evaluated
- Residence Hall recipe with American cheese
- Residence Hall recipe with mozzarella cheese
- Frozen prepared commercial brand

Methodology
- *Preparation of product*
  Two one-half-size counter pans of the Residence Hall recipe were prepared at the same time from identical ingredients and the same recipe except one contained American cheese and one mozzarella cheese. The frozen product was heated according to manufacturer's directions.

- *Test procedure*
  The three products, each identified by a three-digit number, were cut into sample size portions and displayed side by side. Each product was rated on flavor, texture, and appearance by a taste panel consisting of 16 students, 6 foodservice employees, and 4 staff.

Value Analysis Results
- *Cost Evaluation*

Cost Items	Residence Hall Recipe (American cheese)	Residence Hall Recipe (mozzarella cheese)	Commercial Frozen
	cost per portion		
Food	$ .6216	$ .6256	$ 1.165
Labor	.0676	.0676	—
Total	$ .6892	$ .6932	$ 1.165

- *Quality Evaluation*
  The lasagna prepared from the Residence Hall recipe with mozzarella cheese ranked highest in flavor. Both Residence Hall products ranked similar in texture and appearance. The frozen commercial product ranked lowest in all three quality criteria.

Recommendation
- The recommendation based on taste panel evaluations is that the Residence Hall lasagna recipe with mozzarella cheese be used. Also, the made-on-premise product is much less expensive than the purchased frozen. If a decision to serve a frozen prepared product is made because of labor intensity, other brands should be evaluated.

Few foodservice managers consider the first alternative, either because of the expense or the time and labor required. The second two alternatives were compared in the value analysis report on lasagna (Figure 7.5). A product prepared on-site from a recipe using many purchased items (such as lasagna noodles, canned tomatoes, and cheese, which were assembled for baking in the foodservice kitchen) was compared with a commercial frozen product. Based on cost and quality criteria, the alternative of making the lasagna on the premises was recommended rather than that of purchasing a frozen product.

Decision factors in foodservice are quality, quantity, service, and cost. The serving of quality food is a prime consideration in all foodservices, whether menu items are made on the premises or purchased completely prepared from the processor. Quality standards have become well established among most processors of ready-to-serve foods, although variations are evident.

In the lasagna analysis, the testers said the noodles in the convenience product were tender but firm enough to retain their shape and the product held its square form when plated as specified in the standard. The top was covered with meat sauce and cheese with no juices on the plate. The problem was the flavor of the product, which was not acceptable to the students. The standard was that the flavor should be a blend of cheese, meat, and pasta with evidence of Italian seasoning, but should not be overpowering. The taste panel indicated that the frozen product had a strong tomato flavor and was too highly seasoned; however, the made-on-premises product with the mozzarella cheese did satisfy the flavor standard.

Quantity enters into the decision process when the ability to produce in the desired amount is considered. The quantity needed may be too large for a satisfactory "make" decision or too small for the processor to consider. As a result of past make-or-buy decisions, many foodservices have closed bake shops or omitted them from new facilities because acceptable bakery products are commercially available.

Service includes a wide variety of intangible factors influencing the satisfaction of the buyer. Two important factors in the decision process are reliable delivery and predictable service. Foodservices must operate on a rigid time schedule; a late delivery of a menu item cannot be tolerated. The dependability of a supplier in all circumstances must be assured. Clues to suppliers' reliability may be their record of labor relations and reputation in the industry, including the opinions of other customers. The geographic proximity of a supplier also should be considered.

When quality, quantity, and service factors are equal, a make-or-buy decision will be made by a comparison of the known cost from the supplier with the estimated cost of making the product. In this process, the cost of raw food is easily determined, but the costs of labor, energy, equipment depreciation, and overhead may be difficult to calculate.

Set policies for reaching make-or-buy decisions should be avoided because these and other purchasing decisions are influenced by a number of interrelated factors that are extremely variable. The critical factor on one occasion may be noncritical on the next. The buyer must strike a fine balance among all factors and make a decision on the basis of what is best for the foodservice operation.

SPECIFICATIONS

Quality has become the watchword in foodservices. In addition to the assurance of quality in product selection, critical elements in producing quality food and foodservice are the development of and strict adherence to rigorous purchasing specifications. Although people often are surprised when reminded of it, the root of the word *specification* is "specific," which means that some condition or status must be met, definitely and without equivocation. Thus the primary safeguard of foodservice quality is adherence to specifications.

A **specification** has been defined in many different ways, but it is essentially a statement, readily understood by both buyers and suppliers, of the required quality of products, including the allowable limits of tolerance. In the simplest terms, a specification, or spec, may be described as a list of detailed characteristics desired in a product for a specific use.

Types

The three types of specifications applicable to foodservices are technical, approved brand, and performance. Selection of one, or a combination of two or all three, of these types is based on the product being purchased, whether it is food, supplies, or equipment.

Technical specifications are applicable to products for which quality may be measured objectively and impartially by testing instruments. These are particularly applicable to graded food items for which a nationally recognized standard exists. Other examples are parts and metals in fabricated equipment, such as stainless steel and aluminum, which are subject to a thickness requirement that can be measured by gauges. Technical specifications also can be written for detergents and cleaning compounds that can be chemically analyzed.

Approved brand specifications indicate quality by designating a product of known desirable characteristics. Retail brand names on menus have customer appeal. For example, Ocean Spray and Ragu appear on menus for Morrison's Custom Management accounts (Lorenzini & McCarthy, 1992). Specifications are easy to write because the brand name of a product is all that is needed; only the size and quantity must be added. The problem, however, is that purchasing options are limited and competition is eliminated.

In **performance specifications,** quality is measured by the effective functioning of large or small equipment, disposable paper and plastic items, or detergents. For example, in addition to the technical specifications for an insulated beverage carafe for room service in a hotel, the minimum time requirement for maintaining a desired temperature for coffee might be specified.

Of the three types of specifications, the technical is used most often for food products purchased by schools, colleges and universities, large healthcare facilities, and multiunit commercial operations. A specification for a particular product also might be a combination including both technical and performance criteria.

Writing Criteria

Written specifications are necessary for an efficient foodservice operation. When specifications are written, suppliers, the receiving clerk, and the foodservice manager all can determine if the products received are what were ordered. Specification writing can be time consuming and labor intensive, especially for small operations. If, however, specifications for high-priced products, such as meat and seafood, are written first, then writing them for other products will be easier. Specification writing requires a team approach and generally includes the foodservice manager, dietitian, procurement and production unit heads, buyer, cook or chef, and often the financial manager. A specification can be simple or complex, depending on the type used; the brand-name type is the simplest, and the technical type is the most complex.

A good written specification should meet the following criteria. It should be:

- Clear, simple, and sufficiently specific so both buyer and supplier can readily identify all provisions required
- Identifiable with products or grades currently on the market
- Verifiable by label statements, USDA grades, weight determination, and the like
- Fair to the supplier and protective to the buyer
- Realistic and not unattainable for most products
- Capable of being met by several bidders to enable competition

Specific Information

All specifications for food products should include the following information:

- Name of product (trade or brand) or standard
- Federal grade, brand, or other quality designation
- Size of container (e.g., weight, can size)
- Count per container or approximate number per pound (number of pieces per container if applicable)
- Unit on which price will be based

Information that would describe the product in more detail might include:

- Product use (e.g., for salads, soups)
- Product test procedures used by the foodservice operation to determine quality compliance (e.g., degree of ripeness, flavor characteristics)
- Quality tolerance limits (number of substandard products in a container of produce, for example)
- Weight tolerance limits (range of acceptable weight, such as for meat, poultry, or seafood)

Any other information that helps to describe the condition of the product should be included, like the following:

- *Canned goods:* type or style, pack, syrup density, size, specific gravity
- *Meat and meat products:* age, exact cutting instructions, weight tolerance limit, composition, condition upon receipt of product, fat content, cut of meat to be used, market class

- *Fresh fruits and vegetables:* variety, weight, degree of ripeness or maturity, quality tolerance limit, geographical origin
- *Frozen foods:* temperature during delivery and upon receipt, variety, sugar ratio
- *Daily products:* temperature during delivery and upon receipt, milk fat content, milk solids, bacteria content

Were it not for the availability of nationally accepted grades and other criteria developed by the USDA and other organizations, writing a specification would be an almost insurmountable task for the buyer as well as difficult for the supplier to interpret. Buyers can now write definite specifications by citing a known standard, which is in itself a rigorous specification. These referenced standards constitute a common technical language for buyers and suppliers. For reasons of brevity, specifications are limited to meat products in this text because meat is generally the major cost item in foodservice.

Additional Information

The USDA has published a greater volume of specification material pertaining to foodservice than any other agency. As mentioned earlier in this chapter, the USDA has established grading standards for fruits, vegetables, eggs, poultry, dairy, and meat products.

The USDA Institutional Meat Purchase Specifications, commonly referred to as IMPS, simplify specification writing for large-volume users of meat. Use of these specifications is related to the food acceptance service of the USDA, a part of the total service of federal inspection and grading.

An extremely valuable feature of IMPS is the numbering system for the identification of carcass cuts and various cuts or types of meat products. In addition to the benefits of the numbering system, IMPS information on ordering includes USDA quality and yield grades, as well as weight ranges, portion cut tolerances, fat limitations, and refrigeration requirements for beef, pork, veal, and lamb.

One of the most useful guides for writers of meat purchase specifications is the *Meat Buyers Guide*, published by the National Association of Meat Purveyors. This organization, commonly referred to as NAMP, coordinated the publication of the guide with the USDA. The numbering system and the general purchasing information are identical to those of IMPS. The advantage of the NAMP *Meat Buyers Guide* is the arrangement of the text material and the excellent colored illustrations of each numbered item. Most wholesalers of meat products are members of NAMP, so the guide provides an excellent adjunct to communication between buyer and supplier. Also included in the guide is a description of USDA identification marks for quality and yield grades and federal inspection stamps for beef.

Grades and Inspection

The USDA quality grades for beef are U.S. Prime, U.S. Choice, U.S. Select, U.S. Standard, U.S. Commercial, U.S. Utility, U.S. Cutter, and U.S. Canner. These grades pertain to the palatability qualities of beef, namely tenderness, juiciness, and flavor. The

USDA conducts a voluntary meat grading service to identify beef, and when this service is used, a USDA Quality stamp will appear on the carcass (Figure 7.6). Resource material for writing specifications for other food products is included in the references in Appendix A.

Yield grades for beef provide a nationwide uniform method of identifying "cutability" differences among beef carcasses. Specifically, these grades are based on the predicted percentage of carcass weight in closely trimmed, boneless, retail cuts from the chuck, rib, loin, and round. The Yield Grade stamp is shown in Figure 7.6. Because of trimming done by the purveying industry, however, Yield Grade stamps will rarely appear on cuts purchased by the foodservice industry. Beef carcasses, as of April 1989, no longer had to be graded for both quality and yield; they may be graded for quality or yield, or both. Because quality denotes palatability and yield cutability, each quality grade could have a yield from 1, increased muscle and decreased fat, to 5, decreased muscle and increased fat. In Table 7.2 those yield grades indicated by an *X* are in the largest supply for each quality grade.

The mark of federal inspection is a round stamp (Figure 7.6) identifying the slaughterhouse of origin for carcasses and the meat fabricating house for further cuts. This stamp indicates that all processing was done under government supervision and that the product is wholesome.

Packers' Brands

Many food producers use their own brand names and define their own quality standards. The grading system closely follows federal grades and standards but with more flexibility than is allowed in the USDA system. For example, one brand or label may mean the highest quality, while another is used to indicate a lower quality. Canners, packers, and distributors usually put their top-ranking label on the highest-quality products. For example, many packers of canned and frozen fruits and vegetables choose their own quality labels rather than those established by the USDA; some packers also use their own color code for various quality levels. The buyer

Figure 7.6.   USDA meat stamps.

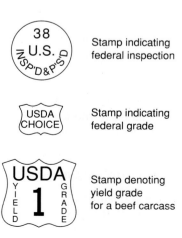

Stamp indicating federal inspection

Stamp indicating federal grade

Stamp denoting yield grade for a beef carcass

Table 7.2. Yield grades for beef carcasses

	USDA Grades				
	Yield Grades[a]				
Quality Grades	**1**	**2**	**3**	**4**	**5**
U.S. Prime			X	X	X
U.S. Choice		X	X	X	
U.S. Select[b]		X	X		
U.S. Standard	X	X	X		
U.S. Commercial		X	X	X	X
U.S. Utility		X	X	X	
U.S. Cutter		X	X		
U.S. Canner		X	X		

[a] The yield grades reflect differences in yields of boneless, closely trimmed, retail cuts. As such, they also reflect differences in the overall fatness of carcasses and cuts. Yield Grade 1 represents the highest yield of retail cuts and the least amount of fat trim. Yield Grade 5 represents the lowest yield of retail cuts and the highest amount of fat trim.

[b] Changed by author from U.S. Good to U.S. Select to conform with current USDA grade designation.

Source: From *Institutional Meat Purchase Specifications for Fresh Beef* by Agricultural Marketing Service, Livestock Division, January 1975, Washington, DC.

must realize that the word *Fancy* as a brand name does not mean that the product meets the requirements of the U.S. Fancy grade. Because not all canned fruits and vegetables are graded by the USDA, the careful buyer will open cans of different brands for comparison before placing major orders. This process is commonly called "can cutting" and constitutes the only means of acquiring firsthand information about drained weight, appearance, texture, flavor, or any scoring factors of particular interest to the buyer. Following selection of a product by this or some other process, the buyer of larger-volume quantities might utilize the USDA acceptance service for inspection at the cannery. This service is provided for a fee, as described in the market regulation section of this chapter.

Sample Specification

The sample specifications shown in Table 7.3 are for steak menu items in three distinct types of restaurants, each with a menu appropriate to its type. The first specification is for a first-class hotel or restaurant, the second for a family-style restaurant, and the third for a budget steak house. Note that, in each case, the NAMP *Meat Buyers Guide* numbers are used to define the meat product. The NAMP identifications define these steaks as follows:

- NAMP #180. Strip loin, short cut, boneless
- NAMP #1179. Strip loin steaks, bone-in, short cut
- NAMP #1177. Strip loin steaks, bone-in, intermediate

The NAMP #180 is a whole loin and will be portion cut on the premises by the hotel chef, as indicated on the specification by the 10- to 12-pound weight for the loin.

In these three specifications, the term *strip steaks* has noticeably different meanings and further distinctions are necessary. The first-class restaurant has specified USDA Prime, Yield Grade 2. The family-style restaurant calls for USDA low Choice, and the budget steakhouse uses USDA Commercial and specifies cows only, no bulls. In each case, the specifications also include trim factor, weight range, method of tenderization, packaging, special considerations, and state of refrigeration. The meat

Table 7.3.   Specifications for steak menu items in three types of restaurants

Specification Item	First-Class Hotel or Restaurant	Family Style Restaurant	Budget Steakhouse
Menu item	New York strip steak, 12-oz. boneless	New York strip steak, 12-oz. with bone	charbroiled strip loin
Product	strip loin	bone-in strip steak	bone-in strip steak
National meat buyers' guide number (NAMP#)	180	1179	1177
Grade	USDA Prime, yield grade two	USDA low Choice (note that this specifies where on the grade)	USDA Commercial (cows only, no bulls)
Trim factor	not to exceed ¼ inch fat with smooth trim	¼ inch to ½ inch fat cover, maximum 2-inch tail	maximum ¾ inch fat, maximum 3-inch tail
Weight range	10 to 12 pounds	12-oz. portion ± half ounce	14 oz. portion ± one oz., minimum ¾ inch thick
Method of tenderization	dry age 14 to 21 days	age whole strips seven to nine days prior to cutting; jackard before cutting	pin whole strip before cutting; individually dip portion steaks in liquid tenderizer
Packaging	wrapped in sanitary paper and tied	individual 12 portions per box, six boxes per case; show date of cutting	multivac, 24 portions per box
Special considerations	eye must be between 2½ and 3½ inches in diameter	¾-inch-thick steaks with vein on both sides are not acceptable	
State of refrigeration	fresh only, ship chilled at temperature between 33 and 40°F	frozen—0°F or less	frozen—0°F or less

Source: From *Meat and Fish Management* (pp. 203–205) by S. A. Mutkoski and M. L. Schuren, 1981, North Scituate, MA: Breton Publishers. Used by permission.

grades selected in each case, along with the additional specifications, reflect the judgment of the management on the desires and spending habits of the diners.

METHODS OF PURCHASING

The buyer who has forecast a need for specific products and written specifications must then decide how to make the purchase. Whatever method is chosen, it will belong to one of two general categories of procedures, identified as either informal or formal, both applicable to independent and organizational buying.

Informal

Informal purchasing, in which price quotations are given and orders made by telephone or personally with a salesperson, is often used when time is an important factor. It is usually done under the following circumstances:

- Amount of purchase is so small that time required for formal purchasing practices cannot be justified.
- An item can be obtained only from one or two sources of supply.
- Need is urgent and immediate delivery is required.
- Stability of market (and prices) is uncertain.
- Size of operation may be too small to justify more formal procedures.

If possible, at least two prices for each item should be obtained because informal quotations, being oral, have little legal force. Any prices quoted by telephone should be recorded by the buyer and then checked on the invoice at the time of delivery. Because of the lack of formal records, federal, state, or local laws often determine conditions under which informal purchasing can be used for tax-supported institutions.

Formal

Tax-supported institutions usually are required to use competitive bidding, which is optional for private institutions and commercial foodservices. Competitive bidding usually culminates in a formal contract between buyer and supplier. Understanding legal implications of contract buying is important for both parties. These legal considerations apply equally to buying for a single independent unit, a department within an organization, several departments of an organization, or a group of organizations.

Bid Buying

Bid buying occurs when a number of suppliers are willing to compete over price quotations on buyers' specifications. Buyers generally use the fixed bid or the daily bid.

Figure 7.7.   Example of a bid request form.

BID REQUEST

Bids will be received until   November 15, 1994

Issued by:	Community Hospital	*Address:*	100 North Street Sunnyvale, OK
Date issued:	September 30, 1994	*Date to be delivered:*	Weekly as ordered between 12/30/94 to 6/23/95

Increases in quantity up to 20% will be binding at the discretion of the buyer. All items are to be officially certified by the U.S. Department of Agriculture for acceptance no earlier than 2 days before delivery; costs of such service to be borne by the supplier.

Item No.	Description	Quantity	Unit	Unit Price	Amount
1	Tomatoes, whole or in pieces, U.S. Grade B, #10 cans, 6/cs.	100	cs.		
2	Sweet Potatoes, vacuum pack, U.S. Grade B, #3 vacuum cans (enamel lined), 24/cs.	50	cs.		
3	Asparagus, all green, cuts and tips, U.S. Grade A, #10 cans, 6/cs.	50	cs.		
4	Corn, Cream Style, golden, U.S. Grade A, #10 cans (enamel lined), 6/cs.	50	cs.		
5	Blueberries, light syrup, U.S. Grade B, #10 cans (enamel lined), 6/cs.	20	cs.		

Supplier_____

Figure 7.8.   Example of a form for telephone bids.

COMMUNITY HOSPITAL
QUOTATION RECORD FORM

Date: _____ Wed., 8/24/94 _____   Delivery Date: _____ Fri., 8/26/94 _____

Circle accepted price quotation.

Item	Specifications	Amount Needed	Amount on Hand	Amount to Order	Supplier Jones per unit	total	Supplier L. & M. per unit	total
Tomatoes	U.S. #1, 20# lug, 5 × 6	4 lugs	1	3	$10.50	$31.50	$10.75	$32.25
Lettuce	U.S. #1, Untrimmed Iceberg, 40% hard head, 24 heads, 1 crate	4	1	3	12.00	36.00	11.70	35.10
Potatoes	U.S. #1, Long Russet Bakers 100 count (50# box)	6	2	4	8.82	35.28	9.07	36.28
Onions	U.S. #1, Yellow, med. size, 50# sack	1	0	1	10.25	10.25	10.00	10.00
Watermelon	U.S. #1, Red flesh	400#	0	400#	.18/#	72.00	.155/#	62.00
						$185.03		$175.63

Price Quotations

The fixed bid is often chosen for large quantities, particularly for nonperishable items, purchased over a long period of time. Suppliers avoid committing themselves to that much time because of potential price fluctuations. The buyer selects a group of suppliers and gives them specifications and bidding forms for desired products, which they complete and return. Figure 7.7 is an example of a bid request form.

The daily bid is often used for perishable products that last only a few days. Sometimes this bid is referred to as *daily quotation buying.* Daily bids usually are made by telephone followed by a written confirmation before the buyer accepts the bid. Bids quoted by telephone are recorded by the buyer on a quotation record form similar to that in Figure 7.8.

Whether fixed bids or daily bids are used, only two basic methods for awarding bids are available.

- *Line-item bidding.* Each supplier bids on each product on the buyer's list, and the one offering the lowest price receives the order for that product. Foodservice buyers must consider the size of orders wholesalers are required to deliver.

According to Gunn (1992), the typical wholesaler needs a guarantee of a $600 minimum order to realize a return on investment. The line-item bid is much more time consuming than the all-or-nothing bid and costs more for both the buyer and supplier.

- *All-or-nothing bidding.* This type of bidding, often referred to as the "bottom-line" approach, requires suppliers to bid the best price on a complete list of items (Gunn, 1992). Milk bids in school foodservice operations have been awarded on a bottomline basis for a long time. Gunn asks if you can imagine having one dairy deliver 2% unflavored milk while another delivers whole and skim milk, and maybe a third dairy delivers cottage cheese and yogurt. The foodservice manager would have to receive three deliveries and process three invoices if a single source wholesaler is not used.

Buyers for federal, state, and local institutions must allow bidding by all qualified distributors, but buyers for private institutions and commercial foodservices may select any supplier they want to submit bids. A bid request form, on which the specification and date for closing bids are indicated, is used to solicit bids from prospective suppliers. Bids usually are required to be submitted in an unmarked, sealed envelope and are opened at a specified time. Government purchasing policies usually require that this be done publicly, and the award should be made to the lowest bidder. No such requirement is dictated to the foodservice industry; however, if buyers select qualified suppliers and bids are responsive to the specifications, the lowest bid should be the one accepted. Computer software programs that allow managers to compare bid prices by food category and suppliers are available to assist with the bid analysis process.

Legal Considerations

Purchasing practices and the buyer/supplier relationship are usually predicated upon good faith and not dependent on legal considerations. The purchase/sale interchange, however, is a legal and binding commitment covered in the Uniform Commercial Code (UCC), which has been passed by all state legislatures except Louisiana's. The code provides uniformity of law pertaining to business transactions in eight areas, which are identified as articles. Article 2, which governs purchase/sales transactions, is of prime concern to buyers because the buyer is protected legally against trickery by the supplier. The major legal areas involve the laws of agency, warranty, and contracts. Buyers, therefore, should have knowledge of the basic principles of each to avoid litigation.

Law of Agency.   Every purchasing transaction is governed by at least one law, the **law of agency**; in most purchases, however, many laws are involved. Buyers may involve the foodservice operation and themselves in expensive legal disputes if they do not understand how these laws affect the purchasing function. The law of agency defines the buyer's authority to act for the organization as well as the obligation each owes the other and the extent to which each may be held liable for the other's actions.

Some terms need to be defined before the law can be understood. An **agent** is an individual who has been authorized to act on behalf of another party, known as the **principal.** The business relationship between the agent and the principal is the **agency.** Agents have the power to commit their principals in a purchase contract with suppliers. In large organizations, the power of the agency is created by actions of the board of directors and delegated to the buyer, but in small organizations, this authority may have developed over years without ever having been recorded in a written agreement.

Often the buyer will deal with a salesperson and not the supplier. Generally, salespeople are considered to be special agents empowered to solicit business for their principal or company, and nothing else. The primary interest of most salespeople, quite rightly, is to make as many sales as possible. Most companies protect themselves and their customers by specifying in sales agreements that they are not bound by a salesperson's promises unless they appear in writing in a contract approved by an authorized person in the supplier's office.

Law of Warranty.   The **law of warranty** defines *warranty* as a supplier's guarantee that an item will perform in a specified way (Kotschevar & Donnelly, 1994). The UCC recognizes three types of warranties (Leenders et al., 1989):

- *Express warranty*—promises, specifications, samples, and descriptions of goods that are under negotiation
- *Implied warranty of merchantability*—suppliers "puff" the virtues of their products for the purpose of making a sale
- *Implied warranty of fitness for a particular purpose*—buyer relies on the supplier's skill or judgment to select or furnish suitable goods

The UCC provides that if the supplier knows the particular purpose for which products are required and the buyer is relying on the supplier's skill or judgment to select or furnish suitable products, the implied warranty is that the goods shall be fit for such a purpose. If, however, the buyer provides detailed specifications for the product, the supplier is relieved of any warranty of fitness for a particular purpose.

Law of Contract.   Under the **law of contract**, a *contract* is an agreement between two or more parties. Because contracts constitute an agent's primary source of liability, buyers should always be certain that each contract bearing their signature is legally sound. A contract must fulfill the following five basic requirements to be considered valid and enforceable: the offer, acceptance of the offer, consideration, competent parties, and legality of subject matter. An example of a contract award form for purchasing by a school district is shown in Figure 7.9.

- *The offer.* The first step in entering into a contract is the making of an offer, usually by the buyer. A purchase order becomes a buyer's formal offer to do business. When the supplier agrees to terms, a contractual relationship will be in force. A contract does not exist until both parties agree to the terms stipulated therein.

Figure 7.9. Example of a contract award form.

Contract Award

Board of Education or School _____   Contract Award No. _____

_____   Date Awarded _____
Address
_____   Date Bid Opened _____

This is a notice of the acceptance of Bid # _____ for the period of _____ 19 ____ to _____ 19 ____.

Delivery
Delivery is to be made in two shipments: Week of _____ and _____ between _____ a.m. and _____ p.m.

Notice to Contractors:
This notice of award is an order to ship. Orders against contract are listed by _____ and invoices shall be rendered direct to the _____. The price basis, unless otherwise noted, _____ includes delivery and transportation charges fully prepaid F.O.B. agency. No extra charge to be made for packing or packages.

Names and Addresses of Successful Bidders

Offer
In compliance with the above award, and subject to all terms and conditions listed on the Bid Request, the undersigned offers and agrees to sell to _____ the items listed on the attached schedule.

Bidder _____

_____
Address
By _____
   Signature of person authorized to sign this contract
Title _____

Accepted as to items numbered _____   Accepted by _____

By _____   Date _____

Title _____

Source: Food Purchasing Pointers for School Food Service (p. 17) by U.S. Department of Agriculture, Food and Nutrition Service, August 1977, Washington, DC: Government Printing Office.

- *The acceptance*. The next step is the acceptance of the offer by the supplier. Generally, a clause is included that requires the supplier to indicate acceptance of the terms in the purchase order. Two copies of the order are sent to the supplier, one for the supplier and the other to be signed and returned to the buyer.
- *The consideration*. *Consideration* is the value, in money and materials, that each party pays in return for fulfillment of the other party's contract promise. Failure to do this usually means that the contract will not be legally enforceable. Agreement must be made on quantity, price, and time of delivery between the buyer and the supplier.
- *Competent parties and legality*. The final two requirements of a valid contract are competency and legality. *Competency* means that the agreement was reached by persons having full capacity and authority to enter into a contract. *Legality* means that a valid contract cannot conflict with any existing federal, state, or local regulations or laws.

Independent and Organizational Buying

Buying methods and legal concerns are just as applicable to an independent unit as they are to central and group purchasing. As mentioned previously, buyers must be legally authorized to act as agents of those for whom they purchase.

Independent Purchasing

Independent purchasing is done by a unit or department of an organization that has been authorized to purchase. In the simplest situation, the owner of a restaurant might be the manager, as well as the buyer, and would have full legal authority to execute binding contracts. In small hospitals, the head of the Food and Nutrition Department may have authority to do the purchasing and thus becomes the agent for the principal, which is the hospital.

Centralized Purchasing

Centralized purchasing is based on the principle that the purchasing activity is done by one person or department. In operations that have centralized purchasing, the head of the department usually reports directly to top management, who has the overall responsibility for making a profit. Advantages of centralized purchasing are many and include the following:

- Better control with one department responsible for purchasing and one complete set of records for purchase transactions and expenditures.
- Development of personnel with specialized knowledge, skills, and procedures that result in more efficient and economical purchasing.
- Better performance in other departments when managers are relieved of purchasing and of the interruptions and interviews incidental to purchasing.
- Economic and profit potentials of purchasing, making it a profit rather than a cost center (Heinritz et al., 1991).

Most hospitals have used centralized purchasing for many years, even though they are part of a group purchasing group. The foodservice department purchases canned and frozen products, staples, paper goods, and detergents from the central storeroom. Foodservice managers, because they know the products that will satisfy the customer, often are responsible for ordering perishable foods, such as fresh fruits and vegetables and meat, and storing them in department refrigerators or freezers.

Widely dispersed units under one central management may also use centralized purchasing. A classic example is a school foodservice operation, in which a central purchasing office, usually with warehouse facilities, meets the requirements for all schools in a district. Warehousing and delivery costs add to the purchasing budget, but these costs may be somewhat offset by lower prices from suppliers who do not have to make individual deliveries to all schools in the district.

The profit center concept is most successful in a centralized purchasing department (Reck & Long, 1983). Profit can be made by the purchasing department by adding a percentage that includes overhead, labor, and profit to the price of the product. As a check, individual departments need to be given the authority to purchase products and services directly from suppliers if significant cost savings can be achieved. A buyer in the centralized purchasing department, however, can negotiate for lower prices, consolidate orders from several departments into one large order, develop long-term supply agreements with suppliers in return for larger discounts, and search for low-cost suppliers and substitute materials. In most cases, the ultimate cost of products is lower than individual departments could achieve.

Group Purchasing

Group purchasing, often erroneously called *centralized purchasing,* involves the bringing together of foodservice managers from different operations, most often noncommercial, for joint purchasing. Many healthcare organizations participate in group purchasing, which is external, but have internal centralized purchasing. The economic advantage of group purchasing is that the volume of purchases is large enough to warrant volume discounts by suppliers.

A site is selected for the group purchasing office, and purchasing personnel are hired by the group of managers and paid from group funds. Warehouse space is not required because products are delivered directly to the foodservice operation by a distributor on a cost-plus basis. The participating managers serve as an advisory committee to the purchasing personnel and also have some decision-making power. An example is their agreement on specifications for each item to be purchased. Obviously, wide variations in specifications defeat the purpose of group purchasing because the quantity of a specific product ordered by one foodservice operation would be low, thus decreasing high-volume savings.

The best example of group purchasing is found in healthcare organizations, particularly hospitals. Because volume purchasing has been shown to bring food costs down anywhere from 3% to 10%, the number of hospitals organizing or joining purchasing groups has been increasing steadily. Some foodservice consultants estimate that close to 80% of hospitals in the United States belong to purchasing groups and that about 160 such groups exist (Lorenzini, 1991).

The earliest group purchasing program was started in 1918 by the Greater Cleveland Hospital Association and was supported by hospitals in a 100-mile radius of Cleveland. Currently, many cities such as Cincinnati, Chicago, Los Angeles, and Philadelphia, and states such as Connecticut, Idaho, Mississippi, and Missouri have group purchasing organizations. Large medical centers as well as small hospitals can profit from group purchasing. Members of healthcare purchasing groups bought $15.4 billion in goods and services in 1991, according to 62 respondents to a 1992 survey (Wagner, 1992). These group purchasing organizations have contracts for products or services with various companies. The most widely used contracts continue to be those for medical/surgical supplies, followed by medical equipment and pharmaceuticals. One of the groups has separate contracts for 200 or more products with Kraft Foods. Contracts for healthcare, total quality management, and waste management services are available in many of the group purchasing organizations. Group purchasing has become so large that it has its own professional organization.

Warehouse Club Purchasing

A growing number of restaurateurs are discovering substantial savings and product offerings available through self-service, cash and carry, and wholesale warehouses (Geisse, 1988). These wholesale units operate on a no-frills approach to purchasing that enables prices to be kept exceptionally low. Products remain in original cartons stacked on pallets or metal shelves separated by wide aisles. A wide variety of brand-name items, including food, cleaning supplies, furniture, electronic equipment, and many others needed for operating restaurants, are stocked in plain and unfinished warehouses. Personnel are not available to assist customers. Markups on most articles average 10% over wholesale cost, and food products have an even lower percentage markup. These clubs, which number more than 200 in the United States, have been a boon to owners of a single restaurant. Owners of more than one restaurant also can benefit from **warehouse club purchasing**.

SUPPLIER SELECTION AND EVALUATION

Supplier selection may be the single most important decision made in purchasing. The trend to fewer suppliers, buying rather than making products, electronic data interchange, and continuing improvement in quality, price, and service requires greater communication between buyers and suppliers than ever before. Improving buyer and supplier relationships, therefore, is one of the key concerns in the overall area of supplier selection (Leenders et al., 1989).

The ideal method of purchasing would be based on price, quality, and delivery from the supplier offering the best possible combination of all three. This ideal method is not always an easy task because salespeople might have their own ideas of right and wrong and buyers might be prejudiced against certain suppliers. A **sup-**

plier, often identified as a seller or vendor, is a person who offers products for sale. *Supplier* is the term used by the National Association of Purchasing Management in the *International Journal of Purchasing and Materials Management* and, therefore, will be used in this text. A **buyer**, often called a purchaser, has charge of the selection and purchasing of products.

Traditionally, the supplier seeks out the buyer, but for the purchasing professional, this leaves too much to chance. Few foodservice managers need to be convinced of the advantages of consolidating most purchases with a primary distributor. Gone are the days when buyers had many suppliers or distributors to supply their needs. According to Goebeler (1989), because of mounting competitive pressures, every manager should pay careful attention not only to selecting a distributor but also to maintaining the relationship that ensues. The selection process consists of the inquiry and survey stages followed by an evaluation.

The Survey Stage

The purpose of a supplier survey is to explore all possible sources for a product. The buyer's experience and personal contacts with various suppliers provide the most valuable and reliable information and should be kept in complete and up-to-date supplier files in the purchasing office. These files should contain the name and address of every supplier with whom a foodservice organization has transacted business, plus information on products purchased from the supplier. Additional helpful information would be reliability in meeting commitment dates, willingness to handle emergency orders, and number of times orders were incorrect. The supplier file can be cross-referenced with a file of products listing from whom they were purchased, the prices paid, and the points of shipment.

Buyers should learn to question salespeople because they have a lot of information even though they could be biased about their own products. If they do not have a product a buyer wants, they often will suggest other companies that carry it.

The Inquiry Stage

In the inquiry stage of the supplier selection process, the field must be narrowed from possible to acceptable sources. In general, this process involves comparing potential suppliers to provide the right quality and needed quantity at the right time, all at the right price with the desired degree of service. Quality, quantity, and price should be compared and balanced against one another. The common factor of service is applicable to all purchases, and although it cannot be quantified, it is an important subjective consideration.

Geographic location is a major concern in evaluating a supplier's service. Certainly, shorter delivery distances offer better opportunities for satisfactory service. The supplier's inventory also is an important consideration as are the kinds and quantities of products available to buyers. The selected supplier should be one who keeps current with technological developments, is capable of providing new and improved products as they become available, and whose quality control standards ensure that inspection and storage methods adhere to FDA and USDA standards.

Warranty and service offered by an equipment supplier also are a vital concern for the buyer.

Another factor to consider is that the financial condition of a supplier is vital for maintaining a satisfactory business relationship. In general, the controller or credit manager of the buyer's organization can provide a credit report. When feasible, a visit to the prospective supplier's warehouse facilities will give an indication of the efficiency and general cleanliness associated with a good business. The visit gives the buyer an opportunity to observe employees, including union relations, and order-handling procedures.

After the survey and inquiry stages of selection, the buyer should have a few suppliers from which to choose one or more. The selection process is relatively subjective; more objective evaluations can be made after an order is placed and the product delivered. An evaluation process needs to be developed that can be used for newly selected suppliers as well as for those who have been servicing the operation for some time.

Supplier Performance Evaluation

The evaluation of suppliers is a continuing purchasing task (Leenders et al., 1989). Current suppliers need to be monitored to be sure they are meeting performance expectations, and new ones need to be screened to determine if they should be seriously considered in the future. Buyers, however, tend to use suppliers who have proven to be reliable. The new supplier should be given the opportunity to make a few deliveries to the foodservice operation before performance evaluation. It is difficult to select a supplier performance evaluation instrument that is fair but really evaluates performance. Many foodservice managers develop their own forms. Figure 7.10 is an example of an evaluation form that could be modified for a specific operation.

Evaluations should be conducted periodically to keep the supplier's record file up-to-date. Suppliers who have good evaluations will appreciate knowing the results. Suppliers who do not rate well should be told why before severing the business relationship.

PURCHASING PROCESS

Procurement in any foodservice operation requires procedures to accomplish the routine purchasing transaction as quickly and accurately as possible. Undoubtedly, each foodservice operation has its own procedures unique to its needs, but very likely these procedures conform to the basic pattern followed by many other foodservices. The adoption of definite purchasing procedures implies utilization of appropriate records for each phase in the purchasing process.

Purchasing Procedures

Several basic procedures appear in some form in every purchasing unit (Figure 7.11). These procedures are simple and should be adapted to the particular needs of

Figure 7.10.   Example of a supplier evaluation form.

SUPPLIER PERFORMANCE EVALUATION

Supplier: _____   Date: _____

Company	Excellent	Good	Fair	Poor
Size and/or capacity				
Financial strength				
Technical service				
Geographical locations				
Management				
Labor relations				
Trade relations				
Products				
Quality				
Price				
Packaging				
Uniformity				
Service				
Delivers on time				
Condition on arrival				
Follows instructions				
Number of rejections				
Handling of complaints				
Technical assistance				
Emergency deliveries				
Supplies price changes promptly				

the various departments of an organization. Buyers can add, delete, or modify these procedures to fit their operation.

Recognition of a Need

The most obvious place for a need to occur is the production unit. The second location is the storage area, in which the objective is to have on hand the right product, in the right quantity, at the right time. A third location exists in larger organizations

Figure 7.10. *continued*

Sales Personnel	Excellent	Good	Fair	Poor
• Knowledge:				
Of company				
Of products				
Of foodservice industry				
• Sales calls:				
Properly spaced				
By appointment				
Planned and prepared				
Mutually productive				
• Sales service:				
Obtains information				
Furnishes quotations promptly				
Follows through				
Expedites delivery				
Handles complaints				
Accounting				
Invoices correctly				
Issues credit promptly				

in which the centralized purchasing department has inventory responsibility. Recognition of a need should be followed by action to remedy the deficiency by preparing a requisition.

Description of the Needed Item

In most organizations, the production unit cooks, having recognized a need, initiated a requisition, preferably written, to the storeroom for the required amount of the product. If the storeroom has an adequate supply of the item, the requisition can be honored. If, however, honoring this requisition brings the inventory stock below the acceptable minimum, storeroom personnel will initiate another requisition to purchasing for replenishment of the product to the desired level.

Computer software programs are available that will generate a list of the quantities of ingredients needed to prepare menu items. The caveat is that many of these programs round quantities to the nearest purchase unit; for example, a teaspoon of salt might become a case of salt.

Figure 7.11. Fundamental
steps in purchasing.

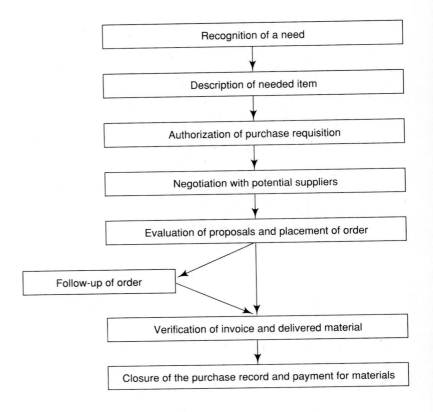

Authorization of the Purchase Requisition

The third procedure in the purchasing process is authorizing the purchasing requisition. In every foodservice organization, a policy should be established to indicate who has the authority to requisition food, supplies, and equipment. No requisitions should be honored unless the person submitting them has the authority to do so. Furthermore, suppliers should know the names of persons authorized to issue purchase orders.

Negotiation with Potential Suppliers

Negotiation in purchasing is the process of working out both a purchasing and sales agreement, mutually satisfactory to both buyer and supplier, and reaching a common understanding of the essential elements of a contract. It is one of the most important parts of the procurement function, if not the most important. Negotiation usually is responsible for vital details such as establishing the qualifications of a particular supplier, determining fair and reasonable prices to be paid for needed products, setting delivery dates agreeable to both the buyer and supplier, as well as renegotiating contract terms when conditions change.

Evaluation of Proposals and Placement of Order

All supplier proposals are evaluated for compliance with the preceding four fundamental steps in purchasing (Figure 7.11). The actual placement of an order follows the evaluation of proposals. Ideally, all orders should be in writing, but if an order is placed by telephone, confirmation in writing should be made promptly. A written record of every purchase should always be on file.

Many software programs will generate a purchase order for the supplier. Also, some suppliers are able to link their computers with those of buyers. Orders can then be placed via computer modem.

Follow-up of Order

Theoretically, buyers should not have to follow-up an order after it has been placed and accepted by a supplier. A follow-up in a foodservice operation is justified, however, when a specific delivery time of certain products is critical to an occasion, such as a major banquet.

Verification of Invoice and Delivered Materials

The **invoice** is the supplier's statement of what is being shipped to the buyer and the expected payment. The invoice should be checked against the purchase order for quality, quantity, and price. Delivery and condition of materials should be in agreement with the purchase order and the invoice. Any differences require immediate action by the buyer. For example, the condition of frozen food should be verified, and any indication of exposure to higher temperatures would be a cause for rejection and immediate communication with the supplier.

Closure of Purchase Record

Closing the purchase record consists of the clerical process of assembling the written records of the purchase process, filing them in appropriate places, and authorizing payment for the goods delivered. The filing system need not be complex; its only purpose is to provide an adequate historical record of these business transactions.

Purchasing Records

The essential records for the purchasing process are the requisition and purchase order, originating with the buyer, and the invoice, prepared by the supplier. These records differ among foodservice operations but all have the same essential information.

Purchase Requisition

The **requisition** is the first document in the purchasing process and may have originated in any one of a number of units in the foodservice operation. An example is shown in Figure 7.12. The following five basic items of information generally are included on all requisitions:

- *Requisition number.* This number is necessary for identification and control purposes and is generally accompanied by a code for the originating department.
- *Delivery date.* This date, on which the product should be in the storeroom for use by the cooks, should always allow sufficient time to secure competitive bids and completion of the full purchase transaction, if at all possible.
- *Budget account number.* This number indicates the account to which the purchase cost will be charged.
- *Quantity needed.* Quantities should be expressed in a common shipping unit, such as cases, as well as the number of items in a unit. For example, the entry on the requisition is for 20 cases of diced tomatoes, with six No. 10 cans to a case.
- *Description of the item.* The description and quantity needed are the two most important pieces of information on the requisition form and, therefore, occupy

Figure 7.12.   Example of a purchase requisition.

COMMUNITY HOSPITAL
PURCHASE REQUISITION

To: _____Purchasing Office_____   Requisition No.: _____FS1201_____

Date: _____August 16, 1994_____   Purchase Order No.: _____1842_____

From: _____Foodservice_____   Date Required: _____September 12, 1994_____

Budget Account No.: _____FS 1101_____

Total Quantity	Unit	Description	Supplier	Unit Cost	Total Cost
20	cases	Tomatoes, diced, U.S. Grade B (Extra Standard) #10 cans, 6/case	L. & M. Wholesale Grocers	$13.86	$277.20

Requested by _____   Approved by _____   Date Ordered _____8/23/94_____ the central space. The information may include the product specification, brand, or catalog number.

In addition to the above information, which is provided by the person responsible for the requisition, the buyer adds the name and address of the supplier and the details of the purchase, including the purchase order number and the price and date ordered.

Purchase Order

A **purchase order** is the document, based on the information in the requisition, completed by the buyer, who gives it to the supplier. It states in specific terms the purchase and sales agreement between the buyer and the supplier. Before acceptance, the purchase order represents an offer to do business under certain terms and, once accepted by a supplier, becomes a legal contract. An example is shown in Figure 7.13.

Format.   The purchase order, like the requisition, exists in a wide variety of formats that have been developed to meet the needs of individual foodservice operations. The principal reason for this variety is that the purchase order is a legal document; the terms presented in it are intended to protect the buyer's interests, which differ from one operation to another. Almost every purchase order, however, includes the following items.

- Name and address of the foodservice organization
- Name and address of the supplier
- Identification numbers (purchase order and purchase requisition)
- General instructions to supplier
- Complete description of purchase item
- Price data
- Buyer's signature

Required Number of Copies.   Three copies of the purchase order are sufficient for basic ordering: the original, which is sent to the supplier; an acknowledgment copy, which is actually a formal acceptance that the supplier returns to the buyer; and a file copy for the buyer's own record. This simplified plan offers only a minimum of control and, therefore, should not be used in large operations.

Ordering for large foodservice organizations requires six copies of the purchase order. The additional three copies are given to the following individuals or departments: receiving copy, which informs receiving personnel that a delivery is scheduled; accounting copy, which is sent to the accounting office; and requisitioner's copy, which notifies the requisitioner that the order has been placed.

Invoice

The invoice prepared by the supplier contains the same essential information as the purchase order; that is, quantities, description of items, and price. When products are delivered, the supplier's invoice must be compared with the purchase order and the quantity received. Any discrepancies or rejections at the receiving point should be noted immediately. In some operations, these problems are handled by the receiving clerk, who notes the discrepancy on the invoice and has the supplier's delivery person initial it.

Figure 7.13.   Example of a purchase order form.

COMMUNITY HOSPITAL
1010 Main Street
Sioux City, Oklahoma

To: L. & M. Wholesale Grocers

200 South Street

Sunnyvale, OK

Purchase Order No.: _____ 1842 _____

Please refer to the above number on <u>all</u> invoices, two copies required.

Date: _____ August 23, 1994 _____

Requisition No.: _____ FS1201 _____

Dept.: _____ Foodservice _____

Date Required: _____ September 12, 1994 _____

Ship to: _____   F.O.B.: _____   Via: _____   Terms: _____

Total Quantity	Unit	Description	Supplier	Unit Cost	Total Cost
20	cases	Tomatoes, diced, U.S. Grade B (Extra Standard) #10 cans, 6/case	L. & M. Wholesale Grocers	$13.86	$277.20

Approved by _____

Title: Director of Purchasing
Community Hospital

If the invoice and the purchase order are in agreement, the invoice will be forwarded to the accounting office for payment. In small operations using only three copies of the purchase order, the receiving clerk only verifies the arrival of the delivered items and sends the invoice to the accounting office for comparison with the purchase order and subsequent payment.

ETHICAL CONSIDERATIONS

Ever since the Watergate scandal resulted in the resignation of a United States president, the public, businesses, and professions have become sensitive to what is and what is not ethical and to the standards by which people are measured. The news media have kept the word *ethics* in public view by reports of congressional hearings on ethical conduct of appointees to important positions. With all of this emphasis on ethics, politicians, appearing almost daily in newscasts, occasionally are accused of lying, and business people, of accepting favors under the table.

Business and professional organizations are updating codes of ethics or standards of practice and are emphasizing them to their members. **Ethics** is defined as the principles of conduct governing an individual or a business. Personal ethics should be distinguished from business ethics. The source of **personal ethics** lies in a person's religion or philosophy of life and is derived from definite moral standards. **Business ethics** may be defined as self-generating principles of moral standards to which a substantial majority of business executives gives voluntary assent. It is a force within business that leads to industrywide acceptance of certain standards of practical conduct (Zenz, 1981). Ethics are created by society; morals are an individual's personal belief about what is right or wrong (Kapoor, 1991).

Code of Ethics

Organizations, to be professional, must have a code of ethics to which its members subscribe. Many businesses also have a code of ethics. A code is a set of rules for standards of professional practice or behavior established by a group. A **code of ethics** is influenced by codes of individuals, but the major emphasis is on the relationships within businesses and professional organizations. A **standard** is the result of the managerial process of planning and is defined as the measurement of what is expected to happen. Many businesses and professional organizations are changing from a code of ethics to standards of practice. For example, the National Association of Purchasing Management (NAPM) changed from a code of ethics to Principles and Standards of Purchasing Practice in January 1992 (Figure 7.14).

The NAPM principles and standards of practice touch on many areas in which the buyer might err ethically. In addition to conforming to the standards, buyers are subject to the law of agency, in which they are the agent and the principal is the employing agency.

A prime example is the code of ethics for the Foodservice Purchasing Managers of the National Restaurant Association, as shown in Figure 7.15. This code was devel-

Figure 7.14.   Principles and standards of purchasing practice.

Principles and Standards of Purchasing Practice

National Association of Purchasing Management ®

LOYALTY TO YOUR ORGANIZATION
JUSTICE TO THOSE WITH WHOM YOU DEAL
FAITH IN YOUR PROFESSION

From these principles are derived
the NAPM standards of purchasing practice.
(Domestic and International)

1. Avoid the intent and appearance of unethical or compromising practice in relationships, actions, and communications.

2. Demonstrate loyalty to the employer by diligently following the lawful instructions of the employer, using reasonable care and only authority granted.

3. Refrain from any private business or professional activity that would create a conflict between personal interests and the interests of the employer.

4. Refrain from soliciting or accepting money, loans, credits, or prejudicial discounts, and the acceptance of gifts, entertainment, favors, or services from present or potential suppliers that might influence, or appear to influence, purchasing decisions.

5. Handle confidential or proprietary information belonging to employers or suppliers with due care and proper consideration of ethical and legal ramifications and governmental regulations.

6. Promote positive supplier relationships through courtesy and impartiality in all phases of the purchasing cycle.

7. Refrain from reciprocal agreements that restrain competition.

8. Know and obey the letter and spirit of laws governing the purchasing function and remain alert to the legal ramifications of purchasing decisions.

9. Encourage all segments of society to participate by demonstrating support for small, disadvantaged, and minority-owned businesses.

10. Discourage purchasing's involvement in employer-sponsored programs of personal purchases that are not business related.

11. Enhance the proficiency and stature of the purchasing profession by acquiring and maintaining current technical knowledge and the highest standards of ethical behavior.

12. Conduct international purchasing in accordance with the laws, customs, and practices of foreign countries, consistent with United States laws, your organization policies, and these Ethical Standards and Guidelines.

Adopted 1/92   2/92/10M

Source: Reprinted with permission from the publisher, the National Association of Purchasing Management, *Principles & Standards of Purchasing Practice,* adopted January, 1992.

Figure 7.15.   Code of ethics of Foodservice Purchasing Managers Group.

FOODSERVICE
PURCHASING MANAGER'S
CODE OF ETHICS

I, _____ , recognized as a
Foodservice Purchasing Professional, hereby subscribe and agree to abide by the following
Principles and Standards of Foodservice Purchasing Practice. To wit my signature of agreement
appears below.

I.   I will consider the interest of my company in all transactions. Acting as its
representative, I will carry out and believe in its established policies.

II. I recognize good business practices can be maintained only on the basis of honest
and fair relationships.

III. I agree to make commitments for only that which I can reasonably expect to
fulfill.

IV. I will provide a prompt and courteous reception to all who request a legitimate
business appointment.

V.   I will avoid comments which may discredit or otherwise harm legitimate
competition. Likewise, information received in confidence will not be used by me
to obtain an unfair advantage in competitive transactions.

VI. I will strive to develop specifications and standards which will enable all qualified
sources to compete for the business without prejudice.

VII. I will not accept, nor encourage, the giving of gifts or entertainment where the
intent is to sway my decision in favor of the donor versus other qualified
competitors.

VIII. I abhor all forms of bribery and will act to expose same whenever encountered.

IX. I will always strive to be up to date on products, materials, supplies,
and manufacturing processes which will ensure my company receives the proper
quality at the most beneficial cost when it is required by the operator.

X.   Recognizing the National Restaurant Association and its Foodservice Purchasing
Managers are engaged in activities designed to enhance the development and
standing of Foodservice Purchasing, I hereby agree to support and participate in
its programs.

I, the undersigned, hereby subscribe,

Witness:
        Chairman, NRA Foodservice Purchasing Managers

Developed by the Foodservice Purchasing Managers

Source: National Restaurant Association. Used by permission.

oped to promote and encourage ethical practices in the industry. To give further credence to the code, the purchasing group agreed that each member should indicate support for the principles by signing the code of ethics document.

Ethical Issues

Buyers face many ethically critical situations in the performance of their duties. Reid and Riegel (1989) refer to such situations as *ethical dilemmas,* and they group them in the following three categories:

- Efforts to gain inside information about competitors that will benefit competition; for example, receiving information about competitors through shared suppliers
- Activities that allow buyers to gain personal benefits from suppliers; for example, free lunches, dinners, entertainment, trips, and gifts
- Activities that manipulate suppliers in such a manner as to benefit the purchasing organization; for example, overstating the seriousness of a problem to obtain concessions from the supplier, threatening the use of a second source, using the organization's economic clout, and permitting information on bids from other suppliers

Ethics Management

Ethical conduct should be an organizational priority. Andrews (1989) asks why business ethics is a problem that snares not just a few criminals in the making but also good people who lead exemplary private lives while concealing information about dangerous products or falsifying costs. By implementing the management functions of planning, organizing, staffing, leading, and controlling, management can be sure that ethics is established formally and explicitly into daily organizational life. Hogner (1987) stated that when planning is applied to managing ethics, management is translating societal expectations for business performance and behavior into policies and procedures. Organizing requires structuring of the organization including a network to ensure that information relevant to ethical behavior will be transmitted.

Staffing requires that recruitment and selection procedures be consistent with the ethical code of the organization. Finally, controlling links all the functions together. To control ethical conduct, performance standards must be established and continually monitored, and appropriate appreciative or corrective action taken.

Leadership has been cited as the principal mechanism for increasing ethical performance in business. Nielsen (1989) concluded that leadership can be used effectively in organizational ethics issues. Depending upon the circumstances, including personal courage, one can choose to act and be ethical both as a leader and an individual. Being part of or leading ethical change is generally the more constructive approach, although many times the only short-term effective approach is for an individual to intervene against others or the organization. Dire consequences for the individual may result, however, from standing up to and opposing unethical behavior.

MATERIALS MANAGEMENT

The concept of **materials management** has been well expressed by Dillon (1973) as the unifying force that gives interrelated functional subsystems a sense of common direction. Its goal is to transform materials or physical resources that enter the system into an output that meets definite standards for quantity and quality. An organization using the concept has a single manager responsible for the planning, organizing, motivating, and controlling of all activities principally concerned with the flow of materials into an organization (Leenders et al., 1989).

Zenz (1981) gave another definition, derived from the basic concept but expressed in operational terms. According to his definition, materials management is an organizational concept of centralized responsibility for those activities involved in moving materials into, and in some cases through, the organization. While the activities vary, they usually include purchasing, receiving, storage, inventory control, and traffic, with production control and related activities sometimes included, depending on the product or process involved. Of all activities performed in the materials management cycle, supplier relationships are perhaps the most susceptible link (Handfield, 1993). The dependency of a foodservice operation on outside sources for products often leaves it at the mercy of suppliers' performance. Buyers should promote an environment of mutual trust in which suppliers can help improve the current purchasing process.

The trend toward the inclusion of materials management in organizational structure has been motivated, primarily, by the need to control the overall cost of materials and to minimize those areas of conflict that have traditionally resulted in excessive inventories. Utilization of the materials management concept, although initiated in industrial manufacturing, has spread to foodservice-related industries, such as major frozen food processors and commercial bakeries. A materials manager has an important role in Stouffer's frozen food division. Moreover, a number of larger hospitals now have a materials manager in a staff position with advisory responsibilities for all activities involving moving materials through the organization.

Fundamentally, a materials manager is a consultant to all departments or units on the movement of materials through the organization. In smaller organizations that cannot justify additional staff persons, the materials management activity could be performed by top-level management. Other similar staff positions in the organization are held by personnel, facility, and operational (financial, time, communication) resource managers with the same type of responsibilities as the materials manager.

SUMMARY

The goal of any foodservice system is to serve quality meals while maximizing value for both the operation and the customer. Value is the perceived relationship between quality and price. Procurement is the first functional subsystem of the transformation element; purchasing, receiving, storage, and inventory control are important functions within procurement. The role of the purchasing manager is changing from just a buyer to that of a communicator, decision maker, negotiator, and problem solver.

The market is the medium through which a change in ownership moves commodities from producer to consumer. Effective purchasing requires knowledge of the food market and the marketing channel. The supplier sends raw food products to processors, who sell the processed items to middlemen, who then sell to the customer. Value added is the increase in value of a product caused by processing and marketing exclusive of the cost of products or overhead. The superdistributor has been added to the list of wholesalers and is the creator of "one-stop shopping."

The food industry is the most controlled industry in the United States today because it is covered by comprehensive and complex federal regulations. The Food and Drug Administration (FDA) ensures that foods other than meat, poultry, and fish are pure and wholesome, safe to eat, and produced under sanitary conditions, and that packaging and labeling are in agreement with the contents. Food irradiation, genetically engineered foods, and nutrition labeling are also under the jurisdiction of the FDA.

The U.S. Department of Agriculture (USDA) has an important role in the food regulatory process. One of the most important functions of the USDA is the grading, inspection, and certification of all agricultural products. The National Marine Fisheries Service, under the Department of Commerce, offers a voluntary inspection program for fish and seafood, which currently is not subjected to mandatory continuous inspection, as are meat and poultry.

Product selection is the responsibility of the buyer, whose primary function is to procure the required products for the desired use at minimum cost. Value analysis, which is an organized technique applicable to cost reduction, and make-or-buy decisions, which are decisions to purchase from oneself (make) or from suppliers (buy), are both continuing processes. Strict adherence to rigorous purchasing specifications is required. Purchasing can be informal or formal, which results in a contract. Supplier selection and evaluation also are very important parts of the buyer's responsibilities.

A well-organized purchasing process should be developed for effective purchasing. The purchasing procedure begins with the recognition of a need and ends with closure of the purchase record. The various phases of the purchasing procedure require written records, including the requisition, the purchase order, and the invoice.

Ethical decisions are very important in buyer and supplier relationships. Many business and professional organizations have changed from a formal code of ethics to standards of practice. Leadership has been cited as the principal mechanism for increasing business ethical performance.

Materials management is the unifying force that gives interrelated functional subsystems a sense of common direction. It is an organizational concept that centralizes responsibility for all activities in moving materials through all the subsystems with concentration in procurement.

BIBLIOGRAPHY

Allen, R. L. (1992). New Senate bill backs seafood inspection plan. *Nation's Restaurant News*, *26*(17), 3, 59.

Amos, S. M. (1992). Looking beyond the "specs" —growing together through profitable partnerships. *The Consultant*, *25*(2), 69.

Andrews, K. R. (1989). Ethics in practice. *Harvard Business Review*, (5), 99.

Avery, A. C. (1978). The art of writing specs. *Food Management*, *13*(3), 40–41, 65, 68.

Bales, W. A., & Fearon, H. E. (1993). *CEOs'/presidents' perceptions and expectations of the purchasing function*. Tempe, AZ: Center for Advanced Purchasing Studies.

Carter, J. R., & Narasimhan, R. (1993). *Purchasing and materials management's role in total quality management and customer satisfaction*. Tempe, AZ: Center for Advanced Purchasing Studies.

Catalog of food specifications: A technical assistance manual. (1992). Vol. I. (5th ed.). Dunnellon, FL: Food Industry Services Group in cooperation with the U.S. Department of Agriculture, Food & Nutrition Service.

Cheney, K. (1993). Tempest in a test tube. *Restaurants & Institutions*, *103*(5), 76–79, 84, 88, 90, 94.

Contract purchasing: A technical assistance manual. (1992). Vol. II. Dunnellon, FL: Food Industry Services Group in cooperation with the U.S. Department of Agriculture, Food & Nutrition Service.

Cook, R. L. (1992). Expert systems in purchasing: Applications and development. *International Journal of Purchasing and Materials Management*, *28*(4), 20–27.

Derr, D. D. (1993). Food irradiation: What is it? Where is it going? *Food & Nutrition News*, *65*(1), 5–6.

Dillon, T. F. (Ed.). (1973). Materials management: A convert tells why. *Purchasing*, *74*(5), 43–45, 47.

Eames, D., & Norkus, G. X. (1988). Developing your procurement strategy. *Cornell Hotel and Restaurant Administration Quarterly*, *29*(1), 30–33.

Ellram, L. M., & Pearson, J. N. (1993). The role of the purchasing function: Toward team participation. *International Journal of Purchasing and Materials Management*, *29*(3), 3–9.

Etherton, T. D. (1993). The new bio-tech foods. *Food & Nutrition News*, *65*(3), 1–3.

FDA issues food biotech guidelines. (1992). *Food Insight*, 6–7.

Fearon, H. E. (1988). Organizational relationships in purchasing. *Journal of Purchasing and Materials Management*, *24*(4), 2–12.

Food facts for food service supervisors: A technical assistance manual. (1992). Vol. III. (2nd ed.). Dunnellon, FL: Food Industry Services Group in cooperation with the U.S. Department of Agriculture, Food & Nutrition Service.

Geisse, J. (1988). Wholesale warehouses offer no-frills shopping. *Restaurants USA*, *8*(11), 14–15.

Goebeler, G. (1989). Distributor viewpoint: Maintaining relationships. *Restaurant Business*, *88*(10), 56.

Gunn, M. (1992). Professionalism in purchasing. *School Food Service Journal*, *46*(9), 32–34.

Handfield, R. B. (1993). The role of materials management in developing time-based competition. *International Journal of Purchasing and Materials Management*, *29*(1), 2–10.

Heinritz, S., Farrell, P. V., Giunipero, L., & Kolchin, M. (1991). *Purchasing: Principles and applications* (8th ed.). Englewood Cliffs, NJ: Prentice-Hall.

Hogner, R. H. (1987). Ethics in the hospitality industry: A management control system perspective. *FIU Hospitality Review*, *5*(1), 34–41.

Kansas State University Cooperative Extension Service. (1990). What is value-added? *Focus on Value-added Agricultural Products, 1*(1), 1.

Kapoor, T. (1991). A new look at ethics and its relationship to empowerment. *Hospitality & Tourism Educator, 4*(1), 21–24.

Keegan, P. O. (1993). 'Healthy' menus may fall victim to labeling act. *Nation's Restaurant News, 27*(24), 1, 100.

Kolchin, M. G., & Giunipero, L. (1993). *Purchasing education and training: Requirements and resources.* Tempe, AZ: Center for Advanced Purchasing Studies.

Konopacki, A. (1989). Power buyers of the 1990s. *Successful Meetings, 38*(5), 131–132.

Kotschevar, L. H., and Donnelly, R. (1994). *Quantity food purchasing* (4th ed.). New York: Macmillan.

Kunkel, M. E. (1993). Position of The American Dietetic Association: Biotechnology and the future of food. *Journal of The American Dietetic Association, 93*(2), 189–192.

Kurtzweil, P. (1993). 'Nutrition facts' to help consumers eat smart. *FDA Consumer, 27*(4), 22–27.

Leenders, M. C., Fearon, H. E., & England, W. B. (1989). *Purchasing and materials management* (9th ed.). Homewood, IL: Irwin.

Leenders, M., & Nollet, J. (1984). The gray zone in make or buy. *Journal of Purchasing and Materials Management, 20*(3), 10–15.

Lewis, H. T. (1975). This business of procurement. *Journal of Purchasing and Materials Management, 11*(2), 7–17.

Liddle, A. (1988). Distribution: A new era dawns. Part 3. *Nation's Restaurant News, 22*(29), F1, F6, F10.

Loecker, K. A., Spears, M. C., & Vaden, A. G. (1983). Purchasing managers in commercial food-service organizations: Clarifying the role. *Professional, 1*(2), 9–16.

Lorenzini, B. (Sept. 18, 1991). Purchasing groups find savings in numbers. *Restaurants & Institutions* (Special ed.). A51–A52.

Lorenzini, B., & McCarthy, B. (1992). The branding evolution. *Restaurants & Institutions, 102*(21), 87, 90, 95, 98, 102, 106, 108.

Martin, R. (1993). Meat inspection proposal spurs price hike fears. *Nation's Restaurant News, 27*(13), 9, 103.

McProud, L. M., & David, B. D. (1976). Applying value analysis to food purchasing. *Hospitals, 50*(18), 109–113.

Miles, L. D. (1972). *Techniques of value analysis and engineering.* New York: McGraw-Hill.

Muller, E. W. (1992). *Job analysis identifying the tasks of purchasing.* Tempe, AZ: Center for Advanced Purchasing Studies.

Mutkoski, S. A., & Schurer, M. L. (1981). *Meat and fish management.* North Scituate, MA: Breton Publishers.

National Association of Meat Purveyors. *The meat buyers guide.* (1992). McLean, VA: Author.

National Center for Nutrition and Dietetics. (Spring 1993). The new food label: Standardizing the terms to dispel confusion. *From the Center, 3.*

National Food Brokers Association. (Undated). *How to select brokers: A guide to interviewing and selecting retail, foodservice, and industrial ingredients brokers.* Washington, DC: Author.

National Food Service Management Institute. (1992). *Impact of food procurement on the implementation of the Dietary Guidelines for Americans in Child Nutrition Programs: Conference proceedings.* University, MS: Author.

National Restaurant Association. (April 1989). *A review of U.S. food grading and inspection programs. Current Issues Report.* Washington, DC: Author.

Nielsen, R. P. (1989). Changing unethical organizational behavior. *Academy of Management Executive, 3*(2), 123–130.

Patterson, P. (1993). Roundtable discussions offer buyers insights, helping hand. *Nation's Restaurant News, 27*(24), 80.

Peddersen, R. B. (1981). *Foodservice and hotel purchasing.* New York: Van Nostrand Reinhold.

Puckett, R. P., & Miller, B. B. (1988). *Food service manual for health care institutions.* Chicago: American Hospital Publishing.

Reck, R. R., & Long, B. G. (1983). Organizing purchasing as a profit center. *Journal of Purchasing and Materials Management, 19*(4), 2–6.

Reck, R. R., & Long, B. G. (1988). Purchasing: A competitive weapon. *Journal of Purchasing and Materials Management, 24*(3), 2–8.

Reid, R. D., & Riegel, C. D. (1989). *Purchasing practices of large foodservice firms.* Tempe, AZ: National Association of Purchasing Management.

Restaurants find bottom-line benefits in pilot voluntary seafood inspection program. (1993), *Washington Weekly, 13*(23), 7.

Riegel, C. D., & Reid, R. D. (1988). Food-service purchasing: Corporate practices. *Cornell Hotel and Restaurant Administration Quarterly, 29*(1), 25–29.

Riegel C. D., & Reid, R. D. (1990). Standards in food-service purchasing. *Cornell Hotel and Restaurant Administration Quarterly, 30*(4), 19–25.

Romeo, P. J. (1988). Distribution: A new era dawns. Part 1. *Nation's Restaurant News, 22*(27), F1, F9, F12, F14, F18.

Rubin, K. W. (1993). The pros & cons: Food irradiation. *FoodService Director, 6*(2), 90.

Ruggless, R. (1993). 'Pharmers' to cultivate more food choices. *Nation's Restaurants News, 27*(24), 62.

Spears, M. C. (1976). Concepts of cost effectiveness: Accountability for nutrition, productivity. *Journal of The American Dietetic Association, 68*(4), 341–346.

Stefanelli, J. M. (1992). *Purchasing: Selection and procurement for the hospitality industry* (3rd ed.). New York: Wiley.

Stephenson, S., & Schuster, K. (1982). Who am I? *Food Management, 17*(4), 44–47, 70, 72, 74, 76, 80, 82.

U.S. Department of Agriculture, Livestock Division, Agricultural Marketing Service. *Institutional Meat Purchase Specifications for Fresh Beef.* (January 1975). Washington, DC: Author.

Van Warner, R. (1988). Distribution: A new era dawns. Part 2. *Nation's Restaurant News, 22*(28), F1, F21, F24.

Virts, W. B. (1987). *Purchasing for hospitality operations.* East Lansing, MI: Educational Institute of the American Hotel and Motel Association.

Wagner, M. (1992). Group purchasing survey: Purchase groups buy goods worth over $15 billion. *Modern Healthcare, 39*, 40, 42, 46, 48–50.

Williams, A. J., Lacy, S., & Smith, W. C. (1992). Purchasing's role in value analysis: Lessons from creative problem solving. *International Journal of Purchasing and Materials Management, 28*(2), 37–42.

Zenz, G. J. (1981). *Purchasing and the management of materials* (5th ed.). New York: Wiley.

Receiving, Storage, and Inventory Control

As stated previously, procurement is the managerial function of acquiring material for production. Food is the material resource that is transformed into quality meals to satisfy the customer of a foodservice operation. Now that purchasing procedures and principles have been described, receiving, storage, and inventory control, the remaining activities in the procurement subsystem, will be examined.

Because the entire procurement process is a profit generator, these other aspects of procurement are as critical to cost control and profit generating as is purchasing. Without proper controls in the receipt and storage of food and supplies, the careful planning for menu design and purchasing would be partially nullified. Planning and controlling, therefore, are the management functions with direct relevance to receiving, storage, and inventory control.

Purchasing was defined in chapter 7 as an activity concerned with the acquisition of goods. **Receiving** can be defined as an activity for ensuring that products delivered by suppliers are those that were ordered in the purchasing activity. After food and supplies have been received properly, they must be placed in appropriate **stor-**

age, which is the holding of goods under proper conditions to ensure quality until time of use. **Inventory** is a record of material assets owned by an organization, and **inventory control** is the technique of maintaining assets at desired quantity levels. The major objectives of this control are to maintain quality of goods and supplies, to minimize inventory costs, and to ensure that adequate quantity is on hand for production and service needs.

RECEIVING

Receiving has become known as the missing link in the procurement of materials because quite often food quality problems are caused by breakdowns in receiving procedures. One of the problems is related to a different product arriving on the receiving dock than what was ordered (Gunn, 1992). The receiving process involves more than just acceptance of and signing for delivered products. It also includes verifying that quality, size, and quantity meet specifications, that the price on the invoice agrees with the purchase order, and that perishable goods are tagged or marked with the date received. Furthermore, the products received should be recorded accurately on a daily receiving record, then transferred promptly to the appropriate storage or production areas to prevent loss or deterioration. Consistent and routine procedures are essential to the receiving process, along with adequate controls to preserve quality of products and prevent loss during the delivery and receipt of products.

In a foodservice establishment, between 30% and 50% of the revenues are spent on food purchases. In far too many operations, however, the receiving of these valuable purchases is entrusted to any employee who happens to be near the unloading or storage area when shipments from suppliers arrive. When responsibility for receiving is not assigned and procedures are not systematized, a number of problems may arise, such as careless losses, failure to assure quality and quantity of goods delivered, and pilferage. These potential losses can cost a foodservice operation more than its net profit each year.

The economic advantages gained by competitive purchasing, based on complete and thorough specifications, can be lost by poor receiving practices. Properly designed and enforced controls in receiving will ensure management that a dollar value in quantity and quality is received for every dollar spent.

Elements of the Receiving Activity

Elements of good receiving practices include competent personnel with specified responsibilities, proper facilities and equipment, well-written specifications, procedures for critical control, good sanitary practices, adequate supervision, scheduled receiving hours, and procedures to ensure security. Receiving is generated in purchasing through development of specifications and issuance of purchase orders.

Competent Personnel

Responsibility for receiving should be assigned to a specific member of the foodservice staff. In a small operation, this individual may have additional duties, but, if possible, the person responsible for the receiving should not be involved in food purchasing or production. Separating the duties of purchasing and receiving is basic to a check-and-balance system for ensuring adequate control. The owner or foodservice manager should assume the responsibility for either purchasing or receiving if sufficient employees are not available.

In addition to being familiar with food products and quality, receiving personnel must be able to detect old products, excess shrinkage, short weights, and products that do not meet specifications (Keiser & DeMicco, 1993). The receiving clerk should be provided with the specifications and purchase order sent to the supplier as a basis for checking all products delivered.

Employees need to be well trained in the skills and knowledge needed to perform receiving tasks competently. Many of the receiving tasks can be explained and demonstrated through on-the-job training; however, learning how to use specifications for evaluating products is more difficult and may require specific training sessions.

Facilities and Equipment

Adequate space and equipment are necessary to perform receiving tasks properly. In many foodservice operations, the receiving area may serve as an entrance for employees and salespeople, a place for general storage, and a passage to where trash is stored, all of which suggest a need for monitoring and good security procedures.

Ideally, the receiving area should be located near the delivery door, storeroom, refrigerators, and freezers to minimize the time and effort in movement of food into appropriate storage. In small operations, a wide hallway may be used as the receiving area; in larger operations, additional space is needed. In either case, the receiving area should be located near the delivery door for two reasons: union contracts of delivery persons may stipulate "inside-door" delivery only, and security and sanitary concerns arise when persons who are not foodservice employees are permitted in back-of-house areas (Dietary Managers Association, 1991).

In large facilities, a receiving office generally is located near the delivery entrance. Enough space should be available to permit all incoming products to be inspected and checked at one time. In smaller facilities, a desk at the receiving entrance facilitates the check-in of products.

Many products require such minimal inspection as checking the package, label, and quantity on boxes of cereal and bags of flour. Deliveries of other products, such as meat, may require opening packages to inspect quality, count, and weight. Storage should not begin until the delivery personnel have left to eliminate confusion for the receiving clerk. Time and money can be saved by providing facilities that require a minimum of handling of products and permit direct transfer to storage and areas of use.

The size of the receiving area for a specific food facility is influenced by the nature and volume of materials received or being transferred out at any one time. For exam-

Figure 8.1. Motorized vehicles for storeroom work.

ple, a hand truck may be sufficient in some operations, but a forklift may be required for pallet deliveries (Figure 8.1). In commissary operations, the space must be sufficient for the various kinds of carts used for transporting foods to satellite service centers. A typical cart for transporting food from a commissary or base kitchen to a satellite school foodservice operation is shown in Figure 8.2. The receiving area in the satellite service center also must be designed to accommodate delivery carts.

The receiving department requires certain equipment, including scales in good working order. Both platform and counter scales should be available (Figures 8.3 and 8.4); portion scales are useful for checking portion cuts of meat. In large operations, the scales print the weight of the product on the reverse side of the invoice or packing slip or print a tape that can be attached to eliminate any doubt in the weighing-in process. Accuracy of all scales should be checked periodically.

Figure 8.2. Bulk food transporter insulated to keep food hot or cold.

Source: Crescent Metal Products, Inc., Cleveland, OH. Used by permission.

Figure 8.3.   Platform scale.

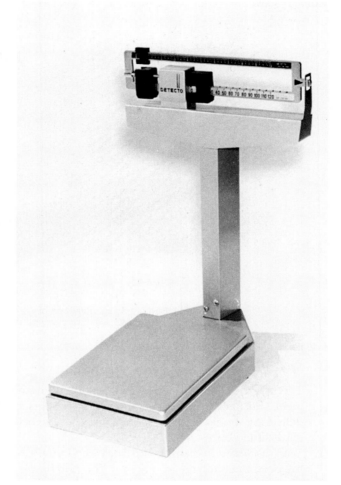

Source: Detecto Scale, a division of Cardinal Scale Manufacturing
Co., Webb City, MO. Used by permission.

An unloading platform of a convenient height for delivery trucks is needed, as well as a ramp to facilitate the unloading of trucks that do not match the platform height. Dollies and hand trucks are important to expedite the movement of products to storage with the least amount of effort.

Other equipment in the receiving area should include a table for inspection of deliveries, and tools, such as a can opener, crowbar, claw hammer, and short-blade knife for opening containers and packages. A thermometer is needed to check if chilled or frozen products are delivered at appropriate temperatures according to specification. Clip boards, pencils, and marking and tagging equipment are also necessary. A file cabinet should be available for storing records and reports as well as a calculator for verifying the computations on the invoice.

Figure 8.4. Counter scale.

Source: Detecto Scale, a division of Cardinal Scale Manufacturing Co., Webb City, MO. Used by permission.

Specifications

The employee who receives orders should know the standards the suppliers must meet and have a notebook or file box of specifications available for reference. All deliveries should be checked against these specifications and nothing below standard should be accepted.

As discussed in chapter 7, a copy of all purchase orders is provided to the receiving personnel in large operations. The purchase order includes a brief specification, data on quantity ordered; pricing information, and general instructions to suppliers such as delivery date and shipping instructions. With this information, receiving personnel are alerted that a delivery is scheduled on or before a certain date and that they must be prepared to count and inspect anticipated shipments.

Critical Control in Receiving

Receiving is a critical control point for many foods, especially perishable ones such as raw meat and vegetables. Setting-up procedures for inspection and standards for acceptance are necessary to prevent foodborne illness. A HACCP flowchart for safe receiving procedures for beef stew ingredients, as shown in chapter 5 (see Table 5.1), should be included in the procedures manual of a foodservice operation for all potentially hazardous foods (Applied Foodservice Sanitation, 1992). Note that the hazard for receiving these foods is contamination and spoilage and the corrective action, if the standard is not met, is rejection of the delivery.

Sanitation

The receiving area should be designed for easy cleaning. The floor should be of material that can be easily scrubbed and rinsed and has adequate drains and a water connection nearby to permit hosing down of the area. Storage for cleaning supplies should be conveniently located.

Because insects tend to congregate near loading docks, adequate screening must be provided. Electrical or chemical devices for destroying insects are often mounted near the outside doors in the delivery area.

Adequate Supervision

The management of a foodservice operation should monitor the receiving area at irregular intervals to check security and ensure that established receiving procedures are being followed. A member of the management staff should recheck weights, quantities, and quality of products received at various times as part of the control system for the foodservice system.

Scheduled Hours

Suppliers should be directed to make deliveries at specified times. This policy reduces the confusion of too many deliveries arriving at one time and ensures that deliveries will not arrive during meals or after receiving personnel are off duty.

In small operations, receiving personnel are frequently assigned other duties, especially during periods around meal service. In many hospitals, for example, a porter may have the responsibility for receiving, for transporting meal carts to patient floors, and for maintenance duties after meal service. When deliveries arrive at inappropriate times, the receiving clerk may be pressured and thus not check in goods properly. Therefore, midmorning and midafternoon deliveries are best in such operations. Another consideration in scheduling deliveries is the nature of the shipment. The receiving schedule in some operations requires that perishable foods be delivered during the morning hours and staple goods during the afternoon.

Security

An owner of an investigation company stated that foodservice managers can reduce internal theft by 60% if employees know that management is watching (Lorenzini, 1992). One practice that should be followed to guard against theft at the time products are received is not to have the same person responsible for purchasing and receiving. Other elements of good receiving, previously discussed, also serve as components of a receiving security system; scheduled hours for receiving and adequate facilities and equipment for performing receiving tasks will prevent many problems. Another important practice is to move products immediately from receiving to storage. In addition, delivery persons generally should not be permitted in storage areas, further protecting the security of the area. Salespeople also should be excluded from the area, but usually for a different reason. Barring them from the storeroom prevents them from checking stock and influencing the purchaser.

Receiving Process

Detailed procedures are important to assure that incoming products are received properly. The steps in the receiving process are outlined in Figure 8.5.

Inspection Against the Purchase Order

A written record of all orders must be kept to provide a basis for checking deliveries. This record, the purchasing order, as was shown in Figure 7.13, becomes the first control in the receiving process by including a brief description of the product, quantity, price, and supplier. In small foodservice operations using informal purchasing procedures, this record may be as simple as a notebook for listing this basic ordering information. In large operations with more formal procedures, however, one copy of the purchase order generally is transmitted to the receiving department as a record of scheduled deliveries.

Incoming shipments should be compared with the purchase order to ensure that the products accepted were in fact ordered. The purchase order also will permit the receiving personnel to determine partial deliveries or omission of ordered products. If a comparison of incoming deliveries with purchase orders indicates that the appropriate products have been delivered, then quality should be checked according to the receiving HACCP flowchart in the procedure manual. In reality, receiving is the initial phase of total quality control in the procurement subsystem.

As stated previously, specialized training of receiving personnel is important to ensure that quality standards are known and can be recognized. All products should be compared with the characteristics in the established specifications. Without exception, the count or weight and quality tolerance levels of all products must be

Figure 8.5.   Steps in the receiving process.

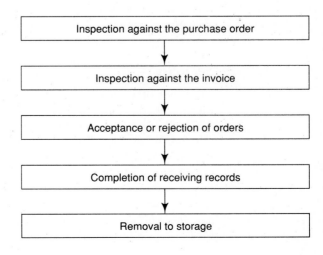

verified before the delivery invoice is signed. When too many products are below the tolerance levels for weight or quality, products should be rejected by the receiving clerk. A printed credit memorandum form may be used; one copy should be attached to the invoice with the discrepancy noted, initialed by the delivery person, and sent to the controller (Keiser & DeMicco, 1993). Another copy should be given to the delivery person to transmit to the supplier, and the third copy should be kept by the receiving department.

For some products, a foodservice operation may have a standing order and, therefore, not maintain a specific purchase order or other record for each delivery. In some operations, an inventory quantity level may have been established for such products as bread, dairy products, coffee, and eggs. Some foodservice operations permit delivery personnel employed by the supplier to check the quantity on hand and deliver enough to restore the inventory level. This practice cannot be recommended for reasons of security and quality control.

Inspection Against the Invoice

After products have been checked against the purchase order and specifications, the delivery should be compared to the invoice prepared by the supplier. The invoice is the supplier's statement of what is being shipped to the foodservice operation and the expected payment.

Obviously, checking the quantities and prices recorded on the invoice is a critical step in the receiving process. Three receiving methods are used in foodservice operations: invoice, blind, and electronic receiving.

- *Invoice receiving.* The receiving clerk checks the quantity of each product against the purchase order. Any discrepancies are noted on both the purchase order and the invoice. This method is quick but can be unreliable if the receiving clerk does not compare the two records and only looks at the delivery invoice.
- *Blind receiving.* The receiving clerk uses an invoice or purchase order with the quantity column blanked out and records on it the quantity of each product received. This method requires that each product be checked because the amount ordered is unknown; it is time consuming for both the receiving clerk and the deliverer and, therefore, expensive in labor costs.
- *Electronic receiving.* Technology is speeding up the receiving process although it is still too expensive for small foodservice operations. Tabulator scales, which weigh and automatically print the weight on paper, are being used in large operations as is the Universal Product Code (UPC). Bar codes are appearing on cartons and packages in foodservice operations. Hand-held scanners that can read a UPC bar speed up the receiving process (Stefanelli, 1992).

If the quantities and prices are correct and the receiving clerk has checked the quality of the products, the invoice should be signed. Generally, two copies of the invoice are required, one for foodservice records and the other for the accounting office. In small operations, only one copy may be required. If errors have been made in the delivery or in the pricing, corrections must be reported on the invoice before it is signed. The delivery person also should initial any correction of errors.

Acceptance or Rejection of Orders

Delivered products become the property of the foodservice operation whenever the purchase order, specifications, and supplier's invoice are in agreement. Payment then will be due at the agreed upon time for products charged on the invoice.

Rejection at the time of delivery is much easier than returning products after they are accepted. If, however, errors are discovered after a delivery has been accepted, the supplier should be contacted immediately. Reputable suppliers generally are willing to correct problems. The foodservice manager should find out why the problem was not detected at the time of delivery and should make changes in receiving procedures. Whether products are returned after acceptance or rejected at the time of delivery, accounting personnel must be sure credit is given by the supplier.

Occasionally, a substandard product may be accepted because it is needed immediately and time does not permit exchanging it or finding an alternate source. The buyer might try to negotiate a price reduction with the supplier. If a product is not available or only a partial amount is delivered, the buyer needs to decide if it should be back ordered.

Completion of Receiving Records

The receiving record provides an accurate list of all deliveries of food and supplies, date of delivery, supplier's name, quantity, and price data. This information is helpful in verifying and paying invoices and provides an important record for cost control of all foods delivered to the kitchen and storeroom.

An example of a receiving record used in a community hospital is shown in Figure 8.6. The columns on the form that show distribution of products to the kitchen or storage are useful in some food cost-control systems. Products delivered directly to the kitchen will be included in that day's food cost, whereas products sent to the storeroom will be charged by requisition when removed from stores for production and service.

This kind of receiving record documents the transfer of products to storage in facilities, in which receiving and storage tasks are performed by different employees. The receiving record provides a checkpoint in the control system. This record usually is prepared in duplicate, with one copy sent with the invoices to the accounting department and the other retained in the receiving department. If the receiving clerk is responsible for verifying price extensions on invoices, this should be done before the invoice is forwarded to the accounting office. Accounting personnel, however, also should verify the arithmetic extensions on invoices.

Removal to Storage

Products should be transferred immediately from receiving to the secure storage area; personnel should not be permitted to wait until they have time to move food

Figure 8.6.   Example of a receiving record.

COMMUNITY HOSPITAL
RECEIVING RECORD

Date: August 26, 1994

Quantity	Unit	Description of Item	Name of Supplier	Inspected and Quantity Verified by	Unit Price	Total Cost	Distribution	
							To Kitchen	To Storage
3	lugs	Tomatoes	L. & M.	JH	$10.75	$32.25		X
3	crates	Lettuce	L. & M.	JH	11.70	35.10		X
4	boxes	Bakers	L. & M.	JH	9.07	36.28	X	
1	sack	Onions	L. & M.	JH	10.00	10.00		X
400	lbs.	Watermelon	L. & M.	JH	.155	62.00	X	

and supplies to storage. Because the products are now the property of the foodservice operation, security measures are important to prevent theft and pilferage. Also, spoilage and deterioration may occur if refrigerated and frozen products are held at room temperature for any period of time.

Foodservice operations may indicate on the receiving record various procedures for marking or tagging products for storage. *Marking* consists of writing information about delivery date and price directly on the case, can, or bottle before it is placed in storage. Daily food cost calculations can be done quickly because prices do not have to be looked up, and fewer products will spoil on the shelves because the ones with the oldest dates will be used first. *Tagging* products also facilitates stock rotation to ensure that older products are used first, which is particularly important with perishable food products such as expensive meat and fish. Data such as date of receipt, name of supplier, brief description of product, weight or count upon receipt, and place of storage may be written on the tag.

STORAGE

Storage is important to the overall operation of a foodservice because it links receiving and production (Figure 8.7). Dry and low-temperature storage facilities should be accessible to both receiving and food production areas to reduce transport time and corresponding labor costs. Proper storage maintenance, temperature control, and cleaning and sanitation are major considerations in ensuring quality of stored foods. Competent employees are as essential for storage positions as they are for all other positions in a foodservice operation. Storage employees check in products from the receiving unit, provide security for products, and establish good material-handling procedures. Only those personnel authorized to store goods, issue foods, check inventory levels, or clean the areas should have access to storage facilities.

Prevention of theft and pilferage in storage facilities is a major concern. The sad commentary is that, once again, the customer pays for the problem. Stefanelli (1992, p. 256) defines **theft** as "premeditated burglary" and **pilferage** as "inventory shrinkage." Theft occurs when someone drives a truck up to the back door and steals expensive products and equipment. Pilferage focuses on the employee who steals a couple of steaks before checking out. In some foodservice operations, locks for storage areas are replaced periodically to prevent entry with unauthorized duplicate keys; only authorized persons should have access to keys.

Figure 8.7. Food flowchart.

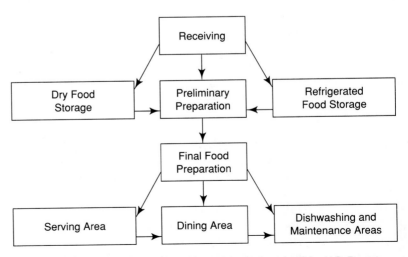

Source: Food Storage Guide for Schools and Institutions (p. 2) by U.S. Department of Agriculture, Food and Nutrition Service, 1975, Washington, DC: Government Printing Office.

Ideally, storage should be located on the same floor as the production area and be visible to the foodservice manager, but this is not always possible. In large operations, the main storage areas often are located on a lower level or in another area of the building, making surveillance by the manager difficult. In institutions such as a large hospital, dry storage for food and supplies may be in the central storeroom for the entire hospital, and security responsibility is assumed by the centralized purchasing department. In multiunit restaurant operations, a central warehouse usually is kept for major storage, and a limited storeroom and small refrigerators and freezers for a few days' supply is provided in the individual restaurant operations.

The type of storage facilities will vary greatly, depending on the type of foodservice (refer to chapter 5 for discussion of types of foodservices). In an assembly/serve foodservice, in which many fully processed frozen food products are used, the need for frozen storage may be much greater than in a conventional operation. In a commissary foodservice, the central production area demands large-storage capability.

Regardless of the type of foodservice, all foods should be placed in storage as soon as possible after delivery, unless they are to be processed immediately. Dry groceries, canned foods, and staples should be placed in dry stores. Perishable foods must be placed in refrigerated or frozen storage promptly. In addition, storage facilities must be available for china, flatware, trays, utensils, and nonfood products, such as paper supplies, detergent, and cleaning products.

In some storage areas, fast-moving products are placed near the entrance and slower-moving ones are stored in less accessible locations. Generally, however, products are categorized into groups, then arranged either alphabetically or according to frequency of use within the groups. Foods that give off odors should be stored separately. New stock should be placed in back of older stock to prevent loss from deterioration. Foods should be checked periodically for evidence of spoilage, such as bulging or leaking cans. Ideally, canned foods should not be kept more than 6 months even when stored under proper conditions (Thorner & Manning, 1983), because it is difficult to determine the length of time foods have been canned prior to delivery.

The advantage of purchasing large quantities to save money must be weighed against the possibility of storage loss and the costs of maintaining inventories. Inventory costs are discussed in more detail later in this chapter.

Dry Storage

The **dry storage** area for food provides orderly storage for foods not requiring refrigeration or freezing and should provide protection of foods from the elements, insects, and rodents as well as theft. Cleaning supplies and pesticides should not be kept in the food storeroom. Instead, a separate locked room should be provided to prevent such items from contaminating food products.

Facilities

The floors in the dry storage area should be slip resistant and easy to clean. Walls and ceilings should be painted light colors and have smooth surfaces that are impervious to moisture and are easy to wash and repair. The number of doors allowing access to the storeroom should be limited. The main door should be heavy-duty and of sufficient width for passage of the equipment used to transport foods from receiving to dry storage or from dry storage to production areas. This door should lock from the outside; however, a turnbolt lock or crash bar should be provided to permit opening from the inside without a key in case people are inadvertently locked in. A Dutch door often is used as the main entrance; the lower half is locked at all times except when accepting large shipments, and the upper door is open during issuing hours.

In large operations, a recording time lock may be used on the door to the storeroom. When the storeroom is opened at an unscheduled time, it is recorded by this type of locking system; the tape from the time lock should go to the controller each day for review.

Storeroom windows should be opaque to protect foods from direct sunlight. For security, most storerooms are designed without windows or have windows protected with grates or security bars. Adequate lighting is needed in storage areas, not only for sanitation purposes but also to prevent theft. Closed circuit television systems are being used more and more in storage areas.

Good ventilation in the dry stores area is essential to assist in controlling temperature and humidity and preventing musty odors. A thermometer should be mounted in an open area for easy reading. Dry storage temperature should be cool, within a range of 50°F to 70°F. Spices, nuts, and raisins, which should be stored in temperatures not over 52°F, quite often are kept in a refrigerator. Humidity often is overlooked in storerooms. For most food products, a relative humidity of 50% to 60% is considered satisfactory. Humidity above this level may result in rusting cans, caking of dry and dehydrated products, growth of bacteria and mold, and infestation of insects in the storeroom (Longrée & Armbruster, 1987).

Dry storerooms frequently are located in a basement and, therefore, have all sorts of pipes along the ceiling, such as water, heating, and sewer pipes. Leakage from these pipes is a common source of trouble in basement storerooms, especially leakage from sewer pipes that causes contamination of food. Hot water and steam pipes may create high temperatures in the storeroom and, therefore, should be well insulated (Longrée & Armbruster, 1987).

Equipment

Sectional slatted platforms, delivery pallets, and metal platforms with wheels provide a useful type of storage for case lots of canned goods or for products in bags. Their distance from the floor must be in accordance with local health department requirements. Shelving is required whenever less-than-case lots must be stored or when management prefers that canned goods be removed from cases for storage. Adjustable metal shelving is desirable because it allows for various shelf heights and is verminproof (Figure 8.8). Shelving must be sturdy enough to support heavy loads without sagging or collapsing, and it should be located at least 2 inches from walls to provide ventilation. If the size and shape of the room permit, shelving should be arranged for accessibility from both sides.

High-density shelving (Figure 8.9) has been designed that permits maximum use of storeroom space. The shelf units are installed in floor runners with shelves that can be moved electrically or mechanically when access to products is needed.

Metal or plastic containers with tight-fitting covers should be used for storing cereal products, flour, sugar, dried foods, and broken lots of bulk foods. These containers should either be placed on dollies or have built-in wheels for ease of movement from one place to another (Figure 8.10).

Figure 8.8.   Metal shelving for storage area.

Source: Crescent Metal Products, Inc., Cleveland, OH. Used by permission.

Figure 8.9.   High-density storage shelves.

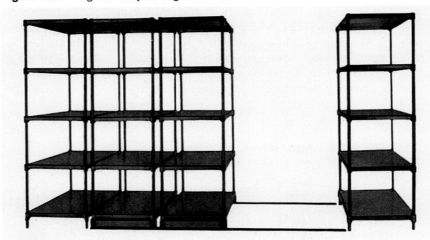

Source: Market Forge, Everett, Massachusetts. Used by permission.

Aisles between shelves and platforms should be wide enough for equipment with wheels. Forklift trucks for moving food and supplies or pallets require much wider aisles than do handcarts.

Products should be arranged in the storeroom according to a plan, and every product should be assigned a definite place. Time can be saved if forms for checking inventory are designed to match the arrangement of products on the shelves.

Low-Temperature Storage

Perishable foods should be held in refrigerated or frozen storage for preservation of quality and nutritive value immediately after delivery and until use. The type and amount of low-temperature storage space required in a foodservice operation will vary with the menu and purchasing policies. An excessive amount of refrigeration and frozen storage increases capital costs and operating expenses. Too much also encourages a tendency to allow leftovers to accumulate and spoil. Astute foodservice managers have found that limiting the amount of low-temperature storage discourages excess purchases and overproduction.

Facilities

In very large organizations, especially commissaries, separate refrigerated units are available for meats and poultry, fish and shellfish, dairy products and eggs, and fruits and vegetables. Freezers are available for frozen foods and ice cream; also, tempering boxes are used for thawing frozen products. Ideal storage temperatures vary among food groups, and the more precise the temperature, the better the quality of the products. Table 8.1 provides a detailed list of recommended storage temperatures and maximum storage times according to type of food, most of which has little processing.

Low-temperature storage units can be categorized into the following three types:

- *Refrigerators.* Storage units designed to hold the temperature for storing meat and poultry at 30°–36°F, fish and shellfish at 30°–32°F, dairy products and eggs at 36°–40°F, and fruits and vegetables at 40°–45°F.
- *Tempering boxes.* Separate units for thawing frozen foods, specially designed to maintain a steady temperature of 40°F regardless of room temperature or product load.
- *Storage freezers.* Low-temperature units for frozen foods that maintain a constant temperature in the range of −10°–0°F.

In many foodservice operations, especially small ones, satisfactory storage of various products can be maintained with fewer units kept at the following temperatures:

- Dairy products, eggs, meat, and poultry at 32°–40°F
- Fresh fruits and vegetables at 40°–45°F
- Frozen foods at −10°–0°F

Figure 8.10. Dollies and containers for use in storage facilities.

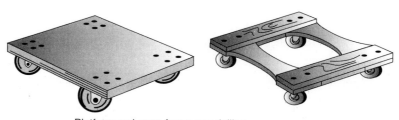

Platform and open-frame can dollies

Container on dolly

Container equipped with casters

Source: From *Food Storage Guide for Schools and Institutions* (p. 11) by U.S. Department of Agriculture, Food and Nutrition Service, 1975, Washington, DC: Government Printing Office.

Table 8.1.   Recommended storage temperatures and times

Food	Refrigerator Storage (32°F–40°F [0°C–4°C])	Freezer Storage (0°F [–18°C] or below)	Dry Storage (30°F–70°F [10°C–21°C])
Roasts, steaks, chops	3–5 days	Beef and lamb: 6 months Pork: 4 months Veal: 4 months Sausage, ham, slab bacon: 2 months Beef liver: 3 months Pork liver: 1–2 months	Never
Ground meat, stew meat	1–2 days	3–4 months	Never
Ham, bake whole	1–3 weeks	4–6 months	Never
Hams, canned	12 months	Not recommended	Never
Chicken and turkey	2–3 days	Chicken: 6–12 months Turkey: 3–6 months Giblets: 2–3 months	Never
Fish or shellfish	30°F–32°F (–1°C–0°C) on ice, 2–3 days	3–6 months	Never
Shell eggs	1–2 weeks	Not recommended	Never
Frozen eggs	1–2 days after thawing	9 months	Never
Dried eggs	6 months	Not recommended	Never
Fresh fruits and vegetables	5–7 days	Not recommended	Never
Frozen fruits and vegetables	—	Variable, depends on kind	Never
Canned fruits and vegetables	—	Not recommended	12 months
Dried fruits and vegetables	Preferred	Not recommended	2 weeks
Canned fruit and vegetable juice	—	—	Satisfactory
Regular cornmeal	Required over 60 days	Not recommended	2 months
Whole wheat flour	Required over 60 days	Not recommended	2 months
Degermed cornmeal	Preferred	Not recommended	Satisfactory
All-purpose and bread flour	Preferred	Not recommended	Satisfactory
Rice	Preferred	Not recommended	Satisfactory

Source: Reproduced with permission from *Food Service Manual for Health Care Institutions, 1994 Edition*, published by American Hospital Publishing, Inc., copyright 1994.

Many times, only one refrigerator is in the facility, and everything but frozen foods is kept in it. In cases like this, either the lettuce and tomatoes will freeze or the dairy products, eggs, meat, and poultry will deteriorate quickly. The quality of products certainly will diminish.

Humidity control also is important for maintaining food quality in low-temperature storage. A humidity range between 75% and 95% is recommended for most foods. Perishable foods, however, contain a great deal of moisture; therefore, a relative humidity of 85% to 95% is recommended. If humidity is not sufficient, evaporation will cause deterioration such as wilting, discoloration, and shrinking.

Low-temperature storage units are designed as walk-in or reach-in refrigerators or freezers. In large operations, walk-in units generally are located in the storage area with separate reach-in units in the production and service areas for dairy products, salad ingredients, or desserts. In small operations, the trend is away from walk-ins and toward reach-ins because less floor space is required and the capital investment generally is less.

Hard-surface, easily cleaned floors, walls, and shelves should be made of smooth, nonabsorbent material. The floor level should be the same as that of the area in which the walk-in is located to permit carts to be rolled in. As with dry storage, cartons of food products should not be placed on the floor. Drains are needed for removal of scrubbing water and condensate, and they should be located inside the low-temperature storage units. Uniform ventilation and adequate lighting are essential in the unit to maintain sanitary conditions. All low-temperature storage units should be cleaned on a regularly scheduled basis according to manufacturers' instructions. Most refrigerators and freezers in foodservice operations today are self-defrosting; if not, periodic defrosting of these units must be scheduled.

All refrigerators and freezers should be provided with one or more of the following kinds of thermometers:

- *Remote reading thermometer.* Placed outside the unit to permit reading the temperature without opening the door
- *Recording thermometer.* Mounted outside the unit and continuously records temperatures in the unit
- *Bulb thermometer.* Mounted or hung on a shelf in the warmest area inside the unit

Temperatures in all units should be checked at least twice a day. An employee should be assigned to check and write down temperatures at specified times as a control measure. Some foodservice operations have an alarm or buzzer that is activated when temperatures rise above a certain level. Employees should be trained to open refrigerator doors as infrequently as possible; obtaining all foods needed at one time keeps temperatures down while conserving energy. If temporary power failures occur, refrigerators and freezers should be opened as seldom as possible.

In both low-temperature and dry storage, foods that absorb odors must be stored away from those that give off odors. Typical foods that emit and absorb odors are listed in Table 8.2.

Frozen food should be wrapped in moistureproof or vaporproof material to prevent freezer burn and loss of moisture. The original packages of most frozen foods

Table 8.2. Foods that give off and/or absorb odors

Food	Gives off Odors	Absorbs Odors
Apples, fresh	Yes	Yes
Butter	No	Yes
Cabbage	Yes	No
Cheese	Yes	Yes
Cornmeal	No	Yes
Eggs, dried	No	Yes
Eggs, fresh shell	No	Yes
Flour	No	Yes
Milk, nonfat dry	No	Yes
Onions	Yes	No
Peaches, fresh	Yes	No
Potatoes	Yes	No
Rice	No	Yes

Source: From *Food Storage Guide for Schools and Institutions* (p. 27) by U.S. Department of Agriculture, Food and Nutrition Service, 1975, Washington, DC: Government Printing Office.

are designed with this in mind. For fresh foods that are to be held in storage, specifications may include special instructions for frozen storage wrapping.

As discussed earlier, precooked frozen foods are being used more frequently in foodservice operations. Because the shelf life varies widely, the quality and stability of these foods require special attention. The list in Figure 8.11 should be useful as a general guideline for storage of precooked frozen foods.

Critical Control in Storage

Storage is the second critical control point for beef stew. The microbiological safety of raw beef and fresh vegetables while in storage before production is critical. The HACCP flowchart for safe storage procedures, as shown in chapter 5 (see Table 5.1), also should be included in the foodservice procedure manual because the products are potentially hazardous foods (Applied Sanitation Manual, 1992). The hazards for storing raw beef include cross-contamination of other foods and bacterial growth and spoilage; for storing vegetables, cross-contamination from raw, potentially hazardous foods is a major concern. The final corrective action, if the standards are not met, is to discard the product.

INVENTORY

As previously defined, *inventory* is a record of material assets owned by an organization. Inventory is supported by the actual presence of products in the storage areas. Materials held in storage represent a significant investment of the organization's assets. Although the monetary value of food and supplies in storage will be clear to management, employees may not always grasp this concept. Foodservice employees who would not steal money from the cash register may see nothing wrong in taking products from storage now and then. They may also not see as money down the drain the losses that result from failure to store foods promptly or rotate stock.

Figure 8.11.   Stability of foods under frozen storage.

Foods with Short Storage Life (2 weeks to 2 months)

Product	Maximum storage Life at 0°F (−17.8°C)
Bacon, Canadian	2 weeks
Batter, gingerbread	3–4 months
Batter, muffin	2 weeks
Batter, spice	1–2 months
Biscuit, baking powder	1–2 months
Bologna, sliced	2 weeks
Cake, sponge, egg yolk	2 months
Cake, spice	2 months
Dough, roll	1–2 months
Frankfurters	2 weeks
Gravy	2 weeks
Ham, sliced	2 weeks
Poultry giblets	2 months
Poultry livers	2 months
Sauce, white (wheat flour base)	2 weeks
Sausage	2 months

Foods with Medium Storage Life (6 to 8 months)

Chicken, fried	Meals on a tray	Pies, meat
Crab	Meat balls	Potatoes, French-fried
Fish, fatty	Meat loaf	Soups
Fruit, purees	Pies, chicken	Shrimp
Ham, baked, whole	Pies, fruit, unbaked	Turkey
Lobster		

Foods with Long Storage Life (12 months or longer)

Applesauce	Candies	Peanuts
Apples, baked	Cherries	Pecans
Bread	Chicken, creamed	Plums
Bread (rolls)	(waxy rice flour-based)	Stew, beef
Blackberries	Chicken à la king	Stew, veal
Blueberries	Cookies	Waffles
Cake, fruit	Fish, lean	

Source: Adapted from pages 84, 85, 86, *Quality Control in Food Service,* Revised Ed. Thorner/Manning, The AVI Publishing Company, Inc., Westport, Connecticut, 1983.

Foods, beverages, and supplies in storage areas must be considered as valuable resources of the operation and treated accordingly. For inventory control to be effective, access to storage areas should be restricted, authorized requisitions for removing goods from storage required, and inventory levels monitored. The inventory control system also requires maintenance of accurate records. Improvements in management effectiveness for inventory control require a plan and timely measures of performance. Issuing products, conducting inventories, and controlling methods all need some type of record keeping.

Issuing Products

Issuing is the process used to supply food to production units after it has been received. Products may be issued directly from the receiving area, especially if they are planned for that day's menu, but, more often, food and supplies are issued from dry or low-temperature storage. The issuing process entails control of food and supplies removed from storage and provides information for food cost accounting and, in some cases, information for a perpetual inventory system (Keiser & DeMicco, 1993).

Direct Issues

Products sent directly from receiving to production without going through storage are usually referred to as *direct purchases* or *direct issues.* To have accurate food cost information and better control, direct issues should be limited to food that actually will be used on the day it was delivered (Keiser & DeMicco, 1993). If the food is not used the same day, the recorded food cost will be unrealistically high on the day of delivery, and the food cost on the day of actual use of the product will be unrealistically low.

Storeroom Issues

Foods that are received but not used the day they are purchased are identified as storeroom issues; these products are issued from a storage area when needed for production or service. Control of issuing from storage has two important aspects. First, goods should not be removed from the storeroom without proper authorization and, second, only the required quantity for production and service should be issued. A requisition is needed to provide these two controls.

The cook or ingredient room supervisor requisitions from the storeroom the desired products and quantity and size for each according to a recipe. The ingredient room concept is discussed in detail in chapter 10. At the time of issue, the storeroom employee completes a daily issue record form, shown in Figure 8.12, that contains the daily issue number, issuing storeroom identification, and date of issue. As on the requisition form (Figure 8.6), the columns consist of the requisition number, quantity, unit, and product description. In addition, columns for unit receiving the issue, unit price and total price of each product, and identification of issuing storeroom employee are included on the daily record form.

In foodservice operations that have an inventory control person, the production or ingredient room requisition would be sent to that person, who would send it to

Figure 8.12.   Daily issue record.

Issue Number: _____92_____

Issuing Storeroom: ___Dry Stores #1___   *Date of Issue:* ___9/14/94___

Req. No.	Quantity	Unit	Description of Item	Issued to	Unit Price	Total Price	Issued by
823	10	#10 cans	Tomatoes, diced	Cook's unit	$ 2.31	$23.10	AV
	1	1½# box	Oregano, dried leaf		10.95	10.95	
	1	3# box	Dried, minced onion		4.93	4.93	
	1	1# box	Dried, diced green pepper		7.24	7.24	
	1	20# box	Spaghetti, long thin		10.75	10.75	
	1	1-gal. bag	Catsup		3.18	3.18	
	8	2-oz. can	Bay leaves		3.59	26.32	
	1	11-oz. can	Thyme, grd.		4.99	4.99	
	1	1 gal.	Worcestershire sauce		4.38	4.38	
	1	26-oz. box	Salt, iodized		.26	.26	
	3	#10 can	Tomato puree		2.07	6.21	

the storeroom employee after checking availability of the product in inventory. If a computer-assisted inventory system is used, an inventory number for each product is required. Costs would not have to be entered, because these data are available from the stored information in the computer. The storeroom requisition may be generated by the computer. The person responsible for ordering food and supplies would only need to review the computerized list of issues and make needed adjustments. An example of a computer-generated storeroom order form is shown in Figure 8.13.

If an ingredient control room has been established, the employees in this unit are responsible for requisitioning supplies from various storage and prepreparation areas, weighing and measuring ingredients for menu items, and providing the needed products to production employees on a scheduled basis. A limited inventory may be maintained in the ingredient room, usually the balance of a product not needed that day, such as the remainder of a bag of flour or sugar.

Inventory Records

Placing products in storage, taking them out when needed, and ordering more when necessary are inadequate for control of valuable resources. Inventory control records

Figure 8.13. Computer-generated order form.

018-HOU-231-01 2.01 83024 ORDER REPORT
DECEMBER 3, 1993 AT 2.23 PM

1281-DERBY FOOD CENTER 010-9
MEAT - ORDER GROUP

01/05/94 - PITTMAN DATE 01/10/94 - DELIVERY DATE MONDAY USAGE PERIOD WED-THU

STORAGE LOCATION	CODE	ITEM NAME	BRAND	BIN QT.	ORDER AMOUNT (IN ORDER UNITS)	DELIVERED AMOUNT	NEEDED AMOUNTS LBS.	NEEDED AMOUNTS ORDER UNITS	SUGG ORDER UNITS	INVENTORY ON HAND
41000	0021010102	EGGS WHOLE FRESH			2 CS		41.70	1.2 CASE	2	
43000	0012110141	BEEF FAJITA MEAT			67 LB		67.00	67.0 POUND	67	
43000	0012113701	GROUND BEEF BULK			206 LB		206.00	206.0 POUND	206	
43000	0012113751	PATTIES GROUND BEEF 5			12 BOX		192.00	12.0 BOX	12	
43000	0012117820	STEAK KC STRIP LOIN BONELESS 8OZ			20 LB		20.00	20.0 POUND	20	
46000	0012149608	PORK STRIPS JULIENNE			34 LB		33.60	33.6 POUND	34	
47000	0013145398	BACON SLICED LAYOUT			30 LB		30.00	30.0 POUND	30	
47000	0013186108	DELI HAM			48 LB		48.00	48.0 POUND	48	
47500	0013113801	DRIED BEEF			10 LB		9.20	9.2 POUND	10	
47500	0013186001	WIENERS BULK			30 LB		30.00	30.0 POUND	30	
48000	0022001000	CHEDDAR CHEESE SHREDDED			47 LB		46.40	46.4 POUND	47	
48000	0022001107	AMERICAN PROCESSED CHEESE LOAF			9 LB		8.10	8.1 POUND	9	
48000	0022001115	AMERICAN PROCESSED CHEESE RIBBON SLI			126 LB		126.00	126.0 POUND	126	
48000	0022001166	CREAM CHEESE			4 LB		3.10	3.1 POUND	4	
48000	0022001409	PARMESAN CHEESE GRATED			8 LB		7.20	7.2 POUND	8	
48000	0022001506	MOZZARELLA CHEESE SHREDDED			42 LB		42.00	42.0 POUND	42	
48000	0022001646	MONTEREY JACK CHEESE			25 LB		24.30	24.3 POUND	25	
49000	0013165852	CHICKEN ASSORTED COUNTRY FRIED			28 CS		500.00	27.8 CASE	28	
49000	0013168002	TURKEY, STAY FRESH BREAST			32 LB		32.00	32.0 POUND	32	
49000	0013168100	TURKEY BREAST SMOKED			0 LB		0.00	0.0 POUND	0	
49000	0013168223	TURKEY ROLLED RAW (RAW ROAST)			29 LB		28.30	28.3 POUND	29	
49500	0013152726	SHRIMP BREADED SMALL			175 LB		175.00	175.0 POUND	175	

Source: Kansas State University Housing and Dining Services. Used by permission.

must include adequate procedures to permit the foodservice manager to have up-to-date and reliable data on costs of operation. Inventory records have four basic objectives:

- Provision of accurate information of food and supplies in stock
- Determination of purchasing needs
- Provision of data for food cost control
- Prevention of theft and pilferage

Issuing procedures are only one component of inventory records. A periodic physical count of food and supplies in storage is another requisite element of any inventory control system. More sophisticated records, such as a perpetual inventory, are maintained in many operations to assist in achieving the objectives.

Physical Inventory

A **physical inventory** is the periodic actual counting and recording of products in stock in all storage areas. Usually, inventories are taken at the end of each month. In large operations, a complete inventory rarely is taken at one time. Instead, inventories often are taken by storage areas, or a section of one, each week with all areas covered by the end of a month.

The process involves two people, one of whom, as a control measure, is not directly involved with storeroom operations. One person counts the products, which should be arranged systematically, and the other person records the data on a physical inventory form, an example of which is shown in Figure 8.14. This form is designed to match the physical arrangement of products on the shelves, thereby greatly facilitating the actual physical count. Space should be included to record quantity in stock, unit size, name of food item, item description, unit cost, and total value of the inventory on hand.

As discussed in receiving, in some operations, the price is marked on a good before it is stored. The unit cost is recorded at the time of the physical count; otherwise, the pricing of the inventory will be completed by the bookkeeper. If the physical inventory is computerized, employees taking the inventory only need to record the amount in stock and enter it into the computer for calculating the beginning inventory value.

In operations in which a monthly inventory is taken, food costs can be determined in one of two ways. The simplest method is to calculate cost of food in the following manner:

> **Beginning inventory**
> $+$ Purchases
> $=$ Cost of food available
> $-$ Ending inventory
> $=$ Cost of food used

Figure 8.14. Physical inventory form.

Quantity on Hand	Unit Size	Food Item	Item Description	Unit Cost	Total Inventory Value
	#10	Asparagus	All green cuts and tips, 6/#10/case		
	#10	Beans, green	Cut, 6/#10/case		
	#10	Beans, lima	Fresh green, small, 6/#10/case		
	#10	Beets	Whole, 6/#10/case		
	#10	Carrots	Sliced, medium, 6/#10/case		
	#10	Corn	Whole kernel, 6/#10/case		
	#10	Peas	Sweet peas, 4 sv., 6/#10/case		
	#10	Potatoes, sweet	Whole, 6/#10/case		
	#10	Tomatoes	Whole peeled, juice packed, 6/#10/case		
	46 oz.	Tomato juice	Fancy, 12/46 oz./case		

PHYSICAL INVENTORY

Date _____　　　Taken by _____　　　Beginning Inventory $_____

For inventories done as infrequently as two or three months, daily food cost is determined by computing the cost of the direct and requisition issues, as shown in Figure 8.15. Direct issues for milk, bread, and produce often are delivered by the supplier directly to the using units. Requisition issues are made only on requisitions from using units and are taken from inventory. The monthly food cost is then calculated by adding the daily food costs for the entire month. The lower part of the daily expense worksheet has space for entry of totals of the direct and requisition issues, total food today, and total food to date. Totals also are recorded for nonfood today and to date. All entries on this form are in dollars.

As a check, the cost of food used may be found by adding the cost of purchases to the beginning inventory cost that yields the cost of food available. Subtracting the

Figure 8.15. Example of a daily expense worksheet.

Date _____

Direct Issues ($)					Requisition Issues ($)				Nonfood ($)	
Dairy	Bakery									
Milk	Other	Bread	Other	Produce	Meat/Fish & Poultry	Frozen	Groceries		Linen	Supplies

TOTAL DIRECT ISSUES _____

TOTAL REQUISITION ISSUES _____

TOTAL FOOD TODAY _____

TOTAL FOOD TO DATE _____

TOTAL NONFOOD TODAY _____

TOTAL NONFOOD TO DATE _____

Figure 8.16. Reconciliation of food cost records.

Beginning inventory	$ 2,000
+ Purchases	+ 8,000
= Cost of food available	= $10,000
− Ending inventory	− 1,925
= Cost of food used	= $ 8,075
− Total direct and requisition issues	− 7,850
= Difference	= $ 225

cost of ending inventory yields the cost of food used. An example of reconciliation of food cost records is shown in Figure 8.16, in which total direct and requisition issues are subtracted from the cost of food used. A zero difference would be quite unlikely. If the difference appears to be significantly large, the foodservice manager will need to analyze the reasons for the discrepancy and implement tighter controls over products in storage.

Perpetual Inventory

A **perpetual inventory** keeps purchases and issues continuously recorded for each product in storage so that the balance in stock is always available (Keiser & DeMicco, 1993). The result is a record of the quantity in stock at any given time as well as the value of food and supplies. An example of a perpetual inventory record is shown in Figure 8.17.

If a perpetual inventory record is used, it is generally restricted to products in dry and frozen storage. Produce, milk, bread, and other fast-moving products usually are not kept on perpetual inventory but are considered to be direct issues. Fresh meats, fish, and poultry delivered on the day of use or 1 or 2 days in advance are not recorded on this record but are charged to the food cost for the day on which they are used. If large quantities of frozen meat are purchased at one time and stored until needed, however, these products will be included on perpetual inventory records. This method of purchasing is a modification of the **just-in-time (JIT)** manufacturing philosophy introduced into the United States by the Japanese in the late 1970s.

JIT actually is a philosophy and strategy that has effects on inventory control, purchasing, and suppliers (Heinritz, Farrell, Guinipero, & Kolchin, 1991). The objective is "to draw material through the production and distribution system on an 'as needed' basis rather than on a forced feed flow driven by an order quantity" (Dion, Banting, Picard, & Blenkhorn, 1992, p. 33). In foodservice operations, the objective of **just-in-time purchasing** is to purchase products just in time for production and immediate consumption by the customer without having to record it in inventory.

Once a product is put into inventory, capital is tied up. If, however, the product is considered part of daily food cost, the money is not "sitting" on storeroom shelves making no interest. Buyers must have a good relationship with local suppliers to keep transportation costs low and delivery time short. Serving high-quality perishable products at the lowest possible cost is a concern for foodservice operators, and using JIT concepts might help.

Perpetual inventories require considerable labor to maintain and are used only in large operations that keep a large quantity of products in stock. The increasing use of computers, however, makes maintaining a perpetual inventory record much easier. After the computerized inventory control system has been established, the perpetual inventory record can be kept up-to-date very simply by recording issues from the storeroom on a daily basis. However it is done, a perpetual inventory record is not sufficient for accurate accounting and control of food and supplies. A physical inventory should be conducted on a monthly basis for verification.

Figure 8.17.   Perpetual inventory record.

Item: Tomatoes, diced   *Purchase Unit:* 6/10 cs   *Issue Unit:* #10 can

Issuing Storeroom: Dry Stores #1

Date	Order No. or Requisition No.	Quantity In (purchase unit)	Quantity Out (issue unit)	Quantity on Hand (issue unit)
8/23/94	PO 1842	20 cs		120 cans
8/27/94	R 823		10 cans	110 cans

Inventory Control Tools

With increased pressures for cost containment in all foodservice operations, the need for inventory control has become more important. The major functions of a control system are to coordinate activities, influence decisions and actions, and assure that objectives are being met. In brief, these functions ensure that adequate products are available for production and service and minimize inventory investment (Flamholtz, 1979). Excess inventory is a nonpaying investment.

In foodservice operations, the activities to be coordinated and integrated are those in the procurement, production, and distribution and service subsystems. For example, if chili is a menu item and canned tomatoes are not available in the inventory, production will be disrupted and, in turn, service will be affected.

The second major function of inventory control, decision making, can be made somewhat routine through the use of computers. The role of the manager as the one responsible for establishing policies and procedures and for monitoring operations to ensure that plans are being implemented appropriately is still paramount, however.

Using tomatoes as an example again, the computer may provide information on when and how many cases to order. The established policies and procedures may detail the processes for ordering, receiving, and storing products. Management, however, still must decide issues such as whether or not to take advantage of forward buying if a tomato crop failure is anticipated. An essential factor in this control system is feedback from the production and service units to procurement, the unit responsible for inventory control.

With the increasing size and complexity of foodservice operations, inventory management and control have become more complicated and critical. A variety of tools is available to assist managers in determining quantities for purchase, inventory levels, and cost of maintaining inventories. Several of these techniques are discussed.

ABC Method

In most foodservice operations, a small number of products account for the major portion of the inventory value; therefore, a method for classifying products according to value is needed. This method is generally referred to as the *ABC analysis* and is shown in Figure 8.18.

The principle of the **ABC inventory method** is that effort, time, and money for inventory control should be allocated among products according to their value. Products should be divided into three groups, as shown in the figure, with the high-value, *A*, and medium-value, *B*, products given priority.

The high-value *A* products represent only 15% to 20% of the inventory but typically account for 75% to 80% of the value of total inventory. The inventory level of these expensive products, for example frozen lobster tails, should be maintained at an absolute minimum.

The medium-value *B* products represent between 10% and 15% of total inventory value and 20% to 25% of the products in inventory. *C* products are those whose dollar value accounts for 5% to 10% of the inventory value but make up 60% to 65% of

Figure 8.18.   ABC analysis.

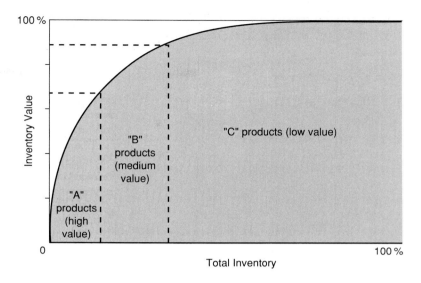

the inventory. Less concern, obviously, should be directed to the proportionally lower value products (Zenz, 1981).

In applying the ABC concept to a foodservice operation, an analysis of inventory products and classification of them into the three categories would assist the food-service manager in deciding the amount of time and effort that should be spent in controlling inventory. In a restaurant with both food and beverage operations, liquor would be classified as an *A* product and controlled very closely; sugar packets would be classified as *C* products and monitored less closely.

Minimum-Maximum Method

An often-used method for controlling inventory involves the establishment of minimum and maximum inventory levels, commonly called the **mini-max method**. Theoretically, the minimum inventory level could be zero if the last product were used as a new shipment arrived. The maximum inventory then would be the correct ordering quantity.

In reality, however, this extreme policy is not practical because it would likely involve running out of products at a critical time. In the mini-max method, a safety stock, which is maintained at a constant level both on the inventory record and in the storerooms, is established. The maximum inventory level consists of the safety stock plus the correct ordering quantity, which is the difference between the safety stock and maximum inventory level. Figure 8.19 is a graphic representation of the mini-max principle.

The **safety stock** is a backup supply to ensure against sudden increases in usage rate, failure to receive ordered products on schedule, receipt of products not meeting specifications, and clerical errors in inventory records. The size of safety stock depends on the importance of the products, value of the investment, and availability of substitutes on short notice. Size is dependent upon lead time and usage rate.

Figure 8.19.   Graphic representation of mini-max principle.

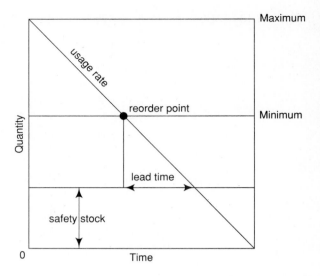

Lead time is the interval between the time that a requisition is initiated and receipt of the product. The shortest lead times occur when material is purchased from a local supplier who carries it in stock, and the longest when products are ordered from out-of-town suppliers. Safety stock of a product is part of the quantity on the shelves in the storeroom and must be rotated; it is noted only on inventory records. Stock rotation, in which the oldest products are used first, should be a policy in the storeroom.

The **usage rate** of a product is determined by experience and forecasts. The **reorder point** is established from the lead time and usage rate, and it is the lowest stock level that safely can be maintained to avoid a stockout or emergency purchasing. The maximum inventory level is equal to estimated usage plus the safety stock.

Economic Order Quantity Method

The total annual cost of restocking an inventory product depends directly on the number of times it is ordered in a year. To decrease these costs, orders should be placed as seldom as possible by ordering larger quantities. The holding cost of an inventory, however, is directly opposed to the concept of large orders.

The **Economic Order Quantity (EOQ)** concept is derived from a sensible balance of ordering cost and inventory holding cost. Ordering cost diminishes rapidly as the size of the orders is increased, and holding cost of the inventory increases directly with the size of the order. In Figure 8.20, the ordering cost is a curve diminishing in ordinates as the abscissae, the order quantities, increase. The holding cost varies directly with the order quantity and, therefore, shows as a straight line. The objective of EOQ is to determine the relationship between the ordering cost and the holding cost that yields the minimum total cost, which is the point at which both costs are equal and the two lines intersect. This relationship expressed mathematically yields the formula for EOQ:

$$EOQ = \sqrt{\frac{2 \times \text{ordering cost} \times \text{total annual usage}}{\text{holding cost}}}$$

In this model, the *ordering cost* includes the total operating expenses of the purchasing and receiving departments, expenses of purchase orders and invoice payment, and data processing costs pertinent to purchasing and inventory. *Total annual usage* is the number of units to be used annually. *Holding cost* is the total of all expenses in maintaining an inventory and includes the cost of capital tied up in inventory, obsolescence of products, storage, insurance, handling, taxes, depreciation, deterioration, and breakage. The EOQ may be determined by solving a mathematical formula or by using tables that are the result of formula calculations.

In the development of the basic EOQ formula, the following several assumptions were made (Montag & Hullander, 1971):

- Total annual usage is known and constant.
- Withdrawals are continuous at a constant rate.
- Quantity purchases are available instantly.

Figure 8.20. Graphic representation of the Economic Order Quantity concept.

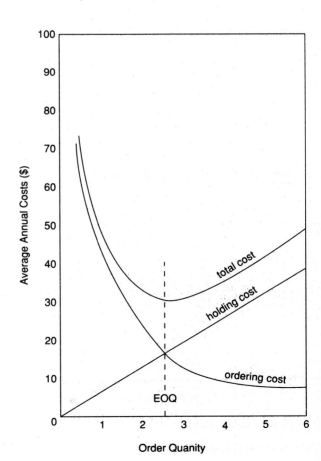

- Shortages are not tolerated.
- Unit cost is constant.
- Ordering cost is constant.
- Unit cost of inventory products is constant.

The disadvantages of the EOQ method include its unreliability when the demand rate is highly variable, the requirement of forecasting all products, and the necessity of past demand data. In foodservice operations, use of the EOQ concept is contraindicated when the following conditions exist (Montag & Hullander, 1971):

- Total annual usage is uncertain.
- Withdrawals are discontinuous or variable.
- Unit cost is dependent on price breaks.
- Lead time is not constant.
- Ordering cost is variable.
- Stock holding cost is variable.

The EOQ procedure is impractical unless computer assistance is available. Furthermore, the expenditure of large amounts of time and money required by this procedure is not justified on inventory analysis of products that account for very little of the total inventory costs. The EOQ method is not suitable for many foodservice operations because of a variable demand for certain food products, seasonal menu changes, and indefinite lead times. EOQ is acceptable, however, for large-scale food processing operations such as preparation of commercial frozen meals and wholesale bakeries. The EOQ method should not be considered a panacea for all inventory problems. The quantitative approach to inventory problems must be augmented by the knowledge, skills, and judgment of the foodservice manager.

Inventory Valuation Methods

Inventories, which are subject to change because usage and product replacement are continual, represent a significant portion of current assets including inventory that are owned by the foodservice operation. In every accounting period of generally 1 year, an accounting cycle usually begins and ends each month. Hence, inventories are taken monthly and the dollar value is included as an asset, or a debit, on the monthly balance sheet. A beginning and ending inventory occurs each month; the ending inventory becomes the beginning inventory for the next month.

Rotation of products, using the oldest first, is imperative if food quality is important. Pricing, or valuing, products does not have to follow this pattern in keeping inventory records. The important factor is that the method of valuing be consistent within an accounting period. The five principal *methods of inventory valuation* are actual purchase price, weighted average, FIFO (first in, first out), LIFO (last in, first out), and latest purchase price (Kim, Finley, Fanslow, & Hsu, 1992).

Actual purchase price involves pricing the inventory at the exact price of each individual product. Marking each product with the purchasing price, as it is received, is necessary. This method requires detailed record keeping and, therefore, is used only in very small foodservice operations.

The second method is **weighted average,** in which a weighted unit cost is used, and is based on both the unit purchase price and the number of units in each purchase. Again, cans or packages must be marked with the purchasing price at the time of receipt.

Inventory valuation using the **FIFO** method means that pricing closely follows the physical flow of products through the operation. The oldest products in the storeroom are used before the newest ones. The ending inventory reflects the current cost of products because inventory is valued at the prices for the most recent purchases.

LIFO is based on the assumption that current purchases are largely, if not completely, made for the purpose of meeting current demands of production. The purchase price of the oldest stock, therefore, should be charged out first. Generally, the value of the inventory will be lowest using LIFO and highest using FIFO based on the assumption that current prices will be higher than older ones. Foodservice managers choose LIFO when determining the value of inventory to reduce profit on financial statements in order to decrease income taxes, particularly in a high-inflation period. It minimizes the value of the closing inventory, which maximizes food cost for the time period.

The **latest purchase price** method uses the latest purchase price in valuing the ending inventory. This method often is used in foodservice operations because it is simple and fast (Kim et al., 1992).

The method chosen for valuing inventory is important because it will affect the determination of price of menu items sold, which in turn will affect the profit or loss figure. The balance sheet or statement of financial condition of a foodservice organization will also be affected because inventories are a current asset. These concepts are discussed in more detail in chapter 21, Management of Financial Resources.

SUMMARY

Receiving, storage, and inventory control are important activities in the procurement subsystem of a foodservice operation. Receiving is a process for ensuring that products delivered by suppliers are those that were ordered in purchasing, and storage is the holding of goods under proper conditions to ensure quality until time of use. Inventory is a record of material assets owned by an organization and is supported by the actual presence of products on the premises; inventory control is the technique of maintaining assets at desired quantity levels.

Elements of good receiving include competent personnel with specified responsibilities, proper facilities and equipment, well-written specifications, good sanitary practices, procedures for critical control, adequate supervision, scheduled receiving hours, and procedures to ensure security. Each received product must be checked against the purchase order and specification, and finally the delivery should be compared to the invoice prepared by the supplier. Products should be accepted or rejected based on checking; a receiving record should document all transactions and serve as a control mechanism.

Products should be transferred immediately to the proper secure storage area or to the production unit. Proper storage maintenance, temperature control, and cleaning and sanitation are major considerations in ensuring quality of stored food. Storage facilities will vary depending on the type of foodservice system. The major types of storage are dry and low temperature; proper facilities and equipment are required for each type. Storage is the second critical control point for microbiological safety of products and should be included in the procedure manual.

Issuing, either direct or storeroom, is the process for supplying products to the production units after they have been received or stored. Issuing provides information for food cost accounting. Inventory, which is a record, is supported by actual products on the shelf. Inventory control must include adequate record-keeping procedures to assure management has recent and reliable data on costs of operation. A physical inventory is a requirement of inventory control; a perpetual inventory, now that it can be computerized, is being used by more foodservice operations. Purchasing products just in time for production and immediate consumption by the customer without putting them in inventory is a concept being looked at by foodservice managers. With the increasing size and complexity of foodservice operations, inventory management and control have become more complicated and critical. ABC analysis, mini-max, and economic order quantity are tools available to assist managers in determining quantities for purchase, inventory levels, and cost of maintaining inventories.

Inventories represent a significant portion of current assets, and several methods for valuation have been developed. The five principal methods of inventory valuation are actual purchase price, weighted average, FIFO (first in, first out), LIFO (last in, first out), and latest purchase price.

BIBLIOGRAPHY

Dietary Managers Association. (1991). *Professional procurement practices: A guide for dietary managers*. Lombard, IL: Author.

Dion, P. A., Banting, P. M., Picard, S., & Blenkhorn, D. L. (1992). JIT implementation: A growth opportunity for purchasing. *International Journal of Purchasing and Materials Management, 28*(4), 32–38.

Educational Foundation of the National Restaurant Association. (1992). *Applied foodservice sanitation* (4th ed.). Chicago: Author.

Flamholtz, E. (1979). Organizational control systems as a managerial tool. *California Management Review, 22*(2), 50–59.

Food Industry Services Group in cooperation with the United States Department of Agriculture, Food and Nutrition Service. (1992). *Guidelines for the storage and care of food products: A technical assistance manual* (3rd ed.). Dunnellon, FL: Author.

Geller, A. N. (1991). Rule out fraud and theft: Controlling your food-service operation. *Cornell Hotel & Restaurant Administration Quarterly, 32*(4), 55–65.

Gunn, M. (1992). Professionalism in purchasing. *School Food Service Journal, 46*(9), 32–34.

Heinritz, S., Farrell, P. V., Guinipero, L., & Kolchin, M. (1991). *Purchasing: Principles and applications* (8th ed.). Englewood Cliffs, NJ: Prentice-Hall.

Keiser, J. R., & DeMicco, F. (1993). *Controlling and analyzing costs in food service operations* (3rd ed.). New York: Macmillan.

Kim, I. Y., Finley, D. H., Fanslow, A. M., & Hsu, C. H. C. (1992). *Inventory control systems in foodservice organizations: Programmed study guide*. Ames: Iowa State University Press.

Lary, B. K. (1991). Employee theft: Tracking and prosecuting perpetrators of on-the-job theft. *Restaurants USA, 11*(7), 18–22.

Longrée, K., & Armbruster, G. (1987). *Quantity food sanitation* (4th ed.). New York: Wiley.

Lorenzini, B. (1992). The secure restaurant. Part II: Internal security. *Restaurants & Institutions, 102*(25), 84–85, 90, 92, 96, 102,

Montag, G. M., & Hullander, E. L. (1971). Quantitative inventory management. *Journal of The American Dietetic Association, 59*(4), 356–361.

Puckett, R. P., & Miller, B. B. *Food service manual for health care institutions*. Chicago: American Hospital Publishing, 1988.

Schwartz, W. C. (1989). Employee theft: Guess who's leaving with dinner. *Restaurants USA, 9*(8), 24–29.

Stefanelli, J. M. (1992). *Purchasing: Selection and procurement for the hospitality industry* (3rd ed.). New York: Wiley.

Thorner, M. E., & Manning, P. B. (1983). *Quality control in food service* (rev. ed.). Westport, CT: AVI.

U.S. Department of Agriculture, Food and Nutrition Service. (1975). *Food storage guide for schools and institutions*. Washington, DC: Government Printing Office.

Zenz, G. J. (1981). *Purchasing and the management of materials* (5th ed.). New York: Wiley.

Purchasing for The Fairfax

Purchasing and receiving, storage, and inventory control activities in the procurement subsystem have been covered in chapters 7 and 8. These activities are important in a senior living services community because the quality of menu items served to the residents is dependent upon the quality of the food purchased.

Background Information

The purchasing agent in The Fairfax and Belvoir Woods Health Care Center has overall responsibilities for purchasing and receiving all foodservice products used in the community (Figure 1). The policy of Marriott Senior Living Services communities is to purchase all food from approved suppliers according to specifications. Currently, the F&B directors use Marriott's Food Standards and Specifications of Food and Services Management in healthcare facilities. These specifications include most products and meet the dietary needs of all residents, especially those eating in the assisted living and skilled nursing dining rooms.

Marriott has a sophisticated and efficient procurement/distribution system that is unparalleled in the lodging and foodservice industries. It serves Marriott hotels, restaurants, and contract foodservices and provides them with purchasing economies of scale, an individualized ordering procedure organized around individual unit needs, a professional staff, and an extensive quality assurance program. These key benefits also are available to Marriott's franchisees, partners, and clients. In one section, *Marriott's Senior Living Services Food and Beverage Policy and Procedure Manual* describes food purchasing, receiving, and storage, all of which are helpful to The Fairfax F&B director and purchasing agent.

Economies of Scale

Each year the Marriott Corporation purchases about $2.5 billion of food, services, supplies, and equipment. Through the company's national and regional purchasing agreements, Marriott utilizes significant economies of scale to control purchasing costs and to standardize quality. Marriott's central purchasing staff negotiates

Figure 1. Job description for the purchasing agent.

JOB TITLE: Purchasing Agent

OCCUPATION CODE:

DEPARTMENT NAME: Food and Beverage

DEPARTMENT NUMBER:

Overall Responsibilities

Purchase and receive all foodservice products used throughout the community so that the community's food production products meet, or exceed, Marriott hospitality and service standards. Maintain food inventory control.

Working Relationships

Reports to: Executive Chef or Director of Food and Beverage
Supervises: Storeroom Attendant/Receiver
Interfaces with: Suppliers, kitchen staff and various community supervisors and managers

Primary Job Duties

Procurement

- Procure all food products, as directed, from Marriott-approved vendors.
- Ensure that all food purchases are made within community and division standards and budgetary guidelines.
- Develop and maintain positive supplier relations.
- Report all purchasing expenditures to the Director, Food and Beverage, on an ongoing basis.

Inventory

- Maintain proper inventory levels to meet expected community needs, emergency requirements, security control, and storage availability.
- Inspect all goods received for quality and quoted price and compliance with Marriott specifications.
- Responsible for proper storage of all food items according to Senior Living Division standards.
- Ensure that all stock is dated and rotated prior to storage.
- Responsible for security of all inventory, particularly high-cost items (i.e. meats, seafood, liquor, etc.).

Administrative

- Complete all purchasing documentation requirements and maintain appropriate purchasing records. (i.e., purchase log, invoices, requisitions).
- Organize and administer weekly inventories.
- Maintain a daily Freezer and Refrigerator Temperature log.
- Maintain a daily Freezer and Refrigerator Pull List.

Figure 1. *continued*

Other

- Assist supervisor with any other reasonable task, as assigned.
- Practice all safety and loss prevention procedures; adhere to universal precautions and all infection control guidelines.
- Attend inservices as required.
- Maintain and protect the confidentiality of resident information at all times.

Job Qualifications

Required

- Ability to use a calculator and perform mathematical calculations.
- Ability to read and write English.
- Be able to physically handle weights up to and including 30 pounds.

Preferred

- High school diploma or equivalent.

OSHA Occupational Exposure Category

After careful analysis, it has been determined that this position falls into OSHA Occupational Exposure Category _____ and requires the following protective equipment be worn by anyone filling this position:

Training will be provided in how to properly and effectively use the equipment listed above, in addition to education regarding precautionary measures, epidemiology, modes of transmission and prevention of HIV/HBV.

I have read and agree that the contents of this job description accurately reflect what is expected of me in my current position.

_____   _____
Employee's Signature   Date

_____
Employee's Printed Name

_____   _____
Immediate Supervisor's Signature   Date

Source: Courtesy of Marriott Corporation.

contracts at the national level for products that are in demand across the country and integrally involves itself in the evaluation and selection of regional suppliers.

In addition to its corporate-owned and contract distribution program, regional programs are developed for perishable products such as bakery, dairy, and meats and for products needed only in selected locales. Product needs in the United States are handled through the company's 11 regional procurement offices, which assist Marriott-related operations in almost every geographical area (Figure 2). These offices are located in San Francisco, Los Angeles, Denver, Boston, Houston, Chicago, Atlanta, Ft. Lauderdale, Seattle, Edison (New Jersey), and Washington, D.C. In addition to regional procurement capabilities in the United States, Marriott has international procurement offices that handle needs outside the United States.

In distributing products, Marriott uses approximately 50 major full-line distribution companies around the country, including Sysco, Kraft Foodservice, and Rykoff-Sexton. Four are owned and operated by Marriott and distribute products to operations on the East Coast and West Coast. Contracts are developed with other distributors for staples, frozen products, and items that cannot be found in Marriott's distribution centers in some geographical locations. Additionally, more specialized distributors that handle truffles, caviar, and saffron are identified and contracted across the country. Both the distributors and specialized distributors are under contract on a competitive cost-plus basis.

Marriott's purchasing and distribution programs are massive but well-organized to provide each local unit, such as schools and hospitals, with Marriott foodservice contracts and Marriott hotels, restaurants, and continuing care retirement communities, a wide variety of products based on the unit's specifications. Unit F&B directors order directly from those suppliers contracted by Marriott, thus offering customers unique local products. Marriott is an equal opportunity buyer. It seeks to ensure that all qualified businesses, including those that are minority owned, have the opportunity to work with the company.

The Senior Living Services Division has developed standard operating procedures for receiving, storage, and inventory control. Tagging all perishable food items, developing receiving schedules, and refusing delivery of products not meeting specifications are examples of general procedures for receiving. Specific procedures also have been developed for receiving meats, poultry, seafood, canned and frozen fruits and vegetables, produce, dairy products, and bread and bakery products. Temperatures for food products in dry and refrigerated storage are listed in a manual of standard operating procedures.

Inventory is taken each week in Marriott operations and entered on an inventory/order form. Figure 3 is a partial inventory list of meat and poultry items. For each item, the form shows an item number, a description, the weight or quantity in each pack, and the price of each pack. (For example, item 01526.1 is 6-ounce boneless chicken breasts, which come 48 in a pack and cost $37.72.) The number of packs of each item in the storeroom is multiplied by the pack price, giving the

Figure 2. Marriott corporate procurement regions.

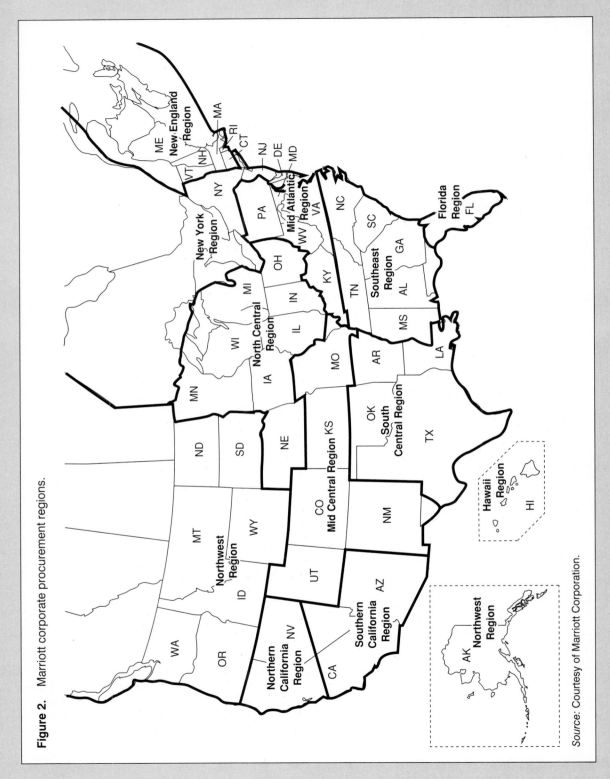

Source: Courtesy of Marriott Corporation.

Figure 3. The Fairfax food and beverage weekly inventory/order form.

PERIOD# ___2___   WEEK# ___3___   DATE: ___2/18/94___

ITEM	ITEM DESCRIPTION	PACK	PRICE	TOTAL INVENT	TOTAL
4100.45	MEAT/SEAFOOD/EGGS & CHEESE				
00240.2	Beef Knuckle (Pot Roast)	24 lbs	$44.40	0	$0.00
00461.8	Flank Steak	80 lbs	$228.00	1	$171.00
00307.7	Ground Beef – Lean 80/20	4/5 lbs	$29.33	1	$29.33
00211.9	Ribeye Roll 6 Pc #112A	70 lbs	$297.36	1	$223.02
00144.9	Steamship Round w/handle	65 lbs	$114.41	0	$0.00
00110.2	Tenderloin Peeled 6 Pc	25 lbs	$215.50	0	$0.00
00156.2	Top Round – No Roll, 3 Piece	65 lbs	$113.75	1	$85.31
01051.1	Ham Fresh Bnls B–R–T	26 lbs	$40.66	0	$0.00
00690.4	Pork Loin Bnls. B–R–T	2/9 lbs	$41.94	8	$335.52
09727.6	Sausage Link 1oz	12 lbs	$14.23	5	$71.15
01230.1	Sausage Sweet Italian 8/1	10 lbs	$15.50	1	$7.75
01306.4	Spare Ribs 3.5 Down	30 lbs	$37.79	1	$37.79
00960.1	Lamb Leg of B-R-T	9.0 lbs	$40.35	0	$0.00
00965.2	Lamb Rack french Cap Off	4/3 lbs	$145.24	14	$1,960.74
00747.0	Veal Osso Boco	17 lbs	$75.89	0	$0.00
00574.6	Strip Steak 6oz #1180	54 each	$142.97	3	$428.91
00716.1	Tender 4oz #1190A	54 each	$144.59	0	$0.00
09376.9	Beef Stew 90% Lean	2/10 lbs	$43.86	0	$0.00
00406.5	Chip Steak 4oz	6/5 lbs	$90.60	1	$67.95
07046.7	Liver Beef	10 lbs	$10.23	1	$10.23
00555.0	Short Rib 3oz Bnls	10 lbs	$53.24	0	$0.00
00710.2	Short Rib 8oz Bone-in	22.5 lbs	$77.03	0	$0.00
01166.5	Swiss Steak 5oz Cubed	20 lbs	$56.67	1	$56.67
00782.0	Tender Medallions 2.5oz	100 each	$66.66	8	$533.28
00766.8	Pork Chop 4oz CC 40 each	10 lbs	$26.24	5	$131.20
00771.4	Pork Diced	20 lbs	$39.72	3	$119.16
00963.6	Lamb Cubes 1 inch	10 lbs	$29.23	0	$0.00
00741.2	Veal Cubes 1 inch	3/5 lbs	$48.09	1	$48.09
00680.5	Veal Cutlets 4oz.	10 lbs	$58.10	1	$29.05
01584.9	Chicken Quarters 2½ lbs	70 lbs	$59.88	0	$0.00
01615.2	Turkey Brst. Ckd. Skinless	2/9 lbs	$37.24	4	$130.34
01471.4	Chicken Breast Cordon Bleu 7.5oz	24 each	$43.67	4	$174.68
01554.7	Chix Brst. 4oz. Bnls. Sknls.	48 each	$28.68	9	$258.12
01526.1	Chicken Breast 6oz Boneless	48 each	$37.72	0	$0.00
01515.6	Chicken Breast 8oz. Boneless	24 each	$37.72	0	$0.00
01545.8	Chicken Cut 8 Piece IQF	96 each	$28.33	0	$0.00
01565.2	Chicken Diced ½" 60/40	10 lbs	$25.19	3	$75.57
01574.1	Chicken Meat-Pulled & Cooked – IQF	10 lbs	$23.47	2	446.94
01506.7	Chix Tender Brd.	4/3 lbs	$37.59	1	$37.59
01501.6	Chicken Patty Breaded 4oz.	60 each	$38.13	0	$0.00
01670.5	Chicken Quarters 8-¾oz.	60 each	$33.34	3	$83.35
01521.1	Cornish Hen – Split 9oz	24 each	$20.46	1	$20.46
01609.7	Duck Breast Split 6-7oz.	32 each	$68.43	3	$205.29
01653.5	Turkey Breast Split S>F>	12 lbs	$33.60	2	$67.20
01618.7	Turkey Ground	2/10 lbs	$18.96	0	$0.00
01619.8	Turkey Honey Smoked Ckd	2/7 lbs	$33.80	9	$304.20
09906.6	Turkey Sausage Links 1oz	12 lbs	$20.60	0	$0.00
01633.4	Turkey Whole Tom Grade A	2/24 lbs	$44.76	0	$0.00
13218.7	Bacon Bits Imitation	12/1 lbs	$13.47	0	$3.37
09527.3	Bacon Canadian 1 Pc/Bx	7.5 lbs	$26.63	1	$26.63
08144.2	Bacon Layout 26/30 Ct	15 lbs	$21.64	3	$64.92
08602.9	Bologna Beef	10 lbs	$14.83	0	$0.00
00074.4	Corned Beef Brisket Raw	2/11 lbs	$41.12	0	$0.00

Source: Courtesy of Marriott Corporation.

total price for each item on the inventory. This total inventory figure becomes the basis for the next week's order. After checking the menu, the projected census, and the amount on hand, the amount to be ordered is determined by using a computer software program.

Professional Staff

Marriott's corporate purchasing and distribution staff is respected within the foodservice industry. The staff is well known for high-quality work; for being strong, but fair, in negotiations; and for unyielding integrity. The staff members understand the foodservice industry because many have been restaurant operators or owners. Their expertise and experience enable them to advise operators on product quality specifications, current buying practices, and cost-cutting alternatives to current programs.

Quality Assurance

Marriott has an excellent corporate quality assurance staff that works with the company's procurement team to ensure that established standards are maintained by suppliers. They continually monitor product quality and make on-site inspections of suppliers' facilities and distribution centers serving the Marriott Corporation.

The Problem

Baked New Zealand lamb chops dijonaisse are on the Today's Menu occasionally, but the residents have expressed discontent with the menu item. In the Dining Journal, in which residents comment about meals, some wrote that the flavor was too strong and others that the meat was too dry. Each of the managers in the Food and Beverage Department is required to read the journal daily for comments pertaining to their area and to respond to the resident by phone or reply in the journal.

The dining room manager discussed the problem of the New Zealand lamb chops with the residents who made comments and related to the F&B director their concerns and suggestions that American lamb would be much better. The manager was not convinced that the origin of the lamb was the problem but thought that baking individual lamb chops could have some effect on the intensity of flavor and dryness. The F&B director decided to purchase American lamb from another Marriott supplier but instead of individual rib lamb chops, he ordered racks of lamb, each with 9 to 11 ribs. The racks of lamb would be roasted and then sliced into chops before serving, which should make the flavor less concentrated and the meat more moist than individual baked chops. Two rib chops are considered a serving but smaller or larger portions are available upon request.

The F&B director followed his standard ordering schedule. He placed the meat order on Friday for a Monday delivery to be served on Tuesday. When the meat order arrived on Monday, the racks of lamb were missing. The director phoned the supplier, who explained that orders for lamb racks have to be placed 7 days in advance of delivery and that his order would arrive on Thursday. A quick decision for Tuesday's menu had to be made by the director and chef. How would you solve the problem?

Points for Discussion

- How did this procurement problem affect the entire foodservice system?
- List all the alternatives the F&B director might have to consider in deciding what to do about Tuesday's menu.
- How could the F&B director incorporate residents and their opinions into the procurement decision-making process?
- Why is knowledge of food preparation and service important in procurement problem solving and decision making?

4

Production

Production follows procurement in the foodservice system. The production unit is where most of the activity happens in the foodservice organization. Food products are converted into menu items through the efforts of many employees to meet the goal of providing quality food in quantity to serve to customers.

- **Chapter 9, Production Planning.**   One of the goals of a foodservice organization is to prepare menu items in the needed quantity and with the desired quality. Forecasting the number of menu items needed for a particular meal and the quantity of food to be purchased for the menu items occurs in the production unit.

- **Chapter 10, Ingredient Control.**   Centralized assembly of ingredients to be used in a recipe often is done in an ingredient room located near the kitchen. Standardized recipes are necessary for an ingredient room to be effective.

- **Chapter 11, Quantity Food Production and Quality Control.**   The foodservice manager must understand the principles of food production in order to perform competently in the foodservice organization. Standards are necessary for all menu items in order to evaluate quality.

- **Chapter 12, Food Safety.**   Many opportunities exist for food contamination causing foodborne illness, which is recognized as a major health problem in the United States today. A well-designed food safety program monitors critical control points for errors in handling food products and eliminates those errors.

- **Chapter 13, Labor Control.**   Foodservice managers are challenged by controlling and reducing labor costs and simultaneously increasing labor productivity. Proper staffing and scheduling of employees are complex tasks, but they are necessary to control labor costs.

- **Chapter 14, Energy Control.**   Energy management is essential for a foodservice organization. Because the foodservice industry is energy intensive, the impact of rapidly increasing energy costs is serious.

Production Planning

In the simplest possible terms, the objective of **food production** is the preparation of menu items in the needed quantity and with the desired quality, at a cost appropriate to the particular foodservice operation. *Quantity* is the element that distinguishes production in foodservices from home or family food preparation. *Quality*, an essential concomitant of all food preparation, becomes an extremely vital consideration in mass food production due to the number of employees involved. Quality includes not only the aesthetic aspects of a food product but also the nutritional factors and the microbiological safety of the product. Cost, of course, determines whether or not a product should be produced for a specific customer. Serving filet mignon as a school lunch item obviously would be too expensive. In the 1990s, foodservice operators are being challenged by rising costs of food, energy, and labor. The key to success is that managers make the effort to change strategies and management styles (Cichy, 1983).

After procurement, production is the next major subsystem in the transformation element of the foodservice system; it is highlighted in Figure 9.1. Because of the increased use of partially processed foods, such as peeled and sliced apples, any prepreparation will be done in the production unit. **Production** in the generic sense is the process by which products and services are created. In the context of foodservice, production is the managerial function of converting food purchased in various stages of preparation into menu items that are served to a customer.

303

Figure 9.1. Foodservice systems model with the production function highlighted.

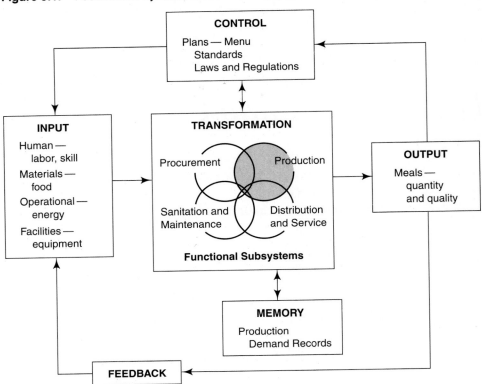

In foodservice operations today, production is no longer considered merely cooking in the kitchen, but it involves planning, controlling ingredients, production methods, food quality, labor productivity, and energy consumption. Each of these controls is discussed in separate chapters of part 4.

In essence, foodservice managers responsible for production are resource managers, and they may be designated as such in some organizations. Those who rely on experience to make decisions could have difficulty surviving in today's competitive market. Innovative approaches to decision making are required for allocation and control of resources (Lambert & Beach, 1980). For example, analytical and computer techniques used in production industries can be adapted by the foodservice manager for determining resource requirements for production.

Planning for production is the establishment of a program of action for transformation of resources into products and services. The manager identifies the necessary resources and determines how the transformation process should be designed to produce the desired products and services. Once this process has been developed, planning must be integrated with the other managerial functions of organizing and controlling.

Planning, organizing, and controlling are overlapping managerial functions, however, and cannot be considered separately. For example, the foodservice manager and the production supervisor might have established a schedule of preparation times to prevent vegetables from being overcooked (controlling), but then suddenly they must revise the production schedule (planning) because an essential employee went home sick and a critical task had to be reassigned. The content of jobs has to be analyzed to be sure all tasks are covered (organizing).

PRODUCTION DECISIONS

As discussed in chapter 3, planning encompasses the setting of goals and objectives by top management and the development of policies and procedures by middle management. Eventually, decisions must be made concerning the necessary quantities to produce and the standards of quality that must be maintained within the limitations of costs. In foodservice operations, as in industry, managers must estimate future events. Thus, forecasting, planning for aggregate or total output, and production scheduling are important elements for decision making.

All these planning decisions must be made within the constraints of the existing facility. Much too often, in a hospital or nursing home, the number of patients or residents is increased, but the capacity of the equipment in the kitchen is not. If the anticipated demand exceeds the present capacity, then the facility must be expanded, future production curtailed, or more ready prepared foods purchased to handle the increased demand.

Production planning primarily is the effective synthesis of quantity, quality, and cost objectives. The objective of the production subsystem is to transform human, material, facility, and operational resources into outputs. The secondary objectives must specify the following (Adam & Ebert, 1992):

- Product/service characteristics
- Process characteristics
- Product/service quality
- Efficiency
     Effective employee relations and cost control of labor
     Cost control of material
     Cost control in facility utilization
- Customer service
     Producing quantities to meet expected demand
     Meeting the delivery date for products or services

These industrial production planning objectives are equally applicable to foodservice operations. The characteristics of the product or menu items depend upon the type of operation, such as short orders in a limited-menu restaurant or hotel coffee shop, individual item selections in a full-service restaurant, or a fixed menu in a school. For example, a ground meat patty would be served as a grilled hamburger on a bun for a limited-menu restaurant, a charbroiled ground steak for a full-service restaurant, or an oven-baked hamburger for a nursing home.

Production process characteristics include the method of food preparation, ranging from grilling to broiling to baking. The process and product characteristics are closely related because both determine the quality of the product and service. Efficiency of the process depends upon the control of costs for labor, material, and facility utilization. All of these secondary objectives lead to customer service.

Obviously, effective planning cannot be done without reasonably accurate forecasting of future demand quantities. Finally, standards of quality must be established for all products that will be produced. Maintenance of quality is a cost factor because of employee training, inventory control of both raw and prepared food items, and sanitation programs.

The cost element in planning is the result of the correlation of food, storage, issue, and production costs with labor, facility, and energy costs. These elements must be considered in all planning. Whenever planning goes beyond a day-to-day basis, forecasting becomes absolutely necessary.

PRODUCTION FORECASTING

Forecasting is the art and science of estimating events in the future and provides the data base for decision making and planning. The art of forecasting is the intuition of the forecaster, and the science is the use of past data in a tested model. Both are required to estimate future needs. Forecasting is described as a function of production and constitutes the basis for procurement.

Production Demand

Forecasting is a function of production but also is needed for procurement. Food products have to be available for producing menu items for the customer. The primary result of forecasting should be customer satisfaction; customers expect to receive what they ordered. In addition, the foodservice manager is concerned with food cost; both overproduction and underproduction affect the bottom line.

Overproduction generates extra costs because the salvage of excess food items is not always feasible. Leftover prepared food spoils easily and requires extreme care in handling and storage. Even though some leftover foods might be salvageable by refrigeration, certain foods may break down and lose quality. An example is a custard or cream pie that must be held under refrigeration for microbiological safety but develops a soggy crust quickly and cannot be served. Policies and procedures for the storage of overproduced food items should be well defined and rigorously enforced.

Attempts to reduce overproduction costs by using a leftover high-priced food as an ingredient in a low-cost menu item reduces profits. For example, using leftover rib roast in beef stew, soup stock, or beef hash, all of which could be prepared with less expensive fresh meat, is difficult to justify. In addition to the higher food cost, planning and carrying out these salvage efforts incur higher labor costs that could have been avoided had overproduction not occurred. Customers often suspect that leftovers are being used, which can be damaging to the image of a foodservice operation.

Underproduction can increase costs as much as overproduction. Customers will be disappointed if the menu item is unavailable, and they often have difficulty in making another selection. Furthermore, underproduction may involve both additional labor costs and often the substitution of a higher-priced item.

A wise manager will insist that a similar backup item be available when underproduction occurs. For example, in a university residence hall foodservice, if the grilled meat patties run out, an excellent replacement would be frozen minute steaks, quickly grilled. Such a substitution certainly would increase customer satisfaction even though it hurts the bottom line.

Quantity Demand

The desire for an efficient foodservice operation requires that the production manager know the estimated number of customers or the number of servings of each menu item in time to order from the procurement unit. Good forecasts are essential for managers in planning smooth transitions from current to future output, regardless of the size or type of the foodservice (i.e., schools, hospitals, or restaurants). Forecasts vary in sophistication from those based on **historical records** and intuition to complex models requiring large amounts of data and computer time. Choosing a **forecasting model** that is suitable for a particular situation is essential.

Historical Records

Adequate historical records constitute the basis for most forecasting processes. This is especially true of foodservices in which past customer counts, number of menu items prepared, or sales records were used to determine the number of each menu item to prepare long before sophisticated forecasting methods were available. In fact, such historical records are the root of most of these methods. Such records must, however, be accurate and complete, or they cannot be extended into the future with any reliability.

Effective production records must be identified by date and day of the week, meal or hour of service, special event or holiday, and even weather conditions, if applicable. An example of a historical record for a catered party to announce an engagement is shown in Figure 9.2. This seasonal menu also could be used for a June wedding reception, a graduation party, or even Father's Day.

Caterers, both social and employed by an organization, must keep accurate records of the amount of food for each event to prevent underproduction or overproduction of menu items on the next similar occasion. Catering is a profit enterprise, and reliable past records are essential because events are not repetitive; elaborate forecasting methods generally are not feasible.

Although the production unit records reveal the vital information on menu items served to customers it is by no means the only organizational unit that should keep records. Only by cross-referencing records of sales with those of production can a

Figure 9.2. Example of a historical record.

ENGAGEMENT ANNOUNCEMENT
COCKTAIL PARTY

Date: June 17, 1994—Saturday   *Weather:* 80°F and sunny
Time: 5:00 P.M.   *Special Notes:* Serve on patio in parents' home

Iced Shrimp (3)* with
Tangy Cocktail Sauce

Guacamole
Cherry Tomatoes (2)*
Tortilla Chips

Thin Slices of Rare Beef Tenderloin (2)*
Cocktail Buns   Horseradish Sauce

Chicken Liver Pâté with Pistachios (2 pâtés)
Tiny Melba Toast

Assorted Raw Vegetables (3)*
Curry Dip

Whole Strawberries with Stems (2)*
Sour Cream and Brown Sugar Dip

Butter cookies (2)*
Champagne

*per person

---

reliable historical basis for forecasting be formalized. In foodservices in which meals are sold, as in restaurants, hospital employee and guest cafeterias, and schools, records of sales will yield customer count patterns that can be useful for forecasting. These data can be related to the number of times customers select a given menu item or the daily variations induced by weather or special events.

Historical records in the production unit provide the fundamental base for forecasting quantities when the same meal or menu item is repeated. These records should be correlated with those kept by the purchasing department, which include the name and performance of the supplier and price of the food items.

Forecasting Models

Selecting a forecasting model for a foodservice operation can be a difficult task for a manager. The easiest method is to guess how many people are expected or how much of each menu item is needed. Amazingly, many chefs and cooks can guess quite accurately, especially if the customer count is approximately the same at each meal or the menu is static. But when customers can choose where they want to eat, guessing does not always work. A more scientific method for forecasting is needed. Forecasting models are available and should be researched before deciding which one to use.

Criteria for a Model.   Numerous forecasting models have been developed during the past three decades, but, as one might expect, the trend has been toward sophisticated models using computer-based information systems. According to Fitzsimmons and Sullivan (1982), the factors deserving consideration when selecting a forecasting model are cost, required accuracy, relevancy of past data, forecasting lead time, and underlying pattern of behavior.

Cost of Model.   The cost of a forecasting model involves the expenses of both development and operation. The developmental costs arise from constructing the model, validating the forecast stability, and, in the case of large operations, writing a computer program. In some cases, educating managers in the use of the model is another cost element. Operational costs, including the cost of making a forecast after the model is developed, are affected by the amount of data and computation time needed. More elaborate models require large amounts of data and thus can be very expensive.

Accuracy of Model.   The quality of a forecasting model must be judged primarily by the accuracy of its predictions of future occurrences. An expensive model that yields accurate forecasts might not be as good a choice as a cheaper and less accurate model. This is a decision the foodservice manager must make.

Relevancy of Past Data.   In most forecasting models, the general assumption is that past behavioral patterns and relationships will continue in the future. If a clear relationship between the past and the future does not exist, the past data will not be relevant in developing forecasts. In these cases, subjective approaches, such as those that rely heavily on the opinions of knowledgeable persons, may be more appropriate.

Lead Time.   Lead time pertains to the length of time into the future that the forecasts are made. Usually, these times are categorized as short-, medium-, or long-term. The choice of a lead time depends on the items being forecast: A short-term lead will be chosen for perishable produce, and a medium- or long-term lead for canned goods.

Pattern of Behavior.   As stated above, many forecasting models depend on the assumption that behavioral patterns observed in the past will continue into the future and, even more basic, that actual occurrences follow some known pattern. These patterns, however, may be affected by random influences, which are unpredictable factors responsible for forecasting errors. Not all forecasting models work equally well for all patterns of data; therefore, the appropriate model must be selected for a particular situation.

Types of Models.   Forecasting models have been classified in numerous ways, but the three most common model categories are time series, causal, and subjective. A model in one of the classifications may include some features of the others. In all methods of forecasting, trends and seasonality in the data must be considered.

Time Series Model.   The frequently used **time series forecasting model** involves the assumption that actual occurrences follow an identifiable pattern over time. Although time series data have a specific relationship to time, deviations in the data make forecasting difficult. To reduce the influence of these deviations, several methods have been developed for smoothing the data curve.

The time series models are the most suitable for short-term forecasts in foodservice operations. Actual data may indicate a trend in a general sense but not give forecast information. To make the past data useful, variations must be reduced to a trend line that can be extended into the future. Two time series models, moving average and exponential smoothing, are used more frequently in rationalizing foodservice data than any other type, although causal models may be used as well.

The most common and easiest of the smoothing procedures is the **moving average forecasting model,** which can be used only on items that are of the same kind. The process begins by taking the average of the number of portions sold for the last five or more times the menu item was offered as the first point on the trend line. The second point on the line is determined by dropping the first number and adding the most recent number of portions sold to the bottom of the list and then taking another average. The repetitive process is continued for all the data.

An example of the moving average model is shown in Table 9.1. Data are for the number of hamburgers sold over the last 10 days. A 5-day moving average is used.

Table 9.1.   Example of a moving average

Day	Number of Hamburgers Sold	5-Day Moving Average
1	150	
2	180	
3	185	176
4	170	186
5	195	187
6	200	182
7	185	183
8	160	180
9	175	
10	180	

Figure 9.3. Graph illustrating moving average smoothing effect.

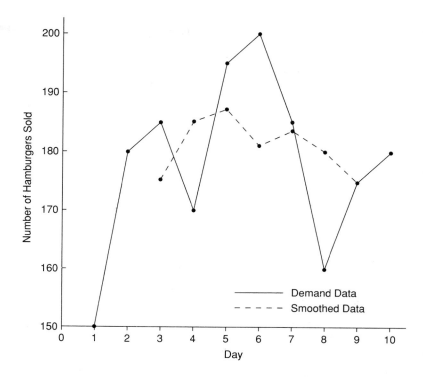

The first 5-day moving average is calculated by adding the number sold for each of those days and dividing by five, giving an average of 176 hamburgers. The next moving average is calculated by adding the number sold for days 2 through 6 and dividing by five. The procedure is repeated by dropping the earliest day's data and adding the next for a total of 5 days.

Demand data and moving average values plotted on the graph (Figure 9.3) illustrate the smoothing effect of the model. Note that the smoothed data curve eliminates the daily variations in demand and thus indicates a trend of the past. This averaging process, when continued, yields data points that smooth out the data to a comparatively constant pattern for use in the forecast.

The **exponential smoothing forecasting model** is a popular time series model that can be set up on a computer spreadsheet. It is very similar to the moving average model except that it does not uniformly weight past observations. Instead, an exponentially decreasing set of weights is used, giving recent values more weight than older ones. Also, the only data required are the weights, the alpha judgment factors, that are applied to the most recent values including the last customer demand and the last forecast, thus eliminating storing historical data (Makridakis & Wheelwright, 1989). Alpha is the judgment factor, or smoothing coefficient, and indicates how well the manager believes the most recent data represent current customer count or number of sales.

The judgment factor, alpha, is a number between 0 and 1 and is used to adjust for any errors in previous forecasts. Alpha is the weight assigned to the most recent customer demand and 1 – alpha is the weight for the most recent forecast. When alpha has a value close to 1, the new forecast will include a substantial adjustment for any error that occurred in the preceding forecast. Conversely, when alpha is close to 0, the new forecast will not show much adjustment for the error from the one before. The value of alpha has been tested in foodservice, and if no major changes occur in the data, customer demand for succeeding weeks is not expected to differ greatly from the past (Messersmith & Miller, 1992). An alpha of 0.3 is commonly used for demand, leaving 0.7 for the forecast. The most recent forecast values are multiplied by the $1 - \alpha$ quantity, which places a greater weight on recent values. The $1 - \alpha$ quantity acquires exponents in increasing order as the forecast is repeated, thus decreasing weights of older values and having a lesser influence on the trend curve than more recent data. The mathematical expression for this smoothing model is

$$F_t = \alpha D_{t-1} + (1 - \alpha)F_{t-1}$$

where

α = a constant, usually between 0.1 and 0.3 (judgment factor)
F_t = smoothed value at time t (new forecast)
D_{t-1} = actual observed value at time $t-1$ (last demand)
F_{t-1} = preceding smoothed value (last forecast)

Stated in words, this forecast equation is

$$\text{New forecast} = \begin{bmatrix} \text{judgment} \\ \text{factor} \end{bmatrix} \begin{bmatrix} \text{last} \\ \text{demand} \end{bmatrix} + \begin{bmatrix} 1 - \text{judgment} \\ \text{factor} \end{bmatrix} \begin{bmatrix} \text{last} \\ \text{forecast} \end{bmatrix}$$

Fitzsimmons and Sullivan (1982) summarized exponential smoothing as a popular technique for short-term demand forecasting for the following reasons:

- All past data are considered in the smoothing process.
- More recent data are given more weight than older data.
- The technique requires only a few pieces of data to update a forecast.
- The model can be easily programmed and is inexpensive to use.
- The rate at which the model responds to changes in the underlying pattern of data can be adjusted mathematically.

Causal Model.   Causal forecasting models, like time series models, are based on the assumption that an identifiable relationship exists between the item being forecast and other factors. These factors might include selling price, number of customers, market availability, and almost anything else that might influence the item being forecast. Causal models vary in complexity from those relating only one factor, such as selling price, to items being forecast to models utilizing a system of mathematical equations that include numerous variables.

The cost of developing and using causal models is generally high, and consequently they are not used frequently for short-term forecasting, such as for perishable produce. They are, however, popular for medium- and long-term forecasts, such as for canned goods.

The most commonly adopted causal models are called **regression analysis fore-casting models**, and they use regression analysis. Following standard statistical terminology, the items being forecast are called dependent variables, and the factors determining the value of the dependent variables are called the independent variables. Regression models require a history of data for the dependent and independent variables to permit plotting over time. Once this is done, the regression process involves finding an equation for a line that minimizes the deviations of the dependent variable from it. Two principal kinds of regression models are linear and multiple.

In linear regression the word *linear* signifies the intent of the analysis to find an equation for a straight line that fits the data points most closely. In conventional statistical terminology, the item being forecast is called the dependent variable (Y), and the factors that affect it are called independent variables (X).

In the analysis, historic demand data for a single variable will result in a derived equation from a linear regression process in the form of a straight line

$$Y = a_0 + a_1 X$$

in which a_0 and a_1 are numerical constants determined by the regression analysis. As shown in Figure 9.4, a_0 is the intercept of the line on the Y axis and a_1 is the slope of the line. In use, X will be a single independent variable quantity. Data points in the figure are the Y (dependent variable) values for specific values of X (independent variable). Preliminary plotting of the variables on graph paper would be advisable to ascertain if they could be represented reasonably by a straight line. The forecasting value consists of the assumption that the linear relationship between the variables will continue for a reasonable time in the future, or quite simply that the line may be extended. Use of the equation requires only substitution of an anticipated future value for X and then solution for Y, which is the forecasting quantity.

Figure 9.4. Typical regression line.

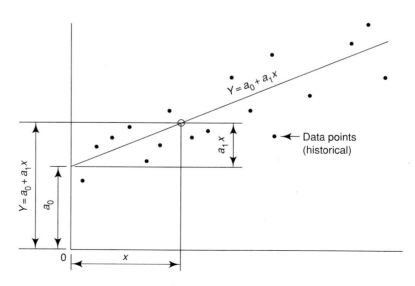

Examples of independent variables in hospital foodservice are total number of patient trays served, patient census, cafeteria customer capacity, number of employees, number of patients on regular diets, and number of patients on each modified diet. For example, roast beef might be a popular item for a foodservice, and the relationship between the historic number of patient trays and the pounds of roast beef could yield a regression equation. To forecast beef demand, an anticipated future count of trays would then be inserted into the equation as X to solve for Y, the pounds of roast beef needed.

If determination of the effects of more than one independent variable ($X_1, X_2, \ldots X_n$) on the dependent one is desired, the process is called multiple regression and the derived equation will have the following form:

$$Y = a_0 + a_1X_1 + a_2X_2 + \ldots + a_nX_n$$

Multiple regression analysis is quite complex, and a good computer program is needed for the solution.

Subjective Model.   A **subjective forecasting model** generally is used when relevant data are scarce or patterns and relationships between data do not tend to persist over time. In these cases, little relationship exists between the past and the long-term future. Forecasters must rely on opinions and other information, generally qualitative, that might relate to the item being forecast.

One of the subjective forecasting models is the **Delphi technique**, which involves a panel of experts who individually complete questionnaires on a chosen topic and return them to the investigator. Results of the first questionnaire are summarized and returned to the panel for revision. Questionnaires are revised successively until a simple majority agreement is reached. The Delphi method can be time-consuming and expensive and is not especially suitable for foodservice forecasting. It is being used, however, by the National Restaurant Association for predicting the status of both the foodservice industry and the foodservice manager in the year 2000.

Other qualitative forecasting techniques include market research, panel consensus, visionary forecast, and historical analogy. *Market research* is a systematic and formal procedure for developing and testing hypotheses about actual markets. *Panel consensus* is based on the assumption that a group of experts can produce a better forecast than one person. This differs from the Delphi method by requiring free communication among the panel members. A *visionary forecast* is characterized by subjective guesswork and imagination. The *historical analogy* involves comparative analysis of the introduction and growth of new items with similar new-product history.

PRODUCTION SCHEDULING

Production scheduling in foodservice operations can be defined as the time sequencing of events required by the production subsystem to produce a meal. Scheduling occurs in two distinct stages—planning and action—and is essential for production control.

In the planning stage, forecasts are converted into the quantity of each menu item to be prepared and the distribution of food products to supervisors in each work center. As an example, 500 servings of grilled marinated chicken breasts, pasta with pesto, asparagus spears in a red pepper ring, green salad with artichoke hearts and diced tomatoes, sour dough rolls, and frozen yogurt with fresh strawberries have been forecast for a special dinner. The foodservice director assigns the production of the 500 servings of chicken, pasta, and asparagus to the supervisor of main production, the green salad to the salad unit, and the rolls and dessert to the waitstaff.

Supervisors in each unit assume the responsibility for the action stage by preparing a production schedule. Each item is assigned to a specific employee and the time to start its preparation is recorded on the schedule. Careful scheduling assures that the food is prepared for service without lengthy holding and deterioration in quality. Supervisors give feedback to the manager by writing comments on the production schedule.

In small operations in which only one cook and perhaps an assistant are on duty at a time, the foodservice manager might also need to assume the responsibility for the action stage. Every foodservice operation, however small, must have a daily production schedule to control both labor and food costs.

Production Schedule

The **production schedule**, frequently called the *production worksheet,* is the major control in the production subsystem because it activates the menu and provides a test of forecasting accuracy. The menu, of course, must be based on standardized recipes, as described in chapter 11.

The production schedule is highly individualized in various foodservices and may vary from a one-page form for manual completion to a computer program printout. Regardless of the form, certain basic information must be included on each schedule. The unit, production date, and meal should be identified as well as such other pertinent information as actual customer count, weather, and special events. In addition, the following information must be included to make it a specific action plan:

- Employee assignments
- Preparation time schedule
- Menu item
- Over- and underproduction
- Quantity to prepare
- Substitutions
- Actual yield
- Additional assignments
- Special instructions and comments
- Prepreparation

The sample production schedule in Figure 9.5 is from a large university residence hall foodservice. Note that the general information previously mentioned is displayed at the top of the worksheet and that the production area is marked as the specific destination for this schedule. The meal count at the conclusion of produc-

Figure 9.5. Residence hall foodservice production schedule.

Unit ___Main Production___

Date ___8/29/94___

Meal Count ___2153___

Weather ___Fair___

Meal _____ Bkf. _____ Lunch ___X___ Dinner

Comments ___Basketball Game___

Employee	Menu Item	Quantity to Prepare	Actual Yield	Instructions	Time Schedule	Left over Amount	Run-out Time	Substitution	Cleaning Assignment
Wege Whatley	Country Fried Steak	1200	1220	Use 2 tilting fry pans and oven number 3	Begin frying 2:15 See Frying Time schedule	35 servings	—	—	Whatley— tilting fry pans
Lundin	Giant Rolled Tostados	1000	1020	Serve open face on cafeteria line and customers will roll their own		50 tortillas 10 lbs meat mixture 1 gal cheese sce.	— — —	— — —	Lundin— slicer and attachments
McCurdy	Whipped Potatoes	1200	1150	If necessary use instant as a back up	Begin steaming potatoes at 3:00	12 lbs	—	—	
McCurdy	Cream Gravy	2000	1600	Make 4 batches 600–600– 400–400 (if needed) Serve over both steak and potatoes		2 gal	—	—	
Wege	Mexican Rice	900	900	Use 12 × 10 × 4 in. pans	See Baking Time schedule	18 lbs	—	—	
Mockery	Refried Beans	850	850	Add bean liquid as needed to maintain a moist product		12 lbs	—	—	Mockery— oven number 1, shelves and doors
Mockery	Broccoli Spears	1000	850	Season with melted margarine	Begin 4:00	0	6:00	2½ lbs Asparagus spears	
Mockery	Yogurt Cup	20	20	Serve whole container— blueberry, cherry, rum raisin, plain	Prepare based on demand	8	—	—	

Prepreparation

Employee	Menu item	Quantity	Instructions	Employee	Menu item	Quantity	Instructions
McCurdy	Roast Beef	600 lbs	Pan beef in baking pans, cover and refrigerate	Lundin	(Omelet) fresh eggs	1 case	Break into 60 qt mixer bowl
Mockery	Hard Cooked Eggs	5 doz.	For garnish on spinach	Wege	Ham	10 lbs	Dice for omelets

Source: Kansas State University Housing and Dining Services. Used by permission.

tion is also recorded, which validates this schedule as part of historical data. Specific headings previously listed constitute the column headings.

The production schedule generally is posted on a bulletin board in the unit. The name of the employee in the left-hand column readily enables personnel to find designated duties. The menu item column identifies the recipe by name. Often the unit supervisor will distribute the recipes, either in card form or as a computer printout, to the appropriate employees at the time the schedule is posted.

The *quantity to prepare* is the forecast amount for each menu item. The *actual yield* is the portion count produced by the recipe. Current practice is to include portion size and count on the recipe. If not contained on the recipe, this information should be placed on the production schedule. Note that the actual yield indicates overproduction for some items and underproduction for others. The *instructions* column, completed by the unit supervisor, gives special information and comments on equipment to be used and service instructions. In addition, this column should contain any specific information not included in the recipe, such as for refried beans, "Add bean liquid as needed to maintain a moist product," and for broccoli spears, "Season with melted margarine."

The *time schedule,* completed by the unit supervisor, is intended to assure that the various menu items will be produced for service at the desired time. It also has references to preparation methods standardized in this main production unit for a particular item, such as country fried steak and Mexican rice.

The *leftover amount* column indicates over- or underproduction. Underproduction is indicated by a zero and a time entry in the *runout time* column. The *substitution* column is used when an item is underproduced, as the broccoli spears were.

In this residence hall foodservice, cleaning assignments are in addition to regularly scheduled cleaning. The name of the person responsible is recorded in the *cleaning assignment* column. Instructions for *prepreparation* for the following meal, whether the same day or next, are in the same form as the production schedule. They begin with the name of the person and are followed by the menu item, quantity, and special instructions. The items listed under prepreparation will be on the schedule for the following meal.

The emphasis in this text on production scheduling is justified by its being an important element of production control with bearing on the cost of materials, labor, and energy. Regardless of the perfection of the schedule and the assignment of employees to implement it, however, the production employees are the ones who make the schedule work. Realization of this simple fact implies the value of production employee meetings.

Batch cooking is a variant of production scheduling but is not always done in foodservice operations. In batch cooking, the total estimated quantity of menu items, often vegetables, is divided into smaller quantities, placed in pans ready for final cooking or heating, and then cooked as needed. The example shown in Table 9.2 is a time schedule developed for steaming rice for the dinner meal in a university residence hall foodservice. It illustrates the way in which production can be scheduled to meet the customer demand throughout the meal service time with assurance that a fresh product is being served.

Table 9.2. Example of time schedule for batch cooking of rice

Buttered Rice (800 servings)	
Steam at 5# pressure, 30 min uncovered	
Time	**Quantity**[a]
3:30 pm	1 pan
3:45 pm	1 pan
4:00 pm	2 pans
4:30 pm	4 pans
4:45 pm	2 pans

[a] 4# rice/12 × 20 × 4 in. pan = 80 servings, 10 pans = 800 servings.

High-speed equipment available today, such as steamers and convection ovens, has made batch cooking feasible for a broad range of menu items. Because many vegetables do not hold up well in a heated service counter, they are frequently prepared using batch cookery. In Figure 9.6, guidelines are given for batch cooking of vegetables. Grilled, deep fried, and broiled items are examples of other products that should be cooked in small quantities to meet service demands.

Production Meetings

Foodservice managers in small operations and unit supervisors in large ones should hold a meeting daily with employees in the production unit. Ordinarily, these pro-

Figure 9.6. Procedure for batch cookery of vegetables.

Procedure

1. Steamers and small tilting trunnion kettles behind the service line are the most useful kinds of equipment for vegetable batch cooking.
2. Divide each vegetable into batches small enough to be served within 20 to 30 minutes. Arrange in steamer pans ready to be placed in steamers, or in containers ready for pouring into the kettles.
3. Keep the prepped vegetables in the cooler until needed.
4. Cook batches as needed. In planning, allow time for loading and unloading the equipment, for cooking, for finishing the product with desired seasoning, sauce, or garnish, and for carrying to the serving line.
5. Undercook slightly if the vegetable must be held before serving.
6. Have all your seasonings, sauces, and garnishes ready for finishing the dish.
7. Do not mix batches. They will be cooked to different degrees, and colors and textures will usually not match.

Source: Professional Cooking (p. 393) by W. Gisslen. Copyright ©1989 by John Wiley. Reprinted by permission of John Wiley & Sons, Inc.

duction meetings can be rather short, but when a menu is changed, more time is required to discuss new recipes and employee assignments. In foodservice operations that serve breakfast, lunch, and dinner, these meetings are generally scheduled after lunch, when activity in the production unit is minimal.

During these meetings, production unit employees can be encouraged to discuss the effectiveness of the schedule just completed. Problems such as underproduction and suggested corrective measures should be recorded for the next time the menu appears in the cycle.

The meeting should conclude with a discussion of the production schedule for the following three meals. At this time, the employees should review recipes for the various menu items, possible substitutions, and prepreparation for the following day. Free discussion of work loads is appropriate for such meetings and can be a morale builder for the employees who really make the schedule work.

SUMMARY

The objective of food production is the preparation of food items in the needed quantity and with the desired quality at a cost appropriate to the particular service. Production is the second subsystem in the transformation element of the foodservice system and generally is the process by which goods and services are created. In foodservice, production is the managerial function of converting food purchased in various stages of preparation into menu items that are served to a customer.

Production planning primarily is the effective synthesis of quantity, quality, and cost objectives that are applicable to foodservice operations in which the product or menu items depend on the type of operation. Forecasting is a function of production and provides the data base for decision making and planning. Production demand has a degree of uncertainty and varies with the type of foodservice operation. Overproduction generates extra costs because salvage of leftover menu items is not always feasible; underproduction also can increase cost because of additional labor for preparation, and often the substitution, of a higher-priced item.

Good forecasts are essential for managers in planning smooth transitions from current to future output, regardless of size or type of foodservice. Historical records constitute the base for most forecasting processes. Forecasting models are categorized as time series, causal, or subjective. The moving average and exponential smoothing models in the time series category currently are the most applicable to foodservice operations because they can be set up on a computer spreadsheet.

Production scheduling in foodservice operations is defined as the time sequencing of events required by the production subsystem to produce a meal. It occurs in two distinct stages. In the planning stage, forecasts are converted into the amount of each menu item to be prepared and the distribution of the food products to supervisors in each work center. Supervisors in each unit assume the responsibility for the action stage by preparing a production schedule on which each item is assigned to a specific employee and the start time for preparation is recorded. Batch cooking is part of production scheduling and is necessary for serving fresh menu items. Production meetings should be held daily by the supervisor with the employees in the unit. The meeting should conclude with a discussion of the production schedule for the following meal or meals, review of recipes, and prepreparation. Employee work loads also should be discussed.

BIBLIOGRAPHY

Adam, E., & Ebert, R. (1992). *Production and operations management: Concepts, models, and behaviors* (5th ed.). Englewood Cliffs, NJ: Prentice-Hall.

Buchanan, P. W. (1993). *Quantity food preparation: Standardizing recipes and controlling ingredients*. Chicago: The American Dietetic Association.

Cichy, R. (1983). Productivity pointers to promote a profitable performance. *The Consultant, 16*(1), 35–36.

Dougherty, D. A. (1984). Forecasting production demand. In J. C. Rose (Ed.), *Handbook for health care food service management* (pp. 193–199). Rockville, MD: Aspen Publishers.

Fitzsimmons, J. A., & Sullivan, R. S. (1982). *Service operations management*. New York: McGraw-Hill.

Georgoff, D. M., & Murdick, R. G. (1986). Manager's guide to forecasting. *Harvard Business Review, 64*(1), 110–112, 119–120.

Konnersman, P. M. (1969). Forecasting production demand in the dietary department. *Hospitals, 43*(18), 85–87.

Lambert, C. U., & Beach. B. L. (1980). Computerized scheduling for cook/freeze food production plans. *Journal of The American Dietetic Association, 77*, 174–178.

Makridakis, S. G. (1990). *Forecasting, planning, and strategy for the 21st century*. New York: Free Press.

Makridakis, S., & Wheelwright, S. C. (Eds.). (1989). *Forecasting methods for management* (5th ed.). New York: Wiley.

Messersmith, A. M., & Miller, J. L. (1992). *Forecasting in foodservice*. New York: Wiley.

Messersmith, A. M., Moore, A. N., & Hoover, L. W. (1978). A. multi-echelon menu item forecasting system for hospitals. *Journal of The American Dietetic Association, 72*, 509–515.

Miller, J. J., McCahon, C. S., & Miller, J. L. (1993). Foodservice forecasting: Differences in selection of simple mathematical models based on short-term and long-term data sets. *Hospitality Research Journal, 16*(2), 95–102.

Miller, J. L., & Shanklin, C. S. (1988). Forecasting menu item demand in foodservice operations. *Journal of The American Dietetic Association, 88*, 443–446, 449.

Miller, J. L., Thompson, P. A., & Orabella, M. M. (1991). Forecasting in foodservice: Model development, testing, and evaluation. *Journal of The American Dietetic Association, 91*, 569–574.

National Restaurant Association. (1988). *Foodservice industry 2000*. Current issues report. Washington, DC: Author.

National Restaurant Association. (1992). *Foodservice manager 2000*. Current issues report. Washington, DC: Author.

Palmer, J., Kasavana, M. L., & McPherson, R. (1993). Creating a technological circle of service. *Cornell Hotel & Restaurant Administration Quarterly, 34*(1), 81–87.

Wood, S. D. (1977). A model for statistical forecasting of menu item demand. *Journal of The American Dietetic Association, 70*(3), 254–259.

10

Ingredient Control

Ingredient control is a major component of quality and quantity control in the production subsystem as well as a critical dimension of cost control throughout the foodservice system. As discussed in chapters 7, 8, and 9, the process of ingredient control begins with purchasing, receiving, and storage of foods and continues through forecasting and production.

Two major aspects of ingredient control are ingredient assembly and utilization of standardized recipes. An **ingredient room,** or ingredient assembly area, is designed for measuring ingredients to be transmitted to the various work centers.

The development and use of **standardized recipes** greatly facilitate purchasing and actual food production. When adjusted to an accurate forecast quantity, these recipes provide assurance that standards of quality will be consistently maintained. If a program for standardization of recipes could be purchased, no foodservice manager would be without one (Buchanan, 1993). Instead, a well-planned program beginning with the standardization of recipes and production procedures needs to be developed individually for each foodservice operation.

INGREDIENT ASSEMBLY

As discussed in chapter 8, concepts related to receiving, storage, and inventory control are important components of ingredient control, particularly issuing from storage. Clear policies and procedures control the issue and assembly of all food and supplies, from delivery to service, by requiring proper authorization for removal of products from storage and by issuing only required quantities for production and service. The ingredient room concept was introduced in chapter 8 as an aspect of issue control used in many foodservices, especially those with computer-assisted inventory systems.

Advantages of Centralized Assembly

Some type of issuing is used in most foodservice operations. The ingredient room contributes to cost reduction and quality improvement by stopping production employees from withdrawing large quantities of products from storage whether or not they are needed. Use of the ingredient room has many advantages including the redirection of cooks' skills away from the simple tasks of collecting, assembling, and measuring ingredients to production, garnishing, and portion control. By limiting access to ingredients, over- and underproduction of menu items can be eliminated, thus controlling costs.

The concept of an ingredient room dates from the late 1950s. Flack (1959), one of the first to implement a central ingredient room, reported reduced labor cost as a major benefit. Dougherty (1984) indicated that approved issue control does not require increased personnel and need not require renovation of facilities; activities being performed in other areas of the foodservice operation can be reassigned to give more emphasis to issue procedures. In recent years, as competition has increased and the available labor pool has shrunk, managers have shown an increased interest in quality, cost controls, and more efficient use of labor. They are adopting, therefore, the ingredient control concept within their foodservice operations (Buchanan, 1993).

Centralized Ingredient Control

Traditionally, in foodservice operations, individual cooks obtain recipe ingredients from storerooms, refrigerators, and freezers. Ingredients are issued in cases, boxes, or bags. In production areas, ingredients that are not currently being used generally are stored in storage bins or on shelves. For example, a cook may keep sugar and flour bins under the work counter and spices and condiments above on a shelf.

Keeping track of unused portions of issue units, especially perishable products, provides a challenge for the cook or foodservice manager and leads to decreased control. For example, 5 pounds of frozen mixed vegetables left over from a 30-pound case must be held and used the next time mixed vegetables are on the menu. With the vast number of ingredients used in foodservice operations today, controlling partially used packages of food items can become a major problem. A cook may decide to add the extra mixed vegetables to the soup kettle rather than return the unneeded amount to storage. Such a practice alters the recipe and adds to cost. This is only one example of problems caused by the traditional method of issuing ingredients.

With centralized ingredient control, the cook is issued only the 25 pounds needed for the forecast production demands on the day of service. The excess of 5 pounds is held in frozen storage in the ingredient room until the next time mixed vegetables are needed for a recipe or as a vegetable on a future menu. Control of unused portions is facilitated because storage is located centrally rather than in various work units throughout the kitchen.

As Buchanan (1993) indicated, food production includes basically two functions: prepreparation and production. Traditionally, both functions have been performed by cooks. As discussed previously, focusing the cooks' efforts and attention on direct production tasks and away from the simple tasks of prepreparation, which can be assigned to less-skilled employees, can lead to operational efficiencies. Observation of critical control points is extremely important in prepreparation of food products. Good sanitation practices are necessary to avoid contamination and cross-contamination in preparing foods for production, as Table 5.1 in chapter 5 shows.

In addition, less-skilled employees can develop skill in performing prepreparation tasks, thereby reducing labor cost. Combining tasks for two or more recipes using similar ingredients is another efficiency. For example, chopped onions may be needed for both meat loaf and a sauce at lunch and for both soup and salad on the dinner menu. By centralizing prepreparation, all the onion that is needed can be chopped at one time and divided into separate batches for each of the four recipes.

The microbiological safety of food products in the ingredient room is very important. Critical control points for prepreparation need to be determined; procedures for corrective action should be in the employees' policy and procedure manual. Standards for preparation of products used in beef stew were identified in the HACCP flowchart shown in chapter 5 (see Table 5.1). Washing hands and sanitizing utensils and equipment are cited often as corrective measures if standards were not met.

Function of the Ingredient Room

The primary function of the ingredient room is to coordinate assembly, prepreparation, measuring, and weighing of the ingredients to meet both the daily production needs and the advance preparation needs of recipes for future meals. Buchanan (1993) indicated that specific activities vary among foodservice operations but that, in general, ingredient rooms operate 24 hours in advance of production needs.

An ingredient room may be limited to premeasuring only dry and room-temperature ingredients, or it may be a center in which all ingredients, whether they are at room temperature, refrigerated, or frozen, are assembled, weighed, and measured. The availability of appropriate equipment, such as prepreparation equipment and low-temperature storage, will help determine the activities to be performed in an ingredient room. A freezer in or near the assembly area is required, for example, if ingredient room employees withdraw and thaw frozen products in advance.

Storeroom employees generally assemble full cases and unopened cans for delivery to production. Partial amounts from cans or cartons should be weighed and measured in the central ingredient assembly area. This practice can eliminate waste in a foodservice operation by avoiding partially used cans or cartons of the same product in several locations. For example, canned tomato sauce may be needed by both the main production and salad units; with a central ingredient room, only one partially used can would need to be stored.

Meat products, which have been ordered according to production demands, may or may not be handled in the ingredient room. For example, preportioned meats that have been ordered in a quantity appropriate for the production demand may go directly from receiving to production. Greater control is possible, however, if these products were distributed through the ingredient room.

Prepreparation tasks, such as cleaning and slicing or dicing vegetables or breading and panning meat items may be done in a prepreparation area or the ingredient room. Today, many of these products, especially produce, are purchased already prepped.

After all ingredients for each recipe have been weighed, measured, chopped, or otherwise prepared, each ingredient is packaged and labeled. Ingredients for each recipe are then transported with a copy of the recipe to the appropriate work unit or held until the scheduled distribution time.

Many computer programs already in existence support the ingredient room (Buchanan, 1993). Recipes that are increased by computer are easier to read than those that are increased manually, and they are correct if correct data are used. This and other types of computer-adjusted recipes increase the speed and productivity of the ingredient room employee and eliminate product errors and costly mistakes. Computers also can produce consolidated ingredient lists by individual ingredients or by production area. In a computer-assisted operation, the recipe can be adjusted and peel-off labels printed out for each ingredient. These labels, which usually have the ingredient name and quantity, facilitate marking the ingredient packages for each recipe.

Ingredient Room Organization

In the design of a new foodservice facility, an ingredient room that can be secured should be located between the storage and production areas. In an existing facility, the ingredient room may be located in or near a storeroom, combined with the prepreparation area, or put in a designated part of the production unit. According to Dougherty (1984), the placement of a central ingredient room should facilitate the flow of materials from storage to production.

Necessary large equipment includes refrigeration, which should be in or near the area, as well as a water supply. Trucks or carts are needed for assembly and delivery of recipe ingredients and portable bins for storing sugar, flour, and other dry ingredients. A worktable or a counter is required, with shelving over or near it for such products as spices. An example of an ingredient room worktable (State of New York, 1981), is shown in Figure 10.1.

Figure 10.1.   Ingredient room worktable arrangement.

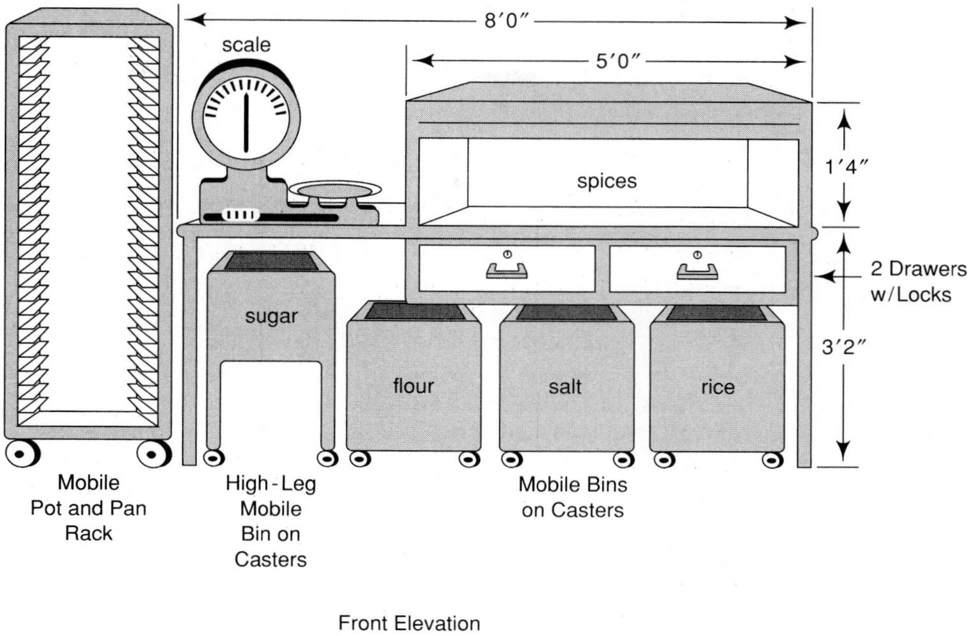

Front Elevation

Scales are the most essential pieces of equipment for an ingredient room. A countertop scale, similar to the one shown in chapter 8, and a portion scale are required for weighing the various types of recipe ingredients. Other equipment will vary depending on the specific functions assigned to the ingredient room, such as a slicer, vertical/cutter mixer, food waste disposal, and mixer.

Ingredient Room Staffing

According to Dougherty (1984), in an operation without an ingredient room, production employees spend about one third of their time determining needs, obtaining supplies, and weighing and measuring ingredients. By centralizing these activities, production employees are free for higher-level skill tasks, which allows management to reassign less-skilled employees from production to the ingredient room. In smaller operations in which a full-time employee is not needed for ingredient assembly, schedules of production employees can be arranged to permit them to weigh and assemble ingredients. As Dougherty (1984) indicated, the more activities are centralized, the greater will be the realized benefits.

Employees assigned to the ingredient room must be literate, able to do simple arithmetic, and familiar with storage facilities. They are often responsible for receiving, storage, and ingredient assembly. Qualifications and training, therefore, must be specific to each of these areas of responsibility. Training should include the following areas:

- Environmental conditions required to store specific foods
- Ventilation and humidity factors in dry storage
- Safety precautions in handling and storage of nonfood items and toxic materials
- Sanitation standards to prevent contamination or deterioration of foods during storage
- Security measures to ensure against pilferage

Figure 10.2.   Job description.

INGREDIENT ASSEMBLY CLERK

Name: _____   Coffee Break: _____

Hours: _____   Lunch Hour: _____

Coffee Break: _____

GENERAL RESPONSIBILITIES AND DUTIES

I. Storeroom Requisition

Pick up storeroom requisition from main kitchen office. This list is a consolidation of all ingredients needed for one day's production of menu items that would be channeled through the ingredient assembly area. Check all items against the inventory on hand in the ingredient assembly area and mark out supplies not needed with a felt pen. (For example, sugar will be issued in a 100# bag. The total amount will not be used in any one day. Annotate this requisition, give it to the storeroom clerk for filling and delivery.)

II. Production Schedule and Production Recipes

Pick up production schedule and production recipes from the main kitchen office.

A. Production recipes. There will be one recipe for each menu item that will be channeled through the ingredient assembly area. These recipes have been adjusted according to the forecast.

B. Production schedule. This form will be prepared by the food production manager and will include the following information:

1. name of employee responsible for product preparation
2. name of recipe or menu item
3. time that ingredients are to be delivered to specified production area

Always check the production schedule to determine how to plan your work. Arrange production recipes according to the times listed on the production schedule so that those to be prepared first are weighed and packaged first.

III. Assembling, Weighing, and Packaging

Accurately assemble, weigh, and measure ingredients according to the amounts on the production recipe printout. Set up ingredients on carts according to the following procedure.

A. Choose a cart large enough to accommodate the ingredients for each recipe.

• Weighing and measuring procedures

The job description for an ingredient room clerk depends on the activities included in the procedures for the ingredient room. An example of a job description is shown in Figure 10.2. The following factors are among those to consider in scheduling personnel for ingredient assembly:

Figure 10.2.   *continued*

B. Using masking tape and felt-tip pen, labe trays and shelves with recipe name. All menu items set up on cart should have approximately the same delivery time.

C. Package ingredients as follows:
1. Paper cups and lids. Use for small amounts of liquid items, such as Worcestershire sauce, lemon juice, salad oil.
2. Plastic bags (2 to 3 sizes). Use for all wet ingredients, such as frozen eggs, fresh vegetables. Fasten bags securely and label with felt marker. Write name of ingredient and weight on bag.
3. Glassine sandwich bags. Use for small amounts of spices or other dry ingredients.
4. Paper bags (assorted sizes). Use for dry ingredients.
5. Counter pans. Use for larger quantities of ingredients, such as bread crumbs, flour, celery. Cover the pan with either a lid or plastic wrap.
6. Original container. Always leave items in the original container whenever possible, such as no. 10 cans or wax paper around margarine.
7. Jars (½ and 1 gal). For vinegar and salad oil.

Put small quantities of ingredients for the same recipe on 18-inch × 26-inch sheet pan. After all items are weighed, labeled, and covered, place them on assigned cart shelves. If the cart is not to be delivered immediately, push it into a walk-in refrigerator if any items require refrigeration. Meat items, such as ground beef, should not be put on the cart until just before delivery.

The production recipe printout should be delivered with the assembled ingredients to the production area. If necessary, inform the cook that the ingredients have been delivered. Ingredients may be transferred from the carts to a table when practical.

IV. General Responsibilities
You are responsible for
A. storing and refrigerating opened cans and using these opened cans first;
B. storing items and rotating stock in the ingredient assembly area;
C. using proper sanitation procedures;
D. washing the equipment that you have used;
E. keeping your work area clean and orderly at all times; and
F. other duties as assigned.

Source: From "Issue Control and Ingredient Assembly System," by D. A. Dougherty. In J. C. Rose (Ed.), *Handbook for Health Care Food Service Management* (pp. 203–204), 1984, Rockville, MD: Aspen Publishers.

- Size of operation
- Frequency and time of deliveries
- Size of ingredient room and location of other storage areas
- Type, number, and complexity of menu items to be assembled
- Number of workstations to be supplied
- Schedule for delivery of ingredients to production and serving areas
- Extent of prepreparation performed in ingredient assembly area

Dougherty (1984) stated that 12-hour coverage by at least one staff member, 7 days a week, is needed in a 300-plus bed healthcare facility. The maximum number of employees needed for even large hospitals is three unless an extremely large amount of prepreparation is required.

RECIPES

A **recipe** is a formula by which weighed and measured ingredients are combined in a specific procedure to meet predetermined standards. The recipe is actually a written communication tool that passes information from the foodservice manager to the ingredient room and production employees. In addition, the recipe is an excellent quality and quantity control tool, constituting a standard for each item on the menu that meets customer and management approval. Cost for each recipe can be easily computed because the ingredients and the amount will be the same each time the recipe is used.

Recipes should have a format that is easily understood by the personnel who are responsible for the production and presentation of menu items to customers. Once a recipe has been tested repeatedly and accepted by management and customers, it becomes a standardized recipe and always gives the same results.

Format

Most recipes are written in a definite pattern or style that is identified as a *format*. For most effective use, all recipes in a particular foodservice operation should be in the same format. This uniformity of style simplifies recipe usage by cooks.

Large quantity recipes generally differ in format from home recipes, which have a list of ingredients followed by procedures. The cook in a foodservice operation is more likely to make errors if required to read alternately from the top and bottom of the recipe. A block format, in which the ingredients are listed on the left side of the recipe and the corresponding procedures directly opposite them on the right, generally is used for quantity recipes. In a complete block format, horizontal lines separate each group of ingredients with procedures from those of the next, and vertical lines separate the ingredient, amount, and procedure columns. A modified version, in which only horizontal lines separate the required ingredients for each procedure, is often used. An example of a modified block format is the recipe for lasagna in Figure 10.3 (Shugart & Molt, 1993). This **recipe format** is suitable for both recipe cards and computer printouts.

Figure 10.3.   Example of the format of a typical quantity recipe.

LASAGNA

Yield: 48 portions or 2 pans 12 × 20 × 2 inches   *Portion:* 6 oz.
Oven: 350°F   *Bake:* 40–45 minutes

Ingredient	Amount	Procedure
Ground beef	5 lb AP	Cook beef, onion, and garlic until meat has lost pink
Onions, finely chopped	12 oz	color.
Garlic, minced	2 cloves	Drain off fat.
Tomato sauce	3 qt	Add tomato and seasonings to meat.
Tomato paste	1 qt	Continue cooking for about 30 minutes, stirring occa-
Pepper, black	1 tsp	sionally.
Basil, dried, crumbled	1 tsp	
Oregano, dried, crumbled	1 Tbsp	
Noodles, lasagna	2 lb 8 oz	Cook noodles according to directions.
Water, boiling	2 gal	Store in cold water to keep noodles from sticking.
Salt	2 oz	Drain when ready to use.
Vegetable oil	2 Tbsp	
Mozzarella cheese, shredded	2 lb 8 oz	Combine cheeses.
Parmesan cheese, grated	6 oz	Arrange in two greased 12 × 20 × 2-inch counter pans in
Ricotta cheese or cottage cheese, dry or drained	2 lb 8 oz	layers in the following order: Meat sauce, 1 qt Noodles, overlapping, 1 lb 12 oz Cheeses, 1 lb 4 oz Repeat sauce, noodles, and cheeses. Spoon remainder of meat sauce on top. Bake at 350°F for 40–45 minutes. Cut 4 × 6.

Approximate nutritive values per portion

Calories (kcal)	Protein (grams)	Carbohydrate (grams)	Fat (grams)	Cholesterol (mg)	Sodium (mg)	Iron (mg)	Calcium (mg)
338	21.4 (25%)	26 (30%)	16.8 (44%)	81	840	3.1	254

Note: 1½ oz (¾ cup) dehydrated onions, rehydrated in 1 cup water, may be substituted for fresh onions.

Source: Food for Fifty (p. 416) by G. S. Shugart and M. Molt. Copyright © 1993 by Macmillan Publishing Company. Reprinted by permission of the publisher.

Specific information should be included on each recipe to simplify its use by those preparing and serving the food. Generally, recipes include the following information:

- Name of food item
- Coded identification
- Total yield
- Portion size and number of portions
- Cooking time, if required
- List of ingredients in order of use
- Amount of each ingredient by weight, measure, or count
- Procedures
- Panning or portioning information
- Serving and garnishing suggestions

The important basic information pertaining to the detailed recipe is shown in the heading of the lasagna recipe (Figure 10.3). It includes the yield, portion size, oven temperature, and baking time. Ingredients are listed in order of use in four procedural groups. Amounts are given to the right of the ingredients. Standard U.S. weights and measures are generally used for all ingredients except small amounts of spices and oil, which are given in the common household units of teaspoons (tsp) and tablespoons (Tbsp).

Recipes for production greater than the 48 portions in the lasagna example could be written entirely in standard measures without the use of teaspoons, cups, and other household units. Many of the computer software programs have the capability of calculating quantities in metric, but, so far, these measures are not commonly used in the United States. Some programs will calculate only in weights, some only in measures, but the majority of the large programs calculate in both. The ability of the metric system to define small amounts in either weight or liquid measures obviates the necessity of resorting to household measures. In small foodservice operations, measures may be used rather than weights, but weighing is more accurate than measuring for determining the proper amount of an ingredient.

Corresponding procedures for each group of ingredients are printed directly opposite them on the right side of the recipe. The layering of the ingredients in lasagna is very important and is detailed in the last procedure. Temperature, baking time, and portioning instructions are repeated for the convenience of the cook. Procedures should be checked for clarity and explicitness, permitting a cook to prepare a perfect product without asking managers for further explanation. In recipes that require the use of equipment such as mixers, time and speed should be specified. For example, the procedure for combining the first three ingredients in a cake might be "cream shortening, sugar, and vanilla on medium speed 10 minutes."

Additional information may be added at the bottom of recipes, such as the approximate nutritive values per portion and variations of the recipe. Special serving instructions, including garnishing and portioning suggestions as well as storage requirements before and after service are often included. If recipe cards are used, these additions can be printed at the bottom or on the back. The ingredient and procedure portion of a recipe, however, should never be printed on the back of the card. For long recipes, a second card or page should be used.

Before adopting a particular format, variations might be tried to give the cooks an opportunity to choose the one they like best. Flexibility in recipe formats is not possible if a computer program is used. The foodservice manager should compare recipe formats in software packages before purchasing one. Once a format is chosen, all recipes should be printed in that style. In converting to a new format, a good method is to adapt the most used recipes first and then gradually extend the conversion to the entire file. The production manager will need to conduct inservice training sessions on the new format for production employees.

Recipes for use at a workstation should be in large print that is easily readable at a distance of 18 to 20 inches; large file cards or 8½-by-11-inch paper must be used. Recipes for the ingredient room or production unit should be in a plastic cover while in use and in some type of rack at the workstations. Because of the larger print, production employees should not have to pick up recipes for a closer look.

The recipe name is generally in bold letters, either in the middle or to the side, at the top of the recipe card. In most operations, a file coding system is established for quick access to each recipe. For example, major categories can be established for menu items, such as beverages, breads, cakes, cheese, cookies, eggs, fish, meat, pies, poultry, salad, sandwiches, soups, and vegetables. Each recipe can be coded with a letter designating the menu item category and a sequential number that identifies the individual recipe. Whatever system is developed should be easy for employees to follow. Today, many foodservices use color coding of major categories to make identification easier.

Methods of maintaining a master file of recipes vary with the foodservice operation. The importance of keeping a backup file of all recipes in a computerized operation cannot be emphasized enough. A minimum of two sets of recipes should be available, one to be kept in a permanent file accessible only to the foodservice manager and the other in a file for use by the cooks. The number of sets depends upon how many employees in the organization need recipes for specific reasons. For example, persons planning menus or purchasing food might save many hours if the recipes are readily available in their office files.

For operations using a computer-assisted system, the recipe file, which permits individual printouts on demand, is the key data base. An example of a computer-generated recipe is shown in Figure 10.4. Because computerized recipes are generated each time an item is on the menu, a protective cover is not needed. The computer-generated menus also may be readily modified, enabling the foodservice manager to maintain an up-to-date file.

Standardization

The ideal of every manager is to have recipes that consistently deliver the same quantity and quality product when followed precisely. Printed recipes from various sources will not guarantee uniform products in every foodservice. Variations in ingredient characteristics, customer demands, personnel, and equipment may require alterations to the recipe or even preclude its successful use. Production procedures are complicated and difficult to establish because many people are involved, each of whom has definite ideas about how a product should be prepared. **Recipe stan-**

Figure 10.4. Example of a developmental computer-generated recipe from a university residence hall foodservice operation using the percentage method for recipe adjustment.

CHOCOLATE WINDSOR CAKE   RECIPE CODE - 1C-C1-C-033-3   STATUS - DEVELOPMENTAL   2

```
/         /   -*******   -*****************************************************************
                                    ***NUMBER OF PORTIONS              120           .133 LBS.  /   $0.0547
                                    PORTION SIZE / CCST                ONE OR TWO
EQUIPMENT-MIXER BOWL - FLAT PADDLE  MEAL PATTERN ALLOWANCE
  18X26X2 IN. PANS GREASED          SUGGESTED SERVING UTENSIL          SPATULA
  OVEN AT 350 DEG. F.               PAN SIZE                           18X26X2
                                    NUMBER OF PANS                     2
                                    WEIGHT PER PAN                     8.00 LBS.
                                    HANDLING LOSS                      3.00 PERCENT
                                    MINIMUM BATCH
RECIPE SOURCE-AIB/MAYFIELD/SEAL     MAXIMUM BATCH
                                    FORECAST UNIT                      60
12/17/92  5.53 PM    900917         *TCTAL RECIPE WEIGHT / COST        16.5 LBS.  /   $6.5640
                                    TCTAL RECIPE VOLUME
```

| | | | WEIGHTS AND | | |
CODE	PERCENT	INGREDIENT	MEASURES	AP/EP	STEP	PROCEDURE
0082021350	17.80	FLOUR CAKE	2.9 LBS		A	1. CCMBINE IN MIXER BOWL. MIX
0082050503	3.36	COCOA	0.55 LBS			ONE MINUTE ON SPEED NO. 1.
0082093008	27.29	SUGAR GRANULATED	4.5 LBS			
0031021000	1.78	NON FAT DRY MILK SOLIDS	0.29 LBS			
0082054100	0.59	SALT	0.10 LBS			
0082050058	0.71	BAKING POWDER	0.12 LBS			
0032050104	0.36	BAKING SODA	0.06 LBS			
0000000001	9.10	WATER	1.5 LBS		B	2. ADD AND MIX 1 MINUTE ON SPEED NO. 1
0041002555	11.08	SHORTENING CAKE	1.8 LBS			AND 3 MINUTES ON SPEED NO. 2.
						3. SCRAPE BOWL AND PADDLE.
0000000001	8.31	WATER	1.4 LBS		C	4. ADD AND MIX 1 MINUTE ON NO. 1 SPEED
0082051003	0.24	EXTRACT VANILLA	0.04 LBS			AND 2 MINUTES ON NO. 2 SPEED.
0031020205	13.05	EGGS WHOLE 30LB FZN	2.2 LBS		D	5. BLEND IN. MIX 2 MINUTES ON NO. 1
0000000001	6.33	WATER	1.0 LBS			SPEED.
						6. SCALE 8.0 LBS PER 18X26X2 INCH
						GREASED PAN.
						7. BAKE AT 350 DEG. F. FOR 25-30
						MINUTES OR UNTIL CAKE TESTS DONE.
						8. CCOL. FROST.
						9. CUT 1C X 6.
					E	10. ONE PORTION IS ONE SQUARE.

Source: Kansas State University Housing and Dining Services. Used by permission.

dardization, or the process of "tailoring" a recipe to suit a particular purpose in a specific foodservice operation (Buchanan, 1993), is one of the most important responsibilities of a production manager.

Standardization requires repeated testing to ensure that the product meets the standards of quality and quantity that have been established by management. Food cost and selling price cannot be correctly calculated unless recipes are standardized to use only specific ingredients in known amounts to yield a definite quantity. A standardized recipe must be retested whenever a small change is made in an ingredient, such as substitution of frozen or dehydrated vegetables for fresh in a recipe for beef stew.

Justification

Standardizing recipes is a time-consuming task, and managers have to be convinced that the process is worthwhile. Conviction follows from the realization that the most successful foodservice operations use only recipes that have been developed specifically for them. Probably the best example of the use of standardized recipes is in the multiunit limited-menu chains. Each batch of hot biscuits, pizza dough and toppings, fried chicken, and french fries is the same in each unit of the chain every day. This essential uniformity often is assured by recipes available to only a few employees in the main ingredient room of the corporate commissary. The packaged ingredients are sent to each unit throughout the region, the nation, and even the world. Without this stringent control, these operations could not maintain national and international reputations for quality.

Among the advantages for using standardized recipes are the following (Buchanan, 1993):

- *Promotes uniform quality of menu items.* All products should be the same high quality.
- *Promotes uniform quantity of menu items.* Recipes should produce a specific, designated quantity.
- *Encourages uniformity of menu items.* Servings of the same menu item should have the same size and appearance.
- *Increases productivity of cooks.* Clarifying procedures on recipes improves the efficiency of cooks.
- *Increases managerial productivity.* Recipes eliminate guesswork in food ordering and questions asked by cooks, thus freeing managers to concentrate on satisfying the customer.
- *Saves money by controlling overproduction.* Waste is controlled if only the estimated number of portions is produced.
- *Saves money by controlling inventory levels.* More just-in-time purchasing is possible, thus increasing cash flow.
- *Simplifies menu item costing.* Precise calculation of serving costs is vital in establishing selling prices.
- *Simplifies training of cooks.* Recipes with detailed procedures provide an individualized training program for new cooks and chefs.

- *Introduces a feeling of job satisfaction.* Cooks know that menu items will always be the same quality.
- *Reduces anxiety of customers with special dietary needs.* A nutrient analysis can be done on each recipe and ingredients can be identified for those with allergies or other health concerns.

Although standardized recipes offer many advantages in a foodservice operation, the key to success is ensuring that recipes are followed carefully and consistently each time an item is produced. Because the human element can be a major variable in product quality and uniformity, employee supervision and training are critical aspects in a quality assurance program.

The use of standardized recipes has some limitations. Even when they are followed, standardized recipes will not improve a product made from inferior ingredients; good specifications for quality products are essential (Mitani & Dutcher, 1992). Standardization also cannot eliminate the variation found in food. For example, climate, degree of maturity, growing regions, and age of products can affect the menu item. Ingredient substitutions also will affect the final product; the recipe has to be standardized again before the menu item is served to the customer. Recipes must be standardized for each foodservice operation. Foodservice managers should review and update previously standardized recipes to reflect changes in the organization. For example, the government's dietary guidelines (see chapter 3) have changed the way food is being prepared; salt, saturated fats, and sugar have been decreased, requiring successful recipes to be standardized again or eliminated.

Method

Standardization is basically a recipe-testing process designed to adapt recipes to a specific foodservice operation. Before any recipe is standardized, it should be thoroughly analyzed, and the relative proportion of ingredients to the menu item should be examined. For example, in a pour-type batter, like that for pancakes or waffles, the ratio of flour to liquid should be 1:1. A recipe varying too much from this would produce a batter too thin or too thick and obviously would need to be modified.

After modification, the recipe should be tested. Careful weighing and measuring of all ingredients are especially important. Notations regarding mixing and combining procedures, preparation and cooking time, temperatures, equipment and utensils, and the method of serving should be made at each stage of the standardization process. The total yield and number of portions must be determined as well as portion size. Special attention should be paid to clarity of procedures for each stage of preparation. Any procedural change also must be noted to ensure that the product is prepared the same way each time.

The final step in recipe standardization is evaluation of the product for acceptability by a taste panel consisting of cooks or chefs, managers, and customers. In many college residence hall foodservices, new products are added to the service line as bonus menu items, and students are asked to indicate on a questionnaire if the items should be included on menus. These panel members understand the product and are quite competent to estimate customer acceptance; they need not have the expertise to conduct a formal analysis. A distinction between this type of panel and a scientific taste panel will be more evident when product evaluation is discussed in chapter 11.

Buchanan (1993) stated that recipes still in a developmental stage may be used in an operation. This status should be clearly indicated on the recipe, however, as on the computer recipe for chocolate Windsor cake in Figure 10.4. She recommended the following procedural steps for standardization of recipes:

- Record amount of ingredients and production procedures cooks currently use or use a file of tested recipes.
- Select 100, 50, or 20 portions as a base size for all recipes in the file.
- Determine, with cooks, order of standardizing recipes.
- Assign one cook to the recipe standardization task.
- Prepare a trial batch of the base number of portions.
- Identify a group of persons to serve on a taste panel for evaluating products in each batch.
- Develop an evaluation form for recipes being tested.
- Make one ingredient change suggested on evaluation forms.
- Revise the recipe, if necessary, after each evaluation.
- Repeat trial batches and evaluations until three batches produce the expected quality and quantity.
- Assign a cook unfamiliar with the recipe to prepare the final recipe.
- Store standardized recipes in a permanent file.

Recipe Adjustment

Computer software programs for recipe adjustments are being used in many food-service operations today to help control food cost. Recipes can be easily adjusted to a specific quantity, for example 160 servings, instead of having to use 150 or 200 portions, which are the base size in many recipes.

Three procedures have been developed for the adjustment of recipes: the factor method, the percentage method, and direct reading measurement tables. Frequently, in many foodservice operations, home-size recipes are used to develop quantity recipes. This process involves special considerations in adjusting recipes designed for a small number of servings to a quantity appropriate for 100 portions or more. In the sections that follow, the three methods of recipe adjustment are described, and a section on adjusting home-size recipes to quantity production is included.

Factor Method

To increase a recipe using the **factor recipe adjustment method**, the ingredients are generally changed from measurements to weights and multiplied by a conversion factor. To simplify the adjustment process, the ingredients should be converted to whole numbers and decimal equivalents, that is, 2 lb 10 oz would be converted to 2.625 lb, but rounded to 2.6 lb for use. Only one decimal place is used in a recipe (e.g., 2.6 lb) unless the original amount is less than 1 pound, in which case two places would be shown (e.g., 0.62 lb). Whenever possible, liquid measures also should be stated in weights; however, liquid measurements may be converted to decimal equivalents of a quart or gallon. Table 10.1 provides data for converting ounces into decimals of a pound and cups and quarts into decimal parts of a gallon.

Table 10.1. Decimal conversions for weights (in pounds) and volume measures (in gallons)

(Pounds) Weight Measure		Decimal Unit	(Gallons) Volume Measure		
oz	lb.		cup	qt	gal
½		.031	½		
1		.063	1	¼	
1½		.093	1½		
2	⅛	.125	2	½	
2½		.156	2½		
3		.188	3	¾	
3½		.218	3½		
4	¼	.250	4	1	¼
4½		.281	4½		
5		.313	5	1¼	
5½		.343	5½		
6	⅜	.375	6	1½	
6½		.406	6½		
7		.438	7	1¾	
7½		.469	7½		
8	½	.500	8	2	½
8½		.531	8½		
9		.563	9	2¼	
9½		.594	9½		
10	⅝	.625	10	2½	
10½		.656	10½		
11		.688	11	2¾	
11½		.719	11½		
12	¾	.750	12	3	¾
12½		.781	12½		
13		.813	13	3¼	
13½		.844	13½		
14	⅞	.875	14	3½	
14½		.906	14½		
15		.938	15	3¾	
15½		.969	15½		
16	1	1.000	16	4	1

Note: This chart *cannot* be used to determine decimal parts of cups or quarts, only gallons . . . and pounds.

Source: ©1993, The American Dietetic Association. *Quantity Food Preparation: Standardizing Recipes and Controlling Ingredients*, 3rd edition. Used by permission.

Table 10.2. Conversion to decimal units

Ingredients	Amount	Decimal Units
Ground ham	4 lb	
Ground beef	4 lb	
Ground pork	4 lb	
Chopped onion	2 oz	.125 lb
Black pepper	½ tsp	
Bread crumbs	1 lb	
Eggs	12	1.5 lb
Milk	1 qt	

After ingredient quantities are converted to the new amounts, they may be changed into units suitable for the weighing and measuring equipment used in the foodservice operation. In most recipes, dry ingredients will be stated in pounds and ounces and liquids in cups and quarts. Because scales may not be accurate for weighing very small amounts, recipe ingredients of 1 oz or less are often stated in tablespoons or teaspoons.

Converting a recipe by using the factor method requires the following:

- Changing ingredient amounts to weight measures using whole numbers and decimals
- Dividing the desired yield by the recipe yield to determine the conversion factor
- Multiplying all recipe ingredients by the conversion factor
- Reconverting the decimal unit back into pounds and ounces or quarts and cups
- Rounding off amounts to quantities simple to weigh or measure and within an acceptable margin of error
- Checking math for possible errors

To illustrate use of the factor method, assume a college residence hall foodservice has ham loaf on the menu. The recipe in the file is for 50 portions; however, the forecast production demand is 250 portions. Using the procedure outlined above, the ham loaf recipe would be adjusted to the desired number of servings by first converting the ingredients, as appropriate, to decimal equivalents, as shown in Table 10.2.

The conversion factor to adjust the base recipe from 50 to 250 portions would be determined by dividing the 250 by 50; the resulting factor is 5. The next step is to multiply the ingredient amounts by the factor. To assist the cooks in using the recipe, ingredients for 250 portions of ham loaf would be stated in pounds and ounces or quarts and cups, as shown in the last column in Table 10.3.

Percentage Method

In the **percentage method**, measurements for ingredients are converted to weights and then the percentage of the total weight for each ingredient is computed. The number of portions is forecast, which provides the basis for determining the ingredient weights from the ingredient percentages. Formulas that have been con-

Table 10.3   Conversion from decimals

Ingredients	50 Portions	50 × 5 = 250 Portions	Conversion from Decimals
Ground ham	4 lb	20 lb	20 lb
Ground beef	4 lb	20 lb	20 lb
Ground pork	4 lb	20 lb	20 lb
Chopped onion	.125 lb	.625 lb	10 oz
Black pepper	½ tsp	2½ tsp	2½ tsp
Bread crumbs	1 lb	5 lb	5 lb
Eggs	1.5 lb	7.5 lb	7 lb 8 oz
Milk	1 qt	5 qt	5 qt

verted to percentages need not be recalculated. This method allows adjustment to the portion size or forecast and permits a shift of ingredients to be done easily.

Percentages can be readily determined using a desk calculator, and with computer assistance they are made even easier. McManis and Molt (1978) described the following step-by-step method for recipe adjustment via the percentage method.

Step 1   Convert all ingredients from measure or pounds and ounces to tenths of a pound. Make desired equivalent ingredient substitutions, such as frozen whole eggs for fresh or powdered milk for liquid.

Step 2   Total the weight of ingredients in a recipe after each ingredient has been converted to weight in the edible portion (EP). For example, the weight of carrots or celery should be the weight after cleaning and peeling. The recipe may show both AP (as purchased) and EP weights, but the edible portion is used in determining the total portion weight.

Step 3   Calculate the percentage of each ingredient in the recipe in relation to the total weight.

Formula:

$$\frac{\text{Individual ingredient weight}}{\text{Total weight}} \times 100 = \text{Percentage of each ingredient}$$

Step 4   Check the ratio of ingredients. Standards of ingredient proportions have been established for many items. The ingredients should be in proper balance before going further.

Step 5   Establish the weight needed to give the desired number of servings, which will be in relation to pan size, portion weight, or equipment capacity.

Examples include the following:

- Total weight must be divisible by the weight per pan.
- A cookie portion may weigh 0.14 lb per serving; therefore, 0.14 times the number of desired servings equals the weight needed.
- Recipe total quantities should be compatible with mixing bowl capacity.

Use the established portions, modular pan charts, or known capacity equipment guides to determine batch sizes. The weight of each individual serving is the constant used in calculating a recipe.

Step 6 Cooking or handling loss must be added to the weight needed and may vary from 1 to 30 percent, depending on the product. Similar items produce predictable losses that with some experimentation can be accurately assigned. The formula for adding handling loss to a recipe is as follows:

$$100\% - \text{handling loss} = \text{yield \%}$$
$$(\text{yield \%})(\text{total quantity}) = \text{desired yield}$$
$$\text{total quantity} = \frac{\text{desired yield}}{\text{yield \%}}$$

Example: Yellow cake has a 1% handling loss. Desired yield is 80 lb of batter for 600 servings.

$$100\% - 1\% = 99\% \text{ or } .99$$
$$.99 \text{ total quantity} = 80 \text{ lb batter}$$
$$\text{total quantity} = \frac{80 \text{ lb}}{.99}$$

total quantity = 80.80 lb of ingredients for 80 lb available batter

Step 7 Multiply each ingredient percentage number by the total weight to give the exact amount of each ingredient needed. After the percentages of each ingredient have been established, any number of servings can be calculated and the ratio of ingredients to the total will be the same. As in the factor method, one decimal place on a recipe is shown unless the quantity is less than 1 pound, in which case two places are shown.

Reviewing the process of expanding a recipe will help illustrate recipe adjustment by this method. Tables 10.4, 10.5, and 10.6 demonstrate how to expand a recipe using the percentage method. The result is a brownie recipe for 60 portions of 0.12 lb each.

Table 10.4. Original recipe, brownies

Ingredients	Amount
Eggs	12
Sugar	2 lb
Fat, melted	1 lb
Vanilla	¼ C
Cake flour	12 oz
Cocoa	8 oz
Baking powder	4 tsp
Salt	2 tsp
Nuts, chopped	12 oz

Source: "Recipe Standardization and Percentage Method of Adjustment" by H. McManis and M. Molt, 1978, *NACUFS Journal*, p. 40. Used by permission.

Table 10.5.   Calculate percent, brownies

Percent*	Ingredients	Measure	Pounds
20.34	Eggs	12	1.32
30.82	Sugar	2 lb	2.00
15.41	Fat, melted	1 lb	1.00
1.70	Vanilla	¼ C	0.11
11.56	Cake flour	12 oz	0.75
7.71	Cocoa	8 oz	0.50
0.49	Baking powder	4 tsp	0.0316
0.41	Salt	2 tsp	0.127
11.56	Nuts, chopped	12 oz	0.75
100%			6.4886

* Individual ingredients weights = Ingredient percents total weight.

Source: "Recipe Standardization and Percentage Method of Adjustment" by H. McManis and M. Molt, 1978, *NACUFS Journal*, p. 40. Used by permission.

Direct Reading Measurement Tables

The third method of recipe adjustment uses **direct reading measurement tables**. These tables have the advantage of being simple and quick to use and require no mathematical calculations. Tables have been developed for both measured and weighed ingredients.

Buchanan (1993) developed tables for adjusting weight and volume of ingredients in recipes that are divisible by 25. Beginning with weights and measures for 25 portions, incremental values are given in these tables for various magnitudes up to 500. Use of these tables allows adjustment of recipes with a known yield in one of the amounts indicated to desired yields divisible by 25. An excerpt from a direct reading table for adjusting weight ingredients of recipes is shown in Table 10.7. For example, the amount of ground beef needed for 225 portions, using a recipe designed to produce 100 portions, can be determined easily using the table. If the 100-portion recipe requires 21 pounds of beef, then reading to the right for the amount for 200 portions and the left for 25 portions, the total of 47 lb 4 oz can be determined quickly.

Adapting Home-Size Recipes

Although good quantity recipes are readily available today, managers often prefer to develop their own formulations, rely on the expertise of cooks who may prepare an item without a written recipe, or adjust a home-size recipe to quantity production. In a residence hall or school foodservice, students may bring recipes from home and request that items be prepared. A nursing home resident might have a favorite item and share the recipe with the cook as a possible selection for the menu.

Special considerations are necessary in adjusting a recipe designed for 6 to 8 servings to an appropriate quantity for 100 servings or more. The following suggestions are given for expanding recipes from home to quantity size:

Table 10.6. Recipe for 60 servings, 0.12 lb each, 3% handling loss

Percent	Ingredients	Pounds
20.34	Eggs	1.51
30.82	Sugar	2.29
15.41	Fat, melted	1.14
I.70	Vanilla	0.13
11.56	Cake flour	0.86
7.71	Cocoa	0.57
0.49	Baking powder	0.04
0.41	Salt	0.03
11.56	Nuts, chopped	0.86
100%		7.42*

* Calculations:

60 × .12 lb each serving = Batter needed (7.2 lb).

3% Handling Loss = 7.2/0.97 = 7.42 lb.

7.42 lb × each individual ingredient percent = Amount of each ingredient.

Source: "Recipe Standardization and Percentage Method of Adjustment" by H. McManis and M. Molt, 1978, *NACUFS Journal*, pp. 40–41. Used by permission.

- Know exactly what ingredients are used and in what quantity.
- Make the recipe in the original home-size quantity following instructions exactly and noting any unclear procedures or ingredient amounts.
- Evaluate the product for acceptability to determine if the recipe has potential for expansion.
- Proceed in incremental stages in expanding the recipe, keeping in mind the quality and appearance of the original product. Evaluate quality at each stage, deciding if modifications are necessary as the recipe is adjusted.
- Determine handling or cooking losses after increasing the recipe to an amount close to 100 servings; usually 5% to 8% loss is typical. The actual yield of the recipe should be reviewed carefully. Mixing, cooking, and preparation times

Table 10.7. Excerpt from a direct reading table for adjusting weight ingredients of recipes divisible by 25

25	50	75	100	200	300	400	500
5 lb	10 lb	15 lb	20 lb	40 lb	60 lb	80 lb	100 lb
5 lb 4 oz	10 lb 8 oz	15 lb 12 oz	21 lb	42 lb	63 lb	84 lb	105 lb
5 lb 8 oz	11 lb	16 lb 8 oz	22 lb	44 lb	66 lb	88 lb	110 lb
5 lb 12 oz	11 lb 8 oz	17 lb 4 oz	23 lb	46 lb	69 lb	92 lb	115 lb
6 lb	12 lb	18 lb	24 lb	48 lb	72 lb	96 lb	120 lb

Source: ©1993, The American Dietetic Association. *Quantity Food Preparation: Standardizing Recipes and Controlling Ingredients*, 3rd edition. Used by permission.

Figure 10.5. Example of an adjusted home recipe using a computer software program.

Cinnamon Biscuits

COOKING TIME: 10-15 minutes   TEMPERATURE: 450°

PORTION: 2.5 oz

PAN SIZE:

INGREDIENTS:

	8 Servings	50 Servings	120 Servings	150 Servings
Flour, white	1½ c 2 T	2 qt 2 c	1 gal 2 qt	1 gal 3½ qt
Baking Powder	1 T ¼ t	⅓ c 2 T	1 c	1 ¼ c
Salt	½ t	2½ t	2 T	2 T 1½ t
Cream of Tartar	½ t	2½ t	2 T	2 T 1½ t
Sugar, granulated	1½ t	3 T 1t	½ c	½c 2 T
Milk, nf inst dry	3 T ½t	1¼ c	3 c	3¾ c
Shortening, veg	½ c	3 c	1 qt 3¼ c	2 qt 1 c
Water	½ c 1 T	3½ c	2 qt ⅓ c	2 qt 2½ c
Sugar, granulated	½ c 2 T	3¾ c	2 qt 1 c	2 qt 3¼ c
Cinnamon	2½ t	¼ c 1 T	¾ c	¾ c 3 T
Margarine, regular	1 T ¾ t	½c 2 T	1½ c	1¾ c 2 T
Milk, fl 2% fat	1 T ½ t	⅓ c 2 T	1 c 1 T	1⅓ c

(HANDLING LOSS 8.0%)

1. Mix flour, baking powder, salt, cream of tartar, sugar, and nonfat dry milk with mixer (low speed).
2. Add shortening. Mix until crumbly and coarse crumbs (low speed).
3. Add water all at once. Mix (low speed) to form soft dough. DO NOT OVERMIX. Dough should be as soft as can be handled.
4. Place dough on lightly floured board or table. (For larger quantities, divide dough in smaller portions). Knead lightly 15-20 times.
5. Roll dough to approximately ¼ inch thickness in shape of rectangle.
6. Combine sugar, cinnamon, melted margarine and milk. Spread on dough - rectangle, going close to edges.
7. With French knife, cut dough into approximately 3" x 3" squares.
8. Bring four corners to center and pinch together. Place in 2-inch deep baking pan, having biscuits close together and touching.
9. Biscuits may be held several hours in refrigerator until time to bake.
10. Bake just before serving until light brown (DO NOT OVERBAKE).
11. Serve warm.

ABBREVIATIONS:

nf = Nonfat   inst = instant   veg = vegetable   fl = fluid

Source: "Mom," Evelyn Larson, from Great-Grandma Larson

Micro Foodcare System
(700-15) 8-24-93

Source: Noaleen Ingalsbe, R.D. Used by permission.

should be noted, especially for producing the item in quantity, because these items increase substantially for quantity production.

- Check ingredient proportion against a standard large quantity recipe for a product of similar type to assess balance of ingredients.
- As with standardized recipes, evaluate products using taste panels and customer acceptance assessments before recipes are added to the permanent file.

The recipe adjustment method varies with the computer software program, which should be checked at the time of purchase. Research, using various computer programs for recipe adjustment, has shown a great variation in the quantity of each ingredient and, therefore, the quality of the product (Lawless & Gregoire, 1987–88; Lawless, Gregoire, Canter, & Setser, 1991). The factor method, however, is the most common for recipe adjustment in a computer program.

Consultant dietitians often have found that cooks in small nursing homes and hospitals prefer to work with recipes adjusted from home-size recipes, rather than use recipes initially designed for quantity. As mentioned previously in this chapter, cooks in these small operations also tend to prefer recipes with measures stated in volume rather than in weights. An example of a home-size recipe for cinnamon biscuits, which was adapted to various yield quantities from an original recipe for eight servings, is shown in Figure 10.5. This recipe was generated from a data base compiled for a computer-assisted foodservice management program developed by a consultant dietitian in Kansas.

Careful evaluation is important in adjusting home-size recipes to quantity production. Some recipes that are suitable for service at home are simply not practical to make in quantity because of the time constraints of large-scale operations. If extensive labor time is required, a product may be too costly for most foodservice operations.

SUMMARY

Ingredient control begins with purchasing, receiving, and storage of food and continues through forecasting and production. With centralized ingredient control, cooks can focus on production tasks and the simple tasks of preparation can be assigned to less-skilled workers, and this leads to operational efficiencies.

The primary function of the ingredient room is to coordinate assembly, preparation, measuring, and weighing of ingredients to meet both daily production and the preparation needs of recipes for future meals. The organization of the ingredient room is essential and should be self-contained in a small amount of space. Equipment should be adequate but not excessive. Employees assigned to the ingredient room should be trained according to a job description.

Standardized recipes are essential for quality and quantity control in a foodservice operation. A recipe is a formula by which measured ingredients are combined in a specific procedure to give predetermined results. The format of a recipe should be consistent for a particular operation; large-quantity recipes generally are done in a block format. Recipe standardization requires repeated testing to ensure that the product meets the standard of quality and quantity that has been established by management.

Three procedures have been developed for the adjustment of recipes: the factor method, percentage method, and direct reading measurement tables. In the factor method, ingredients are changed from measurements to weights and multiplied by a conversion factor determined by the change in yield. In the percentage method, measurements for ingredients are converted to weights, and the percentage of each ingredient of the total weight is computed. To ensure that the correct number and size of portions are prepared using this method, handling loss must be included in the formula. Direct reading measurement tables have been developed for adjusting weight and volume ingredients of recipes that are divisible by 25.

Occasionally a home-size recipe is adjusted for quantity production. The recipe should be expanded in incremental stages and evaluated for quality at each stage; some home recipes simply are not practical to make in quantity.

BIBLIOGRAPHY

Buchanan, P. W. (1993). *Quantity food preparation: Standardizing recipes and controlling ingredients* (3rd ed.). Chicago: The American Dietetic Association.

Dougherty, D. A. (1984). Issue control and ingredient assembly systems. In J. C. Rose (Ed.), *Handbook for health care food service management* (pp. 199–205). Rockville, MD: Aspen Publishers.

Flack, K. E. (1959). Central ingredient room simplifies food preparation and cuts costs. *Hospitals, 33*(17), 125, 128, 132.

Gisslen, W. (1989). *Professional cooking* (2nd ed.). New York: Wiley.

Kotschevar, L. H. (1988). *Standards, principles, and techniques in quantity food production* (4th ed.). New York: Van Nostrand Reinhold.

Lawless, S. T., & Gregoire, M. B. (1987–88). Selection of computer software for recipe adjustment. *NACUFS Journal, 13*(1), 24–27.

Lawless, S. T., Gregoire, M. B., Canter, D. D., & Setser, C. S. (1991). Comparison of cakes produced from computer-generated recipes. *School Food Service Research Review, 15*(1), 23–27.

McManis, H., & Molt, M. (1978). Recipe standardization and percentage method of adjustment. *NACUFS Journal,* 35–41.

Mitani, J., & Dutcher, J. (1992). Standardized recipes: Are they worth the hassle? *School Food Service Journal, 46*(8), 70–72.

Payne-Palacio, J., Harger, V. F., Shugart, G. S., & Theis, M. (1994). *West & Wood's Introduction to foodservice* (7th ed.). New York: Macmillan.

Puckett, R. P., & Miller, B. B. (1988). *Food service manual for health care institutions.* Chicago: American Hospital Publishing.

Shugart, G., & Molt, M. (1993). *Food for fifty* (9th ed.). New York: Macmillan.

State of New York, Office of Mental Retardation and Developmental Disabilities, Office of Mental Health, Division of Alcoholism. *Nutrition services ingredient room manual.* (1981). New York: Author.

Quantity Food Production and Quality Control

Most organizations have adopted the total quality management philosophy, which has the goal of providing customer satisfaction by increasing the quality of products and services. That is also a goal of the foodservice system. The foodservice manager has to train employees to make decisions based on the quality of food served to the customer.

Production of quality food in quantity involves a highly complex set of variables. Quantity food production varies widely with the type of operation and foodservice. Many types of foodservice operations and their different objectives were described in chapter 1. The one-meal-a-day pattern of a school cafeteria, for example, presents a different production challenge than the 24-hour-a-day limited-menu restaurant operation. Because the menu is the basic plan for the foodservice system, planning in the food production subsystem depends on menu item selections.

In chapter 5, four basic foodservices were described: conventional, commissary, ready prepared, and assembly/serve. Differences in processing foods in these various foodservices were indicated. Obviously, those using completely prepared foods requiring only thawing and heating prior to service have different production demands than conventional foodservices in which foods are prepared from scratch.

As stated in chapter 9, quantity food production involves control of ingredients, production methods, quality of food, labor productivity, and energy consumption, all of which are critical to controlling costs. Issues related to planning for production also were discussed. The true test of the production plan, however, is whether the food is acceptable to the customer, produced in the appropriate quantity, microbiologically safe, and prepared within budgetary constraints.

Quantity is the primary element that introduces complexity to food production in the foodservice system. Producing food for 100 people requires much more careful planning and large-scale equipment than does preparing food for a small group at home. In operations serving several thousand, the complexity is drastically increased. Large-scale equipment and mechanized processes are requisites to producing the vast quantities of food to serve such customers.

The foodservice manager must understand the principles of quantity food production. Expertise in food preparation techniques and equipment usage will enable the manager to perform competently in planning and evaluation of foodservice operations. As one expert in foodservice management stated, quantity food production is the nuts and bolts of the foodservice industry.

In this chapter, the basic objectives and methods of food production are described. Development of product standards are then discussed as well as production controls for all phases of operation in the production subsystem.

OBJECTIVES OF FOOD PRODUCTION

Food is cooked for three primary reasons:

- Destruction of harmful microorganisms, thus making food safer for human consumption
- Increased digestibility
- Change and enhancement of flavor, form, color, texture, and aroma

First, cooking at proper temperatures can destroy pathogens. The amount of heat required to kill a particular microorganism depends on such factors as time, method, type of food, and type and concentration of the organism. Adequate cooking is a major factor in foodservice sanitation, but proper handling before and after cooking is also critical. Referring to the HACCP flowchart for beef stew in chapter 5 (see Table 5.1), bacterial survival and contamination are the major hazards in the cooking standards. Actions for correcting these hazards are to control time and temperature and to discard products that have been contaminated by employees' hands or mouths.

Many foods become more digestible as a result of cooking. For example, protein in cooked meat is more digestible than in raw meat. Raw starch in foods such as potatoes and flour will gelatinize during cooking or baking and become more digestible.

Nutrient value of some foods can be decreased if they are improperly cooked. Vegetables, for example, are often handled improperly during prepreparation or production, thus causing vitamin or mineral loss. The amount of cut surface and length of time between prepreparation and production affect nutrient retention. Nutrients also may be leached out in the cooking water or oxidized in cooking. The degree of loss is affected by the amount of water, length of cooking time, and cooking temperature. Proper attention to time and temperature control is important in preserving quality of food, especially the nutrient value. The foodservice manager should be well versed in concepts of food science in order to understand changes that occur in food during production.

The aesthetic quality of food can be enhanced by cooking; food can also be ruined or be made less palatable by improper procedures. The quality of any cooked food depends primarily on the following four variables:

- Type and quality of raw ingredients
- Recipe or formulation for the product
- Expertise of production employees and techniques used in preparation
- Method and duration of holding food items in all stages from procurement through service

The first two variables have been discussed in chapters 7, 8, and 10. The latter two are emphasized in this chapter.

Cooking may enhance, conserve, develop, or blend flavors, as in a sauce or soup. Preservation of color is another aesthetic objective in cooking food; for example, the dull color of overcooked vegetables makes them unappealing. Also, overcooked cabbage, cauliflower, and broccoli become mushy and develop a strong flavor, thus losing appeal.

The contrasts between the qualities of meat prepared properly and improperly provide another vivid example of the effect of cooking on the aesthetic quality of food. Consider the juicy flavorful quality of a properly broiled steak as compared to one that is dry, tough, and overdone. Baked products also are affected greatly by improper preparation and cooking techniques; tenderness is affected by overmixing, and the effects of over- and underbaking are obvious.

These examples all emphasize the importance of adequate controls throughout the prepreparation production and service processes. The holding stage, particularly in cook-chill or cook-freeze foodservices, is another critical component affecting the aesthetic quality and acceptability of food.

METHODS OF PRODUCTION

A variety of processes is involved in production of food for service. Preparation may be as simple as washing and displaying the food, such as fresh fruit, or as complex as the preparation of a lemon meringue pie. Production may include cooking, chilling, and freezing processes or a combination; in this section of the chapter, the emphasis is on cooking methods.

Cooking is scientifically based on principles of chemistry and physics. Properties of many ingredients used in food production cause reactions of various types. For example, baking powder, when exposed to moisture, gives off carbon dioxide, and the combination of egg, liquid, and oil produces an emulsion identified as mayonnaise.

Heat Transfer

Heat is the factor that causes many reactions to occur, and the type and amount of heat greatly affect the resulting product. Heat is transferred in four ways: conduction, convection, radiation, and induction.

Conduction is the transfer of heat through direct contact of one object or substance with another. Transfer can occur in any of the three states of matter: solid, liquid, or vapor. Metals, as a group of solids, are good conductors; however, different metals conduct heat at different rates. For example, copper, iron, and aluminum are effective conductors for cooking vessels; stainless steel developed from iron is not as effective. In cooking by conduction, the heat is first transferred from a heat source, usually gas or electricity, through a cooking vessel to food. Conduction is the dominant means of heat transfer in grilling, boiling, frying, and, to some degree, in baking and roasting. In pan broiling or grilling a steak, for example, the heat is transferred from the source to the pan or grill and then directly to the meat. In pan frying, fat is the transfer agent between the pan and the food.

Convection is the distribution of heat by the movement of liquid or vapor and may be either natural or forced. Natural convection occurs from density or temperature differences within a liquid or vapor. The temperature differences cause hot air to rise and cool air to fall. Thus, in a kettle of liquid or a deep fat fryer, convection keeps the liquid in motion when heated.

Forced convection is caused by a mechanical device. In convection ovens and convection steamers, for example, fans circulate the heat, which is transferred more quickly to the food, causing faster cooking. Another means of achieving the convection effect is to circulate the food, rather than the heat. A reel oven with shelves that rotate much like a Ferris wheel is an example of moving the food rather than the air. Stirring is another form of forced convection in which heat is redistributed to prevent concentration of heat at the bottom of the container. For example, in cooking a sauce or pudding, stirring is important not only to speed up the cooking but also to prevent scorching and burning.

Radiation pertains to the generation of heat energy by wave action within an object. The waves do not possess energy but they induce heat by molecular action upon entering food. Infrared and microwave are the two types of radiation used in food production. **Infrared waves** have a longer wavelength than visible light does. Broiling is the most familiar example of infrared cooking. In a broiler, an electric or ceramic element heated by a gas flame becomes so hot that it emits infrared radiation, which cooks the food. High-intensity infrared ovens are designed to heat food even more rapidly. Infrared lamps are commonly used in foodservice operations for

Figure 11.1. Microwave/convection oven.

holding food at a temperature acceptable for service. For example, in restaurants, infrared lamps are frequently placed over the counter where cooks set the plates of food for pickup by servers.

Microwaves have a very short length and are generated by an electromagnetic tube. In use, microwaves penetrate partway into the food and agitate water molecules. The friction resulting from this agitation creates heat, which in turn cooks the product. Because microwave radiation affects only water molecules, a waterless material will not become hot in a microwave oven. Thus, disposable plastic or paper plates can be used for heating or cooking some foods. Most microwaves penetrate only about 2 inches into food, and heat is transferred to the center of large masses of food by conduction. Microwave cooking is not a predominant method in foodservice operations, but microwave ovens are widely used for heating prepared foods for service. For example, microwave units are frequently available in hospital galleys, and vending operations often have small microwave ovens for heating sandwiches and soups. Cooking units, as shown in Figure 11.1, are now available with microwave and convection functions that can be used singly or together. Microwaves provide the major thawing and heating energy and hot-air jets provide surface color and texture (Lampi, Pickard, Decareau, & Smith, 1990). For example, potatoes may be cooked more rapidly using both microwave and convection, yielding the texture and flavor of a baked potato.

Induction is the use of electrical magnetic fields to excite the molecules of metal cooking surfaces (Riell, 1992). This method is widely used in Japan; it has been used in some other U.S. industries and is just beginning to be used in foodservice operations here. An induction-heat grill that cooks magnetically has been developed. The grill has no open flame and needs no thermostat for control. Temperatures are determined by specific alloys in four removable grills, which also allow a single grill to be divided into four different cooking zones. According to foodservice operators, induction heating is faster and more even than any other method of cooking. It also is cleaner, easier to ventilate, and does not require gas piping.

Cooking Methods

Cooking methods are classified either as moist heat or dry heat. **Moist heat methods** are those in which the heat is conducted to the food product by water or steam. **Dry heat methods** are those in which the heat is conducted by dry air, hot metal, radiation, or a minimum amount of hot fat. Different cooking methods are suitable for different kinds of food. For example, tender cuts of meat should be prepared using a dry heat method, and a tougher cut, such as that used for stew, should be cooked using moist heat.

Quantity food production and equipment are closely related and should be discussed together. Both are important components of cooking technology. Essentially, the equipment available dictates the choice of cooking methods. Almost every technical problem relating to cooking involves either method or equipment. The choices of cooking method and equipment are vital because of their effect on many aspects of daily operations, such as labor scheduling, productivity, product quality, speed of service, sanitation and maintenance, energy conservation, menu flexibility, and cost control.

The technology of cooking has become more complex over the past few decades, primarily because of innovations resulting in increased efficiency and sophistication of equipment. Combination cooking methods apply both dry and moist heat to the menu item (Ryan, 1991). Convection ovens and steamers, for example, are more efficient than their conventional counterparts. Less tender cuts of meat can be braised in a little oil in the oven followed by adding liquid and steaming until tender. These efficiencies are due to the fan circulation of heated air in the ovens and to higher temperatures of steam under pressure. In both types of equipment, cooking time is reduced and less energy is used. New cooking technology has been introduced by the increased sophistication of equipment, such as that in which two types of heat transfer are applied simultaneously. Examples include a convection roasting oven with steam injection and equipment that simultaneously grills and broils. In the following discussion of moist and dry heat cooking methods, equipment is a secondary consideration.

Moist Heat

The most common moist heat methods of cooking are boiling, simmering, stewing, poaching, blanching, braising, and steaming. These methods are similar and have only slight differences.

To boil, simmer, stew, or poach means to cook a food in water or a seasoned liquid. The temperature of the liquid determines the method. To *boil* means to cook in a liquid that is boiling rapidly. Boiling generally is reserved for certain vegetables and starches. The high temperature toughens the proteins of meat, fish, and eggs, and the rapid movement breaks delicate foods.

Simmering or *stewing* means cooking in a liquid that is boiling gently, with a temperature of 185°–205°F. Most foods cooked in liquid are simmered, even though the word *boiled* may be used as a menu term, such as "Boiled Corned Beef and Cabbage." Simmering or stewing is frequently done in a **steam-jacketed kettle** (Figure 11.2). The jacket creates the space surrounding the kettle through which steam is introduced to provide the necessary heat. Most kettles today make their own steam and are identified as *self-contained.*

To *poach* is to cook in a small amount of liquid that is hot but not actually bubbling. The temperature range is about 160°–180°F. Poaching is used to cook delicate food, such as fish, eggs, or fruit. Because poached foods are bland, the chef has to create maximum flavor without using high-calorie, fat, and sodium sauces for healthful cooking (Donovan, 1993).

Blanching is defined as cooking an item partially and briefly, usually in water, although some foods, such as french fries, are blanched in hot fat. Two methods are used for blanching in water. To dissolve blood, salt, or impurities from certain meats and bones, the item is placed in cold water, brought to a boil, simmered briefly, then cooled by plunging in cold water. Blanching is also used to set the color of and destroy enzymes in vegetables or to loosen skins of vegetables and fruits for easier peeling. For this latter purpose, the item is placed in rapidly boiling water, held there until the water returns to a boil, and then quickly cooled in cold water.

Figure 11.2.   Three different types of steam-jacketed kettles. *A,* Floor-mounted stationary self-contained kettle (electric). *B,* Tabletop-mounted, tilting, self-contained kettle (gas). *C,* Floor-mounted, tilting, self-contained cooker/mixer kettle (electric).

A   B   C

Source: Groen, Elk Grove Village, IL. Used by permission.

Figure 11.3.   Tilting fry or braising pan.

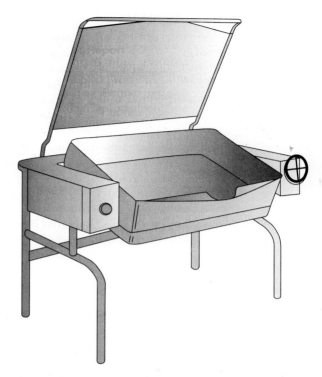

Source: CTX Division, Pet Inc., St. Louis, MO. Used by permission.

Braising involves cooking food in a small amount of liquid, usually after preliminary browning. In today's price-conscious economy, chefs are showcasing the more flavorful, less tender and costly cuts of meat by preparing homey braised beef dishes (Gilleran, 1991). Braised meats usually are browned in a small amount of fat, which gives desirable appearance and flavor to the product and sauce. With today's emphasis on healthful cooking, cubed meat can be quickly seared in hot stock, taking on a brown color without taking on added fats, and then finished at a lower temperature (Donovan, 1993). The liquid often is thickened before service by reducing it over hot heat rather than using a thickening agent. In some recipes for braised items, no liquid is added because the item cooks in its own moisture. Braising may be done on the range or in the oven, although today a covered **tilting fry pan** or braising pan is frequently used, such as the one shown in Figure 11.3. Other terms describing the braising process are *pot roasting, swissing,* and *fricasseeing.*

To *steam* is to cook foods by exposing them directly to steam. In many foodservice operations, compartment and high-pressure steamers are being replaced by the **pressureless convection steamer** (Figure 11.4), in which a fan directs the steam flow throughout the steamer cavity, encircling the food. This eliminates the need for steam pressure to cook the food and is well accepted by cooks because the steamer door can be opened at any time without fearing a gust of steam. Cooking in a steam-jacketed kettle is not steaming, because the steam does not actually touch the food. In foodservice operations, steaming is most commonly used for vegetable preparation.

Figure 11.4.   Pressureless convection steamer.

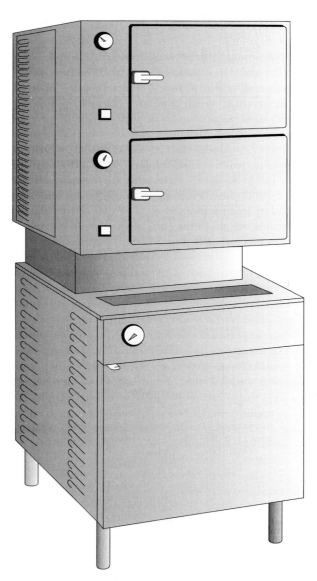

Source: CTX Division, Pet Inc., St. Louis, MO. Used by permission.

Dry Heat

The major methods of cooking without liquid are roasting, baking, oven frying, broiling, grilling, barbecuing, and frying, including sautéing, pan frying, and deep frying. Delmar, the author of *The Essential Cook*, contended that including foods cooked in oil as dry cooking does not make sense because oil is wet (Batty, 1990). He stated that three methods of cooking should be recognized: cooking in liquids, cooking in dry heat, and cooking in fat. In this text frying will be discussed under dry heat.

Roasting and Baking.   Cooking food by surrounding it with hot, dry air, usually in an oven, is referred to as *roasting* or *baking.* Generally, *roasting* refers to oven cooking of meat and poultry, and *baking* to desserts or bread. In roasting, cooking uncovered is essential. If the product is covered, the steam is held in, which changes the process from dry to moist heat cooking, such as braising or steaming.

The term *oven broiling* is sometimes used to refer to high-temperature cooking in an oven. Actually, oven broiling is a misnomer because it is not true broiling. In this process, food such as steaks or hamburger patties is cooked on baking sheets in the oven, usually at temperatures of 400°–450°F. Oven frying is a variation of oven broiling and involves placing food on greased pans and usually dribbling fat over it before baking it in a hot oven. The resulting product is much like fried or sautéed foods. These processes frequently are used when broilers, grills, or deep fat fryers are not available or are inadequate to handle the production demand. These methods are used most often in large healthcare operations or other large noncommercial foodservices.

Roasting or baking uses a combination of all three modes of heat transfer: conduction, convection, and radiation. Heat is conducted through the pan to the food and convection circulates air, either naturally or forced; heat also radiates from the hot walls of the cooking chamber.

A wide range of oven types is used in foodservice operations. The three main categories are hot-air ovens, infrared broilers, and microwave ovens. Several types of hot-air ovens are in use: range, deck, forced convection, conveyor, reel ovens, and low-temperature roasters. Infrared broilers and microwave ovens were described in the heat transfer section of this chapter.

The **range oven**, in use primarily in small operations, is part of a stove, generally called a *range,* and is under the cooking surface. The **deck oven** has traditionally been the standby of hot-air ovens in foodservice operations, so named because the pans of food usually are placed directly on the metal decks. Cooking chambers vary in size, depending on their function, either roasting or baking. Deck ovens may be stacked on each other, up to three sections high, and frequently are identified as *stack* or *sectional ovens.* Most deck or stack ovens have a separate heat source under each cooking chamber. A major problem with deck ovens is fluctuation in temperature.

The convection oven has a fan on the back wall that creates currents of air within the cooking chamber. This process eliminates hot and cold air zones, thereby accelerating the rate of heat transfer. The convection oven has three major advantages over the traditional range oven. It has more space than other ovens and accommodates two to three times as much food, reduces cooking time by 30% to 40% and cooks at 40°–50°F lower temperatures, thus conserving energy.

The standard convection oven is the square cabinet type that holds between 6 and 11 full-size baking pans and can be double-stacked to conserve floor space. An example is shown in Figure 11.5. Another type of convection oven is the roll-in rack oven. This oven operates on the same principle as the cabinet-type convection oven, but accommodates 16 or 20 full-size baking pans and can be rolled into the oven on casters.

Figure 11.5.   Convection oven.

Source: CTX Division, Pet Inc., St. Louis, MO. Used by permission.

Figure 11.6. Convection combo, double stacked, stand mounted (gas).

Source: Groen, Elk Grove Village, IL. Used by permission.

The **combination convection oven/steamer**, referred to as the "combo" or "combi," directs the flow of both convected air and steam through the oven cavity to produce a super-heated, moist internal atmosphere (Beasley, 1991). This is the "hottest" foodservice equipment item on the market today. Combo ovens are considered a revolution in cooking (Spertzel, 1992). Foodservice design consultants predict that it will replace most ovens and steamers in foodservice operations in the future. Meat, seafood, poultry, vegetables, and even delicate meringues, pastries, and breads with a crusty surface can be prepared in this oven. Four cooking methods are combined in one unit: convection, steam, convection plus continuous steam, and convection plus cycled steam. Gas combos (Figure 11.6), with a design certified by the American Gas Association, are much appreciated by foodservice operators who live in areas of the country in which natural gas costs up to one third less than electric power (Bean, 1990).

The combo performs such cooking functions as roasting, baking, steaming, blanching, simmering, braising, poaching, defrosting, broiling, grilling, reheating, proofing, and holding (Lampi et al., 1990). The advantage of this combination is its versatility, which permits menu expansion with a single piece of equipment, conservation of valuable floor space, and faster cooking with minimal shrinkage and maximum retention of flavor, color, and nutrients. Some manufacturers even include altitude sensors in the combi-steamer ovens that are being used in Arizona and Florida (Riell, 1992).

As customers demand fresh and healthful food, foodservice operators are responding with batteries of space-saving ovens capable of turning out small batches of products quickly (Chaudhry, 1993). Schlotzsky's, a Texas-based sandwich chain, has increased sales by baking its own bread for sandwiches and crusts for pizzas by using three different ovens: convection, conveyor, and pizza. Everything is baked daily in steam-injected convection ovens, and the pizzas are finished off in small stack-style pizza ovens, identified as *impingers* (Figure 11.7). The impingers are conveyorized electric or gas-fired ovens that make the bottom of the crust toasted and flaky. Air impingement permits heating, cooking, baking, and crisping of foods, two to four times faster than conventional ovens, depending on the food product. Sandwiches go through the conveyor oven two times; the cheese on a sliced bun is melted first and then after the meat is added, it is heated again.

The increased product yield of meats roasted at low temperatures over longer than usual periods of time has led to the development of *low-temp cooking and holding ovens* (Figure 11.8). Yield of meats can be increased up to 25% and energy consumption reduced by 30% to 50% (Beasley, 1991). Mobile low-temp ovens are flexible because they can also be used as warmers. Cooking temperatures are from

Figure 11.7.   Mobile single, double, and triple impingers.

Source: Lincoln Foodservice Products, Inc., Fort Wayne, IN. Used by permission.

Figure 11.8. Low-temp cooking and holding oven.

Source: Alto-Shaam, Menomonee Falls, WI. Used by permission.

100°F to 325°F, and holding temperatures are from 60°F to 200°F. Stacking low-temp roaster ovens are frequently used in catering operations.

Broiling.   **Broiling** is similar to grilling except that the heat source is above the rack with the food for broiling and below the rack for grilling (Donovan, 1993). Cooking by radiant heat is broiling. Food is placed 3 to 6 inches from the heat source, depending on the type and intensity of the heat. The temperature required depends on the fattiness, tenderness, or thickness of the food. Traditional broilers lack precise temperature controls, and foods must be closely monitored during the cooking process. Some new types of equipment, however, such as conveyorized infrared broilers, give more flexibility and control over the broiling process by varying the conveyor speed.

Charbroiling has become popular in many foodservice operations, especially in steak houses and fast-food hamburger establishments. Most **charbroilers** (Figure 11.9) use either gas or electricity, with a bed of ceramic briquettes above the heat source and below the grid. Because the heat source in charbroiling is from below, it technically is a grilling and not a broiling method.

Grilling.   Grilling, griddling, and pan broiling are all dry heat cooking methods that use heat from below. Grilling is the culmination of a chef's culinary experience (Solomon, 1992). Grilling is taking on an international flavor that is exciting customers. Asian teriyakis, Tex-Mex fajitas, Middle Eastern shish kebabs, and all-American cowboy-grilled steaks are some of the highlights. The fajita grill has been modified to include a rack on one side to hold fajita skillets upright, keeping them hot and easy to reach (Lorenzini, 1992). **Grilling** is done on an open grid over a heat

Figure 11.9. Electric char-broiler.

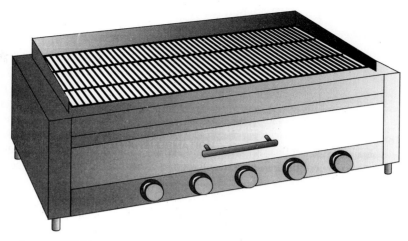

Source: CTX Division, Pet Inc., St. Louis, MO. Used by permission.

source, which may be an electric or gas-heated element, ceramic briquettes, or exotic woods and flavored chips.

Several manufacturers have added to a grill a hinged or removable top with its own heat source that permits cooking both sides of the food at one time (Riell, 1992). Known as **Clamshells**, as shown in Figure 11.10, these grills are popular with limited-menu restaurants serving many hamburgers. The gas Clamshell is easy to

Figure 11.10. Clamshell with gas broiler hood and griddle base.

Source: Lang Manufacturing Co., Redmond, WA. Used by permission.

operate; when the broiler hood is lowered, the burners turn on automatically. Penetrating infrared heat from the broiler and contact heat from the griddle drive the juices to the center of the meat, resulting in a perfectly cooked product in half the normal cooking time.

Mesquite grilling is found in upscale restaurants where there is a demand for fresh seafood, steaks, and gourmet burgers. Restaurants must be willing to support the increased cost of mesquite wood as fuel and the special equipment, as shown in Figure 11.11. Smoke from mesquite wood imparts a smoky and somewhat subtle tangy flavor, which has become a trademark in some restaurants; other popular woods include cherry, maple, oak, hickory, apple, and mango (Solomon, 1992). Because wood burns at a very high temperature, compliance with the safety requirements of building codes in many areas makes this method very costly.

Griddling with or without small amounts of fat requires a solid cooking surface called a **griddle**. Meats, eggs, and pancakes are frequently cooked on a griddle. *Pan broiling* is similar to griddling except that a sauté pan or skillet is used instead of a griddle. If fat is not poured off as it accumulates, the process becomes pan frying.

Barbecuing.   Barbecue-style meats can be cooked in conventional equipment or specialized barbecue equipment as follows ("Cooking Equipment," 1986):

- Basic open pit: steel or concrete wood-burning firebox under a grate, some with rotisserie spits
- Cylindrical or kettle-shaped smokers with domed lids, some with pans
- Large-capacity, closed-pit wood-burning rack or reel ovens with fireboxes to the rear
- Gas or electric upright rack ovens that cook at temperatures from 175° to 250°F. Smoke is dispersed by wood-charring or a smoke-concentrate device.
- Pressurized **barbecue smokers** that combine heat (350°–425°F) with pressure (12–15 psi) and smoke from wood chips

Figure 11.11.   Mesquite grill.

Figure 11.12.   Smoker.

Source: Alto-Shaam, Menomonee Falls, WI. Used by permission.

The smoker shown in Figure 11.12 is an electric, compact-size oven with racks to smoke up to 100 pounds of meat at a time. One and one half pounds of wood chips are used, and the smoke can be mild, medium, or heavy. In many areas, hoods or vents are not required. The dry rub of spices and the barbecue sauce unique to the operation add a piquant flavor to the meat.

Frying.   **Frying** is cooking in a fat or oil. Dry heat methods using fat include sautéing, pan frying, and deep frying. To *sauté* means to cook quickly in a small amount of fat. Sautéed foods are frequently cooked "à la minute" because the chef does not prepare the menu item until an order arrives in the kitchen (Donovan, 1993). The pan is preheated and not overcrowded to prevent food from simmering in its own juices. Meats are often dusted with flour to prevent sticking and help achieve uniform browning. After a food is sautéed, wine or stock is frequently added to dissolve brown bits of food clinging to the sides or bottom of the pan, a process called *deglazing.* Generally, this liquid is used in a sauce served with the sautéed item. Sautéing is commonly used in restaurants featuring French and northern Italian food.

Pan frying involves cooking food in a moderate amount of fat in a pan over moderate heat. While pan frying is similar to sautéing, it uses more fat and requires a longer cooking time. This method is used for larger pieces of food, such as chops or chicken. Pan frying is not used widely in foodservice operations, except as an initial step for a braised product.

Figure 11.13. Fryer with a
built-in filter.

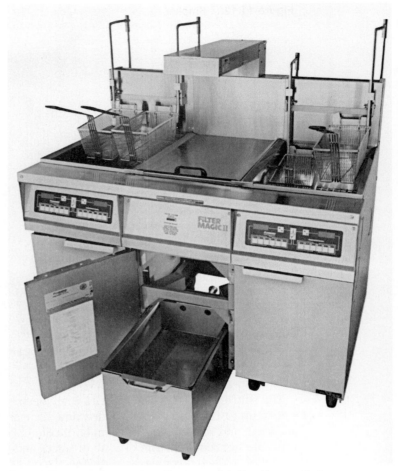

Source: Frymaster Corporation, Shreveport, LA. Used by permission.

In *deep frying,* food is submerged in hot fat. A well-prepared deep fried product should have minimum fat absorption, minimum moisture loss, attractive golden color, crisp surface or coating, and no off-flavors imparted by the frying fat. Foods for deep fat frying are dipped in a breading or batter that provides a protective coating and helps give the product crispness, color, and flavor. Obviously, the type and quality of the breading and batter affect the quality of the finished product.

Tremendous improvement has been made in gas and electric deep fat frying equipment for foodservice operations. Today, solid state electronics monitor the cooking cycle and control cooking temperature in **deep fat fryers**. Two of the most important developments in frying technology are precise thermostatic control and fast recovery of fat temperature, permitting foodservice operations to produce consistent quality fried food rapidly. Improvements for deep fat fryers have centered on fat filtration and automatic controls (Lampi et al., 1990). The filtration operation is an integral part of the equipment and is easy to use (Figure 11.13). A built-in filter features automatic washdown and an instant change filter. Automatic controls minimize the need for the operator to use judgment. Control functions range from a simple indication of elapsed time to presetting times for more than one product per basket, lift controls, holding time indication, adjustment of frying time for lower temperatures, and automatic shut down.

Another development in frying concepts is the *pressure fryer.* A pressure fryer can be described as a fryer with an air-tight lid that fastens securely over the kettle before frying (Figure 11.14). The fryer develops steam from moisture escaping the food. The pressurized steam creates an equilibrium of pressures between the steam within the food and the fat on the outside, thus minimizing moisture loss. The fat is pushed

Figure 11.14.   Pressure fryer.

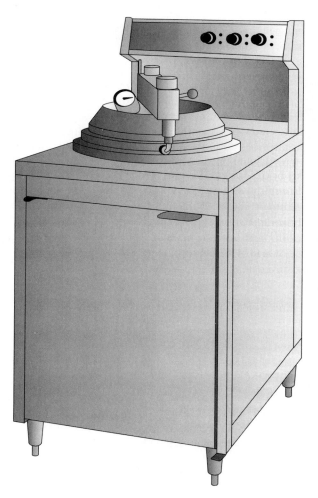

Source: CTX Division, Pet Inc., St. Louis, MO. Used by permission.

into closer contact with the food surface, which reduces cooking time. The general turbulence in the oil during frying also speeds cooking time. Pressurized fryers are used most frequently in restaurants specializing in fried chicken.

PRODUCT STANDARDS

Standards are a result of the managerial process of planning. In the discussion in chapter 3 on the managerial function of controlling, these standards were defined as the measurement of what is expected to happen. They therefore provide the basis for monitoring performance of the organization and taking corrective action deemed necessary.

The word *quality* is often used in combination with standards and is central to the traditional statement of the objective of a foodservice: the production of the highest possible quality food. This leads to the difficult problem of defining quality.

In chapter 2, the objective of a foodservice was defined as the production of food to satisfy the expectations, desires, and needs of a particular customer. The primary **quality attributes** of this food are microbiological, nutritional, and sensory, which require controls throughout the procurement/production/service cycle to maintain them. The increased time lapse between the preparation and distribution of food to the customer in some of the foodservices is critical and requires monitoring for nutritional value. David (1979) defined quality food very simply as that which is satisfying to the customer as well as nutritious and safe.

Although the customer's interpretation of quality is certainly important in the foodservice operation, quality can also be defined from a scientific standpoint. Thorner and Manning (1983) stated that the analyst or technologist refers to quality as an index or measurement obtained by grading or classifying a product's chemical and physical characteristics in accordance with explicit, predetermined specifications. The essential elements that must be evaluated in establishing the quality of a product are flavor, texture, appearance, consistency, palatability, nutritional value, safety, ease of handling, convenience, storage stability, and packaging.

They concluded that there are two dominant factors in the evaluation of quality:

- The actual chemical or physical measurement of the product
- The acceptance of the product by consumers

In addition, management will be concerned with the relationship of quality to cost, profit, and selling price, in particular to cost of the product, profits generated, and consumer acceptance.

Many factors are responsible for poor-quality food, among them poor sanitation, improper handling, malfunctioning equipment, incorrect prepreparation or preparation, and carelessness. Thorner and Manning (1983) compiled the following list of the prime factors responsible for significant quality changes:

- Spoilage due to microbiological, biochemical, physical, or chemical factors
- Adverse or incompatible water conditions

- Poor sanitation and ineffective warewashing
- Improper and incorrect precooking, cooking, and postcooking methods
- Incorrect temperatures
- Incorrect timing
- Wrong formulations, stemming from incorrect weight of the food or its components
- Poor equipment maintenance
- Presence of vermin and pesticides
- Poor packaging

Any of these factors, either singly or in combination, will contribute to poor quality and cause changes affecting the flavor, texture, appearance, and consistency of food.

The management of a foodservice operation should define the quality standards appropriate to a particular operation. In chapter 7, an example of specifications for the type of steak that would be purchased for three different types of restaurants was described. This is an excellent illustration of the need to define quality in relation to the objectives of a specific foodservice.

In Figure 11.15, a typical menu for a university residence hall foodservice is given, together with the quality standards for evaluating each of the menu items. The assumption is made that the items will be prepared using a standardized recipe defining the ingredient amounts, precise production methods, product yield, and serving directions. Too frequently, foodservice managers have failed to define standards of quality for menu items, and therefore the basis for control and evaluation is not available. In Appendix B, product standards for several menu items for a university foodservice are included. These standards are good examples of the efforts of a management team to define quality for foodservice operations.

In the remainder of this chapter, various types of production controls that affect the microbiological quality, nutritional value, and acceptability of foods are discussed. Chapter 12 focuses on sanitation principles in quantity food production.

PRODUCTION CONTROLS

Control is the process of ensuring that plans have been followed. Therefore, the essence of control, as described in chapter 3, is comparing what is set out to do with what was done, and taking any necessary corrective action. Control has not been effective until action has been taken to correct unacceptable deviations from standards.

In essence, then, *quality control* means assuring day-in, day-out consistency in each product offered for service. *Quantity control,* simply stated, means producing the exact amount needed—no more, no less. Each type of control is directly related to control of costs and thus to profit in a commercial operation or to meeting budgetary constraints in a nonprofit establishment. Over- and underproduction create managerial problems and have an impact on cost. Time and temperature, product yield, portion control, and product evaluation all relate directly to quality, quantity, and indirectly to control of costs.

Figure 11.15.   Example of quality standards for a lunch menu in a university residence hall foodservice.

MENU

Cheeseburger on Bun

Hot Dutch Potato Salad

Sliced Tomatoes   Dill Pickle Spears

Fudge Brownie

Fresh Fruit Cup

STANDARDS FOR MENU ITEMS

Cheeseburger. The hamburger patty should be evenly browned, juicy, and glossy with a moist and tender interior. The appearance should be pleasing and the flavor typical of beef. Processed American cheese covering the grilled patty will be melted without becoming stringy. The color will be yellow orange, typical of the type of cheese, and the flavor will be distinctive but will not dominate the meat.

Bun. The top crust should be golden brown and tender. The interior texture should be fine, even grained, free from large air bubbles, and have thin cell walls. The white crumb should be moist and silky. The bun will be fresh.

Hot Dutch Potato Salad. The cooked potato slices will be soft but firm. Diced celery, chopped onion, pimento, parsley, and crisp ⅛-inch-wide bacon pieces will be identifiable both in flavor and color. The tart vinegar sauce will be adequately sweet and thick enough to coat all ingredients. The salad will be served hot.

Fudge Brownie. The exterior should be dark brown and smooth with a darker interior. Texture should be slightly chewy and free of crumbling. The brownie should be slightly moist and tender, not doughy or pasty. Nut pieces should be evenly dispersed and identifiable. Each brownie should be approximately 3 inches square and ¾ inch uniform height.

Fresh Fruit Cup. Fruit should be selected for color, texture, flavor, and shape— e.g., golden peaches, blue plums, red watermelon, green grapes, white bananas. The fruit should be cut into bite-size pieces without predominance of any one type. A simple syrup flavored with lemon, lime, or other fruit juices should slightly coat the fruit.

Time and Temperature Control

Time and temperature are critical elements in quantity food production and must be controlled to produce a high-quality product. Excess moisture loss will occur in most products if the cooking time is extended even with a correct temperature. Most food production equipment today has timing devices as an integral part of construction. Equipment is available with computer controls that can be programmed by the manufacturer or operator to control cooking times (Riell, 1992).

Temperature is recognized as the common denominator for producing the correct degree of doneness. To assure this degree, temperature gradations vary dramatically for different food categories, depending on the physical and chemical changes that occur as food components reach certain temperatures. When foods are subjected to heat, the physical state is altered as moisture is lost. Chemical changes are complex and differ greatly among various food products. An area in which progress is being made is the development of more sophisticated ways of measuring "doneness," such as moisture readings, vapor-content analyzers, and optical sensors to detect color (Riell, 1992).

The proper control of temperature is often dependent on thermostats, which control temperature automatically and precisely. A well-designed thermostat should not be affected by changes in ambient temperatures, but instead should respond quickly to changes on the sensing element and be easily set. The functioning of the thermostat should be checked periodically.

Standardized recipes should state temperatures for roasting and baking. Figure 11.16 shows the temperatures commonly used in food production, in both centigrade and Fahrenheit. Tables should be developed for cooking time and temperature for items on the menu, including end-point temperatures for various products. For example, the degree of doneness for meat is best stated in terms of end-point temperature.

Various types of thermometers are needed for determining end-point temperatures or for checking the temperature of foods being held. A meat thermometer, for example, should be used to determine the internal temperature of meat. Hamburgers are now being cooked thoroughly to 160°F as a result of the *E. coli* breakouts in limited-menu restaurants (Van Warner, 1993). The chart in Table 11.1 shows the internal temperatures of meats at three levels of doneness: rare, medium, and well done. In meat cookery, however, carry-over cooking must be taken into account, especially for large cuts of meat. *Carry-over cooking* means that the internal temperature of the meat will continue to rise even after the meat is removed from the oven. This phenomenon occurs because the outside of the roasting meat is hotter than the inside, and heat continues to be conducted into the meat until the heat is equalized throughout the piece.

Time and temperature are closely related elements in cooking. Some foods, like prime rib roast, can be cooked for a longer period of time at lower temperatures to achieve the desired degree of doneness; others, like strip steak, should be cooked for a shorter time at higher temperatures.

The integral nature of time-temperature relationships is perhaps best illustrated with baked products in which accurate timing and temperature control are critical to producing high-quality products. For example, a cake baked at too high a temperature will cook too quickly on the outside before the baking powder or soda has produced expansion in the batter. As a result, the cake will crack and become too firm before the cake is done internally.

Product Yield

Yield is the amount of product resulting at the completion of the various phases of the procurement/production/service cycles. It usually is expressed as a definite weight, volume, or serving size. For most foods, losses in volume or weight occur in

Figure 11.16.   Centigrade and Fahrenheit thermometer showing temperatures commonly used in the kitchen.

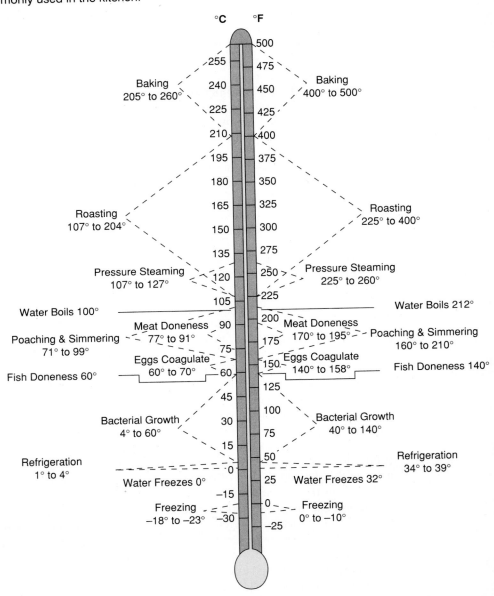

Source: Adapted from *Food Preparation for the Professional* (p. 108), by D. A. Mizer, M. Porter, and B. Sonnier, 1978, New York: John Wiley & Sons.

Table 11.1.   Internal temperatures of cooked meats

Meat	Rare	Medium	Well Done
Beef	120–125°F (49–52°C)	140–145°F (60–63°C)	160°F (71°C)
Lamb	125–130°F (52–54°C)	145°F (63°C)	160°F (71°C)
Veal	—	145–150°F (63–66°C)	160°F (71°C)
Pork	—	—	165–170°F (74–77°C)

Source: Professional Cooking, 2nd ed. (p. 210), by W. Gisslen, Copyright © 1989 John Wiley & Sons, Inc. Reprinted by permission of John Wiley & Sons, Inc.

each phase, even though a few foods, such as rice and pasta, increase in volume during production. The weight of meat, fish, and poultry **as purchased (AP)** decreases before it is served to the customer. This change occurs for many reasons, including the following: skin and bones may be removed, fat trimmed, or juices lost during preparation. Buyers must know how many **edible portions (EP)** of a food are required before deciding on the amount of food to be purchased. Only delicatessen cold meat products are 100% edible; the cost of bones, skin, and cooking loss is included in the purchase price.

The *Food Buying Guide* (U.S. Department of Agriculture, 1984), for use in child nutrition programs, provides data on the edible yield percentages of various foods to assist buyers in planning amounts to purchase. The cooked yields of selected meat, poultry, and fish are shown below.

Product	**Cooked Yield**
Ground beef (less than 30% fat)	42%
Pork ribs	45%
Chicken breast (with skin & bones)	66%
Turkey roll	66%
Fish sticks (raw breaded, 75% fish)	56%

As an example, if a 4-ounce grilled boneless and skinless chicken breast is desired, a 6.1-ounce chicken breast with rib bones and skin would have to be purchased. The 4-ounce EP of the grilled chicken breast is divided by the 66% yield of the chicken breast with rib bones and skin that would need to be purchased (AP).

The EP price is very important to the foodservice manager. The AP price of the chicken breast with bones and skin is $0.126 per ounce or $0.77 for 6.1 ounces. To calculate the EP price of the 6.1-ounce chicken breast, the AP price is divided by the EP yield percentage (66%).

$0.77 AP price ÷ 0.66 EP yield = $1.17 EP price

In addition to losses during prepreparation, losses may occur during actual production, in portioning, or in panning for baking. **Cooking losses** also account for decrease in yield of many foods, primarily because of moisture loss. As discussed previously in this chapter, the time and temperature used for cooking a menu item will affect the yield.

Table 11.2. Percent yield by weight of potato baked by two methods

Cultivar	Method 1[a] (%)	Method 2[b] (%)
Irish cobbler	83	97
Katahdin	77	95
Russet burbank	76	94

[a] Baked in 204°C (400°F) oven, skin rubbed with oil.

[b] Baked in 204°C (400°F) oven, skin covered with aluminum foil

Source: "Guidelines for Determining and Reporting Food Yields," by R. H. Matthews, 1976, *Journal of The American Dietetic Association, 69,* p. 398.

Handling loss occurs not only during production but also during portioning for service. McManis and Molt (1978) stated that handling losses must be considered in determining desired yield. They indicated that cooking and handling loss may vary from 1% to 30%, depending on the product being produced and served. In the brownie recipe included in chapter 4, (see Table 10.4), a 3% handling loss was reported to illustrate the percentage method of recipe adjustment.

When preparing food in quantity, these losses can have a cumulative impact on the number of portions available from a recipe. That loss must be considered when estimating production demand. Total yield and number of portions should be stated in a standardized recipe, taking into account changes in yield that occur from as purchased (AP) to edible portion (EP) to serving yield.

Equipment and cooking procedures also affect food yield. As Table 11.2 shows, the cooking method can have a major impact on resulting yield. A russet Burbank potato baked in foil resulted in 94% yield by weight compared with 76% for one baked without foil.

Less data are available, however, on handling losses during the production and service of food. Studies should be conducted to determine these data in order to have accurate information on production and service yields. In computer-assisted management information systems, data on yields from purchase to service are especially important in order for the computer to produce reliable information for ordering, recipe adjustment, production planning, cost, and nutrient composition of food items.

Portion Control

Portion control is one of the essential controls in production of food in quantity. In essence, it is the achievement of uniform serving sizes, which is important not only for control of cost but also for customer satisfaction. Customers of any foodservice establishment are concerned about value received for price paid, and they may be dissatisfied if portions are not the same for everybody. In a commercial cafeteria, for example, if one person is served a larger portion of spaghetti than another, customers may believe they have been treated unfairly even though the smaller portion may be adequate to satisfy their appetites.

Achieving portion control results from following well-defined steps, beginning with the purchase of food according to detailed and accurate specifications to assure that the food purchased will yield the expected number of servings. Often, products are ordered by count or number, with a definite size indicated. As an example, the food buyer may order 150 4-ounce portions of round steak for the Swiss steak recipe. An alternate way of controlling portion size is buying products in individual serving sizes. Examples are individual boxes of cereal, butter and margarine pats, packaged crackers, and condiment packets.

The next step in portion control is the development and use of standardized recipes, which were discussed in chapter 10. A standardized recipe will include information on total number and size of portions to be produced from the recipe. Following recipe procedures carefully during prepreparation, production, and service is necessary to ensure that the correct number of portions will result. For example, cooking a roast at too high a temperature will result in excess moisture loss and decreased yield.

Another aspect of portion control is knowing the size and yield of all pans, dishers, and ladles and stating the specific production and service equipment for each standardized recipe. Employees should be trained to use the appropriate equipment.

The approximate yield and typical uses for various sizes of dishers are given in Table 11.3. A number is commonly used to indicate the size of a disher, which specifies the number of servings per quart when leveled off. For example, a level measure of a No. 8 disher yields eight servings per quart, each portion measuring about 1/2 cup. Dishers are frequently used during production to ensure consistent size for such menu items as meatballs, drop cookies, and muffins, and during service to ensure correct serving size.

Table 11.3. Disher equivalents

Disher Number[a]	Approximate Measure[b]	Approximate Weight[b]	Suggested Use
6	10 Tbsp (⅔ cup)	6 oz	Entrée salads
8	8 Tbsp (½ cup)	4–5 oz	Entrées
10	6 Tbsp (⅜ cup)	3–4 oz	Desserts, meat patties
12	5 Tbsp (⅓ cup)	2½–3 oz	Croquettes, vegetables, muffins, desserts, salads
16	4 Tbsp (¼ cup)	2–2¼ oz	Muffins, desserts, croquettes
20	3⅕ Tbsp	1¾–2 oz	Muffins, cupcakes, sauces, sandwich fillings
24	2⅔ Tbsp	1½–1¾ oz	Cream puffs
30	2⅕ Tbsp	1–1½ oz	Large drop cookies
40	1½ Tbsp	¾ oz	Drop cookies
60	1 Tbsp	½ oz	Small drop cookies, garnishes
100	Scant 2 tsp		Tea cookies

[a] Portions per quart.
[b] These measurements are based on food leveled off in the disher. If food is left rounded in the disher, the measure and weight are closer to those of the next-larger disher.
Source: Adapted from *Food for Fifty* (p. 47) by G. Shugart and M. Molt, 1993, New York: Macmillan. Used by permission.

Table 11.4.   Ladle equivalents.

Approximate Weight[a]	Approximate Measure[a]	Approximate Portions per Quart	Suggested Use
1 oz	⅛ cup	32	Sauces, salad dressings
2 oz	¼ cup	16	Gravies, some sauces
4 oz	½ cup	8	Stews, creamed dishes
6 oz	¾ cup	5	Stews, creamed dishes, soup
8 oz	1 cup	4	Soup

[a] These weights and measures are based on food leveled off in the ladle. If food is left rounded in the ladle, the weight and measure are closer to those of the next-larger ladle.

Source: Adapted from *Food for Fifty* (p. 48) by G. Shugart and M. Molt, 1993, New York: Macmillan. Used by permission.

The approximate weight and measure for various sizes of ladles are given in Table 11.4. Ladles are used for portioning liquid or semiliquid foods, such as gravies, soups, and cream dishes. Generally, the number of ounces a ladle will hold is marked on the handle. The 1-ounce ladle holds 1/8 cup whereas a 6-ounce ladle holds 3/4 cup.

A spoon is often used for serving various foods, but it makes portion control much more difficult. Servers should be trained to weigh a menu item served with a spoon to establish a visual image of the correct typical portion. Another helpful tool is the spoodle, which is a spoon with a bowl (Figure 11.17). This tool permits better portion control than a regular serving spoon.

A variety of other tools are available to assist in consistent portioning of menu items. The biscuit cutter, in Figure 11.18, is a hollow metal rolling pin with compart-

Figure 11.17.   Spoodles.

Source: Vollrath Co., Sheboygan, WI. Used by permission.

Figure 11.18.   Biscuit Cutter.

Source: Moline Co., Duluth, MN. Used by permission.

Figure 11.19.   Expandable cake cutter.

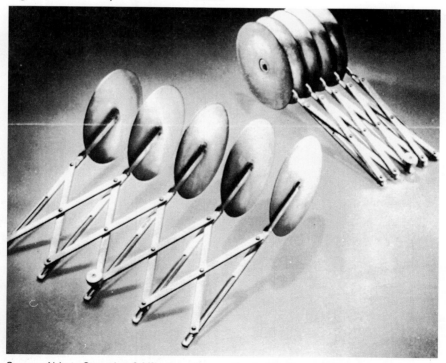

Source: Abbate Stamping & Mfg., Inc., Bensenville, IL. Used by permission.

ments. It produces the desired portion size when rolled through the dough. Expandable cutters (Figure 11.19) are frequently used for cutting foods prepared in large pans, such as gelatin salads, sheet cakes, and bar cookies. These products can be cut by rolling through the food both lengthwise and crosswise to yield an exact number of portions per pan. This technique not only results in standard portion sizes but also speeds up the portioning process. The bun divider (Figure 11.20) is another example of a compartment rolling device; it produces equal-size buns. Another tool is the pie marker (Figure 11.21), which simplifies the cutting of pie into equal-size wedges.

Perhaps the most elementary method of portion control is counting pieces. For example, one Delicious apple (113 count per box) may be the specified serving, or two strips of bacon, or two pancakes dipped onto the grill with a No. 12 disher.

Glass, china, or ceramic service dishes or paper containers of various sizes may be used to control serving size. The number of ounces of a beverage, for instance, can be controlled by the size of a cup.

Scales of many kinds are used throughout a foodservice because weight control is essential from the time the food enters the foodservice operation from the supplier until it is served. The use of scales during receiving and production has been dis-

Figure 11.20. Miller bun divider.

Source: Miller Bun Divider Co., Milwaukee, WI. Used by permission.

Figure 11.21.   Pie markers.

cussed in previous chapters. Portion scales are an important tool in quantity control in both the production and service units. Cooks and servers should have portion scales available for checking serving sizes.

Slicers are a valuable piece of equipment to assist in ensuring portion control because foods can be sliced more evenly and uniformly than can be done by hand. An example of a slicer is shown in Figure 11.22. A device on a slicer permits adjusting the blade for the desired degree of thickness for a product. Portion control guides in a foodservice operation should include the correct setting on the slicer for various food items. Automatic slicers are available that do not require the operator to return the food holding device manually, which frees the worker to continue other tasks while food is being sliced. The slicer is used most commonly for meat and meat products; however, it can also be used for vegetables, cheese, bread, and many other items. In portioning sliced foods for service, a portion scale should be readily available for a periodic check on portion size.

To illustrate how lack of control can create food shortages and drive up food cost, assume the server is not careful in serving the chicken and vegetable stir-fry entrée and heaps, rather than leveling-off, the product in the ladle. The result of yielding only four ladles per quart and not the intended five has several ramifications. The recipe calculated to serve 600 will serve only about 480; the desired food cost of

Figure 11.22.   Food slicing machine.

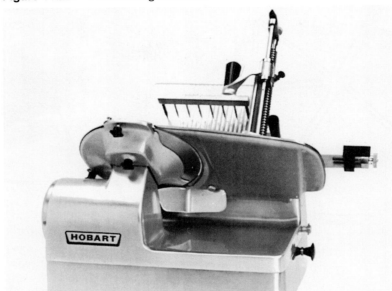

Source: Courtesy Hobart Corporation, Troy, OH. Used by permission.

$1.00 per serving will increase to $1.25. If a product is substituted because of the 120-portion shortage, additional cost may be incurred; the substitution may be a quickly prepared item that is more expensive than the chicken and vegetable stir-fry entrée, resulting in further cost overruns.

Foodservice operations should develop portion control guides to be used by production and service personnel. Table 11.5 shows typical portion sizes for selected menu items served in an employee cafeteria. Training and supervision are necessary, however, to ensure that proper tools and techniques are followed.

Foodservice managers must realize that portion control is essential for operating in the black. All food items must be purchased on the basis of weight and measure, and everything served on the basis of specified weight or portion size. Portion control is critical, not only for cost control but also for customer satisfaction. The key steps in ensuring portion control are purchasing by exact specifications, using standardized recipes, and using proper tools and techniques during all aspects of production and service.

Product Evaluation

Product evaluation is an important component of a foodservice quality control program. As emphasized in chapter 2, quality controls are essential at every step of the

Table 11.5.   Portion control guide for selected menu items in an employee cafeteria

Menu Items	Portion Size
Breakfast Items	
Bacon	2 slices
Cereal, hot	4 oz ladle
Eggs, soft cooked or fried	2
Eggs, scrambled	#16 disher
English muffins	2 halves
Sausage link	2 links
Syrups or sauces	2 oz ladle
Luncheon Entrées	
Baked beans	1 #12 disher
Barbecue meat on bun	1 #16 disher
Meat, egg, or fish salad on sandwich or cold plate	1 #12 disher or 2 #24 dishers
Soups, stews, chili, chowders	1 6 oz ladle
Stuffed green peppers	#8 disher stuffing in half pepper
Dinner Entrées	
Chicken, country fried	1 breast and 1 wing OR 1 thigh and 1 leg
Chicken, curry	4 oz ladle
Fried shrimp	5 if 18–20/lb
Meat balls	2 meat balls, #16 disher
Stroganoff or beef strips in sauce or gravy	4 oz ladle
Potatoes, Vegetables, and Substitutes	
Bread dressing	1 #12 disher
Broccoli, spears	2–3 spears
Okra, whole, french fried	5 pieces
Sweet potatoes, mashed	#12 disher
Tater tots	10–12 pieces
Salads	
Cottage cheese, with fruit or vegetable on plate	#24 disher
Deviled eggs	2 halves
Potato salad	#12 disher
Tomato wedge or slice	3 for salad (6 per tomato)
Desserts	
Cobblers	18" × 26" pan, cut 6 × 10
Ice cream and sherbets	#12 disher
Jumbo cookie	#20 disher (before baking)
Pies	9" tin, 8 per pie

foodservice operations, from development of procurement specifications through distribution and service to the customer. Quality control is a continuous process of checking to determine if standards are being followed, and if not, taking corrective action. In order to evaluate a product, it must be compared to both the specification requirements for the ingredients and the determined standard for the finished product.

Food Quality

Food quality is evaluated by sensory, chemical, and physical methods. Sensory methods are used to determine if foods differ in such qualities as taste, odor, juiciness, tenderness, or texture and to define the extent and direction of the differences (Palmer, 1972). Chemical and physical methods for testing food often are used with sensory analysis to identify reasons for differences in color, texture, and flavor. These methods are usually more reproducible and less time-consuming than sensory analysis, but they are limited to areas in which they have been shown to measure the quality that is apparent to the senses (Palmer, 1972).

In this text, **sensory analysis** is defined as a science that measures the texture, flavor, and appearance of food products through human senses. Commonalities in all the definitions are measurement, food characteristics, and human senses.

In very large foodservice organizations, competent personnel in quality control laboratories routinely conduct sensory, physical, and chemical tests during the initial development of new menu items. As discussed in chapter 5, commissary foodservices frequently employ food technologists and microbiologists as staff specialists with responsibility for regular monitoring of quality control. Large multiunit operations often have a sophisticated quality control program centralized at corporate headquarters.

Physical and chemical testing of foods may be limited in smaller operations in which a full-scale quality control program is not justified. These tests are in the domain of food science and are beyond the scope of this text. Because sensory evaluation of new menu items by panels and consumer testing can be a regular part of any foodservice operation, practical applications will be discussed.

Skelton (1984) contended that to ensure continuing success, foodservice operations should use sensory evaluation for new menu items and for maintaining the quality of existing items. As she stated, the cost of an error in a menu innovation can be extremely high, involving such costs as reprinting menus, training staff, and loss of business from dissatisfied patrons.

The necessary sensory evaluation of new menu items requires a sensory or consumer panel before serving them to the customer. Sensory panels are relatively small, ranging from 6 to 12 persons trained to judge quality characteristics and differences among food items. Panel members must be experienced in the use of score cards and in the vocabulary of food description; they also must be able to distinguish among various levels of the basic tastes (sweet, salt, sour, and bitter) and to repeat their assessments with reasonable precision.

In contrast to the trained panel, the consumer panel usually includes 50 to 100 persons who are reasonably representative of the target market. The objective in using consumer panels is to evaluate acceptance of, or preference for, a menu item.

Analytical and affective sensory tests are used for product evaluation and defined as follows:

- *Analytical sensory test:* differences and similarities of quality and quantity of sensory characteristics are evaluated by a panel of specially trained persons, commonly identified as a *trained panel.*
- *Affective sensory test:* preference, acceptance, and opinions of a product are evaluated by consumers who have no special sensory training, commonly identified as an *untrained panel.*

Skelton (1984) outlined three general purposes for sensory tests in foodservice operations: discriminating among food items, describing characteristics, and determining acceptance and preference.

- *Discrimination sensory test:* This test determines detectable differences among food items. For example, judges may be presented three samples and asked to choose the one that is different. If the panel has difficulty identifying the odd sample, the conclusion can be made that the product was not altered by the ingredient or process change.
- *Descriptive sensory test:* Quality control and recipe development both depend on descriptive tests to provide information about certain sensory characteristics. Adjectives, numerical scales, and rankings are used to evaluate such attributes as taste, aroma, texture, tenderness, and consistency. Figure 11.23 gives a list of terms that are useful in describing food products.
- *Acceptance and preference sensory test:* This test, used with a consumer panel, is intended to answer the questions of whether or not people will like the menu item. Using this test, preference for certain characteristics may be rated, or an overall preference score may be attained. The conclusion can be made that one recipe is not better than the other if consumers indicate no difference in overall preference.

Properly designed, executed, and analyzed sensory tests can be used to assist the foodservice manager in developing products that are more likely to succeed in the marketplace and in setting standards for ingredients and menu items (Setser, 1992). Table 11.6 outlines typical problem areas and the type of panel and category of sensory test appropriate to each.

Sensory Analysis Instruments

Before introducing a new menu item, the recipe must be standardized using the methods described in chapter 10. The final step in recipe standardization is evaluation of the product for acceptability by a taste panel that usually consists of cooks, supervisors, and managers. Standards for evaluating a recipe for blueberry muffins and stir-fried chicken and vegetables are shown in Figure 11.24.

Sensory analysis can be a useful tool for recipe standardization. In the early stages of this process, cooks and managers together can help in determining which sensory attributes are most affected by changes in formulation and ingredients and in indicating general trends in acceptability for those attributes. The two basic tests in this type of sensory evaluation are discrimination and acceptance. For discriminatory tests, such as difference tests and ranking, panelists do not need extensive training and large panels are not required. The paired comparison test can be used to differentiate between a pair of coded samples on the basis of some specified characteristic, for example, sweetness, crumbliness, moistness when chewing, lightness, or

Figure 11.23.   Terms used in judging food products.

Appearance (optical properties)

- Color

blueness	dark	greenness	redness
bright	dull	light	white
browness	grayness	pale	yellowness

- Other optical properties[a]

clear	irridescent	scum	translucent
cloudy	lustrous	sediment	transparent
frothy	muddy	shiny	uniform pigment distribution
glossy	opaque	sparkling	

[a] attributes related to transmission or reflectance of light but not related to pigmentation

Appearance (physical form)[b]

broken	large	rounded	smooth
crumbly	loose particles	shriveled	stringy
curdled	medium	shrunken	uniform shape
flat	rough	small	uniform size
irregular			

[b] visual perceptions related to dimensions and adherence between particles

Aroma and Flavor by Mouth[c]

burned	eggy	medicinal	scorched
buttery	fishy	nutty	spicy
caramelized	floral	putrid	starchy
dairy-like	fruity	rancid	

[c] sensations produced by volatile substances through nasal and oral cavities that provide characteristic flavors perceived in foods and beverages

Quality Judgments

acceptable	fresh	low	optimum
appealing	full-bodied	mellow	pleasing
delicate	good	normal	poor
desirable	high	objectionable	rich
excellent	ideal	obnoxious	stale

degree of browning. A sample evaluation form for a paired comparison test is found in Figure 11.25.

The ranking test extends the paired comparison test to three or more coded samples, and panelists are asked to rank them by intensity of the characteristics that differentiate the products. For example, if moistness of crumb obviously is affected by changing the formulation of the recipe, this characteristic is one that should be ranked. An example of this type of questionnaire for a ranking test is provided in Figure 11.26.

Figure 11.23.   *continued*

Taste and Chemical Feeling Factors

astringent	burning	sharp	sweet
biting	coolness	sour	tart
bitter	salty		

Textural Attributes

- Consistency—resistance to deformation through continuous changes of form

liquid	stiff	thick	viscous
runny	syrupy	thin	watery
slimy			

- Geometrical properties—size, shape, and orientation of particles perceived by tactile nerves

abrasive	crystalline	granular	rough
aerated	even	irregular	sharp
amorphous	fine	lumpy	smooth
beady	flaky	porous	stringy
broken	flat	powdery	uneven
cellular	fluffy	puffy	uniform
coarse	foamy	pulpy	
creamy	grainy	regular	

- Mechanical deformation—reaction of food to stress

adhesive	elastic	mealy	stiff
bouncy	firm	pasty	sticky
brittle	fracturable	plastic	tacky
chewy	friable	rubbery	tender
cohesiveness	gooey	short	toothpacking
crisp	gummy	soft	tough
crumbly	hard	solid	
crunchy	limp	springy	

- Mouth feel—physical feel of moistness and oilness of food as it is broken down

dehydrating	greasy	moist	saliva-inducing
dry	juicy	mouthcoating	soggy

Source: Compiled from many sources.

Table 11.6.   Problems solved by sensory evaluation.

Problem	Type of Panel	Category of Test
(1) Recipe development: Maximizing quality	Trained	Discrimination and Description
(2) Shelf-life: Storage time and temperature	Trained	Discrimination and Description
(3) Acceptance: Likelihood of purchase	Consumer	Acceptance or Preference
(4) Convenience food: Best substitution	Trained or Consumer	Description or Acceptance
(5) Quality control: Product consistency	Trained	Discrimination and Description

Source: "Sensory Evaluation of Food" by M. Skelton, 1984, *Cornell Hotel and Restaurant Administration Quarterly, 24*(4), p. 51. Used by permission.

Panelists should receive the samples for all tests in a random order to avoid any order biases in the testing. Ten to 12 panelists should be used. Results of the discrimination testing can be compared to affective testing of acceptance or preference to determine how the changes are liked. For affective tests, the number and type of panelists are an important issue. Such tests are aimed at determining the response of the consumer to the product. If too few representative or unbiased panelists are used, the results can be questionable. In the laboratory, 20 to 40 panelists generally can establish relative desirability; for hedonic evaluations of products, 50 to 100 judgments usually are necessary. Larger panels are needed only when unusually high precision is required.

For the relative desirability information, the use of the paired comparison or ranking test is also appropriate. In this type of test, the panelists' response to the following statement is requested: "Two samples of stir-fried chicken and vegetables are pre-

Figure 11.24.   Examples of quality standards for standardizing recipes.

Blueberry Muffins

A blueberry muffin should be scored on overall acceptability, including appearance, texture, tenderness, and flavor. The crust should be crisp, shiny, pebbly, and golden brown with a well-rounded top free from knobs. It should be large in volume compared to weight. The interior crumb should be moist, light, and tender with a coarse, even grain and no tunneling. The whole blueberries should be evenly distributed through the muffin. The blueberries should be moist but not discolor the muffin. The muffin flavor should be delicate, not bready or too sweet, and the blueberries should have a natural taste.

Stir-Fried Chicken and Vegetables

The chicken and vegetables should be scored on overall acceptability, including appearance, texture, tenderness, and flavor. The appearance should be pleasing and the chicken and vegetables identifiable. The bite-size pieces of chicken should be tender and juicy. The fresh vegetables should retain some of the original crispness and natural color. Fresh ginger and garlic should enhance, but not overpower, the flavors of the other ingredients. The sauce should be clear and ingredients glazed.

Figure 11.25. Example of paired comparison test form.

Name of Panelist _____ Date _____

Product _____

Evaluate the sweetness of the two samples of blueberry muffins. Taste the muffin on the left first. Indicate which is sweeter.

Code number 581   *Code number 716*

         —   —

Comments:

sented. Taste each sample in the order specified and indicate which saltiness level you prefer." Further information can be obtained by using a hedonic scale for individual attributes. An example of such a scale for three characteristics of interest in the blueberry muffins is given in Figure 11.27. A comparison of the attribute intensity information from the discrimination testing and the relative preference data might suggest changes for the next test run, which likely would involve adjustments in ingredient quantities, cooking times, procedures or serving methods. Generally, only one change should be made at a time in systematically evaluating the effect of the various changes. A cost analysis of different preparation methods for blueberry muffins is shown in chapter 18 (see Table 18.1).

Overall acceptability using the hedonic scale of two or more different muffins could also be obtained after recipe modifications are considered complete. Such evaluations of the product would likely include assessments from customers. To perfect some recipes, many tests might be needed. If test products are acceptable, the

Figure 11.26. Example of ranking test form.

Name of Panelist _____ Date _____

Rank the muffins for crumb moistness as you chew them. The least moist muffin is ranked first and the most moist sample is ranked fourth. Place the code numbers on the appropriate lines. Test the samples of the coded muffins in the following order: *212, 336, 471, 649.*

     ____   ____   ____   ____
      1   **2**   **3**   **4**

Figure 11.27.   Example of a hedonic scale for rating the acceptance of blueberry muffin attributes.

Name of Panel Member: ___John Green___

Date: ___8/29/94___

PRODUCT: ___Blueberry Muffin___

Please rate the muffin by checking one point on the following scale to indicate your evaluation of each attribute.

SWEETNESS	SALTINESS	CRUMBLINESS
☐ Like extremely	☐ Like extremely	☐ Like extremely
☐ Like very much	☐ Like very much	☑ Like very much
☐ Like moderately	☐ Like moderately	☐ Like moderately
☐ Like slightly	☐ Like slightly	☐ Like slightly
☐ Neither like, nor dislike	☑ Neither like, nor dislike	☐ Neither like, nor dislike
☑ Dislike slightly	☐ Dislike slightly	☐ Dislike slightly
☐ Dislike moderately	☐ Dislike moderately	☐ Dislike moderately
☐ Dislike very much	☐ Dislike very much	☐ Dislike very much
☐ Dislike extremely	☐ Dislike extremely	☐ Dislike extremely

COMMENTS: Explain your decisions

Muffin not sweet enough for tart blueberries

recipe is then put into the format being used in the operation and placed in the permanent file.

SUMMARY

The foodservice manager must understand the principles of food production in order to perform competently in the planning and evaluation of foodservice operations. Food should be acceptable to the customer, produced in the appropriate quantity, microbiologically safe, and within budgetary constraints. Food is cooked to destroy harmful microorganisms, increase digestibility, and change and enhance flavor, form, color, texture, and aroma. The importance of adequate controls throughout the preparation, production, and service processes needs to be emphasized.

Cooking is scientifically based on principles of chemistry and physics. Heat causes many reactions to occur, and the type and amount of heat greatly affect the resulting product. Heat is transferred by conduction, convection, radiation, and induction.

Cooking methods are classified either as moist heat or dry heat. Moist heat methods are those in which the heat is conducted to the food by water or steam. Dry heat methods are those in which the heat is conducted by dry air, hot metal, radiation, or a minimum amount of hot fat. Cooking methods and type of equipment are closely related and are important components of cooking technology, which has become more complex, primarily because of innovations resulting in efficiency and sophistication of equipment.

Standards, defined as the measurement of what is expected to happen, provide the basis for monitoring performance of the organization and taking necessary corrective action. The word *quality* often is used in combination with standards and should be defined in relation to the objectives of a specific foodservice. Management should be concerned with the relationship of quality to cost, profit, and selling price.

Standards are a result of the process of planning, and control is the process of ensuring that plans have been followed. Effective control does not occur until action has been taken to correct unacceptable deviations from standards. Quality control means assuring consistency in each product offered for service, and quantity control means producing the exact amount needed. Both are related to control of costs and thus to profit or to meeting budgetary constraints.

Time and temperature are critical elements in quantity food production and must be controlled to produce a high-quality product. Yield is the amount of product resulting at the completion of the various phases of the procurement, production, and service cycles and usually is expressed as a definite weight, volume, or serving size. Portion control is one of the essentials in production of food in quantity and is dependent upon the use of standardized recipes, which include the size and number of portions to be produced.

Product evaluation is an important component of a foodservice quality control program. Food quality is evaluated by sensory, chemical, and physical methods. Sensory analysis is defined as a science that measures the texture, flavor, and appearance of food products through human senses and should be used for new menu items and for maintaining quality of existing items. One type of sensory test is identified as analytical in which differences and similarities of quality and quantity of sensory characteristics are evaluated by a panel of specially trained persons; the other type is affective because preference, acceptance, and opinions of a product are evaluated by consumers.

BIBLIOGRAPHY

American Society for Testing & Materials, *American society for testing & materials: Manual on sensory testing, methods*. (1988). ASTM Special Technical Publication No. 434. Philadelphia: Author.

Batty, J. (1990). Essentials from concept to clean-up. *Restaurants USA, 10*(10), 18–21.

Bean, R. L. (1990). What makes the Groen Gas Combo so special? *The Consultant, 23*(2).

Beasley, M. A. (1991). Improving operational efficiencies—Part II. *Food Management, 26*(8), 52, 57.

Bowers, J. (Ed.). (1992). *Food theory and application* (2nd ed.). New York: Macmillan.

Chaudhry, R. (1993). Small ovens, big impact. *Restaurants & Institutions, 102*(2), 133, 136.

Cooking equipment: Barbecue and smoke ovens. (1986). *Foodservice Equipment Supplies Specialist Magazine, 39*(5), 35.

David, B. D. (1979). Quality and standards—the dietitian's heritage. *Journal of The American Dietetic Association, 75*, 408–412.

Decareau, R. V. (1992). Microwaves in foodservice. *Journal of Foodservice Systems, 6*, 257–270.

Delmar, C. O. (1989). *The essential cook: A practical guide to foods and cooking*. Chapel Hill, NC: Hill House.

Donovan, M. D. (Ed.). (1993). *The professional chef's ® techniques of healthy cooking*. New York: Van Nostrand Reinhold.

Garey, J. G., & Simko, M. D. (1987). Adherence to time and temperature standards and food acceptability. *Journal of The American Dietetic Association, 87*, 1513–1518.

Giese, J. (1992). Advances in microwave food processing. *Food Technology, 46*(9), 118–123.

Gilleran, S. (1991). Braising brings rich flavors to beef. *Restaurants & Institutions, 101*(22), 113–114, 121, 124.

Gisslen, W. (1989). *Professional cooking* (2nd ed.). New York: Wiley.

Jernigan, A. K., & Ross, L. N. (1989). *Food service equipment* (3rd ed.). Ames: Iowa State University Press.

Klein, B. P., Matthews, M. E., & Setser, C. S. (June 1984). *Foodservice systems: Time and temperature effects on food quality*. North Central Regional Research Publication No. 293. Urbana-Champagne: University of Illinois.

Lampi, R. A., Pickard, D. W., Decareau, R. V., & Smith, D. P. (1990). Perspective and thoughts on foodservice equipment. *Food Technology, 44*(7), 60, 62, 64–66, 68–69, 132.

Lorenzini, B. (1992). Cookware for ethnic fare. *Restaurants & Institutions, 102*(22), 153, 156.

McManis, H., & Molt, M. (1978). Recipe standardization and percentage method of adjustment. *NACUFS Journal*, 35–41.

McWilliams, M. (1993). *Foods: Experimental perspectives* (2nd ed.). New York: Macmillan.

Mizer, D. A., Porter, M., & Sonnier, B. (1987). *Food preparation for the professional* (2nd ed.). New York: Wiley.

National Restaurant Association. (1979). *Sanitation operations manual*. Chicago: Author.

Palmer, H. H. (1972). Sensory methods in food-quality assessment. In P. C. Paul, & H. H. Palmer (Eds.), *Food theory and application* (pp. 727–738). New York: Wiley.

Riell, H. (1992). Equipment's cutting edge. *Restaurants & Institutions, 12*(4), 22–23.

Ryan, N. R. (1991). Combination cooking. *Restaurants & Institutions, 101*(7), 73–74.

Scriven, C. R., & Stevens, J. W. (1989). *Manual of equipment and design for the foodservice industry*. New York: Van Nostrand Reinhold.

Setser, C. S. (1992). Sensory evaluation. In B. S. Kamel, & C. E. Stauffer, (Eds.), *Advances in baking technology* (pp. 254–291). New York: VCH Publishers.

Shugart, G., & Molt, M. (1993). *Food for fifty* (9th ed.). New York: Macmillan.

Skelton, M. (1984). Sensory evaluation of food. *Cornell Hotel and Restaurant Administration Quarterly, 24*(4), 51–57.

Solomon, J. (1992). The art of global grilling. *Restaurants USA, 12*(5), 14–17.

Spertzel, J. K. (1992). Combi ovens: A revolution in cooking. *School Food Service Journal, 45*(8), 94, 96.

Stone, H., & Sidel, J. L. (1992). *Sensory evaluation practices* (2nd ed.). New York: Academic Press.

Terrell, M. E., & Headlund, D. B. (1989). *Large quantity recipes* (4th ed.). New York: Van Nostrand Reinhold.

Thorner, M. E., & Manning, P. B. (1983). *Quality control in foodservice* (rev. ed.). Westport, CT: AVI.

U.S. Department of Agriculture, Food and Nutrition Service. (January 1984). *Food buying guide for child nutrition programs*. Program Aid No. 1331. Washington, DC: Author.

Van Warner, R. (1993). Government reaction to food safety should quell fears, but be prudent. *Nation's Restaurant News, 27*(16), 21.

Food Safety

Food spoilage is a difficult term to define because people have different concepts about edibility. Generally, however, **spoilage** is understood to denote unfitness for human consumption due to chemical or biological causes. Longrée and Armbruster (1987) identified the following criteria for assuring foods are fit to eat:

- The desired stage of development or maturity of the food
- Freedom from pollution at any stage in production and subsequent handling
- Freedom from objectionable chemical and physical changes resulting from action of food enzymes; activity of microbes, insects, and rodents; invasion of parasites; and damage from pressure, freezing, heating, or drying
- Freedom from microorganisms and parasites causing foodborne illnesses

Many opportunities exist for food contamination in a foodservice operation. As Minor (1983) stated, food is often prepared far ahead of service, thus permitting time for bacterial growth. The possibility of transmission of disease or harmful substances carried by vectors (flies, cockroaches, weevils, mice, and rats) and by foodservice staff is much greater than in the home. The potential for contamination also is more likely wherever groups of people congregate. Employees may be a hazard if they are not trained in safe food-handling practices.

Today, foodborne illnesses are recognized as a major health problem in the United States. Estimates of the cost of foodborne illnesses to the United States economy range from $1 billion to $10 billion, according to Hecht (1991). According to economists of the Economic Research Service (ERS), the most costly foodborne bacterial pathogens are *Campylobacter*, *Salmonella*, and *Staphylococcus*. Cost estimates include the direct medical costs of food illnesses as well as the indirect cost of time lost from work and other expenditures (Roberts, 1989). The number of Americans estimated to have foodborne illness each year is over 50 million. Adding to the concern is the fact that most incidents of foodborne illness are never reported; symptoms may mimic a flu-type virus that many people call the "24-hour bug" (VanLandingham, 1993). According to the Centers for Disease Control and Prevention (CDC), 77% of foodborne illnesses can be attributed to foodservice operations, 20% to homes, and 3% to food processing plants (Dykstra & Schwarz, 1991). The CDC tracks the epidemiological character of human diseases and has suggested the following to improve food safety:

- More training for food handlers
- Better protocols for investigating foodborne outbreaks
- Hazard analyses in food operations
- Improved data on how pathogenic organisms are carried and spread, how food preparation contributes to proliferation, and how food handling contributes to an outbreak

Microbial agents are not the only cause of foodborne illness in foodservice operations. Chemicals, herbicides, pesticides, antibiotics, metals, and hormones may contaminate food. In addition, some foods can be poisonous, such as certain varieties of mushrooms. Over the past several years, contamination of the food supply by chemicals or other agents has been a major news topic. Although many incidents have been overemphasized, others represent serious potential health hazards.

All foods deteriorate, some more rapidly than others. Degrees of perishability require food-handling practices that will maintain safety for consumption. The clients of a foodservice establishment have every reason to expect that the food served will be safe and wholesome. Customers can initiate legal action against a foodservice for not protecting their safety.

The foodservice organization has a tremendous responsibility for safeguarding its customers. Food-handling practices beginning with procurement and continuing throughout the production and service of food must be designed with the safety of the public in mind.

MAJOR TYPES OF FOOD SPOILAGE

The major types of food spoilage are microbiological, biochemical, physical, and chemical. The extent of contamination of some foods may be difficult to determine from their appearance, odor, and taste, as with inadequately refrigerated potato salad at a church picnic. In other foods, mold, discolored or altered appearance, off-odors, or off-flavors are obvious signs of contamination.

An in-depth discussion of food microbiology is beyond the scope of this text. An overview, however, is presented from the standpoint of controlling practices to ensure production of microbiologically safe food.

Microbiological

Thousands of species of microorganisms have been identified. The three most common forms are bacteria, molds, and yeast. They are found everywhere that temperature, moisture, and substrate favor life and growth. Some species are valuable and useful in preserving food, producing alcohol, or developing special flavors if they are specially cultured and used under controlled conditions. Other microbial activity, however, can be a primary cause of food spoilage. Food spoilage caused by microorganisms is called **microbiological spoilage.**

Certain microorganisms and parasites are transmitted through food and may cause illnesses in people who ingest contaminated items. According to Longrée and Armbruster (1987), microorganisms causing foodborne illnesses include bacteria, molds, yeasts, viruses, rickettsiae, protozoa, and parasites, such as trichinae. They stated, however, that although most microorganisms producing foodborne illness are bacteria, less than 1% of all bacteria can be considered "enemies of man and many are his friends."

Bacteria

Bacteria are microscopic, unicellular organisms of varying size and shape, including spherical, rod, and spiral. In most instances, the presence of bacteria cannot be seen in food, even if the contaminants are present in sufficient number to produce foodborne illness. Sometimes, however, bacterial contamination may cause a turbid appearance or a slime on a food surface that is visible to the human eye.

Four requirements for bacterial growth are time and temperature, food and moisture, acidity (pH), and oxygen (Educational Foundation of the National Restaurant Association [NRA], 1992). Although requirements for growth vary among different types of bacteria, all bacterial cells pass through various phases. When the multiplication of bacteria is steady, the number of cells produced over a certain period of time can be plotted. Figure 12.1 shows a typical bacterial growth curve.

In general, multiplication of bacteria is affected by available moisture in food. Water may become less available through the presence of solutes such as salt and sugar, through freezing, or through dehydration. Microorganisms vary in their response to oxygen; some bacteria require only a very small amount of oxygen.

The degree of a food's acidity or alkalinity, expressed as **pH value**, also affects bacterial growth. The pH value represents the hydrogen ion concentration and is expressed on a scale from 0 to 14, with 7 expressing neutrality. Values below 7 indicate acidity; those above, basic or alkaline materials. Bacteria vary widely in their reaction to pH. Although some are quite tolerant to acid, they generally grow best at a pH near neutral, so acid is frequently used in food preservation to suppress bacterial multiplication.

Longrée and Armbruster (1987) indicated that multiplication of the organisms causing food infections and foodborne illnesses are supported in slightly acid, neutral, and slightly alkaline food materials. The pH of some common foods is listed in Figure 12.2.

Microorganisms have specific temperature requirements for growth. At its optimum temperature, a cell multiplies and grows most rapidly, but a cell will also grow within its minimum and maximum temperatures around its optimum.

Figure 12.1. Stages in the growth curve of bacteria. (1) Lag phase—no multiplication occurs. (2) Log phase—accelerated growth occurs. (3) Stationary phase—competition for nourishment causes slowdown in multiplication and death of some bacteria. (4) Decline phase—bacterial cells die more quickly due to lack of nutrients or their own waste products.

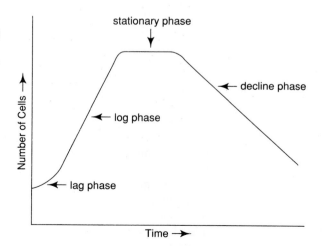

Source: From Applied Foodservice Sanitation (p. 21) by The Educational Foundation of the National Restaurant Association, 1992, Chicago: Author.

Figure 12.2   pH of some common foods.

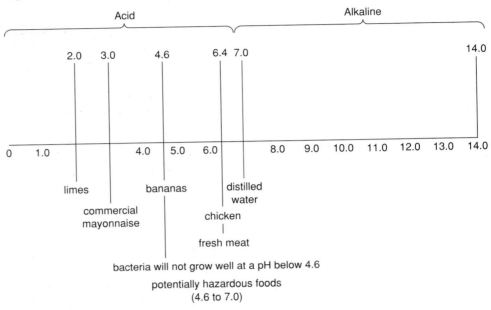

Source: From *Applied Foodservice Sanitation* (p. 26) by The Educational Foundation of the National Restaurant Association, 1992, Chicago: Author.

Various types of bacteria respond differently to temperature. In general, spores of microorganisms are more heat resistant than vegetative mature cells, which are dormant and asexual. Some bacteria form spores inside the wall of their cells when they mature. These spores are more resistant to high heat, low humidity, and other adverse conditions than are vegetative bacteria cells. They may remain dormant for long periods of time and germinate when conditions are favorable into new, sensitive, vegetative cells.

The heat resistance of microorganisms is their thermal death time, or the time required at a specified temperature to kill a specified number of vegetative cells or spores under specific conditions. Thermal death depends on the age of the organism, the temperature to which it is exposed, the length of time for which heat is applied, the presence of moisture, and the nature of the medium. As this discussion indicates, time and temperature are important in preserving microbiological quality in foods.

Various inhibitors have a pronounced effect on bacterial multiplication and death. According to Longrée and Armbruster (1987), inhibitors may be integral in the food, developed during processing as a product of the microorganism's metabolism or added purposely by the processor. The benzoic acid in cranberries and lysozyme in egg whites, for example, are natural inhibitors of these foods. Alcohol produced in the growth and fermentation of yeast, in fruit juices, or in the production of wine is an example of an inhibitory substance that may accumulate and become toxic.

Molds and Yeasts

Molds and yeasts are other common forms of microorganisms important in food sanitation. Molds are larger than bacteria and more complex in structure. In general, they grow on a wide range of substrates—moist or dry, acid or nonacid, high or low in salt or sugar. Molds also grow over an extremely wide range of temperature, although the optimum temperature is between 77° and 86°F. Because mold growth may appear as highly colored, cottony, powdery, or fuzzy tufts and patches, it is probably the most common type of spoilage that can be identified by the naked eye.

Yeasts are not known to cause foodborne illnesses, but they may cause spoilage of sugar-containing foods. They are unicellular plants that play an important role in the food industry, particularly in the fermentation or leavening of beer, wine, and bread. Yeasts can induce undesirable reactions, however, resulting in sour or vinegary taste.

Viruses

Viruses are small pathogens that multiply in the living cells of the host but not in cooked food. They are capable of causing diseases in plants, animals, and humans. Viruses can be carried in food and water, but they multiply only in the living cell. In many respects, viruses resemble bacteria in that the right temperature, nutrients, moisture, and pH are necessary for effective growth and reproduction. Examples of human diseases caused by viruses are influenza, poliomyelitis, chicken pox, and hepatitis, some of which have been associated with foodborne outbreaks. Many viruses are inactivated by high temperatures (149°–212°F) and by refrigeration.

The so-called *Norwalk virus* is rapidly increasing as a health threat. Outbreaks have occurred through water contaminated with sewage, raw shellfish harvested from polluted growing areas, and feces of infected food handlers who have not scrubbed their hands after using the toilet. In 1982, an infected employee in a bakery caused some 3,000 illnesses by mixing uncooked butter cream icing in a giant vat with his bare arms up to the elbows in the mixture. An estimation has been made by a virologist at the University of Wisconsin that Norwalk and other Norwalk-like viruses account for approximately 40% of all serious, nonbacterial foodborne illness. Norwalk multiplies in the cells of the human intestinal tract, passes through, and cannot grow again until it reaches another human. Unable to multiply in food, as bacteria do, it must wait for ingestion by a person.

Other Microorganisms

Rickettsiae include such human diseases as typhus fever, Q fever, and Rocky Mountain spotted fever. Cows infected with the organism causing Q fever may transmit the disease to humans through their milk if it has not been properly heat treated. Like viruses, rickettsiae multiply in living tissues only.

Protozoa may be carried by food and cause illness when ingested. Usually microscopic in size, protozoa are unicellular, animal-like forms that are distributed widely in nature, especially in sea water, but also in lakes, streams, and soil. Amoebic dysentery is caused by one of the pathogenic protozoa that can be spread by water and food. Nearly all animals and humans carry protozoa in their intestinal tracts, which has implications for food-handling practices of foodservice employees.

Parasites include *trichinae, tapeworms,* and *roundworms.* Trichinosis occurs when people are served a trichina-infested product, the main source being pork. This disease is preventable, however, if food is cooked to a proper end-point temperature. The 1991 Pork Industry Group of the National Live Stock and Meat Board recommends that today's lean pork should be cooked to an end temperature of 160°F for medium and 170°F for well done. An internal temperature of at least 150°F is considered a margin of safety to kill any trichinella larvae that may have infested the pork (Education Foundation of the NRA, 1992). If the pork is cooked in a microwave oven, it must be heated to an internal temperature of 170°F because post-cooking standing times are necessary to allow all parts of the meat to reach the temperature. The USDA, however, has developed processing methods for pork products that assure a safe product. Problems with trichinae can be prevented by purchasing pork from approved sources and by adequate cooking.

Tapeworms are parasitic intestinal worms of flattened tapelike form that may cause disease in humans when larva-infested meat, either beef or pork, is ingested. Tapeworm contamination can be prevented by appropriate sanitary measures during agricultural production and by the federally regulated inspection of cattle carcasses for evidence of tapeworms. Proper meat processing procedures—especially during boning, which involves a great deal of hand contact—also are important to prevent contamination.

The current popularity of raw seafood dishes, like sushi, sashimi, and ceviche, and undercooked fin fish, has introduced a new source of tapeworm and roundworm (*Anasakis*) infestation. *Anasakis* can be particularly devastating because the parasite attaches itself to the wall of the digestive organs and requires surgery to dislodge it. These parasites are destroyed by cooking or freezing. The Food and Drug Administration (FDA) in 1987 ruled that fish that is not to be thoroughly cooked must be blast frozen to −31°F or below for 15 hours or frozen to −10°F or below for 168 hours (7 days). In addition, the foodservice operator must keep a record of the process on file for 90 days.

Biochemical

Biochemical spoilage is caused by natural food enzymes, which are complex catalysts that initiate reactions in foods. Off-flavors, odors, or colors may develop in foods if enzymatic reactions are uncontrolled. When fruit is peeled and exposed to air, for example, undesirable browning due to enzyme activation by oxygen occurs. Enzymes may have desirable effects as well, such as the natural tenderizing or aging of meat.

Enzyme formation can be controlled in much the same manner as microorganisms. Thorner and Manning (1983) stated that heat, cold, drying, addition of inhibiting chemicals, and irradiation are the principal means for controlling and inactivating natural food enzymes.

Physical

Temperature changes, moisture, and dryness can cause physical spoilage of food. Excessive heat, for example, breaks down emulsions, dehydrates food, and destroys certain nutrients. Severe cold also causes various kinds of deterioration; certain starches used in preparing sauces or gravies, for instance, may break down at freezing temperatures.

Excessive moisture may lead to various types of spoilage. For example, excess moisture may support the growth of mold and bacteria, as previously mentioned. It also may cause physical changes, such as caking, stickiness, or crystallization, which may affect the quality of a food item. A powdered beverage product, for instance, may not be dispensed properly in a vending machine if the moisture level is too high.

A *physical hazard* is the danger of particles that are not supposed to be in a food product (Educational Foundation of the NRA, 1992). Chips of glass or metal from broken glasses or enamelware dishes are obviously dangerous. Metal curls from a worn-out can opener can fall into the food when the can is being opened. Dangers caused by physical contaminants may result from tampering incidents, particularly with soft-packed food items. Food items delivered to the foodservice operation should be rejected if evidence of tampering is seen.

Chemical

Chemical spoilage may result from interaction of certain ingredients in a food or beverage with oxygen or light. A chemical hazard is the danger posed by chemical substances contaminating food all along the food supply chain (Educational Foundation of the NRA, 1992). The reaction of incompatible substances can lead to chemical spoilage, such as the effect of certain metals on foods. As an example, the zinc often used in galvanized containers may render poisonous such acid foods as fruit juices, gelatins, sauerkraut, and tomatoes. The following four kinds of chemical hazards besides pesticides can occur in a foodservice operation:

- Contamination of food with foodservice chemicals, such as detergents and sanitizers
- Use of excessive quantities of additives, preservatives, and spices
- Acidic action of foods with metal-lined containers
- Contamination of food with toxic metals

Pesticides are chemicals that kill or discourage the growth of pests, which are defined as organisms that cause damage to food, making it inedible, unappealing, or unsafe (Chaisson, Petersen, & Douglass, 1991). Pesticides typically are applied to crops growing in the field but also may be applied after harvest to prevent insect or mold infestation during transport or storage. Much research is being conducted on ways to reduce reliance on applied pesticides.

Integrated plant management (IPM) is the most promising new approach being used in agriculture today. In addition to using pesticides carefully, IPM incorporates the latest agricultural technologies and biological controls, including pest predators and pest diseases, to decrease the amount of pesticide used. Computerized forecasting may be used to predict disease and weather conditions. Plant breeding and genetic engineering may reduce the use of applied pesticides by manipulating genes to "program" plants to make high levels of their own pesticides. The U.S. Department of Agriculture (USDA) and FDA are responsible for monitoring the food supply to ensure that residue levels are within tolerance limits.

Foodservice chemicals, including detergents, polishes, caustics, and cleaning and drying agents, are poisonous to humans and should never come in contact with food. Labels should be read carefully for directions on how to use and store under safe conditions away from food.

Use of additives and preservatives is still being debated by scientists and legislators. Excessive amounts of certain food additives and preservatives have caused illness, but the reasons have not yet been determined. One foodborne illness that remains in question is that which results from the use of too much monosodium glutamate (MSG), a food additive that serves as a flavor enhancer. Because it is often heavily used in Chinese and Japanese foods, apparent reactions to MSG have been called the "Chinese Restaurant Syndrome." MSG also is very high in sodium. Most of the symptoms after ingestion are subjective. They include a feeling of tightening of the face and neck skin, tingling sensations, dizziness, and headache. MSG apparently affects only sensitive persons (Cody & Keith, 1991). Federal law requires that MSG be listed on the label of any product to which it is added. Foodservice managers should avoid adding MSG to recipes.

Food irradiation is classified as a food additive and is regulated by the FDA. It controls microbes responsible for foodborne illness and extends the shelf life of refrigerated foods, such as fresh fruits and vegetables, by delaying ripening, as well as the shelf life of stored foods, like spices and dried herbs. Food irradiation was discussed in more depth in chapter 7.

Preservatives used to preserve the flavor, safety, and consistency of foods have been linked to food contamination. Excessive use of sulfites has affected a number of sensitive individuals, particularly asthmatics. The FDA prohibits the use of sulfites on raw fruits and vegetables that are to be served or sold to customers.

Several food additives or preservatives, when used in excessive amounts, have caused illness. Nitrites, for example, are preservatives used by the meat industry to prevent growth of certain harmful bacteria and as a flavor enhancer (Educational Foundation of the NRA, 1992). Scientists have established a link between cancer and nitrites when meat containing them is overbrowned or burned. As a result, the meat industry has decreased levels of nitrites in meats. In the 1980s, a number of food-related illnesses, allergic in type, were traced to sulfites used for fresh fruits and vegetables, shrimp, and dried fruit. For packaged foods, proper labeling is now required. Restaurants and supermarkets are now using lemon juice or citric acid for preserving color in these foods, especially in salad bars (Longrée & Armbruster, 1987).

Chemical contamination can occur when high-acid foods are prepared or stored in metal-lined containers. Poisoning may result if brass or copper, galvanized, or gray enamelware containers are used. Fruit juices should never be stored in enamelware coated with lead glaze or tin milk cans. Cases of poisoning have been recorded that are attributed to use of improper metal utensils. Sauerkraut, tomatoes, fruit gelatins, lemonade, and fruit punches have been implicated in metal poisonings.

Toxic metals also have been implicated in food poisoning cases. Copper may become poisonous when it is in prolonged contact with acid foods or carbonated beverages. The vending industry recently has voluntarily discontinued all point-of-sale carbonation systems that do not completely guard against the possibility of backflow into copper water lines, which may dissolve the copper. Also, food such as meat placed directly on cadmium-plated refrigerator shelves may be rendered poisonous.

FOODBORNE PATHOGENS

Table 12.1 presents a tabulation of the incubation period, duration of illness and symptoms, as well as the reservoir or source, foods implicated, spore former, and prevention for various bacteria-causing diseases. "Food poisoning," as outbreaks of acute gastroenteritis are popularly called, is caused by microbial pathogens that multiply profusely in food. These attacks are either foodborne intoxications or foodborne infections. Their symptoms, which are frequently violent, include nausea, cramping, vomiting, and diarrhea.

Foodborne intoxications are caused by toxins formed in the food prior to consumption, and **foodborne infections** are caused by the activity of large numbers of bacterial cells carried by the food into the gastrointestinal system of the victim. The symptoms from ingesting toxin-containing food may occur within as short a period of time as 2 hours. The incubation period of an infection, however, is usually longer than that of an intoxication.

Salmonella

Salmonella frequently has been associated with foodborne illnesses and causes salmonellosis. The bacterium does not release toxins into the food in which it multiplies; rather, the ingested cells continue to multiply in the intestinal tract of the victim, causing illness. The incubation period ranges from 6 to 72 hours, but is usually 12 to 36 hours. Individuals differ in susceptibility to *Salmonella* infections; the symptoms include acute gastroenteritis, with inflammation of the small intestine, nausea, vomiting, diarrhea, and frequently a moderate fever, often preceded by headache and chills.

The primary source of *Salmonella* is the intestinal tract of carrier animals. A carrier appears to be well and shows no symptoms or signs of illness but harbors causative organisms. Various insects and pets may be reservoirs of *Salmonella*. Food animals are important reservoirs, especially hogs, chickens, turkeys, and ducks. The disease salmonellosis is spread largely by contaminated food and is believed to be one of the major communicable diseases in this country.

A number of raw and processed foods have been found to carry *Salmonella*, especially raw meat, poultry, shellfish, processed meats, egg products, and dried milk.

Table 12.1. Major foodborne diseases of bacterial origin

	Salmonellosis Infection	Shigellosis Infection	Listeriosis Infection	Staphyloccal Intoxication	Clostridium Perfringens Toxin Mediated Infection	Bacillus Cereus Intoxication	Botulism Intoxication
Bacteria	*Salmonella* (facultative)	*Shigella* (facultative)	*Listeria monocytogenes* (reduced oxygen)	*Staphylococcus aureus* (facultative)	*Clostridium perfringens* (anaerobic)	*Bacillus cereus* (facultative)	*Clostiridium botulinum* (anaerobic)
Incubation Period	6–72 hours	1–7 days	1 day to 3 weeks	1–6 hours	8–22 hours	½–5 hours; 8–16 hours	12–36 hours + 72
Duration of Illness	2–3 days	Indefinite, depends on treatment	Indefinite, depends on treatment, but has high fatality in the immuno-compromised	24–48 hours	24 hours	6–24 hours; 12 hours	Several days to a year
Symptoms	Abdominal pain, headache, nausea, vomiting, fever, diarrhea	Diarrhea, fever, chills, lassitude, dehydration	Nausea, vomiting, headache, fever, chills, backache, meningitis	Nausea, vomiting, diarrhea, dehydration	Abdominal pain, diarrhea	Nausea and vomiting; diarrhea, abdominal cramps	Vertigo, visual disturbances, inability to swallow, respiratory paralysis
Reservoir	Domestic and wild animals; also humans, especially as carriers	Human feces, flies	Humans, domestic and wild animals, fowl, soil, water, mud	Humans (skin, nose, throat, infected sores); also, animals	Humans (intestinal tract), animals, and soil	Soil and dust	Soil, water

Foods Implicated	Poultry and poultry salads, meat and meat products, milk, shell eggs, egg custards and sauces, and other protein foods	Potato, tuna, shrimp, turkey and macaroni salads, lettuce, moist and mixed foods	Unpasteurized milk and cheese, vegetables, poultry and meats, seafood, and prepared, chilled, ready-to-eat foods	Warmed-over foods, ham and other meats, dairy products, custards, potato salad, cream-filled pastries, and other protein foods	Meat that has been boiled, steamed, braised, stewed or roasted at low temperature for a long period of time, or cooled slowly before serving	Rice and rice dishes, custards, seasonings, dry food mixes, spices, puddings, cereal products, sauces, vegetable dishes, meat loaf	Improperly processed canned goods of low-acid foods, garlic-in-oil products, grilled onions, stews, meat/poultry loaves
Spore Former	No	No	No	No	Yes	Yes	Yes
Prevention	Avoid cross-contamination, refrigerate food, cool cooked meats and meat products properly, avoid fecal contamination from foodhandlers by practicing good personal hygiene	Avoid cross-contamination, avoid fecal contamination from foodhandlers by practicing good personal hygiene, use sanitary food and water sources, control flies	Use only pasteurized milk and dairy products, cook foods to proper temperatures, avoid cross-contamination	Avoid contamination from bare hands, exclude sick foodhandlers from food preparation and serving, practice good personal hygiene, practice sanitary habits, proper heating and refrigeration of food	Use careful time and temperature control in cooling and reheating cooked meat dishes and products	Use careful time and temperature control and quick chilling methods to cool foods, hold hot foods above 140°F (60°C), reheat leftovers to 165°F (74°C)	Do not use home-canned products, use careful time and temperature control for sous-vide items and all large, bulky foods, keep sous-vide packages refrigerated, purchase garlic-in-oil in small quantities for immediate use, cook onions only on request

Source: From *Applied Foodservice Sanitation* (p. 35) by The Educational Foundation of the National Restaurant Association, 1992, Chicago: Author.

399

Meat mixtures, dressings, gravies, puddings, and cream-filled pastries are among the menu items frequently indicated in salmonellosis. Food handlers and poor sanitation practices are often associated with outbreaks. Care must be exercised in production, storage, and service to ensure that food is not held for long periods at warm temperatures, cooled slowly, or cut on contaminated surfaces.

The organism grows best in a nonacid medium, although under certain conditions it may grow at low pH levels. The optimum temperature for growth of *Salmonella* is the temperature of the human body, or 98.6°F, but different strains of the bacterium may grow at many other temperatures. Organisms also have been shown to survive freezing and freezer storage.

Shigella

Shigella is another bacteria that causes foodborne illness, shigellosis, sometimes called bacillary dysentery. It is an infection that occurs 1 to 7 days after the ingestion of the bacteria. *Shigella* causes diarrhea, cramps, and chills and usually a fever. Humans are the prime reservoir for *Shigella*. Carriers excrete *Shigella* in their feces and transmit the bacteria to the food if they do not wash their hands properly. Flies also are thought to carry the bacteria. Foods involved are raw produce and moist-prepared foods, such as potato, tuna, turkey, and macaroni salads that have been handled with bare hands during preparation. Shigellosis could be prevented if employees wash their hands after using the toilet, food is rapidly cooled, and flies are controlled.

Listeria monocytogenes

Listeria monocytogenes is the bacterium responsible for listeriosis and is widely distributed in nature. It has been isolated from feces of healthy human carriers as well as sheep, cattle, and poultry. It has been detected in normal and mastitic cow's milk and has been isolated from improperly fermented silage, unwashed leafy vegetables and fruit, and soil. The bacterium has been found in dairy and meat processing factories with some degree of frequency.

The listeriosis disease has been linked to consumption of contaminated delicatessen food, milk, soft cheeses (like Mexican-style feta, Brie, Camembert, and blue-veined cheeses), and undercooked chicken. The bacteria occurs in a neutral or slightly alkaline medium. It grows at refrigeration temperatures and survives the minimum high-temperature and short-time treatment for pasteurizing milk and the standard heat treatment for the manufacture of cottage cheese. Postprocessing contamination must be minimized to assure that consumers have a safe food supply (Ryser & Marth, 1989).

Staphylococcus aureus

Staphylococcus aureus, a bacterium commonly referred to as *staph* or *S. aureus*, is one of the principal causative agents in foodborne illness. *S. aureus* grows in food, developing a toxin that, when ingested, causes an inflammation of the lining of the stomach and the intestinal tract, or gastroenteritis. Many strains of *S. aureus* are known, some of which are extremely pathogenic. Their symptoms include nausea, cramps, vomiting, and diarrhea, and they usually appear within 2 to 3 hours after ingestion of toxic food, but the time may vary up to 6 hours. The duration of the illness may be from 24 hours to 2 days; mortality is low.

Distribution of *S. aureus* is widespread, as it is relatively resistant to adverse environmental conditions. Cells may survive for a long time, even months, in dust, soil, frozen foods, or on various surfaces, such as floors or walls. The human body is one of the most important sources of the organism. Even healthy people harbor *S. aureus* on their skin and in their mouths, throats, and nasal passages. *S. aureus* especially thrives in infected skin abrasions, cuts, and pimples. Although *S. aureus* is heat-sensitive and can be controlled and destroyed by heat, it produces a heat-stable toxin that persists long after the cells are destroyed.

Staphylococcal intoxication is a fairly frequent cause of foodborne illness, with foods high in protein the usual culprits. Cream pies, custards, meat sauces, gravies, and meat salad are among the products most likely to be involved in foodborne intoxication. The appearance, flavor, or odor of the affected food items are not noticeably altered.

Temperatures must be carefully controlled to prevent multiplication of staphylococci in food. The organism multiplies even under refrigeration if temperatures are not sufficiently low or if the cooling process does not proceed rapidly enough. Longrée and Armbruster (1987) outlined the following precautions for controlling foodborne illnesses caused by *Staphylococcus aureus*:

- Sanitary precautions in connection with the food handler
- Proper measures to preclude multiplication of the organism during preparation, service, and storage of food
- Proper use of heat for the destruction of the organisms before toxins are formed

Clostridium perfringens

Clostridium perfringens is considered the third most common cause of foodborne illness in the United States, following *S. aureus* and *Salmonella*. This bacterium is an anaerobic, gram-positive, spore-forming rod. Several toxigenic strains that are associated with gastrointestinal illness have been differentiated. The bacterial cells and, more recently, the toxins have been implicated in the *C. perfringens* gastroenteritis. It is a common inhabitant of the intestinal tract of healthy animals and human beings and occurs in soil, sewage, water, and dust.

The infected food has invariably been held at room temperature or refrigerated in a large mass for several hours. Symptoms usually begin between 8 and 15 hours after ingestion and continue for 6 to 12 hours. Symptoms are similar to staph intoxication but milder, and they include nausea, cramping, and diarrhea. Meats, meat mixtures, and gravies are frequently implicated. Vegetative cells succumb through cooking, but spores may survive. Overnight roasting of meat is, therefore, not recommended because of the low temperatures often used, and careful reheating of leftover meat is important. Prevention of *C. perfringens* multiplication can be achieved by refrigerating foods at 40°F or below or holding them at 145°F or higher. In addition, rapid cooling of cooked foods is an important practice.

Bacillus cereus

Foodservice managers are beginning to be concerned about the *B. cereus* toxin, which is found in soil and, therefore, gets into many foods once thought to be safe (Educational Foundation of the NRA, 1992). Two toxins formed by *B. cereus* have different times of onset and symptoms. One form causes diarrhea and abdominal pain and has an onset time of 8 to 16 hours; the other form causes vomiting that usually occurs 30 minutes to 5 hours after ingestion of contaminated food.

The *B. cereus* bacteria are found in grains, rice, flour, spices, starch, and in dry mix products such as those used for soups, gravies, and puddings. They also have been found in meat and milk. Time and temperature are very important in preventing rapid increase in the vegetative bacteria and development of spores. Foods should not be held at room temperature for any period of time, but should be chilled to at least 45°F within 4 hours after preparation. Foods stored in the refrigerator need to be in shallow pans with a food depth of less than 2 inches and should be used as quickly as possible after preparation.

Clostridium botulinum

Clostridium botulinum produces a toxin that affects the nervous system and is extremely dangerous. The disease called *botulism* is the food intoxication caused by bacteria. Several types of the pathogen have been identified, each producing a specific toxin. Toxins are very potent and may persist in food for long periods of time, especially under low-temperature storage.

The illness is treated by administering an antitoxin specific to the particular toxin involved. Therapy may not be satisfactory, however, because of the lag time between ingestion of the food, appearance of symptoms, diagnosis, and administration of an antitoxin. Improved food processing techniques have led to greatly reduced incidence of botulism, although inadequately processed home-canned foods are still frequently associated with botulism.

Toxin production accompanies growth of the organism, both of which are affected by the composition, moisture, pH, oxygen, and salt content of the food. The temperature and length of food storage also are factors influencing toxin production and growth. The toxin is stable in acid, but unstable in alkaline. Meats, fish, and low-acid vegetables have been found to support toxin formation and growth.

Temperature requirements of the strains vary. The optimum temperature for toxin formation and growth of the most common types of *C. botulinum* in this country is about 95°F. Freezing prevents growth in toxin formation but does not kill the organism. Destruction of the spores, however, may require several hours of boiling or a shorter heating period at higher temperatures. The toxins are less resistant to heat and are denatured readily at temperatures above 176°F; about 15 minutes at boiling temperatures should deactivate them. Precautions for avoiding botulism include procuring foods from safe sources, rejecting home-canned products and low-acid products, destroying canned goods with defects such as swells or leaks, storing foods under recommended conditions, and using appropriate methods for thawing frozen foods.

In addition to improperly processed products, other suspicious foods include smoked, vacuum-packed fish; garlic products packed in oil; grilled onions; baked potatoes; turkey loaf; and stew. Sous vide products offer a potential risk because they are vacuum packaged. Soil-grown vegetables, particularly potatoes, can be prime carriers of this toxin. A recent incident involved a potato wrapped in foil and boiled in a saucepan and afterward left out at room temperature overnight in 95°F weather. Subsequently, the foil was removed and the potato heated in the microwave oven and eaten for supper. On the following morning, the person who consumed the potato reported not feeling well, and at noontime he was unconscious and not breathing. The final report was that the victim was completely paralyzed and unable to blink his eyes, but made some response to stimuli to his feet. Scrapings from the foil confirmed that the potato had been laced with *Clostridium botulinum*. The toxin was so potent that mere traces from the foil were enough to kill the mice in the health department laboratory in less than an hour. Authorities reported that wrapping a potato in foil is not necessary to establish an anaerobic condition. The potato flesh itself would be anaerobic if the spores got in through the eyes of the potato.

Emerging Foodborne Pathogens

Liston (1989) reported that from 1982 to 1986, 31% of all foodborne illness outbreaks reported to the CDC were of unknown etiology. The reason for this is still unclear, but perhaps centralized production of food products and the widespread use of refrigeration to hold fresh produce for long periods could have contributed to the emergence of bacteria able to grow at low temperatures. According to Ryser and Marth (1989), within the past 15 years, several pathogenic bacteria-causing foodborne illnesses have emerged, including *Campylobacter jejuni, Vibrio cholerae, Escherichia coli,* and the Norwalk virus, as shown in Table 12.2.

Campylobacter Jejuni

Campylobacter jejuni was a well-known pathogen in veterinary medicine before it was considered a human pathogen. It is now recognized as one of the most common causes of gastroenteritis in humans, with a ranking similar in importance to *Salmonella*. A pathogen of cattle, sheep, pigs, and poultry, it is present in the flesh of these food animals and thus may be introduced into the kitchen with the food supply. A mode of transmission to humans closely resembles that of *Salmonella*. The incubation time varies from 2 to 20 days, usually 4 to 7 days. Gastrointestinal symptoms are abdominal pain; slight to severe bloody diarrhea is common; and the symptoms may last from 2 to 7 days.

This pathogen must be expected to be present in areas where raw meats and poultry are handled even when the techniques of sanitary handling are practiced. In addi-

Table 12.2.   Emerging pathogens that cause foodborne illness

	Campylobacteriosis Infection	E. Coli 0157:H7 Infection/Intoxication	Norwalk Virus Illness
Pathogen	*Campylobacter jejuni*	*Escherichia coli*	Norwalk and Norwalk-like viral agent
Incubation period	3–5 days	12–72 hours	24–48 hours
Duration of illness	1–4 days	1–3 days	24–48 hours
Symptoms	Diarrhea, fever, nausea, abdominal pain, headache	Bloody diarrhea; severe abdominal pain, nausea, vomiting, diarrhea, and occasionally fever	Nausea, vomiting, diarrhea, abdominal pain, headache, and low-grade fever
Reservoir	Domestic and wild animals	Humans (intestinal tract); animals, particularly cattle	Humans (intestinal tract)
Foods implicated	Raw vegetables, unpasteurized milk and dairy products, poultry, pork, beef, and lamb	Raw and undercooked beef and other red meats, imported cheeses, unpasteurized milk, raw finfish, cream pies, mashed potatoes, and other prepared foods	Raw vegetables, prepared salads, raw shellfish, and water contaminated from human feces
Spore former	No	No	No
Prevention	Avoid cross-contamination, cook foods thoroughly	Cook beef and red meats thoroughly, avoid cross-contamination, use safe food and water supplies, avoid fecal contamination from foodhandlers by practicing good personal hygiene	Use safe food and water supplies, avoid fecal contamination from foodhandlers by practicing good personal hygiene, thoroughly cook foods

Source: From *Applied Foodservice Sanitation* (p. 48) by the Educational Foundation of the National Restaurant Association, 1992, Chicago: Author.

tion to flesh foods, raw milk has been implicated in outbreaks of gastroenteritis caused by *Campylobacter.* The variety of menu items that have been involved in foodborne outbreaks attests to the fact that multiplication of the organism is possible in many different foods and food mixtures including cake icings, eggs, poultry, and beef.

Additional research is needed to determine the time-temperature relationships for cooking, using a variety of food items and preparation procedures. According to Ryser and Marth (1989), food handlers must take appropriate precautions to prevent cross-contamination between raw meat (e.g., meat, poultry, seafood) and cooked or ready-to-eat foods (e.g., fruits and vegetables that are frequently consumed without additional cooking).

Esberichia Coli

Esberichia coli 0157:H7, or *E. coli,* caused the January 1993 outbreak of food poisoning in western states. *E. coli,* a bacteria that can be transmitted by eating raw or undercooked ground beef, was linked to hamburger patties served in one limited-menu restaurant chain (McLauchlin, 1993). This bacteria is most likely found on the surface of meat and is killed when cuts such as steaks are grilled (McCarthy, 1993). Grinding meat, however, transfers the bacteria from the surface to the inside of the product, making *E. coli* more difficult to kill. The FDA issued guidelines to federal, state, and local food safety regulators recommending that restaurants cook ground beef products to 155°F until the juices run clear and all pink color is gone from the inside (McLauchlin, 1993).

Symptoms of the infection begin with the sudden onset of severe abdominal cramps followed by watery and grossly bloody diarrhea. The organism damages the mucosal cells of the colon, causing bleeding, which is why the illness is called *hemorrhagic colitis.* Raw or undercooked ground beef and red meats including pork and lamb are the major vehicles of transmission. It has also been found in prepared foods such as cream pies and mashed potatoes as well as in untreated water, unpasteurized milk, imported cheeses, and finfish. The FDA and the National Restaurant Association recommend that restaurateurs use a "thermocouple-type" wire inserted into hamburger patties to determine the cooked temperature of the meat because the probes are much thinner than a regular thermometer and read more accurately on an electronic or digital display (McLauchlin, 1993). Thorough cooking and reheating, good sanitation, and refrigeration at 40°F or below are all important ways to prevent growth of the bacteria ("Facts about E. coli," 1993). Avoidance of cross-contamination of cooked meat by contaminated equipment, water, or food handlers also is an important preventive measure.

Norwalk Virus

Norwalk is a viral illness caused by poor personal hygiene among infected foodhandlers (Educational Foundation of the NRA, 1992). This virus is characterized by nausea, vomiting, diarrhea, abdominal pain, headache, and low-grade fever. The incubation period is 24 to 48 hours followed by an additional 24 to 48 hours of illness. Because it is a virus, it does not reproduce in food but remains active until the food is eaten. The Norwalk virus is found in the feces of humans. It is transmitted through contaminated water and human contact, raw vegetables fertilized by manure, salads, raw shellfish, eggs, icing on baked pastries, and in manufactured ice cubes and frozen foods.

CONTROLLING MICROBIOLOGICAL QUALITY OF FOOD

The ultimate goal of the sanitation program in a foodservice operation is to protect the customer from foodborne illness. The role of the foodservice manager is to take responsibility both for serving safe food to customers and for training employees on a continual basis.

Control of Food Quality

Control of the microbiological quality of food must focus on the food itself; the people involved in handling food, either as employees or customers; and the facilities and equipment, including both large and small equipment. The legal fees, medical claims, lost wages, and loss of business associated with foodborne illness can be overwhelming (Educational Foundation of the NRA, 1992). Good sanitation will pre-

Figure 12.3.   Transmission of a foodborne illness from an intermediate source to food and on to humans.

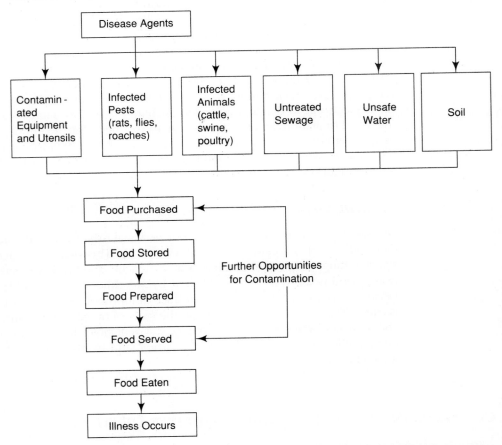

Source: Reproduced by permission from *Applied Foodservice Sanitation,* third edition (p. 18), 1985, by the Educational Foundation of the National Restaurant Association, Chicago: Author. Used by permission.

Figure 12.4. Transmission of a foodborne illness from infected human beings to food and back to other human beings.

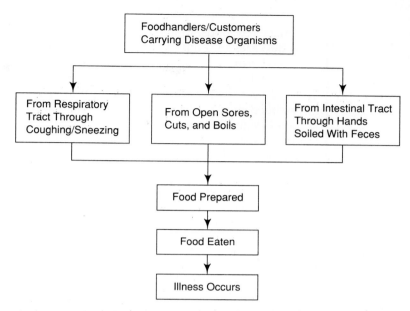

Source: From *Applied Foodservice Sanitation* (p. 9), by the Educational Foundation of the National Restaurant Association, 1992, Chicago: Author.

vent foodborne illness breakouts, maintain customer goodwill, and keep the bottom line from bottoming out.

Quality of Food

Clearly, the condition of the food brought into the facilities is a critical aspect to consider. Time-temperature control also is critical during storage, production, and service.

Condition of Purchased Food.   Possibilities for contamination of food before it is purchased include contaminated equipment, infected pests and animals, untreated sewage, unsafe water, and soil, as shown in Figure 12.3. After purchase, possibilities of contamination exist in storage, preparation, and service of food. Following human consumption, illness occurs. Figure 12.4 illustrates the possible transmission routes from infected persons through respiratory tract discharges, open sores, cuts, and boils, or through hands soiled with feces into food being prepared. The consumed food then completes the transmission to other persons.

Foods traditionally sold in the grocery refrigeration case have had a good safety record until lately when they have been implicated in a number of serious foodborne disease outbreaks, such as salmonellosis from pasteurized milk and listeriosis from a specific brand of Mexican-style cheese. Uncertainty caused by these episodes is intensified by the emergence of newly recognized pathogens that grow at refrigeration temperatures.

Concern is growing about the microbiological safety of the new generation of refrigerated foods, which include such foods as frozen or restaurant dinners; frozen or delicatessen entrées; dry, frozen, or canned pasta; fresh, refrigerated salads; canned, frozen, or dry gravies; canned or dry soups; and frozen, canned, or refrigerated cooked meat. Unlike traditional refrigerated foods, new generation products do not have a readily apparent preservation system. Many of these products have an extended shelf life, but the packaging may enhance growth of pathogens. Other risks are inadequate temperature control for safety and extended shelf life, a poor distribution system, and partial processing, which prevents warning of hazards by destruction of spoilage flora.

Relationship of Time and Temperature.   Reducing the effect of contamination is largely a matter of temperature control in the storage, production, and service of foods. As discussed previously, the growth of harmful organisms can be slowed or prevented by refrigeration or freezing, and the organisms themselves can be destroyed by sufficient heat. According to the Educational Foundation of the NRA, failure to cool food properly and to thoroughly heat or cook food are the first two most frequently cited factors in causing a foodborne illness outbreak (Educational Foundation of the NRA, 1992). Minimum, maximum, and optimum temperatures vary for the various pathogenic microorganisms; in general, however, they flourish at temperatures between 40° and 140°F (Figure 12.5). This temperature range is commonly called the *food danger zone* because bacteria multiply rapidly within it. The longest period that food may safely remain in this zone is 4 hours, although food should not be in the 60°–100°F range longer than 2 hours. Safe temperatures, then, as applied to potentially hazardous food, are those of 40°F and below and 140°F and above. As implied by the data given above, both time and temperature are important in handling food to preserve microbiological quality.

In foodservice operations, ingredients and partially and fully prepared menu items are subjected to a wide range of temperatures. In relating food handling practices to the time and temperature effects on microorganisms in foods, the following are temperature zones at which foods are handled:

- *Freezing, defrosting, and chilling zone.* Temperatures should prevent multiplication of organisms causing foodborne intoxications and infections over an extended storage period.
- *Growth or hazardous zone.* Temperatures allow bacterial multiplication.
- *Hot holding zone.* Temperatures are aimed at preventing multiplication, but usually do not kill the organism.
- *Cooking zone.* Temperatures should be sufficiently high to destroy bacterial cells within a short period of time.

Employee and Customer

Employee personal hygiene and good food handling practices are basics of a sanitation program in a foodservice facility. One major risk is that unsanitary employees can contaminate food products that go from receiving to service (Beasley, 1993).

Figure 12.5. Important temperatures in sanitation and food protection.

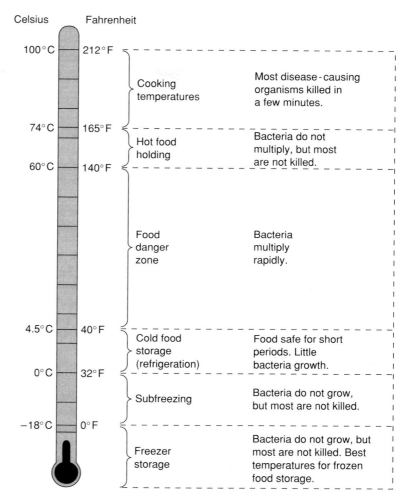

Celsius   Fahrenheit

100°C — 212°F

Cooking temperatures — Most disease-causing organisms killed in a few minutes.

74°C — 165°F

Hot food holding — Bacteria do not multiply, but most are not killed.

60°C — 140°F

Food danger zone — Bacteria multiply rapidly.

4.5°C — 40°F

Cold food storage (refrigeration) — Food safe for short periods. Little bacteria growth.

0°C — 32°F

Subfreezing — Bacteria do not grow, but most are not killed.

−18°C — 0°F

Freezer storage — Bacteria do not grow, but most are not killed. Best temperatures for frozen food storage.

Source: Adapted from *Keeping Food Safe to Eat,* Home and Garden Bulletin No. 162, 1970, Washington, DC: U.S. Department of Agriculture.

When interviewing prospective employees, managers should note their personal grooming habits and appearance. Employees who practice poor hygiene at home and at work are the third most frequently cited cause of outbreaks of foodborne illnesses, according to the Educational Foundation of the National Restaurant Association (1992). Foodservice managers, therefore, should emphasize the importance of food safety and sanitation to the employee before hiring and then after hiring, when an educational process should begin. Because the most critical aspect of personal cleanliness is frequent and thorough hand washing, proper methods should be emphasized in training programs. Government statistics show improper hand washing accounts for 25% of all foodborne illnesses (Weinstein, 1991b). Using gloves can give a false sense of security because the foodhandler might not change the gloves after handling a contaminated product, resulting in cross-contamination of other products. Managers need to conduct daily inspections of all employees to ensure that proper sanitation practices are being followed.

Control of contamination from customers is more difficult, but various aspects of facility design or policies and procedures can assist in this arena. Sneeze guards on a cafeteria service counter or salad bar can prevent the spread of bacteria as can isolation procedures for a patient hospitalized with a highly communicable disease.

Facilities and Equipment

Restaurants need to be the cleanest places on earth. If not, the potential for the transmission of disease is great (Weinstein, 1991a). The design and construction of foodservice facilities and equipment have a major impact on the effectiveness of the sanitation program; maintenance of the facilities to ensure a sanitary environment is also critical and is discussed in depth in chapter 16. Weinstein (1991a) said that no computer program for a restaurant's physical plant can replace a broom, a mop, some detergent and water, a scrub brush, a pair of hands, and some old-fashioned elbow grease for cleaning and sanitation.

Food Safety Program

A well-designed food safety program monitors all food production activities for errors in handling and eliminates those errors. Critical control points first have to be identified before a Hazard Analysis Critical Control Point model can be selected.

Critical Control Points

Based on the food product flow shown in the models in chapter 5, each type of foodservice has between four and nine critical control points. The type of foodservice influences the number of areas requiring precise monitoring (Table 12.3). For example, the commissary foodservice requires monitoring at all nine points, whereas the assembly/serve has only four critical control points. In the conventional foodservice, in which food is served after heat processing, monitoring at control points concerned with storage following processing and the subsequent heating of precooked items generally is not applicable. The discussion of each of the nine critical control points requiring monitoring within various foodservices follows.

Food Procurement.   With the exception of certain cultured foods, such as cheeses, only pathogen-free ingredients with low levels of microbial count should be purchased. Verification of suppliers' sanitary practices in the warehouse and in transportation vehicles is in order.

Food Storage.   Food is stored immediately after receipt in both conventional and assembly/serve foodservices. In commissary and ready prepared systems, foods as purchased are stored upon receipt, and prepared menu items are stored after food

Table 12.3. Location of nine critical control points applicable in four alternative foodservice systems

Critical Control Points	Alternate Foodservice Systems			
	Commissary	Conventional	Ready Prepared	Assembly/Serve
Food procurement	X	X	X	X
Food storage	X	X	X	X
Food packaging	X		X	
Preprocessing	X	X	X	
Heat processing	X	X	X	
Food storage following heat processing	X		X	
Heat processing of precooked menu items	X		X	X
Food product distribution	X			
Foodservice	X	X	X	X

Source: Foodservice Systems: Product Flow and Microbial Quality and Safety of Foods, North Central Regional Research Bulletin, No. 245, by N. Unklesbay, R. B. Maxcy, M. Knickrehm, K. Stevenson, M. Cremer, and M. E. Matthew, 1977, Columbia: Missouri Agricultural Experiment Station.

production. Foodservice managers should establish and monitor controls for the temperature and length of storage for each menu item.

Food Packaging.   Proper packaging has an important role in both the commissary and ready prepared foodservices by facilitating handling and protecting the food from contamination and unwanted changes in color, texture, and appearance. Packaging materials vary, depending on the type of foodservice, but they must all provide an effective moisture barrier; controls are needed to ensure that the food product is able to be chilled sufficiently in a short time. Prepared menu items are packaged either in individual portions or in bulk. In a hospital with a cook-freeze ready prepared foodservice operation, items for patient service may be packaged individually, and those for the employee cafeteria may be packaged in bulk.

Preprocessing.   In commissary, conventional, and ready prepared foodservices, gross contamination of food products is eliminated during prepreparation. For example, dirt that may cling to fresh fruits or vegetables is washed off. Opportunities for cross-contamination are possible, however, between raw and processed food products; strict separation in storage and food preparation areas is, therefore, necessary.

Heat Processing.   The commissary, conventional, and ready prepared foodservices all involve heat processing of menu items. Effectiveness of heat processing in destroying microorganisms is influenced by food mass and the capacity and type of equipment.

As indicated in the discussion on time-temperature relationships, appropriate heat processing is critical to maintaining microbial quality in foods. If consumed soon after preparation, as in a conventional foodservice, the food should be safe if it was not held under inappropriate conditions or for long periods of time. In a ready prepared foodservice, heat processing temperatures become even more important, because items are cold-stored, then reheated.

Food Storage Following Heat Processing.   Depending on the type of foodservice and the attributes of menu items, foods may be stored heated, chilled, or frozen. Because high temperatures and humidity cause loss of food quality, prolonged hot-holding is not recommended. In commissary and conventional foodservices, the hot-holding period should be as short as possible because of the adverse effects on nutritional and sensory qualities. Temperature control during this period is critical to ensure that food is not held in the hazardous zone for an extended period.

In school foodservice, this concern is of particular relevance because prepared food frequently is transported from a central production facility or a base kitchen to satellite service centers. The time between food production and end of service may be as long as 4 hours, which creates a potentially hazardous condition.

Chilled storage in commissary and ready prepared foodservices represents another critical control point. The necessity for rapid cooling after cooking, assuring that food does not remain in the danger zone for extended periods, has been well documented (Longrée & Armbruster, 1987). Monitoring of temperatures during the cooling period should be part of a quality control program. Types of containers for storing foods affect cooling rate significantly; specialized chilling equipment has been designed for use in cook-chill systems. For example, chili stored in a 2-inch cafeteria counter pan will chill more quickly than if it is stored in a stock pot.

Both commissary and ready prepared foodservices utilize the cook-freeze concept. Heat processed menu items must be frozen as rapidly as possible, and precautions must be taken to control microbial growth during both freezing and thawing processes. Blast freezing has been designed to rapid-freeze food items for later storage at usual freezer temperatures of 0°F. Periodic checks should be conducted to determine the bacterial count of menu items because of the increased potential for microbial growth in prepared food items.

Heat Processing of Precooked Menu Items.   Effective managerial monitoring of product temperatures and storage times is needed to ensure the sensory and microbiological quality of foods when served. The reheating process from the chilled, frozen, tempered (30°F or slightly below), or thawed stage should be rapid, reaching an internal temperature of 165°F. Convection and microwave ovens are most often used for heat processing of prepared menu items.

Food Product Distribution.   In commissary foodservices, chilled and heated menu items are transported to satellite service centers that may be widely dispersed from the central facilities, increasing the opportunities for contamination. Menu items should be transported at temperatures either below 40°F or above 140°F.

Service of Food.   During service, foods must be monitored for contamination and temperature control. Menu items should be served either at 40°F or above 140°F. Thus, the old adage "hot foods hot and cold foods cold" is an important principle for both the sensory and microbiological quality of food. Covering food aids in protecting against contamination and surface evaporation. Managerial monitoring must be effective at point of service, or the effectiveness of all previous controls throughout the flow of food products from procurement to consumption may be nullified (Unklesbay, 1977).

Hazard Analysis Critical Control Point Model

The Hazard Analysis Critical Control Point (HACCP) concept refers to a model developed initially for quality control in the food processing industry, with special emphasis on microbial control. The most common misconception about HACCP is that if managers are using a TQM philosophy, they have a HACCP program in place (King, 1992). HACCP is part of a TQM program, but it is only the part that is concerned with food safety. It has its own rules and standards that are specific to food safety and sanitation. Critical control points are those steps in production processing in which loss of control would result in an unacceptable safety risk. HACCP is a preventive approach to quality control, identifying potential dangers for corrective action.

The original HACCP model was modified by Bobeng and David (1978) to include not only microbiological but also nutritive and sensory quality. They applied the model to quality control of entrée production in conventional, cook-chill, and cook-freeze hospital foodservice operations. HACCP models were developed during three phases: selection of control points, using flow diagrams; identification of critical control points; and establishment of monitors for control.

Figure 12.6 shows the flow of entrée production in hospital foodservice operations and the various control points. Twelve control points were established and four critical control points identified: ingredient control and storage, equipment sanitation, personnel sanitation, and time-temperature relationships (Table 12.4).

The flow of food is best illustrated by designing a flowchart such as the one for beef stew in Table 5.1 (Educational Foundation of the National Restaurant Association, 1992). All potentially hazardous foods on the menu should have a flowchart beginning with receiving and finishing with service. Because of food safety concerns of foodservice managers and customers, the HACCP concept was introduced in chapter 5 and then emphasized in the following chapters throughout the text: chapter 8, receiving and storage; chapter 10, prepreparation; chapter 11, cooking; chapter 15, holding and service; and chapter 5, cooling.

The National Advisory Committee for Microbiological Criteria for Foods has identified seven major principles involved in establishing a HACCP program in a foodservice organization. The principles include the following:

- Assessing hazards at each step in the flow of food and developing procedures to lower the risk for each
- Identifying critical control points
- Setting up control procedures and standards for critical control points

Figure 12.6. Control points for entrée production in convention-al, cook-chill, and cook-freeze hospital services.

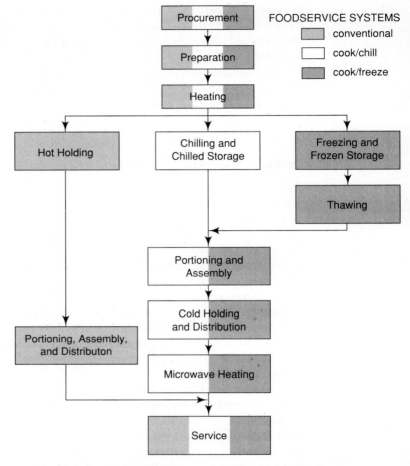

Source: B. J. Bobeng and B. D. David: "HACCP Models for Quality Control of Entrée Production in Hospital Foodservice Systems." Copyright The American Dietetic Association. Reprinted by permission from *Journal of the American Dietetic Association,* Vol. 73: 524, 1978.

- Monitoring critical control points
- Taking corrective action if a deviation from procedures occurs at a critical control point
- Developing a record-keeping system that documents the HACCP plan
- Verifying that the HACCP program is working

Procedures for Complaints

No foodservice manager is immune to outbreaks of foodborne illness (Cheney, 1993). A cook might fail to heat up the grill to the correct temperature, a refrigerator

Table 12.4. Critical control points during entrée production in conventional, cook-chill, and cook-freeze hospital foodservices.

Control Points	Foodservice Type	Ingredient Control and Storage	Equipment Sanitation	Personnel Sanitation	Time-Temperature Relationship
		Critical Control Point			
Procurement	1, 2, 3[a]	X			
Preparation	1, 2, 3		X	X	X
Heating	1, 2, 3				X
Hot holding	1				X
Chilling and chilled storage	2		X		X
Freezing and frozen storage	3				X
Thawing	3		X		X
Portioning and assembly	2, 3		X	X	X
Portioning, assembly, and distribution	1		X	X	X
Cold holding and distribution	2, 3		X		X
Heating	2, 3				X
Serving	1, 2, 3		X	X	X

[a] 1 = Conventional; 2 = Cook–chill; 3 = Cook–freeze.
Source: B. J. Bobeng and B. D. David: "HACCP Models for Quality Control of Entrée Production in Hospital Foodservice Systems." Copyright The American Dietetic Association. Reprinted by permission from *Journal of the American Dietetic Association,* Vol. 73: 524, 1978.

might go on the blink, employees might forget to wash their hands before cutting meat or produce, and a supplier might deliver a contaminated product. Customers who believe their health has been harmed by food eaten in the foodservice establishment have a right to take the manager to court (Educational Foundation of the National Restaurant Association, 1992). The customer might have a legitimate grievance, and managing this crisis correctly could be difficult.

When someone complains of foodborne illness, it is a good idea to have that person complete a complaint report similar to the one shown in Figure 12.7. This will ensure that the right questions are asked even if the business is hectic at the time. Cheney (1993) condensed the information in *Applied Food Service Sanitation* (Educational Foundation of the National Restaurant Association, 1992) into the following steps for receiving a complaint:

• Obtain all the pertinent information including the names and addresses of all party members, the employee who served the meal, the date and time of the customer's visit, and the suspect meal.

Figure 12.7. Foodborne illness/complaint report.

Foodborne Illness/Complaint Report

Complainant name: _____ Phone _____ Work

Address: _____ Phone _____ Home

_____

Others in party? _____ (Get names and addresses) Use back of form if additional space is needed.

_____

_____

_____

Onset of symptoms: Date _____ Time _____

Symptoms:
- ☐ Nausea ☐ Diarrhea ☐ Fever ☐ Blurred vision
- ☐ Vomiting ☐ Dizziness ☐ Headache ☐ Abdominal cramps

Other _____

Medical treatment: Doctor (Hospital)   Name _____   Address _____   Phone number _____

Suspect meal: _____

Location: _____

Time & date: _____

Identification (brand name, lot number): _____

Description of meal: _____

Leftovers: _____ (Refrigerate, *do not freeze*)

Other foods or beverages consumed before or after suspect meal: Date   Time   Location   Description

_____

_____

Other agencies notified: Agency   Person to contact   Phone

_____

_____

Remarks:

Report received by: _____ Date: _____ Time: _____

Referred to: _____

Source: Make a S.A.F.E. Choice: A New Approach to Restaurant Self-Inspection (p. 17) by the National Restaurant Association, 1991, Washington, DC: Author.

- Remain concerned and polite, but do not admit liability or offer to pay medical bills.
- Never suggest symptoms, but let the complainant tell his or her own story.
- Record the time that the symptoms started, which will help in identifying the disease and determining the foodservice operation's responsibility.
- If possible, try to get a food history of all the meals and snacks eaten before and after the person has eaten the suspect meal.
- Never offer medical advice; gather information but do not interpret symptoms.

The NRA diagrammed a step-by-step procedure for crisis management of foodborne illness complaints, which is shown in Figure 12.8.

Figure 12.8. Crisis management—foodborne illness complaints.

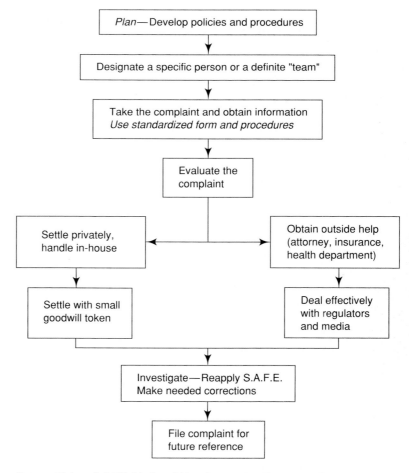

Source: Make a S.A.F.E. Choice: A New Approach to Restaurant Self-Inspection (p. 19) by the National Restaurant Association, 1991, Washington, DC: Author.

SANITATION REGULATIONS AND STANDARDS

The protection of the food supply available to the consumer is the responsibility of governmental agencies at the federal, state, and local levels. Trade associations and institutes, professional societies, and private associations and foundations are especially concerned about microbiological standards of food products.

Regulations

In chapter 7, key federal agencies involved in the wholesomeness and quality of food from producer to purchaser were discussed. The sanitation and service of food after it is purchased is controlled largely by state and local agencies and private organizations.

Role of Governmental Agencies

The U.S. Public Health Service (PHS) and its subdivision, the FDA, both of which are agencies within the U.S. Department of Health and Human Services, are charged specifically with promoting the health of every American and the safety of the nation's food supply. The PHS identifies and controls health hazards, provides health services, conducts and supports research, and develops training related to health.

Two agencies within the PHS specifically related to sanitation regulations and standards are the CDC and the FDA. The CDC investigates and records reports of foodborne illness and is charged with protecting public health by providing leadership and direction in the control of diseases and other preventable hazards. The Bureau of Training within the CDC develops programs for control of foodborne diseases in the foodservice industry. The CDC is responsible for providing assistance in identifying causes of disease outbreaks, including foodborne illnesses.

The FDA directs its efforts toward protecting the nation's health against unsafe and impure foods, unsafe drugs and cosmetics, and other potential hazards. The FDA has been established as the regulatory agency with responsibility for food safety in the United States, and the Food Safety and Inspection Service (FSIS) has the same responsibility under the USDA. The FDA uses laboratory analysis of foods to detect microbiological and chemical toxicants, adulteration, and incorrect labeling; the USDA does on-site inspections of meat and poultry operations and checks animal health, sanitation, and product labeling (Cody & Keith, 1991).

The FDA is responsible for regulating interstate shipment of food, as discussed in chapter 7, and for inspecting foodservice facilities on interstate carriers such as trains, planes, and ships operated under the U.S. flag. In addition, although local agencies have the primary responsibility for inspecting foodservice establishments, the FDA assists these agencies by developing model codes and ordinances and providing training and technical assistance. Amendments to the Federal Food, Drug, and Cosmetic Act include the Pesticide, Food Additives, Color Additives, and Labeling Acts. The FDA's jurisdiction has been redefined as new concerns have arisen. The FDA also administers the Fair Packaging and Labeling Act and the Tea Importation

Act. The FSIS administers the Wholesome Poultry Products, Egg Products Inspection, Federal Meat Inspection, and Wholesome Meat Acts (Cody & Keith, 1991). The General Accounting Office (GAO) prepared a report on the proliferation of federal food safety laws and overlapping responsibilities together with resource constraints leading to duplication of services ("GAO Calls," 1993). The greatest problem has been in clarifying the responsibilities of the USDA, which oversees meat and poultry, and the FDA, which oversees most other products. The GAO suggests that the president appoint a blue-ribbon panel to develop broad-based agreement on food safety standards before revising them.

Many state and local governments have adopted the U.S. Public Health Service codes in establishing standards of performance in sanitation for foodservice establishments. These codes generally require that employees have medical examinations to determine their qualifications to handle food safely and have a food handler's permit, which usually requires a short training program on sanitation practices.

State and local health agencies act to ensure that foodservice establishments (Longrée & Armbruster, 1987)

- Are equipped, maintained, and operated to offer minimal opportunities for food hazards to develop
- Use food products that are wholesome and safe
- Are operated under the supervision of a person knowledgeable in sanitary food handling practices

State or local agency officials make periodic inspections to foodservice operations to compare their performance with standards of cleanliness and sanitation. Any deficiencies must be corrected before the next inspection. The agency generally has the authority to close an operation that has an inordinate number of deficiencies in meeting the sanitation standards.

The Environmental Protection Agency (EPA), another agency within the U.S. Department of Health and Human Services, also has responsibility in certain areas related to sanitation in foodservice establishments. The EPA endeavors to comply with environmental legislation and control pollution systematically by a variety of research, monitoring, standard setting, and enforcement activities. Of particular interest to the foodservice industry are programs on water standards, air quality, pesticides, noise abatement, and solid waste management.

Role of Other Organizations

A number of other organizations are active in upgrading and maintaining the sanitary quality of various food products and establishing standards for foodservice operations. For example, many trade and professional organizations serving various segments of the food industry have established sanitary standards for food processing operations.

The **Educational Foundation of the National Restaurant Association** has exerted aggressive leadership in developing standards and promoting training in foodservice sanitation. It has developed a crisis management program, identified as SERVSAFE®, that concentrates on three areas of potential risk—food safety, responsible alcohol service, and customer safety (NRA, 1991; Educational Foundation of the National Restaurant Association, 1992). SERVSAFE courses focus on the manager's role in assessing risks, establishing policies, and training employees. SERVSAFE programs include employee training materials for each subject, such as study guides, videos, and an employer's kit with a leader's guide and other teaching aids. The *Applied Food Service Sanitation* book is the major component of the food safety course. Employees who purchase the book and register with the Educational Foundation may take the course, either in a classroom or at home. Everyone who completes the course and satisfactorily completes a certification examination administered by the foundation is eligible for an Educational Foundation SERVSAFE certificate.

In the healthcare industry, the **Joint Commission on Accreditation of Health-care Organizations (JCAHO)** has encouraged high standards of sanitation by including assessment of sanitary practices in its accreditation standards and visits. The **American Dietetic Association (ADA)** has emphasized food safety for many years and has published *Food Safety for Professionals: A Reference and Study Guide* (Cody & Keith, 1991) and *Pesticides in Food: A Guide for Professionals* (Chaisson, Petersen, & Douglass, 1991). Dietitians can receive continuing education hours toward dietetic registration and use the legally protected professional designation Registered Dietitian (RD) by completing a self-study program that uses the Cody and Keith publication as well as four specially prepared ADA pamphlets.

The **National Association of College and University Food Services (NACUFS)** published a standards manual in 1991. These standards were designed to be used as a self-monitoring program for improving operations and as part of a voluntary peer review program. The **American School Food Service Association (ASFSA)** has provided leadership in the school foodservice segment of the industry in promoting good sanitation practices and has been active in providing employee training.

The **American Public Health Association (APHA),** a professional society representing all disciplines and specialties in public health, has a standing committee on food protection that establishes policies and standards for food sanitation. Several other associations for professionals in the area of food protection and sanitation also promote standards and enforcement procedures in food safety and sanitation.

Members of the **National Council of Chain Restaurants** have formed an ad hoc group, the Food Safety Council, charged with helping formulate government policies and alleviating customer concerns (Allen, 1993). The council wants to convey to the customer the seriousness with which restaurateurs view the hamburger mishap and project a positive active role in shaping food safety for the future. The group also will participate in developing new federal food safety policies. The food safety committee is planning to participate in the public hearings being held by USDA's Food Safety and Inspection Service concerning the development of a new method to reduce pathogens in meat and poultry products. Finally, the group will serve as a watchdog for the foodservice industry by collecting and disseminating information on food safety problems at state and local levels.

The **National Sanitation Foundation (NSF)** is one of the most influential semi-private agencies concerned with sanitation. The NSF, organized by a group of industrial leaders and public health officials, is a nonprofit, noncommercial organization that seeks solutions to problems involving cleanliness and sanitation. On the basis of research results, the NSF develops minimum sanitation standards for equipment, products, and devices. Manufacturers may request that the NSF evaluate their equipment, and they will receive an NSF Testing Laboratory Seal of Approval for equipment meeting NSF standards.

Microbiological Standards

The increased number of highly processed foods now being used in some foodservice operations has caused concern about new potential hazards to the public health. In more recent years, the FDA has been concerned with the establishment of microbiological quality standards, setting acceptable microbial levels for food products to control abuses at the retail or customer level. Microbiological standards are being developed cooperatively by governmental agencies, professional societies, and industry because of concerns over the safety of the food supply.

The HACCP model, currently being used by most foodservice organizations, actually started in 1971 when the head of quality control at Pillsbury introduced the concept at a national conference for food protection (Stern, 1990). The HACCP model identifies which specific foods are at risk of meeting standards and the locations in the process line where mishandling is likely to occur. NRA has developed a *Sanitary Assessment of the Food Environment* (S.A.F.E.) model based on HACCP that is being used by restaurant health inspectors.

Foodservice Operation Inspections

State and local health inspections of foodservice operations have changed in the last several years (Stern, 1990). The trend is changing from floors, walls, and ceilings to hazard analysis. Inspectors are still concerned about the cleanliness of the facility, but they are really looking at those areas where hazards exist and are trying to decide if those hazards will cause foodborne illness. Critical control points need to be identified in the food preparation process, in which the food might become contaminated if bacteria survive cooking temperatures and grow if food is held at incorrect temperatures.

A self-inspection program, an **internal audit**, should not be developed primarily for the purpose of preparing for regulatory agency inspections, an external audit, but it should be a continual check on the adequacy of the sanitation program in a foodservice operation. Managers, supervisors, and employees should understand that the purpose of self-inspection is to promote positive sanitation practices, not to take punitive action.

Audits of Sanitation Standards

Evaluation of the maintenance of foodservice sanitation standards is accomplished in two ways: external and internal audits of facilities and practices. An **external audit** may be performed by governmental or nongovernmental agencies; an **internal audit** is the responsibility of the management of a foodservice organization and may be part of a self-inspection program for sanitation or a component of a broader total quality management program.

External Audits

External audits are performed by federal, state, and local governmental agencies, as indicated in the previous discussion of the role of these agencies in monitoring sanitation in foodservice establishments. Inspections are conducted by a local or state sanitarian every month or annually, at any time during a calendar year. The *sanitarian*, often referred to as the inspector or health official, represents the state or local health department and is trained in sanitation principles and methods and in public health. The local health code serves as the sanitarian's guide.

In some areas, HACCP principles have replaced traditional health department inspections that stress the appearance of the facility and spot-checking temperatures. Inspectors trained in these principles examine the procedures related to the flow of food from receiving to service and may verify critical control points for each step. Many state and local ordinances are patterned on the Model FDA Food Service Establishment Inspection Report, shown in Figure 12.9. Note the item numbers that are marked with an asterisk on the left. These items are weighted at 4 or 5 points and are similar to critical control points in the HACCP program. The foodservice operation receives a score after violations are subtracted from the perfect score of 100.

A good sanitation program and well-trained employees result in safe food and are reflected in a good sanitation report (Educational Foundation of the National Restaurant Association, 1992). The foodservice manager should check the credentials of the sanitarian to be sure the visit is a routine inspection and not the result of a customer complaint. The manager should accompany the inspector during the inspection and take advantage of his or her expertise. Most foodservice managers become defensive with the inspector. Instead the manager should cooperate with the inspector and instruct employees to do the same. Problems should be discussed with employees and corrected immediately. According to the FDA model code, if the operation receives a score of less than 60, corrective actions must be taken on all violations within 48 hours. The larger violations should have a time frame approved by the inspector. The sanitarian usually has the authority to close the operation if violations are excessive and dangerous to the public health.

The inspection process in a foodservice facility may begin before a facility is built, as many jurisdictions require a review of plans and specifications for new construction or extensive remodeling. Once a facility is completed, inspection visits are usually conducted prior to issuance of permits to operate. After a foodservice operation has opened, inspections will occur periodically, depending on the workload of the responsible agency and severity of violations at previous inspections. The growth of the foodservice industry has not been matched by expansion of the capacity of health agencies to monitor operations.

Figure 12.9. Foodservice establishment inspection report.

TIME INSPECTION STARTED	COUNTY	CITY	YEAR	IDENT NO.	TYPE	FEE CODE	SANIT CODE	MO.	DAY	YEAR	TRAVEL TIME	INSP TIME

PURPOSE

Regular 1
Follow-up 2
Complaint 3
Investigation 4
Other (Pre.) 5

RATING SCORE

100 less weight of items violated →

OWNER'S NAME
ESTABLISHMENT'S NAME
STREET ADDRESS
CITY
STATE AND ZIP

KANSAS DEPARTMENT OF HEALTH AND ENVIRONMENT FOOD ESTABLISHMENT INSPECTION REPORT

Received by: Inspected by:

NAME   TITLE   NAME   TITLE

FOOD

*01	Source; sound condition, no spoilage	5
02	Original container; properly labeled	1

FOOD PROTECTION

*03	Potentially hazardous food meets temperature requirements during storage, preparation, display, service, transportation	5
*04	Facilities to maintain product temperature	4
05	Thermometers provided, conspicuous, and accurate	1
06	Potentially hazardous food properly thawed	2
*07	Unwrapped and potentially hazardous food not re-served	
*07 A	Cross contamination prevented; damaged/detained food segregated	4
08	Food protection during storage, preparation, display, service, transportation	2
09	Handling of food (ice) minimized	2
10	In use, food (ice) dispensing utensils properly stored	1

PERSONNEL

*11	Personnel with infections restricted	5
*12	Hands washed and clean, good hygienic practices	5
13	Clean clothes, hair restraints	1

FOOD EQUIPMENT & UTENSILS

14	Food (ice) contact surfaces: designed, constructed, maintained, installed, located	2
15	Non-food contact surfaces: designed, constructed, maintained, installed, located	1
16	Dishwashing facilities: designed, constructed, maintained, installed, located, operated	2
17	Accurate thermometers, chemical test kits provided, gauge cock (¼″ IPS valve)	1
18	Pre-flushed, scraped, soaked	1
19	Wash, rinse water: clean, proper temperature	2
*20	Sanitization rinse: clean, temperature, concentration, exposure time, equipment, utensils sanitized	4
21	Wiping cloths: clean, stored, use restricted, sanitizer	1
22	Food-contact surfaces of equipment and utensils clean, free of abrasives, detergents	2
23	Non-food contact surfaces of equipment and utensils clean	1
24	Storage, handling of clean equipment/utensils	1
25	Single-service articles, storage, dispensing, used	1
26	No re-use of single service articles	2

WATER

*27	Water source, safe: hot & cold under pressure	5

SEWAGE

*28	Sewage and waste water disposal	4

PLUMBING

29	Installed, maintained	1
*30	Cross-connection, back siphonage, backflow	5

TOILET & HANDWASHING FACILITIES

*31	Number, convenient, accessible, designed, installed, maintained	4
32	Toilet rooms enclosed, self-closing doors, fixtures, good repair, clean: hand cleanser, sanitary towels/tissue/hand-drying devices provided, proper waste receptacles	2

GARBAGE & REFUSE DISPOSAL

33	Containers or receptacles, covered: adequate number, insect/rodent proof, frequency, clean	2
34	Outside storage area enclosures properly constructed, clean; controlled incineration	1

INSECT, RODENT, ANIMAL CONTROL

*35	Presence of insects/rodents - outer openings protected, no birds, turtles, other animals	4

FLOORS, WALL & CEILINGS

36	Floors: constructed, drained, clean, good repair, covering installation, dustless cleaning methods	1
37	Walls, ceiling, attached equipment: constructed, good repair, clean surfaces, dustless cleaning methods	1

LIGHTING

38	Lighting provided as required, fixtures shielded	1

VENTILATION

39	Rooms and equipment vented as required	1

DRESSING ROOMS

40	Rooms clean, lockers provided, facilities clean, located, used	1

OTHER OPERATIONS

*41	Necessary toxic items properly stored, labeled, used	5
42	Premises maintained, free of litter, unnecessary articles, cleaning maintenance equipment properly stored. Authorized personnel	1
43	Complete separation from living/sleeping quarters. Laundry.	1
44	Clean, soiled linen properly stored	1

* Critical Items Requiring Immediate Attention.

Failure to comply with applicable requirements may result in license suspension, license revocation or civil penalty.

Source: U.S. Food and Drug Administration.

Internal Audits

A foodservice organization should have its own program of self-inspection as a means of maintaining standards of sanitation. In organizations with a TQM program, an audit of sanitation practices should be one of its major components. Employees can be an important part of an internal audit when they are empowered to take corrective action if a critical control point is violating the code. A voluntary food safety program of self-inspection will assure the government and the public that the foodservice operation is protecting the safety of food in each step of production.

EMPLOYEE TRAINING IN FOOD SANITATION

Trained personnel who have good personal hygiene habits and follow recommended food handling practices are critical to an effective sanitation program. Equally important are strong leadership by management, provision of appropriate tools and equipment, and continual follow-up. The time, money, and effort that go into a sanitation program are wasted if the foodservice staff is not knowledgeable about appropriate sanitation practices (Longrée & Armbruster, 1987). Although training principles and processes are discussed in chapter 20, the critical importance of protecting customers from foodborne illness justifies mention here.

The education of foodservice managers should include microbiology and sanitation to equip them for leadership in design and implementation of inservice sanitation programs. Continuing education is then needed to maintain competency.

Professional associations have an important role in upgrading sanitation practices in the foodservice industry. As mentioned earlier in this chapter, several professional associations are concerned with training standards and programs for foodservice employees. The certification programs of several organizations, like the American School Food Service Association and the Dietary Managers Association, include knowledge and competency in sanitation practices as a major component. A major contributor to the upgrading of sanitation practices is the Educational Foundation of the NRA through its national uniform sanitation, training, and certification plan for foodservice managers.

In 1984, the Educational Testing Service developed a national testing program for the Food Protection Certification Program that is voluntary and not required by state and local regulatory agencies. The certificate has national recognition and certifies that the holder has qualified by passing a rigorous examination.

Training of nonmanagerial employees is the responsibility of the foodservice manager. Initial training in safe food handling practices and personal hygiene is needed for new employees.

Many teaching aids are available to assist the manager in conducting training programs on safe food handling. Federal and state health agencies, as well as commercial and private organizations, have videotapes, films, slides, posters, and manuals available free or at low cost. Many foodservice organizations also have developed comprehensive manuals for training and as a reference for their own personnel. The *Sanitation Operations Manual* (National Restaurant Association, 1984) published by the

Figure 12.10.   Sanitation mini-posters.

Source: National Restaurant Association Catalog of Publications. Used by permission.

NRA is an excellent sanitation resource. The NRA also periodically distributes brochures, mini-posters, and other aids, such as those shown in Figure 12.10, that are eye-catching and useful in reminding personnel to use good food handling practices.

SUMMARY

Many opportunities exist for food contamination in a foodservice operation, and foodborne illness is being recognized as a major health problem in the United States today. The major types of food spoilage are microbiological, biochemical, physical, and chemical. Microbiological organisms causing foodborne illness include bacteria, molds, yeasts, viruses, rickettsiae, protozoa, and parasites. Biochemical spoilage is caused by natural food enzymes, which are catalysts that initiate reactions in foods, such as off-flavors, odors, and colors. Physical spoilage of food may be caused by temperature changes, moisture, and dryness. Chemical spoilage may result from interaction of certain ingredients in a food or beverage with pesticides, foodservice chemicals such as detergents, additives and preservatives, and toxic metals such as zinc, copper, or cadmium.

Foodborne illnesses, which are outbreaks of acute gastroenteritis, are caused by microbial pathogens that multiply profusely in food. These illnesses are either foodborne intoxications caused by toxins formed in the food prior to consumption or foodborne infections caused by the activity of large numbers of bacterial cells carried by the food into the gastrointestinal system of the victim. *Staphylococcus aureus, bacillus cereus*, and *clostridium botulinum* are foodborne intoxications. *Salmonella, shigella, listeria monocytogenes, staphylococcus aureus*, and *clostridium perfringens* are bacteria that multiply in the intestinal tract and cause foodborne infections. Emerging foodborne pathogens include *campylobacter jejuni, escherichia coli,* and the Norwalk virus.

Control of the microbiological quality of food must focus on the food itself, employees or customers handling the food, and facilities and equipment. The quality of food depends upon its condition when purchased and the time-temperature control during storage, production, and service. Employee personal hygiene and cleanliness of the facility and equipment also contribute to food safety.

A well-designed food safety program monitors food production activities in which errors in handling occur and eliminates those errors. Critical control points have to be identified before a Hazard Analysis Critical Control Point (HACCP) model can be selected. The type of foodservice influences the number of areas, or critical control points, that need monitoring. HACCP is a preventive approach to quality control, identifying potential dangers for corrective action.

The protection of the food supply available to the consumer is the responsibility of governmental agencies at the federal, state and local levels. Two agencies within the Public Health Service specifically related to sanitation regulations and standards are the Centers for Disease Control (CDC) and Food and Drug Administration (FDA). The CDC investigates and records reports of foodborne illness and is charged with protecting the public health. The FDA protects the nation's health against unsafe and impure foods and other potential hazards. The U.S. Department of Agriculture (USDA) oversees meat and poultry, and the FDA oversees most other products. A number of other organizations are active in upgrading and maintaining the sanitary quality of various food products and establishing standards for foodservice operations.

Microbiological standards are being developed cooperatively by trade associations and institutes, professional societies, and private associations and foundations because of concerns over the safety of the food supply. Sanitation inspections have changed from checking only the cleanliness of the facility to also checking areas where hazards exist that could lead to foodborne illness. Evaluation of foodservice sanitation standards is accomplished through external audits performed by governmental and nongovernmental agencies and internal audits by management that may be part of a self-inspection program for sanitation or a component of a broader total quality management program. Continuing education in food safety, therefore, is necessary for foodservice managers because they are responsible for training employees in safe food handling practices and personal hygiene.

BIBLIOGRAPHY

Allen, R. L. (1993). NCCR creates group to battle food fears. *Nation's Restaurant News, 27*(24), 7.

American Home Economics Association. (1980). *Handbook of food preparation* (8th ed.). Washington, DC: Author.

Beasley, M. A. (1993). Food safety through hygiene. *Food Management, 28*(5), 36.

Bobeng, B. J., & David, B. D. (1977). HACCP models for quality control of entree production in foodservice systems. *Journal of Food Protection, 40*(9), 632–638.

Bobeng, B. J., & David, B. D. (1978). HACCP models for quality control of entree production in hospital foodservice systems. I: Development of Hazard Analysis Critical Control Point models. II: Quality assessment of beef loaves utilizing HACCP models. *Journal of The American Dietetic Association, 73*(5), 524–535.

Chaisson, C. F., Petersen, B., & Douglass, J. S. (1991). *Pesticides in food: A guide for professionals*. Chicago: The American Dietetic Association.

Cheney, K. (1993). Managing a crisis. *Restaurants & Institutions, 103*(13), 51, 56, 58, 62, 66.

Cody, M. M., & Keith, M. (1991). *Food safety for professionals: A reference and study guide*. Chicago: The American Dietetic Association.

Dkystra, J. J., & Schwarz, A. R. (1991). The race against germs. *School Food Service Journal, 45*(6), 92, 94, 98.

Educational Foundation of the National Restaurant Association. (1992). *Applied foodservice sanitation* (4th ed.). Chicago: Author.

Facts about E. coli. (1993). *Nutri-facts* (Kansas Beef Council), *19*(Spring), 1.

GAO calls for a single government agency to police food-safety. (1993). *FoodService Director, 6*(2), 36.

Gebo, S. (1992). *What's left to eat?* New York: McGraw-Hill.

General Accounting Office. (1990). *Food safety and quality: Who does what in the federal government*. GAO/RCED-91-91B. Washington, DC: Author.

Government regulation of food safety: Interaction of scientific and societal forces. A scientific status summary by the Institute of Food Technologists' expert panel on food safety and nutrition. (1992). *Food Technology, 46*(1), 73–80.

GRAS status: What's in a name? (1992). *Food Insight*, (March/April), 6–7.

Hecht, A. (1991). The unwelcome dinner guest: Preventing food-borne illness. *FDA Consumer Magazine*, (January/February), DHSS Publication No. (FDA) 91-2244.

King, P. (1992). Implementing a HAACP program. *Food Management, 27*(12), 54, 56, 58.

Liston, J. (1989). Current issues in food safety—especially seafoods. *Journal of The American Dietetic Association, 89*(7), 911–913.

Liston, J. (1990). Microbial hazards of seafood consumption. *Food Technology, 44*(12), 56, 58–62.

Longrée, K., & Armbruster, G. (1987). *Quantity food sanitation* (4th ed.). New York: Wiley.

Longrée, K., & Blaker, G. G. (1982). *Sanitary techniques in foodservice* (2nd ed.). New York: Wiley.

Martin, R. (1993). Push for safety: A quiet crusade. *Nation's Restaurant News, 27*(25), 1, 75.

McCarthy, B. (1993). All about burgers. *Restaurants & Institutions, 103*(7), 89.

McLauchlin, A. (1993). Restaurants urged to cook ground beef thoroughly. *Restaurants USA, 13*(3), 7.

Minor, L. J. (1983). *Sanitation, safety & environmental standards.* L. J. Minor Foodservice Standards Series, Vol. 2. Westport, CT: AVI.

National Advisory Committee for Microbiological Criteria for Foods. (1989). *HACCP principles for food production.* Washington, DC: U.S. Department of Agriculture, Food Safety and Inspection Service.

National Association of College and University Food Services. (1991). *National Association of College and University Food Services professional standards manual* (2nd ed.). East Lansing: Michigan State University.

National Restaurant Association. *Sanitation of restaurants.* (1975). Chicago: Author.

National Restaurant Association. *Sanitation operations manual.* (1984). Chicago: Author.

National Restaurant Association. (1986). *A restaurateur's guide to consumer food safety concerns.* Current issues report. Washington, DC: Author.

National Restaurant Association. (1991). *Make a S.A.F.E. choice: Sanitary assessment of food environment.* Washington, DC: Author.

Roberts, T. (1989). Human illness costs of foodborne bacteria. *American Journal of Agricultural Economics, 71*(2), 468–474.

Ryser, E. T., & Marth, E. H. (1989). "New" food-borne pathogens of public health significance. *Journal of The American Dietetic Association, 89*(7), 948–954.

Snyder, O. P. (1986a). Applying the hazard analysis and critical control points system in foodservice and foodborne illness prevention. *Journal of Foodservice Systems, 4*(2), 125–131.

Snyder, O. P. (1986b). Microbiological quality assurance in foodservice operations. *Food Technology, 40*(7), 122–130.

Stern, G. M. (1990). Goodbye to the white-glove test: Restaurant inspections are changing. *Restaurants USA, 10*(8), 32–34.

Thorner, M. E., & Manning, P. B. (1983). *Quality control in foodservice* (rev. ed.). Westport, CT.: AVI Publishing.

U.S. Department of Agriculture. *Keeping Food Safe to Eat,* Home and Garden Bulletin No. 162, 1970, Washington, DC: Author.

U.S. Department of Health, Education, and Welfare, Public Health Service, Food and Drug Administration. (1978). *Food service sanitation manual: 1976 recommendations of the Food and Drug Administration.* DHEW Publication No. (FDA) 78-2081. Washington, DC: Government Printing Office.

Unklesbay, N. (1977). Monitoring for quality control in alternate foodservice systems. *Journal of The American Dietetic Association, 71*(4), 423–428.

Unklesbay, N., Maxcy, R. B., Knickrehm, M., Stevenson, K., Cremer, M., & Matthews, M. E. (1977). *Foodservice systems: Product flow and microbial quality and safety of foods.* North Central Regional Research Bulletin No. 245. Columbia: University of Missouri-Columbia Agricultural Experiment Station.

VanLandingham, P. G. (1993). On foodborne illnesses: It's time to yell 'fowl!' *Nation's Restaurant News, 27*(17), 32.

Weinstein, J. (1991a). The clean restaurant. I: Physical plant. *Restaurants & Institutions, 101*(12), 90–91, 94, 96, 98, 100, 106–107.

Weinstein, J. (1991b). The clean restaurant. II: Employee hygiene. *Restaurants & Institutions, 101*(13), 138–139, 142, 144, 148.

Wolf, I. D. (1992). Critical issues in food safety, 1991–2000. *Food Technology, 46*(1), 64–70.

13

Labor Control

Controlling and reducing labor costs and simultaneously increasing labor productivity have been growing challenges for many years among managers of all types of foodservice operations. Many different approaches have been used, a number of which have been helpful, such as increasing the use of convenience foods, decreasing the number of items on a menu, improving the efficiency of the layout and equipment, and increasing employee benefits (Keiser & DeMicco, 1993). Also needed for streamlining a labor force are the following seven steps: personnel policies, job analysis, work simplification, work production standards, forecasting work load, scheduling, and control reports. According to Keiser and DeMicco (1993), however, labor cost recently has emerged as the predominant cost in hospitality operations. In the past, when labor was abundant and cheap, and there were few government controls, food was considered the highest-cost item. That period has passed, and as Keiser and DeMicco stated, if management is to generate a profit for commercial operations or provide the best services resources permit for noncommercial enterprises, control of labor costs is inevitable.

Bernstein (1992) said he has a brilliant idea that will overcome the skyrocketing labor costs that may be ruining operators' bottom lines and raising havoc with the industry. He suggested that perhaps robots could run a foodservice operation. Computerized french fry equipment and self-service ordering technology already are being tested by limited-menu chains. Bernstein concluded that robots must never be a substitute for people; the industry is asking for trouble if robots replace beginning workers. The industry must continue to provide employment and work toward a balance between automation and limited robotics and the dignity of individuals.

The foodservice industry has some unique labor control problems. The most competent personnel in the work force will choose industries in which financial compensation is higher than in foodservice. According to the 1992 Bureau of Labor Statistics, foodservice employees are more likely than other workers to be young, female, and unmarried (Iwamuro, 1993). They also are more likely to work on a part-time basis. The typical foodservice employee is

- Female (58%)
- Under 30 years of age (58%)
- A high school graduate (65%)
- Single—includes never married, divorced, and widowed (64%)
- Living in a household with relatives (82%)
- Living in a household with two or more wage earners (77%)
- Living in a household similar to the U.S. average in terms of income
- A part-time employee with weekly hours averaging 25.0
- An individual with relatively short tenure

In addition, the foodservice industry is an important employer of new workers entering the labor force. Nearly 5% of the work force was employed in food preparation and service occupations during 1992. Eating and drinking places are the nation's largest retail employer; 3 out of 10 retail employees work in these establishments.

For many years, the foodservice industry has been the biggest employer of teenagers, who consider their jobs temporary; even that source of supply is diminishing because of population trends. In addition, foodservice employees' work hours may be long and scheduled at times that teenagers study or have social activities. The prediction is that the elderly and handicapped might take up this employment slack. Scheduling and staffing become top priorities for many foodservice managers. Lowering labor costs by improving productivity and providing satisfactory wages that attract and retain competent employees is a monumental challenge for the manager.

Ninemeier (1986) offered a number of points that justify efforts to control labor cost.

- Direct labor costs are increasing due to elevation of minimum wage scales and competition for qualified people.
- Increases in labor costs often are not balanced by increases in productivity.
- Fringe benefit packages are often very creative and unfortunately expensive.
- People with different attitudes, beliefs, problems, goals, and personalities, as are commonly found among hospitality staff, are difficult to manage.
- Managers often recognize the need to help employees find meaning in their jobs and get more than the salary or wage from their work.

- Foodservice is labor intensive, and generally food industry technology has not been able to replace people with equipment.

Cost-containment pressures and the prospective payment plans in the healthcare segment of the industry, for example, have caused a major emphasis to be placed on improved productivity. All segments of the industry have been faced with inflationary pressures with a concomitant rise in costs. Because labor represents a major component of the total operational cost in a foodservice establishment, controlling labor cost is a key aspect of attaining financial objectives in foodservice operations.

Effective utilization of labor in the foodservice industry is especially difficult because of the following unique characteristics:

- Many 7-day-a-week operations, some of which may require staff coverage 24 hours a day
- Peaks in service demand requiring additional staff at those times
- Seasonal variations in patronage of establishments
- The highly perishable nature of food items both before and after production
- The labor intensive aspect of most production and service activities
- The large number of unskilled and semiskilled personnel in the industry

These and other characteristics emphasize problems facing the foodservice manager in utilizing the work force to its fullest capacity. In this chapter, staffing and scheduling and improving productivity are discussed.

STAFFING AND SCHEDULING

The terms *staffing* and *scheduling* are sometimes used interchangeably; in fact, they refer to separate but interrelated functions. **Staffing** concerns the determination of the appropriate number of employees needed by the operation for the work that has to be accomplished. Job analyses and work production standards provide the bases for determining staffing needs.

Scheduling, however, means having the correct number of workers on duty, as determined by the staffing needs. Scheduling involves assignments of employees to specific working hours and workdays. The challenge of scheduling is having sufficient staff for busy meal periods without having excess staff during slack periods between meals.

Variables

Staffing and scheduling employees depend on many factors. Operational differences, such as which meals are served or where the foodservice operation is located, has a great effect on the number of employees and the time they will work. Also, the type of foodservice, whether it is conventional, commissary, ready prepared, or assembly/serve, affects staffing and scheduling.

Operational Differences

In foodservice operations, staffing and scheduling can become extremely complex because of the highly variable nature of the foodservice business. In a commercial foodservice, for example, the weekend dinner meal is often a peak time, and Tuesday or Wednesday night might attract only a small number of diners. An operation serving primarily a lunch crowd in a business area, however, may have very low volume in the evening. Hospital censuses generally tend to vary seasonally; patient loads are frequently at a low point during the summer vacation months and the November and December holiday season.

In a university residence hall foodservice, a school lunchroom, and some other foodservice operations, customer participation is much more predictable. In these operations, however, a number of other factors affect participation, such as scheduled campus events or what is on the menu. These examples illustrate the staffing and scheduling demands on foodservice managers to ensure adequate personnel without overstaffing.

Scheduling is further complicated by absenteeism, labor turnover, vacations and holidays, days off, and differing skills of employees. In addition to all these considerations, special events and catering functions usually will have an impact on the schedule. Determination of minimal staffing requirements is based primarily on providing coverage for all events in the operation.

Type of Foodservice

The type of foodservice is a major determinant of staffing needs in the operation. A distinctive example is provided by contrasting the conventional and the assembly/serve foodservices discussed in chapter 5. Obviously, the staffing needs are much different in a 200-bed hospital with a conventional foodservice operation, in which all or most foods are prepared from scratch, from those in an assembly/serve foodservice, in which foods require little or no processing and fewer and less highly skilled employees.

One of the advantages of the ready prepared foodservice operations using cook-chill or cook-freeze methods is that the majority of the employees, especially those in production, can be scheduled on a regular 40-hour, 5-day, Monday-through-Friday basis. This schedule is possible, of course, because production is for inventory, rather than for immediate service. As a result, only a skeleton staff is needed at mealtime, during weekends, and on holidays, primarily for service, dishwashing, and cleanup. Fewer employees have to work the undesirable early morning, late evening, weekend, or holiday schedules.

Relief Employees

Scheduling only full-time employees to do all the work in a foodservice operation could create some problems. During rush hours, customers would complain that the operation is understaffed, but during slack times the foodservice manager would be concerned that employees are sitting around with nothing to do while increasing labor costs. A solution for many service industries is to hire part-time or temporary employees to help out in rush hours.

Part-Time Employees

In some foodservice operations, most of the staff are part-time employees, a practice particularly prevalent in the limited-menu restaurant. Part-time employees quite often are not eligible for many benefit programs, such as vacation and sick leave time, holidays, or insurance. In some organizations, part-time employees receive these benefits when the hours of employment reach a specified level. Benefits such as vacation and sick leave time may be prorated according to the number of hours worked.

Teenagers or college students traditionally have made up the bulk of the part-time work force in limited-menu operations; changes in the national birthrate, however, have created a short supply of this age group. Employers need to be aware of labor laws affecting teenagers and understand the following federal restrictions when hiring ("Employers Should," 1993).

- Workers must be at least 14 years old to work in foodservice jobs.
- Minors aged 14 and 15 are limited to working 8 hours a day and 40 hours a week. During weeks when school is in session, workers under 16 may not work before 7:00 A.M. or after 9:00 P.M.
- 14- and 15-year-olds may not hold jobs in which they do repair or maintenance work on machines or equipment; they may not cook (other than at soda fountains, lunch counters, snack bars, or cafeteria serving counters) or bake; and they are not permitted to load or unload goods to and from trucks.
- Finally, no worker under 18 is permitted to operate power-driven meat and food slicers.

All states, many of which have higher standards than federal law, have teenage employment laws as well. Any employer who is covered by both laws must observe the higher standard. A telephone survey among 398 National Restaurant Association members was conducted to determine employment practices and policies for hiring teenagers (Gordon, 1992). Approximately 32% said they were visited by a federal or state labor inspector and 20% were visited by both. Fines were imposed on those cited.

Other foodservice operations that regularly utilize part-time student workers are college and university foodservices. In these operations, most of the servers are college students. Healthcare institutions commonly use student employees for service and cleanup during the evening meal. Part-time employees also make up the majority of the work force in school foodservice operations, where most of the workers are women who want to work outside the home for only a few hours during the day.

In foodservice operations in which the patronage varies greatly within the week, or in establishments involved in catered events, adequate staffing is provided by scheduling persons to work only on busy weekends or to assist with special events. **Split shift scheduling**, in which employees are scheduled to work during peak hours only, is another way in which foodservice managers attempt to have adequate staffing when they need it and minimal staffing during between-meal low-volume times. Dining room hostesses, waiters, waitresses, and other service personnel are frequently scheduled to work during the noon meal, take a break during the afternoon when the dining room may be closed, and return for the evening meal. Such scheduling must be done carefully, however, because many states have work span laws that require the hours worked to fall within a given span of time. Instead of using split shift scheduling, most operations now use part-time personnel.

Temporary and Leased Employees

A number of creative ideas are being tried in foodservice operations for dealing with labor shortages caused by the shortage of young, entry-level workers and more stringent immigration laws (Coppess, 1988). By the year 1995, the 15- to 24-year-olds in the work force will decline 14%; the 45- to 64-year-olds will increase by 17%; and the over-60-year-olds will increase by 18% (Humes, 1988). Typical entry-level school-age workers are going to be much harder to find and keep in the foodservice work force. Much emphasis is being placed on hiring unconventional workers, such as the handicapped and elderly, but more needs to be done to alleviate the labor shortage problem.

The latest entrepreneurial effort to solve staffing dilemmas is the private agency that offers temporary employees (Coppess, 1988). Depending on the agency, employees are available with skills for jobs ranging from dishwasher to executive chef, from hotel administrator to controller or marketing director. This service is new to the foodservice industry, and some managers question the quality of temporary employees and the cost of using them on a regular basis. A foodservice operation that needs temporary help calls an agency, explains its requirements, and agrees to an hourly fee. The agency reviews its database of available employees and makes the closest match possible. The employee reports for work, completes the assignment, and is paid by the agency, which then bills the company. Hourly rates range from 50% to 150% above the wage a company would pay a permanent employee. Most of the agencies provide employee benefits, workers' compensation, and liability insurance.

Another idea that was started in Chicago is leasing staff. More than 1,500 leasing companies nationwide employ more than 1 million leased employees in all kinds of businesses (Resnick, 1993). Skyrocketing insurance costs and the cost of complying with government regulations make taking care of employees a difficult task. Restaurant owners can cut costs and headaches and concentrate on operating a restaurant by leasing employees from companies that provide everything from payroll processing to discounted healthcare coverage and workers' compensation insurance. The leasing company hires all the employees away from the restaurant and leases them back for a monthly percentage. Under this arrangement, the leasing company handles employee-related tasks, and the client maintains supervisory control of the employees, sets the pay rate, and does scheduling. Some leasing companies even will screen and hire employees if desired by the manager. The cost to the client often is lower than it would be if the foodservice operation were responsible. Foodservice managers should choose a leasing company carefully and do a thorough check of its financial health. Should the leasing company fail, foodservice operations are not free of liability for paying healthcare and workers' compensation claims.

Table 13.1.   Staffing for kitchen

Jobs to Be Filled	For 0–49 Customers	For 50–99 Customers	For 100–175 Customers	For 175+ Customers
Chef	1	1	1	1
Cook	1	2	3	4
Salads—pantry	1	2	2	3
Dishwasher	1	2	3	3
Potwasher	1	1	1	1
Cleaner	0	1	1	1
Storeroom person	0	1	1	1
Baker	0	1	1	1

Indices for Staffing

Several indices have been developed to assist managers in various segments of the foodservice industry in determining their staffing needs. Lundberg and Armatas (1980) proposed using staffing tables for commercial operations based on the number of customers expected for a given meal on a given day. These tables can be used to take into account day-to-day and meal-to-meal fluctuations. In Tables 13.1 and 13.2, staffing is shown for a restaurant kitchen and dining room at various patronage levels.

Fairbrook (1979) indicated that industrywide standards for staffing have not been developed to the level of sophistication of those in manufacturing industries, such as the automobile industry. Fairbrook (1979) proposed a rule of thumb of approxi-

Table 13.2.   Staffing for dining room

Jobs to Be Filled	Number of Customers								
	0–37	38–58	59–75	76–95	96–112	113–129	130–145	146–166	167+
Hostess	1	1	1	1	1	1	1	1	1
Waiter/ Waitress	2	3	4	5	6	7	8	9	10
Bus person	1	2	2	3	3	3	3	4	5
Bar waiter/ waitress	1	1½	1½	2	2	2½	2½	2½	2½

Figure 13.1.   Staffing guide.

Staffing guide for residence hall kitchen serving 750 students

Regular Employees

A.M. Shift
(5:00 a.m.–2:30 p.m.)*

1 Lead Cook
1 Second Cook
2 Salad and Line Workers
1 Dishwasher
1 Potwasher/Porter
1 Hostess/Cashier

P.M. Shift
(10:00 a.m.–7:00 p.m.)

1 Lead Cook
1 Second Cook
1 Salad and Line Worker
1 Dishwasher
1 Potwasher/Porter

*Regular employees are normally scheduled for a nine-hour period, which in residence halls, allows time for two 30-minute meal periods. Those working other shifts take only one 30-minute meal period.

Student Employees

Breakfast (7:30–9:30)

Dining Room Runner 1
Serving Line 1
Serving Line Runner 1
Dishroom 1
Total Hours: 8

Lunch (11:15–1:30)

Dining Room Runner 1
Serving Lines (2) 4
Serving Line Runners 2
Dishroom 3
Total Hours: 22½

Dinner (4:15–6:30)

Dining Room Runner 1
Serving Lines (2) 4
Serving Line Runners 2
Dishroom 3
Dining Room Porters 2
Hostess/Cashier 1
Total Hours: 29¼

Please note: The above-cited staffing example gives a productivity of approximately ten (10) meals per labor hour:

750 students × 3 meals/day = 2,250 meals paid for per day
less: 29% missed meals factor = 1,600 meals eaten per day

Total hours worked: Regular Employees   96 per day
Total hours worked: Student Employees   59¾ per day
TOTAL 155¾ per day
Meals per Labor Hours: 1,600 ÷ 155¾ = 10.3

Source: Fairbrook, P.: *College and University Food Service Manual*. Stockton, CA: Colman Publishers, 1979, p. 125.

mately 10 meals per labor hour for a college and university residence hall, utilizing a combination of regular and student employees (Figure 13.1).

Rose (1980) indicated that an index of 17 minutes per meal, or 3.5 meals per labor hour, is frequently used as a staffing standard for a conventional foodservice in an acute care facility. These standards are based on recommendations published by the U.S. Public Health Service (Coble, 1975).

Sneed and Kresse (1989) summarized **productivity levels**, the meals per labor hour, for various foodservice operations by meals per labor hour as follows:

Limited-menu restaurant	9.5
Fine dining restaurant	1.4
Family restaurant	4.8
Cafeteria	5.5
Acute care facility	3.5
Extended care facility	5.0
School foodservice	13 – 15.0

These productivity levels reflect industry averages and should serve only as a guide for determining staffing needs and evaluating employee performance. Sneed and Kresse (1989) stated that although productivity levels can be compared to industry averages, comparison of an operation with itself over time could be more valuable. Deviations in productivity can be determined and improvement identified by self-comparison.

Approximately 1.55 employees are necessary for everyday coverage of full-time positions (Keiser & DeMicco, 1993). The number of positions to be filled should be multiplied by 1.55 employees per position. Although meals must be served 365 days per year in many foodservice operations, full-time employees are generally available an average of only 236 days a year because of days off and benefit days, as shown below:

365 days/year
104 days off/year (52 weeks × 2)
 10 vacation days/year
  8 holidays
  7 sick leave days
---
129 total days not worked
236 days worked full time

To determine the actual number of employees needed, the number of full-time positions needs to be multiplied by 0.55 to find the number of relief employees necessary in addition to the full-time staff.

Staffing needs are often stated in **full-time equivalents (FTEs)**. The absolute FTEs indicate the minimum number of employees needed to staff the facility; the adjusted FTEs take into account the benefit days and days off. A minimum number of FTEs is computed by dividing the total number of hours for operating the foodservice for a period of time by the normal work load hours of one employee. For example, if a restaurant requires 200 hours of labor a day and the normal working period is 8 hours, 200 divided by 8 would give 25 as the FTEs. Because this calculation does not include any time off, the actual number will be more than 25 employees.

A **meal equivalent** has the same effect as a meal in any situation involving meal count. In determining staffing needs in relation to meals produced and served, a healthcare institution serving many nourishments must reduce the count to meal equivalents. Ho and Matthews (1978) suggested that dividing the number of nourishments by 6 yields a satisfactory number of meal equivalents. In other words, 6 nourishments require as much staffing as one meal; the total staffing needs then are the sum of regular meals plus the meal equivalents. In a commercial operation, the meal equivalent may be determined by dividing the total dollar sales for a specific time period by the cost of a prototype meal. For example, a prototype meal might consist of a salad, entrée, bread, dessert, and beverage.

Issues in Employee Scheduling

Some unusual problems occur in scheduling for a foodservice operation (Keiser & DeMicco, 1993). The hours between breakfast and dinner in a three-meal-a-day operation do not lend themselves to two full shifts. The manager needs to determine the type of work schedule that would be best for the operation. Overtime increases labor costs and should be carefully investigated before approval. Finally, alternative work schedules need to be examined.

Types of Schedules

Three types of work schedules—master, shift, and production—must be made by the foodservice manager. The master schedule will show days on and off duty, and vacations. The shift schedule will indicate the position and hours worked and may indicate the number of days worked per week as well as relief assignments for positions when regular workers are off. The production schedule was discussed in chapter 9.

Master Schedule.   In most foodservice facilities, a **master schedule**, which includes days off, serves as an overall plan for employee scheduling. Generally, some type of rotation is used for scheduling days off, especially in 6- or 7-day-a-week operations, permitting employees to have some weekend time off on a periodic basis. A policy of every other weekend or every third weekend off is not uncommon. The master schedule provides the basis for developing the weekly, biweekly, or monthly schedule.

Figure 13.2 depicts a master schedule for the cooks' unit in a university residence hall foodservice serving approximately 2,200 meals daily. The schedule is based on a 2-week cycle for the 1995 spring semester. The calendar dates for week I and week II are shown at the bottom of the schedule. The specific hours of employment for each position are in the first column with the name of the assigned employee in the second column. The assignment for these duties is indicated for each day of week I and week II; days off are indicated by an *X*. Note that each full-time employee has one weekend off during the 2-week cycle, either Saturday of the first week and Sunday of the second or Saturday of the second week and Sunday of the first. The word *lunch* on an employee's line means that the cook must help with lunch preparation because the cook assigned to that duty has the day off. A similar master schedule is made for each work unit in the foodservice operation.

A paid vacation of two weeks is common in most organizations and may be extended to three weeks or longer for employees after a defined number of years of service. Holidays generally vary between 8 and 10 days per year. Because many food-

Figure 13.2. Example of a master schedule for a cook's unit.

HOURS	NAME	*WEEK I SUN	MON	TUE	WED	THU	FRI	SAT	**WEEK II SUN	MON	TUE	WED	THU	FRI	SAT
5:30 – 2:10	King, JoAnn	X				X					X				X
5:30 – 2:10	Lewis, Jane			X				X	X				X		
6:00 – 2:40	North, Edra	X				X						X			X
6:00 – 2:40	Evans, Cindy	X			X		X				X				
6:00 – 2:40	Roberts, Jerry				X			X	X					X	
5:30 – 2:10	Vail, Gloria	X	X							X					
10:20 – 7:00	Mack, Thelma	X		X				X	X				X		
10:20 – 7:00	Schartz, Phyllis	X				X					X				X
10:20 – 7:00	Long, Peggy	X	***Lunch	X				X	X				X	Lunch	X
10:20 – 7:00	Adams, Anna Mae	X			X		Lunch	X	X					X	
10:20 – 7:00	Harris, Alberta	X	X					X	X					X	
10:20 – 7:00	Watt, Wilhelmine	X				Lunch	X	X	X	Lunch	Lunch	X			X
10:20 – 7:00	Mann, Maxine	X					X	X	X	Lunch	Lunch				X

* Week I Calendar

10/02/94 – 10/08/94	12/11/94 – 12/17/94	
10/16/94 – 10/22/94	12/25/94 – 12/31/94	
10/30/94 – 11/05/94	1/08/95 – 1/14/95	
11/13/94 – 11/19/94	1/22/95 – 1/28/95	
11/27/94 – 12/03/94	2/05/95 – 2/11/95	

** Week II Calendar

10/09/94 – 10/15/94	12/18/94 – 12/24/94	
10/23/94 – 10/29/94	1/01/95 – 1/07/95	
11/06/94 – 11/12/94	1/15/95 – 1/21/95	
11/20/94 – 11/26/94	1/29/95 – 2/04/95	
12/04/94 – 12/10/94	2/12/95 – 2/18/95	

*** = Help with preparation

Source: Kansas State University Housing and Dining Services, Manhattan, KS. Used by permission.

service operations continue to conduct business during holidays and vacation time, scheduling of vacations and holidays can present a special challenge for managers. In fact, in some commercial operations, holidays may be among the highest volume days of the year and require extra staff rather than only a skeleton staff. Employees may be compensated by being paid at a double-time rate instead of being given scheduled time off.

Careful vacation planning is needed to avoid higher labor costs in the form of overtime or the excessive use of replacement workers. Managers should attempt to schedule vacations at low-volume periods throughout the year and to stagger them in order that only minimal relief staffing will be required. Effective scheduling of vacations can be accomplished by requesting each employee to submit preferences with alternates early in the year. A master vacation schedule can then be prepared that will permit the most effective labor utilization.

Shift Schedule.   The **shift schedule** shows the staffing pattern of the operation. In the schedule shown in Figure 13.3, the staffing for a dishroom operation for three meal periods covering 16 hours has been divided into two basic shifts. For the most part, this rigid shift scheduling is not the most effective approach to scheduling in foodservice operations. For instance, in the example in Figure 13.3, all six dishwashers come on duty at 7:00 A.M., and soiled dishes in any quantity may not come into the dishroom until 7:30 or 7:45 A.M. One or two of the workers may be required to fill the dishwasher and prepare it for use for the breakfast dishes; the other workers would probably have time that would be difficult to use efficiently in some other way.

A variation of a shift schedule is a **staggered schedule** (Figure 13.4), which provides for employees to begin work at varying times, generally resulting in better utilization of the labor force. Staggered scheduling will usually lead to reduction in idle time and is more adaptable to the fluctuating pattern of activity in a foodservice operation.

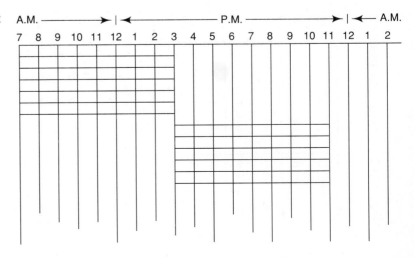

Figure 13.3.   Example of a shift schedule for a dishroom operation.

Figure 13.4. Example of a staggered schedule for a dish-room operation.

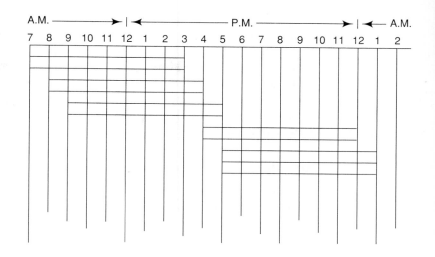

Computer software is available for automated scheduling, but the complex nature of the process and variations in the program often require close monitoring by management and an occasional override. Kasavana (1986) named the following functions that can be included in a labor scheduling format: identification of employee, shift, workstation, overtime, temporary worker files, and job skill banks. A list of backup personnel also can be developed. Adaptation of software to an operation generally requires the services of a computer system specialist.

Control of Overtime

Uncontrolled overtime is a key factor in driving up labor cost. In some instances, employees may need to work beyond their normal hours; in other cases, however, supervisors may use overtime as a substitute for proper scheduling and planning. Moreover, employees may try to create opportunities for overtime because of the time-and-a-half wage rate they may receive.

As Fairbrook (1979) indicated, if the basic staffing pattern is the first determinant of labor cost, then control of overtime is the second, and the key to the overtime problem is tight control. With proper staffing and realistic work schedules, Fairbrook said, overtime becomes necessary only in emergencies.

Policies regarding time cards must be formulated to ensure that employees check out at scheduled times, unless authorized for a later time. Most organizations today have a computerized timekeeping system instead of the older clock-in and clock-out system. Employees only have to pass a personal identification card through the recognition slot of a machine. Managers may then get daily printouts of the number of hours each employee worked the day before. If overtime was recorded, managers are alerted to the labor cost changes in adequate time to take corrective measures. Controls are needed for legal reasons as well as for control of labor cost. Labor laws define when overtime pay is due to an employee, and the recorded time from the clock usually provides the official record of time worked for employees covered by overtime provisions.

Overtime that can be anticipated may be controlled by requiring overtime authorization, as shown in Figure 13.5. Approval of overtime is indicated by the signature of designated members of the managerial staff. Unpredictable overtime required by a breakdown of equipment or unexpected business should require an overtime report filed within 24 hours. The amount of overtime should be shown and the reason stated. The completed form should be signed by the supervisor.

Alternate Work Schedules

Alternatives to the standard workweek have been a topic of interest and experimentation in recent years. Trends have included the introduction of several discretionary time work schedules and new forms of part-time employment. These new approaches can be grouped into three categories:

- Compressed workweek
- Discretionary working time
- Part-time employment

Three work patterns characterize the *compressed workweek* trend. A change in days holds the total hours constant, but reduces the number of days worked so that employees may work, for example, a 4-day week, 10 hours per day. Some foodservice operations have experimented with this approach, but it has not gained widespread popularity in the industry. The second pattern, a change in hours worked, shortens the number of hours in the week, but the number of days worked remains the same. A 7-hour day, for instance, is characteristic of some organizational scheduling patterns. The third pattern is a change in days and hours, in which both dimensions are changed.

Discretionary working time modifications to the standard workweek include the staggered start system and flexible working hours. In using the staggered start, the

Figure 13.5.   Example of an overtime authorization form.

Overtime Authorization

Date _____

Name _____   Unit _____

Reason for Overtime _____

_____

_____

Amount of Overtime: _____ Hrs.

_____
Signature of Unit Manager

organization or the employee chooses, from a number of management-defined options, when they wish to start their fixed-hour working day.

Flexible working hours, or *flexitime* as the system is sometimes called, is the second major variation of discretionary working time schedules. Generally, the organization defines a range of hours within which employees may select their starting time. A number of variations to flexitime systems have been developed. Flexitime is impractical for many foodservice operations, however, because of the demands of production and service and scheduled mealtimes.

Task contracting is another alternative to flexible work schedules in which the employee contracts to fulfill a defined task or piece of work. This approach has been used by some foodservice caterers who may contract an individual for preparation of particular food items or for serving a particular function.

Part-time employment is being used more widely in many different organizations today. As discussed earlier in this chapter, foodservice operations commonly use part-time employees as a way of controlling labor costs. Two new variations in part-time employment, job sharing and job splitting, have emerged. In job sharing, a single job is divided and shared by two or more employees, each of whom must be capable of performing the entire range of tasks in the job description. In job splitting, the tasks that constitute a single job are divided, with subsets of differentiated tasks assigned to two or more employees.

Several of these alternative work schedules have been tried or are now being utilized in the foodservice industries. Others are less applicable because of the unique characteristics of the industry. When experimenting with new scheduling methods, foodservice managers should use a careful process that involves gaining employee acceptance, pilot testing, and carefully evaluating outcomes before making a change.

PRODUCTIVITY IMPROVEMENT

Productivity improvement is a phrase widely used in newspapers, on television, in news magazines, and in the trade and professional literature of almost all fields. The decline in productivity in this country has been a widespread concern in many organizations. Low productivity has been characteristic of the foodservice industry: the popularly accepted statistic is that its productivity level is only 40% to 45% (Sneed & Kresse, 1989). Productivity concerns the efficient utilization of human, equipment, and financial resources, and it is often expressed mathematically as the ratio of output to input (Olsen & Meyer, 1987). Productivity measures, principles of work design, work measurement, and the use of **quality circles** have much effect on the improvement of productivity in a foodservice organization.

Productivity Measures

Measurement of productivity is important for foodservice managers. Productivity can be evaluated by means of the ratio of one or more inputs compared with a variety of outputs (Lieux & Winkler, 1989). Formulas for frequently used measures of labor

Figure 13.6.   Selected productivity measures.

$$\text{Meals/labor hour} = \frac{\text{total meals served/day}}{\text{labor hours/day}}$$

$$\text{Minutes/meal} = \frac{\text{labor minutes/day}}{\text{total meals served/day}}$$

$$\text{Payroll cost/day} = \Sigma \text{ of hourly rate of each employee} \times \text{hours worked for all employees}$$

$$\text{Payroll cost/meal served} = \frac{\text{total daily payroll cost}}{\text{meals served/day}}$$

$$\text{Labor cost/day} = \text{total payroll cost/day} + \text{total of all other direct labor costs (fringe benefits, and so on)/day}$$

$$\text{Labor cost/meal served} = \frac{\text{total labor cost/day}}{\text{meals served/day}}$$

productivity are shown in Figure 13.6. Record-keeping systems should be established to record data on a systematic basis to determine the productivity measures that the management of the foodservice organization has selected for analysis. These measures can be used to examine trends over time within a particular operation, to compare various operations within an organization, or to compare the results from a specific foodservice operation with available industry data.

Principles of Work Design

Konz (1990) proclaimed that the key to productivity improvement is to assist the worker to "work smart and not hard." *To work smart* means to work more efficiently; to *work hard* is to exert more effort. He contended more potential for improvement exists in reducing excess work than through making the staff work harder. Konz also asserted that people do not like to work hard and, therefore, they resist efforts to make them do so.

Productivity is the ratio of output to input, or the ratio of goals to resources of the foodservice system. Productivity can be increased by reducing input, by increasing output, or by doing both at the same time. As indicated previously, productivity in the foodservice industry tends to be low. High turnover, rising costs, and the relatively high percentage of the revenue dollar devoted to labor indicate that foodservice managers should focus their efforts on increasing productivity.

Labor inefficiency may be the result of several factors: poor product design, work methods, management, or workers (Konz, 1990). Material waste, improper tools or methods, inadequate maintenance, poor production scheduling, absences without cause, and carelessness are examples.

Work design refers to a program of continuing effort to increase the effectiveness of work systems (Kazarian, 1979). Industrial engineers, for example, have applied work analysis and design techniques in the manufacturing industries for many years. More recently, these principles have been applied to the service industries.

Principles of work analysis and design have been developed from several different fields of study over time. Among the major contributions to the study of scientific management is the classic work of Frederick Taylor and the Gilbreths. Their principles of materials handling and motion economy are the ones most directly related to labor productivity and will be discussed here.

Materials handling refers to the movement and storage of materials and products as they proceed through the foodservice system. Good design of materials movement will lead to increased efficiency and decreased activities that do not add appreciable value to the end product. The amount of materials handling often depends on the location and arrangement of storage areas, preparation and production areas, and equipment. The key principles of materials handling, outlined in Figure 13.7, should be used to develop a system for moving materials efficiently within the foodservice facility.

The *principles of motion economy*, primarily from the early work of the Gilbreths, relate to the design of work methods, of the workplace, and of tools and equipment. These principles, summarized in Figure 13.8, specify that movement should be simultaneous, symmetrical, natural, rhythmic, and habitual.

The *principles of motion economy* that pertain to the human body—the use of both hands, coordination of hands and eyes, and continuous motion—are aimed specifically at reducing the effort and energy required to do a job. Principles related to the design of the workplace and of tools and equipment identify situations that

Figure 13.7.   Principles of materials handling.

1. Minimize all material movements and storages.
2. Use the shortest and straightest routes for the movement of materials across the workplace.
3. Store materials as close to the point of first use as possible.
4. Minimize handling of materials by workers unless absolutely necessary.
5. Preposition all materials at the workplace as much as possible to reduce handling effort.
6. Handle materials in bulk if at all possible.
7. Provisions should be made to remove scrap, trash and other wastes at the point of creation.
8. Take advantage of gravity to move materials when feasible.
9. Use mechanical aids to lift heavy materials that are frequently used at workplaces.
10. Built-in leveling devices can be used to keep materials at a convenient working height.
11. Use mechanized conveyors to move materials that follow a fixed route across the workplace if they do not interfere with the work.
12. Use well-designed containers and tote pans that are easy to pick up and move.
13. Consider the use of interlocking containers for moving greater loads with ease and safety.
14. Consider changing design of products involved to improve their materials-handling characteristics.

lead to easy body motions, such as locating tools within easy reach and placing objects in fixed positions. The graphics in Figure 13.9, which have been adapted from employee training materials on work simplification, illustrate several of the key principles.

Figure 13.8.   Principles of motion economy.

A. Use of the human body
1. The number of motions required to complete a task should be minimized.
2. The length of necessary motions should be minimized.
3. Both hands should be used for work, and they should begin and end their activities simultaneously.
4. Motions of hands and arms should be in symmetrical and opposite directions.
5. Both hands should not be idle at the same time except for rest.
6. Motions should be confined to the lowest possible classifications needed to perform the task satisfactorily.
7. Smooth curved motions should be developed in preference to straight-line or angular motions.
8. Motion patterns should be developed for rhythmic and habitual performance.
9. The motions should be arranged to take advantage of momentum.
10. The number of eye fixations required for the task should be minimized.
11. Intermittent use of the different classifications of movements should be provided to combat fatigue.

B. Design and layout of the workplace
1. Materials, tools, and controls should be located within the normal working area.
2. Materials and tools should have a fixed location.
3. Work requiring the use of eyes should be done within the normal field of vision.
4. Tools and materials should be prepositioned to facilitate picking up.
5. Gravity feed bins or containers should be used to deliver incoming materials close to the point of use.
6. Gravity should be used to deliver outgoing materials.
7. The height of the working surface should be designed to allow either a standing or sitting position.
8. The environment of the workplace should be conducive to productive motions.

C. Design of tools and equipment
1. Tools, hand equipment, and controls should be designed for easy grasp.
2. Two or more tools should be combined if possible.
3. Jigs, fixtures, or foot-operated devices should be used to relieve the work of the hands.
4. Equipment should be designed so the inherent capabilities of the body members are fully utilized.
5. Levers and controls should be designed to make maximum contact with the body member.

Figure 13.9. Illustrations of work design principles in foodservice operations.

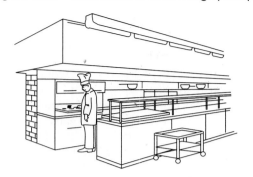

Proper equipment arrangement helps you "work smart."

Adjust work heights to your elbow by:

1. changing height of work on the table. (Use different thicknesses of cutting boards or platforms to adjust the level of the work.)

2. adjusting equipment height. (Use a table to suit your height and job.)

3. adjusting the chair. (Use adjustable-height chairs.)

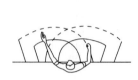 

Maximum work areas, based on your reach distance, affect tiring.

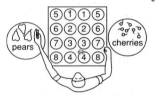

Arranging work and using both hands in rhythm and order improves your methods.

YES	NO

Planned arrangements save time, motion, and effort.

Drop delivery using proper heights of equipment helps you "work smart."

Using continuous curved motions avoids unneeded starts and stops.

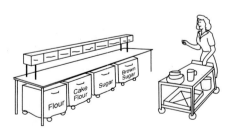

Equipment on wheels can help organize work.

Using muscles smoothly when lifting prevents strain.

Source: The Basic Four of Work by S. Konz and J. Maxwell, 1980. Unpublished manuscript. Manhattan, KS: Kansas State University.

Work Measurement in Foodservice

Work measurement is a method of establishing an equitable relationship between the amount of work performed and the human input used to do that work. Several productivity measures were outlined previously in this chapter. In any production operation, work measurement is necessary for effective use of human resources.

Data from work measurement studies can aid in evaluating alternative production/service subsystems, determining and controlling cost, staffing, scheduling work, deciding whether to make or buy, planning facilities and layout, identifying needs for changes in employee assignments, and timing or sequencing of tasks. David (1978) indicated that work measurement data are also needed for developing useful managerial aids, such as production time standards. Activity analysis, activity or occurrence sampling, elemental standard data, and predetermined motion time are the primary techniques of work measurement for analysis in foodservice operations.

Activity analysis involves continuous observation for a chronological record of the nature of activities performed by individual workers, work performed at one workstation, work units produced, or the amount of time that equipment is used and for what purpose. Data are used to establish standards for short- or long-cycle work by persons in motion. A simplified technique has been developed that involves employee recording of activities at periodic intervals, usually between 5 and 15 minutes. This technique has been referred to as an employee time log reporting system (DenHartog, Carlson, & Romisher, 1978). Figure 13.10 shows a time sheet and the codes for reporting foodservice activities on an employee time log in a school foodservice operation. Employees are asked to enter data on the form every 10 minutes during each day of a time study, according to the function and project from the code sheet. Data permit an analysis of time devoted to various operations within the school foodservice unit.

Occurrence sampling, or activity sampling, are terms used in the literature to describe a method for measuring working time and nonworking time of people employed in direct and indirect activities, and to measure operating time and downtime of equipment. The term *work sampling* also may be used to refer to this technique. Konz (1990) stated that occurrence sampling is a more accurate term than work sampling, however, because what is being sampled is the occurrence of the various events that may invoke direct work, indirect work, and delays. By means of intermittent, randomly spaced, instantaneous observations, estimates can be made of the proportion of time spent in a given activity over a specified time.

Occurrence sampling has been widely used for studying work in foodservice operations. Several of these studies are cited in the reference list at the end of this chapter. Much of the classic work has been done at the University of Wisconsin, where the *Methodology Manual for Work Sampling: Productivity of Dietary Personnel* (University of Wisconsin-Madison, 1967) was developed. In the manual, methods for conducting activity sampling studies in foodservice operations are described. Work functions and classifications, including direct work, indirect work, and delays, are defined in the manual (Figure 13.11).

The number of observations required in occurrence sampling depends on the type of study, the type of operation, and the number of personnel. Data from an occurrence sampling study are used to calculate labor minutes per meal equivalent

Figure 13.10. Form and codes for employee time log system.

TIME REPORTING SHEET FOR THE LINCOLN PUBLIC SCHOOLS' FOODSERVICE STUDY

LINCOLN PUBLIC SCHOOLS
DIVISION OF BUSINESS AFFAIRS
TIME MANAGEMENT REPORTING FORM

_____
DEPARTMENT

• EMPLOYEE NAME _____   _____   WORK WEEK

Day	Unit	Function	Project	Day	Unit	Function	Project	Day	Unit	Function	Project

CODES FOR REPORTING FOODSERVICE ACTIVITIES: EMPLOYEES, FUNCTIONS, AND PROJECTS

Coding for Foodservices Work Analysis

Code/Function		Code/Project	
1	Washing	1	Salad Fruit/Veg
2	Chopping	2	Fruit Can/Fresh
3	Dipping	3	Dessert
4	Stirring	4	Vegetables
5	Mixing	5	Entree
6	Panning	6	Juice
7	Shaping	7	Milk
8	Weighing	8	Breads
9	Buttering	9	Restroom
10	Cutting	10	Dishes
11	Sorting	11	Trays
12	Opening	12	Silverware
13	Peeling	13	Carts
14	Supplying	14	Tables
15	Cleaning	15	Refrig./Freezers
16	Scraping	16	Pots & Pans
17	Shredding	17	Steam Tab/Count
18	Pouring	18	Stove
19	Planning	19	Ovens
20	Ordering	20	Mixer
21	Counting	21	Work Schedules
22	Wrapping	22	Qualify Planner
23	Accounting	23	Deposit Slip
24	Deliming	24	Floors
25	Putting Away	25	Steam Equipment
		26	Dish Machine
		27	Dish Ally
		28	Food Order
		29	Nonfood Order
		30	Clothes

Source: "Employee Logs as a Basis for Time Analysis" by R. DenHartog, H. Carlson, and J. M. Romisher, 1978, _School Food Service Research Review, 2_(2), p. 99. Used by permission.

Figure 13.11.   Work function classification and definitions.

Direct Work Functions

Any essential activity contributing directly to the production of the end product (end product is total number of meals served per day).

- **Processing**
  Act of changing the appearance of a foodstuff by physical or chemical means.
  - **Preparation or preliminary processing**
    Preliminary act or process of making ready for preparation, distribution, or service.
  - **Preparation or cooking**
    Final act or process of making ready for distribution or service.
- **Service**
  Act of preparing facilities for distribution and of portioning and assembling pre-pared food for distribution to patients and to cafeteria customers (to coffee shop also if dietary is responsible for operation of coffee shop).
- **Transportation**
  Act of transporting food, supplies, or equipment from a location in one function-al area to a designated location in another area within the department or to patients' wards.
  - **Transportation of food**
    Act of moving food from a location in one functional area to a designated location in another area within the department.
  - **Transportation of equipment, supplies, and other**
    Act of moving equipment, supplies, and other items from a location in one functional area to a designated location in another area within the department.
  - **Delivery of trays to patients** (if this function is performed by dietary)
    Act of removing patients' trays from food trucks, dumbwaiter or trayveyor, and carrying to patients' bedside.
  - **Return of trays from patients** (if this function is performed by dietary)
    Act of removing trays from patients' bedside to food trucks; dumbwaiter on the ward.
  - **Transportation empty**
    Act of moving without carrying or guiding anything from a location in one func-tional area to a designated location in another area within the department.
- **Clerical (routine)**
  Act of receiving, compiling, distributing, and storing of routine records of data and information necessary for operation of the department.
- **Cleaning**
  Act of removing soil or dirt to provide sanitary conditions for the use of equip-ment, facilities, and supplies.
  - **Pot and pan washing**
    Act of scraping, washing, or rinsing quantity food containers and cooking utensils.

Figure 13.11.   *continued*

- **Dishwashing**
  Act of preparing for or removal of soil or dirt to provide sanitary conditions for use of tableware (china, silverware, glassware, and trays).
- **Housekeeping**
  Act of removing soil or dirt to provide sanitary conditions for the use of installed and mobile equipment and facilities.
- **Receiving**
  Act of acquiring, inspecting, and storing food and/or supplies from an area outside the department.

Indirect Work Functions

Any catalytic activity which contributes to production of the end product.

- **Instruction or teaching**
  Act of directing or receiving direction by oral or written communication in a training or classroom situation or on the job.
- **Appraisal**
  Act of judging or estimating the value or amount of work in order to make decisions for future planning.
- **Conference**
  Act or oral communication with one or more persons in the form of a scheduled meeting.
- **Clerical (original or non-delegable)**
  Act of compiling and formulating management control records of data and information necessary for the operation of the department.

Delays

All time when an employee is scheduled to be working and is not engaged in either a direct or an indirect work function.

- **Forced delay**
  The time an employee is not working due to an interruption beyond his control in the performance of a direct or an indirect work function.
- **Personal and idle delays**
  The time an employee is not working due to personal delays or avoidable delays.
  - **Personal delays**
    The time an employee is not working due to time permitted away from the work area.
  - **Idle time**
    Any avoidable delay (other than forced or personal delay) that occurs for which the employee is responsible.

Source: Adapted from *Methodology Manual for Work Sampling, Productivity of Dietary Personnel* by the University of Wisconsin-Madison, 1967.

or labor minutes for some other specific activity. For example, Block, Roach, and Konz (1985) used occurrence sampling to study cleaning times for vegetables in a university residence hall foodservice. One advantage of occurrence sampling is that several workers in a specific area can be studied simultaneously by a single observer.

Elemental standard data are time values that have been determined for many elements and motions common to a wide variety of work (David, 1978). From these values, total times for specific tasks can be synthesized. David (1978) stated that job variables significantly affecting normal time for a given type of operation must first be hypothesized, then time data collected on the number and variety of jobs in that operation. Data are used to determine the relationship between normal time and each of the variables believed to affect normal time significantly.

Predetermined motion time includes techniques in which tasks are broken down into basic motions with known normal time values (David, 1978). The purpose is to establish cycle time for a specific operation without actually performing the task. Instead, the predetermined time for the basic motions that make up the cycle are synthesized. One technique, Methods Time Measurement (MTM), is widely used in industry but is time-consuming; David (1978) concluded that MTM is usually not applicable to long-cycle work or work with limited repetition, such as that in foodservice operations. An alternative technique has been developed, called **Master Standard Data (MSD)**, in which seven basic elements of work are combined into larger, more condensed elements.

Montag, McKinley, and Kleinschmidt (1964) were among the first to apply MSD to foodservice operations. They concluded that the method was applicable for developing coded standard elements with universal application in foodservice operations. Several studies listed in the references for this chapter cite other studies that used MSD for examining production times in foodservice facilities. One of the studies (Ridley, Matthews, & McProud, 1984) used MSD to develop labor times for the assembling and microwave heating of menu items in a hospital galley. They also found that the technique could be used effectively for developing standard labor times because data from their study indicated that total labor time under actual conditions in a hospital galley was similar to MSD-predicted time.

David (1978) concluded that progress is being made in foodservice systems toward developing standards for labor time using techniques of work measurement. She asserted, however, that application of the more complex measures requires a combination of the expertise of the foodservice operation and the systems analyst. Each foodservice operation should establish its own standards of productivity because of unique differences among operations. She emphasized that work measurement in the industry should continue, but cautioned that quantitative productivity standards should not be the only index for measuring the effectiveness of a foodservice operation. At some point, increased productivity can be achieved only by sacrificing the quality of food and service or the level of employee satisfaction (Ruf & David, 1975).

Productivity and Quality

After World War II, a time when U.S. manufacturers were producing high-quality war products with unskilled labor, service men and women returned to the work force, as discussed in chapter 2. With an immediate absence of foreign competition, U.S. manufacturers returned to prewar production methods (Beasley, 1991). At the same time, Japanese industrialists changed customers' perceptions from one of poor-quality products to one of high-quality, high-tech products. The Japanese had to learn to manufacture superior products in a cost-effective manner. The quality circle concept evolved from quality training initially provided to managerial personnel and first-line supervisors, which was then applied to employee involvement in quality improvement. Konz (1990) described a quality circle as a small group of employees, ranging from 3 to 25 members.

American industry executives, who tried to learn from the Japanese, decided to use the concept of quality circles. These executive did not look, however, at the whole concept and principles of the quality improvement process and the role management must have in the transformation process. They did not understand that quality circles only work when corporate hierarchies are replaced by participative management that gives workers the power to implement their ideas even if they must bypass managers. Quality circles failed miserably in the United States primarily because the infrastructure was not created ("Quality Circles Anyone"? 1991). Employees did not know what the themes were or what would be done with suggestions for implementation. U.S. management did not follow through and, therefore, it let the employees down.

Executives of American companies finally realized that if manufacturers were ever going to reclaim market share from Japanese competitors, top managers would have to lead efforts to improve productivity and overall efficiency. Also, many noncommercial organizations, including those in healthcare and education, began to be challenged by customers to provide higher-quality products and services. As a result, the total quality management (TQM) philosophy is being implemented in commercial and noncommercial organizations. For TQM to be successful, managers need to be completely committed to the philosophy before they involve employees. Today, foodservice managers are listening to the customers, empowering employees to meet customers' expectations, and treating employees as internal customers as well as increasing productivity of quality products and services while containing costs. TQM is here to stay!

SUMMARY

Managers of foodservice operations are challenged by controlling and reducing labor costs and simultaneously increasing labor productivity. Labor rather than food cost has emerged as the predominant cost in hospitality industries. Labor control is necessary if management is to generate a profit for commercial operations or provide the best services that resources permit for noncommercial foodservices. Effective utilization of labor is difficult because many operations are open 7 days a week and have peaks in service demands and seasonal variations in patronage. Also, the highly perishable nature of food items, labor-intensive aspect of activities, and large number of unskilled personnel contribute to labor utilization.

Proper staffing and scheduling of employees are responsibilities of managers. Both can become extremely complex because of the highly variable nature of the business. Adequate numbers of employees are needed, but overstaffing should be avoided to prevent increased labor cost. Scheduling is complicated by absenteeism, labor turnover, vacations and holidays, days off, and differing skills of employees. Special events and catering functions also have an impact on the schedule. Use of part-time employees is one way that foodservice operations can provide adequate personnel at peak periods without having excessive numbers of personnel during slower periods.

Several indices have been developed for various segments of the foodservice industry to assist managers in determining staffing needs, which are often stated in full-time equivalents (FTEs). Also, meal equivalents are used to measure productivity.

Three types of work schedules need to be made by the foodservice manager: the master schedule, the shift schedule, and the production schedule. Computer software is available for automated scheduling, but the complex nature of the process requires close monitoring by management. Alternatives to the standard workweek are grouped into three categories: the compressed workweek, discretionary working time such as flexitime, and task contracting.

Productivity improvement includes productivity measures, principles of work design, work measurement, and the philosophy of total quality management. Productivity concerns the efficient utilization of resources—human, equipment, and financial—and often is expressed mathematically as the ratio of output to input. The principles of materials handling and motion economy are the ones most directly related to labor productivity. Work measurement is a method of establishing an equitable measurement for analysis in foodservice operations.

Quality circles, a Japanese management technique, failed in the United States primarily because employees did not understand the concept and managers did not explain it to them. Finally, executives realized that if manufacturers were ever going to reclaim market share from Japanese competitors, top managers would have to lead efforts to improve productivity and overall efficiency. As a result, foodservice managers are adopting the total quality management philosophy.

BIBLIOGRAPHY

Beasley, M. A. (1991). The story behind quality improvement. *Food Management, 26*(5), 52, 56, 58, 60.

Bernstein, C. (1992). Robots to run foodservice? *Restaurants & Institutions, 102*(19), 20.

Block, A. A., Roach, F. R., & Konz, S. A. (1985). Occurrence sampling in a residence hall foodservice: Cleaning times for selected vegetables. *Journal of The American Dietetic Association, 85*(2), 206–209.

Chew, W. B. (1988). No-nonsense guide to measuring productivity. *Harvard Business Review, 88*(1), 110–118.

Coble, M. C. (1975). *A guide to nutrition and food service for nursing homes and homes for the aged* (rev. ed.). Washington, DC: U.S. Department of Health, Education, and Welfare.

Coppess, M. H. (1988). Temps: Can they help your labor shortage? *Restaurants USA, 8*(1), 16–18.

David, B. D. (1978). Work measurement in food service operations. *School Food Service Research Review, 2*(1), 5–11.

DenHartog, R., Carlson, H., & Romisher, J. M. (1978). Employee logs as a basis for time analysis. *School Food Service Research Review, 2*(2), 98–101.

Drucker, P. F. (1992). *Managing for the future: The 1990s and beyond.* New York: Truman Talley Books/Dalton.

Employers should brush up on teen-labor laws for summer hire. (1993). *Washington Weekly, 13*(25), 4.

Fairbrook, P. (1979). *College and university food service manual.* Stockton, CA: Colman Publishers.

Freedman, D. H. (1992). Is management still a science? *Harvard Business Review, 70*(6), 26–28, 30, 32–38.

Gordon, E. (1992). How restaurants deal with teen employees. *Restaurants USA, 12*(9), 36–38.

Ho, A. K., & Matthews, M. E. (1978). Activity sampling in two nursing home foodservice systems. *Journal of The American Dietetic Association, 73*(6), 647–653.

Humes, S. J. (1988). Viewpoint: "Special" hires. *Foodservice Director, 1*(4), 16.

Iwamuro, R. (1993). Foodservice employee profile. *Restaurants USA, 13*(6), 37–41.

Kasavana, M. (1986). Automated labor management. *Restaurant Business, 85*(12), 160–162.

Kazarian, E. A. (1979). *Work analysis and design for hotels, restaurants and institutions* (2nd ed.). Westport, CT: AVI.

Keiser, J. R., & DeMicco, F. (1993). *Controlling and analyzing costs in food service operations* (4th ed.). New York: Macmillan.

Konz, S. (1990). *Work design: Industrial ergonomics* (3rd ed.). Worthington, OH: Publishing Horizons.

Lieux, E. M., & Winkler, L. L. (1989). Assessing productivity of foodservice systems in nutrition programs for the elderly. *Journal of The American Dietetic Association, 89*(6), 826–829.

Lundberg, D. E., & Armatas, J. P. (1980). *The management of people in hotels, restaurants, and clubs* (4th ed.). Dubuque, IA: Wm. C. Brown.

Mayo, C. R., & Olsen, M. D. (1987). Food servings per labor hour: An alternative productivity measure. *School Food Service Research Review, 11*(1), 48–51.

Montag, G. M., McKinley, M. M., & Kleinschmidt, A. C. (1964). Predetermined motion times: A tool in food production management. *Journal of The American Dietetic Association, 45,* 206–11.

National Restaurant Association. (1988). *Foodservice industry 2000.* Current Issues Report. Washington, DC: Author.

Newstrom, J. W., & Pierce, J. L. (1979). Alternative work schedules: The state of the art. *Personnel Administrator, 24*(10), 19–23.

Ninemeier, J. D. (1986). *Planning and control for food and beverage operations* (2nd ed.). East Lansing, MI: Educational Institute of the American Hotel and Motel Association.

Olson, M. D., & Meyer, M. K. (1987). Current perspectives on productivity in food service and suggestions for the future. *School Food Service Research Review, 11*(2), 87–93.

Palmer, J., Kasavana, M. L., & McPherson, R. (1993). Creating a technological circle of service. *Cornell Hotel and Restaurant Administration Quarterly, 34*(1), 81–87.

Pickworth, J. R. (1987). Minding the Ps and Qs: Linking quality and productivity. *Cornell Hotel and Restaurant Administration Quarterly, 8*(1), 40–45.

Quality circles anyone? (1991). *Training, 28*(Suppl. 3), 8.

Resnick, R. (1993). Leasing workers. *Restaurants USA, 13*(6), 14–17.

Ridley, S. J., Matthews, M. E., & McProud, L. M. (1984). Labor time code for assembling and microwave heating menu items in a hospital galley. *Journal of The American Dietetic Association, 84*(6), 648–654.

Rose, J. C. (1980). Containing the labor costs of foodservice. *Hospitals, 54*(6), 93–94, 96, 98.

Ruf, K. L., & David, B. D. (1975). How to attain optimal productivity. *Hospitals, 49*(24), 77–81.

Ruf, K. L., & Matthews, M. E. (1973). Production time standards. *Hospitals, 47*(9), 82, 84, 86, 88, 89, 90.

Sneed, J., & Kresse, K. H. (1989). *Understanding foodservice financial management*. Rockville, MD: Aspen Publishers.

University of Wisconsin–Madison. (1967). *Methodology manual for work sampling: Productivity of dietary personnel*. Madison: Author.

14

Energy Control

World energy markets are comparatively stable because of the relatively inexpensive oil resulting from the allied victory in the Persian Gulf War (Dale & Kluga, 1992). This trend could be quickly overturned if members of the Organization of Petroleum Exporting Countries (OPEC) disagree on imported oil issues in industrial economies. The Gulf War alerted many countries that energy conservation is a good practice to protect industry against disruption by distant, isolated events.

Before the 1970s, foodservice managers had little concern for energy costs, viewing them as a minor portion of overall operational costs. In most noncommercial foodservices, utility costs were not considered important because they were generally absorbed by management and not charged directly to the operation. As a result, poor practices developed during the many years when energy costs were very low. One such practice, for example, is that the first cook who arrives in the morning routinely turns on all the lights and ovens.

The oil crisis, however, forced managers in all industries to reexamine practices and develop energy control programs. The cost of energy in a hotel is between 2% and 6% of the total operating budget (Dale & Kluga, 1992). Natural gas is the major source of energy supplying nearly 60% of a hotel's requirements. The kitchen, laundry, and banquet departments are the major users; the kitchen uses natural gas for cooking, serving, and storing food and for dishwashing. Because the foodservice industry is energy intensive, the impact of changes in energy costs is noticeable. As a result, this industry has become increasingly conscious of the amount of energy and the kinds of energy sources needed to maintain its operations.

The foodservice industry is as dependent on steady sources of energy as are many manufacturing industries. The final product, food that is ready to eat, depends greatly on energy-consuming equipment: refrigerators, freezers, ovens, ranges, fry-ers, holding equipment, dish machines, and water heaters. The space in which per-sonnel work and customers are served must be lighted, heated, cooled, and venti-lated. In many foodservice operations, transportation may also be a major consumer of energy. The commissary foodservice, for example, is highly dependent on trans-portation, and the recent tax of 4.3 cents per gallon of gasoline will increase the total meal cost.

In the foodservice systems model in chapter 2, the elements of the foodservice system were outlined, identifying utilities as an operational resource or input into the system. Obviously, heating, cooling, and lighting are important in each of the functional subsystems: procurement, production, distribution and service, and sani-tation and maintenance. The quality of the products and services, or the outputs, requires energy dependent equipment and processes. As with any resource, control measures are needed to ensure effective utilization of energy in the foodservice sys-tem (the control element). As discussed in this chapter, various records (the mem-ory element) are needed to monitor energy utilization.

To illustrate the dependence of the foodservice system on energy, consider the cri-sis a manager faces when confronted with a power failure at a critical time during production or service. Loss of power, if prolonged, also has tremendous implications for loss of food in refrigerators or freezers.

In this chapter, energy utilization in foodservice operations and energy conservation are discussed. In addition, an energy management system is outlined, including meth-ods and forms for analyzing utility consumption and cost and equipment operation.

ENERGY UTILIZATION

An abundance of energy resources have been available in the United States through most of its history. But the ever-increasing demand for energy has led to increased reliance on an imported supply, particularly for crude oil.

Oil and gas constitute almost three fourths of the energy consumption; both are fossil fuels and thus nonrenewable resources. One reason for the U.S. energy prob-lem is that the United States relies heavily on oil and gas as sources of fuel, but, as shown in Table 14.1, it has less than 12% of the world's oil and natural gas reserves. Projections suggest that this reliance on oil and gas will decrease. These fuels are expected to continue to be the major sources of energy in this country, however, still constituting almost 50% by the year 2000.

The cost of foodservice energy, like other costs, has increased dramatically during the past several years. To illustrate this point, Borsenik (1983) stated that in 1983 the foodservice industry spent more than $3 billion on new equipment, but more than $6 billion on energy, a cost projected to grow at ever-increasing rates in the future. He reported that, traditionally, only 0.25% to 0.5% of total food and beverage sales was devoted to energy costs. Limited-menu restaurants in 1992 paid more for utili-ties than for any other foodservice segment, according to the National Restaurant Association's (NRA) *Restaurant Industry Operations Report, 1993.* Based on a

Table 14.1. Projected consumption of energy in the U.S. for selected years

	1985	1990	2000
Domestic oil	26%	22%	19%
Imported oil	20	13	12
Natural gas	26	22	16
Coal	20	30	32
Nuclear	4	9	11
Other	4	4	11

Source: Energy Policy: Choices for the Future by the Energy Research and Development Administration, 1976, Washington, DC: Office of Public Affairs.

national survey of restaurants, the following information regarding utilities, including electricity, gas, water, and waste removal, was found.

Restaurant Type	Median per Seat
Limited-menu (fast food)	$236.00
Full-menu table service	$175.00
Limited-menu table service	$160.00
Cafeterias	$ 91.00

According to Pannell, Kerr, Foster Consulting (1993), the accounting firm that regularly tracks cost in the lodging industry, the percentage of total sales dollars expended for energy increased from 4.4% in 1990 to 4.6% in 1992. The annual energy cost per room for various types of lodging properties, including foodservice, was as follows for 1992:

Type	Cost
Full service	$1,394.00
Limited service	$ 582.00
Resort properties	$2,056.00
Suite	$1,213.00

These costs are expected to remain stable; costs that do increase may be attributable to the way the property is managed.

A 1984–85 study of seven East Coast limited-menu to full-service restaurants found the following energy use (Burman, 1988):

28.0% for climate control
17.8% for hot water
13.5% for lighting
 5.8% for refrigeration
34.9% for cooking equipment

Refrigeration and cooking equipment accounted for 40% of the restaurants' energy usage.

The United States uses a mix of primary energy sources to heat, cool, illuminate, and power its commercial, industrial, residential, and transportation sectors (Dale & Kluga, 1992). Oil provides more than 40% of the U.S. primary energy needs, coal about 23%, and natural gas more than 20%. Electricity is the most expensive energy and requires more than 36% of the primary resources in the United States. Energy content and cost of fuel sources should be considered by foodservice managers. Each resource is sold and billed by a different unit: electricity in kilowatt-hours (kwh), natural gas in therms or thousands of cubic feet (MCF), and fuel oil by the gallon or barrel. To compare these units for energy content and cost, the British thermal unit (Btu) is used. One Btu contains enough energy to heat 1 pound of water 1 degree Fahrenheit. A common yardstick measures in millions of Btus (MMBtus). The approximate current average cost is as follows (Dale & Kluga, 1992):

Energy Source	Cost per MMBtu
Natural gas	$ 4.82
Fuel oil (number two)	$ 5.12
Electricity	$21.10

Table 14.2 provides a list of direct and indirect energy expenditures within a foodservice operation. **Direct energy** refers to energy expended within the foodservice operation to produce and serve menu items at safe temperatures. It is required for any storage, heating, cooling, packaging, reheating, distributing, or serving functions to be performed for any menu item prepared for service within a facility.

Indirect energy refers to energy expended to facilitate functions that use energy directly. It supports the other necessary functions, like waste disposal, sanitation, and maintenance of optimal work environment, that are involved with the production and service of menu items.

Unklesbay and Unklesbay (1982) named three vital aspects of food quality that should be assured by the effective use of direct energy:

- Sensory quality
- Food safety
- Nutrient retention

Table 14.2.   Direct and indirect energy expenditures within foodservice operations

Direct Energy Expenditures	Indirect Energy Expenditures
Food storage	Food waste disposal
—dry, refrigerated, frozen	Sanitation procedures
Preprocessing	—personnel, equipment
Heat processing	Optimal working environment
Food packaging	—light, heat, air-conditioning
Food storage following heat processing	—physical facilities
—heated, chilled, frozen	
Heat processing of precooked menu items	
Food distribution to units	
Food service	

Source: Unklesbay, N.: Energy consumption and school foodservice systems. *School Food Service Res Rev 1*(1):9, 1977.

Table 14.2 identifies eight categories of activities in which energy is directly used to ensure food quality. As stated previously, however, the activities occurring in a specific operation depend upon the market form of foods and the type of foodservice. Thomas and Brown (1987) studied the amount and cost of electricity for processes specific to the cook-freeze assembly/serve foodservice operation in a 500-bed hospital. The amount and cost of electricity needed to chill or freeze, hold under refrigeration, and heat the food were minimal.

The need for energy conservation should be obvious. In many businesses, however, management simply has not given sufficient priority to reducing energy usage. Several reasons may be responsible: first, energy costs are currently relatively minor compared to food and labor costs; second, managers have tended to increase menu prices rather than attempt to reduce cost; and third, energy costs are of limited concern if profit margins are at projected levels. Projected trends emphasize the fact, however, that foodservice managers can no longer be complacent about controlling energy costs.

ENERGY CONSERVATION

Energy Edge, a research project promoting superefficient commercial buildings, was conducted by the largest electrical wholesaler in the Pacific Northwest (Burman, 1988). Twenty-nine new commercial buildings including four restaurants were selected for the study, and design modifications, energy saving features, and monitoring of the electricity usage were funded by the wholesaler. Restaurants were actively recruited, even though they are the smallest buildings in the program, because they are high-intensity energy users and have been overlooked in previous energy studies. Although energy is only 5% of the operating budget for a limited-menu restaurant, any increased efficiency can boost profit margins. For the Energy Edge restaurants, heat recovery systems, which use waste heat from refrigeration and ice machine compressors to preheat air for the dining area, are the largest energy saving factor. Heat recovery is expensive, but the payback is rapid. In addition to this system, the other Energy Edge components are a super ventilator for kitchen cooling, a heat pump hot water heater, and photocells to control outside lights.

Computer modeling, a series of mathematical calculations, is often used in designing large commercial buildings. It is a relatively new application for restaurants and generally costs from $3,000 to $5,000 for a small facility. Energy usage of a building can be simulated by determining the square footage of glass areas, wall insulation value, amount of heat from the grill, number of people in and out, and the hours of use. Results of this study could change the attitude of foodservice operators as they become cognizant of continuing increases in energy cost.

The National Restaurant Association has endorsed an Environmental Protection Agency (EPA) program called "Green Lights." The EPA is asking businesses to volunteer to survey lighting and install energy-efficient lighting in 90% of their facilities within 5 years. Participating businesses will receive EPA guidance, financing recommendations, and national and regional recognition. The EPA estimates that the national energy bill would be reduced by $15.8 billion annually if all of corporate America participates in the program. In addition, the reduction in carbon dioxide emissions would be equivalent to removing one third of the nation's cars from streets and highways ("Association Endorses," 1992).

Data in the energy utilization section of this chapter demonstrate the need for energy conservation in foodservice operations. In the guide for energy conservation, published by the Federal Energy Administration (FEA) on the recommendation of the FEA Food Industry Advisory Committee, the following suggestions were given for initiating an energy conservation program (National Restaurant Association, 1976):

- *Assign responsibility.* A manager, assistant manager, or energy committee should be appointed to take responsibility for developing and putting into action an energy conservation program.
- *Conduct a survey of the facility.* A survey should be conducted to identify the equipment that requires the most energy. A worksheet can be used for recording the kwh of electricity per year, the MCF of natural gas per year, the Btus in millions per year as well as the total energy cost in dollars per year. This information is on the equipment nameplate. Equipment that requires high levels of energy and is used extensively during the day should be priority targets for conservation measures.
- *Perform an energy audit.* Energy bills from the prior year should be reviewed to forecast energy use for the current year. Procedures for conducting such an audit are discussed later in this chapter.
- *Establish energy conservation goals.* After a review of energy bills from the prior year and the needs of the facility, goals should be established for reducing the amount of energy and the costs.
- *Consult with utility company representatives and equipment manufacturers.* Utility companies can install meters on kitchen equipment for monitoring and assessing effectiveness of conservation programs. Rates at various usage levels also should be reviewed.
- *Develop a program for the operation.* Data from the survey, recommendations from utility company representatives, and a review of literature and guidelines will assist in developing a program for energy reduction.
- *Conduct a training program for employees.* An energy conservation program will be effective only to the degree that employees are trained.
- *Assess effectiveness of the program.* As with any program, a follow-up and evaluation should be conducted and changes made if necessary.

The energy conservation checklist in Figure 14.1 (Jernigan, 1981) in the manual produced for Maryland institutions, as well as other references at the end of the checklist, outlines measures that can be utilized within the operation to reduce energy. These measures are related to administrative policies and practices, food production techniques, and equipment utilization. Periodic use of the checklist by the individual or committee assigned responsibility for the energy conservation program will provide data for assessing its effectiveness.

Figure 14.1.   Energy conservation checklist.

ADMINISTRATION

_____ 1. Evaluate and revise cycle menus to use foods that require shorter cooking times and thus save energy.

_____ 2. Prepare a schedule distributing energy intensive operations (electric heat-producing equipment and compressors) throughout the day and night in order to limit the number of appliances used at one time, thus reducing peak demand charges.

_____ 3. Prepare a schedule of correct preheating times based on manufacturers' instructions for kitchen equipment (4).*

_____ 4. Prepare a schedule showing when backup units (such as fryers, ovens, and broilers) will be needed (i.e., during "rush" hours) (4).

_____ 5. Prepare a schedule of regular cleaning and maintenance of all equipment based on manufacturer's recommendations.

_____ 6. Keep records of equipment breakdowns, repairs, parts replacements, and maintenance.

_____ 7. Prepare a schedule of food deliveries so that cold and frozen food storage areas will not be either overloaded or underutilized.

_____ 8. Specify energy-saving models when ordering new equipment.

_____ 9. Request that the utility company check for proper gas pressure on lines coming into the building (4). Check air/gas mixture on gas appliances to ensure proper combustion.

_____ 10. Develop guidelines for overhead exhaust hood operation based on minimum health department requirements. Do not operate hoods continuously (5).

FOOD PRODUCTION

_____ 1. Prepare food in quantities for optimum utilization of equipment based on manufacturer's recommendations (4.5). This may require cooking some foods partially or completely in advance.

_____ 2. Set cooking dials to the lowest safe temperature which gives a satisfactory product (4.6). Dialing higher does not reduce preheating time, but a lower temperature results in lower energy consumption because less heat is lost to the surrounding air.

_____ 3. Consider using steam cookers to begin cooking some foods, to fully cook foods such as rice, pasta, and vegetables, and to maintain temperatures of all foods.

_____ 4. Use flat-bottomed pots and kettles when surface cooking for maximum transfer of heat.

_____ 5. Cover pots, pans, and kettles when cooking to keep heat in and decrease cooking time (5).

_____ 6. Do not wrap potatoes or other foods in aluminum foil before baking (4).

_____ 7. Trim food products and discard unusable parts before storing to reduce volume and quicken chilling (7).

_____ 8. Allow hot foods to cool a few minutes (but not below 140°) before refrigerating or freezing them.

_____ 9. Be sure cooking and heating units (ovens, fryers, dishwashers) are turned off when not in use (4).

_____ 10. Direct fans to cool workers, not hot food and cooking equipment (5).

_____ 11. Unpack cold or frozen food from cartons upon delivery (7).

_____ 12. Place cold or frozen food in proper storage after delivery to eliminate rechilling (7).

_____ 13. Take empty racks out of cold or frozen food storage units to eliminate the cost of cooling them (7).

_____ 14. Consolidate frozen and refrigerated foods where possible. Full units use energy more efficiently than partially full ones.

_____ 15. Thaw frozen foods in refrigerated units both to conserve energy and for proper sanitation (4,5,7). As food thaws, it cools surrounding air, placing less of a power demand on the refrigerator.

* See references at end of figure.

Figure 14.1.   *continued*

EQUIPMENT
Cold and frozen food storage units

_____ 1. Avoid frequent, lengthy openings of freezers and refrigerators (5.7).

_____ 2. After entering cold and frozen food walk-ins, close the door behind you. (Be certain all walk-ins will open from the inside.)

_____ 3. Install pilot lights, which light up when lights inside have been left on, on switches of walk-in units (7).

_____ 4. Use calibrated thermometers to check temperatures in all refrigerated and frozen units (7).

_____ 5. Set temperature controls for proper storage of particular foods in each unit (7).

_____ 6. Level free-standing upright units so that doors automatically swing shut from an open position.

_____ 7. Be sure that items do not jam against doors, damaging the gasket and the door. Check gaskets and seals on doors for cold air leakage (7).

_____ 8. Keep an area 4 feet wide in front of compressors and coils clear. Check compressors for leakage and level of refrigerant (7).

_____ 9. Maintain proper tension on compressor belts and replace when worn or damaged.

_____ 10. Clean coils of ice and dust build-up (5,7).

_____ 11. Check refrigerators and freezers daily for loss of temperature and short cycling times. Refrigerant level may be low.

_____ 12. Check outside walls of refrigeration and freezer units for cold spots, which may indicate insulation shift or water logging.

_____ 13. Defrost refrigerators and freezers on a regular basis (7).

_____ 14. Check for correct cycling of automatic defrosters (7).

_____ 15. Close ice-maker storage bins when they are not in use.

Ovens

_____ 1. Determine the cooking capacity for each oven and select the oven with the least amount of waste space and the most energy efficient oven able to do the job (4,7).

_____ 2. Schedule baking and roasting to fill ovens to capacity (4,7).

_____ 3. Begin to cook foods (except for those that will overcook or dry out) while pre-heating the oven (4). Do not preheat ovens sooner than is necessary (4,5).

_____ 4. Load and unload ovens quickly to avoid unnecessary heat loss (4).

_____ 5. Avoid opening ovens to check foods. Keeping doors closed allows foods to cook faster and lose less moisture. Rely on timers to tell when a product is done.

_____ 6. Calibrate oven thermostats often (7).

_____ 7. Turn off oven, allowing receding heat to finish cooking (6,7).

_____ 8. Level oven and oven racks often.

_____ 9. Clean up spills as they occur to avoid build-ups of carbon deposits.

_____ 10. Clean and repair oven doors and hinges to ensure a snug fit so that heat will not escape (6).

_____ 11. Consider using a microwave for thawing (5), cooking, and reheating. A microwave may use less energy as it requires less time to cook foods.

_____ 12. Contact a serviceman to check microwave timer for accuracy and to check oven for radiation leakage.

_____ 13. Clean interior surfaces of convection ovens using a nonabrasive material to avoid damage.

_____ 14. Check fan motors of convection ovens for proper performance.

_____ 15. Clean fan blades of convection ovens following manufacturers' instructions.

_____ 16. Use proper utensils and pans for convection ovens. Do not damage door seals or interior surfaces.

Deep fat fryers

_____ 1. Set fryer thermostat only as high as necessary to reach optimum frying temperatures (8).

_____ 2. Check temperatures of cooking oil with a reliable thermometer. Smoking fat in fry-

Figure 14.1.   *continued*

ers indicates that the temperature is too high or that the fat has broken down (6).

____ 3. Load frying baskets only to stated capacity (8). Overloading increases cooking time, using more energy.

____ 4. Drain and clean fryer after each use (4).

____ 5. Filter particles from oil at least once a day and replace oil often (4,8).

____ 6. Make sure the level of oil is high enough to cover foods at all times (8).

____ 7. Clean heating elements and interiors weekly. Remove burned foods, grease, and carbon deposits (8).

Surface cooking units

____ 1. Group kettles and pots close together on closed top ranges to minimize heat loss (4).

____ 2. Burners should be one inch smaller in diameter than the pots placed on them (4).

____ 3. Reduce heat when food begins to boil, maintaining liquids at a simmer.

____ 4. Operate gas burners only when ready to cook.

____ 5. Clean and adjust gas burners so that flame is blue, with a distinct, white center cone.

____ 6. Clean pilot lights, orifices, and burners often with a stiff brush (consult manufacturer's instructions) (9).

____ 7. Remove grease and dirt build-up from range top, burner plates, and coils to allow maximum heat transfer (4,9).

____ 8. Clean all cooking surfaces and wipe up spills as they occur.

Griddles

____ 1. Group foods close together on griddle top to minimize heat loss (4).

____ 2. Heat only the portion of the griddle being used (4).

____ 3. Check thermostats against a reliable thermometer and recalibrate if necessary (9).

____ 4. Clean griddle surfaces of food and fat particles after each use with a spatula and grill brick (4,9).

____ 5. Clean and wipe out grease troughs daily (9).

Broilers

____ 1. When briquets become hot, turn broiler heat to medium (4) or low.

____ 2. Load heated broilers to maximum capacity, utilizing entire surface area.

____ 3. Infrared broilers should be turned off when not in use as they can be reheated quickly.

____ 4. On infrared broilers, check ceramic refactory units for cracks, blackening, or crumbling (9), and replace if units show damage.

____ 5. Keep briquets clean (4).

____ 6. Clean broiler drip pans. Wash grids and grease troughs after each use with a mild soap (9).

____ 7. A clear blue flame with a distinct center cone is a sign of proper burner operation on broilers (6).

____ 8. Clean burner orifices.

____ 9. Check air shutters and openings to ensure that they are clear.

____ 10. Clean and adjust pilot lights on gas broilers.

Steam cookers and tables

____ 1. When using a steam cooker, fill cooking vessels to maximum capacity according to manufacturer's instructions.

____ 2. Check steamer and steam table thermostats for proper calibration.

____ 3. Clean debris from steamer and door seals so that steam will not escape when door is closed. Replace gaskets if necessary.

____ 4. Use steam tables to keep foods hot but not allow clouds of steam to escape.

____ 5. Check condensate traps for steam leakage.

____ 6. Maintain proper insulation on steam lines.

____ 7. Repair all steam leaks, no matter how small, promptly.

____ 8. Turn off steam tables and cookers when not in use.

Figure 14.1.   *continued*

Hoods

_____ 1. Have an adequate air supply to the hood to make exhausted air without exhausting cooled air from public spaces.

_____ 2. Turn off hood fans when equipment underneath is not in use.

_____ 3. Adjust oversized exhaust hoods in the food preparation area so that no more air than necessary is exhausted (4,5).

_____ 4. Clean grease from exhaust hoods and filters regularly (4,8).

Dishwashing

_____ 1. Scrape and stack dishes before washing. Avoid using hot or warm water for this task (5).

_____ 2. Do not turn on booster heaters on dishwashers until they are needed.

_____ 3. Be certain that dishwashers and booster heaters are turned off when they are not in use.

_____ 4. Wash dishes only when a dishwasher is full (5).

_____ 5. Completely fill dishracks before sending them through flight-type and conveyor dishwashers.

_____ 6. Check rinse water temperature, flow controls, and pressure regularly.

_____ 7. Check power rinse for adequate but not excessive water flow and for automatic shut-off when a rack has gone through the machine.

_____ 8. Adjust power dryers to shut off when dishes are dry.

_____ 9. Clean dishwashers daily.

_____ 10. De-lime dishwashers when lime deposits are present.

_____ 11. Check for proper temperatures in dishwashers (140°F to 160°F for wash water, 180°F for rinse water).

_____ 12. Consider using chemical sanitizing agents instead of excessively hot water in dishwashers.

_____ 13. Inspect pumps and feed and drain valves for leakage.

_____ 14. Lubricate the speed reducer on conveyor-type dishwashers.

_____ 15. In the pot and pan washing area, fill sinks for washing, rinsing, and sanitizing instead of using continuously running water.

Hot water

_____ 1. Do not leave faucets running.

_____ 2. Use hot water only when cold water will not do.

_____ 3. Eliminate the use of hot water for hand-washing wherever possible.

_____ 4. Maintain insulation on hot water pipes. Check for missing or damaging insulation.

_____ 5. Install flow restrictors on faucets. They are particularly effective when water duration, not amount of water, is the important factor.

_____ 6. Use hot tap water for cooking when possible. A water heater uses less energy than a range top to heat the same amount of water.

_____ 7. Repair leaky faucets immediately.

References

(1) Mueller Associates, Inc.: Maryland Energy Audit Training Manual for Schools, Hospitals, Local Government, and Public Care Institutions. Baltimore: Maryland Energy Office, 1979.

(2) Department of Natural Resources: Energy Savings Workbook. Baltimore: Maryland Energy Policy Office, 1979.

(3) Moulton, C. C.: Sources to tap for energy information. Restaur. Hospitality 62:6 (Sept.), 1978.

(4) Energy Audit Workbook for Restaurants. Washington, DC: U.S. Department of Energy, 1978.

(5) Schools and Hospitals Energy Conservation Workbook. Publ. No. 2099. Milwaukee, WI: Johnson Controls, Inc., 1979.

(6) Operations/technology: Good equipment maintenance saves energy. Institutions 84:99 (April 1), 1979.

(7) West, B., Wood, L., Harger, V., and Shugart, G.: Food Service in Institutions. 5th ed. New York: John Wiley & Sons, 1977.

(8) Energy tips: Deep fat frying. Restaur. Bus. 77:2 (Dec. 1), 1978.

(9) Energy tips: Maintenance saves energy. Restaur. Bus. 77:2 (Sept. 15), 1978.

Source: B. S. Jernigan: Guidelines for Energy Conservation. Copyright The American Dietetic Association. Reprinted by permission from *Journal of The American Dietetic Association,* Vol. 79: 459–462, 1981.

ENERGY MANAGEMENT

Control of energy costs requires maintaining a record-keeping system for tracking utility costs and monitoring equipment utilization. Energy utilization for lighting, heating, ventilating, and air conditioning of the facilities must also be monitored and controlled.

Preparing and maintaining appropriate records will facilitate identification of the amount and cost of energy use as well as of any trends that may be developing. Energy use can then be related to the number of customers, and energy cost can be examined in relation to sales volume, which will provide indicators of the success or failure of the energy management program. Monitoring the use and cost of water can also be accomplished, thus providing an analysis of all utilities (Thompson, 1992).

Dollar costs of energy are, of course, important, and comparisons of current costs with those of prior years will provide one basis of comparison. Because of rising utility rates, however, those figures may not be as meaningful as utilization data. Actual utilization, therefore, should also be analyzed. An analysis of energy usage and cost is an important management function. The glossary in Figure 14.2 lists some of the key terms related to energy consumption.

The three forms in Figures 14.3, 14.4, and 14.5, developed by the National Restaurant Association, are examples of records that should be maintained regularly to track energy use and costs. Typical data for a restaurant have been entered on each form along with explanations of the data, which provide an illustration for a straightforward method of tracking utility consumption and cost. Electricity and gas utilization have been converted to MBtus to provide a standard base for comparison; one MBtu is equal to 1,000 Btus.

As emphasized in the technical bulletin on energy management prepared by the NRA (1982), the information provided by these records will point out any "leaks" in the utility system of an operation, such as:

- Development of poor operating practices
- Malfunctioning of equipment
- Structural damage to the building
- Misreading of meters
- Improper billing by utility companies

In recording data each month, comparable periods should be used to permit comparisons with prior years. For example, if the reading date for the electric meter is on the fifth day of the month, the consumption for a particular month should be for the period from that day to the fifth of the following month.

One factor to be taken into account in analyzing utility utilization is the extremities in the weather. Some winters are colder and some summers warmer than others; deviations in utilization may be explained by the additional demand on heating or air conditioning systems. One way of measuring these variances is the use of degree days, which are deviations of the mean daily temperatures from 65°F. In comparing

Figure 14.2.   Glossary of energy-related terms.

Glossary

Ampere—a unit of measure of an electric current.

British thermal unit (Btu)—a unit of heat energy. The amount of heat required to raise 1 lb. of water 1 degree Fahrenheit.

Demand charge—based on the highest electric energy use over some fixed period of time during the billing cycle, usually 15 or 30 minutes. As the peak demand determines the size of power generating plants, transmission lines and other equipment needs, it is designed to pay for the utility company's investment.

Demand charge discount—rewards a customer who has a reasonable constant monthly electrical demand. For example, a customer may earn a discount for maintaining a billing demand in excess of 75 percent of his highest maximum demand in the preceding months.

Energy charge—determined by the electric energy used (kilowatt hours). It is designed to cover the base fuel expenses necessary in the generation of electrical energy.

High load factor discount—rewards a customer who uses his demand for many hours of operation. It is a percentage index comparing the actual kilowatt hours used with the maximum billing demand times the total hours in a billing period.

Horsepower—a unit of power equal to 746 watts.

Kilowatt—one thousand watts. Most electrical foodservice equipment is rated in kilowatts.

Kilowatt hour—a measure of work (kilowatt × hours). The unit recorded by electric meters and utilized for billing.

Purchased energy or fuel adjustment—an incremental adjustment to rates reflecting changes in costs of purchased energy or fuel since current rates became effective. It is usually expressed as a fractional amount added to the current rate.

Therm—a unit used for measuring natural gas equal to 100,000 Btu.

Volt—the push that moves the electric current.

Watt—a unit of power. One watt equals the flow of one ampere at a pressure of one volt (watts = volts × amperes).

Source: Energy Management System, 1982, Chicago: National Restaurant Association. Used by permission.

the prior year's utility use with the current year's, a comparison in the number of degree days would perhaps explain deviations.

The form in Figure 14.3 is a worksheet designed to develop information required for the forms shown in Figures 14.4 and 14.5. After utility bills have been received for the current month, the Figure 14.3 worksheet should be completed. The form in Figure 14.4 will assist in monthly tracking of utility consumption as well as in comparing the current year's consumption of gas, electricity, and water with the prior year's. The form in Figure 14.5 will enable the manager to track total energy utilization and utility cost as well as to compute ratios that will assist in analyzing utilization and cost in relation to volume and sales. These ratios are: MBtus per customer, and utility cost as a percentage of sales.

Careful planning and analysis of equipment operation are important aspects of energy control in a foodservice operation. The first planning step is to list all the equipment in an operation requiring energy, as in the form shown in Figure 14.6. Data shown on this sample form are for a restaurant that serves from 11:00 A.M. to 9:00 P.M. Each piece of equipment should be listed in the appropriate column, along with its location and energy source. In some operations, the manager may wish to

Figure 14.3.   Form for recording monthly utilities.

FORM A: MONTHLY UTILITY WORKSHEET

Month _____ January _____ ,19 __ 94 __

				MBTU[a]	Cost

Gas _____ _____ cubic feet × _____ _____ _____

____ 1937 ____ therms   × 100   193,700 $900.71[b]

Electricity __ 20,833 __ kWhr   × 3,412   71,082 $1470.81[c]

Fuel Oil #2 _____ gals.   × 140   _____ _____

Propane _____ gals.   × 91.6   _____ _____

Steam _____ lbs.   × 1   _____ _____

Water __ 67,200 __ gals.   _____ $100.80[d]

TOTAL   264,782 MBTU $ 2472.32

Customer count for month _____ 16,800 _____

MBTUs   Total MBTUs   264,782
──────   =   ──────────   =   ──────   =   15.8
per customer Customers   16,800

Sales for month ____ $45,360 ____

Utility costs   Total utility cost $2,472.32
as percent   =   ─────────────   =   ─────────   =   5.4%
of sales   Sales   45,360

[a] MBTU = 1000 BTU   [b] 1000 cubic feet (MCF) = $4.65
[c] 1 kilowatt hour (kWH) = $0.0706   [d] 1 gallon water = $0.0015

1. Enter the month and year. In this example the information is for January 1994.
2. From your gas bill, obtain the consumption figure. It may be expressed in therms, cubic feet or 100 and 1000 multiples of cubic feet. If gas is metered in 100 cubic feet (CCF) the multiple for conversion to MTBU is 100. If gas is metered in 1000 cubic feet (MCF) the multiple for conversion is 1000. In this example the gas is metered in therms and 1937 therms is equal to 193,700 MBTU. The cost, also obtained from the bill, is $900.71. The conversion values given are average values and satisfactory for these calculations.
3. The total number of kilowatt hours used during the month was 20,833 and the conversion to MBTU provides a value of 71,082. The total cost (including demand charges and any other miscellaneous charges) is $1470.81.
4. Water is metered in gallons or cubic feet. The total consumption during the month was 67,200 gallons at a total cost of $100.80. This includes all other charges as leak insurance and sewage.
5. These are the total figures for MBTU and dollar costs obtained by adding each of the two columns. In this example the totals are 264,782 MBTUs and $2472.32.
6. The customer count for this period is 16,800. It may be more appropriate in your operation to use number of transactions, number of parties or whatever unit is most suitable for your restaurant.
7. MBTUs per customer is obtained by dividing the total MBTUs by the customer count. In the example, dividing 264,782 by 16,800 results in 15.8.
8. Sales for the month were $45,360 obtained from bookkeeping records. The total cost of utilities as determined on this form was $2472.32. Dividing $2472.32 by $45,360 results in a utility cost percentage of 5.4%.

Source: Adapted from *Energy Management System*, 1982, Chicago: National Restaurant Association. Used by permission.

Figure 14.4. Form for analyzing and tracking utility consumption.

FORM B: UTILITY CONSUMPTION AND TRACKING

Year _1994_

Month	Gas therms ①			Electricity							Water gallons ③			④		
				Use KWH			Demand KW ②									
	1994	1995	Percent Change	1994	1995	Percent Change	1994	1995	Percent Change		1994	1995	Percent Change	1994	1995	Percent Change
JAN	2041	1937	−51	23,617	20,833	−118	612	595	−27		73,600	67,200	−87			
FEB																
MAR																
APR																

1. Enter your gas consumption figures in this section. In the example, consumption for January 1994 was 2041 therms while 1937 was the figure for January 1995. This is a 5.1 percent reduction which is indicated in the third column.

2. Electricity use is recorded in this section. Kilowatt hours were 23,617 and 20,833 for January 1994 and January 1995 respectively which represents a 11.8 percent reduction. Demand expressed in kilowatts decreased from 61.2 to 59.5 for a downward change of 2.7 percent. Significant increases in demand may indicate the need to stagger equipment operation or may also point to malfunctioning electrical devices and controls.

3. Water consumption is entered in this portion. Water consumption was 73,600 gallons in January 1994 and 67,200 gallons in January 1995 for a 8.7 percent reduction. Increases in water consumption may indicate system leaks, improperly operating faucets and faulty valves.

4. This section is to be used for other energy sources as propane, fuel oil or steam.

Source: Adapted from *Energy Management System*, 1982, Chicago: National Restaurant Association. Used by permission.

Figure 14.5. Form for computing cost and use indicators for utility consumption.

FORM C: UTILITY CONSUMPTION, TRACKING AND COST

Month	Total MBTU ①			Total Utility Cost ②			MBTU per customer ③			Utility Cost as Percent of Sales ④		
	1994	1995	Percent Change	1994	1995	Percent Change	1994	1995	Percent Change	1994	1995	Percent Change
JAN	284,681	264,782	-7.0	2162.57	2472.32	14	16.6	15.8	-4.8	5.1	5.4	-0.3
FEB												
MAR												
APR												

1. The information for this section is obtained from the monthly utilities worksheet (Form A) where all energy used is converted to the common unit of BTUs. The 264,782 MBTU figure for January 1994 of the example is from Form A while the 284,681 MBTU is the like figure for January 1994. There has been a 7.0 percent reduction in the total number of BTUs from one January to the next.

2. Total utility costs, including water for January 1995, was obtained from Form A and is $2472.32. The prior January figure was $2162.57 which indicates a change of 14 percent.

3. This section relates energy consumption to number of customers. The 15.8 MBTU per customer figure was calculated in Form A and from historical data 16.6 was the prior year figure indicating a reduction of 4.8 percent.

4. Utility costs as a percent of sales were 5.4 for January 1995. This figure was obtained from Form A. For January 1994, the figure was 5.1 which indicates a change of -0.3 percent. While total utility costs remained almost constant, utility costs as a percentage of sales decreased because of larger sales volume.

Source: Adapted from *Energy Management System*, 1982, Chicago: National Restaurant Association. Used by permission.

complete a separate form for each unit in the production area. Next, preheat times should be established for each piece of equipment; the menu should then be examined to estimate processing and cooking times for each menu item. From this information, the "on" time for each piece of equipment can be established.

In a commercial operation with a set menu, a master schedule that could be used from day to day would suffice. If operating hours or meal periods are different on certain days, a variation in the master schedule would be needed to accommodate the differences. In a foodservice operation with a menu that varies from day to day, an equipment schedule should be prepared on a daily basis.

Figure 14.6.   Example of an equipment schedule for energy consumption.

EQUIPMENT SCHEDULING

Equipment	Location	Utility*	AM 6 7 8 9 10 11	PM 12 1 2 3 4 5 6 7 8 9 10 11	AM 12 1 2 3	AM 4 5 6
Proof Box	Bakery	G				
Oven	Bakery	G				
Donut Fryer	Bakery	G				
Exhaust Hood #1	Kitchen	E				
Exhaust Hood #2	Kitchen	E				
Fryer #1	Kitchen	G				
Fryer #2	Kitchen	G				
Griddle #1	Kitchen	E				
Griddle #2	Kitchen	E				
Range #1	Kitchen	G				
Range #2	Kitchen	G				
Convection Oven	Kitchen	G	Schedule daily – if needed 15 minutes preheat			
Broiler	Kitchen	G				
Deck Oven	Kitchen	E	Schedule daily – if needed 40 minutes preheat			
Tilting Skillet	Kitchen	E	Schedule daily – if needed 10 minutes preheat			
Steam Table	Kitchen	G				
Coffee Urn	Kitchen	G				
Warming Lamps	Kitchen	E	High energy user – turn on only when needed			
Dish Machine	Kitchen	G				
Coffee Warmer Station #1	Dining Rm	E				
Hot Chocolate Dispenser	Dining Rm	E				
Soup Warmer	Dining Rm	E				
Bun Warmer Station #1	Dining Rm	E				
Coffee Warmer Station #2	Dining Rm	E				
Bun Warmer Station #2	Dining Rm	E				

* G - Gas   E - Electricity

Source: Adapted from *Energy Management System*, 1982, Chicago: National Restaurant Association. Used by permission.

To illustrate the use of such a schedule, based on the data in Figure 14.6, assume that the baking is done during the morning hours before the restaurant opens at 11:00 A.M. The bakers are scheduled to report to work at 6:00 A.M. With mixing and other preparation time, the proof box for yeast breads will not be needed until 7:30 A.M., and the oven will not be needed until 8:00 A.M. Therefore, with preheat time of 15 minutes for the proof box and 40 minutes for the oven, turn-on times at 7:15 A.M. and 7:20 A.M., respectively, will be sufficient for the equipment to be ready at the required times.

Before the days of rapidly rising energy costs, bakers would have turned on the proof box and oven shortly after arriving at work at 6:00 A.M. This example shows how energy savings are possible by altering this type of practice with careful planning and scheduling.

As mentioned earlier, however, employee training is one of the keys to a successful energy conservation program, and the employees are the key to the success of an equipment operation schedule. Ultimately, they are the ones responsible for turning equipment on and off at appropriate times. Before initiating an equipment operation schedule, employee training sessions are needed to present information on the objectives of the program and on its importance of controlling operational costs as well as to elicit cooperation and support.

To assist in executing the equipment scheduling aspect of the energy management program, the NRA has developed a series of labels that can be mounted on or close to various pieces of equipment to remind employees of various energy conservation issues (Figure 14.7). Data presented earlier in this chapter indicated that refrigeration and cooking equipment are responsible for approximately 40% of the energy use in restaurants. Therefore, the importance of equipment control in these operational areas should be obvious.

Energy conservation tricks for reducing overhead expenses should be explored by foodservice managers, who have been dealing with increasing energy usage over the past several decades. The first big item has been air conditioning dining rooms because customers demand comfort in both summer and winter (Patterson, 1993). The drive for comfort also has extended to the kitchen and is becoming a union bargaining position in most areas. Many electric utility companies are providing analyses of air conditioning demand that involve monitoring temperatures and compressors and cutting off units during slow periods while keeping temperatures within the comfort zone.

Refrigerator compressors and condensing coils generate a lot of heat. The amount of air conditioning can be decreased by 5% to 10% in the kitchen by placing them in remote locations outside the building or in the basement. The same procedure can be used for the serving and dining areas. A heat pump water heater uses not only heat generated by cooking equipment but also returns cool air to the kitchen. Managers are always looking for ways to reduce costs of energy in cooking equipment; manufacturers are responding by beginning to build insulated equipment. The surprise is that manufacturers have found that keeping heat inside the equipment makes cooking more efficient. High-efficiency gas burners in energy efficient equipment produce more Btus from the same amount of gas; infrared heating in fryers has the same results. Induction heating in electric equipment promises not only more

Figure 14.7. Examples of energy conservation reminders.

Source: National Restaurant Association Catalog of Publications. Used by permission.

efficiency but also lower kitchen temperatures. Induction fryers and grills are also being developed by manufacturers (Patterson, 1993).

Water conservation is not only good for the environment, but also it is good for saving money (Thompson, 1992). The problem is how to conserve water without cutting back on customer service. Many restaurateurs in California and Nevada have experienced droughtlike conditions in recent years and have had to conserve water by serving it only if requested by the customer. Employees have to be trained to conserve water in cooking and in cleaning. It is not necessary to use 15 gallons of water to cook 4 pounds of pasta. Hoses for cleaning floors or sidewalks can be replaced by buckets of water and a deck brush. Check dripping faucets; a little drip could flush $1,000 down the drain in a year. The objective is to use water as efficiently as possible, while maintaining high sanitary standards.

During the past decade, concern about energy costs has stimulated research in the energy utilization of foodservice operations; a number of studies are listed in the references for this chapter. Much of the work has been conducted at the University of Missouri-Columbia and at the University of Wisconsin-Madison. Research has focused on energy modification of recipes, energy utilization of various equipment, energy demand with differing oven loads, and energy usage in the different foodservice operations. This research is yielding valuable data to assist foodservice managers in understanding energy utilization in foodservice operations and designing energy control programs.

A recent study on energy consumption and cost for the production of school foodservice meals was funded by the Division of Applied Research of the National Food Service Management Institute (Messersmith, Wheeler, & Rousso, 1994). An energy audit was conducted to determine the cost of energy used by foodservice equipment to produce one meal and the amount of energy used per square foot of production space in four school foodservice operations, each in a different district and with a different foodservice operation. The following types of foodservice operations were used in the study:

- Conventional on-site production and delivery
- Conventional on-site production and one satellite service center
- Ready/prepared cook-chill on-site production and delivery

The fourth operation, located in the district administration building, is a ready/prepared cook-chill commissary with 90 satellite centers in which approximately 29,000 meals are prepared each day. Energy consumption in the commissary and one satellite operation was used in the study.

Each operation was audited 2 days in the spring and 2 days in the fall. The schools' utility meters were read at half-hour intervals during the same period that energy equipment data were collected in the kitchens. Calculations for electric, gas, and steam energy were converted to British thermal units for comparison. The overall average per meal was 2,590 Btus; the average energy cost per meal was $0.13. Energy per square foot of production space averaged 432 Btus. These results indicate that an energy conservation program could help the bottom line of a school foodservice operation.

SUMMARY

Since the early 1970s, the oil crisis has resulted in inflated prices and forced the United States to look to conservation and alternate sources of energy. The oil crisis forced foodservice managers and others in the industry to reexamine practices and develop energy control programs. Because the foodservice industry is energy intensive, the impact of rapidly increasing energy cost is especially severe. The industry depends on energy-consuming equipment: food storage, production, and holding equipment in addition to dishwashers, water heaters, and air conditioners.

Energy conservation programs should be initiated in all industries including foodservice. A manager or committee should be assigned responsibility for energy conservation. Equipment requiring energy should be identified, energy conservation goals established, and an educational program developed for employees. Follow-up and evaluation should be a component of an energy conservation program to provide an assessment of its effectiveness.

Energy management is essential for a foodservice operation. Regularly tracking energy costs and monitoring equipment utilization will assist in energy control. Comparisons of current costs of energy with those of prior years are important, but utilization data are far more important. In addition, current research is yielding valuable data to assist foodservice managers in understanding energy utilization in foodservice operations and designing energy control programs.

BIBLIOGRAPHY

Association endorses new EPA "green" utility program. (1992). *Washington Weekly, 12*(36), 2–3.

Aulbach, R. E. (1988). *Energy and water resource management* (2nd ed.). East Lansing, MI: Educational Institute of the American Hotel and Motel Association.

Borsenik, F. D. (1983). Energy and foodservice equipment. *The Consultant, 16*(1), 12–20, 24.

Burman, J. (1988). Keeping an edge on energy costs. *Restaurants USA, 8*(7), 20–23.

Dale, J. C., & Kluga, T. (1992). Energy conservation: More than a good idea. *Cornell Hotel and Restaurant Administration Quarterly, 33*(6), 30–35.

Energy Research and Development Administration. (1976). *Energy policy: Choices for the future*. Washington, DC: Office of Public Affairs.

Federal Energy Administration, Food Industry Advisory Committee. (1977). *Guide to energy conservation for foodservice*. Washington, DC: Government Printing Office.

Federal Energy Administration, Office of Industrial Programming. (1976). *Energy use in the food system*. Washington, DC: Author.

Harrington, R. E. (1991). A kilowatt saved . . . *Restaurants USA, 11*(2), 11.

Jernigan, B. S. (1981). Guidelines for energy conservation. *Journal of The American Dietetic Association, 79*(4), 459–462.

McProud, L. M. (1982). Reducing energy loss in food service operations. *Food Technology, 36*(7), 67–71.

Messersmith, A. M., Rousso, V., & Wheeler, G. (1993). Energy management in three easy steps. *School Food Service Journal, 47*(9), 41, 42, 44.

Messersmith, A. M., Rousso, V., & Wheeler, G. (1994). School food service energy. *School Food Service Research Review, 18*(1), 38–44.

Messersmith, A. M., Wheeler, G., & Rousso, R. (1994). Energy used to produce meals in school food service. *School Food Service Research Review, 18*(1), 29–36.

National Restaurant Association. (1981). *Efficient use of energy: Goal of NRA.* Chicago: Author.

National Restaurant Association. (1982). *Energy management system.* Chicago: Author.

National Restaurant Association. (1988). *Foodservice and energy to the year 2000.* Washington, DC: Author.

National Restaurant Association. (1993). *Restaurant industry operations report.* Washington, DC: Author.

Pannell, Kerr, Foster Consulting. (1993). *Trends in the hotel industry.* USA edition. Philadelphia: Author.

Patterson, P. (1993). Energy-saving tricks reduce overhead expenses. *Nation's Restaurant News, 25*(17), 56, 65.

Riehle, H. (1991). Restaurateurs are conserving energy. *Restaurants USA, 11*(4), 36–37.

Shanklin, C. W., & Hoover, L. (1993). Position of The American Dietetic Association: Environmental issues. *Journal of The American Dietetic Association, 93,* 589–591.

Thomas, C. J., & Brown, N. E. (1987). Use and cost of electricity for selected processes specific to a hospital cook-chill/freeze food-production system. *Journal of Foodservice Systems, 4*(3), 159–169.

Thompson, P. K. (1992). Saving water can save you money. *Restaurants USA, 12*(4), 10–11.

Thorner, M. E., & Manning, P. B. (1983). *Quality control in foodservice* (rev. ed.). Westport, CT: AVI.

Unklesbay, N. (1982). Overview of foodservice energy research: Heat processing. *Journal of Food Protection, 45*(10), 984–992.

Unklesbay, N. (1983). Integration of energy data into food industry decision making. *Food Technology, 37*(12), 55–59, 110.

Unklesbay, N., & Unklesbay, K. (1982). *Energy management in foodservice.* Westport, CT: AVI.

Managing Production at The Fairfax

Production is the second subsystem in the foodservice system, but it is probably the most active of all the subsystems. It is the nucleus of all activities that occur in the Food and Beverage Department because raw food products are converted into menu items to satisfy the residents. Quantity and quality control are important as well as labor and energy control. Knowledge gained from the six chapters in part 4 is necessary to solve the case problem.

Background Information

Most menu items served in The Fairfax and Belvoir Woods Health Care Center are produced in the kitchen on the upper level of the Community Center. Some baked goods are purchased, such as a premixed muffin batter that is baked in the kitchen and certain bread items and cakes. Cobblers and special rolls are baked in the kitchen.

The F&B director has contracted with Dunkin' Donuts to supply the coffee shop/country store with donuts every day. Marriott has been successful in using national brands in foodservice operations. It offers an array of established brands that customers know and prefer. Marriott began offering branded products 10 years ago in its Travel Plazas and found that retail brands consistently increase sales, profits, and customer counts. Today, the company offers national retail brands in stadiums and arenas, businesses, hospitals, colleges, and secondary schools as well as in plazas at toll roads and airports.

Most menu items, however, are produced in the kitchen. Recipes are from the master file in the corporate office. Figure 1 shows a recipe for Cajun Beef Kabobs and a photograph of the plated menu item. The recipe includes ingredients for 6 and 72 kabobs, the procedure, how it is to be served, garnishes, and suggested accompaniments. In Figure 2, the procedure for marinating beef cubes with marinades for each type of diet is given, followed by modified recipes, first for low-fat and low-sodium beef kabobs (I) and then for soft and bland beef kabobs (II) modifications. A good understanding of the production system requires an understanding of the kitchen design, production procedures, menus, and staffing.

Figure 1. Cajun beef kabobs and a recipe for the menu item.

CAJUN BEEF KABOB

Ingredients:

Tenderloin tails, 1 oz cubes	6 ea	72 ea
(weight)	(6 oz)	(4½ lb)
Cajun Spice Mix	1 t	¼ C
(Recipe FD 699)		
Green peppers, 1¼" squares	3 ea	36 ea
(weight)	(½ oz)	(6 oz)
Onions, ½" wedges	2 ea	24 ea
(weight)	(½ oz)	(6 oz)
Mushroom caps	2 ea	24 ea
(weight)	(½ oz)	(6 oz)
Cherry tomatoes, whole	2 ea	24 ea
(weight)	(1 oz)	(12 oz)
Skewers, 12"	1 ea	12 ea
Oil	1 t	¼ C
Rice, cooked, hot	¾ C	2¼ qt
(weight)	4½ oz)	(3 lb 6 oz)
(F&SM Recipe 4300)		
Yield:	1 serving	12 servings

Figure 1. *continued*

Procedure:
1. Toss tenderloin cubes with Cajun Spice Mix to coat lightly. Refrigerate until needed.
2. Blanch green peppers, onions and mushroom caps in boiling water for 20 seconds. Drain well. Refrigerate until needed.
3. Assemble skewers as follows: cherry tomato, mushroom cap, meat, green pepper, meat, onion, meat, green pepper, meat, onion, meat, green pepper, meat, mushroom cap, cherry tomato. **Note:** Evenly space ingredients on skewer to allow for even heat penetration.
4. Cover and refrigerate until ready to cook.
5. As needed for service, brush each kabob with oil and place on preheated medium broiler or char-broiler. Cook, turning to brown evenly on all sides, to desired degree of doneness.
6. Serve immediately.

Service:
Place ¾C (4½ oz) hot rice along center of oval entrée plate. Top with 1 Cajun Beef Kabob (6 oz cooked wt).

Plate Garnish:
Parsley sprig

Suggested Accompaniments:
1. Substitute Dirty Rice (Recipe FD 710) for plain rice
2. Fresh Okra
3. French Bread

Source: Courtesy of Marriott Corporation.

Kitchen Equipment

The kitchen is designed for efficient use. It contains production, service, and warewashing areas. The salad area, which is part of the production unit, contains a pantry for dry goods and a walk-in refrigerator for fresh produce and other ingredients. The major equipment in the salad area is the following:

- 1 2-compartment sink with counters on either side and an overhead shelf
- 1 food processor on a mobile table

The production area contains two back-to-back sections. The equipment in the first section includes the following:

- 1 hand-washing sink
- 1 four-burner range
- 2 5-gallon steam-jacketed kettles on a stand
- 1 30-gallon tilting steam-jacketed kettle on a stand
- 1 30-gallon tilting fry pan on a stand
- 2 double-unit convection ovens
- 1 15-foot worktable with sink and overhead shelf in front of equipment

Figure 2. Modified recipes for beef kabobs.

7652   BEEF KABOBS

Ingredients	24 Servings	96 Servings	Procedure
MARINATED BEEF CUBES:			
REG DIETS:			
SOFT AND BLAND DIETS:			
LOW FAT AND CALC DIETS:			
LOW SODIUM AND LOW SODIUM CALC DIETS:			
Meat Marinade (Recipe 8468)	1 qt	1 gal	Prepare marinade for each diet according to recipe. Pour marinade over beef cubes for each diet. Place in suitable covered containers in refrigerator. **Marinate overnight.**
Top Sirloin, 1¼" cubes	72 ea	288 ea	
(weight)	(5 lb 10 oz)	(22½ lb)	

7652   BEEF KABOBS I   Portion: Serve 1 kabob (5½ oz by wt.)

Ingredients	24 Servings	96 Servings	Procedure
REG DIETS:			
LOW FAT AND CALC DIETS:			
LOW SODIUM AND LOW SODIUM CALC DIETS:			
Skewers, wood or bamboo	24 ea	96 ea	Place ingredients on skewer in the following order:
Marinated Beef Cubes (see above)	72 ea	288 ea	• green pepper • beef cube
Green peppers, 1½" × 1" chunks	48 ea	192 ea	• onion wedge • green pepper
(weight)	(13 oz)	(3¼ lb)	• beef cube
Onions, 1¼" wedges	48 ea	192 ea	• onion wedge
(weight)	(1½ lb)	(6 lb)	• mushroom
Mushrooms, fresh, whole	24 ea	96 ea	• beef cube
(weight)	(12 oz)	(3 lb)	• cherry tomato
Cherry tomatoes, fresh	24 ea	86 ea	Brush kabobs with remaining marinade. Bake in a 350°F standard oven for 15 to 20 minutes. Serve immediately.
(weight)	(1 lb)	(4 lb)	

7652   BEEF KABOBS II   Portion: Serve 1 kabob (5½ oz by wt.)

Ingredients	24 Servings	96 Servings	Procedure
SOFT AND BLAND DIETS:			
Skewers, wood or bamboo	24 ea	96 ea	Place ingredients on skewer in the following order:
Marinated Beef Cubes (see above)	72 ea	288 ea	• mushroom • beef cube
Carrots, frozen, baby, whole, thawed	48 ea	192 ea	• zucchini • carrot
(weight)	(1 lb)	(4 lb)	• beef cube
Zucchini, slices, 1¼" thick	48 ea	192 ea	• zucchini • carrot
(weight)	(3 lb)	(12 lb)	• beef cube
Mushrooms, fresh, whole	48 ea	192 ea	• mushroom
(weight)	(1½ lb)	(6 lb)	Brush kabobs with remaining marinade. Bake in a 350°F standard oven for 15 to 20 minutes. Serve immediately.

Source: Courtesy of Marriott Corporation.

- 1 15-foot worktable with overhead shelf in front of equipment
- l slicer on mobile stand in front of equipment
- mobile heated carts and a roll-in refrigerator, which are stored against a wall for transportation of menu items in bulk to the various foodservice operations

A wall separates the main production area from the serving area, which contains the following equipment for menu items that need to be prepared after the order is placed:

- 1 reach-in refrigerator/freezer
- 1 broiler
- 1 2-burner range top
- 1 griddle with oven base
- 1 compartment steamer with capacity to hold 12 4-inch hotel pans
- 1 hand-washing sink

Production Procedures

Policies and procedures for the production unit are specific and clear in the *Marriott's Senior Living Services Food and Beverage Policy and Procedure Manual.* Examples of selected production procedures are

- Daily meetings with the production staff are essential to critique the day's production and plan for tomorrow's meals. A printed production sheet for each work area must be issued to the associates for each day's menu.
- Completed production sheets are to be kept on file for 1 year as historical reference documents.
- Production forecasting should be completed a minimum of 1 week in advance of ordering.
- Correct handling of leftover menu items is an essential factor of safe food practices.
- Maintaining the correct temperatures (listed in the procedure) for storage, holding, and serving will ensure optimum quality at time of service.
- The entire serving area must be checked by management or unit supervisors for adequate supplies, cleanliness, correct serving temperatures, and the like.
- A taste panel monitors the daily quality of menu items for all meals and records results on a Taste Panel Control Record. Figure 3 is a copy of the taste panel procedures, and Figure 4 is the taste panel evaluation form.

Forecasting

Forecasting quantity demand for menu items is not as difficult as it is for restaurants because the number of residents is constant. Also, many of the menu items

Figure 3. Taste panel procedure.

The Taste Panel monitors the daily quality of menu items. All meals must be tasted, and results must be recorded on the Taste Panel Form (Figure 4).

Purpose

A Taste Panel is conducted in order to assure our customers of a fresh, hot, appetizing, and attractive menu item prepared according to the Senior Living Services Division recipe.

When

Conducted *at least* 15 minutes prior to opening for business for every meal.

Where

Conducted at the chef's counter after the serving line has been completely set up for service.

Who

Three participants with *at least* one member of management.

What

All hot menu items, including grill items, and as many cold menu items prepared in the kitchen as necessary. Cold menu items that should be tested are salad mixtures (chicken, tuna, ham, potato, cole slaw), home made dessert items such as custard and chocolate pudding.

How

Using a spoon, take a bite-size portion of each menu item to be tested and place it on your plate. Use a different spoon for each food product. *Never* dip spoon used to taste one product into pan containing another product. Taste each food product and using the criteria outlined on the Taste Panel form, evaluate each product and enter appropriate comments on the form. Be as *objective* as possible, basing your comments on recipe compliance. Comments such as "not salty enough" or "too spicy" are invalid unless judged against the standard. Taste all products even if you do not like them. Remember, you are only being asked to taste for evaluation purposes.

Source: Courtesy of Marriott Corporation.

are prepared after the order is brought to the kitchen; thus, leftovers of expensive menu items are minimal. Family and friends are welcome to dine with residents in the main dining room, private dining room, or coffee shop. Residents, however, are asked to notify the dining room manager of guests at least 24 hours in advance. The number of residents eating in the dining rooms usually is as follows:

- Main dining room: 70–100 for lunch, 300–350 for dinner
- Coffee shop and bistro: 70–100 for dinner
- Assisted-living dining room: 45
- Skilled-nursing dining room: 55, of whom 15–20 residents suffering from a stroke are fed by the nursing staff in the restorative dining room until they can feed themselves.

Figure 4. Taste panel evaluation form.

MARRIOTT CORPORATION
SENIOR LIVING SERVICES
TASTE PANEL

EVALUATE:
TEMPERATURE
APPEARANCE
SEASONING
TEXTURE

CODE: ACCEPTABLE - A
UNACCEPTABLE - UA

ACTION: REJECTED - R
CORRECTED - C

UNIT # _____ DATE _____

Minimum Serving Temperatures

Sauces and Gravies 160°–180°	Meats 150°	Dip Entrées 160°	Vegetables 160°–180°	Soups 160°–180°	Coffee 180°	Cold Foods 40°

Breakfast: Time _____

ITEM	Temp.	Testers			ACTION
		1	2	3	

Lunch: Time _____

ITEM	Temp.	Testers			ACTION
		1	2	3	

Dinner: Time _____

ITEM	Temp.	Testers			ACTION
		1	2	3	

484

More than one seating is required at dinner in the main dining room, which has only a 236-seat capacity, and in the 62-seat coffee shop and bistro.

Associate Staffing

Staffing for the kitchen and main dining room (independent living) requires planning. A weekly staffing pattern for associates in the production unit for 1 week is shown in Figure 5, as well as the number of daily and weekly hours for each associate. One-half hour is subtracted from the cook's hours each day for meals. Total hours per week for the category 09 supervisors and the category 06 cooks also are recorded. Production staff includes cooks, prep cooks, a cafeteria and salad bar attendant, and an expeditor. The waitstaff includes a host/hostess, servers, dining room and coffee shop/country store attendants, aisle persons, and an expeditor. Full-time staffing equivalents for the foodservice operation are

Production	13.5
Utility (janitors and dishwashers)	7.9
Waitstaff	29.7
Supervisors (including purchasing agent)	7.0
Managers (including F&B Director)	4.0
Total FTEs	62.1

The Problem

Virginia is known for its winter ice storms. During one such storm, The Fairfax community had a power outage at 4:00 P.M., and of course, the kitchen was affected. Dinner preparations had already started because the main and healthcare dining rooms and coffee shop open for serving at 5:00 P.M. Even though most of the kitchen equipment depends upon gas for operation, the fire extinguishers require electricity and were affected by the outage. Whenever an electrical outage occurs, the gas shuts off in the kitchen and the cooking equipment becomes inoperable to prevent a fire from occurring. With no electricity, extinguishing a fire would be impossible. In addition to the outage problem, a number of associates called in to report they would not be able to travel to work because highways were closed due to the heavy ice storm.

The F&B director and chef knew they had less than an hour to find a solution. If the dining rooms were not open at 5:00 P.M., many residents would be unhappy, which could lead to complaints. Policies and procedures are available for disasters such as a power outage (Figure 6). After a quick conference, the director and chef

Figure 5. Weekly staffing pattern for main kitchen supervisors (category 09) and cooks (category 06).

Main Kitchen NAME	SAT	SUN	MON	TUES	WED	THUR	FRI	HOURS
Supervisor — 09								
AM	6—2	5—1	—	—	6—2	6—2	6—2	40
PM	—	1—9	1—9	1—9	1—9	1—9	—	40
Purchasing Agent	—	—	8—4	8—4	8—4	8—4	8—4	40
Total — 09								120
Cooks — 06								
Lead	1—9	7—3	6—2	6—2	—	—	1—9	37.5
#1 — Cook	1—9	1—9	1—9	1—9	1—9	1—9	1—9	52.5
#2 — Cook	1—9	1—9	1—9	1—9	1—9	1—9	1—9	52.5
#3 — Cook	12—8	1—9	1—9	1—9	1—9	1—9	1—9	52.5
#4 — Cook	8—4	7—3	—	8—4	8—4	8—4	8—4	45.0
#5 — Cook	9—5	9—5	9—5	9—5	9—5	9—5	9—5	52.5
#6 — Cook	—	7—3	—	—	—	—	—	7.5
#7 — Cook	—	11—3	—	—	—	—	—	4.0
#1 — Prep Cook	8—4	8—4	8—4	8—4	8—4	8—4	8—4	52.5
#2 — Prep Cook	10—6	10—6	10—6	10—6	10—6	10—6	10—6	52.5
#3 — Prep Cook	12—8	7—3	8—4	8—4	8—4	8—4	8—4	52.5
Cafeteria	—	—	8—4	8—4	8—4	8—4	8—4	37.5
Salad Bar	—	—	4—9	4—9	4—9	4—9	4—9	20
Expeditor	—	—	3—10	3—10	3—10	3—10	3—10	32.5
Total — 06								551.5
TOTAL								671.5

Source: Courtesy of Marriott Corporation.

Figure 6. Policy and procedure for developing a disaster plan.

POLICY: The Food and Beverage Department of a Marriott property shall have an exclusive disaster and fire plan that is related to the plans of other departments.

Note: Each community has its own disaster plan as written by management and staff. The Food and Beverage Director should write the departmental procedure based on the following factors:

- Duration of the plan (community should be self-sustaining for a minimum of one week)
- Estimated number of people to be served:

  Residents in the community
  Disaster victims
  Staff
  Volunteer workers
  Police, firefighters, and others

- Quantity of food in storage
- Types and quantity of disposable ware needed and in storage
- Menu for the duration of the disaster
- Do's and don'ts during the disaster
- Telephone numbers of key food and beverage associates
- Responsibilities of key associates during the disaster
- List of equipment and power supply for each. Is a hook-up to emergency power available?
- Location of emergency lighting such as flashlights
- Identification of the emergency water supply
- List of emergency suppliers and telephone numbers
- Notation of any other pertinent information

Source: Courtesy of Marriott Corporation.

came up with a solution that enabled them to serve dinner on time. How would you have handled the problem?

Points for Discussion

- Using the foodservice systems model, analyze how the environmental effect of an ice storm made an impact on The Fairfax foodservice system.
- What kinds of disaster plans should a foodservice manager have ready for use? Give some examples of other kinds of disasters for which the manager should prepare. Who should be involved in the disaster planning process?
- How can a foodservice manager utilize resources in the outside community to respond to emergency situations? What are some other groups, organizations, agencies, or foodservice operations that a manager might network with in a crisis situation?

5

Distribution, Service, Sanitation, and Maintenance

Distribution and service is the third subsystem, and sanitation and maintenance is the fourth in the foodservice system. The goal of each foodservice operation is to provide quality food and service that will satisfy the customer. For complete customer satisfaction, the equipment and facility need to be clean and well maintained.

- **Chapter 15, Distribution and Service.**   Distribution of menu items to customers can be complex, depending upon the type of foodservice operation. Total quality service is a paradigm shift from just product and service quality to a total customer value quality.

- **Chapter 16, Sanitation and Maintenance of Equipment and Facilities.**   Sanitation and maintenance are important throughout all the subsystems and are critical in a foodservice organization. A properly designed facility is basic for sanitation that could not be accomplished without a preventive maintenance program.

15

Distribution and Service

Distribution and service is the third subsystem in the transformation element of the foodservice system (Figure 15.1). Service is a major component of distribution of menu items to customers and, therefore, covers all types of foodservice operations. Vending machines serve customers, who want a snack or a quick meal, by meeting their needs as does a waitperson in a fine dining restaurant under leisurely conditions.

Depending on the type of foodservice operation, distribution may or may not be a major function. Regardless of the type of foodservice, holding and serving hot food can be hazardous if critical control points are not established. Referring to the HACCP model in Table 5.1 in chapter 5, the hazard for holding and service is contamination because of bacterial growth, and the corrective action, if the standard is not met, is sanitizing the equipment and keeping the food at 165°F.

Distribution is a major concern in hospital foodservices in which patients are served in individual rooms located on many floors and perhaps in separate buildings. Ensuring that the appropriate food is sent to the appropriate place for service to a

Figure 15.1.   Foodservice systems model with the distribution and service subsystem highlighted.

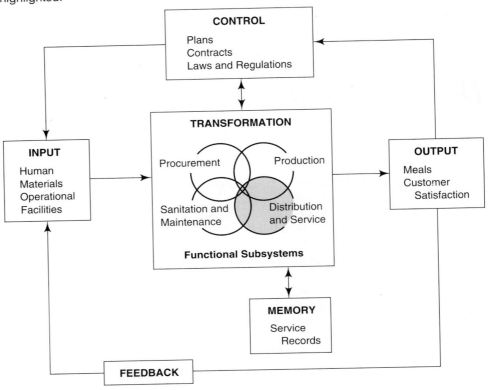

particular patient is a complex process, which is further complicated by the need to ensure that the food is at the right temperature and aesthetically appealing. In contrast, in limited-menu restaurant operations, where customers pick up the menu items directly after production and either go off premises for consumption or to a table in the facility, distribution is relatively simple. In fact, distribution and service become the responsibility primarily of the customers and not the employees.

More than $62 billion is spent annually on take-out food, according to Alabama-based Morrison's Custom Management (Ryan, 1991). Two thirds of that total is spent at limited-menu, full-service, and other restaurants, and one third is spent at retail food stores like delis and supermarkets. Morrison's estimate is that 64% of food-to-go is take-out, 29% is drive-thru and 75% is delivery. To speed up service, many limited-menu restaurants have installed a double drive-thru, which has two windows and two lanes for traffic (Follin, 1991). The 90s really can be called the decade of take-out and delivery as operators continue to develop new technology for delivery of products to consumers who want restaurant food at home. Pizza, chicken, and hamburgers are the top three menu items customers are likely to purchase from table service restaurants for take-out (Iwamuro, 1992). Experts also have predicted that customers in the future will order meals via television and a computer modem tuned into a home-delivery channel.

Service takes many forms in a foodservice establishment, from that in the upscale fine dining restaurant involving several highly trained employees to that in the many self-service operations—cafeteria, vending, or buffet. The method, speed, and quality of the services provided can make or break a foodservice establishment. The quality of food may be excellent, the sanitation of the establishment above reproach, the procurement and storage of food ideal, but if service is lacking, the operation will be rated poor by the customers.

This chapter covers distribution and service in the four basic types of foodservices, which were introduced in chapter 5. Different types of service in foodservice operations are then compared. Also, emphasis is put on service management, which is being discussed a lot in the industry today.

DISTRIBUTION IN FOODSERVICE TYPES

The four types of foodservices are conventional, commissary, ready prepared, and assembly/serve. Although distribution and service have some characteristics in common, their differences lead to different demands in some aspects of the distribution function.

Conventional

In the conventional operation, most menu items are produced on the premises and distributed for service to a serving area or areas close to the production facilities. Hot- and cold-holding equipment is needed to maintain the proper temperature for various menu items between the time of production and service. Adherence to critical control points for holding hot food and serving it is emphasized in the HACCP model in Table 5.1. If standards are not met, utensils and equipment need to be washed, rinsed, and sanitized, and the beef stew reheated to 165°F. Depending on the service areas, this holding equipment may be stationary or mobile. Some equipment is versatile and can be used for distribution, holding, and service. For example, the mobile modular serving units in Figure 15.2 could be used for transporting food for a catered function in a dining room away from the main kitchen and for holding the food until time of service. The units then provide a service counter for self- or waitstaff service.

In a healthcare facility, however, patient service may take place throughout the facility, requiring a more complex distribution process than in other foodservice operations. Meal assembly, for example, may be centralized or decentralized. In a facility with **centralized meal assembly,** the time between production, assembly, distribution, and service can be minimal. The trays are first assembled for service at a central location in or close to the main production facilities. The assembled trays are then distributed to the patient units using a variety of carts. Some institutions use

Figure 15.2. Examples of distribution and service equipment: *top,* hot serving unit with heated storage base; *bottom,* cold serving unit with refrigerated storage base.

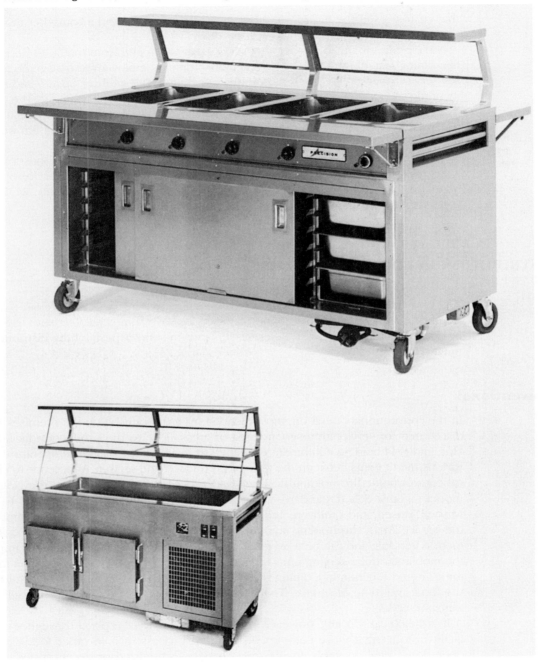

Source: Precision Metal Products, Inc., Miami, FL. Used by permission.

heated and refrigerated tray carts, which may be motorized or pushed manually by hospital employees.

The high initial and maintenance costs of heated and heated/refrigerated carts have led to development of other methods for maintaining proper temperatures on assembled trays. One of these methods uses specially designed dishes that have been preheated in an infrared oven, then transferred to an insulated base. The hot menu items are portioned onto the plate, which is covered by a dome designed to fit the base container, thus keeping food warm until service to the patient. This unit is placed on the individual patient's tray and other menu items that have been individually wrapped are added. The assembled trays are then transported in an unheated cart to patient units for service.

The process of meal distribution in a centralized tray operation includes activities relating to the movement of assembled trays from the point of assembly to the patient area. A method for thermal retention is needed in larger operations in which the time between meal assembly and service to the patient is too long to maintain proper temperatures. Refrigerated support for cold foods may also be needed. Figure 15.3 outlines methods for thermal retention support in conventional foodservice operations (Hysen & Harrison, 1982). These categories include hot thermal retention, hot and cold thermal retention, and no thermal support. Benefits and constraints of these methods of distribution are described in Table 15.1.

An example of a centralized tray assembly unit is shown in Figure 15.4. The layout uses mobile equipment, which has been widely accepted because of its flexibility and the ease of facility maintenance that it provides. This setup can be readily rearranged or moved for cleaning.

Each assembly area requires support equipment to assemble the trays, such as that shown in Figure 15.4. The size of the area and the number of trays to assemble will determine the type of conveyor required to provide support. Hysen and Harrison (1982) outlined options for tray assembly equipment (Figure 15.5), including

Figure 15.3. Methods of thermal retention support in conventional foodservice.

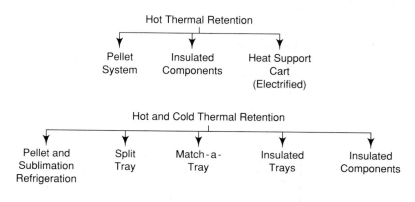

Source: Adapted from "State-of-the-Art Review of Health Care Patient Feeding System Equipment" in *Hospital Patient Feeding Systems* (p. 172) by P. Hysen and J. Harrison, 1982, Washington, DC: National Academy Press.

Table 15.1.   Benefits and constraints of various methods of meal distribution

Type of Meal Distribution	Benefits	Constraints
Hot Thermal Retention		
Pellet system	Support equipment and system operation are conventional and uncomplicated. No requirement for a special plate: any standard-size china. No special insulated delivery cart is required.	Provisions for maintenance of cold items such as milk, salads, jello, ice cream are not made. Hot food cannot be held for a long period of time (more than 45 minutes). Additional serviceware pieces need to be inventoried, stored, transported, and washed.
Insulated components	Only the dinner plate and food are heated; there are no pellet bases to heat. Simple in operation requiring no special pellet dispensers to purchase. No burn hazard to the attendant or patient because there is no hot pellet base or pellet disk. No special insulated delivery cart is required.	Additional serviceware pieces need to be inventoried, stored, transported, and washed. Attractive insulated components are often taken home by patients as useful memento of their hospital experience.
Heat support cart	Thermal energy can be controlled to plate and/or bowl as required. The cart allows for food to remain heating until tray is removed for service to the patient. Each chart has an insulated drawer for ice cream and other frozen desserts. Heat energy continues to be supplied to food during the transportation process.	Special sophisticated motorized carts and special trays with heaters are required. The potential for maintenance/repair problems is high. Cart and the trays are dependent on the use of disposable dishes. Disposable dishes could be uneconomical from an operational cost standpoint and could be considered unacceptable from an aesthetic perspective. No provisions are made for maintenance of cold food items at proper temperatures except ice cream.
Hot and Cold Thermal Retention		
Pellet and sublimation refrigeration	A synergistic heat maintenance effect is achieved. Simplicity of cart construction and ease of sanitation. Cart is lightweight, which provides for ease of mobility.	Operational cost and complexity of the required carbon dioxide cooling system is a consideration. Patient trays are not completely assembled at a central assembly point. Final assembly occurs in patient areas.
Split tray	Centralized supervision and control of the total assembly process. No reassembly of tray components is required in the patient areas. Good temperature retention of both hot and cold items. System accommodates late trays within a reasonable period.	Cart is heavy and bulky. A motorized version may be required if any ramps are to be negotiated. Carts are difficult to sanitize. Initial cost of the cart is high and maintenance costs can be high. Due to the relatively heavy weight and limited maneuverability, carts and wall surfaces are subject to damage.

Table 15.1. *continued*

Type of Meal Distribution	Benefits	Constraints
Match-a-tray	Same as described for split tray except that consolidation is required on the patient level.	Same as described for split tray. Additional labor must be applied at the patient area to reassemble the complete patient meal.
Insulated trays	Maintains hot and cold zones well without external heat or refrigerant sources. Simplicity of transport is achieved. Does not require a heavy, enclosed delivery cart. Stacked trays protect and insulate food. Less load on the dishwashing facility due to disposables. No complex components to repair, replace, or maintain.	Purchase of special disposable dishes results in higher operational costs. Food holding time is limited to 45 minutes. Long-range cost could be substantially higher than other systems due to disposable and lease costs. Hot foods may take on a "steamed" appearance in the hot compartment due to relatively small volume and lack of venting. Possible adverse patient reaction to eating from a compartmentalized tray. Trays can be difficult to sanitize completely due to deep cavity construction. Top and bottom tray compartments do not nest; more storage area required. Rigid presentation and placement of dishes is a limitation of the system.
Insulated components	Only the dinner plate and food are heated. No pellet bases to heat. Hot and cold foods are placed in insulated containers. There is no burn hazard to the attendant or patient because there is no hot pellet base or pellet disk. Cold food items can be held longer than 30 minutes. No special insulated delivery cart is required.	Additional serviceware pieces need to be inventoried, stored, transported, and washed. Attractive insulated components are often taken home by patients as useful mementos of their hospital experience. Hot food holding time is limited to 30 minutes.
No Thermal Support		
Covered tray	Tray is a simple standard unit. Equipment cost of the system is low.	Requires an immediate and responsive transportation system. High labor component is required for transportation process. No thermal support is available for entrée and other food items.

Source: Adapted from "State-of-the-Art Review of Health Care Patient Feeding System Equipment" in *Hospital Patient Feeding Systems* (pp. 168–172) by P. Hysen and J. Harrison, 1982, Washington, DC: National Academy Press.

Figure 15.4.   Centralized tray assembly unit.

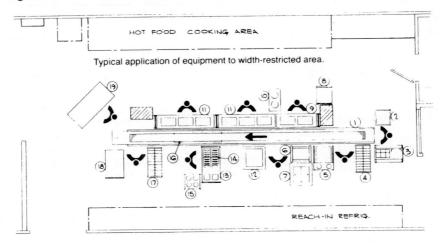

Typical application of equipment to width-restricted area.

Source: ©1984, Caddy Corporation of America, Pittman, NJ. Used by permission.

manual conveyors, such as a trayslide; simple mechanical conveyors, such as a roller type; and motorized conveyors, which may be straight-line or circular. The straight-line layout shown in Figure 15.4 illustrates other equipment needed for holding and dispensing food and other items needed for tray assembly.

In **decentralized meal assembly,** the food products are produced in one location and transported to various locations for assembly at sites near patients. Equipment to maintain proper temperatures—food warmers, hot food counters, and/or

Figure 15.5.   Types of conveyors for meal assembly process.

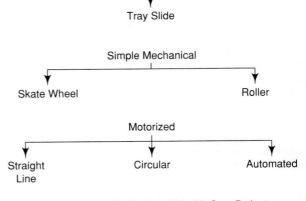

Source: "State-of-the-Art Review of Health Care Patient Feeding System Equipment" in *Hospital Patient Feeding Systems* (p. 172) by P. Hysen and J. Harrison, 1982, Washington, DC: National Academy Press.

refrigerated equipment—must be provided at each location. Because some foods, such as grilled or fried menu items, do not transport or hold well, some cooking equipment may be available in the service units for these difficult-to-hold foods. Even in centralized meal assembly, a few items such as coffee and toast may be prepared on the patients' floors.

Since the early 1950s, healthcare institutions have moved toward centralized tray assembly patterned after airline foodservices. Centralized assembly has the advantages of eliminating double handling of food and facilitating supervision of meal assembly because the activity takes place in one location rather than throughout the facility. In addition, it allows for standardization of portions, uniformity in presentation, and decreased waste. Finally, less staff time is needed, and the space occupied by floor kitchens can be used for other purposes. Decentralized meal assembly is still used in some institutions, however, because it offers the advantage of less time between assembly and service to patients, allowing for potentially higher-quality food. Decentralized facilities also offer greater flexibility in providing for individual patient needs and in making last-minute substitutions and changes.

Depending on the layout and design of the healthcare facility, a combination of meal assembly and distribution methods may be used. Some facilities may even serve groups of patients in the dining room, while other patients are provided tray service in their hospital rooms. Group service is especially common in nursing homes and other extended care facilities, such as psychiatric hospitals.

Commissary

Commissary foodservice operations are characterized by centralized production with distribution of prepared menu items to several remote areas for service and possibly final production. Service at these areas may be self-service, cafeteria service, or tray service. The centralized production facility is referred to as the *commissary;* in this book, the service areas are called *satellite service centers.*

As discussed in chapter 5, the commissary foodservice operation can take many forms. In school foodservice operations, for example, a central kitchen may be a conventional foodservice for a secondary school, but it may also provide food to several satellite service centers for elementary schools in the district.

The unique feature of distribution in the commissary foodservice is that food must be transported to remote locations of the satellite service centers. As discussed in chapter 5, these facilities may be relatively close, within the same city or county, as is the case for most school foodservice operations, or they may be located great distances away from the central production unit, which is typical of many large commercial chain operations.

For this reason, a commissary foodservice requires specialized distribution equipment, tailored to its particular needs. Menu items produced in the central commissary may be transported either frozen, chilled, or hot and in bulk or in individual portions. As discussed in chapter 8, the receiving area of the commissary and the satellite service centers must be designed to accommodate the distribution equipment. Also, as pointed out in chapter 12, special precautions are necessary to preserve the microbiological quality of foods because of the length of time between production and final service to the customer. In evaluating commissary foodservices, transportation costs must be considered because they may add materially to the total cost of the operation, involving purchase, operation, and maintenance of trucks for distribution.

Ready Prepared

In ready prepared foodservice operations, menu items are produced and held, either frozen or chilled, for service at a later time. They may be packed in bulk, in individual portions, or in combination containers. For example, in airline foodservice, two or three menu items may be portioned onto an individual serving dish.

The distribution equipment needed by ready prepared operations depends on whether foods are in bulk quantities or individual portions and if a cook-chill or cook-freeze method is used. A unique characteristic is the heat processing of prepared items before service. Microwave, convection, and infrared ovens are commonly used in the service unit for final heat processing of chilled or frozen foods. Cold temperature support also is required during the distribution process.

Two types of carts are used predominantly for cold temperature support. One type is insulated to maintain temperature during distribution to remote service areas where the carts are connected to wall-mounted or floor-borne refrigeration units. Hysen and Harrison (1982) described these carts as lightweight and thus easy to transfer; they are also easy to sanitize. In more common use is the roll-in refrigerator cart. If the cart is the enclosed type, its doors should be opened before it is placed in the refrigerated unit to permit proper circulation of chilled air.

In addition to the equipment that is used for heat processing before service, some institutions use contact plate heater carts and integral heat ovens and carts for patient tray service. The benefits and constraints of the various methods of heat processing for patient service are enumerated in Table 15.2 (Hysen & Harrison, 1982).

The place on the patient floors where final heat processing and meal assembly occurs is generally referred to as a *galley*. The equipment in a galley includes cold temperature support equipment, equipment needed for final heat processing of menu items, some small equipment such as a coffee maker and toaster, a sink, a small storage area, and a desk area for the dietetic personnel.

Assembly/Serve

The assembly/serve foodservice operation generally uses foods that are ready to serve or that require little or no processing prior to service. As pointed out in chapter 5, bulk, preportioned, and preplated frozen foods are the three market forms of foods that fit into this category.

When foods are served cafeteria style, the bulk form is used, and a means of heating before service is required. This heat processing can be done in the service unit or an auxiliary area, using one of the methods described in the discussion of the ready prepared operation. If preportioned or preplated items are purchased, heat processing similar to the methods described for ready prepared foodservices can be used. Cold temperature support equipment may be needed for distributing foods to service areas and for holding prior to heat processing for service.

Table 15.2.   Benefits and constraints of various methods of heat processing

Method	Benefits	Constraints
Microwave ovens	Food is cooked very rapidly. "On-demand" patient feeding can be achieved.	Food is easily overcooked, and some foods tend to rethermalize unevenly, leaving hot and cold spots. Food does not brown, causing some foods to have an unnatural appearance. Trained operator is required to rethermalize all food products. Employee training is essential to the success of the program. Maintenance of microwave ovens can be a significant cost factor.
Convection ovens	Oven cavities can accommodate 12 to 30 meals at a time; thus higher efficiency can be achieved in the rethermalization and reassembly process as compared to a microwave system.	Speed is increased as compared to a conventional still air oven; however, the process is not as fast as a microwave oven. Some food products experience excessive cooking losses; in others, there is a thickened surface layer on the food from the rethermalization process. Some food products do not rethermalize to a uniform temperature.
Infrared ovens	Food is rethermalized at a faster speed than conventional still air ovens. Oven cavities can accommodate 16 to 24 meals at a time; thus higher efficiency can be achieved in the rethermalization and reassembly process as compared to a microwave system.	Energy consumption for rethermalization is comparatively high. Soups are not accommodated by the infrared equipment and must be separately handled. Dishes and covers become very hot in the rethermalization process. Food products may burn to or stick to the heated dish.

CATEGORIES OF SERVICE

The foodservice systems model, the organizing framework for this book, defines quality meals and customer satisfaction as its primary outputs or outcomes. As the process that provides for the culmination of these outcomes, service of food to the customer is both a subsystem of the model and the ultimate objective of a foodservice organization.

Quality food poorly served will often result in complaints and loss of customer goodwill; mediocre food well served may meet with satisfaction. Complaints about poor service rank high in the foodservice industry. People become frustrated and discontented if they have to stand in a long service line or wait for food to arrive, or

Table 15.2.   *continued*

Method	Benefits	Constraints
Integral heat ovens and carts	Minimum intervention by employees is required to rethermalize foods. Efficiency and speed of service is enhanced due to multiplicity of meals rethermalized at the same time. Integrally heated dish acts as "pellet" system to continue to provide thermal support to hot food after service to patient.	Certain food items, such as soup or hot breakfast cereals, are difficult to rethermalize. Dishes must be sprayed with a release agent to prevent sticking when using certain food items. Warewashing time is increased, particularly for the breakfast service, because of the food that sticks to dishes. Ongoing operation costs are comparatively high due to replacement and lease costs. Inflexible presentation of the tray and rigid placement of items when employing the cart-borne system.
Contact plate heater carts	Reduced pantry labor due to rethermalizing and refrigerating patient trays in the delivery cart. Allows pantry to be reduced in size and lowers equipment cost by eliminating need for reheating ovens. Minimum intervention by employees after assembled tray has been dispatched from main tray assembly location.	Cart maintenance may be a problem due to complex electrical components. Special trays and dishes are required—usually disposable dishes—which can increase operating costs. Rethermalization can only be done from the chilled state, not from the frozen state. Cart is presently being field-tested; its performance has not been proven. Operating cost appears to be high, based on preliminary data available. An inflexible presentation of the tray and rigid placement of items are aesthetic limitations of the system.

Source: Adapted from "State-of-the-Art Review of Health Care Patient Feeding System Equipment" in *Hospital Patient Feeding Systems* (pp. 178–179) by P. Hysen and J. Harrison, 1982, Washington, DC: National Academy Press.

if they are served the wrong menu item or one prepared differently from what was ordered.

Service can be categorized in a variety of ways and, in fact, a number of combination services exist. In this chapter, three types of service are discussed: table and counter service, self-service, and tray service. The table service restaurant with the self-service salad bar is an example of a combination of service types.

Service of food and beverages is one of the most diverse activities imaginable, assuming many forms and occurring in a wide variety of places, at all hours of the day and night. Because of today's life-styles, it can range from fine service with tableside preparation, to coffee and doughnuts in the factory, to hot dogs at the beach.

Table and Counter Service

Table and counter service have traditionally been the most common forms of service in the commercial segment of the industry. Table service can be very simple or extremely elaborate; its distinguishing characteristic is service by a waitperson. In most table service operations, a hostess, host, or maître d'hôtel is responsible for seating guests in the dining room.

The most common method of table service in the United States, often referred to as *American-style* service, involves the plating of food in the kitchen or service kitchen and then presenting it to the guest. In more elaborate service, food is often prepared at the table—as with bananas Foster or steak Diane. Another type of table service is called *family-style,* in which food is brought to the table on platters or in bowls by the waitstaff and then passed around the table by the guests. Restaurants featuring country fried chicken or barbecue ribs will frequently feature family-style service, as do some elementary schools.

Counter service is often found in diners, coffee shops, drug store fountains, and other establishments in which patrons are looking for speedy service. People eating alone can join others at a counter and enjoy the companionship. The most common counter arrangements are shown in Figure 15.6. These arrangements provide not only fast service for a customer but also efficiency for the establishment. The counter attendant is usually responsible for taking the orders, serving the meals, busing dishes, and cleaning the counter, and may even serve as cashier except at peak periods.

A well-trained and courteous waitstaff and other service employees are the keys to successful table service operations. In upscale restaurants offering sophisticated service, the job of the waitstaff is highly specialized and truly an art.

Tipping waitstaff has been a common practice in table and counter service restaurants in the United States, as is charging for service in Europe. Tips are voluntary, and service charges are not. The service charge is a predetermined amount added to each customer's check and is considered part of the restaurant's gross receipts subject to income tax; it is not a tip. The employer is under no obligation to give the service charge to the waitstaff. In most cases, however, all employees benefit from it by increases in salaries. Tips left in addition to service charges belong to the waitstaff and are treated the same as all voluntary tips by wage and tax laws.

The foodservice manager must be aware of federal and state legislation and regulations that have provisions covering tipping and tipped employees. The Internal Revenue Service (IRS) designed the form 8027 in 1983 after Congress passed a law to obtain more information from restaurants about employee tip reporting ("IRS says 75%," 1993). The law requires restaurants to give information to the IRS each February on the 8027 form including their annual sales, the amount of tips reported by employees, the amount of sales charged by customers, and the amount of tips charged to credit cards.

Figure 15.6.   Various layouts for counter service.

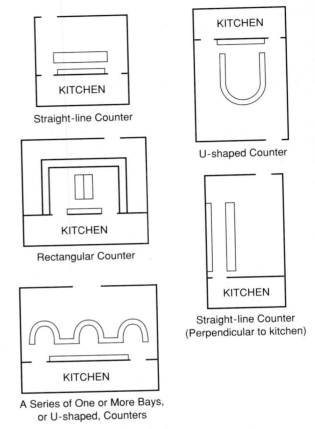

Straight-line Counter

U-shaped Counter

Rectangular Counter

Straight-line Counter
(Perpendicular to kitchen)

A Series of One or More Bays,
or U-shaped, Counters

Source: How to Manage a Restaurant or Institutional Food Service (p. 105) by J. W. Stokes, 1982, Dubuque, IA: Wm. C. Brown.

Approximately 25% of the estimated number of restaurants file an 8027. The amount of charged tips that have been reported by employers has been greater than the total amount of tips reported by employees for the same time. Even though tip reporting is the responsibility of the employee and employers cannot do anything about it, federal agents could go back to employers and ask them to pay their share of FICA (Federal Insurance Contribution Act) payroll taxes that the IRS says employees earned but did not report. The fact that the IRS does not have the staff to do this in all district offices could preclude the restaurateur from being billed for back taxes.

Handling and controlling guest checks is another concern of managers in utilizing a waitstaff. The first element of effective control is to ensure that all menu items are charged to the customer. A duplicate check procedure is the most widely used; the waitperson writes the order on a customer check, simultaneously preparing a carbon duplicate that is submitted to the production area to obtain menu items. The second element of control is to charge proper prices for menu items. Electronic cash registers with preprogrammed prices are widely used for this reason. The assurance that all checks are accounted for is the third element of control. This can be accomplished by having all customer checks sequentially numbered, keeping reserve stocks of checks in locked storage, issuing checks to waitstaff in numerical order, and then sorting the used customer checks periodically into numerical order to determine if any are missing.

A large number of top restaurant chains are accepting credit cards as a means of offsetting slow growth and gaining new customers. Besides, both chains and credit card companies are eager to open new markets. New technology is ultimately responsible for creating the biggest symbol of American consumer culture—the credit card (Bertagnoli, 1989). Most restaurant chains use the "Swipe" method of credit card transactions. The credit card number is used to authorize the transaction and check the card against a computer file of lost or stolen cards within seconds. Customers ordering home or office delivery of food by telephone can give the order taker their card number, and authorization is obtained while they wait. Delivery employees for some restaurants carry a portable addressograph to imprint card information on the check.

Self-Service

Today, self-service foodservice operations cover a wide spectrum, cafeteria service being one of the most commonly used forms. Self-service is characteristic of the limited-menu restaurant industry, with counter pickup service, take-out, and drive-thru window service the most common approaches. Other self-service operations include buffets, vending machines, refreshment stands in recreational and sports facilities, and mobile foodservice units that range from the small, hot dog cart rolled down the street by the operator, to sophisticated operations in motorized vans equipped for preparing a variety of menu items. Today's round-the-clock eating patterns in every imaginable place have created a demand which self-service satisfies.

Cafeteria service is characterized by advanced preparation of most menu items and a maximum of self-service except for hot food. It is the predominant form of service used in noncommercial foodservice and employee-feeding operations. Self-busing of trays and dishes is also a common practice as a means of reducing labor costs. Traditionally, the commercial cafeteria has placed a great deal of emphasis on food display, merchandising, and marketing of menu items. Managers of noncommercial foodservices now realize the profit possibilities in a cafeteria and are including service as a regular part of the business.

The straight-line counter, which may vary greatly in length, is the most common cafeteria counter arrangement, for self- and waitstaff service. Generally, the length varies with the quantity and variety of menu items, instead of being dependent on the number of persons to be served.

An alternative arrangement to the straight-line counter is the hollow square, sometimes called the scramble system. In this layout, the various stations or food counters are positioned to form three or four sides of a square, with space between the counters and perhaps a center island. This layout allows customers to move from one station to another without being held up by the entire line. The hollow square layout not only decreases lines but also permits more people to be served in a smaller space.

Buffet service has enjoyed increasing popularity in recent years in all types of food-service. Periodic scheduling of buffets in a college residence hall foodservice, an employee cafeteria in a hospital, or an industrial foodservice operation can serve as a monotony breaker and a means of creating goodwill.

Buffet service enables a facility to serve more people in a given time with fewer employees. The usual procedure in commercial operations is for guests to select the entrée, vegetables, and salad from the buffet before going to the dining table set with flatware, napkins, and water. Quite often, table service is used for the main course and beverages, and they are supplemented by salad and dessert bars.

Vending machines, dubbed the silent salesman with a built-in cash register, annually move billions of dollars of products and services to customers around the world. Selling items from machines is nearly as old as recorded history, but the impact of vending machines on the U.S. economy was not recognized before the middle 1940s. Even though the external customer communicates only with the vending machines, employees, the internal customer, work behind the scenes to ensure that customer needs are being met. Temperatures must be recorded daily on all perishable food, and strict adherence to the coding, product handling, and rotation procedures must be maintained. Sanitation procedures and schedules must be developed and checked. Employees servicing the machines also must be trained in customer relations because they represent the vending operation when interacting with the customer.

Many organizations use a contract company rather than setting up their own vending operation. The contract should be reviewed periodically as should the accident, liability, and hazard insurance carried by the vending supplier (Beasley, 1990). Also, the supplier's compliance with city, county, state, and other regulatory agency standards should be checked. Before selecting a supplier, it can be helpful to visit the headquarters to see how and under what conditions the food is prepared. Competitive bidding has proven beneficial to the purchaser of the service who can negotiate commission rates as well as replacement of equipment when necessary.

However, many foodservice directors operate their own vending business, giving them the opportunity to tailor a vending program to their customers' needs. These directors should respond to new market trends and technological advances in vending, as they do in their cafeteria and catering programs (Beasley, 1993). They have to be innovative and creative in finding ways to make vending a profit center. Using focus groups of customers from various demographic groups to suggest new items and for taste-testing will make the vending machines customer-friendly.

Healthful food products, including those that are lower in calories, fat, cholesterol, and sodium should be stocked in vending machines (Beasley, 1993). Salads and fruit plates packaged in upscale, deli-type containers with colored, rigid plastic plate-bases and clear domes have customer appeal. One vending machine can dispense espresso, three strengths of regular coffee, decaffeinated, gourmet coffees, cappuccino, fresh-brewed leaf tea, hot chocolate, and soup. Retail branded foods, which generally sell at higher prices than nonbranded items, are gaining popularity. Retail branded entrées are beginning to appear because operators believe this will upgrade the perceived quality of menu items.

Vending machines have been modified to accept credit cards, which is appreciated by cashless customers (Beasley, 1990). Other new payment options include the use of charge cards that permit customers or a department to be billed. The use of a debit card permits the customer's balance on hand to be reduced after each transaction.

Tray Service

Tray service, in which food is carried to a person by a foodservice employee, is used primarily in healthcare institutions and for in-flight meal service in the airline industry. Room service, in which food is served on a tray or on a cart in a customer's hotel or motel room, is a variation of tray service. In airline service, food is produced in a commissary by a food contractor who provides meals or snacks according to the specifications of the airline. Specialized tray assembly equipment is tailored to the needs of the operation. On airplanes, thermal support is needed for heat processing of menu items and cold support for chilled items. Controls are required to ensure that the proper number of meals are provided and to monitor other items, ranging from dishes and flatware to individual tea bags. As an illustration of the complexity of airline service, the food may be loaded onto a plane at one location and the empty trays unloaded at a location across the world.

SERVICE MANAGEMENT

Service management is a philosophy, a thought process, a set of values and attitudes, and a set of methods, according to Albrecht and Zemke (1985). To transform an organization to a customer-driven one takes time, resources, planning, imagination, and tremendous commitment by management. Mill (1986) stated that many customer complaints in restaurants seem to be the result of overly high expectations. James L. Heskett, a professor at Harvard University's Graduate School of Business Administration, said that complaints are like golden nuggets that give insight to a business (Herlong, 1991). For every complaint heard, about five go unheard. A complainer is basically saying, "Give me a reason to come back."

Boorstin (1992), a professor at the University of Chicago, stated that the challenge that faces the foodservice industry today is not so much food as it is service. He believes that *service* is a word like *quality*, and it means different things to different people at different times. Service should be thought of as an American way of life, and in America each individual holds the power. The word *service* comes from *servile*, the Latin word for *slave*.

Total Quality Service

Albrecht (1992) is concerned that many service organizations are adopting the total quality management (TQM) philosophy, which is based on numbers and work processes with little emphasis on customer value. He suggests that the emphasis be changed from management to service to create a **total quality service (TQS)** model. The emphasis then is to take the long view and focus on the reason the organization exists, which is to serve. The TQS philosophy emphasizes that all quality standards and measures should be customer referenced and should help people guide the organization to deliver outstanding value to its customers. Quality standards should be means to an end but not ends in themselves.

Since World War II, the world has been embracing a quality paradigm. Albrecht (1992) defines a **paradigm** as a mental frame of reference that dominates the way people think and act. Quality has been focused primarily on zero defects in a product and very little on service quality. As the 21st century approaches, the distinction between product and service will become obsolete. They will be replaced by **total customer value**, the combination of the tangible and intangible experienced by customers at the various moments of truth that become their perception of doing business with an organization (Albrecht, 1992). Quality must start with the customer, not with the product or work process that creates it. A **paradigm shift**, defined as a new set of rules by Barker (1992), from a quality to a total customer value paradigm is occurring in organizations.

Eating a meal in a restaurant often is a one-time, special event for a customer, but serving it is a repetitive, mundane occurrence for an employee. Customers cannot forget a bad experience; employees can agree to do better the next time, which is too late. Good service, which satisfies a customer, does not always give total customer value. Exemplary service, however, delights customers by totally exceeding their expectations (Marvin, 1992). Customers keep a mental score and assign a subconscious point value to their experience. The more positive the experience, the higher the score. If the score is higher for one restaurant than for its competitors', the high score operation becomes the restaurant of choice. If a competitor, however, has a higher score, the original restaurant is in trouble. Any foodservice operation and its service are only as good as its staff.

The U.S. Department of Commerce has found that more than 90% of dissatisfied customers will drift over to the competition, but not always silently (Bode, 1993). This customer will tell as many as nine other people about the bad experience. To prevent this from happening, more and more restaurant owners are hiring **mystery shoppers** who not only record their own experiences but also those of other diners. The mystery shopper must be able to turn a narrative into a snapshot so the management can picture what the dining experience was like. Mystery shopping services are available in many cities. A representative visits the restaurant and then develops a mystery shopper evaluation form. The representative hires shoppers to pose as ordinary customers, but they evaluate the restaurant as they dine. An excerpt on guest services from a four-page evaluation form for Showbiz Pizza/Chuck E. Cheese's is shown in Figure 15.7.

One service may have as many as 5,000 shoppers consisting of persons from the age of 18 to 85 who usually have full-time positions elsewhere. These shoppers must complete a vigorous training program, much of which is done with audio-video tapes. In the beginning, they visit the operation a number of times a week until once or twice a week is sufficient. They record the perception of their own dining experi-

Figure 15.7.   Excerpt from a mystery shopper evaluation form.

McBIZ CORPORATION
SHOWBIZ PIZZA/CHUCK E. CHEESE'S
MYSTERY SHOPPER EVALUATION FORM

Shopper Code: _____
Male:_____   Female:_____
Points Possible: 1,000
Points Earned: _____

Location: _____

Date of Visit: _____   Time Entered: _____   Time of Exit: _____   Day: _____

Level of Business: Full: _____   3/4 Full: _____   1/2 Full: _____   1/4 Full: _____

Office use only		
		Please circle "Y" for yes and "N" for no.
		GUEST SERVICES (20%)
		Server's Name:
		Upon delivery of food did the employee:
	Y N 47.	Ask if anything else was needed
	Y N 48.	Deliver plates with pizza
	Y N 49.	Employee checked back to inquire about satisfaction or talk with you while you were seated during your meal
	Y N 50.	Employee offered to take away dirty plates, empty trays, etc. or box leftovers while you were seated
	Y N 51.	Tables were cleared and wiped quickly after other guests left table **(Do not penalize if guests are in other areas of the restaurant)**
	Y N 52.	Manager was visible to you during your visit
	Y N 53.	Manager talked or interacted with guests
	Y N 54.	Manager was professional and well groomed
	Y N 55.	Employees appeared to stay focused on their job responsibilities and the needs of guests
	Y N 56.	Employees were helpful, pleasant, courteous, and friendly to guests
	Y N 57.	Your child had an opportunity to hug Chuck E. Cheese (Live, walk-around costumed/mascot)
	Y N 58.	An employee at any time during your visit mentioned the fan club sign-up program to you
	Y N 59.	All employees were wearing name tags
	Y N 60.	All employees appeared neat, well groomed and in uniform
	Y N 61.	Red or blue shirt
	Y N 62.	Khaki pants or shorts
	Y N 63.	Red or blue visor
	Y N 64.	White tennis shoes

Comments _____
_____

Total Possible: 80   Multiply __× 2.__   Points Scored: _____   Guest Points: _____

Source: Adapted from McBiz Corporation, Topeka, KS. Used by permission.

ence but also monitor and later record experiences of other diners. The mission of the mystery shopper can be divided into three phases: monitoring, motivation, and maintenance. *Monitoring* occurs because employees never know who the shopper is and when she or he is in the restaurant. *Motivation* of employees is much more simple if the continual flow of feedback from the mystery shopper leads to higher salaries for good performance. Finally, *maintenance* of superior customer service is possible with fewer visits from the mystery shopper.

The Staff

Until managers begin to treat employees as if they are customers, service will never reach the exemplary stage. Employees are internal customers and are very important team members. *Empowerment* will continue to be a key buzzword. If it works, it will result in a more efficient and customer-responsive organization (Stephenson & Weinstein, 1992). Many foodservice managers are empowering employees to make decisions that will contribute to total customer value. Although every employee is not comfortable with this philosophy of management, those who are will blossom and do a better job.

Cross-training is another technique being used by foodservice managers to involve employees in the total customer value concept. It usually results in a loyal staff because employees have the opportunity to understand how the foodservice operation works and to find out what is happening in each unit (Weinstein, 1992). Some operations have established a cross-training program in which a front-of-the house employee starts as a buser at a minimum wage and progresses rapidly to a runner and finally a head runner at a higher salary. Then the employee works in the kitchen at yet a higher salary to see how it operates, learns the computer program, and observes how management handles relationships between front- and back-of-the-house employees. Kitchen employees are given the same opportunity in the dining areas. This cross-training can break down barriers between employees in the front- and back-of-the-house, creating a climate that adds to total customer value. For example, cooks, who have been cross-trained, begin to realize that demands of the waitstaff are not personal demands but demands of customers.

In some of the fine dining restaurants patterned after those in Europe, service is considered an honorable profession and a career (Ryan, 1993). All new staff, regardless of experience, must go through an apprenticeship program that might last a year to become a fully qualified waitperson. Other operations often have a rigorous and lengthy training program that uses written tests covering a general knowledge of the restaurant and its foods and wines and essays on hypothetical situations that might happen in the restaurant.

Employees no longer can be told to assimilate into the corporate culture of the organization (Schuster, 1992). Instead managers are asking employees to share their individual culture with them and other employees. Managing cultural diversity can be a challenge for a foodservice manager. According to Boss (1992), managing cultural diversity is about recognizing, accommodating, and even cultivating differences in the workplace. Rewards await managers who can learn about cultural differences from employees. The first is a motivated work force, which causes productivity to increase, which will enhance the foodservice operation's competitive advantage in the market. The second is that the responsiveness to an increasingly diverse work force by the manager will urge employees to be more responsive to customers, who are becoming more diverse. The work force of the future will consist of an increase in women, minorities, and foreign-born immigrants with an even greater variety of backgrounds, values, cultures, and attitudes than it has today (Schuster, 1992). Managers of noncommercial foodservice operations are among the first to face the impact of these demographic changes as well as to manage a multicultural work force. This will occur because prospective employees will come from culturally diverse labor pools to fill the large number of entry-level, less-skilled positions in the industry.

Burnout, emotional exhaustion, is common among many employees in the work force. Satisfaction of employees with their jobs should be thoroughly examined by management because it can have a great effect on the quality of service given to the customer. If good employees begin to be absent or are late more than usual, have lost enthusiasm, or have decreased productivity, something is wrong (Lorenzini, 1992). Perhaps the employee has burnout, which is a form of silent stress. Oftentimes managers take advantage of good employees and overload them with tasks that temporary help could do. The cost for temporary help is much less than the cost of losing a good employee. Employees like to be noticed and should be praised when they do a good job. The manager should treat employees as they want to be treated.

The Special Customer

Customers have been mentioned many times in this text—their demographics, lifestyles, and menu preferences. The importance of pleasing customers has been discussed in every chapter, but what about customers who don't fit the typical pattern but fit into special groups and have to, or choose to, eat away from home. How is the foodservice industry taking care of steady customers, solo diners, customers with small children, and disabled customers? Are restaurants, school and college foodservices, and hospitals treating them the same as the regular customer?

Restaurant customers often are considered transient, especially during vacation times when they stop to eat in a restaurant on the way to their destination. Most restaurants could not survive if customers only visit during vacations or for special events. Restaurateurs are thanking customers for being loyal and steady customers (Reichheld, 1993) and telling them how much their business is appreciated (Fintel, 1991). A chain of restaurants in California rewards its customers who dine regularly at their restaurants with priority reservations by giving them a plastic card with numbers to a special telephone. When that telephone rings, the front-of-the-house employee knows a steady customer is calling for a reservation, usually for the same day. Top priority is given to that customer.

A restaurant in Virginia Beach has created a T.L.C. (The Local Customer) program. Locals qualify as frequent diners by eating at the restaurant 10 times during off-season, which runs from September through February. They become card-bearing regulars, which qualifies them to reserve a table during peak season, when reservations are not accepted. They also receive bonuses including a complimentary special, such as an appetizer or a glass of wine. The secret of getting customers to return is to make them feel special.

Traditionally, restaurateurs have welcomed groups of two or more customers and shunned solo diners, who typically avoid dining at fine restaurants because they feel unwelcome (Stern, 1990). Many times they eat at limited-menu and other lower-priced operations because they are embarrassed by having the maître d', usually in a loud voice, ask if they are alone. Several demographic trends, however, speak to the need to recognize single diners as important customers (Prewitt, 1992). Among them are the increasing assertiveness and financial clout of working women, increased business travel, and the rapid growth of affluent senior citizens and retirees. Staff in many operations are being trained to be more sensitive to solo diners. Dining room designers have come up with ideas to make these customers comfortable. Booths are desired by many of these customers, who usually come prepared with a book or magazine and appreciate more space. Another design idea that gives flexibility to single diners is a banquette, which is an upholstered sofa that runs along a wall and forms a long seating space. Small tables for one to four customers are placed in front of the banquette. Tables for one are no longer being placed in corners or outside the kitchen door, but are being distributed throughout the area. Single-dining areas that cater to travelers and local singles are being incorporated into some main dining rooms. Also, place settings often are laid out at the bar, especially at lunchtime, for business people dining alone.

A communal table to accommodate singles who enjoy the fellowship of other solo diners has been added to dining areas in some restaurants. This has to be carefully considered because singles often are not extremely social and would prefer to be left alone with their books or newspapers. Why should restaurants bother with singles? Studies by the National Restaurant Association revealed that single diners, often professionals in their 20s and 30s, spend more than half of their food budget eating out; the number of solo diners has increased to about 20 million people.

Many restaurants now have children's menus and games to keep children busy while parents eat (Crosby, 1992). Some foodservice operations are even offering child care to customers. The greatest increase in birthrates since the baby-boom ended in 1964 occurred in the 80s and 90s. Because finding babysitters is a challenge to parents, foodservice managers are trying to solve the problem by providing child care within or near their facilities.

At the A Piece of Quiet restaurant in Denver, child care has been placed on its menu. The Kids Cafe, in a separate room with a one-way soundproof window between the cafe and the restaurant, allows parents to watch their children. The child-to-staff ratio is five to one. Staff members who work with children are trained in CPR, the Heimlich Maneuver, and dealing with parents. Because parents and children are both on the premises, the restaurant does not need a Colorado child-care license. The cost is $5 per child and includes a meal for the child.

The Rockwells chain in New York has converted empty dining room space on slow nights into a supervised play area for children. This service is free to customers unless the child eats. Then parents are billed for the food. Play area supervisors are good babysitters and not certified child-care professionals because parents and children can see each other at all times. The Bistro Banlieue in Illinois opted to use a nearby Kinder-Care facility for supervising children off-site. Parents drop their children off at the center on Monday, Tuesday, and Wednesday evenings, and the restaurant picks up the tab.

McDonald's, after testing its indoor playground concept for children, is currently opening 25 to 30 units of Leaps & Bounds in 100-seat foodservice facilities across the country (Keegan, 1993). Counselors go through a rigid training program, and the safety and security of children are strictly monitored. Entry fees for children range from $4.95 to $5.95 depending on location; adults are free. Food is not allowed in the play area but can be purchased from the "Great Eats" snack bar that serves "kid friendly" food to be eaten at tables outside the play area.

People with disabilities often were not able to eat away from home because restaurants and transportation were not accessible to them, especially if they were in wheelchairs. Hearing and visual impairments also are considered disabilities. With the passage of the Americans with Disabilities Act (ADA) on January 26, 1992, public accommodation rules took effect. On that day, a sign was put on the door of each of the 4,000 company-owned Pizza Huts announcing to customers and the public that "Pizza Hut is committed to implementing the American with Disabilities Act and pledges to take all reasonable steps to accommodate our customers and employees" (Weinstein, 1992). One week before the act took effect, a conference was held for regional managers to explain the company's commitment to removing barriers and changing service practices that prevent disabled customers from enjoying all the services. Time limits for improvement were established, and employees were trained to be sensitive to the special needs of disabled customers.

Front door, aisle, table, self-service, and restroom access must be available to disabled customers. ADA rules apply not only to restaurants but also to all noncommercial foodservice operations. Hospital and nursing home foodservice managers have had more experience in feeding disabled persons than have any other managers, but they still have to comply with the new rules.

Managing Service

Romm (1989) stated that the business of the restaurant takes place in the social space created between the guest and the server. Improving the consistency and quality of the service staff and treatment of customers can be a problem facing the foodservice industry. Managers would like to generate friendly behavior that will be perceived as authentic between the waitstaff and the customer. This process requires two steps. First, more attention needs to be focused on service employees, and their place in the business structure should be reexamined; and second, techniques being used to change employees' behavior need to be examined. Service employees are expected to nurture and entertain customers, but management seldom nurtures employees. The most common response to a need for increasing profits is to reduce payroll expenses by reducing labor hours; yet the employee is expected to deliver good service to the customer. Wallace (1988) commented that the norm today for businesses is to provide more physical facilities for service rather than flesh-and-

Figure 15.8.   The service profit chain.

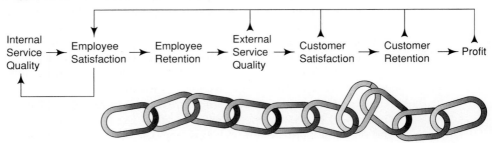

Source: Reprinted by permission of *Harvard Business Review.* An exhibit from "Customer Satisfaction Is Rooted in Employee Satisfaction," by Leonard A. Schlesinger and James L. Heskett, in "How Does Service Drive the Service Company?" by Michael R. Quinlan (November-December 1991). Copyright © 1991 by the President and Fellows of Harvard College; all rights reserved.

blood servers to make them operative. Therein lies one reason for the public's growing discontent with American service.

If a major goal of a foodservice operation is to provide excellent service to the customer, management styles will need to change. The foodservice industry has a dual problem—customers and staff. A new way of looking at the business must be developed that will change the distribution of resources so they will better support the providing of a crafted guest-customer experience and the recognition of the emotional and social components in the work of the staff (Romm, 1989).

Ideas have to change and managers need to develop what Heskett (1987) called a *new service concept,* which is the way an organization would like to have its service perceived by its customers and employees. He stated that executives who have had careers in both service and manufacturing agree that reaching a high quality level is more difficult in services.

Schlesinger and Heskett (cited in "How Does Service," 1991) developed a model that emphasizes a **service profit chain** with a goal of profit rather than customer satisfaction. Profit is most closely linked to customer retention that results from customer satisfaction, as shown in Figure 15.8. Value, service quality, and costs are driven in turn by employee retention, employee satisfaction, and the quality of the internal service support that employees use to help customers. Employee satisfaction is especially high in service organizations that not only give high value to customers but also select service employees carefully, train them well, and empower them to solve customer problems. In this model, managers are coaches rather than supervisors, and spans of management control can be increased to 20, sometimes even 40, employees to 1 manager.

SUMMARY

Distribution and service is the third major subsystem in the foodservice system, and with procurement and production, it meets the goal of providing quality food and services that result in customer satisfaction. Distribution of food to the customer can be complex, as in hospitals, or simple, as in limited-menu restaurants.

Distribution and service differs in the four foodservice types. In the conventional foodservice, holding equipment is needed to maintain the proper temperature of food between production and service. Tray service, generally in healthcare organizations, requires complex distribution and service because food must be delivered to people in many different areas.

Commissary foodservice is characterized by centralized production with distribution of the food to several remote areas designated as satellite service centers. In ready prepared foodservices, menu items are produced and held, either frozen or chilled, for service at a later time. The assembly/serve foodservice uses foods that are ready to serve or that require little or no processing before service.

Table and counter, self-, and tray service are the three most common categories of service in the United States. Tipping waitstaff is a common practice in table and counter service restaurants in the United States; federal and state legislation have provisions covering tipping. Handling and controlling guest checks is a concern of managers in an operation utilizing waitstaff.

Self-service operations include cafeterias, buffets, and vending service. Vending is becoming more sophisticated each year because of advances in technology and a service mandate. Tray service, in which food is carried to a person by a foodservice employee, is used primarily in healthcare institutions and for in-flight meal service in airplanes.

Service management is a philosophy, a thought process, and a set of values and attitudes. Changing an organization to one that is customer driven takes time and requires a commitment from management. Total quality service is a paradigm shift from product and service quality to a total customer value quality. Employees must be treated as customers and empowered to make decisions that will contribute to total customer value. Special customers such as steady customers, solo diners, those with small children, and disabled customers have special needs that need to be recognized by foodservice managers. A model has been developed that links profit to customer retention resulting from customer satisfaction.

BIBLIOGRAPHY

Albrecht, K. (1988). *At America's service*. Homewood, IL: Dow Jones-Irwin.

Albrecht, K. (1992). *The only thing that matters*. New York: HarperCollins.

Albrecht, K., & Zemke, R. (1985). *Service America! Doing business in the new economy*. Homewood, IL: Dow Jones-Irwin.

Anticipating paradigm shifts. (1993). *Food Management, 28*(1), 92.

Barker, J. A. (1992). *Future edge: Discovering the new paradigms of success*. New York: William Morrow.

Beasley, M. A. (1989). *Opportunities in vending*. Chicago: American Society for Hospital Foodservice Administrators of the American Hospital Association.

Beasley, M. A. (1990). Ways vending can work for you. *Food Management, 25*(8), 42, 46.

Beasley, M. A. (1993). Vending: New paths to success. *Food Management, 28*(3), 32.

Bertagnoli, L. (1989). Credit cards spark chain reaction. *Restaurants & Institutions, 99*(16), 44–45, 50, 56, 58.

Bode, D. (1993). The secret service of professional mystery shoppers. *Nation's Restaurant News, 27*(35), 66.

Boorstin, D. J. (1992). Service in America. *Restaurants & Institutions, 102*(11), 70, 74, 82.

Boss, D. (1992). A diversity of individuals. *Food Management, 27*(9), 16.

Brault, D. (1993). Customer time at table unchanged. *Restaurants USA, 13*(3), 37–38.

Crosby, M. A. (1992). Bouillabaisse, burgers, and babysitting. *Restaurants USA, 12*(6), 13–16.

Fintel, J. (1991). How to keep them coming back. *Restaurants USA, 11*(5), 14–15.

Follin, M. (1991). Two for the road: Double drive-thru restaurants. *Restaurants USA, 11*(3), 26–28.

Frinstahl, T. W. (1989). My employees are my service guarantee. *Harvard Business Review, 67*(4), 28–32.

Gotschall, B. (1989). Catering prevails when funding fails. *Restaurants & Institutions, 99*(14), 148–152.

Herlong, J. E. (1991). There is no such thing as a complaint. *Restaurants USA, 11*(5), 16–17.

Heskett, J. L. (1986). *Managing in the service economy*. Boston: Harvard Business School Press.

Heskett, J. L. (1987). Lessons in the service sector. *Harvard Business Review, 65*(2), 118–126.

How does service drive the service company? (1991). *Harvard Business Review, 69*(6), 146–150, 154, 156–158.

Hysen, P., Boehrer, J., Greenberg, A., Noseworthy, E., Prentkowski, D., Wilson, T., & Boss, D. (1993). Anticipating paradigm shifts in foodservice. *Food Management, 28*(1), 100–103, 106–108.

Hysen, P., & Harrison, J. (1982). State-of-the art review of health care patient feeding system equipment. In *Hospital patient feeding systems* (pp. 159–192). Washington, DC: National Academy Press.

IRS says 75% of restaurateurs covered by tip-reporting rules are not filing required forms. (1993). *Washington Weekly, 13*(31), 3.

Iwamuro, R. (1992). Carryout and delivery from tableservice restaurants. *Restaurants USA, 12*(10), 48–51.

Keegan, P. O. (1993). McD's playground concept grows by 'Leaps & Bounds.' *Nation's Restaurant News, 27*(11), 3, 65.

Lorenzini, B. (1992). Fire up burnt-out employees. *Restaurants & Institutions, 102*(7), 27–28.

Martin, W. B. (1986a). Defining what quality service is for you. *Cornell Hotel and Restaurant Administration Quarterly, 26*(4), 32–37.

Martin, W. B. (1986b). *Quality service: The restaurant manager's bible*. Ithaca, NY: Cornell University, School of Hotel Administration.

Marvin, B. (1992). Toward exemplary service. *Restaurants & Institutions, 102*(11), 86–87, 93, 95, 97.

Mill, R. C. (1986). Managing the service encounter. *Cornell Hotel and Restaurant Administration Quarterly, 26*(4), 39–46.

Miller, R. P. (1984). Vending: Contract or self-operated. In J. C. Rose (Ed.), *Handbook for health care food service management* (pp. 265–270). Rockville, MD: Aspen Publishers.

National Restaurant Association. (1988). *Waitstaff Compensation: Tips vs. Service Charges*. Current Issues Report. Washington, DC: Author.

National Restaurant Association and National Center for Access Unlimited. (1992). *Americans with Disabilities Act: Answers for foodservice operators*. Washington, DC: Authors.

Pickworth, J. R. (1988). Service delivery systems in the food service industry. *International Journal of Hospitality Management, 7*(1), 43–61.

Prewitt, M. (1992). Wanted: Singles. *Nation's Restaurant News, 26*(50), 27–28.

Reichheld, F. F. (1993). Loyalty-based management. *Harvard Business Review, 71*(2), 64–73.

Romm, D. (1989). "Restauration" theater: Giving direction to service. *Cornell Hotel and Restaurant Administration Quarterly, 29*(4), 31–39.

Ryan, N. R. (1991). Food to go. *Restaurants & Institutions, 101*(6), 84–85, 88, 92, 98.

Ryan, N. R. (1993). They give service with a style. *Restaurants & Institutions, 103*(13), 71, 74, 76.

Schuster, K. (1992). Managing cultural diversity. *Food Management, 27*(9), 118–119, 122–123, 128–129.

Shapiro, B. P., Rangan, V. K., & Sviokla, J. J. (1992). Staple yourself to an order. *Harvard Business Review, 70*(4), 113–122.

Sherer, M. (1993). Anticipating paradigm shifts in institutions. *Food Management, 28*(1), 93–96.

Stephenson, S., & Weinstein, J. (1992). Training feeds fighting spirit. *Restaurants & Institutions, 102*(1), 80–81, 84, 88.

Stern, G. M. (1990). Table for one. *Restaurants USA, 10*(3), 14–16.

Stern, G. M. (1991). Service in the 1990s: The way to grow your business. *Restaurants USA, 11*(5), 11–13.

Stokes, J. W. (1982). *How to manage a restaurant or institutional food service* (4th ed.). Dubuque, IA: Wm. C. Brown.

Wallace, J. (1988). Viewpoint. *Restaurants & Institutions, 98*(23), 14.

Weinstein, J. (1992). The accessible restaurant. *Restaurants & Institutions, 102*(9), 96–98, 102, 104, 110, 117.

16

Sanitation and Maintenance of Equipment and Facilities

Sanitation and maintenance is the last major functional subsystem in the foodservice system, and it permeates all other subsystems (Figure 16.1). The receiving area needs to be checked after each delivery and thoroughly cleaned daily. Storage areas also should be put on a regular cleaning schedule to prevent vermin and rodent infestation. Foodservice utensils, dishes, equipment, and facilities require continuous cleaning. Every time a meal is produced and served, cleaning and sanitizing must be done again. Today, cleaning is complicated. Employees have to know what cleaning agent works on what type of soil and whether or not a brush, mop, sponge, scouring pad, or rag should be used (Patterson, 1993).

The two major components of a sanitation program are the foodservice operation and equipment and facilities maintenance. Equipment and facilities are important factors in any HACCP program because they make sanitation procedures more effective (Educational Foundation of the National Restaurant Association [NRA], 1992).

Figure 16.1. Foodservice systems model with the maintenance subsystem highlighted.

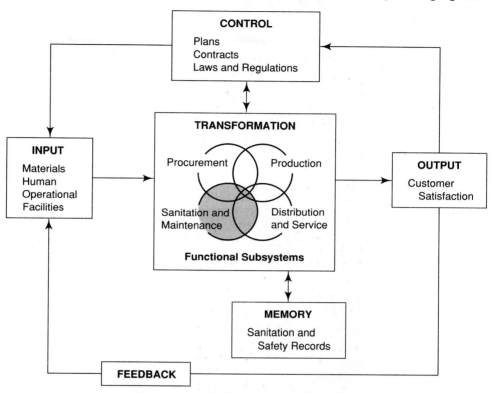

Cleanliness of foodservice facilities ranks high among the concerns of customers and often influences their decision to return to a restaurant. Maintenance of equipment and facilities is important for reasons other than sanitation too. The safety of surroundings is often related to cleaning and maintenance practices. Two examples are spills that are not cleaned up properly, which may cause people to fall, and grease buildup in the hoods over the production equipment, which is a major cause of fires in foodservice operations.

SANITATION

A properly designed foodservice facility is basic for maintenance of a high standard of sanitation. The first requirement for a sanitary design is cleanability, which means the facility has been arranged so that it can easily be cleaned. Equipment and fixtures should be arranged and designed to comply with sanitation standards, and trash and garbage isolated to avoid contaminating food and attracting pests.

For a facility to be clean is not enough; it must also be sanitary. Although the two words are often used synonymously, **clean** means free of physical soil and with an outwardly pleasing appearance—a glass that sparkles, silver that shines, and a floor that is free from dust and grime. These objects may look clean on the surface but may harbor disease agents or harmful chemicals. **Sanitary** means free of disease-causing organisms and other contaminants. Cleaning and sanitizing are both issues of concern in the maintenance of foodservice facilities and equipment, and together they form the basis for good housekeeping in foodservice operations. As stated cogently in the National Restaurant Association's (1979) *Sanitation Operations Manual*, the foodservice manager should operate an establishment by this rule: "Look clean—Be sanitary."

Sanitization

Sanitization is critical for any surface that comes in contact with food, which includes, of course, all dishes, utensils, pots, and pans. Sanitizing is accomplished by heat or a chemical sanitizing compound. In either case, the object must be thoroughly clean and completely rinsed in order for the sanitizing process to work. Caked-on soils not removed by cleaning, for example, may harbor bacteria even after the use of a sanitizing solution.

Exposing a clean object to sufficiently high heat for an adequate length of time will sanitize it. In general, the higher the heat, the shorter the time required. The most common method of heat sanitizing is immersion of an object in water at 170°F for no less than 30 seconds. The temperature must be increased if the time is decreased; the final sanitizing rinse in dishwashing machines, for example, is 180°F for 10 seconds. Another means of heat sanitization is through the use of live, additive-free steam, from water at a temperature of 212°F. The temperature at the surface of the object being sanitized is the critical one, not the temperature as the water or steam leaves its source.

The other method, using chemical compounds capable of destroying disease-causing bacteria, is widely used in the foodservice industry. These chemicals are rigorously regulated by the Environmental Protection Agency (EPA) to assure public safety. Chemical sanitizing is accomplished either by immersing an object in the correct concentration of sanitizer for 1 minute or by rinsing, swabbing, or spraying twice the usual recommended concentration of sanitizer on its surface. Sanitizers are often blended with detergent so the same product can be used for cleaning and sanitizing. Many detergents and sanitizers have been specifically designed for various cleaning and sanitizing tasks within a foodservice operation. Excellent resource materials are available from chemical companies, the major suppliers of these detergents and agents.

Some sanitizing agents are toxic to humans as well as to bacteria and are, therefore, acceptable for use only on nonfood contact surfaces. Other agents may not be toxic but may have undesirable flavors or odors, which make them unacceptable for use in foodservice operations. The three most common chemicals used in sanitizing are chlorine, iodine, and quaternary ammonia. Table 16.1 summarizes the properties of these sanitizers and outlines procedures for their use.

Table 16.1.   Chemical sanitizing agents

	Chlorine	**Iodine**	**Quaternary Ammonium**
Minimum concentration			
—For immersion	50 parts per million (ppm)	12.5 ppm	200 ppm
—For power spray or cleaning in place	50 ppm	12.5 ppm	200 ppm
Temperature of solution	75°F/23.9°C+	75°–120°F/ 23.9°–48.9°C Iodine will leave solution at 120°F/48.9°C	75°F/23.9°C+
Time for sanitizing			
—For immersion	1 minute	1 minute	1 minute; however, some products require longer contact time; read label
—For power spray or cleaning in place	Follow manufacturer's instructions	Follow manufacturer's instructions	
pH (detergent residue raises pH of solution so rinse thoroughly first)	Must be below 10	Must be below 5.5	Most effective around 7 but varies with compound
Corrosiveness	Corrosive to some substances	Noncorrosive	Noncorrosive
Response to organic contaminants in water	Quickly inactivated	Made less effective	Not easily affected
Response to hard water	Not affected	Not affected	Some compounds inactivated but varies with formulation; read label. Hardness over 500 ppm is undesirable for some quats
Indication of strength of solution	Test kit required	Amber color indicates effective solution, but test kits must also be used	Test kit required. Follow label instructions closely

Source: From *Applied Foodservice Sanitation,* Fourth Edition (p. 195), by The Educational Foundation of the National Restaurant Association, 1992, Chicago: Author. Used by permission.

Warewashing

Warewashing is the process of washing and sanitizing dishes, glassware, flatware, and pots and pans either manually or mechanically. Sinks, dishmachines, and pot and pan washing machines are the most common equipment for this process. Specialized equipment, such as flatware washers and glassware washers, also are available.

Manual

Manual warewashing is used primarily for pots and pans but may be used for dishes in small facilities such as day-care centers or nursing homes. Figure 16.2 illustrates a typical setup of a manual dishwashing operation. A scraping and prerinse area with a three-compartment sink is needed, one for washing, the second for rinsing, and the third for sanitizing. Counter space must be provided next to the sanitizing sink to permit dishes and utensils to air dry; health regulations prohibit towel drying because of the potential sanitation hazard. Application of effective sanitizing procedures destroys disease organisms present on equipment and utensils after cleaning and prevents transfer by toweling to consumers. In most operations, a garbage disposal unit is installed in the scraping and prerinsing area.

The procedure outlined in Figure 16.3 delineates the five steps involved in manual dishwashing. Although heat sanitizing is specified in both the illustration and the procedure, a chemical sanitizer may be used. Dishes and utensils must stand in a chemical sanitizer for at least 1 minute; generally, the sanitizing solution is mixed to twice the recommended strength because water carried over from the rinse tank dilutes the effectiveness of the sanitizer.

The facilities and procedure for washing pots and pans manually are similar to those for manual dishwashing. In most operations, these two processes are carried

Figure 16.2.   Setup of three-compartment sink for manual dishwashing.

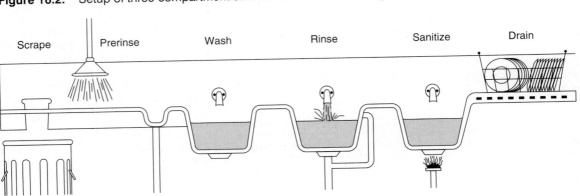

Scrape   Prerinse   Wash   Rinse   Sanitize   Drain

Source: Professional Cooking (p. 17) by W. Gisslen, Copyright © 1989 by John Wiley & Sons, Inc. Reprinted by permission of John Wiley & Sons, Inc.

Figure 16.3.   Procedure for manual dishwashing.

1. *Scrape and prerinse.* The purpose of this step is to keep the wash water cleaner longer.
2. *Wash.* Use warm water at 110° to 120°F (43° to 49°C) and a good detergent. Scrub well with a brush to remove all traces of soil and grease.
3. *Rinse.* Use clean, warm water to rinse off detergent residues. Change the water frequently, or use running water with an overflow, as in Figure 16.2.
4. *Sanitize.* Place utensils in a rack and immerse in hot water at *170°F (77°C) for 30 seconds.* (A gas or electric heating element is needed to hold water at this temperature.)
5. *Drain and air dry.* Do not towel dry. This may recontaminate utensils. Do not touch food contact surfaces of sanitized dishes, glasses, and silverware.

Source: Adapted from *Professional Cooking* (p. 17) by W. Gisslen, Copyright © 1989 by John Wiley & Sons, Inc. Reprinted by permission of John Wiley & Sons, Inc.

out in different areas of the foodservice operation. Separate sink setups are provided in each area, with the pot and pan area usually located adjacent to production areas and the dishwashing facilities in close proximity to the dining room.

Mechanical

The dishwashing operation must include areas for accumulating soiled tableware from dining areas; in a hospital, space must be sufficient for carts holding used trays from patient floors. Space is also needed for the following: for removal of dishes and flatware from baskets or a conveyor; for scraping, prerinsing, and stacking before washing; for stacking and storing dishes and utensils; and for the dishwashing machines. Ideally, dishes should be routed from the dining area in a manner that creates the least noise, confusion, and unsightliness.

Dishwashing machines are generally classified by the number of tanks. A single-tank, stationary-rack machine with doors has the smallest capacity and is used most often in small institutions. It holds a rack of dishes that does not move. Dishes are washed by a detergent and water from below, and sometimes from above, the rack. The wash cycle is followed by a hot water rinse cycle.

Rack conveyor machines with two or three tanks, such as the one in Figure 16.4, are used in larger operations that have greater loads of dishes to be washed. After dishes are scraped and sorted, they are placed in racks designed for plates, cups, or glasses. An overhead hand-operated sprayer is generally used to rinse dishes with warm water. Racks are then pushed onto the conveyor, which moves the racks through the machine at a uniform preset speed. These machines usually have a power wash cycle, a power rinse cycle, and a final rinse cycle. Larger machines may have a prewash tank. The final sanitizing rinse must be 10 seconds in duration and not less than 180°F. The primary purposes for the final rinse are to heat-sanitize the eating utensils and to ensure that detergent is completely rinsed off.

Figure 16.4. Rack conveyor dishwashing machine.

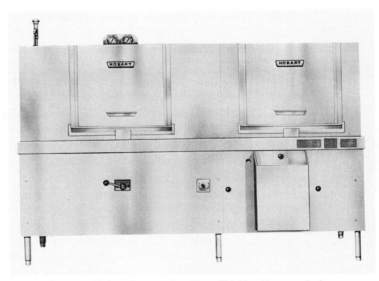

Source: Courtesy Hobart Corporation, Troy, OH. Used by permission.

The flight-type, or rackless, machines have become especially popular in very high-volume operations (Figure 16.5). With these machines, plates and trays are placed directly on a continuous peg or pin conveyor; smaller items such as flatware, glasses, and cups are racked before they are sent through the machine.

Specialized machines are available for washing glass and silverware because of special problems in ensuring their cleanliness and sanitization. Glasswashers are com-

Figure 16.5. Flight-type dishwashing machine.

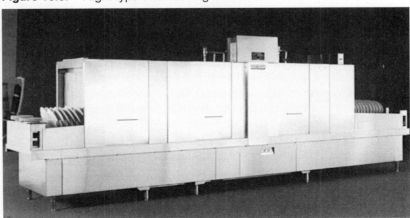

Source: Courtesy Hobart Corporation, Troy, OH. Used by permission.

mon in establishments where large numbers of glasses are washed daily; in bars and lounges, for example, small under-the-counter glasswashers are frequently used.

For sanitizing dishes and utensils, it is critical to ensure that the final rinse reaches the required 180°F but no more than 195°F. At higher temperatures, the water will vaporize before it sanitizes the dishes. General purpose hot water heaters, as a rule, bring the temperature to no higher than 140°F. A booster water heater with sufficient capacity to supply the final rinse water at the required temperature will usually be needed; most dishwashing machines are equipped with booster heaters.

Although dishwashing machines are the most reliable way to clean and sanitize dishes and utensils, many problems can occur if machines are not installed or operated correctly. Table 16.2 enumerates some of the typical problems that occur, along with possible causes and suggested solutions to the problems.

Table 16.2. Dishwashing problems and cures

Symptom	Possible Cause	Suggested Cure
Soiled Dishes	Insufficient detergent	Use enough detergent in wash water to ensure complete soil removal and suspension.
	Wash water temperature too low	Keep water temperature within recommended ranges to dissolve food residues and to facilitate heat accumulation (for sanitation).
	Inadequate wash and rinse times	Allow sufficient time for wash and rinse operations to be effective. (Time should be automatically controlled by timer or by conveyor speed.)
	Improperly cleaned equipment	Unclog rinse and wash nozzles to maintain proper pressure-spray pattern and flow conditions. Overflow must be open. Keep wash water as clean as possible by prescraping dishes, etc. *Change water in tanks at proper intervals.*
	Improper racking	Check to make sure racking or placement is done according to size and type. Silverware should always be presoaked, placed in silver holders without sorting. Avoid masking or shielding.
Films	Water hardness	Use an external softening process. Use proper detergent to provide internal conditioning. Check temperature of wash and rinse water. Water maintained above recommended temperature ranges may precipitate film.
	Detergent carryover	Maintain adequate pressure and volume of rinse water, or worn wash jets or improper angle of wash spray might cause wash solution to splash over into final rinse spray.
	Improperly cleaned or rinsed equipment	Prevent scale buildup in equipment by adopting frequent and adequate cleaning practices. Maintain adequate pressure and volume of water.
Greasy Films	Low pH. Insufficient detergent. Low water temperature. Improperly cleaned equipment	Maintain adequate alkalinity to saponify greases; check detergent, water temperature. Unclog all wash and rinse nozzles to provide proper spray action. Clogged rinse nozzles may also interfere with wash tank overflow. Change water in tanks at proper intervals.

Table 16.2.   *continued*

Symptom	Possible Cause	Suggested Cure
Streaking	Alkalinity in the water. Highly dissolved solids in water	Use an external treatment method to reduce alkalinity. Within reason (up to 300–400 ppm), selection of proper rinse additive will eliminate streaking. Above this range external treatment is required to reduce solids.
	Improperly cleaned or rinsed equipment	Maintain adequate pressure and volume of rinse water. Alkaline cleaners used for washing must be thoroughly rinsed from dishes.
Spotting	Rinse water hardness	Provide external or internal softening. Use additional rinse additive.
	Rinse water temperature too high or too low	Check rinse water temperature. Dishes may be flash drying, or water may be drying on dishes rather than draining off.
	Inadequate time between rinsing and storage	Allow sufficient time for air drying.
Foaming	Detergent: dissolved or suspended solids in water	Change to a low sudsing product. Use an appropriate treatment method to reduce the solid content of the water.
	Food soil	Adequately remove gross soil before washing. The decomposition of carbohydrates, protein, or fats may cause foaming during the wash cycle. Change water in tanks at proper intervals.
Coffee, tea, metal staining	Improper detergent	Food dye or metal stains, particularly where plastic dishware is used, normally requires a chlorinated machine washing detergent for proper destaining.
	Improperly cleaned equipment	Keep all wash sprays and rinse nozzles open. Keep equipment free from deposits of films or materials which could cause foam buildup in future wash cycles.

Source: Recommended Field Evaluation Procedures for Spray-Type Dishwashing Machines, 1982, Ann Arbor: National Sanitation Foundation. Used by permission.

Machines should be equipped with dispensers for detergents and chemical rinses to ensure that the proper amounts are used during the dishwashing cycle. Rinse aids are recommended to speed drying time and permit dishes to be stored more quickly and without moisture.

In some foodservices, high-temperature dishwashing machines have been replaced with low-temperature models that operate at between 120°F and 140°F. In these machines, sanitizing is accomplished by chemicals, not heat. The recommended sanitizing chemical should be used in the proper concentration, and it must be dispensed automatically into the final rinse water. The resulting energy and water savings have led to reduced operational costs using these low-temperature machines. Among their disadvantages are the lower number of dishes per hour and longer drying time than high-temperature machines. In addition, chemical costs may offset a portion of energy savings if the water hardness is excessive.

Many smaller pots and pans can be washed in a dishwashing machine. The scraping and soaking required for burned-on food particles are usually done at the pot and pan sink close to the production areas. A common procedure is to transport pots and pans that have been prerinsed to the dishroom for washing, after the bulk of the dishwashing has been completed.

In large-volume operations, specialized pot and pan washing machines, such as the one shown in Figure 16.6, are being used more frequently to assist in this labor-intensive task. The machines being designed today for use in foodservice operations are heavy duty and capable of cleaning cooked-on foods off of pots and pans. Pot-washing is quite different from dishwashing because it requires pressurized hot water sprayed directly on the soiled surface. Commissary foodservices are installing conveyor-type washers to handle the large volume of pots and pans in these operations. As a rule of thumb, an operation serving 1,000 meals or more per day can justify the investment in a potwashing machine.

FACILITY DESIGN

The physical plant of a foodservice operation includes all areas in which food is received, stored, prepared, and served; dishes, utensils, and equipment are washed and stored; and garbage and trash are assembled for subsequent disposal. Locker rooms for employees and restrooms for employees and customers, which may be considered part of the physical plant, frequently are overlooked by management, but they should not be. Sanitation and maintenance of all these facilities and equipment require constant diligence on the part of the foodservice staff and management. The management of a foodservice operation must exercise constant supervision because cleaning and maintenance tasks generally are considered unpleasant but they are extremely important in maintaining standards of sanitation. Regularly scheduled training programs on proper cleaning procedures should be established.

Kitchen and Dining Areas

Design for sanitation must begin when the facility is being planned. As discussed in chapter 8, floors, walls, and ceilings must be constructed for easy maintenance, and the arrangement and design of the equipment and fixtures should facilitate cleaning. Facilities for proper disposal of trash and garbage are necessary to avoid contaminating food and attracting pests.

Covering materials for floors, walls, and ceilings should be selected for ease of cleaning and maintenance as well as for appearance. In back of the house areas, covering materials are generally very sturdy and are selected primarily on the basis of cleanability and durability. For the front of the house, appearance will generally have greater priority, and as a result, dining areas may present some unique cleaning problems. Customers are sensitive to dusty drapes, streaked or grimy windows, smudged or cracked walls, cobwebbed ceilings, loose floor tiles, and a heavily soiled carpet. Dining room floors need continuous care and in many establishments need to be swept, vacuumed, or mopped when customers have departed after a meal.

Figure 16.6.   Pot, pan, and utensil washer-sanitizer.

Source: Alvey Washing Equipment, Cincinnati, OH. Used by permission.

Once-a-day cleaning by the night porter is generally not adequate for heavy-use dining areas.

A variety of flooring materials is used in construction of foodservice facilities, and each type presents a different cleaning challenge. The material description, advantages, and disadvantages are outlined in Table 16.3. Wood floors and carpeting, often used in dining rooms, are not acceptable in production areas because of problems with cleaning and sanitation. In general, the following procedures should be followed regularly in maintenance of floors:

- Spills should be wiped up promptly to avoid tracking and to eliminate a safety hazard.
- Regular schedules for cleaning floors should be established. Floors subjected to heavy traffic and food spills, such as in the production areas, must be scrubbed daily and hosed and steamed periodically for more thorough cleaning.
- Floor care equipment, including brooms, mops, and vacuums, should be cleaned regularly.

Dish and Equipment Storage

Handling and storage of clean dishes are important aspects of a sanitation program in a foodservice operation. All dishes and utensils must be stored dry and in clean, dust-free areas above the floor and protected from dust, mop splashes, and other forms of contamination. Mobile equipment designed for storage of various types of dishes and glassware is ideal. Dishes and glasses can be placed directly into the storage dolly and moved readily to an area where service or tray setup occurs. Some examples of mobile storage dollies are shown in Figure 16.7. Pots and pans should be stored in stationary or mobile racks with removable shelves. Equipment also is available for washing and sanitizing mobile racks, carts, and shelves.

Garbage and Trash Disposal

Garbage and trash must be handled carefully in a foodservice operation because of the potential for contaminating food, equipment, and utensils and for attracting insects and other pests. The manager needs to establish procedures for handling garbage and trash within the operation and then disposing of the solid waste into the environment.

Procedures for Handling

The following general rules apply to trash and garbage handling in the foodservice operation:

- Garbage and trash containers must be leakproof, easily cleanable, pestproof, and durable with tight-fitting lids. Today, plastic bags are frequently used for lining containers to facilitate disposal.
- Garbage and trash should not be allowed to accumulate anywhere but in containers.

Table 16.3   Description, advantages, and disadvantages of common flooring materials

Material	Description	Advantages	Disadvantages
Asphalt	Mixture of asbestos, lime rock, fillers, and pigments, with an asphalt or resin binder	Resilient inexpensive, resistant to water and acids	Buckles under heavy weight; does not wear well when exposed to grease or soap
Carpeting	Fabric covering	Resilient, absorbs sound, shock; good appearance	Not to be used in food preparation areas. Sometimes problems with maintenance elsewhere
Ceramic tiles	Clay mixed with water and fire	Nonresilient; useful for walls; nonabsorbent	Too slippery for use on floors
Concrete	Mixture of portland cement, sand, and gravel	Nonresilient; inexpensive	Porous; not recommended for use in food preparation areas
Linoleum	Mixture of linseed oil, resins, and cork pressed on burlap	Resilient; nonabsorbent	Cannot withstand concentrated weight
Marble	Natural, polished stone	Nonresilient; nonabsorbent; good appearance	Expensive; slippery
Plastic	Synthetics with epoxy resins, polyester, polyurethane, and silicone	Most resilient; nonabsorbent	Should not be exposed to solvents or alkalies
Rubber	Rubber, possibly with asbestos fibers. Comes in rolls, sheets, and tiles	Antislip; resilient	Affected by oil, solvents, strong soaps, and alkalies
Quarry tiles	Natural stone	Nonresilient; nonabsorbent	Slippery when wet, unless an abrasive is added
Terrazzo	Mixture of marble chips and portland cement	Nonresilient; nonabsorbent. If sealed properly, good appearance	Slippery when wet
Commercial-grade vinyl tiles	Compound of resins and filler and stabilized	Resilient; resistant to water, grease, oil	Water seepage between tiles can lift them, making floor dangerous and providing crevices for soil and pests
Wood	Maple, oak	Absorbent; sometimes inexpensive; good appearance	Provides pockets for dust, insects. Unacceptable for use in food preparation areas. Sealed wood may be used in serving areas

Source: From *Applied Foodservice Sanitation,* Fourth Edition (pp. 161–162), by The Educational Foundation of the National Restaurant Association, 1992, Chicago: Author. Used by permission.

Figure 16.7.   Mobile storage dollies.

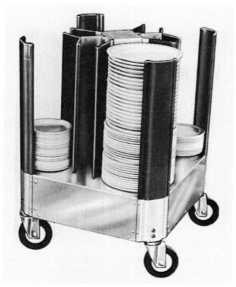

Dish dolly

Tray dolly

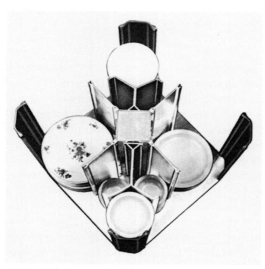

Top view of dish dolly

Cup and glass dolly

Source: Cres-Cor Crown-X, Cleveland, OH. Used by permission.

- Garbage and trash should be removed from production areas on a frequent basis for appropriate disposal.
- Garbage storage areas should be easily cleanable and pestproof. If long holding times for garbage are required, these areas should be refrigerated to prevent decomposition, odor, and infestation by vermin.
- A garbage can washing area with hot water and a floor drain should be located away from food production and storage areas.

Mechanical devices are used in most foodservice facilities to assist in garbage and trash disposal. At a minimum, garbage disposal units should be available in preprepa- ration, dishwashing, and pot and pan washing areas. **Pulpers** are replacing garbage disposal units if water consumption and sewage use are concerns in the community. A pulper works somewhat like a garbage disposal except it dehydrates the product (Riell, 1993). Solid waste is fed into a water-filled pulping tank, where it is pulped into a slurry by a shredding device, and then water is pressed out of it. The waste becomes a semi-dry, degradable pulp ready for disposal. Excess water is recycled in the pulping tank for reuse. Solid waste can be reduced by 85%, which means less space is used in a landfill. Pulpers can handle paper trays, foam, foil, corrugated boxes, bones, food scraps, and some plastics. Pulpers are expensive, but manufactur- ers are beginning to downsize them for use in restaurants and small institutions. Pulpers require a lot of maintenance and cleaning. At the end of the night, the pulper has to be broken down and cleaned properly; otherwise the wad of garbage would be harder than a rock.

Mechanical trash compactors (Figure 16.8) are used for dry bulky trash, such as cans and cartons. Compacting reduces the volume of trash to one fifth of its original bulk.

Solid Waste

In chapter 7, ethical concerns between suppliers and buyers were discussed, but a much broader area of ethics involves responsibility to society, often termed *social responsiveness*. Most foodservice managers are concerned about environmental issues, or the greening of America, and are establishing programs to manage solid waste in their operations. Kleiner (1991) stated that pollution prevention is good for the planet and, thus, for business. To be good for the environment, managers must know how to be very good at production. Professional foodservice organizations have taken a stand on solid waste. The National Restaurant Association (1989) issued a report called *The Solid Waste Problem*. The American Dietetic Association issued a position statement in 1993 to promote environmentally responsible practices that conserve natural resources, minimize the quantity of wastes generated, and mini- mize the adverse impact on human health and the environment (Shanklin & Hoover, 1993). An integrated waste management program to assess the following alternatives for solid waste disposal is being encouraged by both associations:

- Source reduction including reuse of materials
- Recycling of materials including composting
- Combustion with or without energy recovery

Figure 16.8.   Trash compactor.

Source: Precision Metal Products, Inc., Miami, FL. Used by permission.

- Landfilling
- Chemical or biological solution

Restaurants & Institutions conducted two studies on the waste issue; one study looked at the top 60 "institutional giants" (Lorenzini, 1992), and the other, at the top 50 "restaurant chain giants" (Weinstein, 1992a). Hospitals, schools, colleges and universities, and other noncommercial operations have had to overcome hurdles to establish solid waste programs (Lorenzini, 1992). At the University of California, Los Angeles, the Associated Students Organization tried to decide if the university food-service operations should convert to mostly polystyrene disposables; other students were adamant about using only recycled paper. To make the decision, a public forum was established at which students, faculty, and environmental experts could discuss the issue. After hearing the evidence, the students decided that using polystyrene disposables was the best solution to the solid waste problem. Even though hurdles have to be overcome, institutions are beating commercial operations in cutting down on solid waste. Institutions are clearly leading the green revolution (Quinton & Weinstein, 1991).

Solid waste has been defined by the United States Environmental Protection Agency (EPA) as the products and materials discarded after use in homes, commercial establishments, and industrial facilities. Waste is everywhere! If we are not careful, we may be consumed by it (Pence, 1992). Restaurants, take-out operations in particular, have been targeted as the cause of the solid waste problem even though they produce no more trash than other industries. The misconception occurs partly because of the high visibility of restaurants and the fact that customers throw away disposable cups, plates, and napkins with logos after eating either in or outside the restaurant. Foodservice disposables, however, account for a surprisingly small share of landfill volume. If all the paper, plastic, and aluminum from delis, salad bars, schools, households, hospitals, and restaurants were combined, these disposables would account for only 1.84% of landfill volume; 98.16% of what is in landfills is something other than foodservice disposables (Foodservice & Packaging Institute, 1991). A typical limited-menu restaurant meal consisting of hamburgers, fries, and a medium-size drink produces 3.5 ounces of packaging weight (Foodservice & Packaging Institute, Inc., 1991). Even though foodservice disposables are a minor factor in the waste stream, the industry is a major user of recycled products, such as paper napkins and shipping boxes. Source reduction and recycling are the major methods for controlling waste in foodservice operations.

Minimal packaging combined with reduction in cost comes to mind when source reduction is mentioned. The EPA defines **source reduction** as the design and manufacture of products and packaging with minimum toxic content, minimum volume of material, and a longer useful life (Environmental Protection Agency, 1989). The thickness of a 12-ounce beverage can has been decreased by using new designs that retain strength while lowering the average weight by 26% (Testin & Vergano, 1990). Technological changes in packaging materials have been responsible for this decrease in weight. Foodservice managers can practice source reduction through selective buying habits and reuse of products. They should also reexamine specifications to decide if products with lighter weight and less bulky packaging could be purchased (Shanklin, 1991).

Recycling is the second method for decreasing waste in foodservice operations. The Food Service and Packaging Institute (1991) defines **recycling** as the act of removing materials from the solid waste stream for reprocessing into valuable new materials and useful products. Traditional packaging materials, including paper, metals, and glass, have been recycled in varying degrees for many years. Recently, the technology for recycling plastic has improved enough that plastics recycling programs are being developed (Testin & Vergano, 1990). *Composting* is the controlled application of the natural process of organic degradation, according to the Composting Council (Crosby, 1993). Because of its contact with food, foodservice paper is ideal for composting. A commercial composting plant accelerates natural biodegradation, converting mixed organic waste to a nutrient-rich soil conditioner in great demand in agriculture and horticulture.

Combustion, or incineration, is a form of solid waste recycling in which the energy value of combustible waste materials is recovered (Council on Plastics and

Packaging in the Environment, 1991). Modern waste-to-energy plants reduce the volume of waste going to landfills by 80% to 90% while generating electricity and revenue for users. According to the EPA, combustion is a method of waste disposal that should be considered as an alternative to landfills. The estimate is that by the year 2000 as much as 40% of the nation's garbage will be burned to produce energy, a 14% increase over the present level.

The most recent statistics from the EPA indicate that Americans are showing some improvement in decreasing the growth of landfill waste (King, 1993). As of 1990, 17% of municipal solid waste was either recycled or composted, compared to 10% in 1988, and 67% of all waste was sent to landfills, down from 80% in 1988. In a study conducted by *Food Management* for the Society for Foodservice Management, only 51% of the respondents to a questionnaire indicated that they dumped waste from their operations into a landfill (Boss & King, 1993).

Bioremediation companies take advantage of a technological breakthrough to offer a biological solution to an old and nagging restaurant problem, clogged drains and grease traps (Yaffar & Dibner, 1992). Restaurant managers can use a drain cleaning company or a plumber whenever an emergency occurs or can hire a bioremediation company that can establish a chemical or biological **preventive maintenance** program. The **biological solution** is to use bacteria to break down animal fats and food products that clog up drains. Naturally occurring organisms that grow on grease, flour, and other foods have been isolated. Once the food source is depleted, the organisms die and the environment returns to its original natural state. Engineering bacteria in the future is a possibility, but first the EPA will have to decide if releasing a new organism into a foodservice operation is acceptable. Using chemical or biological solutions on a regular basis provides preventive maintenance for a serious problem and also is a known fixed cost in the budget.

Employee and Guest Facilities

Locker rooms should be provided for employees to change clothes. Individual lockers with locks are needed for storing street clothes and personal effects when employees are working. Adequate space and good lighting are necessary for changing clothes as well as for employee safety. Floors should be tile laid in cement or other nonabsorbent materials, especially in the toilet and handwashing sink areas. Employee facilities should be located near the work area.

Guest restrooms should be easily accessible from the dining room. Of all its customer accessibility guidelines, the Americans with Disabilities Act's restroom requirements are the toughest and most expensive to meet (Weinstein, 1992b). In most commercial foodservice operations, the restroom foyer is decorated in harmony with the dining room. Walls might be papered and the floor might be carpeted in this area, but the toilet and handwashing sink areas should have tile floors and walls.

Keeping both employee and guest restrooms clean can be a major management problem. Frequent management inspections are required, especially during changing of employee shifts or guest meal times, when traffic volume is the highest. These inspections are too important to delegate to an employee. Many customers will not return to a foodservice operation if they are dissatisfied with the cleanliness of a restroom. Employee restrooms should be maintained at the same quality as those for guests.

MAINTENANCE

Facilities and equipment are important factors in any HACCP-based program. Poorly designed facilities and equipment make cleaning and sanitization difficult (Educational Foundation of the NRA, 1992). Public health, building, and zoning departments have the power to regulate the building of a facility and to approve plans before new construction or remodeling can be done. The U.S. Department of Justice and the Equal Opportunity Commission, in accordance with the Americans with Disabilities Act, have guidelines for facility design that include access to facilities by disabled customers and employees.

Foodservice managers should look for the National Sanitation Foundation's (NSF) International mark or the **Underwriters Laboratories' (UL)** sanitation classification listings of commercial foodservice equipment that comply with those of NSF International (Educational Foundation of the NRA, 1992). One of the standards of NSF International is that equipment is easy to clean. The UL mark indicates compliance of equipment with electrical safety standards. Local authorities also provide checklists of desirable or necessary features for good sanitation. Well-planned preventive maintenance and pest control, therefore, are essential to keep the facilities and equipment free of contamination.

Preventive Maintenance

Preventive maintenance has two aspects: regular cleaning schedules and standard procedures, and the preventive and corrective maintenance of foodservice equipment and facilities.

Some cleaning should be performed every day and included in the daily tasks of specific employees. Other cleaning tasks may be scheduled on a weekly, monthly, or less frequent basis, as appropriate to the operation, but must be done regularly for proper maintenance of the facilities. In some instances, specific scheduling of additional employees or perhaps specialized cleaning crews may be required. Whatever the schedules, however, proper tools, equipment, and cleaning materials are basic to an effective facility maintenance program and must be on hand as required.

All cleaning tasks should be combined in a master schedule that includes a list of what is to be cleaned, when each task should be done, how the task should be performed, and who has the assigned responsibility. Table 16.4 illustrates a partial master cleaning schedule for a food production area.

Specific cleaning procedures need to be developed to supplement the master schedule. Employees should be instructed in the procedures for cleaning foodservice equipment and in what the proper cleaning devices and materials are. Equipment cleaning procedures must be sufficiently detailed and presented in a step-by-step procedure to ensure that the correct process is followed and that any special

Table 16.4.   Sample cleaning schedule (partial) for food preparation area

Item	What	When	Use	Who
Floors	Wipe up spills	As soon as possible	Cloth, mop and bucket, broom and dustpan	_____
	Damp mop	Once per shift, between rushes	Mop, bucket	_____
	Scrub	Daily, closing	Brushes, squeegee bucket, detergent (brand)	_____
	Strip, reseal	January, June	See procedure	_____
Walls and ceilings	Wipe up splashes	As soon as possible	Clean cloth, detergent (brand)	_____
	Wash walls	February, August		
Work tables	Clean and sanitize tops	Between uses and at end of day	See cleaning procedure for each table	_____
	Empty, clean and sanitize drawers, clean frame, shelf	Weekly, Sat. closing	See cleaning procedure for each table	_____
Hoods and filters	Empty grease traps	When necessary	Container for grease	_____
	Clean inside and out	Daily, closing	See cleaning procedure	_____
	Clean filters	Weekly, Wed. closing	Dishwashing machine	_____
Broiler	Empty drip pan; wipe down	When necessary	Container for grease; clean cloth	_____
	Clean grid tray, inside, outside, top	After each use	See cleaning procedure for each broiler	_____

Source: Applied Foodservice Sanitation, Fourth Edition (p. 217), by The Educational Foundation of the National Restaurant Association, 1992, Chicago: Author. Used by permission.

precautions are heeded. Table 16.5 presents the procedure for cleaning a food slicer along with important safety precautions. Similar procedures need to be developed for all pieces of equipment. Manufacturers' instructions can be useful in developing these procedures.

Bacteriological counts on dishes, utensils, and equipment need to be performed on a regular basis as a check on sanitization. In some localities, health inspectors may perform these tests. In most noncommercial foodservices, these tests are performed by personnel in the facility. In a healthcare facility, for example, laboratory personnel usually have responsibility for bacteriological examinations in the dietetic services department on a regular basis without advanced notice.

Table 16.5.   Sample cleaning procedures for a food slicer

When	How	Use
After each use	1. Turn off machine.	
	2. Remove electric cord from socket.	
	3. Set blade control to zero.	
	4. Remove meat carriage.	
	Turn knob at bottom of carriage.	
	5. Remove the back blade guard.	
	Loosen knob on the guard.	
	6. Remove the top blade guard.	
	Loosen knob at center of blade.	
	7. Take parts to pot-and-pan sink, scrub.	Manual detergent solution, gong brush.
	8. Rinse.	Clean hot water, 170°F (76.7°C) for 1 minute. Use double S hook to remove parts from hot water.
	9. Allow parts to air dry on clean surface.	
	10. Wash blade and machine shell by swabbing. CAUTION: PROCEED WITH CARE WHILE BLADE IS EXPOSED.	Use brush dipped in detergent solution or use a bunched cloth, folded to several thicknesses. Wear steel-reinforced gloves.
	11. Rinse by swabbing.	Clean hot water, clean bunched cloth.
	12. Sanitize blade, allow to air dry.	Clean water, chemical sanitizer, clean bunched cloth.
	13. Replace front blade guard immediately after cleaning shell.	
	Tighten knob.	
	14. Replace back blade guard.	
	Tighten knob.	
	15. Replace meat carriage.	
	Tighten knob.	
	16. Leave blade control at zero.	
	17. Replace electric cord into socket.	

Source: From *Sanitary Techniques in Food Service* (pp. 102–103) by K. Longrée and G. G. Blaker, 1982, New York: John Wiley.

Maintaining equipment and facilities in a good state of repair is important for both sanitation and efficiency in the foodservice operation. To initiate a preventive and corrective maintenance plan, a list of all foodservice equipment should be compiled. In addition, the date of purchase and installation and maintenance information for each piece of equipment should be included. A record of repairs and service and a file of equipment manuals and warranties with the name of the service representative should be available. In some foodservice operations, a computer database has been compiled and a program developed that produces maintenance printout reports to assist managers in scheduling equipment service and maintenance. Corrective maintenance should be scheduled promptly and not put off until there is a major breakdown, which could be much more expensive.

Pest Control

Controlling pests in a foodservice operation is of critical importance. The presence of rodents and insects can be a serious problem because both are sources of contamination of food, equipment, and utensils. Many pests carry disease-causing organisms and cause considerable spoilage and waste. Also, nothing will turn off customers more than the sight of a mouse or cockroach running across the floor in the dining room. Pests can be controlled by following these three basic rules (Educational Foundation of the NRA, 1992):

- Deprive pests of food, water, and shelter by following good sanitation practices.
- Keep pests out of the facility by verminproofing the building.
- Work with a licensed pest control operator (PCO) to rid the operation of pests.

A HACCP-based program will help ensure that potential contamination from pests does not threaten food safety. The foodservice manager should develop an integrated pest management (IPM) program that combines preventive tactics and control methods to reduce pest infestations (Educational Foundation of the NRA, 1992). Although roaches, flies, rats, and mice are the most common pests in a foodservice facility, beetles, moths, and ants may be a problem as well. Cockroaches head the list of the most frequent and bothersome insect pests and can infest any part of a foodservice operation. They carry disease-causing organisms such as *Salmonella*, fungi, and viruses. A PCO should be called immediately when cockroaches are seen because only a trained person can handle the situation. Cockroaches hide and lay eggs in places that are dark, warm, moist, and hard to clean, such as behind stoves and refrigerators and in cracks between ceiling tiles. All incoming shipments of foodstuffs should be inspected and any roach-infested goods refused.

Flies are a greater menace to human health than cockroaches. They transmit foodborne illnesses because they feed on human and animal wastes and garbage. Flies enter the facility primarily through outside doors or other external openings. Control can be facilitated by having tight-fitting and self-closing doors, closed windows, and good screening. Screened or closed storage for garbage is also important. Control, however, should be done by a licensed PCO.

All openings to the outside must be protected against the entrance of rats and mice. Again, constant surveillance for signs of the presence of rodents is needed, and control of rats and mice should be left to a licensed professional PCO. These are formidable pests. *Their control must be entrusted to professionals only* (Educational Foundation of the NRA, 1992).

Pesticides can be toxic, and some may cause fire or explosions if not handled properly. They must be used with extreme care and labeled and stored properly, away from food storage areas. Because pesticides can be hazardous to humans, the foodservice manager should rely on a PCO's knowledge as part of an IPM solution to a consistent pest problem. In the long run, good sanitation practices are the best form of pest control.

SAFETY IN FOODSERVICE OPERATIONS

In comparison with many industrial jobs, those in foodservice are relatively safe occupations. According to the Bureau of Labor Statistics, the number of accidents per 100 full-time employees in eating and drinking places in 1990 was 8.4, which is slightly lower than some other private industries with an accident rate of at least 100,000 for 1990. Motor vehicle manufacturing plants recorded 18.9 injuries per 100 full-time employees; grocery stores, 12.1; and hotels and motels, 10.4 ("Foodservice Injury," 1992). A foodservice facility, however, has many potential hazards; minor injuries from cuts and burns are common, and more serious injuries occur all too frequently. The quantity of hot foods handled, type of equipment used, and the frequently frenetic pace of a foodservice operation require that safety consciousness has a high priority. Accident prevention must be a priority for the management of foodservice operations because accidents may involve injury or even death of employees or customers.

Employee Safety

An **accident** is frequently defined as an event that is unexpected or the cause of which was unforeseen, resulting in injury, loss, or damage. An accident is also an unplanned event that interrupts an activity or function. Although they may or may not be the result of negligence, many accidents can be prevented. Safety is every employer's responsibility. Accidents do not just happen—something causes them, and the majority are controllable (Somerville, 1992).

A foodservice operation should have an accident prevention program that seeks to eliminate all accidents, not just those resulting in personal injury. Accidents are expensive and can result in increased insurance premiums, lost productivity, wasted time, overtime expenses, workers' compensation claims, potential lawsuits, and human suffering. According to Spertzel (1992b), accidents are the thieves of profit. The cost of one accident can wipe out the difference between profit and breaking even. Somerville (1992) considers safety a profit center because when a loss occurs, the cost is great. How many customers would it take to generate that kind of money? Accidents can also result in a fine or legal action if provisions of the Occupational Safety and Health Act are violated.

Many aspects of safety are related to construction and maintenance of the structure and equipment. For example, floors and wiring should be in good repair, and adequate lighting should be provided in work areas, corridors, and the outside of the facility. Exits should be clearly marked, nonslip flooring materials used, and all equipment supplied with necessary safety devices. Also, fire extinguishers of the appropriate type should be readily available throughout the foodservice facility. The basic traffic flow should be designed to avoid collisions.

Most accidents are the result of human error. Employees may lift heavy loads incorrectly, leave spills on the floor, walk across freshly mopped floors, fail to use safety devices on foodservice equipment, block passageways, or fail to clean greasy filters regularly. Many other unsafe practices can be added to this list. Obviously, then, training is an important part of a safety program. Employees should be taught to prevent accidents by learning to recognize and avoid or correct hazardous conditions. The first day on the job is the best time to start educating a new employee about safety procedures and equipment handling (Spertzel, 1992b).

Congress legislated the Occupational Safety and Health Act in 1970. The purpose of the act is "to assure, so far as possible, every working man and woman in the Nation safe and healthful working conditions, and to preserve our human resources."

OSHA allows a compliance officer to enter a facility to determine adherence to standards and to determine if the workplace is free of recognized hazards. During an OSHA inspection, some of the specific conditions for which the compliance officer will be searching include the following:

- Accessibility of fire extinguishers and their readiness for use
- Guards on floor openings, balcony storage areas, and receiving docks
- Adequate handrailings on stairs
- Properly maintained ladders
- Proper guards and electrical grounding for foodservice equipment
- Lighted passageways, clear of obstructions
- Readily available first-aid supplies and instructions
- Proper use of extension cords
- Compliance with OSHA posting and record-keeping requirements

Citations are issued by an OSHA area director upon a review of the compliance officer's inspection report if standards or rules have been violated. Several kinds of violations are possible, which may involve fines or legal action if the violation is sufficiently serious. A task force consisting of representatives from OSHA and the American Insurance Services Groups, Inc., created "Project Safe Georgia," a pilot program to distribute safety knowledge to a variety of often overlooked small industries including restaurants throughout Georgia (Way, 1992). The task force identified the three areas where most restaurant employees are injured: lifting, falling, and cutting. These industries cannot afford to hire full-time safety personnel, but they can develop safety programs, which were piloted in 12 Waffle House restaurants in Georgia for a period of 3 months. Preliminary results indicated a more than 78% decrease in the number of occupational accidents from the same period of the previous year.

More fires start in foodservice than in any other kind of business operation (Educational Foundation of the NRA, 1992). Managers must check their operations regularly and must establish procedures for handling any hazards that could start fires. Hot oil in fryers can burst into flames at its flammable limit of between 425°F and 500°F and be the source of a fire, or it can increase the severity of a fire that is started another way. Also, oil in ventilation systems and on walls, equipment, and other surfaces is highly flammable. Hoods over ranges and filters that are not cleaned regularly provide an ideal environment for a grease fire (Bendall, 1992). A good solution for high-volume restaurants is the extractor ventilator, which is a series of baffles on the hood to extract grease through a centrifugal action. Some have an automatic wash-down feature to clean the inside of the hood with detergent and hot water at scheduled times. Tests have shown that some of these ventilators can remove more than 90% of the grease from the air.

The National Fire Protection Association has identified ABC classes of fire, which are described in Figure 16.9 along with the types of extinguishers for each. The foodservice manager should know the differences among the extinguishers and purchase the proper kind. The local fire department usually is very willing to demonstrate extinguishers. The fire department, however, should always be called before an extinguisher is used. Of course, if the fire is severe, first the building should be evacuated and then the fire department called from another building.

In addition to fire extinguishers, heat and smoke detection devices and some form of fire protection, such as water mist or dry/wet chemicals, should be installed over cooking equipment (Bendall, 1992). Water mist operates from the building's water sprinkler, which has an unlimited supply, and is effective in suffocating all types of fires. Dry/wet chemicals in containers are piped to outlet nozzles above each piece of equipment; once the chemicals are discharged, they have to be replaced immediately to provide continuing fire protection. In many states, a state fire marshal has responsibility for approving the design and construction of buildings from a safety and fire protection standpoint. Also, OSHA inspections are designed to assure safe and healthful working conditions. The foodservice manager should augment any external inspections with a periodic internal review, or self-inspection, of safety conditions and practices, followed by any necessary in-service training. Although small foodservice operations cannot afford to hire a safety coordinator, most can appoint a staff member to coordinate safety efforts.

Insurance companies can be an important safety resource in which the service is either included in the premium or is available at a small charge (Somerville, 1992). This service includes establishing a safety program or reinforcing an existing one. Some insurers conduct audits for the operation and help with employee training by providing safety manuals or films and videos. A comprehensive safety audit includes a thorough inspection of the facility from the sidewalks in. Almost every restaurant that has installed a facility self-inspection report has reduced accidents significantly and has had sizable savings on insurance premiums. Godfather's Pizza saved $500,000 in 1 year.

Although many aspects of safety are concerned with construction and design of facilities, safe practices of employees are also a critical element in a safety program. Ergonomics is another factor of work safety. Ergonomics examines how workers interact with their work environment including equipment, the workstation, and climate, and influences such factors as lighting and footwear, which in turn influence safety. Equipment manufacturers have developed equipment with built-in features such as safety valves on pressure steamers and guards on slicing and chopping machines ("DOL Officially Declares," 1991). The Department of Labor went one step further and issued regulations to officially prohibit 16- and 17-year-olds from using

Figure 16.9.   Description of the classes of fires and fire extinguishers.

CLASSES OF FIRES

Class A fires include: wood, paper, cloth, cardboard, plastics
Examples: a fire in a trash can; a fire that catches drapes or tablecloths in the dining room

Class B fires include: grease, liquid shortening, oil, flammable liquids
Examples: a fire in a deep fryer; spilled flammable chemicals in kitchen

Class C fires include: electrical equipment, motors, switches, and frayed cords
Examples: a fire in a toaster; a fire in the motor of a grinder

CLASSES OF FIRE EXTINGUISHERS

A	AB	BC	ABC
Use on Class A fires only	**Use on Class A and Class B fires only**	**Use on Class B and Class C fires only**	**Use on Class A, Class B, or Class C fires**
Water, stored pressure	Aqueous film-forming foam (AFFF), stored pressure*	Dry chemical, stored pressure	Multipurpose dry chemical, stored pressure*
Water pump tank		Dry chemical, cartridge operated	Multipurpose dry chemical, cartridge operated*
Multipurpose dry chemical, stored pressure		Carbon dioxide, self-expelling*	Halon, stored pressure
Multipurpose dry chemical, cartridge operated		Halon, stored pressure*	
Aqueous film-forming foam (AFFF), stored pressure			
Halon, stored pressure	*Not recommended for deep fryer fires	*Not recommended for deep fryer fires	*Not recommended for deep fryer fires

Source: Adapted in *Applied Foodservice Sanitation* (p. 266) by The Educational Foundation of the National Restaurant Association, 1992, Chicago: Author.

Figure 16.10.   Examples of mini-posters on safe practices in foodservice operations.

Source: National Restaurant Association, Chicago, IL. Used by permission.

power-driven food slicers in restaurants, especially limited-menu operations. Obviously, safety training must have major emphasis in both initial and in-service employee training. Many resource materials on safety and accident prevention are available, such as those of the National Safety Council and the American Red Cross. Also, personnel from state and local fire prevention agencies are often available as speakers. The National Restaurant Association (1988) has published an excellent Safety Operations Manual and mini-posters to serve as employee reminders of safe practices (Figure 16.10).

Customer Safety

Many of the factors discussed for employee safety also apply to customer safety. A crack in the sidewalk, an exit door that does not open, grease on the dining room floor, or a cup of hot coffee that is dropped can cause customers to have serious accidents that end in litigation. Customer safety is the responsibility of the foodservice manager and employees. Emergency action procedures should be included in the employee manual and during employee training sessions (Educational Foundation of the NRA, 1992). A foodservice operation always should have a complete first-aid kit. Some states also have laws specifying the supplies that must be included in this kit. OSHA requires that a restaurant either have a kit equipped according to the advice of a company physician or have physical or telephone access to community emergency services. Ideally, a foodservice operation should have present at all times an employee who is trained and certified in first aid, including how to administer the Heimlich maneuver and how to give cardiopulmonary resuscitation (CPR).

Approximately 60% of all choking incidents occur in restaurants (Herlong, 1991). Prior to 1974, when the Heimlich Maneuver was introduced, about 20,000 choking fatalities occurred each year; currently about 2,000 to 3,000 occur. Laws on first-aid training requirements vary by state. Some states require only that restaurants post Heimlich Maneuver instructional diagrams where all employees can see them, but others require formal training for foodservice employees as well as posting of instructions. The National Restaurant Association, the American Red Cross, and the Heimlich Institute all provide charts and instructional materials on the Heimlich Maneuver.

SUMMARY

Sanitation and maintenance of foodservice equipment and facilities are critical in a foodservice operation. A properly designed foodservice facility is basic for maintenance of a high standard of sanitation. A facility should be clean, free of physical soil, and sanitary, free of disease-causing organisms and contaminants. Sanitization is critical for any surface that comes in contact with food, including dishes, utensils, pots, and pans, and can be accomplished by heat or a chemical sanitizing compound.

Warewashing is the process of washing and sanitizing dishes, glassware, flatware, and pots and pans either manually or mechanically. Manual warewashing is used primarily for pots and pans, but may be used for dishes and flatware in a small operation. Mechanical warewashing requires a dishwashing machine capable of washing, rinsing, and drying dishes, flatware, and glassware. In large operations, heavy-duty pot and pan washing machines have been designed to remove cooked-on food.

Design for sanitation must begin when the facility is being planned. The floors, walls, and ceilings in the kitchen and dining areas should be constructed for easy maintenance, and the design and arrangement of the equipment and fixtures should facilitate cleaning. Handling and storage of clean dishes and pots and pans are important aspects of a sanitation program in a foodservice operation. Garbage and trash must be handled carefully because of the potential for contaminating food, equipment, and utensils and for attracting insects and other pests. Garbage disposal units should be available in preparation, dishwashing, and pot and pan washing areas. Pulpers and trash compactors also reduce the volume of garbage and trash in the operation.

Increased attention is being focused on the solid waste problem in the United States because many landfills are reaching capacity. Foodservice managers are concerned about environmental issues and are establishing programs to manage solid waste in their operations. Source reduction and recycling are the major methods for controlling waste in foodservice operations. Locker rooms for employees and restrooms for guests should be designed to make cleaning and sanitation easy, and frequent management inspections are necessary.

Preventive maintenance has two aspects: regular cleaning schedules and standard procedures and the preventive and corrective maintenance of foodservice equipment and facilities. All cleaning tasks should be combined in a master schedule and specific cleaning procedures should be developed to supplement the master schedule. Maintaining equipment and facilities in a good state of repair is important for both sanitation and efficiency in the foodservice operation.

Controlling pests in a foodservice operation is critical. Rodents and insects can be a serious problem because both are sources of contamination of food, equipment, and utensils. Regular services of a licensed pest control operator should be used for effective control of roaches, flies, and rodents.

A foodservice operation has many potential safety hazards both for the employee and the customer. Safe and healthful working conditions were mandated by the Occupational Safety and Health Act (OSHA) in 1970 to assure that every worker has safe and healthful working conditions, and to preserve human resources. Fire is a particular hazard in the foodservice industry because of the nature of the business. The fire department should be called before using an extinguisher. Safety training must have major emphasis in both initial and in-service employee training programs. Customer safety is the responsibility of the foodservice manager and employees. Ideally, foodservice operations have present at all times an employee who is trained and certified in first aid, including how to administer the Heimlich maneuver and how to give resuscitation.

BIBLIOGRAPHY

Educational Foundation of the National Restaurant Association. (1992). *Applied Foodservice Sanitation* (4th ed.). Chicago: Author.

Bendall, D. (1992). Keeping your kitchen safe and secure. *Restaurant Hospitality*, 76(11), 124, 126.

Boss, D., & King, P. (1993). SFM's waste management survey. *Food Management*, 28(3), 42, 44.

Council on Plastics and Packaging in the Environment. (1991). *Questions and answers on plastics, packaging and the environment*. Washington, DC: Author.

Crosby, M. A. (1993). Composting: Recycling restaurant waste back to its roots. *Restaurants USA*, 13(1), 10–11.

DOL officially declares meat slicers off-limits to 16–17-year-olds. *Washington Weekly*, 11(47), 4.

Environmental Protection Agency. (1988). *EPA Report to Congress: Solid Waste Disposal in the United States*. Vols. 1 and 2. Washington, DC: Government Printing Office.

Environmental Protection Agency, Office of Solid Waste. (1989). *The solid waste dilemma: An agenda for action. Final report of the Municipal Solid Waste Task Force*. Washington, DC: Government Printing Office. EPA/530-SW-89-019.

Foodservice & Packaging Institute, Inc. (1991). *Foodservice disposables: Should I feel guilty?* Washington, DC: Author.

Foodservice injury rates hover near average. (1992). *Washington Weekly, 12*(3), 4.

Gisslen, W. (1989). *Professional cooking* (2nd ed.). New York: Wiley.

Herlong, J. E. (1991). The Heimlich maneuver: Part of a new standard of service? *Restaurants USA, 11*(2), 17–19.

King, P. (1991a). Garbage wars '91: Report from the front. Part two. *Food Management, 26*(2), 40–43.

King, P. (1991b). Garbage wars '91: Report from the front. Part three. *Food Management, 26*(4), 76–81.

King, P. (1993). Recycling & source reduction. *Food Management, 28*(1), 54–55, 58, 60.

Kleiner, A. (1991). What does it mean to be green? *Harvard Business Review, 69*(4), 38–42, 44, 46–47.

Longrée, K., & Blaker, G. G. (1982). *Sanitary techniques in foodservice* (2nd ed.). New York: Wiley.

Lorenzini, B. (1992). Institutional giants face the waste issue. *Restaurants & Institutions, 102*(7), 80–81, 88–89, 92, 96.

Minor, L. J. (1983). *Sanitation, safety & environmental standards*. Westport, CT: AVI.

National Restaurant Association. (1979). *Sanitation operations manual*. Chicago: Author.

National Restaurant Association. (1981). *Safety operations manual*. Chicago: Author.

National Restaurant Association. (1989). *The solid waste problem. Current issues report*. Washington, DC: Author.

National Sanitation Foundation. (1982). *Recommended field evaluation procedures for spray-type dishwashing machines*. Ann Arbor: Author.

Opitz, A. (1992). Recycling: Making a world of difference. *School Food Service Journal, 46*(2), 24–25, 28, 30.

Patterson, P. (1993). Clean it or close it: Business demands strong sanitary focus. *Nation's Restaurant News, 27*(9), 44.

Pence, J. (1992). The greening of food service in North America. In *Food service and the environment: Conference proceedings* (pp. 1–12). Wellington, New Zealand: Victoria University of Wellington.

Puckett, R. P., & Miller, B. B. (1988). *Food service manual for health care institutions*. Chicago: American Hospital Publishing.

Quinton, B., & Weinstein, J. (1991). Operations: Who's leading the green revolution? *Restaurants & Institutions, 101*(30), 32–35, 38, 44, 45, 54.

Riell, H. (1993). Pulpers get a new life at age 40. *Restaurants USA, 13*(3), 9.

Saving the planet. What school food service needs to know about going green. (1992). *School Food Service Journal, 46*(2), 20–21.

Shanklin, C. (1991). Solid waste management: How will you respond to the challenge? *Journal of The American Dietetic Association, 91*, 663–664.

Shanklin, C. W., & Hoover, L. (1993). Position of The American Dietetic Association: Environmental Issues. *Journal of The American Dietetic Association, 93*, 589–591.

Solid waste management resource guide. (1992). *School Food Service Journal, 46*(2), 42.

Somerville, S. R. (1992). Safety is no accident. *Restaurants USA, 12*(7), 14–18.

Spertzel, J. K. (1992a). The great dishroom debate: Permanentware vs. disposables. *School Food Service Journal, 46*(2), 36–38.

Spertzel, J. K. (1992b). Safety is no accident. *School Food Service Journal, 46*(4), 50–53.

Testin, R. F., & Vergano, P. J. (1990). *Packaging in America in the 1990s: Packaging's role in contemporary American society: The benefits and challenges*. Herndon, VA: Institute of Packaging Professionals.

Way, L. (1992). OSHA test promotes foodservice safety. *Restaurants USA, 12*(4), 7.

Weinstein, J. (1992a). Greening of the giants. *Restaurants & Institutions, 102*(19), 146–150, 152, 156.

Weinstein, J. (1992b). The accessible restaurant. *Restaurants & Institutions, 102*(9), 96–98, 102, 104, 110, 117.

Yaffar, E. A., & Dibner, M. D. (1992). Grease-eating microbes: A high-tech solution to a low-tech problem. *Cornell Hotel and Restaurant Administration Quarterly, 33*(6), 84–90.

Serving the Needs of Fairfax Residents

Distribution and service is the third subsystem in the foodservice system, and sanitation and maintenance is the fourth. Serving a varied, well-balanced, and nutritious menu in a prompt, courteous, and sanitary manner is part of the mission statement for the food and beverage department at The Fairfax community. Chapters 15 and 16 are helpful in solving the two problems in this case study.

Background Information

Distribution and Service

The waitstaff pick-up station, part of the kitchen, leads into the main dining room and also is accessible to the private dining room. A wall separates the main production area from the serving area, which contains the following equipment for menu items, like a broiled pork chop or grilled lemon chicken breast, which need to be prepared after the order is placed:

- 1 reach-in refrigerator/freezer
- 1 broiler
- 1 2-burner range top
- 1 griddle with oven base
- 1 steamer
- 1 handwashing sink

Other menu items like chicken parmesan or pasta shells are baked in pans in the kitchen that are transferred to hot food wells with heat lamps in the serving area. Other equipment in this section includes soup wells, a microwave oven, refrigerator, toaster, ice cream freezer, dipper well, and fudge warmer.

This area is separated from the beverage counter by two tables that hold trays used for delivering filled plates to the main dining room. The waitstaff writes down the resident's order on a ticket that is put on the "order wheel" in the kitchen. The menu item is cooked, if needed, and plated by the cooks in the order the ticket was received. The waitstaff also is responsible for portioning and serving beverages, soups, and desserts. A milk dispenser, iced tea machine, juice dispenser, water station with ice bin, coffee urn, cutting board, toaster, and roll warmer are on the beverage counter. A salad bar is available in the main dining room for all lunch and dinner meals and Sunday brunch.

The coffee shop's waitstaff pick-up station for plates filled by an attendant is similar to the station in the main dining room. The coffee shop also has a smaller salad bar. Food for the healthcare center is sent in bulk in heated and refrigerated carts to be plated by a dietary aide and passed to residents by waitstaff in the assisted-living dining room and by certified nursing assistants in the skilled-nursing dining room.

The employees', or associates', self-serve cafeteria is located in the lounge next to the kitchen in the upper level of the community center. Associates may pay for their meals on an à la carte basis or have a complete meal, consisting of soup, salad, entrée, starch, vegetable, dessert, and beverage, for $1.75.

China, bar-style glasses, silver-plated flatware, and cloth tablecloths and napkins are used in all dining rooms. Even the few residents requiring tray service in their rooms receive the same tableware. The one exception is the substitution of special flatware with large handles for the regular flatware for residents in the restorative dining room.

The Fairfax residents are complimentary about the food and service in the elegant restaurant-style dining room, the healthcare dining rooms, informal coffee shop, private dining room, and terrace when weather permits. Personally planned catering services also are available to make residents' special events perfect. The waitstaff in the main dining room and healthcare center dining room wear black

Figure 1. Questions for waitstaff to ask residents.

1. Would you like your beef prepared Rare, Medium, or Well Done?
2. Would you like a breast or leg quarter of chicken? (Breast portion is white meat and leg portion is dark meat.)
3. Would you like white or dark turkey?
4. Would you like gravy with (the entrée)?
5. Would you like that with or without sauce?
   Example: Broccoli with Hollandaise sauce.
6. What type of dressing would you like on your salad?
7. What beverage would you like to order?
8. When would you like your beverage served?
9. What flavor ice cream would you like?
10. Would you like cocktail or tarter sauce with your fried fish?
11. What would you like with your sandwich — ketchup/mayonnaise/mustard?
12. Breakfast questions:
    How would you like your eggs prepared?
    What type of toast would you like — white/wheat/rye/etc.?
    What flavor jelly would you like?

Source: Courtesy of Marriott Corporation.

tuxedos, with skirts instead of trousers for the women. The uniforms in the coffee shop and the associates' cafeteria are casual. Slacks and a good-quality polo shirt with a vest added in the winter are worn by all waitstaff.

The waitstaff concentrates on making dining a pleasant experience for the residents. Their philosophy is that they don't want to say no to any requests of the residents. One Valentine's Day, a special dinner was served, and red roses were flown into Virginia from Oregon for the occasion. The menu featured a choice of grilled filet mignon or poached fresh salmon fillets for the entrée. One of the residents ordered and received a club sandwich and another a broiled lemon chicken breast.

The new director of F&B operations for all Marriott Senior Living Services Communities, who was promoted from the position of F&B director in The Fairfax, has been given the task of writing a service training manual for associates in senior living services. With the emphasis on customer satisfaction, the need for the manual is tremendous. Training programs for waitstaff currently are conducted by the dining room manager, who uses a Marriott video program with handouts on pertinent topics. The training programs are very practical and cover daily problems that could occur in the dining room such as questions waitstaff needs to ask when taking an order (Figure 1) or responses to difficult questions posed by residents (Figure 2).

Figure 2. Responses by waitstaff to residents' difficult questions.

1. *Question:* I have a lactose intolerance. Is the cake made with butter or margarine?
   Response: Just a moment. I'll check with the cook or manager.

2. *Question:* The chicken was awful last time. I guess I'll have the beef.
   Response: I'm sorry you weren't satisfied with the chicken. The beef looks very good this evening.

3. *Question:* I need to get through as quickly as possible.
   Response: I'll try to serve you as quickly as possible.

4. *Question:* My beef is cold.
   Response: I'm sorry it is not as you expected. Let me take it back to the kitchen. I'll just be a moment.

5. *Question:* May I have my vegetables in a side dish, please?
   Response: Yes.

6. *Question:* The portions of beef are always too large. I'd like a smaller portion.
   Response: Would half of a regular portion be fine?

7. *Question:* Those cookies you had last evening were delicious. Could I have a couple of them if you can find any leftover?
   Response: I'll have to ask the manager.

Always remember if you don't know the answer or how to handle the situation, see the manager or dining room supervisor.

Source: Courtesy of Marriott Corporation.

Sanitation and Maintenance

A section on sanitation and preventive maintenance is included in the *Marriott's Senior Living Services Food and Beverage Policy and Procedure Manual*. All foodservice associates are expected to maintain sanitary conditions and implement and practice the "Clean As You Go" program to keep their area clean and safe. Inservice classes on sanitation and personal hygiene are scheduled for associates at regular intervals (Figure 3). These mandatory classes are held during working hours.

Tableware from all dining areas is washed and sanitized in the main kitchen. Dining rooms are cleaned by the waitstaff in each. The service area in the coffee shop is cleaned by the attendant who serves the food and by dietary aides in the healthcare center. The restorative dining room doubles as the activity room during nondining hours. The F&B staff resets the room three times a day. With the exception of coffee and toast, cooking is not done in the healthcare center because the building has no ventilation hoods.

The Problem

This problem has two parts. One is on service and the other on sanitation and preventive maintenance.

Service

Residents were complaining frequently in the Dining Journal about slowness of service. They were unhappy because they had to wait too long for a table, especially at dinner time. This led to other complaints such as having to wait too long before giving an order to the waitstaff or waiting too long for the soup, salad, entrée, dessert, or coffee. Residents liked the food but not the service. The F&B director knew that something had to be done about the problem because if it were not corrected, it would become worse. Hiring more associates was not the answer because of budgetary constraints. He had to work with what was available. He finally came up with a solution that speeded up the service and made the residents happy. How would you as a F&B director handle the problem?

Control and Preventive Maintenance

A horrible odor was noticed in the hall outside the food storeroom in the lower level of the Community Center. The odor was similar to that of spoiled food but was only in the hall, not the storeroom. The kitchen is on the upper level but not over the storeroom. Word began to spread around the center that something was spoiled, and residents were afraid they might get sick if they ate spoiled food. The F&B director tried to determine what caused the odor. Could it be old mop water, spoiled food, or drippings from fish when it was delivered? He then remembered that when The Fairfax was opened, water backed up through the floor drain located beneath the tilting fry pan in the kitchen. He thought the contractor had taken care of the problem by sealing the kitchen floor to prevent water from coming up through the floor. He decided to have the maintenance crew open up the ceiling outside the storage area in the lower level, and they found that the drain

Figure 3. Announcement of an inservice class for associates on personal hygiene.

INSERVICE CLASS
WEDNESDAY, JAN. 19th

ALL HEALTHCARE FOOD AND BEVERAGE STAFF MUST ATTEND

MEETING WILL TAKE PLACE ON THE 2nd FLOOR DINING ROOM

CLASS TOPIC: EMPLOYEE HEALTH AND HYGIENE

Source: Courtesy of Marriott Corporation.

had not been installed correctly when the center was built. The drain pipe was not connected correctly to the drain pan, and the contaminated water was filling up the ceiling space on the lower level, causing the building materials to rot and give off a strong odor. When the problem was corrected, the odor disappeared. If you had been the current F&B director, could you have prevented this from happening?

Points for Discussion

The following questions pertain to service.

- If satisfying the residents is the primary goal of the F&B Department, how can you as the director respond to their demands that are beyond your budget limitations?
- What can you do to make the residents part of the solution rather than part of the problem?
- Utilizing what you have learned about management functions and philosophies, seek a solution that is based on these various theories.

The questions below require an understanding of sanitation and maintenance.

- If you had been the original F&B director, could you have prevented this from happening?
- Develop a step-by-step procedure for handling problems of this type.
- How would you involve the associates in solving the problem?

Management of Foodservice Organizations

The world is living in a period of transformation; organizations, including those in foodservice, are changing from a traditional form of management to a new type that is identified as total quality management. In addition to pleasing the customer, the new organization is using the team approach for solving problems and empowering employees to make decisions.

- **Chapter 17, The Organization Structure.** Layers of middle management are being eliminated and replaced by teams of employees. Jobs are being redesigned to improve organization efficiency and employee job satisfaction.

- **Chapter 18, Linking Processes.** Linking processes are needed to coordinate activities of the foodservice system and include decision making, communication, and balance. The outcome is the accomplishment of goals.

- **Chapter 19, Leadership and Organizational Change.** Trends in leadership power indicate an emphasis in communication and collaboration across functions and units. Organizational change is occurring in many industries as managers are adopting the total quality management or service philosophy.

- **Chapter 20, Human Resources Management.** Employment practices are established by the federal government, but moving people into, within, and out of an organization is the responsibility of human resources management. Training programs are essential to keep employees efficient. Labor unions are undergoing changes in both operation and philosophy similar to those facing organizations.

- **Chapter 21, Management of Financial Resources.** Effective management of financial resources is critical to the success of both profit and not-for-profit foodservice organizations. Control of costs is essential to keep an organization viable.

Designing the Organization

The world is in a period of transformation. Some people attribute the transformation to the emergence of Japan, a non-Western country, as a great economic power, and some others attribute it to the birth of the computer, an information distributor. Peter Drucker stated that he credits the beginning of this transformation to the G.I. Bill of Rights, which gave every American soldier returning from World War II the money to return to school. This was the beginning of a shift to a knowledge society that also is an organization society. The purpose of every organization, commercial or noncommercial, is to integrate knowledge into a common task. Drucker believes that this transformation to a knowledge society will not be complete until 2010 or 2020.

The greatest challenge of managers is to design more flexible organizations (Hirschhorn & Gilmore, 1992). Managers of industries are replacing vertical hierarchies with horizontal networks, linking traditional functions with interdisciplinary teams. They also are developing alliances with suppliers, customers, and even competitors and insisting that employees follow the company's strategic mission without distinction of title, function, or task. The traditional organization chart describes a world that no longer is. Technology, fast-changing markets, and global competition are revolutionizing business relationships.

Foodservice managers, like all other managers, are having a difficult time in making the transformation from the old, complex organization to a networking one. Kanter (1992) says management is a balancing act that includes juggling contradictions to choose the best of opposing alternatives. New organizational models have the best of both worlds—enough structure for continuity but not so much that creative thinking about problems is ignored. All managers must include change in their organization's structure (Drucker, 1992). Change requires continually abandoning what the organization has done and creating the new. If this process is not followed, the organization will become obsolete.

In chain restaurants, in hospitals, and in universities decisions affecting change will be made at the corporate level, and foodservice managers probably will attend many training sessions about implementing the plans. The mission of the organization will be clear and the strategy for implementation will be well defined. Managers of single unit operations, however, could have an even more difficult task in trying to decide if they should stay with the old organizational structure or begin to ease into a new one. In this chapter, the traditional structure will be discussed for those managers who have not started a change, and the new structure will be discussed for those who are thinking about or implementing change.

ORGANIZATION STRUCTURE

An **organization** is defined as a group of people working together in some form of coordinated effort to attain objectives. An ideal organization results in the most efficient use of resources.

The organization structure is based on the objectives that management has established and on plans and programs to achieve these objectives. As indicated previously, different types of structures will be required for traditional and new organizations, each with different objectives.

The Traditional Organization

Kast and Rozenzweig (1985) stated that the **traditional organization** frequently is defined in terms of the following:

- *Organization chart and job descriptions or position guides.* Pattern of formal relationships and duties

- *Differentiation or departmentalization.* Assignment of various activities or tasks to different units or people of the organization
- *Integration.* Coordination of separate activities or tasks
- *Delegation of authority.* Power, status, and hierarchical relationships within the organization
- *Administrative systems.* Guidance of activities and relationships of people in the organization, through planned and formalized policies, procedures, and controls

One of the primary reasons for organizing in the traditional organization is to establish lines of authority, which create order. Without delineation of authority, there may be chaos, in which everyone is telling everyone else what to do.

The New Organization

In the **new organization**, employers are challenged to improve the quality of work life and to develop a corporate, or organizational, culture (Sherman & Bohlander, 1992). Employees today are concerned about having a full-time job, but they are also concerned about their personal lives. Day care, pregnancy leave, and family leave have become important to employees.

Quality of Work Life

To improve the **quality of work life (QWL)** in the organization, managers need to look at the way work is organized and the way jobs are designed. Each organization has special problems and designs jobs to solve its own problems. Some general guidelines, adapted from the Ontario, Canada, Quality of Life Centre (cited in Sherman & Bohlander, 1992), include the following:

- Decisions are made at the lowest possible level.
- Teams of employees are responsible for a complete job.
- Technical and social potential of employees and the organization is developed.
- Quality and quality control are components of production.
- Safety and health of employees are emphasized.
- Immediate feedback of information required to perform a job is available.
- Problems are solved by teams, but responsibility for decisions is shared by all levels of the organization.

Participative management is essential to QWL and is the wave of the future (Sherman & Bohlander, 1992). It involves empowering employees to participate in decisions about their work and employment conditions. The concept is not new, but it had not been acted upon until recently, primarily because managers feared they would lose their authority. Both managers and employees must realize that if the organization is to survive, they must work together to reduce costs and avoid becoming victims of foreign and domestic competition.

Corporate Culture

Culture has become a buzzword in business, but its contribution to the success or failure of a foodservice operation is so important that every owner or manager should learn what culture is and how to use it (Fintel, 1989). **Corporate culture**, or organizational culture, is defined as the shared philosophies, values, assumptions, beliefs, expectations, attitudes, and norms that knit an organization together (Sherman & Bohlander, 1992). Every company has a culture, but it is not always well defined. The most successful companies have adopted a positive culture, one that values its employees and treats them as part of a team. Positive cultures have the following qualities in common (Fintel, 1989):

- Integrity, which is at the top of almost everybody's list, involves trust between people in the organization.
- Bottom-up style of management that gives employees a feeling of being on the team.
- Having fun is important for both managers and employees.
- Community involvement in social service programs is appreciated by employees.
- Physical health and fitness is a belief that a sound mind goes along with a sound body.

Although culture starts at the top, to mean anything it must be passed on to employees at all levels. Managers are encouraged to talk about the organization's culture constantly. Some organizations post signs containing the mission statement or their culture principles for all employees to see. Pizza Hut's culture is called "ownership," the feeling that comes from knowing an employee can affect the company's direction through expertise, innovative ideas, and hard work.

Culture can be used as a hiring tool (Fintel, 1989). Most managers report that they prefer to hire someone who already fits into their culture because those employees tend to be happier and less likely to leave. Rewards are important, too. Rewarding employees by promotion reinforces a culture because those employees provide tangible role models for other employees to follow. Defining an organization's culture and acting on it may take a large investment of time and perhaps some money. The return on the investment is worth this effort; turnover will decrease and customers will be better served.

Division of Labor

Organizing is basically a process of division of labor, which can be divided either vertically or horizontally. Organizing also improves the efficiency and quality of work, as the coordinated efforts of people working together begin to produce a synergistic effect. As defined in chapter 2, *synergy* means that the units or parts of an organization acting in concert can produce more impact than by operating separately. Synergism can result from division of labor and from increased coordination, both of which are products of organization. Improved communication also can be a product of organization and its structurally defined channels. In the traditional organization, employee positions can be discussed as either vertical or horizontal. In new management organizations, the division of labor lines are not as clear-cut because employees work as teams and are empowered to make decisions.

Vertical

Vertical division of labor is based on the establishment of lines of authority. In addition to establishing authority at various levels of the organization, vertical division of labor facilitates communication flow.

The **chain of command** has clear and distinct lines of authority that need to be established among all positions in the organization (Griffin, 1993). The chain of command has two components: the unity of command and the scalar principle. *Unity of command* means that the employee reports to only one manager. The *scalar principle* indicates that a clear and unbroken line of authority extends from the bottom to the top position in the organization.

The 1991 organization chart (Figure 17.1) for the Department of Food and Nutrition Services in the Rush-Presbyterian-St. Luke's Medical Center in Chicago, which is licensed for 1,100 patients, illustrates the vertical division of labor. The department serves between 6,500 and 7,500 meals per day, in the employee cafeteria, the Atrium Court Café for visitors, and a private club, Room 500, which is often used for special events, as well as to patients. The main kitchen is responsible for patient foodservice, and the employee cafeteria kitchen also prepares menu items for the Atrium Court Café. The Room 500 has its own kitchen. Note the levels of management, especially middle management from the associate director of Foodservice Administration through the four operation managers, required for this conventional foodservice in a traditional organization.

Authority, the right of a manager to direct others and to take action, is delegated down the hierarchy of the organization. The tapered concept of authority (Figure 17.2) holds that the breadth and scope of authority become more limited at the lower levels of an organization. To apply this concept to the Rush-Presbyterian-St. Luke's Medical Center depicted in Figure 17.1, the manager with the broadest scope of authority is the associate director for Foodservice Administration; each succeeding level has a narrower scope.

Through the process of delegation, the authority and responsibility of organization members are established. Delegation is the process of assigning job activities and authority to a specific individual within the organization.

Responsibility is the obligation to perform an assigned activity. Because responsibility is an obligation a person accepts, it cannot be delegated or passed to a subordinate. Managers can delegate responsibilities to subordinates in the sense of making subordinates responsible to them; however, this delegation does not make managers any less responsible to their superiors. When a manager delegates responsibility, the manager does not abdicate any responsibility. For example, Joe, the head cook, cannot say to the production supervisor, "It's all his fault," regarding a product failure of one of the assistant cooks, and not bear responsibility himself.

Authority once delegated, however, is given up by the person who delegated it. According to a principle of organization called the *parity principle*, authority and

Figure 17.1. 1991 organization chart.

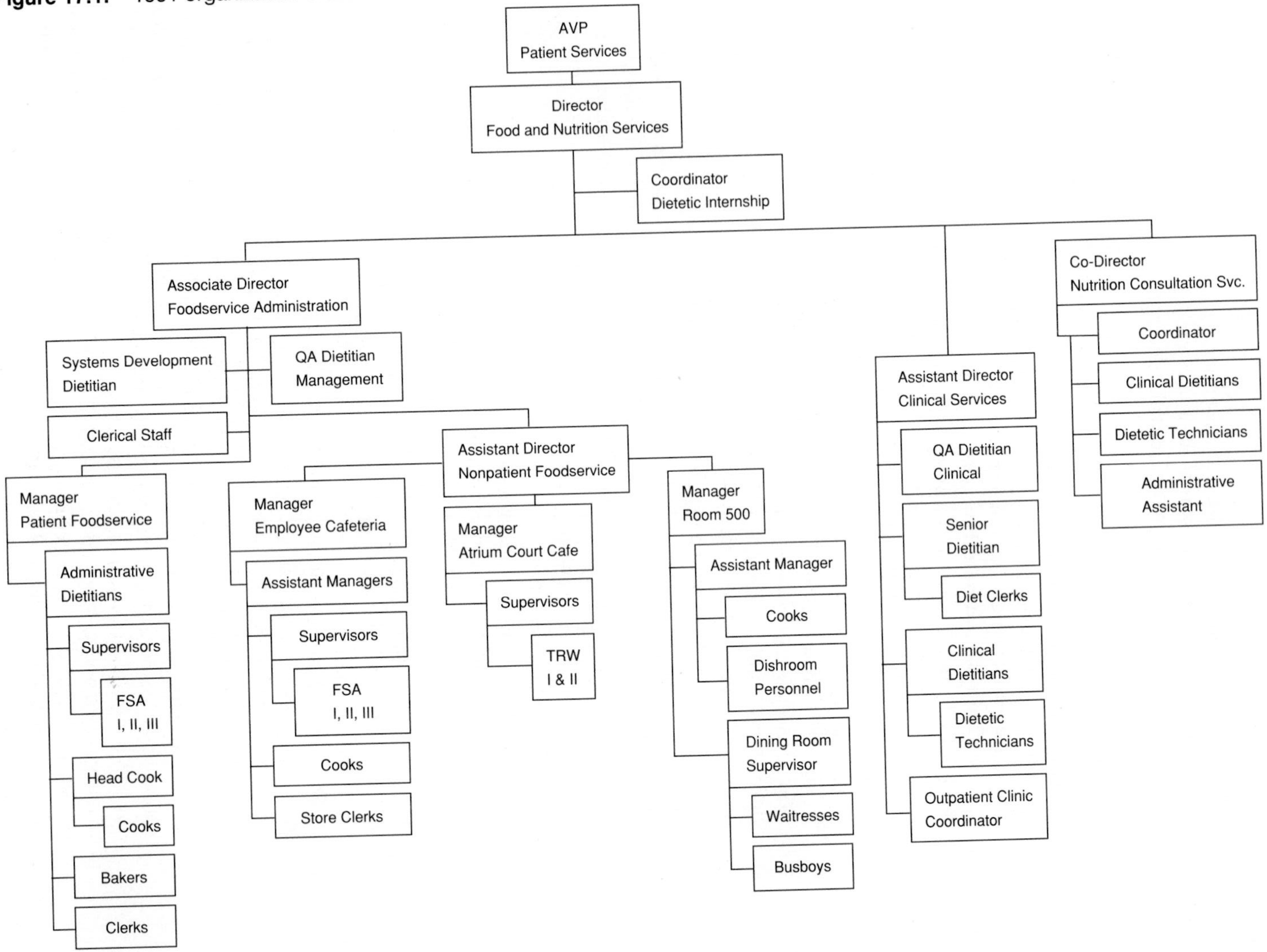

Source: Department of Food and Nutrition Services, Rush-Presbyterian-St. Luke's Medical Center, Chicago, IL. Used by permission.

Figure 17.2.   Tapered concept of authority.

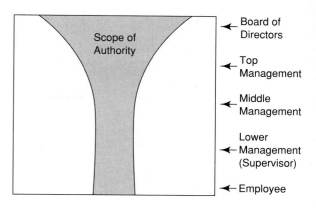

responsibility must coincide; that is, management must delegate sufficient authority so subordinates can do their jobs. At the same time, subordinates can be expected to accept responsibility only for those areas within their authority.

One of the major considerations affecting delegation of authority is decentralization. The key question is: How much of what authority should be granted to whom and for what purpose? The degree to which an organization is centralized or decentralized is basic to this question. These characteristics are at opposite ends of a continuum.

In a centralized organization, most decisions are made at the top, and lower level managers have limited discretion in decision making. The degree of centralization/ decentralization is related to the number of decisions made at lower levels of the organization, the importance of those decisions, and the amount of checking required for decision making by lower level managers.

The degree of decentralization varies widely in large organizations. In some organizations, a high degree of decentralization may exist in major functions, but the auxiliary functions of purchasing, accounting, or personnel may be centralized. In a large hospital, for example, the director of dietetics may have authority over production and service functions but limited authority for purchasing because a purchasing department has procurement responsibility for the entire hospital.

Horizontal

The new organizational philosophy rejects the vertical hierarchy that has been popular in American companies for years. Instead, the companies with the new philosophy are using a **horizontal division of labor** with an emphasis on encouraging employees to share ideas across all levels and departments (Van Warner, 1993). The term *empowerment* has been used for years in foodservice organizations. It has been misused by some managers who think it means that fewer employees can do more work for less money than in the vertical organization, which is a dictatorial environment. Downsizing, or the better term **rightsizing**, is a good idea. In fact, the rightsizing of corporate management has increased productivity in a time of economic slowdown. Empowerment alone cannot compensate for the layers of middle management that have been eliminated and probably will never return.

American executives are re-engineering their organizations in a way that could revolutionize how foodservice managers do business (Van Warner, 1993). Employee teams are revising departments, policies, procedures, and job functions to focus on the customer rather than the process. Managers and employees are being cross-trained to handle multiple jobs, with the entire organization flattened to develop a coaching environment. Cross-training decreases boredom, which could affect the quality of work (Griffin, 1993). The ultimate goal is to create a flexible, quicker-reacting organization that is less distracted by internal problems and bureaucracy. The re-engineered organization focuses on the customer, and every person needs to have the goal of providing the customer with quality, value, and service.

The major changes required for becoming a new organization are evolutionary and occur over years, not weeks or months. An excellent example of this evolution is the Rush-Presbyterian-St. Luke's Medical Center, which was introduced earlier to illustrate the vertical division of labor. In order to stay within the budget, a consultant company was hired to eliminate levels of management and, at the same time, increase customer satisfaction in both products and services. The decision was to reduce the number of levels in each department to three. This required changing the medical center from a traditional to a new organization.

In the Department of Food & Nutrition Services, for example, the first change, in April 1993, was to organize the department by functions, each with its own manager: production, service, patient tray assembly, and materials management/computer systems (Figure 17.3). Because the foodservice operation is decentralized and has many locations, this change was not practical and was rescinded. The second change occurred in September 1993, when the decision was made to go back to the natural divisions of the department (Figure 17.4). In this change, functions, such as production and service, are performed within each division—patient, nonpatient, and Room 500 foodservice operations. Employees in each division were encouraged to cooperate with those in others as well as to share their resources with divisions having similar functions. The goal is to consolidate menus and establish pools of employees—cooks, for example—who have been cross-trained to work in all divisions. Because this plan did not really use the team approach, the hospital initiated another change in November 1993 (Figure 17.5).

Patient foodservices, the largest kitchen in the hospital (Figure 17.4), was chosen to pilot the team approach to decision making. The manager of the patient foodservice kitchen, with coaching from the director of Food and Nutrition Services and with input from the employees, redrew the organization chart to reflect realistically the ways in which teams of employees solve problems. The kitchen currently is divided into four interrelated functional units—production, trayline, sanitation, and materials management (Figure 17.5). Of the three assistant managers in production, two in trayline, two in sanitation, and one in materials management, the manager designated one in each unit to serve as team coordinator. In this most recent revision, the manager serves as a facilitator for the four teams to make sure everyone understands the goals of the organization and the need for effective communication among team members in the patient foodservice division and other divisions. As the

Figure 17.3. April 1993 organization chart.

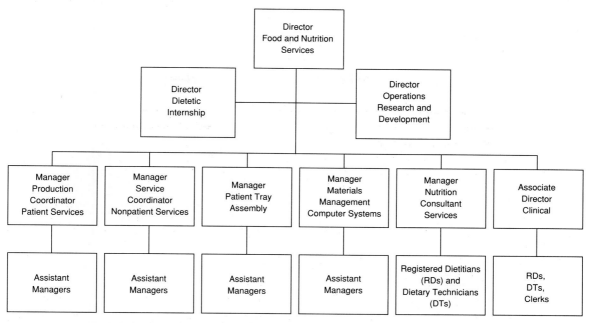

Source: Department of Food and Nutrition Services, Rush-Presbyterian-St. Luke's Medical Center, Chicago, IL. Used by permission.

Figure 17.4. September 1993 organization chart.

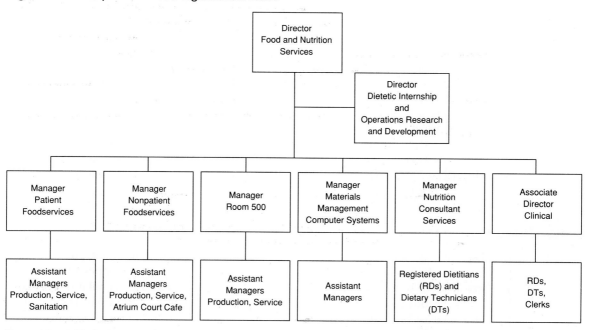

Source: Department of Food and Nutrition Services, Rush-Presbyterian-St. Luke's Medical Center, Chicago, IL. Used by permission.

Figure 17.5.   November 1993 organizational chart.

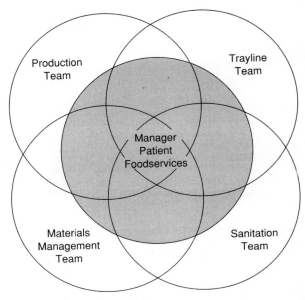

Source: Department of Food and Nutrition Services, Rush-Presbyterian-St. Luke's Medical Center, Chicago, IL. Used by permission.

teams become self-sufficient, the manager will be given other responsibilities, thus requiring the teams to manage the patient foodservice kitchen. The director of the department is convinced that this evolution will continue for many years.

Labor trends indicate that efforts to make changes will be worthwhile (Van Warner, 1993). Qualified employees are decreasing in numbers, probably because of the nation's declining educational programs. Also, people who do not speak English are becoming a bigger part of the work force. For foodservice operations, these labor trends mean that more training is needed in spreading around decision making instead of depending on a few highly skilled managers. The new organizational structure will need to minimize language, cultural, and educational barriers and not emphasize them. Reshaping organizations with empowerment and technology as a base is in order.

Underlying Concepts of Organization

Establishing departments and creating different levels within management are not ends in themselves. As pointed out by Haimann, Scott, and Conner (1985), each department requires a manager and a staff, thereby increasing overall administrative costs. Furthermore, the creation of departments and levels increases coordination and control problems. Why, then, should organizations be departmentalized, and how can they function? The answers can be found in the two underlying concepts of organization: span of management and authority.

Because managers cannot supervise an unlimited number of subordinates, different areas of organizational activity must be defined, with someone placed in charge of each area. Span of management in the traditional organization refers to the number of subordinates who can be supervised effectively by one manager. This concept, also called *span of control,* is the basis for the departmentalization process.

The second concept underlying organization, authority, provides the basis through which managers command work. The source of authority and how it is used are fundamental to the effectiveness of the organization in accomplishing its goals.

Span of Management

The problem of managers being able to manage only a limited number of workers is not unique to any type of organization or any industry. What is the appropriate span? As indicated in chapter 3, a number of factors must be considered when answering this question. One response could be, "It depends on the situation." This situational approach, also called the *contingency approach,* is applicable in answering many questions about organizations. Organizational policies, availability of staff experts, competency of workers, existence of objective work standards, technology, and nature of the work are among the factors affecting the span of management in specific situations.

The narrower the span of management, the more levels needed in the organization. Because each level must be supervised by managers, the more levels that are created, the more managers are needed. Conversely, with a wider span, fewer levels and fewer managers are required. Thus, the resulting organizational shape is a tall, narrow pyramid or a shallow, flat, broad pyramid, as illustrated in Figure 17.6. The leadership style and personality of the manager are other influences on span of management.

Early management theorists attempted to define the appropriate number of superior-subordinate relationships in terms of a mathematical formula. In 1933, Graicunas published a classic paper presenting a formula for analyzing the potential number of these relationships. Based on the work of Graicunas and others, Urwick (1938) stated the concept of span of management as follows: "No superior can supervise directly the work of more than five or, at most, six subordinates whose work interlocks."

During the years since those early publications, researchers have found that a variety of factors influence the appropriate span of management and that this span is not strictly a function of the number of relationships. In addition to those listed previously, the complexity, variety, and proximity of jobs and the abilities of the manager are other factors related to span of management. Today, span of management is a crucial factor in structuring organizations, but no universal formula has been developed for an ideal span (Griffin, 1993).

In a foodservice organization in which many of the workers have low educational levels and limited training, a narrower span of management may be appropriate. If workers are well trained, a wider span is possible and fewer supervisors are needed.

Figure 17.6. Tall versus flat organizations.

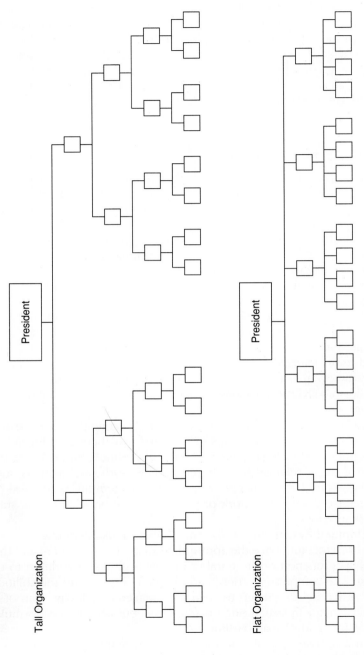

Tall Organization

Flat Organization

Source: Griffin, Ricky W., *Management*, Fourth Edition. Copyright © 1993 by Houghton Mifflin Company. Used with permission.

Formal Versus Acceptance Authority

Up to this point in the book, *authority* has been defined as the right of the manager to direct others and to take action. Actually, this definition deals only with one view of authority, that is, *formal authority.* This view of authority has its roots in the writings of Weber (1947), a German sociologist, who in the late 1800s and early 1900s influenced the development of management thought. He viewed authority as "legitimate power," which involves the willing and unconditional compliance of subordinates based on their belief that it is legitimate for the manager to give commands and illegitimate for them to refuse to obey.

Today, two views of authority are generally recognized in the management literature, formal authority and acceptance authority. Formal authority is considered a top-down theory because it traces the flow of authority from the top to the bottom of the organization (Figure 17.7). As Haimann et al. (1985) indicated, society is the ultimate source of formal authority in the United States because of the constitutional guarantee of private property. Obviously, this statement is directly applicable to public sector institutions, such as public schools, state universities, and hospitals operated by the city, county, state, or federal government, but it applies to organizations in the private sector as well.

Figure 17.7.   Flow of formal authority in organizations.

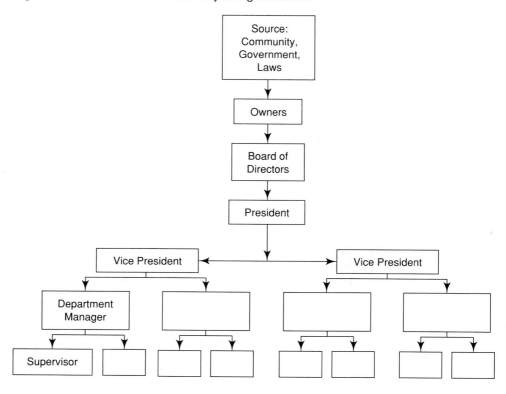

Formal authority is also referred to as *positional authority,* meaning that authority is derived from the position or office. For example, cooks and other production workers recognize the production manager as having certain authority because of the position he or she holds in the organization.

Acceptance authority is based on the concept that managers have no effective authority unless and until subordinates are acquiescent. Although they may have formal authority, this authority is effective only if subordinates accept it. The accept-comply process is related to a subordinate's so-called zone of acceptance. In other words, a manager's order will be accepted without conscious question if it falls within the range of job duties anticipated when the subordinate accepted employment. The employee, however, may refuse to perform a task he or she considers to be outside this range. For example, a cook would willingly prepare any item on the menu, even items added at the last minute, but may be unwilling to help out in the dishroom, feeling that dishwashing is not part of the job.

Acceptance authority is also related to a manager's expertise and personal attributes. *Authority of competence* or expertise is based on technical knowledge and experience. A command may be accepted, not because of organizational title, but because the employee believes the person giving the command is knowledgeable.

This concept underscores the importance of foodservice managers possessing technical skills. If the foodservice manager understands technical operations, such as production methods and equipment operation, foodservice employees view the manager as knowledgeable and generally will be more willing to accept direction.

Subordinates may also accept the authority of the manager because they want to please or help the person giving the command. The charismatic leadership of many well-known historical figures is the epitome of this personal authority.

Departmentalization

One of the first things that happens when people create an organization is that they divide up their work to allow specialization. As the organization grows and tasks become more numerous and varied, this division of labor is formalized into jobs and departments.

Departmentalization, which is the process of grouping jobs according to some logical arrangement, is the most frequently used method for implementing division of labor (Griffin, 1993). Although work units can be structured in a number of ways, all units divide the work and thus establish a pattern of task and authority relationships. This pattern becomes the organizational structure.

In a small delicatessen, for example, a husband and wife may informally share the tasks of preparing sandwiches, salads, and drinks; serving customers; collecting money; wiping tables; washing dishes and utensils; and performing other maintenance duties. They will probably find that each of them will take on the principal responsibility for certain tasks; however, as the business grows, they may need to hire part-time workers to assist at peak periods. These workers will probably be assigned specific duties rather than be responsible for the wide range of duties performed by the husband-wife team. Thus, jobs are created around specialized tasks.

This small business could eventually be the basis for development of a large multi-unit national chain of delicatessens throughout the United States. Additional levels of management would be needed, highly specialized jobs created, and formalized relationships required. At the corporate level, departments focusing on specific functions, such as marketing, procurement, and finance, would be created. This illustration is an example of the development of an organization into jobs, levels of management, divisions, and departments.

Departments are commonly organized by function, product, geography, customer, process, equipment, or time. As indicated earlier, the type and size of the organization are key factors influencing the form an organization structure will take.

Functional

Functional departmentalization occurs when organization units are defined by the nature of the work. (The word *function* is used here to mean organizational functions, such as finance and production, rather than basic managerial functions, such as planning and controlling.) All organizations create some product or service, market these products or services, and finance their ventures. Therefore, most organizations have three basic functions: production, sales, and finance. In the nonprofit foodservice operation, the sales function may be one of clientele service and creation of goodwill, and the finance function may be considered business affairs. Even in these organizations, however, the need to apply marketing concepts is widely recognized.

The primary advantage of such departmentalization is that it allows specialization within function and provides for efficient use of equipment and other resources. It provides a logical way of arranging activities because functions are grouped that naturally seem to belong together. Each department and its manager are concerned with one type of work.

Product and Service

Under departmentalization by a product or a service, all activities required in producing and marketing them are usually under the direction of a single manager. *Product departmentalization* allows workers to identify with the particular product and encourages expansion, improvement, and diversification. Duplication of functions may be a problem, however, because each division or department may be involved in marketing, production, and so forth. This pattern of departmentalization is not common in the foodservice industry, except perhaps in large conglomerate corporations.

Geography

Departmentalization by territory is most likely to occur in organizations that maintain physically dispersed and autonomous operations or offices. *Geographic departmentalization* permits the use of local personnel and may help create customer goodwill and a responsiveness to local customs.

National restaurant chains are often divided into regional areas, with a regional manager and staff responsible for all the operations in a particular area. For example, several of the large contract foodservice companies are divided into several geographic regions, such as East, Midwest, Northeast, Northwest, and South.

Customer

Another type of departmentalization is based on division by type of customers served. A contract foodservice company, for example, that has divisions for schools, colleges, and healthcare is departmentalized by type of customer. A wholesale firm, which distributes products to grocery stores and to hotels, restaurants, and institutions, may be subdivided into two corresponding divisions. This approach to departmentalization permits the wholesaler to serve the specialized needs of both the grocer and the foodservice operator.

Other Types

Process, equipment, and time are other bases for departmentalization. Process and equipment are closely related to functional departmentalization. In large foodservice operations, a deep fat frying section within the production unit would be an example of *process/equipment departmentalization*. A food factory, such as that in a commissary foodservice operation, might be divided into units based on process or equipment because of the specialization needed for the large volume produced in the operation.

Time or *shift* is also a common way of departmentalizing in some organizations. Organizations such as hospitals that function around the clock often organize activities on this basis. Usually, activities grouped this way are first departmentalized on some other basis, perhaps by product or function. Then, within that category, they are organized into shifts. For example, a hospital is departmentalized by functions, such as dietetics services and nursing services; the various departments may then have shifts with a supervisor in charge of each, such as the late shift or early shift.

Departmentalization is practiced as a means not only of implementing division of labor but also of improving control and communications. Typically, as the organization grows, it adds levels and departments. Coordination is another key objective in departmentalization. The type of departmentalization that is best for an organization depends on its specific needs.

Line and Staff

In addition to vertical division of labor through delegation and horizontal division of labor through departmentalization, labor may be divided into line and staff. A **line position** is a position in the direct chain of command that is responsible for the achievement of an organization's goals (Griffin, 1993). Line positions include a procurement, production, or service manager. A **staff position** is intended to provide expertise, advice, and support for line positions. A materials or human resources manager is an example of a staff position.

As indicated in chapter 3, line employees are responsible for production of products and services in the organization, and staff personnel may function in assisting or advising roles. Staff work revolves around the performance of staff activities, the utilization of technical knowledge, and the creation and distribution of technical information to line managers.

Authority

One important difference between line and staff is authority (Griffin, 1993). Line authority is referred to as formal authority created by the organizational hierarchy. Staff authority, however, is based on expertise in specialized activities. Generally, staff personnel provide expert advice and counsel to line managers but lack the right to command them, with two exceptions. First, staff managers exercise line authority over employees in their own departments; second, staff may have functional authority over the line in restricted areas of activity. This functional authority is delegated to an activity and gives managers performing the activity the right to command. Authority granted in this manner, however, is confined to the specialized area to which it was delegated.

The quality control manager, for example, may have functional authority over the work of supervisors in other departments. If inspectors find a product quality problem, they may require the supervisor to suspend production until the problem is corrected.

This example applies directly to a commissary foodservice operation. The microbiologist on the quality control staff may identify a problem with microbial count in a product being produced in a food factory and require that production be curtailed until the source of contamination is identified.

Administrative Intensity

Administrative intensity is the degree to which managerial positions are concentrated in staff positions (Griffin, 1993). Organizations sometimes try to balance their emphasis in terms of administrative intensity. A high administrative intensity organization is one with many staff positions relative to the number of line positions; low administrative intensity emphasizes more line positions. Some organizations are top heavy with staff positions. Organizations, however, would rather spend their money on line managers because by definition they are contributing to basic goals. A surplus of staff positions is a financial drain and an inefficient use of resources. Many organizations have reduced their administrative intensity in recent years by eliminating many staff positions.

Organization Chart

The **organization chart** graphically portrays the organization structure. It depicts the basic relationships of positions and functions while specifying the formal authority and communication network of the organization. The title of a position on the chart broadly identifies its activities; distance from the top indicates the position's relative status. Lines between positions are used to indicate the prescribed formal interaction.

The organization chart is a simplified model of the structure. It is not an exact representation of reality and therefore has limitations. For example, the degree of authority a superior has over a subordinate is not indicated. The chart, however, assists employees of the organization in understanding and visualizing the structure. Charts should be revised periodically because organizations are dynamic and undergo many changes over time.

Responsibility and authority for the preparation, review, and final approval of the organization chart generally lie with top management, although approval may be the responsibility of the board of directors. At the departmental level, the chart may be the responsibility of the department head, although approval may be required from the next level up in the organization.

Vertical organization charts have been the most conventional type, but currently the horizontal chart is being used in the new organizations. In the vertical chart, the levels of the organization are depicted in a pyramid form, with lines showing the chain of command. Special relationships may be indicated by the positioning of functions and lines on the chart. Dotted lines are often used to indicate communication links in an organization. Staff functions may be depicted by horizontal placement between top and department managers. Referring again to the Rush-Presbyterian-St. Luke's Medical Center traditional organization chart (Figure 17.1), the associate director of Foodservice Administration and the co-director of Nutrition Consultation Services are on the same level, which indicates equal responsibility and authority in the department.

Coordination

Coordination is a major component of organization structure (Griffin, 1993). **Coordination** is the process of linking activities of various departments in the organization. Job specialization and departmentalization involve breaking jobs down into small units and then combining the units to form a department. The activities in the department then need to be linked and focused on organizational goals. Some of the most useful means for maintaining coordination among interdependent units are horizontal interaction; policies, procedures, and rules; standards; communication; and committees and task forces.

In a large medical center, for example, *horizontal interaction* is required among departments. Nursing service and dietetics service staff often communicate directly rather than through the vertical organization. Such lateral relationships facilitate communication in an organization.

Policies, procedures, and rules ensure consistency in operations and are an important method of coordination. Managers may also establish schedules and other plans to coordinate action. Events are often unpredictable, however, and must be coordinated by managers using their judgment. Overreliance on rules and regulations can create problems in organizations.

Standards need to be developed before good coordination occurs in a foodservice operation. For example, large limited-menu restaurants usually have specific standards regarding production and service of products. One doughnut chain, for instance, requires that all products not sold within 4 hours after frying must be discarded. Specific formulations and frying procedures must also be followed.

Another way in which managers act to coordinate activities in an organization is in a so-called linking role. Communication is the responsibility of managers for linking with managers at higher levels in the organization and with others at their own level. This concept was derived from Likert's work (1961) in which he described managers as "linking pins."

Appointment of committees and task forces is a mechanism used in organizations for coordination. These groups serve an important role when problem solving must involve several departments. Problems involving half a dozen departments, for example, can be dealt with efficiently by such groups; otherwise, problems have to be referred upward through the chain of command.

Committees and task forces are common in all foodservice operations. Many foodservice organizations in metropolitan centers have formed task forces to recommend ways to feed the needy and homeless, and one solution many find is for the organization to contribute to municipal soup kitchens.

JOB ANALYSIS

Job analysis often is referred to as the base of human relations management because the information collected serves so many functions. Job analysis is the process of obtaining information about jobs by determining what the duties and tasks or activities of those jobs are (Sherman & Bohlander, 1992). In larger organizations, the job analysis often is performed by a specialist retained for that particular purpose. These professionals, identified as job analysts, most often have an industrial engineering background and have access to many tools for the analysis. The job analyst can observe the employee performing the job, have the employee complete a questionnaire, or keep a log. In smaller organizations, the analysis can be conducted by the supervisor of the job or the employee who holds the job. This analysis, however, will not be as sophisticated as that performed by a professional.

The procedure involves a systematic investigation of jobs by following a number of predetermined steps, such as collecting data on tasks, performance standards, and skills required. When completed, job analysis results in a written report summarizing the information from analyzing 20 or 30 individual job tasks or activities. Job analysis is concerned with objective and verifiable information about the actual requirement of a job. Job descriptions and job specifications developed through job analysis should be as accurate as possible to make them valuable to managers who make human resources management (HRM) decisions. The end products of a job analysis are job descriptions and specifications and performance standards.

Job Description

The **job description** lists the duties of a job; the job's working conditions; and the tools, materials, and equipment used to perform it (Griffin, 1993). Job descriptions are gaining importance because they serve as proof that indicates the foodservice operation is in compliance with the Americans with Disabilities Act (Weller, 1992). Most job descriptions include at least three sections: the job title, identification, and duties. If job specifications are not separate, they are usually placed at the end of the job description. A sample job description for a bartender is shown in Figure 17.8.

Job Title

Selection of a **job title** is important to employees because it gives them status as well as indicating their level in the organization. The title may be used to indicate, to a limited extent, the degree of authority the job possesses. The title "sanitation supervisor" indicates that the job involves more authority than "sanitation worker," and the title "head cook" indicates that the job is higher in the organization than "cook" or "assistant cook." Until recently, the titles of some jobs indicated that the job was for a "man" or a "woman," which implies that the job can be performed only by members of one gender. Thus, a waiter or waitress is now a waitperson.

The *Dictionary of Occupational Titles*, commonly referred to as the DOT and compiled by the U.S. Department of Labor (1991), contains standardized and comprehensive descriptions for about 20,000 jobs. An example of a foodservice job description is shown in Figure 17.9. These descriptions have helped bring about a greater degree of uniformity in the job titles and descriptions used by employers in different sections of the nation. The DOT code number facilitates the exchange of statistical data about jobs.

Job Identification

The **job identification** section of a job description usually follows the job title (Sherman & Bohlander, 1992). It includes such information as the departmental location of the job, the person to whom the jobholder reports, and often the number of employees in the department and the DOT code number. This information is found at the top of the sample job description for a bartender (Figure 17.8).

Job Duties

Statements covering **job duties** are usually arranged in their order of importance. These statements should indicate the weight or value of each duty generally measured by the percentage of time devoted to it, as shown in Essential Functions in the bartender job description (Figure 17.8). The description should also state responsibilities, identified as Qualification Standards in the same job description, what the duties entail, and the equipment used by the employee.

Figure 17.8.   Sample job description for a bartender.

SAMPLE JOB DESCRIPTION

Job Title: _____ Bartender _____   **Date:** _____ 1/15/94 _____

Department: _____ Bar _____   **Reports To:** _____ Lead Bartender _____

Prepared By: _____ Hunt (Lead Bartender) _____ **Approved By:** _____ Toms (Mgr) _____

JOB SUMMARY:
Describe the general purpose of the job or why the job exists.

Mixes and serves alcoholic and nonalcoholic drinks to patrons of bar and service bar following standard recipes.

ESSENTIAL FUNCTIONS:
List clear and precise statements of the responsibilities, job duties and major tasks performed—and percentage of time spent on each.

- Mixes ingredients such as liquor, soda, water and sugar to prepare cocktails and other drinks 30%
- Serves wine, draft beer or bottled beer 30%
- Collects money for drinks served 20%
- Orders liquors and supplies 5%
- Arranges glasses and bottles to make an attractive display 10%
- Slices and peels fruits for garnishing drinks 5%

ACCOUNTABILITIES:
Summarize the major results and key outcomes.

Bar and restaurant patron drink orders are filled quickly (within two minutes) and according to recipe. Daily bar inventories correspond with bar cash-register totals. Bar area is always neat and clean.

QUALIFICATION STANDARDS:
List the personal and professional qualifications required, including skill, experience, education, physical, mental safety, and other requirements.

- Able to operate a cash register and make change
- Stands during entire shift
- Reaches, bends, stoops, shakes, stirs and wipes
- Lifts 30-pound cases about 10 times per shift and carries cases up one flight of stairs
- Frequent immersion of hands in water (every 5 minutes)
- Hazards may include but are not limited to cuts from broken glass, burns, slipping, tripping
- Requires one year experience as a bartender

Source: "Preparing Written Job Descriptions" by N. Weller, 1992, *Restaurants USA, 12*(2), p. 21. Used with the permission of the National Restaurant Association.

Figure 17.9.   Job description for fast-foods worker.

311.472–010   FAST-FOODS WORKER (hotel & rest.) alternate titles: cashier, fast foods restaurant

Serves customer of fast food restaurant: Requests customer order and depresses keys of multicounting machine to simultaneously record order and compute bill. Selects requested food items from serving or storage areas and assembles items on serving tray or in takeout bag. Notifies kitchen personnel of shortages or special orders. Serves cold drinks, using drink-dispensing machine, or frozen milk drinks or desserts, using milkshake or frozen custard machine. Makes and serves hot beverages, using automatic water heater or coffeemaker. Presses lids onto beverages and places beverages on serving tray or in takeout container. Receives payment. May cook or apportion french fries or perform other minor duties to prepare food, serve customers, or maintain orderly eating or serving areas.
GOE: 09.04.01 STRENGTH: L GED: R2 M2 L2 SVP: 2 DLU: 86

Source: Dictionary of Occupational Titles (p. 241) by the U.S. Department of Labor, 1991, Washington, DC: Government Printing Office.

Job Specification

The **job specification** lists the abilities, skills, and other credentials needed to do the job (Griffin, 1993). The job specification, popularly referred to as the *job spec,* contains a statement of job conditions relating to the health, safety, and comfort of the employee, including any equipment and any job hazard. Knowing about job content and job requirements is necessary to develop selection methods and job-related performance appraisals and to establish equitable salary rates. In the bartender job description (Figure 17.8), the Qualification Standards are really a job specification that was included in the job description rather than making it a separate document.

One of the problems with job descriptions and specifications is that they are often very poorly written (Sherman & Bohlander, 1992). Job duties often are written in vague rather than specific terms. In today's legal environment, federal guidelines and court decisions require that specific performance requirements are based on valid job-related criteria.

Performance Standards

Performance is the attainment of a desired result and standard at a definite level of quality for a specified purpose; **performance standards**, therefore, define desired results at a definite level of quality for a specified job. Specific details in the standards that can be measured objectively are quality, quantity, and time factors, and they can be grouped under productivity. In many organizations, productivity standards are based on past performance.

In Figure 17.10, an excerpt from a job description for a lead cook is shown. The responsibility and four relevant performance standards have been identified by management with input from the cooks. Standards should be based on work-related behaviors and clearly stated in writing. Activities to be observed and measured

Figure 17.10.   Excerpt from a job description with performance standards for a lead cook.

COMMUNITY HOSPITAL

DEPARTMENT OF FOOD AND NUTRITION SERVICES

JOB DESCRIPTION

JOB TITLE: ___Lead Cook___   DEPARTMENT: ___Dietary___   DATE: ___September 1994___

JOB TITLE OF PERSON TO WHOM REPORTING: ___Production Mgr.___   JOB CODE #: ___116___

JOB PAY: REVISED: ___FEBRUARY 1991___

NO. OF PERSONS SUPERVISED: ___N/A___   EDUCATION REQUIREMENTS: ___High School___
___or Equivalent. Education in Food Service or___
___Cooking Desired.___

PRIOR EXPERIENCE REQUIREMENTS: One year experience or equivalent in institutional cooking, preferably in a healthcare setting.

OTHER COMMENTS: Exposed to heat, humidity, steam, cooking odors, refrigerator temperatures and wet floors. Possible job related injuries include serious cuts from knives or power equipment, burns from cooking equipment and strains, sprains, or falls. Work is performed while standing or walking. Occasionally exerts considerable physical effort in moving or lifting of supplies and/or hot food items.

JOB SUMMARY: Works as a team member with one or more cooks in the daily production requirements for patient and/or employee foodservices. Prepares meats, fish, fowl, vegetables, gravies, sauces, soups, salad ingredients and baked goods according to standardized recipes. Assures freshness, proper serving temperatures, and the minimization of food waste.

Responsibilities	Performance Standards
3 Maintains standards of quality as specified by the Department of Dietetics and all basic food handling guidelines as specified by local, state and federal health agencies.	**A** All foods are to be stored at proper temperatures. Cold foods at or below 45°F, hot foods at 140°F or above. Holding and processing temperatures between these ranges should not exceed 4 hours.
	B All foods are to be covered, labeled and dated when stored.
	C All foods are to be rotated on a first in first out basis in accordance with the department's standards for holding and storing foods.
	D All foods are to be served at the appropriate serving temperature as specified on the steam-table layout diagrams.

should be obvious in the standard. Measuring performance of a standard becomes an unavoidable challenge to the manager in evaluating employees.

According to Puckett and Miller (1988), in most organizations, the performance evaluation system is developed by the human resources department. The mechanics of developing an evaluation instrument and determining how often employees will be evaluated, however, is the responsibility of the manager. The rating scale for each performance standard should be the same as the one used elsewhere in the organization. Generally, descriptive terms such as *always, almost always, sometimes, rarely,* and *never* are assigned numbers to make the scores as objective as possible. The employee receives a score for each performance standard and a total score for each responsibility.

The percentage of time spent by the employee in each responsibility gives the weight of relative importance and is determined by the manager or the employees in the unit. A performance evaluation score is determined by multiplying the total score for each responsibility by its weight. Pesci, Spears, and Vaden (1982) developed a methodology for devising performance standards for employees in a university residence hall foodservice. Their work involved defining major responsibilities and related standards for the position of Cook II. Table 17.1 shows the resulting weights for the various areas of responsibility, as based on the priority ratings provided by the cooks themselves.

Performance evaluations generally are used for salary increases and promotions. Unquestionably, each employee should be given a fair and just performance evaluation.

Table 17.1.   Weights for each major responsibility for Cook II

Area of Responsibility	Foodservice Unit			Overall	Final Weights
	1	2	3		
	← % priority[a] →				%
1. Food production	48	47	43	47	45[b]
2. Equipment care	5	5	6	5	5
3. Storage and handling of food	3	3	4	3	5
4. Employee training	15	16	17	16	15
5. Personal hygiene	3	2	3	3	5
6. Work habits	26	27	27	26	25

[a] Based on priority ratings of tasks within each area of responsibility provided by cooks in each of the food centers.

[b] Weights proposed after review of reliability data and % priority from each food center. Overall % priority data were adjusted to reflect 5% increments for each major responsibility as suggested by Kansas State University, Personnel Services.

Source: "A Method for Developing Major Responsibilities and Performance Standards for Foodservice Personnel" by P. H. Pesci, M. C. Spears, and G. Vaden, 1982, *NACUFS Journal*, pp. 17–21. Used by permission.

JOB DESIGN

Job design is an outgrowth of job analysis and is concerned with structuring jobs to improve organization efficiency and employee **job satisfaction** (Sherman & Bohlander, 1992). Work needs to be divided into manageable units and eventually into jobs that can be performed by employees. A job is a set of all tasks that must be performed by a given employee (Chase & Aquilano, 1992). Each job should be clear and distinct to prevent employees from misunderstanding the job and to help them recognize what is expected of them (Sherman & Bohlander, 1992).

Job design is a complex function because of the variety of factors that enter into arriving at the ultimate job structure, as shown in Figure 17.11 (Chase & Aquilano, 1992). Decisions have to be made about who will perform the job and where and how it is to be performed. One of the considerations in job design is the quality of work life of employees. Job enlargement, job enrichment, job characteristics, and employee work teams are all programs that increase the QWL in organizations.

Job Enlargement

Job enlargement was developed to increase the total number of tasks that employees perform. Tasks are the individual activities that make up a job. Peeling vegetables and measuring or weighing ingredients, for example, are two tasks that are part of an ingredient room worker's job. The rationale for enlarging jobs is that employees become bored if they have to do the same job day after day.

Figure 17.11. Factors in job design.

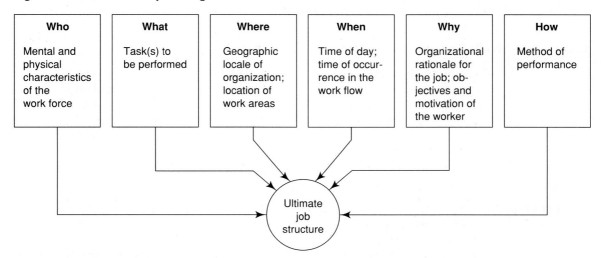

Source: Production and Operations Management, A Life-Cycle Approach (p. 495) by R. B. Chase and N. J. Aquilano, 1992, Homewood, IL: Richard D. Irwin.

Job Enrichment

Job enrichment assumes that increasing the variety of tasks is not enough to improve employee motivation. Thus, job enrichment attempts to increase the number of tasks as well as the control the employee has over the job. Frederick Herzberg was a proponent of job enrichment, which he believed motivated employees while achieving job satisfaction and performance goals. He concluded that the five factors for enriching jobs and motivating employees are achievement, recognition, growth, responsibility, and performance of the complete job rather than parts of it. These factors allow employees to assume more decision-making power and become more involved in their own work. Job enrichment could be a component of TQM because employees, who are internal customers, are empowered to make decisions, which leads to job satisfaction.

Job Characteristics

A new area for job design studies was researched when behavioral scientists focused on identifying various job dimensions that would improve the efficiency of organizations and the job satisfaction of employees (Sherman & Bohlander, 1992). The classic studies on job characteristics by Hackman (1977) resulted in the definition of five job dimensions or task characteristics: skill variety, task identity, task significance, autonomy, and feedback. If these are present to a high degree, Hackman observed that meaningfulness of work, responsibility, and knowledge of actual results of work activities could contribute to work performance and job satisfaction. Hackman concluded that the greater the extent of all five task characteristics in a job, the more likely it is that the job holder will be highly motivated and experience job satisfaction. The **job characteristics** model works when employees have the desire for the autonomy, variety, responsibility, and challenge of enriched jobs. Otherwise, redesign of jobs almost always fails and employees are frustrated.

Employee Work Teams

The outgrowth of the job enlargement, enrichment, and characteristics models is *employee work teams,* in which two or more employees interact regularly to accomplish a narrow set of goals within a certain time period. Groups then design and assign jobs consisting of interrelated tasks to each member (Griffin, 1993). These employee work teams are created by management to solve specific problems, such as serving a customer 15 minutes after the order for pizza is placed.

SUMMARY

The world is in a period of transformation, and organizations are changing from being traditional to much more participative. The organization structure is based on the objectives that management has established and on plans and programs to achieve these objectives. Different types of structures will be required for traditional and new organizations. Quality of work life and corporate culture are being emphasized in the new organization.

Organizing basically is a process of division of labor, either vertically or horizontally. Vertical division of labor, which has been used by traditional organizations, is based on authority that flows from the highest to the lowest ranks of employees. Organizations with the new philosophy are using a horizontal division of labor with an emphasis on encouraging employees to share ideas across all levels and departments. By empowering employees, layers of middle management have been eliminated. Departmentalization is the most frequently used method for implementing division of labor. Labor also may be divided into line and staff. The organization chart graphically portrays the organization structure and shows the relationship of positions and functions while specifying the formal authority and communication network. Coordination is a major component of organization structure because it links activities in various departments.

Job analysis is a procedure involving a systematic investigation of jobs to determine the duties and tasks of each job. The end products are job descriptions and specifications and performance standards. Job design is an outgrowth of job analysis and is concerned with structuring jobs to improve organization efficiency and employee job satisfaction. One of the considerations in job design is the quality of work life of employees.

BIBLIOGRAPHY

Chase, R. B., & Aquilano, N. J. (1992). *Production and operations management: A life-cycle approach* (4th ed.). Homewood, IL: Richard D. Irwin.

Drucker, P. F. (1992). The new society of organizations. *Harvard Business Review, 70*(5), 95–104.

Fintel, J. (1989). Restaurant cultures: Positive cultures can keep companies healthy. *Restaurants USA, 9*(10), 12–16.

Ghorpade, J., & Atchison, T. J. (1980). The concept of job analysis: A review and some suggestions. *Public Personnel Management, 9*, 134–143.

Graicunas, V. A. (1973). Relationship in organization. In L. Gulick & F. L. Urwick (Eds.), *Papers on the Science of Administration* (pp. 181–187). New York: Institute of Public Administration. Original work published 1933.

Griffin, R. W. (1993). *Management* (4th ed.). Boston: Houghton Mifflin.

Hackman, J. R. (1977). Work design. In J. R. Hackman & J. L. Suttle (Eds.), *Improving life at work: Behavioral science approaches to organizational change* (pp. 96–162). Glenview, IL: Scott, Foresman.

Haimann, T., Scott, W. G., & Conner, P. E. (1985). *Managing the modern organization* (4th ed.) Boston: Houghton Mifflin.

Hirschhorn, L., & Gilmore, T. (1992). The new boundaries of the "boundaryless" company. *Harvard Business Review, 70*(3), 104–115.

Kanter, R. B. (1992). The best of both worlds. *Harvard Business Review, 70*(6), 9–10.

Kast, F. E., & Rosenzweig, J. E. (1985). *Organization and management: A systems and contingency approach* (4th ed.). New York: McGraw-Hill.

Katzenbach, J. R., & Smith, D. K. (1993). The discipline of teams. *Harvard Business Review*, *71*(2), 111–120.

Likert, R. (1961). *New patterns of management*. New York: McGraw-Hill.

Mintzberg, H. (1981). Organization design: Fashion or fit? *Harvard Business Review*, *59*(1), 103–116.

Pesci, P. H., Spears, M. C., & Vaden, A. G. (1982). A method for developing major responsibilities and performance standards for foodservice personnel. *NACUFS Journal*, *4*, 17–21.

Puckett, R. P., & Miller, B. B. (1988). *Food service manual for health care institutions*. Chicago: American Hospital Publishing.

Sherman, A. W., & Bohlander, G. W. (1992). *Managing human resources* (9th ed.). Cincinnati: South-Western Publishing.

Urwick, L. F. (1938). *Scientific principles and organizations*. Institute of Management Series No. 19. New York: American Management Association.

U.S. Department of Labor, Employment and Training Administration. (1991). *Dictionary of occupational titles* (4th ed.). Washington, DC: Government Printing Office.

Uyterhoeven, H. E. R. (1992). General managers in the middle. *Harvard Business Review*, *70*(2), 75–85.

Van Warner, R. (1993). Reinventing the organization. *Nation's Restaurant News*, *27*(26), 29.

Weber, M. (1947). *The theory of social and economic organizations*. New York: Free Press.

Weller, N. (1992). Preparing written job descriptions. *Restaurants USA*, *12*(2), 20–23.

18

Linking Processes

Linking processes are needed to coordinate the activities of the system so they can accomplish the goals and objectives. These linking processes are decision making, communication, and balance (Figure 18.1). In chapter 2, the critical role of these processes in the transformation element of the foodservice system was illustrated.

Decision making is the selection of a course of action from a variety of alternatives, and communication is the vehicle whereby decisions and other information are transmitted. Balance concerns management's ability to maintain organizational stability, which is related to effective decision making and communication.

Linking processes are integral to management's effectiveness in performing the functions of planning, organizing, staffing, leading, and controlling. In fact, decision making is sometimes described as the essence of management because managers are making decisions in almost every aspect of their jobs. According to Etzioni (1989), managers of today and tomorrow face continuing information overloads but

Figure 18.1. Foodservice systems model with the linking processes highlighted.

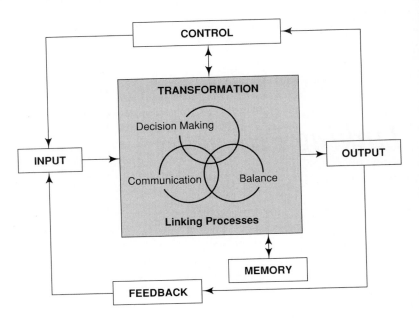

little growth in the amount of knowledge they can use for most complex decisions. Decision makers need to know more than ever before. With the advent of computers, the capacity to collect and process information has grown, but information is not the same as knowledge. Decision making requires seeking the facts from all the information and relating them to current situations so they become usable knowledge.

The decision and communication networks overlay and connect the managerial functions. For example, decisions about standards of control are made during planning and must be communicated to those responsible for controlling activities. To illustrate this point, consider the standards for portion sizes developed in the production planning process and used as a base for service. If periodic checks of roast beef indicate that portions are too large, the service supervisor would communicate the need for corrective action to the cafeteria server.

The linking processes govern the flow of resources or system inputs. Through the networks of decision making and communication, the movement of people, money, and equipment through the system is delineated. These networks also focus resources on the organization's purpose.

In chapter 3, decisional roles of the manager as initiator, disturbance handler, resource allocator, and negotiator were described. Within each of these roles, the manager may be responsible for a range of decisions. The authority delegated to a manager will permit him or her to commit the organization to new courses of action or determine the organization's strategy. As Mintzberg (1975) indicated in describing managerial roles, communication and information come into play by providing input for decision making. Key elements of decision making and communication are discussed in this chapter.

DECISION MAKING

Managers make decisions for the purpose of achieving individual and organizational objectives. Effective managers must be good decision makers. Decision making involves three primary stages—first, definition of the problem; second, identification and analysis of possible courses of action; and third, actual selection of a particular course of action.

Analyzing the decision processes by these stages illustrates the difference between management and nonmanagement decisions. Managerial decisions encompass all three stages; nonmanagerial decisions are concentrated in the last, or choice, stage. For example, when a customer complains that the steak is too well done, the waitperson may follow the traditional course of action and exchange it for a satisfactory one. An employee who is empowered to make decisions under the TQM philosophy might decide that that remedy doesn't go far enough, however, and remove the steak from the check as well. Management, however, must place great emphasis on the problem identification and design phase of decision making.

Types of Decisions

Foodservice managers must make many different types of decisions. Most decisions, however, fall into one of two categories: programmed and nonprogrammed.

Programmed Versus Nonprogrammed

Decisions are often classified as programmed or nonprogrammed. **Programmed decisions** are reached by following established policies and procedures. Normally, the decision maker is familiar with the situation surrounding a programmed decision. These decisions also are referred to as *routine* or *repetitive* decisions. Limited judgment is called for in making programmed decisions. These decisions are made primarily by lower-level managers in an organization. At the doughnut chain that discards products four hours after frying, referred to in chapter 17, supervisors are faced with programmed decisions. These are relatively simple routine decisions regarding whether to retain or discard products, and they are based on the company's rules.

Nonprogrammed decisions are unique and have little or no precedent. These decisions are relatively unstructured and generally require a more creative approach on the part of the decision maker than programmed decisions. Often when dealing with nonprogrammed decisions, the decision maker must develop the procedure to be used. Naturally, these decisions tend to be more difficult to make than programmed decisions. Deciding on a location for a new unit of the doughnut chain and selecting a new frying process and equipment are examples of nonprogrammed decisions.

Because programmed decisions are concerned mostly with concrete problems that require immediate solutions and are frequently quantitative in nature, they tend to be reached in a short time. Nonprogrammed decisions usually involve a longer time horizon because they tend to focus on qualitative problems and require a much greater degree of judgment. When making such judgmental decisions, managers must frequently rely on wisdom, experience, and philosophic insight rather than on established policies and procedures.

Kinds of decisions are also related to level of management in the organization. In chapter 3, the tasks of higher-level managers were described as having a longer time frame and involving more judgment than those of lower-level managers. Conversely, the time perspective of the lower-level manager is shorter, and the tasks need less judgment. Therefore, higher-level managers tend to make mostly nonprogrammed and lower-level managers mostly programmed decisions (Figure 18.2). The foodservice supervisor, for example, is concerned with day-to-day operational decisions, whereas the director of foodservice is responsible for decisions concerned with the future of the overall organization.

Organizational Versus Personal

Nonprogrammed decisions are of two general kinds—organizational and personal. Organizational decisions relate to the purposes, objectives, and activities of the organization, and personal decisions are concerned with the manager's individual goals. These personal decisions, however, may affect the organization, and vice versa.

A manager may decide to resign from a position in a university foodservice, for example, to take a job in a large commercial foodservice organization. After the move from the university, the commercial company may decide to transfer the manager from one regional office to another. The first decision is a personal one affecting two organizations and the second, an organizational decision affecting the individual.

The Decision-Making Process

Foodservice managers make decisions constantly and seldom have time to do a scientific analysis of every problem. The steps outlined in Figure 18.3 show the steps a manager should follow to make rational and logical decisions. These steps in rational

Figure 18.2.   Types of decisions according to levels of management.

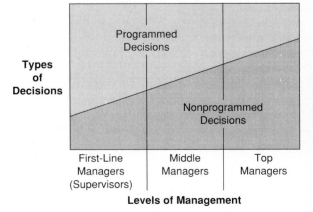

Figure 18.3. Steps in the rational decision-making process.

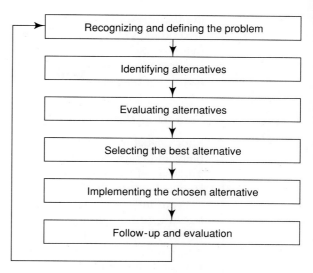

Recognizing and defining the problem

Identifying alternatives

Evaluating alternatives

Selecting the best alternative

Implementing the chosen alternative

Follow-up and evaluation

Source: Griffin, Ricky W., *Management,* Fourth Edition (p. 208). Copyright © 1993 by Houghton Mifflin Company. Adapted with permission.

decision making will keep the decision maker focused on facts and guard against inappropriate assumptions and pitfalls.

Recognizing and Defining the Situation

A stimulus, which can be either positive or negative, is necessary for someone to recognize that there is a problem and a decision is necessary. The problem has to be clearly defined before action is taken. The foodservice manager should understand the problem and its relationship to other factors in the system.

Identifying Alternatives

Once the decision is recognized and defined, alternatives should be identified. Obvious alternatives should be examined as well as those that are creative. The more important the decision, the more attention should be placed on selecting alternatives. When selecting creative alternatives, constraints (legal restrictions, moral and ethical norms, or the power and authority of the manager, for example) might restrict which ones are chosen.

Evaluating Alternatives

Foodservice managers should evaluate all alternatives to determine if they meet the needs of the operation and the feasibility and consequences of using them. For example, in selecting a supplier, a foodservice manager might consider such criteria as dependability, price, delivery time, and variety of products offered. After identifying the possible choices and collecting information, each supplier should be evaluated on the various criteria and then ranked. This evaluation must be conducted carefully to be sure that the selected alternative is successful. How the decision will affect the operation and what the cost will be are major considerations in selecting an alternative. For big decisions, managers should estimate the prices of the alternatives.

Selecting the Best Alternative

After evaluating the alternatives for meeting the needs of the operation and the feasibility and consequences of using them, probably most alternatives will be rejected. The crux of decision making is choosing the best out of those remaining. The decision maker often has to make a subjective choice among the final choices.

Even under the best of conditions, all feasible alternatives cannot be developed because of human or situational limitations. Rarely can managers make decisions that maximize objectives because complete information is never available about all the possible alternatives and their potential results. Complete knowledge of alternatives and consequences would allow the best possible alternative to be chosen; managers often have to make what March and Simon (1958) call a "satisficing," or satisfactory but not optimal decision, because information is lacking. Also, managers may tend to choose the first satisfactory alternative and discontinue their search for additional alternatives. Another pitfall is to continue the search for alternatives when, in fact, the manager may be avoiding making a decision.

Figure 18.4 represents the satisficing approach to decision making. As shown in the model, if the decision maker is satisfied that an acceptable alternative has been found, that alternative is selected. Otherwise, the decision maker searches for an additional alternative. If the value of the new alternative is greater than that of the previous one consistent with the current level of aspiration, it will be chosen. The double arrows indicate a two-way relationship between value and aspirations. The net result determines whether or not the decision maker is satisfied with the alternative. Thus, the manager selects the first alternative that meets the minimum satisfaction criteria and makes no real attempt to optimize; that is, the manager makes a "satisficing" decision, rather than the best possible decision.

Implementing the Chosen Alternative

After an alternative is chosen, it should be implemented into the organization, but that can be difficult to do. When implementing a decision, managers must consider people's resistance to change, which might include insecurity, inconvenience, and fear of the unknown. The decision-making process does not end when the decision is made. The decision must be implemented, and the manager must monitor results to ascertain that the selected alternative solves the problem.

Follow-up and Evaluation

Managers finally have to evaluate the effectiveness of their decision. They need to decide if their chosen alternative was the correct one. If it is not, maybe the second

Figure 18.4.   Model of the satisficing approach.

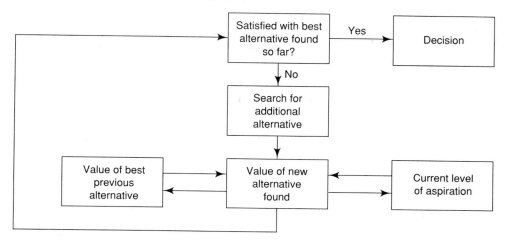

Source: Adapted from *Organizations* (p. 49) by J. G. March and A. Simon. Copyright © 1958 John
Wiley & Sons, Inc. Reprinted by permission of John Wiley & Sons, Inc.

or third alternative would be better or maybe the problem was not correctly defined.
Perhaps the chosen alternative is the best, but more time may be needed or perhaps
the implementation needs to be revised.

Monitoring outcomes results in feedback that should be considered in future
determination of objectives and in decision making. Many good decisions fail
because of poor implementation. The decision must be communicated properly, and
support among employees must be organized. For example, a computer program for
inventory control in a foodservice will have no value until personnel implement it. In
addition, the necessary resources, especially personnel, must be assigned for implementation of the program. Managers often assume that once they make a decision,
their role is over. This is far from true. Proper implementation of the decision is a
critical component in decision making.

Conditions for Making Decisions

Regardless of the approach used, decisions are frequently made at one time for
events that will occur at another, and the conditions are seldom identical. Take, for
instance, the decision to produce sufficient entrées for the usual 2,000 students who
come to lunch on Wednesday in a university residence hall foodservice. This decision may result in excessive overproduction if many students choose not to dash
across campus for lunch because of an unexpected downpour occurring between
11:30 A.M. and 12:30 P.M. Although nature obviously is not under the control of the
decision maker, it affects the outcome of the decision.

The environment within which the decision maker operates, therefore, affects the
decision-making process. Conditions in the environment change and predictions are
difficult; yet managers must make decisions based on the information available, even

Figure 18.5. Certainty-uncertainty continuum in decision making.

though it may be incomplete or involve factors outside their control. These conditions under which decisions are made are referred to as *certainty, risk,* and *uncertainty* (Figure 18.5). They tend to vary with the time frame that encompasses the decision. The longer the future time period involved in the decision, the less certain are the environmental conditions. The various degrees of certainty in relation to time frame are illustrated in Figure 18.6. As shown, the tendency is to move on the certainty/uncertainty continuum into conditions of risk and uncertainty as the time frame becomes longer.

Conditions of Certainty

Under **conditions of certainty**, a decision maker has adequate information to assure results. A decision under conditions of certainty involves choosing the alternative that will maximize the objective. For example, in school foodservice, the number of children scheduled to eat lunch in an elementary school is usually determined by teachers soon after school convenes each morning. These data are transmitted to the school lunch manager, who can then determine production needs based on a known number of lunch participants. Guesswork is not involved in estimating the number of students who will eat. Under conditions of certainty, management science

Figure 18.6. Effect of time and level of knowledge on decision-making situations.

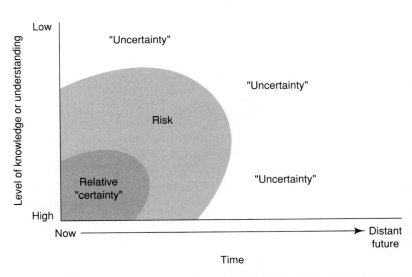

Source: Introduction to Management (p. 170) by E. H. Burack and N. J. Mathys, Copyright © 1983 by John Wiley & Sons, Inc. New York: John Wiley & Sons, Inc.

techniques, such as linear programming, break-even analysis, and inventory control models, have been used effectively.

Conditions of Risk

Because conditions of certainty are becoming less common in today's complex and rapidly changing world, estimating the likelihood or probability of various events occurring in the future is often the only possibility for planning. This condition is called *risk*. Under **conditions of risk**, various probability techniques are helpful in making decisions. In decision making under risk, managers are faced with the possibility that any one of several things may occur. The assumption is that based on the manager's research, experience, and other available information, the probabilities are known about each of these various states.

In the earlier example of the university foodservice and the effect of rain on lunch participation, at the time of production forecasting, the manager might have information about the potential for rain from a weather forecast. The manager can also review past records to determine the predicted effect on weekday meal attendance when rain occurs over the lunch hour. Therefore, while the outcome cannot be known with certainty and the manager is taking a risk in terms of possible over- or underproduction, the weather forecast and prior records may assist in making more accurate projections about participation should an unexpected downpour occur.

Probabilities enable the calculation of an expected value of given alternatives. The expected value of a particular decision is the average of returns that would be obtained if the same decision were made in the same situation over and over again.

Conditions of Uncertainty

When the occurrence of future events cannot be predicted, a state of uncertainty exists. Many changes or unknown facts can emerge when decision time frames are long. To predict what is likely to occur with any degree of certainty, therefore, is quite difficult. In these situations, foodservice managers frequently apply their experience, judgment, and intuition to narrow the range of choices. Input from others may help reduce some of the uncertainty. Involvement of knowledgeable people in the decision process, therefore, may be beneficial. Under **conditions of uncertainty**, some managers will delay decisions until conditions stabilize or will take a path of least risk.

If a decision must be made, however, even though the decision maker has little or no knowledge about the occurrence of various states of nature, one of three basic approaches may be taken. The first is to choose the alternative that is the best of all possible outcomes for all alternatives. This is an *optimistic approach* and is sometimes called the *maximum* approach. A second approach in dealing with uncertainty is to compare the worst possible outcome of each of the alternatives and select the alternative with the least possible negative outcomes. This is the *pessimistic approach*, sometimes called the *maximin* approach. The final approach is to choose the alternative that has the least variation among its possible outcomes. It is a *risk-*

averting approach and may make for effective planning; however, the payoff potential is less than with the optimistic approach.

Decision-Making Techniques

Various techniques have been developed to assist managers in making decisions. Some of these techniques are highly complex and quantitative in nature. In this section, selected techniques are described briefly; however, the mathematics and computations are beyond the scope of this book. Those techniques that have specific applications to procurement or production were discussed in chapters 7 through 11.

Decision Trees

Decision trees allow management to assess the consequences of a sequence of decisions with reference to a particular problem. The approach involves linking a number of event "branches" graphically, which results in a schematic resembling a tree. The process starts with a primary decision that has at least two alternatives to be evaluated. The probability of each outcome is ascertained, as well as its monetary value.

A simplified example of a decision tree is shown in Figure 18.7. The decision in this example concerns the expansion of a restaurant's services to include take-out food. The two alternatives are to expand or not to expand. The restaurateur is faced with the probability of the competitor on the next block also introducing take-out service. To assess the decision in question, the probabilities for all these occurrences should be determined, as well as the effect on net income if each were to occur. This determination will permit the restaurant owner to make the decision with the best potential payoff.

Cost-Effectiveness

Cost-effectiveness is a technique that provides a comparison of alternative courses of action in terms of their cost and effectiveness in attaining a specific objective. It is customarily used in an attempt to minimize dollar cost, subject to some goal requirement that may not be measurable in dollars, or in a converse attempt, to maximize some measurement of output subject to a budgetary constraint.

Cost-effectiveness has been given a great deal of emphasis in recent years in public programs and in public institutions. Concern over increasing costs in tax-supported services has led to the use of this type of analysis.

A structure-of-choice model can be used to illustrate the cost-effectiveness concept. Once the desired goal is determined, alternatives for meeting it are listed. The alternatives are then evaluated for both cost and effectiveness and then ranked in order of desirability by comparison with defined criteria. The alternative selected may not always be the least costly because potential effectiveness is a major consideration.

Quite often, the final decision is based on selecting the best of two alternatives, either of which is considered acceptable. A simplified comparison of alternatives is

Figure 18.7. Example of a decision tree for making a decision on expansion of restaurant services.

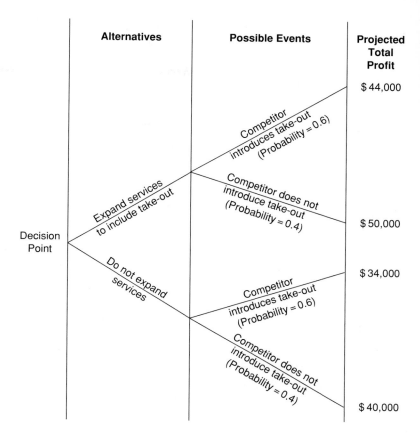

Alternatives	Possible Events	Projected Total Profit
		$ 44,000
	Competitor introduces take-out (Probability = 0.6)	
Expand services to include take-out	Competitor does not introduce take-out (Probability = 0.4)	$ 50,000
Decision Point		$ 34,000
Do not expand services	Competitor introduces take-out (Probability = 0.6)	
	Competitor does not introduce take-out (Probability = 0.4)	$ 40,000

illustrated in Figure 18.8. The cost-effectiveness curves for alternatives *A* and *B* intersect at a point where, for the same cost, the effectiveness is identified. At a cost less than that at the point of intersection, alternative *B* is given greater effectiveness; if costs higher than that at the point of intersection can be tolerated, alternative *A* is preferable. The judgment of the decision maker comes into play in answering the question of whether or not the gain in effectiveness of *A* over *B* at a higher cost is worthwhile.

The cost-effectiveness concept can be illustrated by using a product cost analysis for 48 blueberry muffins done at Pillsbury's Foodservice Technical Center (Table 18.1). Four methods for preparing blueberry muffins are analyzed for labor and ingredient cost and for meeting the needs of a specific foodservice operation. The goal, which is shown on the structure-of-choice model (Figure 18.9), is to serve a high-quality product that meets customers' expectations at the lowest possible cost. The four alternatives are:

- Scratch recipe (Product 1)
- Pillsbury Basic Muffin Mix (Product 2)
- Pillsbury TubeSet® Frozen Batter (Product 3)
- Pillsbury Frozen Baked Muffins (Product 4)

Figure 18.8. Comparison of alternatives in decision making.

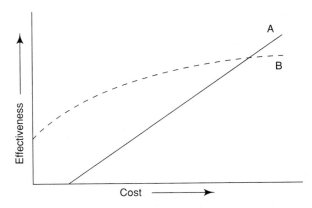

Source: M. C. Spears: "Concepts of Cost Effectiveness: Accountability for Nutrition, Productivity." Copyright The American Dietetic Association. Reprinted by permission from *Journal of the American Dietetic Association*, 68, p. 343, 1976.

The criteria for the selection are that the ingredients or muffins are available on the market, the final product looks good, the customer is pleased, and the food and labor costs are minimum. The effectiveness of each of the four types of muffins will be determined by a taste panel that includes chefs, managers, and customers; a questionnaire for customers to complete; and customer comments gathered by waitpersons. The cost analysis done by Pillsbury's Foodservice Technical Center is used for the cost component of the cost-effectiveness model. The kitchen for the coffee shop is small but has a convection combo (combination convection steamer–convection oven) but does not have a trained baker. After conducting a cost-effectiveness analysis, the alternatives are ranked and the final decision is to use the Pillsbury Muffin Mix. Muffins can be baked according to demand during the breakfast hours, giving a home-made touch to the product, and the cost of each muffin is low enough to make a good profit for the operation. A small supply of the second alternative, Pillsbury Frozen Baked Muffins, however, would be kept in the freezer for emergencies.

Networks

The Program Evaluation and Review Technique (PERT) and Critical Path Method (CPM) are networks for decision making. A **network** is a graphic representation of a project, depicting the flow as well as the sequence of defined activities and events. An *activity* defines the work to be performed; an *event* marks the beginning or end of an activity.

According to Leitch (1989), PERT and CPM currently are the two most widely known and used management science techniques for planning, scheduling, and controlling large projects. For example, a catering manager could use these techniques to improve the planning and control of a complex function such as a wedding. These two techniques have been combined into a PERT-type system that uses a network to

Table 18.1. Product cost analysis for 48 2-oz blueberry muffins

Production Step	Scratch Recipe	Prep Time		
		Pillsbury Basic Muffin Mix[a] (6/5 lb. cartons)	Pillsbury TubeSet® Frozen Batter (6/3 lb. tubes)	Pillsbury Frozen Baked Muffins (96 muffins)
Pan prep (paper liners)	2 min.	2 min.	2 min.	0 min.
Ingredient prep/weighing	6¼ min.	1½ min.	0 min.	0 min.
Mixing and scraping bowl	5½ min.	1¾ min.	0 min.	0 min.
Portioning batter into pans	3 min.	3 min.	4 min.	0 min.
Loading oven	¾ min.	¾ min.	¾ min.	0 min.
Total preparation time	17½ min.	9 min.	6¾ min.	0 min.
Baking time	20 min.	20 min.	20 min.	0 min.
Total prep and bake time	37½ min.	29 min.	26¾ min.	0 min.
Labor cost per batch[b]	$2.14 ($0.044 each)	$1.10 ($0.022 each)	$0.83 ($0.017 each)	$0.00 ($0.00 each)
Clean-up time (and cost)[b]	3 min. ($0.0044 each)	1½ min. ($0.0022 each)	0 min.	0 min.
Ingredient cost per muffin[c]	$0.06 each	$0.11 each	$0.147 each	$0.245 each
Total cost per muffin	$0.1084 each	$0.134 each	$0.164 each	$0.245 each

a Prepared with 12 oz. of frozen blueberries.

b Preparation labor costs based on wages of $7.35/hr.

Labor costs calculated by rounding total preparation time to the nearest minute.

Baking times not included. Clean-up costs based on wages of $4.25/hr.

c Based on anticipated distributor price and case yield information.

Source: Grand Metropolitan Foodservice USA, Minneapolis, MN. Used by permission.

Figure 18.9. Structure-of-choice model for selecting one type of blueberry muffin.

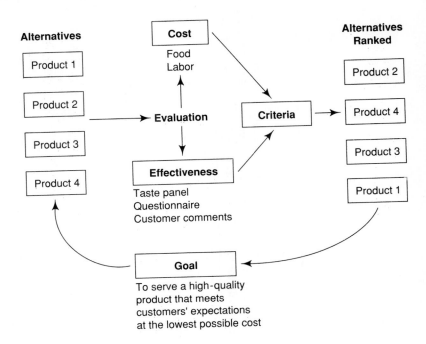

depict the sequence of activities in a particular project. A *project* is defined as a group of tasks that are performed in a certain sequence to reach an objective. In a network diagram, an event is represented by a node (circle), and each activity by a line that extends between two events. Directional arrows are drawn on each line to indicate the sequence in which events must be accomplished.

The first step in managing a project is to specify all the activities required to complete it. The example used by Leitch is the activity sequence in setting up a banquet and serving guests. In Table 18.2, the sequential activities, each identified by a letter, are listed and described. The immediate predecessor for each activity also is listed. For example, activity I, plate salad and serve, cannot begin until activity G, prep salad and store, is completed. Activity A, however, can begin anytime because it does not depend on the completion of other activities.

The next step is to construct the network diagram that connects all these activities. Figure 18.10 is the network representation of the project described in Table 18.2. Because the PERT-type system is typically used for projects in which the time when activities are finished is uncertain, the following three time estimates need to be identified for each activity:

- Most optimistic time, or shortest time, assuming most favorable conditions
- Most likely time, which implies the most realistic time
- Most pessimistic time, or the longest time, assuming the most unfavorable conditions

In the example, a time is estimated for each of the three possible outcomes—optimistic, likely, and pessimistic times—and the mean time in hours is shown on the

Table 18.2.   PERT activities

Activity	Description of Activity	Immediate Predecessor
A	Bring tables and chairs up from basement and arrange in hall	—
B	Pick up tablecloths from laundry and place on tables	A
C	Arrange place setting and decorations	B
D	Fill water glasses	C
E	Turn ovens on and perform equipment check	—
F	Prepare and cook main course	E
G	Prepare salad and store	—
H	Seat guests	D
I	Plate salad and serve	G,H
J	Plate dinner and serve	F,I

Source: "Applications of a PERT-Type System and 'Crashing' in a Food Service Operation" by G. A. Leitch, 1989, *FIU Hospitality Review, 7*(2), p. 67. Used by permission.

Figure 18.10 diagram. Once the times have been plotted for each activity, the minimum time to set up the banquet and serve the guests can be determined. This is done by adding the amount of time, identified as *time duration,* for each path through the network. The path with the longest duration also is known as the *critical path* because if any activities are delayed, the completion of the project also will be delayed. In the example, the manager should not be concerned with the time required to prep the salad, turn on the ovens, or prepare and cook the main course but instead should be concerned with activities A, B, C, D, H, I, and J. These activities lie on the critical path and could delay dinner if they are not completed on time. The

Figure 18.10.   Example of a PERT network chart.

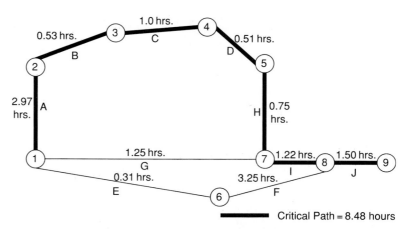

Source: Adapted from "Applications of a PERT-Type System and 'Crashing' in a Food Service Operation" by G. A. Leitch, 1989, *FIU Hospitality Review, 7*(2), pp. 66–67. Used by permission.

expected completion time for this event is 8.48 hours (Leitch, 1989). The foregoing is based on the Leitch article, and no effort was made to disclose the calculations. Currently, computer software is commercially available for such calculations.

Linear Programming

Linear programming is a technique useful in determining an optimal combination of resources to obtain a desired objective. Loomba (1978) contended that linear programming is one of the most versatile, powerful, and useful techniques for making managerial decisions. This concept has been used in solving a broad range of problems in industry, government, healthcare, and education. A few examples of the kinds of decisions that can be determined by linear programming are the optimal product mix, transportation schedule, plant location, and assignment of personnel and machines.

The objective may be to find the lowest cost or highest profit possible from given resources. Linear programming must be considered with recognition of the limitations on its use, however. A general prerequisite is that a linear or straight line relationship exists among the factors involved. In any linear programming problem, the manager must identify a measurable objective or criterion of effectiveness. The constraints must also be specified.

Balintfy, Rumpf, and Sinha (1980) used linear programming techniques to plan the best menus at the lowest cost for school lunches. Balintfy is the pioneer in computer menu planning based on linear programming. In his work, nutrient constraints are specified and aesthetic factors quantified in a model designed to plan least-cost menus. His later work includes food preferences in the menu planning model.

Other Techniques

A variety of other quantitative decision-making techniques have been developed, including game theory, queuing, and simulation models. In the foodservice field, many of these applications have been used in research instead of operational decision making, and most require complex computer models. As the field develops, however, managers will probably be applying these techniques, as have managers in other industries, to improve decision-making effectiveness.

Game theory introduces a competitive note in decision making by bringing into a simulated decision situation the actions of an opponent. Competition for market share is an example of a problem in which game theory might be used. The assumptions are that all competitors, or players, have the objective of winning the game and that they are capable of making independent and rational decisions. Competitors are presumed to be interested in maximizing gains and minimizing losses. Game theory will show the highest gain with the smallest amount of losses, regardless of what the competitor does.

Queuing theory develops relationships involved in waiting in line. Customers awaiting service or work awaiting inspection in a production line are typical of the problems that may be approached by the methods of queuing theory. The theory balances the cost of waiting lines against the cost of preventing them by expanding facilities. The problem is figuring out the cost of total waiting—that is, the cost of tolerating the queue—and weighing it against the expense of constructing enough facilities to decrease the need for the queue. Sometimes, eliminating all delay is more costly than keeping some. The basic framework of queuing is shown in Figure 18.11. Queuing problems can be solved by analytical procedures or simulation.

The concept of **simulation** is to utilize some device for imitating a real-life occurrence and studying its properties, behavior, and operating characteristics. The device can be physical, mathematical, or some other model for describing the behavior of an occurrence that a manager wishes to design, improve, or operate. For example, operating characteristics of a new piece of equipment can be simulated in a laboratory. Similarly, behavior of an occurrence also can be simulated by experimenting on a mathematical model that represents it.

Artificial intelligence is a computer program that attempts to duplicate the thought processes of experienced decision makers (Griffin, 1993). **Expert systems** are artificially intelligent computer software that show significant promise for improving the effectiveness and efficiency of management decision making (Cook, 1992). These systems solve problems by emulating the problem-solving behavior of human experts. Decision making starts with exploring "if . . . then" situations that could occur in solving a problem, and these situations become the knowledge base

Figure 18.11.   Basic framework of a queuing system.

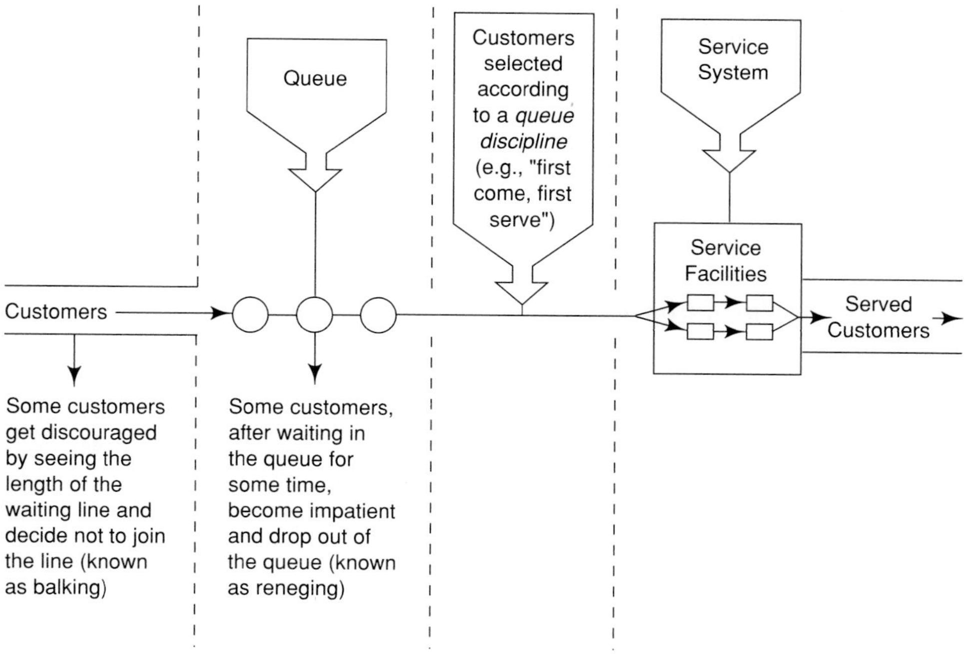

Source: Management: A Quantitative Perspective (p. 429) by N. P. Loomba, 1978, New York: Macmillan.

for the system. The examples below are "if . . . then" statements for a hypothetical steak house's pricing policy. The boldface statement is the current pricing policy for the restaurant.

- If the profit on a steak is high and demand is weak, then price policy to the customer is decrease price.
- If the profit on a steak is normal and demand is steady, then price policy to the customer is maintain price.
- **If the profit on a steak is low and demand is strong, then price policy to the customer is increase price.**

The foodservice manager could ask the system to explain price policy. The response would be that "price policy is increase price." When the manager asks why, the system responds that "price policy is increase price because profit is low and demand is strong." This is a simple example of one component of an expert system, which usually is very complex. By using many "if . . . then" rules, an expert system could be created that can mimic what is actually occurring in an operation.

In any simulation model, the goal is to understand the behavior of a system by testing the model under a variety of operating conditions. The need to experiment on a real-life system thus is eliminated. Sophisticated computer programs have been developed for mathematical simulations. These models require careful delineation of the components of the system and translation into mathematical terms.

Group Decision Making

The idea that a manager may seek input from others in making decisions or may wish to involve them in decision making was discussed previously. Thus far, the focus has been on individual decision makers, but a great number of decisions in most organizations are made by groups. These groups may be standing or specially designated committees, teams, task forces, or project groups. Often, informal groups may be called together to assist with a particular decision.

Individual Versus Group

When should a decision be made by a group rather than an individual? This issue has been greatly debated. According to Burack and Mathys (1983), individual versus group decision making largely depends on factors such as complexity and importance of the problem, time available, degree of acceptance required, amount of information needed to make a decision, and the usual manner in which decisions are made in an organization. They describe three possibilities for managerial decision making.

- *Individual decision.* Managers can make decisions themselves using information available to them.
- *Combination decision.* Managers can make decisions after consulting with others.
- *Group decision.* Managers can allow decisions to be made by the group, of which the manager is usually a member.

Group decision making, then, is utilized because managers frequently confront situations in which they must seek information and elicit judgments of other people—this is especially true for nonprogrammed decisions. In most organizations, rarely does an individual consistently make this type of decision alone. Problems are usually complex, and solutions require specialized knowledge from several fields. Usually, no one manager possesses all the kinds of knowledge needed.

Group decision making also may be used when two or more organizational units will be affected by the decisions. Because most decisions must eventually be accepted and implemented by several units, involving the groups affected in the decision making may be helpful.

Methods of Group Decision Making

Important decisions are being made in organizations by groups rather than by individuals. Group decision making is most often accomplished within interacting, Delphi, nominal, and focus groups.

Interacting Groups. An **interacting group** is a decision-making group in which members discuss, argue, and agree upon the best alternative (Griffin, 1993). Existing groups may be departments, work groups, or standing committees. New groups can be ad hoc committees, work teams, or task forces. An advantage of this method is that interacting promotes new ideas and understanding.

Delphi Groups. A **Delphi group** is used for developing a consensus of expert opinion (Griffin, 1993). A panel of experts, who contribute individually, makes predictions about a specific problem. Their opinions are combined and averaged and then returned to the panel for a second prediction. Members who made unusual predictions may be asked to justify them before sending them to the other members of the panel. When the predictions stabilize, the average prediction represents the decision of the group of experts. The Delphi method is good for forecasting technological breakthroughs but takes too much time and is too expensive for everyday decision making.

Nominal Groups. The **nominal group** method is a structured technique for generating creative and innovative alternatives or ideas (Griffin, 1993). Members of the group meet together but do not talk freely among themselves like members of interacting groups. The manager presents the problem to them and asks them to write down as many alternatives for solutions as possible. They then take turns presenting their ideas, which are recorded on a flip chart. Members then vote by rank-ordering the various alternatives. The top-ranking alternative represents the decision of the group, which can be accepted or rejected by the manager.

Focus Groups. A **focus group** is a qualitative marketing method. It has been successfully used for many years by large, multiunit chains and independents looking for customer feedback (Dee, 1990). The focus group consists of 10 to 20 people brought together for a one-time meeting of about 2 hours to discuss some predetermined aspect of a particular establishment. Men and women are selected to participate if they meet certain criteria, such as being a frequent customer in the restaurant or in a competitive one. Focus groups examine the motivation behind human behavior and, therefore, examine why people act the way they do, not what they do. Bob Evans Farms, Inc., used focus groups to correct sluggish sales in some of its Florida operations, and Burger King used focus groups to introduce its BK broiler.

Advantages and Disadvantages

One advantage of group decision making is that more information and knowledge are available. A group can generate more alternatives than can an individual and also can communicate the decision to employees in their work group or department.

One of the disadvantages is that group processes sometimes prevent full discussion of facts and alternatives. Group norms, member roles, established communication patterns, and cohesiveness may deter the group and lead to ineffective decisions. Also, group decision making takes time and is expensive. One person may dominate the group. Two phenomena often occurring in decision-making groups that may interfere with good decisions are groupthink and risky shift.

Groupthink occurs when reaching an agreement becomes more important to group members than arriving at a sound decision. In cohesive groups, members often want to avoid being too harsh in judging other members. They dislike bickering and conflict, perceiving them as threats to "team spirit." Janis (1983), who popularized the concept of groupthink, suggested that faulty decision making may occur because group members do not want to "rock the boat."

Risky shift is the tendency of individuals to accept or take more risk in groups than they would individually. Riskier decisions may result from group decision making because group members share the risk with others rather than having to bear responsibility individually.

COMMUNICATION

Communication has a major role in determining how effectively people work together and coordinate their efforts to achieve an organization's objectives. All organizations have recognized the importance of communication and have made major expenditures to improve it. Managers must communicate with superiors, other managers, and employees to convey their vision and goals for the organization (Griffin, 1993). These others have to communicate with managers to let them know what is happening in their work life and how they can be more effective.

Early in this chapter, communication was identified as one of the linking processes in organizations that is critical to managerial effectiveness and to the effective functioning of the foodservice system. Mintzberg (1975) found that managers spend a majority of their time communicating, much of which involves verbal communication. He described the informational roles of managers as monitor, disseminator, and spokesman (refer to chapter 3). The remaining seven roles, though not as explicitly, still demand skillful communication.

Because of the importance of communication to organizations and to the personal effectiveness of managers, persons in leadership positions must be well versed on the basics of communication and apply good communication techniques in all their activities. Breakdowns in the communication process may lead to employee dissatisfaction, customer dissatisfaction, misunderstanding, misinterpretation, and a whole range of other problems.

Poor communication, although often cited as the reason for an organizational problem, may be merely a symptom of a more serious situation. Good communication is not a panacea for all organizational problems. It will not compensate for poor planning or poor decisions, although plans and decisions must be communicated to a variety of individuals in an organization for implementation. Thus communication is an extremely important skill for managers, and its significance cannot be overstated.

Communication Defined

Communication is the transfer of information that is meaningful to those involved. It also is defined as the transmittal of understanding. *Effective communication* is the process of sending a message in such a way that the message received is as close in meaning as possible to the message intended (Griffin, 1993). It occurs in many forms, ranging from face-to-face conversation to written memoranda, and involves verbal, nonverbal, and implied messages. Communication in organizations often is viewed from two perspectives: communication between individuals (interpersonal) and communication within the formal organizational structure (organizational). These two basic forms of communication are obviously interdependent and interrelated.

The simplest model of communication is as follows:

Sender ⟶ Message ⟶ Receiver

Regardless of the type of communication, it includes these three elements.

Communication Process

The simple model does not show the complexity of the communication process. A more sophisticated model is shown in Figure 18.12. The process of communication starts when the sender wants to transmit information to the receiver. The sender has a message, an idea, a fact, or some other information to transmit to someone or some group.

For example, the manager of a privately owned restaurant decides to convert the traditional management style currently being used to the new style and has developed a plan to eliminate some middle management positions. The manager is eager to tell the restaurant owner about her plan. This idea may have simple or complex

Figure 18.12. The communication process.

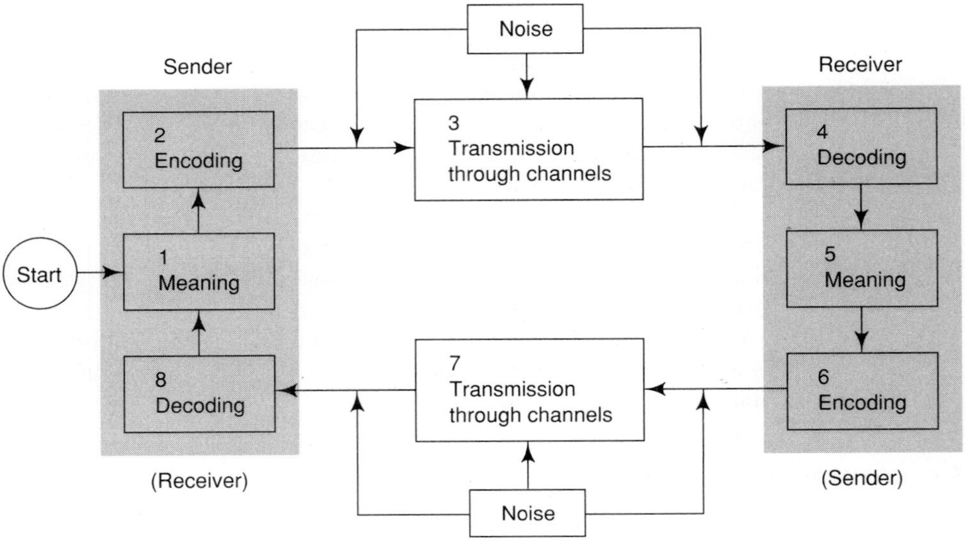

The numbers indicate the sequence in which steps take place.

Source: Griffin, Ricky W. *Management,* Fourth Edition (p. 447). Copyright © 1993 by Houghton Mifflin Company. Reprinted with permission.

meaning to the sender. Meaning is an abstract concept that is highly personal. No direct relationship exists between the symbols and gestures used in communication and meaning, as shown in Figure 18.13. A common problem in communication is the misinterpretation that may result from the receiver not understanding the message in the way the sender intends.

The sender must encode the information to be transmitted into a series of symbols or gestures. For example, the manager might have said to the owner, "I will fire middle managers," or "We will save money," or "I will retrain middle managers for other positions." The manager chose to say "We will save money by eliminating middle management positions." The encoding process is influenced by the content of the message and the familiarity of the sender and receiver.

After the message is encoded, it is transmitted through the appropriate channel. Channels in an organization include meetings, memos, letters, reports, and telephone calls. Because the owner is located many miles away from the restaurant, the manager talked to him over the telephone. After the message is received, it is decoded into a form that has meaning to the receiver. The owner might have thought, "Great, now I will be able to invest in another restaurant with the money saved," or "This is great news for the company," or "She is trying to get her salary increased." This meaning can vary from concrete to abstract. The owner's actual feeling was that this is great news for the company. Often the meaning requires another response and the cycle from sender to receiver starts all over again when a new message is sent back to the original sender. The owner told her immediately that this is great news for the company and made an appointment to meet with the manager in the restaurant within a week to discuss the plan.

Noise refers to all the types of interference that may distort or compete with the message during transmission of messages and is shown at the top and bottom of Figure 18.12. Examples of noise are the inability to hear the sender, who is speaking too softly; distortion of the message by extraneous sounds; and inattention of the receiver. If the owner had been thinking about something else and told the manager that they would talk about the plan on his annual visit to the restaurant, she might wonder why she even bothered to try to save money for him. As it was, his decision to make a special trip to discuss the plan with her positively reinforced her idea as well as her effort to keep him informed.

Communication may be one-way or two-way. In *one-way communication,* the sender communicates without expecting or getting feedback from the receiver. *Two-way communication* exists when feedback is provided by the receiver. Feedback enhances the effectiveness of the communication process and helps to ensure that the intended message is received by allowing the receiver to clarify the message and permitting the sender to refine the communication. One-way communication takes considerably less time than two-way communication, but it is less accurate. If communication must be fast and accuracy is easy to achieve, one-way communication may be both economical and efficient. If accuracy is important, however, and the message is complex, two-way communication is almost essential.

The communication skills, attitudes, knowledge, and the social system or culture of both the sender and receiver affect the communication. Differences in these elements between the sender and receiver may lead to communication problems. An obvious example is when the language is different for the sender and receiver. The foodservice manager responsible for supervising employees who speak little or no English would certainly understand this concept, which also has applicability to

Figure 18.13.   Triangle of meaning.

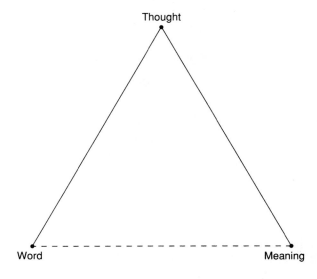

other types of communication. For example, when electronic communication systems are incompatible, communication breakdown also occurs.

Interpersonal Communication

A common and often incorrect assumption made in communicating with other persons is that the message was transmitted and received accurately. The assumption frequently is made that the message was transmitted effectively and the receiver understood it. This assumption is often incorrect and can be the source of many communication problems between individuals.

Interpersonal communication flows from individual to individual in face-to-face situations. The objective in interpersonal communication should be to increase the area of understanding (Figure 18.14). Ideally, the maximum overlap of "what was meant" and "what was perceived" is desired. In the following pages, some of the common barriers to effective interpersonal communications are discussed.

Barriers to Communication

"I didn't understand" is a common reply supervisors hear from employees when discussing why results were different from those expected. The dining room may be set up incorrectly, the wrong number of portions may be prepared, or the dietary aide may come to work at the wrong time. All these problems may be the result of communication breakdown between the supervisor and the foodservice employee.

Sayles and Strauss (1966) identified the following common barriers in interpersonal communication:

- *Hearing an expected message.* Past experience leads one to expect to hear certain messages that may not be correct in some situations.
- *Ignoring conflicting information.* A message that disagrees with one's preconceptions is likely to be ignored. For example, the foodservice manager who

Figure 18.14.   Areas of understanding.

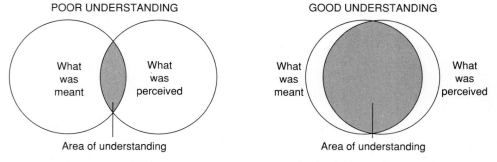

Source: Introduction to Management: A Career Perspective (p. 379) by E. H. Burack and N. J. Math-ys, Copyright © 1983 by John Wiley & Sons, Inc. Reprinted by permission John Wiley & Sons, Inc.

believes the operation has good food and service may ignore customer complaints.

- *Differing perceptions.* Words, actions, and situations are perceived in accordance with the receiver's values and experiences. Different people react differently to the same message.
- *Evaluating the source.* The meaning applied to any message is influenced by evaluation of the source. For example, if the sender is viewed as knowledgeable, the message is interpreted differently than if the sender's knowledge on a particular topic is questioned.
- *Interpreting words differently.* Because of the complexity of language, words have many different meanings.
- *Ignoring nonverbal cues.* Tone of voice, facial expressions, and gestures may affect communication. In an attempt to convey a message, the sender may ignore the fact that the receiver seems preoccupied, and the message will not be heard.
- *Becoming emotional.* Emotion will affect transmission and interpretation of messages. For example, if the person being talked to is perceived as hostile, the response may be defensive or aggressive, creating a negative effect on the communication.

Inference may be a barrier or facilitator in communication. As Haney (1986) discussed, inferences are constantly being drawn in communication with other people; that is, conclusions may be drawn based on incomplete information and action taken as a result. If these inferences are incorrect, problems may occur; but if correct, the inference may lead to efficiency in communication because needless information is avoided. A risk is involved, however, because of the potential of an inference leading to an incorrect conclusion.

Haney identified another barrier in communication that he referred to as "allness." It occurs when one unconsciously assumes that it is possible to know or say everything about something. Arrogance, intolerance of other viewpoints, and closed-mindedness are consequences of allness. Everyone can identify an individual whom we might call a "know-it-all." This person is the epitome of the allness concept.

Ability to interact effectively with other individuals is affected negatively by allness. Professionals, in particular, need to be sensitive to developing "allness" in dealing with nonprofessionals. A common fault among persons with highly specialized knowledge is the feeling that they "know what's best" for the other person.

All these barriers impede interpersonal communication and may lead to more serious problems in relationships. Sensitivity to and awareness of these barriers may assist in improving communication; a number of techniques can also be used to enhance interpersonal communication.

Techniques for Improved Communication

Techniques for improved communication are summarized in Figure 18.15. Using feedback can result in more effective communication because it allows the sender to

Figure 18.15. Techniques for improved communication.

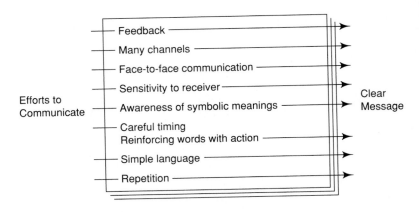

Efforts to
Communicate

Feedback

Many channels

Face-to-face communication

Sensitivity to receiver

Awareness of symbolic meanings

Careful timing
Reinforcing words with action

Simple language

Repetition

Clear
Message

search for verbal and nonverbal cues from the receiver. Questions can be encouraged and areas of confusion clarified as a result of two-way communication in which effective feedback is involved. Also, face-to-face communication may encourage feedback. People generally express themselves more freely in face-to-face situations or in other two-way communication, such as telephone conversations.

Using several channels will improve the chances that a proper message is communicated. For example, following up a verbal message with a written note will serve as a reinforcement.

Sensitivity to the receiver will enable the communicator to adapt the message to the situation. Individuals differ in their values, needs, attitudes, and expectations. Empathy with these differences will facilitate interpersonal communication.

Awareness of symbolic meanings can be particularly important in communication. For example, sensitivity to negative connotations and thus avoiding red flag words, which excite people to anger, is a way to improve interpersonal communication.

"It's not what you say, but what you do" is another tenet for improving communication. This concept, of course, is concerned with timing of the message and reinforcing words with action.

Using direct, simple language and avoiding jargon is another rule for improving communications. Especially important is the need to limit the message to the knowledge level of the receiver. This concept becomes vividly apparent in communication with children, but it may also be important in communicating with adults who may be less willing than children to say "I don't understand." Using the correct amount of repetition can serve as a reinforcement to the receiver's understanding of the message. Unnecessary redundancy or the overuse of cliches, however, may dull the receiver's attention.

Listening is basic to effective communication because receiving messages is as much a part of the process as sending them. Daydreaming and preoccupation with other matters or mentally arguing with points made by the speaker before the talk is finished may preclude individuals from listening. Impatience and lack of interest in the message are other impediments to listening. Rue and Byars (1989) contended that effective listening habits can be developed.

Organizational Communication

Factors discussed in relation to interpersonal communication apply to communication within organizations as well. Effective organizational communication involves getting an accurate message from one person to another. The effectiveness of organizational communication, however, is also influenced by several characteristics unique to organizations.

Lesikar (1977) identified four factors that influence the effectiveness of organizational communication: the formal channels of communication, the authority structure, job specialization, and a factor he calls "information ownership." Formal channels influence communication effectiveness in space. First, as organizations grow, the channels cover an increasingly larger distance. Consider, for example, the small delicatessen that developed into a large national chain of delicatessens, which was described in chapter 17. There, the development of levels of management, divisions, and departments was described in relation to organization growth. With this development, effective communication is more difficult to achieve because of the much greater complexity of the organization.

Second, the formal channels of communication may inhibit the flow of information between levels in the organization. The cook, for example, will almost always communicate problems to the foodservice supervisor rather than to the unit manager. Although this accepted channel of communication has advantages, such as keeping higher-level managers from being overloaded with too much information, it also has disadvantages. Higher-level managers may not always receive all the information they need for staying on top of operations. With additional levels in an organization, the accuracy of messages up and down the organization will also be affected.

The authority structure affects communication because of status and power differences among organizational members. The content and accuracy of communication will be affected by authority differences. For example, conversation between the director of foodservice and a foodservice employee may tend to be a strained politeness or somewhat formal. Neither one is likely to say much of importance.

Job specialization can be both a help and hindrance in communication. It facilitates communication within a work group because members speak the same jargon and frequently develop a group cohesiveness. Communication between work groups, however, may be inhibited. Cafeteria employees may have free-flowing communication, as may also be true of the production employees. Communication between these two groups may not be as effective as that within each group.

The term *information ownership* means that individuals possess unique information and knowledge about their jobs. Such information is a form of power for those who possess it, making them unwilling to share the information with others.

Overcrowded work areas, too many individuals competing for the attention of a manager, the improper choice of a communication channel, and time pressures are other factors that impede organizational communication. For example, a typed letter rather than face-to-face conversation for handling a touchy matter may lead to hurt feelings.

Directions of Internal Communication

Within the organization, managers must provide for communication in four distinct directions: downward, upward, horizontal, and diagonal. Although discussed only briefly in this chapter, managers also must communicate effectively with individuals and groups outside the organization.

Downward.   Individuals at higher levels of the organization communicate downward to those at lower levels. The most common forms of communication are job instructions, policy statements, procedure manuals, and official publications of the organization. In addition, middle- and lower-level managers are usually contacted by written memos or with some other official directive. A summary of various channels available to carry information downward is listed in Figure 18.16.

External means, such as radio, television, and newspaper press releases, may be used to communicate not only with employees but also with the general public. For example, a school district closing because of snow is usually announced over the radio and television as a way of communicating to the school district staff, parents, and students.

Upward.   An effective organization needs open channels of upward communication as much as it needs downward communication. In large organizations, good upward communication is especially difficult. Suggestion boxes, special meetings, and grievance procedures are devices used for upward communication. Some channels for upward flow of information are summarized in Figure 18.17.

Effective upward communication is important because it provides employees with an opportunity to have a say in what happens in the organization. Equally important, however, is the need for management to receive vital information from lower levels in the organization.

Horizontal.   Provision for horizontal flow of communication will enhance organizational effectiveness. This issue was touched on in chapter 17, under Coordination, in the example of nursing service and dietetic service employees who communicate with one another directly. Within the foodservice operation, effective horizontal communication between production and service is critical to ensure that quality food is available at the right time in the right place.

Department head meetings are one way to facilitate interdepartmental communication. The more interdependent the work of organizational units or departments, the greater the need for horizontal communication.

Diagonal.   The use of diagonal channels of communication is a way of minimizing time and effort expended in organizations. Having reports and other information flow directly between departments or units that have a diagonal placement in the organization may result in more effective flow of information.

For example, the ordering clerk in a foodservice operation may send requests directly to the purchasing department rather than through the channels of the food-

Figure 18.16.   Formal downward channels of communication.

Chain of Command
Orders and information can be given personally or in written fashion and transmitted from one level to another. This is the most frequently used channel and is appropriate on either an individual or a group basis.

Posters and Bulletin Boards
Some workers may not read them. This is especially true when the posters or bulletins are not kept current. Thus, this channel may be useful only as a supplementary device.

Company Newsletters or Newspapers
A great deal of information about the company, its products, and policies can be communicated in this way. Readership is increased if some space is allocated to personal items of interest to employees.

Letters and Pay Inserts
Direct mail may be used when top management wants to communicate matters of importance. Since the letter is sent directly to the employee from the company, a reasonable chance exists that it will be read. Inserting a letter with the paycheck ensures that each worker receives a copy.

Employee Handbooks and Pamphlets
Handbooks are frequently used during the hiring and orientation process as an introduction to the organization. Too often, however, they are unread even when the firm demands a signed statement that the employee is acquainted with their contents. When special systems are being introduced, such as a pension plan or a job evaluation system, concise, well-illustrated pamphlets are often prepared to facilitate understanding and stimulate acceptance.

Annual Reports
Annual reports are increasingly written not only for the stockholders but also for the employee. A worker may be able to obtain information about the firm in this way.

Loudspeaker Systems
The loudspeaker system is used not only for paging purposes but also to make announcements while they are "hot."

Source: Adapted from *Management: Concepts and Practices* (pp. 430–432) by R. W. Mondy, A. Sharplin, and E. B. Flippo, 1988, Boston: Allyn & Bacon. Used by permission.

service department. As long as procedures for this direct communication exist, the function of ordering and receiving goods is facilitated. Communication can be either informal or formal.

Informal

Informal networks develop in organizations to supplement formal communication channels. Oral communication usually occurs in face-to-face conversations, telephone calls, and group discussions (Griffin, 1993). The primary advantage of oral

Figure 18.17.   Formal upward channels of communication.

Open-Door Policy
An open-door policy is an established guideline that allows workers to bypass immediate supervisors concerning substantive matters without fear or reprisal. Managers are encouraged to create an environment in which subordinates will feel free to come to them with problems and recommendations.

Suggestion Systems
Many companies have formal suggestion systems. Some have suggestion boxes. When a suggestion system is used, every suggestion should receive careful consideration. Workers should be promptly informed of the results of the decision on each suggestion.

Questionnaires
Anonymous questionnaires sometimes are given to workers in an attempt to identify problem areas within the organization.

The Grievance Procedure
The grievance procedure is a mechanism that gives subordinates the opportunity of carrying appeals beyond their immediate supervisor.

Ombudsperson
An ombudsperson is a complaint officer with access to top management who hears employee complaints, investigates, and sometimes recommends appropriate action.

Special Meetings
Special employee meetings to discuss particular company policies or procedures are sometimes scheduled by management to obtain employee feedback.

Source: Adapted from *Management: Concepts and Practices* (pp. 434–436) by R. W. Mondy, A. Sharplin, and E. B. Flippo, 1988, Boston: Allyn & Bacon. Used by permission.

communication is that it promotes quick feedback to verbal questions or agreements and facial expressions. It is easy to use, and paper and pencil or computers are not needed. The major drawback of using oral communications is that distortion of the message can result.

The **grapevine** is another type of informal communication and may facilitate organizational communication and meet the social needs of individuals within the organization. Although the grapevine may filter or distort messages and occasionally transmit rumors and gossip, its speed and accuracy in getting messages through an organization are often very useful.

The effective manager is one who has learned to use the grapevine to advantage and recognizes its importance in fulfilling the communication needs of members of the organization. The grapevine can and should work for the manager. Informal communication systems can help to speed the work-related flow of information by making use of the natural interaction among people in organizations. Ignoring the grapevine and rumors is a dangerous alternative because a rumor usually has some element of truth and may be a symptom of a larger problem.

Formal

Written communication usually gives the appearance of being more formal and authoritative than oral communication. Because the same words are communicated to all who receive these messages, it tends to be interpreted more accurately and is often used when consistent action is required and a record of the communication is necessary. It also provides a permanent record of the exchange, and the receiver can refer to it as often as needed. A common problem with written communication, however, is proliferation of paperwork that is of questionable importance. Excessive red tape in organizations is often the result of excessive requirements for written communication.

Today most managers prefer not to use written communications because it inhibits feedback and interchange (Griffin, 1993). When a manager sends a letter, it must be written or dictated, typed, mailed, received, opened, and read. If the message is misunderstood, several days may pass before it is rectified. A telephone call could settle the whole matter in just a few minutes. Setting up a meeting by a quick telephone call followed by a written memo is a good practice.

Recent breakthroughs in electronic communication have changed organizational communication drastically (Griffin, 1993). Electronic typewriters, photocopying machines, and computers were early breakthroughs that have given rise to a new version of communications. Today, teleconferences, in which managers stay at their own locations, are taking the place of face-to-face meetings. Cellular telephones and facsimile (FAX) machines make communications easier. Many people use cellular telephones to make calls while commuting back and forth to work. FAX machines permit people to use written communications and receive quick feedback. Psychologists are beginning to associate some problems with these communication breakthroughs. People are no longer part of the organizational grapevine and miss out on the informal communications that take place at work. Also, the use of electronic communications makes it difficult to build a strong culture, develop working relationships, and create a mutually supportive atmosphere of trust and cooperatives.

BALANCE

Balance, a linking process mentioned in chapter 2, refers to managerial adaptations to changing economical, political, social, and technological conditions. Foodservice organizations, as discussed in chapter 17, are the products of the organizing function of management, which provides the mechanism for coordinating and integrating all activity toward accomplishment of objectives. According to Stoner and Wankel (1986), the organizing process involves balancing a company's needs for both stability and change. An organization's structure gives stability and reliability to the actions of its members; both are required for an organization to move coherently toward its goals. Altering the structure of an organization can be a means of adapting to and bringing about change, or it can be a source of resistance to change.

Conditions both outside and inside organizations are changing rapidly and radically today. Externally, environmental conditions generally are becoming less stable and even turbulent (Hampton, 1986). Economic conditions, availability, and cost of materials and money, technological and product innovations, and government regulations all can shift rapidly. A great variety of external forces including competitive actions can pressure organizations to modify their structure, goals, and methods of operation. Pressures for change also may arise from a number of sources within the organization, especially from new strategies, technologies, and employee attitudes and behavior. Internally, employees are changing by placing greater emphasis on human values and questioning authority. They assert their rights under affirmative action, safety, compensation, and other legislation, and they question the fairness of management decisions and actions. Employees desire to improve the quality of their working lives. External and internal forces for change are often linked, especially when changes in values and attitudes are involved.

Organizational design is affected by three types of external environments: stable, changing, and turbulent (Stoner & Wankel, 1986). A stable environment has experienced little or no change; product changes seldom occur and modifications are planned in advance. Market demand is predictable, and laws affecting the organization have remained the same for many years. Stable environments are rare due to the advent of many technological changes. Environmental changes, however, can occur in the product, market, law, or technology. These changes do not surprise managers because trends are apparent. Many service industries function in a changing environment. The organization is in a turbulent environment when competitors market new, unexpected products, when laws are passed that affect businesses in radical ways, and when technology suddenly changes product design or production methods. Few organizations function in a continuously turbulent environment, however; if a radical change does occur, only a temporary period of turbulence is endured before they make an adjustment. For example, hospitals experienced a sudden decrease in demand for their services under the impact of DRGs (diagnosis related groups), which determined cost and length of stay.

Even though many forces for change can be found in organizations, other forces serve to maintain a state of balance. Forces opposing change also are supporting stability or status quo. According to the force-field theory of Lewin (de Rivera, 1976), any behavior is the result of a balance between driving and restraining forces. Driving forces push in one direction while restraining forces push in the other, and the result is a reconciliation of the two sets of forces. An increase in the driving forces might increase performance, but they also might increase the restraining forces. For example, the manager of a foodservice in a very busy theme park decided that employees would have to work every weekend to increase profits. This change might result in a decrease of profits, however, if employees' hostility, distrust, and greater resistance cause additional declines in productivity. Employees who want change push, and those who do not push back. Driving forces generate restraining forces. Decreasing restraining forces, therefore, is usually a more effective method of encouraging change than increasing driving forces. The balance concept suggests that organizations have forces that keep performance from falling too low, as well as forces that keep it from rising too high.

Every organization needs to be committed to its product and service (Peters & Waterman, 1982). In addition, it needs to have a strong set of values and a firm vision of the future. They described the properties of a potential "structure of the eighties" that responds to three needs: for efficiency around the basics, for regular innovation, and to avoid calcification by ensuring responsiveness to major threats. The structure of the organization is based on three pillars, each of which responds to these needs.

- *Stability pillar.* Responds to the need for efficiency and is found in a departmentalized structure
- *Entrepreneurial pillar.* Keeps the structure small, which is a requisite for continual adaptiveness and innovation
- *Habit-breaking pillar.* Includes a willingness to reorganize frequently, adjusting to various forces

SUMMARY

Linking processes are needed to coordinate the activities of the foodservice system in the accomplishment of goals. These processes are decision making, the selection of a course of action from alternatives; communication, the vehicle whereby decisions and other information are transmitted; and balance, management's ability to maintain organizational stability.

Decisions often are classified as programmed, which are reached by following established policies and procedures, and nonprogrammed, which are relatively unstructured and generally require a creative approach. Organizational decisions relate to the objectives and activities of the organization and personal decisions are concerned with the manager's individual goals. The decision-making process has five steps, from defining the situation to evaluating the effectiveness of the decision.

Decisions are seldom made under the same conditions of available information. Environmental conditions under which decisions are made are defined as certainty, risk, and uncertainty. A wide variety of techniques has been developed to assist managers in making decisions, including decision trees and cost effectiveness. Networks for decision making include the Program Evaluation and Review Technique (PERT) and the Critical Path Method (CPM), which are the two most widely used management science techniques for planning, scheduling, and controlling large projects. Linear programming is useful in determining an optimal combination of resources to obtain a desired objective. Group decision making often is used when managers need other people's information and judgment, especially in nonprogrammed decisions.

Communication is the transfer of information that is meaningful to those involved. Communication in organizations can be between individuals or within the formal organizational structure. Regardless of the type of communication, it includes the sender, message, and receiver. An incorrect assumption in communicating with other persons is that the message is transmitted and received accurately. Interpersonal communication flows from individual to individual in face-to-face situations and has a number of barriers. Using feedback can result in more effective communication because it allows the sender to search for verbal and nonverbal cues from the receiver.

Four factors influence the effectiveness of organizational communication: formal channels of communication, authority structure, job specialization, and information ownership. Communication within the organization can be downward, upward, horizontal, or diagonal. Oral communications and the grapevine are considered informal communications; written and electronic communications are considered formal.

Balance, the third linking process, refers to managerial adaptations to changing economical, political, social, and technological conditions. The organizing function of management provides the mechanism for coordinating and integrating all activity toward accomplishment of objectives, and the organizing process involves balancing a company's need for stability and change.

BIBLIOGRAPHY

Balintfy, J. L., Rumpf, D., & Sinha, P. (1980). The effect of preference-maximized menu on consumption of school lunches. *School Food Service Research Review, 4*(1), 48–53.

Batty, J. (1993). Communication between the sexes. *Restaurants USA, 13*(6), 31–34.

Berger, E., Ferguson, D. H., & Woods, R. (1987). How restaurateurs make decisions. *Cornell Hotel and Restaurant Administration Quarterly, 27*(4), 49–57.

Brownell, J. (1987). Listening: The toughest management skill. *Cornell Hotel and Restaurant Administration Quarterly, 27*(4), 65–71.

Burack, E. H., & Mathys, N. J. (1983). *Introduction to management: A career perspective.* New York: Wiley.

Cook, R. L. (1992). Expert systems in purchasing: Application and development. *International Journal of Purchasing and Materials Management, 28*(4), 20–27.

de Rivera, J. (Ed.). (1976). *Field theory as human-science: Contributions of Levin's Berlin group.* New York: Gardner Press.

Dee, D. (1990). Focus groups: Finding out why customers act the way they do. *Restaurants USA, 10*(7), 30–34.

Etzioni, A. (1989). Humble decision-making. *Harvard Business Review, 67*(4), 122–126.

Griffin, R. W. (1993). *Management* (4th ed.). Boston: Houghton Mifflin.

Hampton, D. R. (1986). *Management* (3rd ed.). New York: McGraw-Hill.

Haney, W. V. (1986). *Communication and interpersonal relations: Texts and cases* (5th ed.). Homewood, IL: Richard D. Irwin.

Hersey, P., & Blanchard, K. (1993). *Management of organizational behavior: Utilizing human resources* (6th ed.). Englewood Cliffs, NJ: Prentice-Hall.

Janis, L. L. (1983). Groupthink. In Staw, B. M. (Ed.), *Psychological foundations of organizational behavior* (2nd ed., pp. 514–522). Glenview, IL: Scott, Foresman.

Leitch, G. A. (1989). Application of a PERT-type system and "crashing" in a food service operation. *FIU Hospitality Review, 7*(2), 66–76.

Lesikar, R. V. (1976). *Communication theory and application.* Homewood, IL: Richard D. Irwin.

Lesikar, R. V. (1977). A general semantics approach to communication barriers in organization. In Davis, K., (Ed.), *Organizational behavior: A book of readings* (5th ed., pp. 336–340). New York: McGraw-Hill.

Loomba, N. P. (1978). *Management: A quantitative perspective*. New York: Macmillan.

March, J. G., & Simon, H. A. (1958). *Organizations*. New York: Wiley.

Mintzberg, H. (1975). The manager's job: Folklore and fact. *Harvard Business Review*, *53*(4), 49–61.

Mondy, R. W., Sharplin, A., & Flippo, E. B. (1988). *Management: Concepts, and Practices* (4th ed.). Boston: Allyn & Bacon.

Peters, T. J., & Waterman, R. H. (1982). *In search of excellence: Lessons from America's best-run companies*. New York: Harper & Row.

Rogers, C. R., & Roethlisberger, F. J. (1991). Barriers and gateways to communication. *Harvard Business Review*, *69*(6), 105–111.

Rue, L. W., & Byars, L. L. (1989). *Management: Theory and application* (5th ed.). Homewood, IL: Richard D. Irwin.

Sayles, L. R. (1989). *Leadership: Managing in real organizations* (2nd ed.). New York: McGraw-Hill.

Sayles, L. R., & Strauss, G. (1966). *Human behavior in organizations*. Englewood Cliffs, NJ: Prentice-Hall.

Spears, M. C. (1976). Concepts of cost effectiveness: Accountability for nutrition, productivity. *Journal of The American Dietetic Association*, *68*(4), 341–346.

Stoner, J. A. F., & Wankel, C. (1986). *Management* (3rd ed.). Englewood Cliffs, NJ: Prentice-Hall.

Leadership and Organizational Change

What makes a leader effective? This question has been asked numerous times over the years in attempts to understand the concept of leadership. In spite of intensive research, knowledge of what it takes to be an effective leader is still limited. Leaders play a critical role in helping groups, organizations, and even societies achieve their goals. Also, a number of factors that affect managers' effectiveness in leading organizations have been identified. Leadership is important in all organizations. The focus in this chapter, however, is on the leadership process within the work organization.

Leading is one of the functions that managers perform in the process of coordinating activities of the foodservice system. The leading function involves directing and channeling human effort for the accomplishment of objectives. Therefore, leading is concerned with creating an environment in which members of the organization are motivated to contribute to organizational goals and changes. In other words, leadership is the process of creating a work environment in which people can do their best work.

In this chapter, motivation at work, factors affecting leadership effectiveness, and organizational change are discussed. Several other issues related to organizational behavior are examined, such as job satisfaction, power, and philosophies of human behavior.

MOTIVATION AND WORK PERFORMANCE

Motivation is concerned with the causes of human behavior. An understanding of human behavior is important to managers as they attempt to influence this behavior in the work environment. The study of motivation and behavior is a search for answers to perplexing questions about human nature.

Because of the importance of human resources in organizations, managers must have an understanding of behavior, not only to understand the past but also to predict or change the future. Highly motivated employees in a foodservice or any other organization can elicit substantial increases in performance and decreases in such problems as absenteeism, turnover, grievances, low morale, and tardiness.

Meaning of Motivation

Definitions of motivation usually include such words as *aim, desire, intention, objective, goal*, and *purpose*. A definition commonly quoted is that of Berelson and Steiner (1964):

> **Motivation** is all those inner striving conditions described as wishes, desires, and drives, and is an inner force that activates or moves a person.

The process of motivation can be viewed as a causative sequence:

Needs $\longrightarrow$ Drives or motives $\longrightarrow$ Achievement of goals

In the motivation process, needs produce motives that lead to the accomplishment of goals or objectives. Needs are caused by deficiencies that may be physical or psychological. *Physical needs,* also called innate or primary needs, include food, water, and shelter. *Psychological needs,* also referred to as acquired needs, are those we learn in response to our culture or environment. They include esteem, affection, and power.

Motives are the "whys" of behavior. They arouse and maintain activity and determine the general direction of an individual's behavior. Hersey and Blanchard (1993) explained a motive as something within an individual that prompts a person to action.

Achievement of the goal in the motivation process satisfies the need and reduces the motive. When the goal is reached, balance is restored; however, other needs arise, which are then satisfied by the same sequence of events.

A distinction should be made between positive and negative motivation. Motivation can be either a driving force toward some object or condition (positive motivation) or a driving force away from some object or condition (negative motivation).

Goals can also be positive or negative. A *positive goal* is desirable and the object of directed behavior—an employee's desire to do the best job possible; a *negative goal* is undesirable and behavior is directed away from it—an employee wishes to avoid censure. Goals depend on an individual's subjective experiences, physical capacity, prevailing norms and values, and the potential accessibility of the goal. Furthermore, an individual's self-perception also serves to influence goals.

Both needs and goals are interdependent, and individuals are not always aware of their needs. In addition, needs and goals are constantly changing. As individuals attain goals, they develop new ones; if certain goals are not attained, they develop substitutes. Individuals who are blocked in attempts to satisfy their needs or achieve goals may become frustrated and exhibit dysfunctional or defensive behavior. On occasion, all of us employ defensive behavior as a protective function in our attempts to cope with frustration. Some kinds of defensive behavior have limited negative consequences in the organization, although others may become quite destructive.

Withdrawal, which may be exhibited by apathy, excessive absences, lateness, or turnover, is one mechanism used to avoid frustrating situations. Aggression is a common reaction to frustration that involves a direct attack on the source of frustration or on another object or party. For example, a foodservice employee who is upset with his or her supervisor may slam and bang the pots and pans as a way of venting frustration.

Substitution occurs when an individual puts something in the place of the original object. For example, a foodservice employee bypassed for promotion may seek leadership positions in organizations outside the workplace.

When a person goes overboard in one area or activity to make up for deficiencies in another, the defense mechanism is compensation. Other individuals may revert or regress to childlike behavior as a way of dealing with an unpleasant situation. For example, horseplay in the dishroom is an example of regression.

Some individuals repress frustrating situations and problems; in repression, an individual loses awareness of or forgets incidents that cause anxiety or frustration. Projection is another coping behavior in which an individual attributes his or her own feelings to someone else. For example, a foodservice employee who is displeased about a rule or policy may tell the supervisor how upset another employee is rather than admit personal dissatisfaction.

Rationalization is probably one of the most common reactions to frustration. This behavior enables an individual to present a reason that is less ego deflating or more socially acceptable than the true reason. A baker who blames the oven for poor bakery products is using this defense mechanism.

To some extent, everyone relies on defense mechanisms. Excessive defensive behaviors by employees may be minimized, however, if supervisors encourage constructive behavior. Also, managers who understand defensive behavior should have greater empathy and realize that such behaviors are methods of coping with frustration.

Theories of Motivation

A number of theories of motivation have been developed: need hierarchy, achievement-power-affiliation, two-factor, expectancy, and reinforcement. These theories, described briefly in the following pages, are all different constructs that may prove useful in understanding behavior.

Need Hierarchy

One of the most popular theories of motivation was proposed by Maslow (1943) in the 1940s. This theory, frequently referred to as the **need hierarchy** theory, states that people are motivated by their desire to satisfy specific needs, which are arranged in the following ascending hierarchical order (Figure 19.1).

- *Physiological.* Needs of the human body that must be satisfied to sustain life
- *Safety.* Needs concerned with the protection of individuals from physical or psychological harm

Figure 19.1.   Maslow's need hierarchy and methods of satisfying needs in organizations.

MASLOW'S HIERARCHY	EXAMPLES OF METHODS FOR SATISFYING NEEDS
Self-Actualization Needs (realizing one's potential growth using creative talents)	Challenging work allowing creativity, opportunities for personal growth, and advancement
Esteem Needs (achievement recognition and status)	Title and responsibility of job, praise and rewards as recognition for accomplishments, promotions, competent management, prestigious facilities
Social Needs (love, belonging, affiliation, acceptance)	Friendly associates, organized employee activities such as bowling or softball leagues, picnics, parties
Safety Needs (protection against danger, freedom from fear, security)	Benefit programs such as insurance and retirement plans, job security, safe and healthy working conditions, competent, consistent, and fair leadership
Physiological Needs (survival needs, air, water, food, clothing, shelter, and sex)	Pay, benefits, working conditions

Source: Maslow's hierarchy of needs data from *Motivation and Personality,* 3rd ed., by Abraham H. Maslow. Revised by Robert Frager, James Fadiman, Cynthia McReynolds, and Ruth Cox. Copyright 1954, © 1987 by Harper & Row, Publishers, Inc. Copyright © 1970 by Abraham H. Maslow. Reprinted by permission of Harper Collins Publishers, Inc.

- *Social.* Needs for love, affection, belonging
- *Esteem.* Needs relating to feelings of self-respect and self-worth, along with respect and esteem from one's peers
- *Self-actualization.* Needs related to one's potential or to the desire to fulfill one's potential

According to this theory, each need is prepotent or dominant over all higher-level needs until it has been partially or completely satisfied. A *prepotent need* is one that has greater influence over other needs. Also, according to this theory, a satisfied need is no longer a motivator. A prepotent lower-order need, however, might not have to be completely satisfied before the next higher one becomes potent or dominant. For example, the safety need may not have to be completely satisfied before social needs become motivators.

In our society, the physiological and safety needs are more easily and frequently satisfied than other needs. Many of the tangible rewards, such as pay and fringe benefits, that organizations offer are primarily directed to physiological and safety needs.

The strengths of an individual's needs may shift in different situations. During bad economic times, for example, physiological and safety needs may dominate an individual's behavior, whereas higher-level needs may dominate in good economic times. Also, different methods may be used by different individuals to satisfy particular needs.

Interesting work and opportunities for advancement are means that organizations use to appeal to higher-order needs. Obviously, determining the need level of each individual foodservice employee can be a difficult process, but managers may find this theory useful in attempting to understand their employees' motivations.

Achievement-Power-Affiliation

In his writing on motivation, McClelland (1967) emphasized needs that are learned and socially acquired as individuals interact with the environment. The **achievement-power-affiliation** theory holds that all people have three needs:

- A need to achieve
- A need for power
- A need for affiliation

Achievement Motive.   The need for achievement is a desire to do something better or more efficiently than it has been done before. An individual with a high need for achievement tends to have the following traits:

- Responds to goals
- Seeks a challenge but establishes attainable goals with only a moderate degree of risk
- Exhibits greater concern for personal achievement than rewards of success
- Desires concrete feedback on performance
- Takes personal responsibility for finding solutions to problems
- Maintains a high energy level and willingness to work hard

Persons high in the need for achievement tend to gravitate toward managerial and sales positions. In these occupations, individuals are often able to manage themselves and thus satisfy the basic drive for achievement. Individuals with high achievement needs tend to get ahead in organizations because they are producers—they get things done. These individuals are task oriented and work to their capacity, and they expect others to do the same. As a result, the foodservice manager who has a high need for achievement may sometimes lack the human skills and patience necessary to manage employees with lower achievement motivation.

Power Motive.   The need for power is basically a concern for influencing people. An individual with a high-power need tends to exhibit the following behavior:

- Enjoys competition with others in situations allowing domination
- Desires acquiring and exercising power or influence over others
- Seeks confrontation with others

McClelland and Burnham (1976) identified two aspects of power: positive and negative. Positive use of power is essential for a manager to accomplish results through the efforts of others in an organization. The negative aspect of power is when an individual seeks power for personal benefits, which may be detrimental to the organization.

Affiliation Motive.   The need for affiliation is characterized by the desire to be liked by others and to establish or maintain friendly relationships. A person with a high need for affiliation tends to be one who

- Wants to be liked by others
- Seeks to establish and maintain friendships
- Enjoys social activities
- Joins organizations

McClelland (1967) maintained that most people have a degree of each of these needs, but the level of intensity varies. For example, an individual may be high in the need for achievement, moderate in the need for power, and low in the need for affiliation. Managers should recognize these differing needs in their employees when dealing with them. A foodservice employee with a high need for affiliation, for instance, would probably respond positively to warmth and support, whereas an employee with a high need for achievement would tend to respond to increased responsibility or feedback.

Two-Factor

Herzberg (1966) developed the **two-factor** theory of work motivation, which focuses on the rewards or outcomes of performance that satisfy needs. Two sets of rewards or outcomes are identified: those related to job satisfaction and those related to job dissatisfaction. Those factors related to satisfaction, called *motivators,* include achievement, recognition, responsibility, advancement, the work itself, and the potential for growth. All these factors are related to the environment or content of the job. Those factors related to dissatisfaction, called *maintenance,* or *hygiene,* factors, include pay, supervision, job security, working conditions, organizational policies, and interpersonal relationships on the job. These factors are related to the environment or context of the job.

Based on his research, Herzberg concluded that although employees are dissatisfied by the absence of maintenance factors, the presence of those conditions does not cause motivation. The maintenance factors, he says, are necessary to maintain a minimum level of need satisfaction.

In addition, the presence of some job factors cause high levels of motivation and job satisfaction, but the absence of these factors may not be highly dissatisfying. This second group of factors, which are internal to the job, are a major source of motivation.

Herzberg's research often has been criticized on the basis that the results of his initial interviews can be interpreted in several different ways, that his sample of 200 accountants and engineers is not representative of the general population, and that subsequent research often has not supported his theory. Although Herzberg's theory is not held in high esteem by researchers in the field, it has had a major impact on managers' awareness of the need to increase motivation in the workplace.

An examination of Herzberg's and Maslow's theories of motivation reveals some similarities. Maslow's theory is helpful in identifying the needs or motives; Herzberg's theory provides insights into the goals and incentives that tend to satisfy these needs. The two theories are compared in Figure 19.2. Maslow's physiological, safety, and social needs are motivation factors. The esteem needs, however, involve both status and recognition. Because status tends to be a function of the position a person occu-

Figure 19.2.   A comparison of Maslow's and Herzberg's models of motivation.

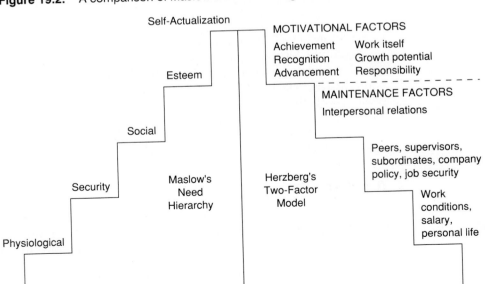

pies and recognition is gained through competence and achievement, esteem needs are related to both maintenance and motivation factors. Self-actualization needs in Maslow's conceptualization are related to Herzberg's motivational factors.

Expectancy

Managers should develop an understanding of human needs and the variety of organizational means available to satisfy employees' needs. The needs approach to motivation, however, does not account adequately for differences among individual employees or explain why people behave in many different ways when accomplishing the same or similar goals.

To explain these differences, several expectancy approaches to motivation have been advanced in the past several years. Two of the most prominent ones were developed by Vroom (1964) and Porter and Lawler (1968). **Expectancy** theory attempts to explain behavior in terms of an individual's goals, choices, and expectations of achieving these goals. This theory assumes people can determine the outcomes they prefer and make realistic estimates of their chances of attaining them.

Expectancy theory is based on the belief that people act in such a manner as to increase pleasure and decrease displeasure. According to this theory, people are motivated to work if they believe their efforts will be rewarded and if they value the rewards that are being offered. The first requirement can be broken down into the following two components:

- The expectancy that increased effort will lead to increased performance
- The expectancy that increased performance will lead to increased rewards

These expectancies are developed largely from an individual's experiences.

The second part of expectancy theory is concerned with the value the employee places on the rewards offered by the organization, also referred to as *valence*. Organizations have tended to assume that all rewards will be valued by employees. Obviously, however, some rewards are more valued than others, and certain rewards may be viewed negatively by some employees.

According to this theory, the factors of expectancy and valence determine motivation, and both must be present for a high level of motivation. In other words, high expectancy or high valence alone will not ensure motivation. If a foodservice employee places a high value on money (high valence) but believes there is little chance of receiving a pay increase (low expectancy), for example, the employee will probably not be highly motivated to work hard because of the low probability of receiving the pay increase.

All employees in an organization do not share the same goals or values regarding pay, promotions, benefits, or working conditions. Managers must consider an employee's goals and values in attempting to create a motivational climate. A key factor in expectancy theory is what the employee perceives as important or of value—"what's in it for me"—not what the manager believes the employee should value.

A major contribution of expectancy theory is that it explains how the goals of employees influence behavior at work. Employee behavior is influenced by assessments of the probability that certain behavior will lead to goal attainment.

Reinforcement

Reinforcement theory, which is associated with Skinner (1971), is often called "operant conditioning" or "behavior modification." Rather than emphasize the concept of a motive or a process of motivation, these theories deal with how the consequences of a past action influence future actions in a cyclical learning process.

According to Skinner, people behave in a certain way because they have learned at some previous time that certain behaviors are associated with positive outcomes and that others are associated with negative outcomes. Further, because people prefer pleasant outcomes, they are likely to avoid behaviors that have unpleasant consequences. For example, foodservice employees may be likely to follow the rules and policies of the organization because they have learned during previous experiences—at home, at school, or elsewhere—that disobedience leads to punishment.

The general concept behind reinforcement theory is that reinforced behavior will be repeated, and behavior that is not reinforced is less likely to be repeated. Reinforcers are not always rewards and do not necessarily have to be positive in nature. The example cited above of the foodservice employee wishing to avoid disciplinary action is an avoidance reinforcer. The current emphasis in management practices is on the use of positive reinforcers, including both tangible and intangible rewards, such as pay increases, promotions, or recognition for good performance.

Job Satisfaction

Closely related to motivation is the concept of job satisfaction. In fact, many managers view motivated employees as being synonymous with satisfied employees; however, important differences should be noted. *Job satisfaction* pertains to individuals' general attitudes toward their occupations. Smith, Kendall, and Hulin (1969) identified five components of job satisfaction: satisfaction with opportunities for promotion, pay, supervision, the work itself, and co-workers. Other factors, such as attitudes toward life in general, health, age, level of aspiration, social status, self-concept, and other activities, also affect job satisfaction.

Job satisfaction refers to the individual's mind-set about the job, which may be positive or negative. It is not synonymous with organizational morale. Morale is related to group attitudes, and job satisfaction is concerned with individual attitudes.

Traditionally, managers have believed that satisfied workers would be good workers; the goal, then, was to keep workers "happy." The satisfaction-performance relationship is much more complex, however. Schwab and Cummings (1970) delineated these three dominant points of view:

Satisfaction $\longrightarrow$ Performance
Performance $\longrightarrow$ Satisfaction
Satisfaction $+ x + y \longrightarrow$ Performance

In other words, some theories hold that satisfaction leads to performance, and others contend that performance leads to satisfaction. The third view is that the relationship between satisfaction and performance is moderated or influenced by a number of variables.

Research generally rejects the more popular view that satisfaction causes performance. According to Greene (1972), some evidence provides support for the view that performance causes satisfaction and that the satisfaction-performance relationship is affected by a number of factors. Furthermore, rewards constitute a more direct cause of satisfaction than does performance, and rewards that are clearly related to current performance influence subsequent performance as well.

The complexity of the job satisfaction–performance relationship is illustrated in the model in Figure 19.3, which was developed from research on job satisfaction and job performance of foodservice employees (Hopkins, Vaden, & Vaden, 1979). Results of the study showed that higher performers had higher job satisfaction and higher organizational identification than low job performers, and their performance outcomes may, in turn, affect future job performance. The researchers also showed that expectations and opportunities arise from the job situation, and they may affect job performance and satisfaction.

Hopkins, Vaden, and Vaden (1979) found that foodservice workers with high performance ratings on several dimensions (quality of work, quantity of work, initiative, judgment, following directions, attendance, and personal relations) were more satisfied with all aspects of their job except pay and promotion. The high-performing group had longer tenure on the job and were mostly full-time workers, not part time.

Job satisfaction does have a positive impact on turnover, absenteeism, tardiness, accidents, and grievances (Schwab & Cummings, 1970). Research among foodservice employees indicates that job satisfaction should be a concern of managers, even though the relationship to job performance may not be clear. In general, low job satisfaction and high employee turnover have characterized the foodservice industry. Poor working conditions, boredom, limited job opportunities, no recognition for performance, low wages, and poor fringe benefits are among factors believed to contribute to the problems of low job satisfaction and low productivity in the industry.

Recently, turnover rates in restaurants have decreased, perhaps reflecting increased efforts to train and motivate employees (Riehle, 1993). The recession in the early 1990s also has made finding jobs more difficult. The annual turnover rates for hourly employees remains noticeably higher than for salaried employees. In a 1991 survey, full-menu table service restaurants reported a median annual turnover rate of 102% for hourly employees, compared with 20% for salaried employees. In limited-menu restaurants, the median annual turnover rate for hourly employees was 122% compared to 25% for salaried employees.

Foodservice managers have to learn to deal with turnover by thinking about the reasons behind it (Bailey, 1992). Money is a big motivator, especially when an employee feels underpaid. Giving raises to employees who are deserving is a morale booster. Burnout is common in foodservice operations. Managers often are so concerned about getting the job done that they forget to challenge employees by giving them more variety in their work. Staff conflicts, which can cause turnover, probably could be resolved if the manager were to serve as a mediator. Employees frequently leave jobs because of personal problems. A sensitive manager often can help an

Figure 19.3. A conceptual model for analyzing work performance in foodservice organizations.

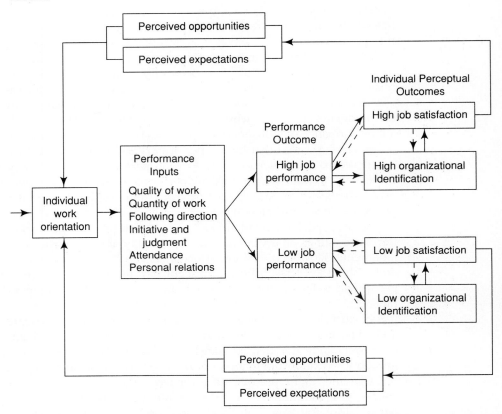

Source: "Some aspects of Organization Identification Among School Foodservice Employees" by D. E. Hopkins, A. G. Vaden, and R. E. Vaden, 1980, *School Food Service Research Review, 4*(1), p. 40. Used by permission.

employee solve these problems just by listening. Employee turnover, however, happens to be part of the restaurant business, and learning to cope with turnover is part of being a restaurant manager.

Rue and Byars (1989) stated that satisfied employees are preferred if for no other reason than that they make the work situation more pleasant. Also, individual satisfaction tends to lead to organizational commitment. Foodservice employees who like their jobs, supervisors, and other job-related factors, for example, will probably be more loyal employees, and those who dislike their jobs will probably be disgruntled and manifest their discontent by being late or absent, behaving in a disruptive way, or quitting.

Before leaving this topic, an important point must be reiterated: Motivation and satisfaction are different concepts. *Motivation* is a drive to perform, but *satisfaction* reflects an individual's attitude in a situation. Factors that determine satisfaction with a job differ from those that determine an individual's motivation. Satisfaction is largely determined by conditions in the environment and in the situation; motivation is determined by needs and goals that may be related to expectations and rewards. The result of motivation is increased effort, which will, in turn, increase performance of individuals who have the ability and direct their efforts appropriately.

LEADERSHIP

Leadership is the process of influencing the activities of an individual or group in efforts toward goal achievement. As Hersey and Blanchard (1993) stated, the leadership process (L) is identified as a function of the leader (l), the follower (f), and other situational (s) variables. This relationship can be expressed as follows:

$$L = f(l,f,s)$$

This definition makes no mention of any particular type of organization because leadership occurs in any situation in which someone is trying to influence the behavior of another individual or group. Thus everyone attempts leadership at one time or another in his or her activities at work, in social settings, at home, or elsewhere. As stated previously, the emphasis in this chapter is on leadership in the work organization.

Dynamic and effective leadership is one major attribute that distinguishes successful from unsuccessful organizations. Leaders are those who are willing to assume significant leadership roles and who have the ability to get the job done effectively. The effective organizational leader is one who is able to influence people to strive willingly for group objectives.

In chapter 17 formal authority and acceptance authority were examined. Leadership is related to acceptance authority. Managers are increasingly finding that subordinates are less willing to follow without question; therefore, a leader today must depend on the acceptance of that leadership by those being led.

Leadership and Power

Leaders possess **power,** and power has various dimensions. The concept of power is closely related to the concepts of leadership and authority because power is one of the means by which a leader influences the behavior of followers.

Traditional Power Structure

Although leadership can be viewed as any attempt to influence, power can be described as a leader's influence potential (Hersey & Blanchard, 1993). Power is the resource that enables a leader to induce compliance or influence others. Some authors distinguish between position power and personal power. *Position power* is derived from a person's official position in an organization, whereas *personal power* comes from personal attributes and expertise.

Several other authors have developed more specific power-base classifications. The one devised by French and Raven (1959) is probably the most widely accepted. They proposed five bases of power: legitimate, reward, coercive, expert, and referent power. Information power and connection power are two additional bases identified by other authorities (Hersey & Blanchard, 1993; Raven & Kluganski, 1970). These seven bases of power, or potential means of influencing the behavior of others, are defined as follows:

- *Legitimate power.* Comes from the formal position held by an individual in an organization; generally, the higher the position, the higher legitimate power tends to be. A leader high in legitimate power induces compliance from others because the followers believe this person has the right to give directions by virtue of his or her position.
- *Reward power.* Comes from a leader's ability to reward others. Examples of formal rewards are increases in pay, promotions, or favorable job assignments.
- *Coercive power.* Comes from the authority of the leader to punish those who do not comply. A leader with coercive power can fire, demote, threaten, or give undesirable work assignments to induce compliance from others.
- *Expert power.* Held by those leaders who are viewed as being competent in their job. Knowledge gained through education or experience and a demonstration of ability to perform are sources of expert power. A leader high in expert power can influence others because of their respect for his or her abilities.
- *Referent power* (sometimes called *charisma*). Based on identification of followers with a leader. A leader high in referent power is generally well liked and admired by others; thus, the leader can influence others because of this identification and admiration.
- *Information power.* Based on the leader's possession of or access to information that is perceived as valuable by others. This power influences others either because they need the information or want to be a part of things.
- *Connection power.* Based on the leader's connections with influential or important persons inside or outside the organization. A leader high in connection power induces compliance from others who aim at gaining the favor or avoiding the disfavor of the influential connection.

These concepts have application to understanding the power of leaders in any organization, including those in the foodservice industry. An example of this power is a hypothetical but typical situation involving a chef in a country club operation. David Lott, the chef at Stoneyville Country Club, is a highly skilled chef who has responsibility for directing the foodservice operations at the club. Mr. Lott is considered fair and consistent but an all-business type of supervisor. He is responsible for hiring and firing all foodservice staff and for recommending pay increments and promotions to the club manager.

Mr. Lott's potential means for influencing the foodservice staff, or, in other words, his power, is derived from the position he holds in the organization. That is, because he is in charge of the foodservice and has authority for punishment and rewards, he has legitimate, coercive, and reward power. Also, from the description of him as a "highly skilled chef," the inference is that he has expert power; from the "all-business" classification, the inference could be made that his referent power is limited.

Although all these dimensions of power are important to leaders, expert and referent powers tend to be related to subordinates' satisfaction and performance, and coercive is the most negative. Expert and legitimate powers appear to be the most important for compliance. In general, depending on the situation, any one of these dimensions is of value to the leader.

Evolving Power Structure

The newer concepts and trends in the power of leadership are exceptionally well presented by Kanter (1989). She stated that competitive pressures are forcing corporations to adopt new, flexible strategies and structures. According to a general in the 1871 French-Prussian War, strategy is applied common sense and cannot be taught (Hinterhuber & Popp, 1992). Profiles of outstanding entrepreneurs indicate that they create a corporate culture in which their vision, philosophy, and business strategies are implemented by employees who think independently.

Managerial authority is diminishing and new tools of leadership are emerging (Kanter, 1989). Managers whose power came from hierarchy and who were accustomed to limited personal control are learning to shift their perspectives and widen their scopes. The new managerial work consists of looking outside an area of responsibility to find opportunities and of forming teams from relevant disciplines to address them. This work involves communication and collaboration across functions, units, and other operations with overlapping activities and resources. Title and rank will be less important factors in success in the new managerial work than having the knowledge, skills, and sensitivity to mobilize people and motivate them to do their best. A prime example of this shift may be seen in a hospital in which the patient tray delivery supervisor discusses the means of improving patient service with the nursing supervisor on the unit. Under the older, more autocratic system, the tray delivery supervisor would have discussed improvements with the foodservice manager, who would have followed through with a decision. Direct communication between the two supervisors involves shifting power and produces more immediate effective action.

Philosophies of Human Nature

Assumptions that managers have regarding other people are major factors determining the climate for motivation. In order to learn how to create an environment conducive to a high level of employee motivation, managers should develop an understanding and philosophy of human nature and its influence on leadership style. A review of the work of McGregor (1960) and Argyris (1957) is useful in gaining this understanding.

McGregor's Theory X and Theory Y

McGregor stressed the importance of understanding the relationship between motivation and philosophies of human nature. In observing the practices and approaches of managers, McGregor concluded that two concepts of human nature were predominant. He referred to these as **Theory X** and **Theory Y**. These two distinct concepts—the negative Theory X, and the positive Theory Y—relate to basic philosophies or assumptions that managers hold regarding the way employees view work and how they can be motivated. These philosophies are summarized in Figure 19.4.

Theory X suggests that motivation will be primarily through fear and that the supervisor will be required to maintain close surveillance of subordinates if the organizational objectives are to be attained. Furthermore, the manager must protect the employees from their own shortcomings and weaknesses and, if necessary, goad them into action. Although Theory X is by no means without its supporters, it is not in keeping with more current concepts of behavioral science. Theory Y, in contrast, emphasizes managerial leadership by permitting subordinates to experience per-

Figure 19.4.   McGregor's Theory X and Theory Y.

Theory X
1. The average human being has an inherent dislike of work and will avoid it if he/she can.
2. Because of this human characteristic of dislike of work, most people must be coerced, controlled, directed, and threatened with punishment to get them to put forth adequate effort toward the achievement of organizational objectives.
3. The average human being prefers to be directed, wishes to avoid responsibility, has relatively little ambition, and wants security above all.

Theory Y
1. The expenditure of physical and mental effort in work is as natural as play or rest. Depending on controllable conditions, work may be a source of satisfaction (and will be voluntarily performed) or a source of punishment (and will be avoided if possible).
2. External control and the threat of punishment are not the only means for bringing about effort toward organizational objectives. People will exercise self-direction and self-control in the service of objectives to which they are committed.
3. Commitment to objectives is a function of the rewards associated with their achievement. The most significant of such rewards, the satisfaction of ego and self-actualization needs, for example, can be direct products of effort directed toward organizational objectives.
4. The average person learns, under proper conditions, not only to accept but to seek responsibility.
5. The capacity to exercise a relatively high degree of imagination, ingenuity, and creativity in the solution of organizational problems is widely, not narrowly, distributed in the population.
6. Under the conditions of modern industrial life, the intellectual potentialities of the average human being are only partially utilized.

sonal satisfaction and to be self-directed. These contrasting sets of assumptions lead to different leadership styles among managers and different behaviors among employees. Managers who hold to the Theory X view tend to be autocratic, and those with a Theory Y philosophy tend to be more participative.

Mondy, Sharplin, and Hippo (1988) suggested that if a manager accepts the Theory Y philosophy of human nature, the following practices might be considered seriously: flexible working hours, job enrichment, participative decision making, and abandonment of time clocks. One should not conclude, however, that McGregor advocated Theory Y as the panacea for all management problems. Although Theory Y is no utopia, McGregor argued that it provides a basis for improved management and organizational performance.

Argyris's Maturity Theory

Review of the work of Argyris also will assist managers in developing an understanding of human behavior. According to Argyris, a number of changes take place in the personality of individuals as they develop into mature adults over the years (Table 19.1). Further, these changes reside on a continuum and the "healthy" personality develops along the continuum from immaturity to maturity.

Argyris questioned the assumption that widespread problems of worker apathy and lack of effort in organizations are simply the result of individual laziness. He suggested that this may not be the case. When people join the work force, he contended, many jobs and management practices are not designed to support their mature personality. Employees who have minimal control over their environment tend to act in passive, dependent, and subordinate ways, and, as a result, to behave immaturely.

According to Argyris, treating people immaturely is built into traditional organizational principles such as task specialization, chain of command, unity of direction, and span of control. He stated that these concepts of formal organization lead to assumptions about human nature that are incompatible with the proper development of maturity in human personality. In colloquial terms, the Argyris theory can be summarized as follows: Management creates "Mickey Mouse" jobs and then is surprised with "Mickey Mouse" behavior.

Argyris challenged management to provide a work climate in which individuals have a chance to grow and mature as individuals while working for the success of the

Table 19.1. Argyris's maturity theory

Immaturity	vs.	Maturity
Passive		Increased activity
Dependence		Independence
Behave in a few ways		Capable of behaving in many ways
Erratic, shallow interests		Deeper, stronger interests
Short time perspective		Long time perspective (past and future)
Subordinate position		Equal or superordinate position
Lack of awareness of self		Awareness and control over self

organization. He contended that giving people the opportunity to grow and mature on the job allows employees to use more of their potential. Furthermore, although all workers do not want to accept more responsibility or deal with the problems that responsibility brings, the number of employees whose motivation can be improved is much larger than many managers suspect.

The Argyris theory has application to understanding jobs and work behavior in the foodservice industry. Although essential to operations, the industry has many jobs that are routine, repetitive, unchallenging, and boring. The foodservice manager should not be surprised to find that employees are not turned on to such jobs as potwasher, dishwasher, or general kitchen worker. The challenge to the foodservice manager is to enrich jobs to the maximum extent possible. One must acknowledge, however, that the potential is limited for making some jobs in the foodservice operation highly motivating and exciting.

Leadership Effectiveness

For many years, the study of leadership concentrated on traits or characteristics essential for effective leadership, particularly focusing on physical attributes and personal qualities. According to Hersey and Blanchard (1993), a review of the research on this trait approach to leadership reveals few significant or consistent findings. The best that can be said is that intelligence, self-confidence, empathy, emotional stability, motivational drive, and the ability to solve problems tend to be associated with effective leaders more often than they are with those who follow.

As suggested by the empirical studies already cited in this chapter, effective leadership is viewed as a dynamic process that varies from situation to situation with changes in leaders, followers, and conditions, rather than as a function of certain traits or characteristics of leaders. These leader behavior or situational approaches are concerned with the behavior of leaders and their group members.

With this emphasis on behavior and environment, the possibility exists for training individuals to adapt styles of leader behavior to varying situations. Most people can generally increase their effectiveness in leadership roles through education, training, and development.

Leadership has been defined as the process of influencing activities of an individual or a group in efforts toward achievement of objectives in a given situation. In essence, then, leadership involves accomplishing goals with and through people; therefore, a leader must be concerned with both tasks and human relationships. This two-dimensional view of leadership is the basis of current leadership theories. Also, the situational dimension is a third aspect of leadership models.

Basic Leadership Styles

Early studies on leadership identified three basic styles: autocratic, laissez-faire, and democratic (Rue & Byars, 1989). Responsibility for decision making is the key factor differentiating these leadership styles. Generally, the *autocratic leader* makes most decisions, the *laissez-faire leader* allows the group to make the decisions, and the *democratic leader* guides and encourages the group to make decisions.

In the early work on leadership styles, democratic leadership was considered the most desirable and productive. Current research does not necessarily support this conclusion. Instead, various styles of leadership have been found to be effective in different situations, which are discussed in more detail later in this chapter. The primary contribution of this early research was the identification of the three basic styles of leadership.

Behavioral Concepts of Leadership

When research shifted from an emphasis on personality and physical traits to an examination of leadership behavior, the focus was on determining the most effective leadership style. Many of the studies were conducted at the University of Michigan and Ohio State University. Building on the work at Michigan and Ohio State, and on results from their own research, Blake and Mouton (1964) proposed another leadership behavior model, the Managerial Grid®.

University of Michigan Leadership Studies.   Leadership studies conducted at the Institute for Social Research at the University of Michigan were designed to characterize leadership effectiveness. These studies isolated two major concepts of leadership: employee orientation and production orientation (Kahn & Katz, 1960).

Employee-centered leaders were identified by their special emphasis on the human relations part of their job, and production-oriented leaders emphasized performance and the more technical characteristics of work. Results of the Michigan studies showed that supervisors of high-producing sections were more likely to have the following traits:

- Receive general rather than close supervision from their superiors
- Spend more time in supervision
- Give general rather than close supervision to their employees
- Be employee oriented rather than production oriented

Likert (1967), a former director of the Institute of Social Research at the University of Michigan, developed a continuum of leadership styles, from autocratic to participative, based on a summary he prepared of the research on leadership. He proposed the following four basic management styles:

- Exploitive autocratic
- Benevolent autocratic
- Consultative
- Participative

In the *exploitive autocratic* management style, employees are motivated by fear, threats, and punishment and seldom by reward. Almost all decisions are made by top management and only occasionally does communication move up from employees to the manager. The *benevolent autocratic* style indicates that only certain minor decisions are made by employees, and communication moving upward is generally ignored. Small rewards are given, but threats and punishment are the norm. In *consultative* style, employees gain some confidence. Information flows up and down, but all major decisions come from the top. The *participative* management style operates on a basis of trust and responsibility. Employees discuss the job with their superiors, and communication flows up, down, and laterally; decision making is spread evenly through the organization.

Likert examined the characteristics of communication flow, decision-making processes, goal setting, control mechanisms, and other operational characteristics of organizations and assessed managerial and leadership styles. Results of these studies indicated that participative was the most effective style of management. Its emphasis is on a group participative role with full involvement of the employees in the process of establishing goals and making job-related decisions.

Ohio State Leadership Studies.   Beginning in the 1940s, researchers at Ohio State University started a series of in-depth studies on the behavior of leaders in a wide variety of organizations. These studies were conducted about the same time as those at the University of Michigan and used similar concepts. Two dimensions of leadership behavior emerged from those studies (Stogdill, 1974): consideration and initiating structure.

Consideration indicates behavior that expresses friendship, develops mutual trust and respect, and develops strong interpersonal relationships with subordinates. Leaders who exhibit consideration are supportive of their employees, use their employees' ideas, and allow frequent participation in decisions.

Initiating structure indicates behavior that defines work and establishes well-defined communication patterns and clear relationships between the leader and subordinate. Leaders who initiate structure emphasize goals and deadlines, give employees detailed task assignments, and define performance expectations in specific terms.

In studying leader behavior, Ohio State researchers found that consideration and initiating structure are separate and distinct dimensions; a high score on one dimension does not necessitate a low score on the other. In other words, the behavior of a leader can be described as a mix of both dimensions. In depicting various patterns of leader behavior, the two dimensions were plotted on two axes, which resulted in four quadrants describing four leadership styles (Figure 19.5).

In general, they found that leaders who showed high consideration for people and initiating structure tended to have higher-performing and more satisfied subordinates than did others. They also concluded that the relationship between these dimensions and leadership effectiveness depends on the group.

The two-dimensional view of leadership is the basis for the Managerial Grid developed by Blake and Mouton (1964, 1978). The two dimensions of the grid are concern for people and concern for production, which are similar to the dimensions of the Ohio State model.

Leadership Grid.   The Blake and Mouton (1964, 1978) Managerial Grid then became the basis for the Leadership Grid® developed by Blake and McCanse (1991), shown in Figure 19.6. It has 9 possible positions along the vertical and horizontal axes for a total of 81 possible leadership styles, although 5 basic styles similar to those of Blake and Mouton are generally discussed. The Leadership Grid was

Figure 19.5. The Ohio State leadership quadrants.

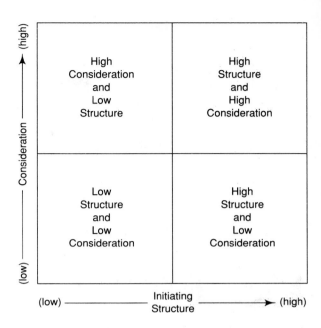

expanded following years of research to include two additional styles: 9 + 9 Paternalistic Management, in which reward is promised for compliance and punishment threatened for noncompliance; and Opportunistic Management, which depends on the style the leader believes will return him or her the greatest personal benefit. A third dimension that was revealed in the research explores the positive and negative motivations underlying the styles.

The Managerial Grid and now the Leadership Grid have been used widely in organization development and have become so popular among some managers that they refer to the styles by number. For example, 9,9, rather than team management, is often used to refer to the leader who has high concern for both people and work.

Situational and Contingency Approaches

Situational and contingency approaches emphasize leadership skills, behavior, and roles thought to be dependent on the situation. These approaches are based on the hypothesis that behavior of effective leaders in one setting may be substantially different from that in another. The current emphasis in leadership research, which is largely focused on the leadership situation, has shifted because previous attempts to determine ineffective characteristics and behaviors were inconclusive.

One of the first situational approaches to leadership was developed by Tannenbaum and Schmidt (1958). Fiedler (1967) has also made significant contributions to understanding contingency or situational approaches to leadership. Reddin (1967a) and, more recently, Hersey and Blanchard (1993) have developed situational models of leadership. The path-goal leadership model is another contingency approach.

Figure 19.6.   The Leadership Grid®.

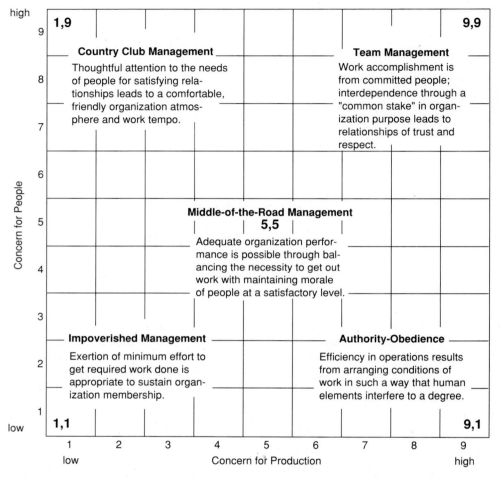

Source: From *Leadership Dilemmas—Grid Solutions* (p. 29), by Robert R. Blake and Anne Adams McCanse (formerly the Managerial Grid Figure by Robert R. Blake and Jane S. Mouton). Houston: Gulf Publishing Company, Copyright © 1991, by Scientific Methods, Inc. Reproduced by permission of the owners.

Leadership Continuum.   Tannenbaum and Schmidt (1958) developed a continuum, or range, of possible leadership behaviors. Each type of behavior is related to the degree of authority used by the manager and the amount of freedom available to subordinates in reaching decisions. The actions range from those in which a high degree of control is exercised to those in which a manager releases a high degree of control.

This continuum was revised and published again in 1973. The revised model, shown in Figure 19.7, reflects two major changes. The interdependencies between the organization and its environment were acknowledged, as shown by the circular addition to the diagram depicting the organizational and societal environment. Also, originally, the terms *boss centered* and *subordinate centered* were used. In revising the model, the terms *manager* and *nonmanager* were substituted to denote functional rather than hierarchical differences.

Three forces were identified as affecting the leadership appropriate in a given situation:

- Forces in the manager
- Forces in subordinates or nonmanagers
- Forces in the situation

In Table 19.2, the factors within each of these forces are listed. These forces differ in strength and interaction in different situations; therefore, one style of leadership is not effective in all situations. In fact, the underlying concept of the continuum is that

Figure 19.7.   Continuum of manager-nonmanager behavior.

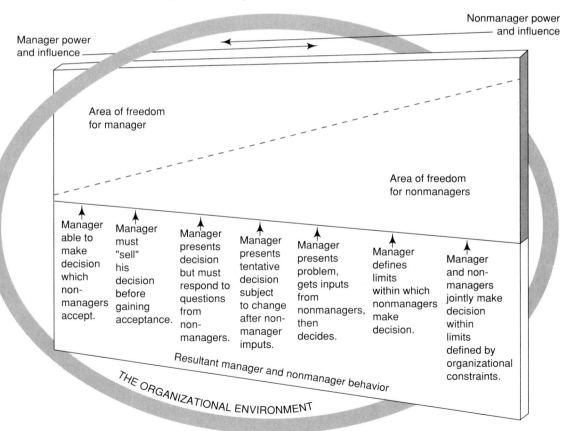

Table 19.2.   Forces in the leadership situation

Forces in the Manager	Forces in Nonmanagers	Forces in the Situation

Value system:
  How the manager personally feels about delegating
  Degree of confidence in staff
Personal leadership inclinations:
  Authoritarian versus participative
Feelings of security in uncertain situations
	Need for independence:
	  Some people need and want direction while others do not
	Readiness to assume responsibility:
	  Different people need different degrees of responsibility
	Tolerance for ambiguity:
	  Specific versus general directions
	Interest and perceived importance of the problem:
	  People generally have more interest in and work harder on important problems
	Degree of understanding and identification with organizational goals:
	  A manager is more likely to delegate authority to an individual who seems to have a positive attitude about the organization
	Degree of expectation in sharing in decision making:
	  People who have worked under subordinate-centered leadership tend to resent boss-centered leadership
		Type of organization:
		  Centralized versus decentralized
		Work group effectiveness:
		  How effectively the group works together
		The problem itself:
		  The work group's knowledge and experience relevant to the problem
		Time pressure:
		  It is difficult to delegate to subordinates in crisis situations
		Demands from upper levels of management
		Demands from government, unions, and society in general

Source: Reprinted by permission of *Harvard Business Review.* An exhibit from "How to Choose a Leadership Pattern" by R. Tannenbaum and W. H. Schmidt, 1973, *51*(3), p. 162. Copyright © 1973 by the President and Fellows of Harvard College; all rights reserved.

the manager may employ a variety of approaches, which are dependent on the forces operating in a particular situation.

For example, a foodservice manager who is generally democratic and involves employees in decision making has to take charge in a crisis situation, such as a kitchen fire, that requires directive and authoritarian leadership.

Successful leaders are keenly aware of the forces most relevant to their behavior at a given time and are able to act appropriately in relation to other individuals involved in a situation and to the organizational and social environmental forces. In general, however, Tannenbaum and Schmidt encouraged managers to shift toward more participative approaches to decision making. Some of the benefits of more participative styles are that they:

- Raise employees' motivational level
- Increase willingness to change
- Improve quality of decisions
- Develop teamwork and morale
- Further the individual development of employees

A criticism of the continuum is that managers will be viewed as inconsistent or wishy-washy if they react differently in every situation. The key to effective leadership using the continuum, however, is that the leader will behave consistently in situations in which similar forces are operating.

Contingency Approach.   Fiedler (1967) developed a leadership contingency model in which he defined three major situational variables. In his theory, Fiedler proposed that these three variables seem to determine if a given situation is favorable to leaders:

- *Leader-member relations.* Personal relations with members of the group
- *Task structure.* Degree of structure in the task assigned to the group
- *Position power.* Power and authority a leader's position provides

Fiedler defined the favorableness of a situation as the degree to which the situation enables the leader to exert influence over the group. The most favorable situation for leaders is one in which they are well liked by the members (good leader-member relations), have a powerful position (high position power), and are directing a well-defined job (high task structure). He examined task-oriented and relationship-oriented leadership to determine the most effective style.

In both highly favorable and highly unfavorable situations, a task-oriented leader seems to be more effective. In highly favorable situations, the group is ready to be directed and is willing to be told what to do. In highly unfavorable situations, however, the group welcomes the opportunity of having the leader take the responsibility for making decisions and giving directions. In moderately favorable situations, however, a relationship-oriented leader tends to be more effective because cooperation is more successful than task-oriented leadership for this particular group. Fiedler's work has been important in demonstrating particular styles of leadership that are most effective in given situations.

Leader Effectiveness Model.   Hersey and Blanchard (1993) developed a leadership model that has gained considerable acceptance. In their model, task behavior and relationship behavior are used to describe concepts similar to those of consideration and initiating structure in the Ohio State studies. They define the terms in the following ways:

- *Task behavior.* The extent to which the leader engages in spelling out the duties and responsibilities of an individual or group. It includes telling people what to do, how to do it, when to do it, where to do it, and who is to do it.
- *Relationship behavior.* The extent to which the leader engages in two-way or multi-way communication. It includes listening, facilitating, and supportive behaviors. (Hersey & Blanchard, 1993, pp. 185, 186, 187)

Recognizing that the effectiveness of leaders depends on how their styles interrelate with the situations in which they operate, an effectiveness dimension was added to the two-dimensional model based on the Ohio State studies (Figure 19.8). The effectiveness dimension was drawn from the work of Reddin (1967a), who contended that a variety of leadership styles may be either effective or ineffective.

By adding the effectiveness dimension to the task and relationship behavior dimensions, Hersey and Blanchard (1993) attempted to integrate the concepts of leadership style and situational demands of a specific environment. When the leader's style is appropriate to a given situation, it is termed *effective;* when the style is inappropriate, it is termed *ineffective.* The difference between effective and ineffective styles is often not the actual behavior of the leader but the appropriateness of the behavior to the environment in which it is used.

As illustrated in their model (Figure 19.9), leadership style in a particular situation can fall somewhere between extremely effective and extremely ineffective; effectiveness, therefore, is a matter of degree. As summarized in Table 19.3, the appropriateness of a leader's style in a given situation is related to the reactions of followers, superiors, and associates. The issue of consistency, which was pointed out with the Tannenbaum and Schmidt leadership continuum, might also be raised with the effectiveness model. Hersey and Blanchard (1993) concluded that consistency is not using the same style all the time; instead, consistency is using the same style for all similar situations and varying the style appropriately as the situation changes.

Figure 19.8.   The effectiveness dimension of leadership.

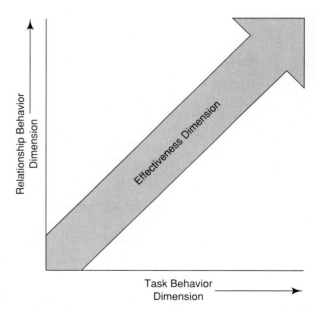

Source: Paul Hersey and Kenneth H. Blanchard, *Management of Organizational Behavior: Utilizing Human Resources,* 6th ed., © 1993, p. 129. Reprinted by permission of Prentice-Hall, Englewood Cliffs, New Jersey.

Figure 19.9.   Leader effectiveness model.

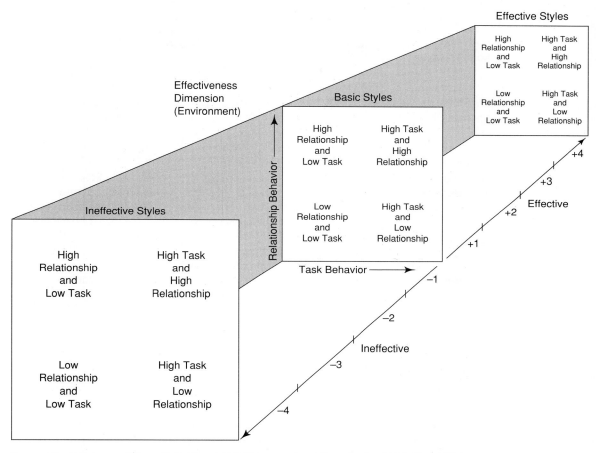

Source: Paul Hersey and Kenneth H. Blanchard, *Management of Organizational Behavior: Utilizing Human Resources,* 6th ed., © 1993, p. 131. Reprinted by permission of Prentice-Hall, Englewood Cliffs, New Jersey.

The task of diagnosing a leadership environment is complex because the leader is the focal point around which all the other environmental variables interact (Hersey & Blanchard, 1993). The leader has a style and expectations and has to interact with the following variables: followers, superiors, associates, and organizations; other situational variables; and job demands.

The Hersey and Blanchard (1993) leadership model also is based on the assumption that the most effective style varies with the followers' level of task-relevant readiness as well as with the demands of the situation. *Readiness* in the work situation is defined as a desire for achievement based on challenging but attainable goals, willingness and ability to accept responsibility, and education or experience and skills relevant to a particular task. Hersey and Blanchard (1993) defined four leadership styles appropriate to various readiness levels of followers in each of their various

Table 19.3.   Effective and ineffective leadership styles

Basic Styles	Effective	Ineffective
High Task and Low Relationship Behavior	Seen as having well-defined methods for accomplishing goals that are helpful to the followers.	Seen as imposing methods on others; sometimes seen as unpleasant and interested only in short-run output.
High Task and High Relationship Behavior	Seen as satisfying the needs of the group for setting goals and organizing work, but also providing high levels of socioemotional support.	Seen as initiating more structure than is needed by the group and often appears not to be genuine in interpersonal relationships.
High Relationship and Low Task Behavior	Seen as having implicit trust in people and as being primarily concerned with facilitating their goal accomplishment.	Seen as primarily interested in harmony; sometimes seen as unwilling to accomplish a task if it risks disrupting a relationship or losing "good person" image.
Low Relationship and Low Task Behavior	Seen as appropriately delegating to subordinates decisions about how the work should be done and providing little socioemotional support where little is needed by the group.	Seen as providing little structure or socioemotional support when needed by members of the group.

Source: Adapted from *The 3-D Management Style Theory*, Theory paper #2—Managerial Styles, by W. J. Reddin, 1967, Canada: Social Sciences Systems.

areas of responsibility (Figure 19.10). A person's readiness level gives the manager a good clue about where to begin further development of that individual. A manager desiring to influence an employee in an area in which the employee is unable and unwilling, indicating a low readiness level (R1), must develop the person by closely supervising the staff member's behavior utilizing the "telling" leadership style (S1), as shown in Figure 19.10.

Once this manager has diagnosed the readiness level of the follower as "low," the appropriate style can be determined (Figure 19.11). The manager would construct a right angle from a point on the readiness continuum to the point where it meets the curved line in the effective styles portion of the model. The leader should use a telling style (S1) in which the staff member must be told what to do and then shown how to do it.

In summary, Hersey and Blanchard's leadership effectiveness model provides a useful framework for leadership (Figure 19.11). Their model suggests that no one best leadership style is appropriate to meet the needs of all situations. Instead, leadership style must be adaptable and flexible enough to meet the changing needs of employees and situations. The effective leader is one who can modify styles as employees develop and change or as required by the situation.

Figure 19.10.   Relationship of leadership style and readiness of followers.

Leadership Styles	**Readiness Level of Followers**
S1 Telling High task and low relationship	**R1** Low Readiness Unable and unwilling or insecure
S2 Selling High task and high relationship	**R2** Low to Moderate Readiness Unable but willing or confident
S3 Participating Low task and high relationship	**R3** Moderate to High Readiness Able but unwilling or insecure
S4 Delegating Low task and low relationship	**R4** High Readiness Able/competent and willing/confident

Source: Paul Hersey and Kenneth H. Blanchard, *Management of Organizational Behavior: Utilizing Human Resources,* 6th ed., © 1993, p. 195. Reprinted by permission of Prentice-Hall, Englewood Cliffs, New Jersey.

Figure 19.11.   Leadership styles appropriate to various readiness levels of followers.

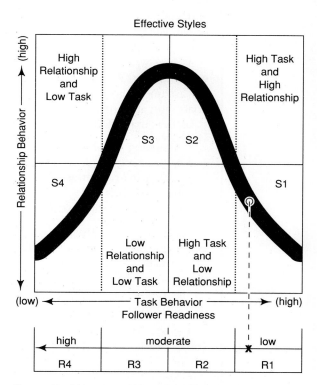

Source: Paul Hersey and Kenneth H. Blanchard, *Management of Organizational Behavior: Utilizing Human Resources,* 6th ed., © 1993, p. 257. Reprinted by permission of Prentice-Hall, Englewood Cliffs, New Jersey.

This leadership model has direct application to leadership in foodservice organizations. Situational leadership contends that strong direction with low-readiness followers is appropriate if they are to become productive. As followers reach high levels of readiness, however, the leader should respond not only by decreasing control but also by decreasing relationship behavior.

Consider this example of applying these concepts. If a foodservice manager is responsible for opening a new unit with a new group of employees who are unaccustomed to working together, the manager should exhibit a directing, or telling, style of leadership. After the unit has been in operation several years and the work group has become accustomed to working as a team, assuming turnover is low and each employee knows his or her job well, a change in leadership style is indicated. A participative, or delegating, style would probably be more appropriate. The foodservice manager, as with any other person in a leadership role, must be flexible in adapting his or her leadership style to changing situations and conditions.

Path-Goal Leadership Model.   Another important contingency leadership concept, which focuses on the leader's effect on the subordinate's motivation to perform, was developed by Evans (1970) and House (1971). The model is based on the expectancy concept of motivation, which emphasizes expectancies and valences, previously discussed in this chapter. The path-goal concept focuses on the leader's impact on the subordinate's goals and the paths to achieve those goals.

As you will recall, expectancies are beliefs that efforts will be rewarded, and valences are the value or attractiveness of those rewards. Basically, the path-goal theory assumes that individuals react rationally in pursuing certain goals because those goals ultimately result in highly valued payoffs to the individual.

Leaders may affect employees' expectancies and valences in several ways:

- Assigning individuals to tasks for which they have high valences
- Supporting employee efforts to achieve goals
- Tying extrinsic rewards (pay increases, recognition, promotion) to accomplishment of goals
- Providing specific extrinsic rewards that employees value

These actions on the part of the leader can increase effectiveness because employees reach higher levels of performance through increased motivation on the job. Additionally, the path-goal theory implies that the degree to which the leader can be effective in eliciting work-goal directed behavior depends on the situation.

The path-goal leadership concept focuses on four types of leader behavior and two situational factors (House & Mitchell, 1974):

- *Directive.* Leadership behavior characterized by providing guidelines, letting subordinates know what is expected, setting definite performance standards, and controlling behavior to ensure adherence to rules
- *Supportive.* Leadership behavior characterized by being friendly and showing concern for subordinates' well-being and needs
- *Achievement oriented.* Leadership behavior characterized by setting challenging goals and seeking to improve performance

- *Participative.* Leadership behavior characterized by sharing information, consulting with employees, and emphasizing group decision making

The *situational factors* include subordinates' characteristics, such as locus of control, and characteristics of the work environment, such as structure and complexity of the task. *Locus of control* refers to the tendency of people to rely on internal or external sources. Internal locus of control is operational with people who attribute task success or failure to their own strengths and weaknesses, whereas external focus of control is characteristic of people with a tendency to attribute success or failure to the nature of the situation around them.

In Table 19.4, effective leader behaviors in relation to various situational factors are summarized. As an application of the path-goal concepts, foodservice employees with a high need for affiliation will be more satisfied with a supportive leader; those with a high need for security will be more satisfied with a directive leader. As indicated in the chart, the nature of the task also influences the leadership behavior appropriate to a given situation.

Implications of Leadership Theories

From this discussion of leadership theories, one might believe that being a truly effective leader is almost impossible. Effective leadership is difficult to achieve, but managers have succeeded in the effort.

A common thread emerges from observations of effective leaders in all types of organizations. Successful leaders have either analyzed situational factors and adapted their leadership style to them or have altered the factors to match their style.

This overview of the leadership literature should make one major conclusion obvious: No one best style of leadership exists. Leadership should be seen as a function

Table 19.4.   Interaction of leader behavior and situational factors

Situational Factors		Effective Leader Behaviors
Subordinate Characteristics	**Characteristics of the Work Environment**	**Effective Leader Behaviors**
High need for affiliation		Supportive
High need for security		Directive
Internal locus of control		Participative
External locus of control		Directive
	Structured task	Directive
	Unstructured task	Supportive
High growth need strength	Complex task	Participative and achievement oriented
Low growth need strength	Complex task	Directive
High growth need strength	Simple task	Supportive
Low growth need strength	Simple task	Supportive and directive

Source: Organizational Behavior: Applied Concepts (p. 362) by R. D. Middlemist and M. A. Hitt, 1981, Chicago: Science Research Associates. Used by permission.

of forces in the leader, in the followers, and in the situation. A range of leader behaviors can be effective or ineffective, depending on styles and expectations of superiors, followers, and associates as well as job demands and organizational characteristics.

The successful leader must have a concern for both tasks and people in the work situation, because leadership is defined as a process of influencing activities of individuals in efforts toward goal achievement. Thus, flexibility and adaptability to changing conditions and situations are underlying concepts of effective leadership in all organizations.

Trends in Organizational Leadership

Is it possible to enhance employee motivation, increase job satisfaction, and improve productivity all at the same time? This question poses a significant challenge for foodservice managers.

Increasingly, America's most highly educated work force has expectations for jobs that not only provide for basic needs but also allow for satisfaction of needs for achievement, growth, recognition, and self-fulfillment. Another important challenge is increasing the productivity of the American worker by finding ways to unlock potential that exists in the overwhelming majority of the work force.

During the past decade, the decline in the productivity growth rate in the United States compared to most other industrialized nations has been a concern. In the foodservice industry, low productivity has been a long-standing problem. Job satisfaction of workers in all organizations has declined over the past couple of decades. As a result, managers are searching for ways to deal with these problems.

Since World War II, Japan has experienced increased prosperity, job security, and successful and stable business organizations. Seeking to discover the secrets of Japan's high productivity has become almost an obsession in U.S. organizations. Surprisingly, some of the more modern elements of Japanese businesses originated in the United States (Griffin, 1993). Dr. W. Edwards Deming went to Japan after World War II to help rebuild the Japanese economy. He convinced business managers to focus on continuously improving their operations by eliminating waste. The United States with government financial support has excelled in basic research while the Japanese have concentrated on applied research by making products better and more efficiently.

Among those looking to the Japanese model for answers to problems of American organizations have been Ouchi (1981) and Pascale and Athos (1981). Ouchi's studies of Japanese organizations led him to the Theory Z approach to management, which suggests that involved employees are the keys to increased productivity.

In contrasting Japanese and American organizations, Ouchi found many differences (Table 19.5). He proposes that American organizations seeking to improve productivity and quality of output would do well to incorporate some of the Theory Z characteristics. The most important change would be the involvement of employees at all levels of the organization.

In their discussion of lessons from the Japanese, Pascale and Athos (1981) described well-managed companies as having superordinant goals expressed in

Table 19.5. Comparison of Japanese and American companies

Japanese Organizations	vs.	American Organizations
Lifetime employment		Short-term employment
Slow evaluation and promotion		Rapid evaluation and promotion
Non-specialized career paths		Specialized career paths
Implicit control mechanisms		Explicit control mechanisms
Collective decision making		Individual decision making
Collective responsibility		Individual responsibility
Wholistic concern		Segmented concern

Source: William Ouchi, *Theory Z: How American Business Can Meet the Japanese Challenge*, ©1981, Addison-Wesley, Reading, Massachusetts, p. 58.

terms of the organization's responsibility to its employees, to its customers, and to the surrounding community. The most effective firms link their purposes and ways of realizing them to human values as well as to economic measures. Schonberger (1992) stated that executives in superior Japanese companies seem to do comparatively little strategic planning. Instead, they spend more time overseeing organizational dedication to "the basics" of competitive advantage.

Common threads that seem to emerge from the discussion of current approaches to leadership, Japanese management, and characteristics of excellently operated companies are the need for innovation and for dynamic, flexible leadership in organizations. Leaders create excitement and work from high-risk positions, especially when opportunity and reward appear high. Zaleznik (1977) insisted that leaders search out opportunities for organizational change.

Maccoby (1981) proposed some similar ideas of leadership for today's organizations. In his view, leadership is needed to develop consensus and revitalize organizations in democratic ways. He described successful leaders as intelligent, ambitious, optimistic, and persuasive communicators who tend to have critical views of traditional authority. But, Maccoby said, they are also caring and flexible, willing to share power, and concerned about self-development and development of others.

Research on chief executive officers and presidents in the noncommercial foodservice industry discovered four foundations of effective leadership (Cichy, Sciarini, Cook, & Patton, 1991). According to them, effective leaders do the following:

- Develop and provide a complete vision
- Earn and return trust
- Listen and communicate effectively
- Persevere when others give up

Cichy, Sciarini, and Patton (1992) surveyed the top 100 foodservice leaders in the United States for their opinions about those personal qualities associated with effective leaders. The leaders agreed or strongly agreed that the following six of the seven keys to effective leadership were important to their leadership style:

- Develop a vision
- Trust your subordinates
- Encourage risk

- Simplify
- Keep your cool
- Invite dissent

The respondents did not attach a high level of importance to the seventh key, "being perceived as an expert." They then ranked the four foundations of leadership from the previous study as follows: (a) trust, (b) vision, (c)communication, and (d) perseverance.

ORGANIZATIONAL CHANGE

Organizational change is any substantive modification to some part of the organization (Griffin, 1993). Over time, the world has changed, but organizations have not. A change is not an extension of what has been going on but is an entirely new way of thinking. Organizations only survive if they can provide value to the customer who believes that what they receive is worth the cost. Every activity in the organization adds cost but few add value. Being a good manager does not mean a person is a good leader, and being a good leader does not necessarily translate into being a good manager. Management is about coping with complexities, and leadership is about coping with change (Kotter, 1990). Part of the reason leadership has become important is that business has become more competitive and volatile. Technological changes are faster, demographics of the work force have changed, and international competition has increased.

Major changes are necessary if an organization is to survive and compete in this environment. More change always requires more leadership. Leaders are needed to develop a vision of the future along with strategies for producing changes needed to achieve that vision. Drucker (1992) stated that leadership is work, which includes setting goals, priorities, and standards. The leader sees leadership as responsibility and is not afraid of strength in associates and subordinates. Finally, effective leadership is to earn trust and is not based on being clever but rather on being consistent. The major change in organizations for many decades has been the conversion of a traditional philosophy of management to a philosophy of total quality management or service.

Converting from a traditional organization to a total quality one requires strong leadership. Total quality management (TQM) is based on zero defects in a product and very little on customer value and service quality. The total quality service (TQS) philosophy, however, emphasizes that all quality standards and measures should be customer referenced and should help people guide the organization to deliver outstanding value to the customers. Foodservice organizations are really using the TQS philosophy. Albrecht (1992) predicted that as the 21st century approaches, distinction between product and service will be obsolete and replaced by total customer value.

Top-level leaders cannot be expected to have detailed knowledge about existing products and those under development (Schonberger, 1992). They really cannot do well at setting goals and policies, but they can exercise vigorous leadership including cheerleading and encouraging implementation in companywide total quality improvement. Their role shifts toward facilitating changes necessary to make TQM and TQS everybody's business.

Organizational Culture

If total quality is the agenda for the organization, then the organizational culture must be redefined, followed by a sound strategy for daily and long-term actions, decisions, and plans that should be incorporated into the culture (Schonberger, 1992). Strategies for creating a corporate culture are discussed in chapter 17. Just as more people are needed to provide strong leadership in complex organizations, more people also are needed to develop cultures that will create leadership.

Institutionalizing a leadership-centered culture is the ultimate act of leadership (Kotter, 1990). Recruiting young people with leadership potential and then managing their careers is the first step in developing this culture. Leaders almost always have had the opportunity to try to lead, take risks, and learn from both success and failure early in their careers. These experiences teach people about the difficulty of leading and also that leadership has potential for providing change. In many organizations that have a leadership-centered culture, managers are encouraged to recognize and reward people who develop leaders, perhaps with a promotion.

Recent studies emphasize the difficulty involved in changing corporate culture (Wilhelm, 1992). If corporate culture is defined but is not working, perhaps changing employee behavior rather than the culture is the answer. Culture should provide the framework of stability, consistency, and behavioral norms required to operate a successful business. Perhaps employee behavior set against the corporate culture framework needs to be changed. Organizational strategy must be clearly defined before any change in employee behavior is attempted. Leaders must have a vision about where the organization is going and how fast. Once the vision has been clearly defined, the organizational strategy with input from the board of directors, management, and employees is outlined to accomplish the vision.

Total Quality Impact on Strategy

Weaknesses of the traditional organization, which appear to be incompatible with the TQM philosophy that includes employee-driven and customer-centered business practices, indicate that a strategy for change needs to be added to the culture (Schonberger, 1992). Principles of TQM can replace much of the strategic planning done in the traditional organization, and any issues that are not clearly covered can be resolved by organizationwide planning with executive oversight. According to Schonberger (1992), the following are the basic principles for continuous improvement that can be incorporated into the management and policies of organizations that adopt the TQM philosophy:

- Ever better, more appealing, less variable quality of the product or service
- Ever quicker, less variable response from producers through suppliers to the final users of products

- Ever greater flexibility in adjusting to variations in the number of customers and the customer mix
- Ever lower cost through quality improvement, use of labor, and non-value-added waste elimination

Each of these principles should be supported by specific operational guidelines. For example, the role of the internal and external customer should be identified in each.

The mental image of the strategic process should be changed from decision makers sifting through volumes of information received from organization managers to leaders seeing that good decisions are being made. The time has come to change the mental image. Effective leadership has been obvious in many organizations that have made changes such as going from a traditional style of management to a total quality philosophy. Human values are being emphasized in these changes.

Human Values in TQM

If employees are empowered to make decisions, they should be considered very important people in the organization. Management has a responsibility to make conditions in their work life pleasant and also to respect the pressures of home life. By mentoring or coaching prospective leaders and placing women in leadership positions, organizational change might become much easier for top management. Also, employees might begin to think that they are very important if top management makes organizational changes that would ease the work/home transition.

Mentoring

A **mentor** can be a teacher, tutor, or coach, depending upon the relationship between the mentor and protégé. A formalized procedure for seeking a mentor has not been developed. Usually ambitious, young employees pursue their own models (Coppess, 1989). Three types of relationships seem most common within the restaurant industry:

- *Model/admirer.* A leader provides an example of professional behavior with or without direct contact with the admirer.
- *Teacher/pupil.* A professional version of a parent/pupil relationship, in which one person directly guides another.
- *Co-mentor.* Peers share guidance and advice with one another.

A consultant who advises corporations on human resource matters stated that the protégé receives the more obvious benefits of guidance, knowledge, and a safety net, and the mentor benefits by contributing to the young person's career development. The mentor must be able to evaluate the protégé to determine when to let go instead of acting like an overprotective parent who never lets go.

One of the strongest selling points to future foodservice leaders is the open ladder to opportunity brought about by the fast growth of the industry (Slater, 1989). By presenting themselves and their employees as role models, dedicated foodservice operators are encouraging young people to stay in school and are showing them that foodservice can be the source of diverse, rewarding careers. Foodservice employs more minority managers and leaders than any other retail industry. The industry really is gender-blind and color-blind, as evidenced by the number of women and minority group members in positions of responsibility.

Women in Leadership Positions

Women are using classic techniques such as mentoring and business networking to overcome barriers to success (Batty, 1992). An unspoken, invisible barrier, identified as the *glass ceiling,* is used to explain that glass can be easily broken and women are beginning to climb through. Former Labor Secretary Lynn Martin wrote that few companies can afford glass ceilings. In a global marketplace that is becoming more competitive, companies need to promote the best people regardless of gender, race, color, or national origin.

Until recently, many male employers have felt that it is a greater expense to have women in management than it is to have men because of career interruptions. Employers today are reluctant to discuss this view because of Equal Employment Opportunity requirements (Schwartz, 1989). Career interruptions, plateauing while raising a family, and turnover are expensive. Gender differences include those related to maternity and those related to traditions and expectations of the sexes. Maternity is biological rather than cultural and cannot be altered.

Women who compete like men may be considered unfeminine; women who emphasize family may be considered uncommitted. Interestingly, a woman often is put down for having the very qualities that would send a man to the top. Male and female roles are beginning to expand and merge. As the socialization of boys and girls and expectations of young men and women become more similar in the future, differences in the workplace will begin to fade. According to Schwartz (1989), access to the most talented human resources is not a luxury in this age of explosive international competition but rather the barest minimum that prudence and national self-preservation require.

Foodservice directors employed in hospitals accredited by the Joint Commission on Accreditation of Healthcare Organizations (JCAHO) were profiled in a study that examined the effect of gender and work-related characteristics on salary (Barrett, Nagy, & Maize, 1992). Results indicated that the 663 directors in the sample are predominantly white female college graduates, and more than half are registered dietitians. Salaries, however, are significantly higher for men than women despite the fact that the majority of the female foodservice directors had completed college and were registered dietitians.

Work/Home Transitions

Kanter (1977) used the phrase *the myth of separate worlds* to describe a state in which an employee's home does not exist or the work world is everything. Many managers value the importance of balancing work life and home life but often do not know how to handle conflicts between the two (Hall & Richter, 1988). At a time when companies are streamlining their work force and employees are expected to do more work, employees are experiencing difficulties in meeting demands of their jobs and managing homes and rearing children. This is no longer only a woman's problem. Concerns with legal issues related to sex discrimination often prevent employers from asking employees about their marital status and family life.

Hall and Richter (1988) believed that the best way to study work/home relationships is to look at the daily transitions that people make as they cross boundaries. Transitions across the boundaries between work and home can be either physical or psychological. In the morning, people begin to think about work before leaving home, but in the afternoons, they do not think about home until they leave work. Reentering the home has been seen as the most hectic time for women because they are involved immediately in household chores. Men, however, go through an unwinding period, perhaps reading a newspaper or watching television, before entering into household activities. Research interviews indicated that home boundaries are much more permeable than work boundaries.

Many organizations are now trying to find ways to help employees and reduce these conflicts. An important step is to legitimize and respect work/home boundaries; breakfast meetings infringe on personal transitions. Meetings scheduled during working hours are much fairer to employees. Some responsibility for reducing conflicts, however, has to be placed on employees. They should, for example, let the family know in advance when peak times will occur at work and home events cannot be scheduled.

Organizational change might have to occur to facilitate balance in daily transitions. Flexibility in work hours, whereby employees may come in late or stay late, for example, might be a solution. Another form of flexibility is flexplace, under which individuals are allowed to do certain work at home. Managers need to be sensitive to individual differences among employees in their ways of making transitions. Finally, providing some kind of child-care benefits, such as child-care vouchers or information about child-care services, as alternates to on-site child care, could reduce stress for employees with children.

SUMMARY

Leadership is concerned with creating an environment in which members of the organization are motivated to contribute to organizational goals. Managers must have an appreciation of human behavior of employees to understand the past and to predict the future. Motivation is an inner force that activates or interacts with the environment. A number of theories of motivation have been developed. Job satisfaction pertains to individuals' general attitude toward their occupations; it may be positive or negative. A negative attitude adversely affects turnover, absenteeism, tardiness, accidents, and grievances. A high job satisfaction has a positive affect.

Leadership is the process of influencing the activities of an individual or group in efforts toward goal achievement, and it is identified as a function of the leader, follower, and other situational variables. Leaders possess power, a concept closely related to leadership, and authority, by which a leader influences the behavior of followers. In the traditional structure, power has many bases or potential means of influencing the behavior of others. The newer concepts and trends in the power of leadership involve communication and collaboration across functions, units, and other operations with overlapping activities and resources.

To create an environment conducive to a high level of employee motivation, managers need to develop an understanding and philosophy of human nature and its influence on leadership style. Situational or contingency approaches emphasize leadership skills, behavior, and roles that are thought to be dependent on the situation.

During the past decade, the decline in the productivity growth rate in the United States compared to most other industrialized nations has been a concern. American organizations are examining Japan's high productivity. The most important Japanese concept that Americans have adopted is the involvement of employees at all levels of the organization in decision making.

Organizational change is any substantive modification to some part of the organization and is an entirely new way of thinking. Leaders are needed to develop a vision of the future with strategies for producing changes to achieve the vision. The major change in organizations has been to convert from a traditional to a total quality management or service philosophy. To accomplish this change, the corporate culture must be redefined, followed by a strategy for daily long-term actions, decisions, and plans. If employees are being empowered to make decisions, management has a responsibility to make conditions in their work life pleasant and also to respect the pressures of home life.

BIBLIOGRAPHY

Albrecht, K. (1992). *The only thing that matters*. New York: HarperCollins.

Argyris, C. (1957). *Personality and organization*. New York: Harper & Row.

Badaracco, J. L., & Ellsworth, R. R. (1988). *Leadership and the quest for integrity*. Boston: Harvard Business School Press.

Bailey, A. D. (1992). Coping with employee turnover. *Restaurants USA, 12*(5), 18–20.

Barrett, E. B., Nagy, M. C., & Maize, R. S. (1992). Salary discrepancies between male and female foodservice directors in JCAHO-accredited hospitals. *Journal of The American Dietetic Association, 92,* 1078–1082.

Batty, J. (1992). Breaking through the glass ceiling. *Restaurants USA, 12*(10), 37–41.

Batty, J. (1993). Finding the balance between work and family life. *Restaurants USA, 13*(3), 26–31.

Bennis, W. (1989). *Why leaders can't lead: The unconscious conspiracy continues*. San Francisco: Jossey-Bass.

Berelson, B., & Steiner, G. A. (1964). *Human behavior: An inventory of scientific findings*. New York: Harcourt, Brace, & World.

Blake, R. R., & Mouton, J. S. (1964). *The managerial grid*. Houston: Gulf Publishing.

Blake, R. R., & Mouton, J. S. (1978). *The new managerial grid*. Houston: Gulf Publishing.

Blake, R. R., & Mouton, J. S. (1984). *The managerial grid III* (3rd ed.). Houston: Gulf Publishing.

Burack, E. H., & Mathys, N. J. (1983). *Introduction to management: A career perspective*. New York: Wiley.

Cichy, R. F., Sciarini, M. P., Cook, C. L., & Patton, M. E. (1991). Leadership in the lodging and non-commercial food service industries. *FIU Hospitality Review, 9*(1), 1–10.

Cichy, R. F., Sciarini, M. P., & Patton, M. E. (1992). Food-service leadership: Could Attila run a restaurant? *Cornell Hotel and Restaurant Administration Quarterly, 33*(1), 47–55.

Coppess, M. H. (1989). Mentoring: Learning the ropes from a pro. *Restaurants USA, 9*(6), 12–14.

Drucker, P. F. (1992). *Managing for the future: The 1990s and beyond*. New York: Truman Tally Books/Dutton.

Evans, M. G. (1970). The effects of supervisory behavior on the path-goal relationship. *Organizational Behavior and Human Performance, 5*(3), 277–298.

Fiedler, F. E. (1967). *A theory of leadership effectiveness*. New York: McGraw-Hill.

French, J. R. P., & Raven, B. (1960). The bases of social power. In Cartwright, D., & Zander, A. (Eds.), *Group dynamics: Research and theory* (2nd ed., pp. 607–623). Ann Arbor: University of Michigan, Institute for Social Research.

Greene, C. N. (1972). The satisfaction-performance controversy. *Business Horizons, 15*(5), 31–41.

Griffin, R. W. (1993). *Management* (4th ed.). Boston: Houghton Mifflin.

Hall, D. T., & Richter, J. (1988). Balancing work life and home life: What can organizations do to help? *Academy of Management Executive, 2*(3), 213–223.

Hersey, P., & Blanchard, K. H. (1993). *Management of organizational behavior: Utilizing human resources* (6th ed.). Englewood Cliffs, NJ: Prentice-Hall.

Herzberg, F. (1966). *Work and the nature of man*. Cleveland: World.

Herzberg, F. (1974). The wise old Turk. *Harvard Business Review 52*(5), 70–80.

Herzberg, F. (1987). One more time: How do you motivate employees? *Harvard Business Review, 65*(5), 109–120.

Hinterhuber, H. H., & Popp, W. (1992). Are you a strategist or just a manager? *Harvard Business Review, 70*(1), 105–113.

Hopkins, D. E., Vaden, A. G., & Vaden, R. E. (1979). Some determinants of work performance in foodservice systems: Job satisfaction and work values of school foodservice personnel. *Journal of The American Dietetic Association, 75*(6), 640–647.

Hopkins, D. E., Vaden, A. G., & Vaden, R. E. (1980). Some aspects of organization identification among school food service employees. *School Food Service Research Review, 4*(1), 34–42.

House, R. J. (1971). A path-goal theory of leader effectiveness. *Administrative Science Quarterly, 16*(3), 321–338.

House, R. J., and Mitchell, T. R. (1974). Path-goal theory of leadership. *Journal of Contemporary Business*, Autumn, 81–97.

Kahn, R., & Katz, D. (1960). Leadership practices in relation to productivity and morale. In Cartwright, D., & Zander, A., (Eds.), *Group dynamics: Research and theory* (2nd ed., pp. 554–570). Elmsford, NY: Row Peterson.

Kanter, R. M. (1977). *Work and family in the United States: A critical review and agenda for research and policy*. New York: Russell Sage.

Kanter, R. M. (1989). The new managerial work. *Harvard Business Review, 67*(6), 85–92.

Kotter, J. P. (1990). What leaders really do. *Harvard Business Review, 68*(3), 103–111.

Likert, R. (1967). *Human organization: Its management and value*. New York: McGraw-Hill.

Maccoby, M. (1981). *The leader: A new face for American management*. New York: Simon & Schuster.

Maccoby, M. (1989). *Why work: Motivating and leading the new generation*. New York: Simon & Schuster.

Maslow, A. H. (1943). A theory of human motivation. *Psychology Review, 50*, 370–396.

Maslow, A. H. (1970). *Motivation and personality* (2nd ed.). New York: Harper & Row.

McClelland, D. C. (1967). *The achieving society*. New York: Free Press.

McClelland, D. C., & Burnham, D. H. (1976). Power is a great motivator. *Harvard Business Review*, *54*(2), 100–110.

McGregor, D. (1960). *The human side of enterprise*. New York: McGraw-Hill.

Middlemist, R. D., & Hitt, M. A. (1981). *Organizational behavior: Applied concepts*. Chicago: Science Research Association.

Mondy, R. W., Sharplin, A., & Flippo, E. B. (1988). *Management: Concepts and practices* (4th ed.). Boston: Allyn & Bacon.

Naisbitt, J., & Aburdene, P. (1990). *Megatrends 2000: Ten new directions for the 1990's*. New York: William Morrow.

Ouchi, W. G. (1981). *Theory Z: How American business can meet the Japanese challenge*. Reading, MA: Addison-Wesley.

Pascale, R. T., & Athos, A. (1981). *The art of Japanese management: Applications for American executives*. New York: Simon & Schuster.

Peters, T. J., & Austin, N. A. (1985). *Passion for excellence: The leadership difference*. New York: Random House.

Porter, L. W., & Lawler, E. E. (1968). *Managerial attitudes and performance*. Homewood, IL: Irwin-Dorsey.

Raven, B. H., & Kluganski, W. (1970). Conflict and power. In Swingle, P. G. (Ed.), *The structure of conflict* (pp. 69–109). New York: Academic Press.

Reddin, W. J. (1967a). The 3-D management style theory. *Training and Development Journal*, *21*(4), 8–17.

Reddin, W. J. (1967b). The 3-D management style theory. *Theory paper No. 2—Managerial styles*. Fredericton, NB, Canada: Social Sciences Systems.

Riehle, H. (1993). Employee turnover rates dropped in 1991. *Restaurants USA*, *13*(6), 42–43.

Rue, L. W., & Byars, L. L. (1989). *Management: Theory and application* (5th ed.). Homewood, IL: Richard D. Irwin.

Schonberger, R. J. (1992). Is strategy strategic? Impact of total quality management on strategy. *Academy of Management Executive*, *6*(3), 80–87.

Schwab, D. P., & Cummings, L. L. (1970). Theories of performance and satisfaction: A review. *Industrial Relations*, *9*(4), 408–430.

Schwartz, F. N. (1989). Management women and the new facts of life. *Harvard Business Review*, *67*(1), 65–76.

Schwartz, F. N. (1992). Women as a business imperative. *Harvard Business Review*, *70*(2), 105–113.

Skinner, B. F. (1971). *Beyond freedom and dignity*. New York: Knopf.

Slater, D. (1989). Role models: Inspiring by example. *Restaurants USA*, *9*(6), 15–17.

Smith, P. C., Kendall, L. M., & Hulin, C. L. (1969). *The measurement of satisfaction in work and retirement: A strategy for the study of attitudes*. Chicago: Rand McNally.

Stogdill, R. M. (1974). *Handbook of leadership*. New York: Free Press.

Tannenbaum, R., & Schmidt, W. H. (1958). How to choose a leadership pattern. *Harvard Business Review*, *36*(2), 95–101. Retrospective commentary (1973). *Harvard Business Review*, *51*(3), 162–164, 166–168, 170, 173, 175, 178–183.

Vroom, V. H. (1964). *Work and motivation*. New York: Wiley.

Wilhelm, W. (1992). Changing corporate culture—or corporate behavior? How to change your company. *Academy of Management Executive*, *6*(4), 72–77.

Zaleznik, A. (1977). Managers and leaders: Are they different? *Harvard Business Review*, *55*(3), 67–78.

Zaleznik, A. (1993). *Learning leadership: Cases and commentaries on abuses of power in organizations*. Chicago: Bonus Books.

Human Resources Management

The most important resources in an organization are its human resources. Selection, training, and compensation of employees were considered basic functions of personnel management for many decades. These functions were considered independent with no regard for their interrelationships. Human resources management emerged from this traditional viewpoint recently, and today, **human resources management** (HRM) represents the extension rather than the rejection of the traditional requirements for managing personnel effectively (Sherman & Bohlander, 1992). Group effort in producing products was the beginning of HRM.

Personnel management went from the factory, in which power-driven equipment and improved techniques made manufacturing products cheaper, to mass production, to HRM. Factory jobs were monotonous and often unhealthy and dangerous. In the late 1880s, state legislation and collective bargaining appeared, and working conditions began to improve. Mass production followed in the early 1890s with the use of labor-saving equipment that further improved production techniques. **Scientific management**, defined as the systematic approach to improving worker efficiency based on the collection and analysis of data, appeared in manufacturing (Sherman & Bohlander, 1991). Frederick W. Taylor, considered the father of scientific management, Frank and Lillian Gilbreth, and Henry L. Gantt all contributed to this era in which performance standards and time studies were emphasized.

HRM began in the 1920s when the human element was added to management functions. The Hawthorne studies, the human relations movement, and behavioral sciences led up to HRM as it is today. Along with HRM came the increase in both federal and state legislation, starting with the National Labor Relations Act in 1935. Human resources managers are responsible for compliance with all state and local laws and regulations that govern work organizations, including foodservice operations. These managers also will need to assume a broader role in organizational strategy and change and at the same time remember the bottom line of the operation.

Moving from a traditional organization to a total quality management culture demands much from the human resources function. Blackburn and Rosen (1993) interviewed human resources professionals at organizations that had received the Baldrige Award to determine what kind of contribution they made to the company's effort to receive the award. Established in 1987, the Baldrige Award recognizes U.S. organizations that excel in quality management. Criteria for the award require major shifts in management philosophy, practices, and policies related to the pursuit of improved quality. If being a recipient of the award is an indication of excellence in quality management, then the human resources management policies in these organizations should support and sustain a total quality culture. A paradigm shift in the human resources policies of the organizations should occur. Traditional human resources policies in command and control cultures of award-winning companies have given way to new policies that are supportive of cultures characterized by employee commitment, cooperation, and communication.

The evolution from traditional HRM practices to new HRM policies indicates the changing role of HRM from a support to a leadership function in the organization, as shown in Table 20.1 (Blackburn & Rosen, 1993). HRM policies in TQM organizations should agree with a corporate culture that is based on employee dedication to quality and customer service. Policies such as communication, job design, conditions of employment, training, evaluation measures, and rewards also must agree with TQM objectives.

Outcomes of building a TQM culture in the Baldrige award organizations indicate that measurable results in the bottom line have been generally impressive. Increased on-time deliveries, reduced customer complaints, and reduction in new product development time were some of the benefits of a total quality program. Not all

Table 20.1.   Evolution of a total quality human resources paradigm

Corporate Context Dimension	Traditional Paradigm	Total Quality Paradigm
Corporate culture	Individualism Differentiation Autocratic leadership Profits Productivity	Collective efforts Cross-functional work Coaching/enabling Customer satisfaction Quality

Human Resources Characteristics	Traditional Paradigm	Total Quality Paradigm
Communications	Top-down	Top-down Horizontal, lateral Multidirectional
Voice and involvement	Employment-at-will Suggestion systems	Due process Quality circles Attitude surveys
Job design	Efficiency Productivity Standard procedures Narrow span of control Specific job descriptions	Quality Customization Innovation Wide span of control Autonomous work teams Empowerment
Training	Job related skills Functional, technical Productivity	Broad range of skills Cross-functional Diagnostic, problem solving Productivity and quality
Performance measurement and evaluation	Individual goals Supervisory review Emphasize financial performance	Team goals Customer, peer and supervisory review Emphasize quality and service
Rewards	Competition for individual merit increases and benefits	Team/group based rewards Financial rewards, financial and nonfinancial recognition
Health and safety	Treat problems	Prevent problems Safety programs Wellness programs Employee assistance
Selection/promotion career development	Selected by manager Narrow job skills Promotion based on individual accomplishment Linear career path	Selected by peers Problem-solving skills Promotion based on group facilitation Horizontal career path

Source: "Total Quality and Human Resources Management: Lessons Learned From Baldrige Award Winning Companies" by R. Blackburn and B. Rosen, 1993, *Academy of Management Executive, 7*(3), p. 51. Used by permission.

Baldrige award recipients have continued to be successful, but most agreed that problems would have been much worse much sooner without their TQM efforts. Human resources executives, once viewed by some executives as having perfunctory roles in the organization, are now welcomed to positions on strategic policy committees (Blackburn & Rosen, 1993).

HUMAN RESOURCES PLANNING

Human resources planning is the process of anticipating and making provision for the movement of people into, within, and out of an organization (Sherman & Bohlander, 1992). The objective is to use these people as effectively as possible and to have available the required number of people with qualifications for positions when openings occur. A strategic plan has to be developed, followed by forecasting future employee needs. Finally, a supply and demand analysis is required. Of course, all human resources planning must be done within the legal environment of the United States.

Strategic Plan

The human resources plan has to fit into the strategic plan of the organization. When the goals of the organization are established, availability of both the internal and external human resources need to be determined. The human resources manager needs to be on the strategic planning team. Members of the team should be aware of current employees' skills, abilities, and knowledge as well as their and future employees' training needs. They also should know what wages and benefits are required to attract qualified employees.

The human resources manager can be proactive by scanning environmental trends such as economic factors, technological changes, political and legislative issues, social concerns including child-care and educational priorities, and demographic trends of the work force (Sherman & Bohlander, 1992). In a booming economy, fewer job candidates may be available because unemployment may be low; in a depressed economy, organizations may have to cut back on operations and consequently on the number of employees. Technological changes will affect the type of specialized personnel an organization will require. Thus, the organizational environment will define the limits within which a human resources plan must operate. After managers fully understand the jobs that are needed in the organization, they can start planning by assessing trends in employee usage and future organizational plans.

Forecasting Supply and Demand

Human resources forecasting involves determination of the number, type, and qualifications of individuals who will be needed to perform specific duties at a certain time. A good sales forecast is often the basis, especially for small organizations (Griffin, 1993). Historical data can be used to predict demand for employees, such as chefs and waitstaff. Large organizations will require a special planner who uses mathematical models for forecasting employment needs. The internal supply of labor, consisting of the number and type of employees who will be in the operation at some future date, needs to be determined first. Then the external supply, the number and type of people who will be available for hiring in the labor market, needs to be forecast. The simplest method is to adjust the present staffing levels for anticipated turnovers and retirements.

Supply Analysis

Once future requirements for employees have been forecast, the manager then has to determine if numbers and types of employees are adequate to staff anticipated openings. Supply analysis uses both internal and external sources. An internal supply analysis begins with the preparation of **staffing tables,** which are pictorial representations of all organizational jobs with the numbers of employees in those jobs and future employment requirements (Sherman & Bohlander, 1992). In conjunction with staffing tables are **skills inventories,** which contain information, usually computerized, on each employee's education, skills, experience, and career aspirations (Griffin, 1993). Forecasting the external supply of labor can be a problem. How does a foodservice manager, for example, predict the number of cooks who will be looking for work in Montana 3 years from now? This information can be calculated from such sources as state employment commissions, government reports, and figures supplied by educational institutions on the number of students in the field.

Balancing Supply and Demand

After comparing the internal demand for employees with the external supply, managers can make plans to deal with internal shortages or with overstaffing. Demand is based on the forecast, and supply is based on finding employees that have the required qualifications to fill vacancies. If shortages are predicted, new employees can be hired, present employees can be retrained, those retiring can be asked to stay on, or labor-saving methods, like using more ready prepared food, can be introduced into the organization. If employees need to be hired, the external labor forecast can help managers plan a recruitment program.

Legal Environment

Union contracts influence human resources planning if clauses on regulating transfers, promotions, and discharges are included. Government legislation also has a vital role. Significant pieces of legislation that affect human resources planning are the Equal Pay Act, Title VII of the Civil Rights Act, Age Discrimination and Employment Act, Pregnancy Discrimination Act, Americans with Disabilities Act, Equal Employment Opportunity Act, and Family and Medical Leave Act, as explained in Table 20.2. Most of these acts have criteria for implementation, such as the number of employees in the organization. Human resources managers in large organizations keep up-to-date on this legislation, but in small operations, which have only one or

Table 20.2.   Major federal laws affecting equal employment opportunity

Date	Act	Explanation
1963	Equal Pay	Equal pay for equal work regardless of sex
1964	Title VII of the Civil Rights Act (amended in 1974)	Forbids discrimination on the basis of race, color, religion, sex, or national origin
1967	Age Discrimination in Employment (amended in 1968)	Forbids discrimination against employees 40 to 69 years of age
1972	Equal Employment Opportunity (amended Title VII of the Civil Rights Act)	Extends coverage of Title VII to include government employees, faculty in higher education, and other employers and employees
1978	Pregnancy Discrimination	Broadens sex discrimination to include pregnancy, childbirth, or related medical conditions
1990	Americans with Disabilities	Prohibits discrimination against persons with physical or mental disabilities or the chronically ill
1993	Family and Medical Leave	Gives employees up to 12 weeks unpaid and job-protected leave per year for themselves or a spouse, parent, or child with a serious health condition

two managers, managers need to be aware of the legislation that has an impact on their operation. Sexual harassment, covered in Title VII of the Civil Rights Act, the Immigration Reform and Control Act, and the Americans with Disabilities Act, currently is receiving the most attention.

Sexual Harassment

The Equal Employment Opportunity Commission (EEOC) has declared **sexual harassment** to be a form of sex discrimination and thus in violation of Title VII of the Civil Rights Act of 1964 (Batty, 1993). The Supreme Court ruled unanimously in 1986 that sexual harassment violates Title VII (Hamilton & Veglahn, 1992). Two recent incidents brought sexual harassment to the forefront (Aaron & Dry, 1992). First, the allegations of sexual harassment brought against Clarence Thomas by Anita Hill during his Supreme Court confirmation hearings were unprecedented. Second, President Bush signed civil rights legislation that allows an employee in a sexual harassment case to recover damages.

The EEOC has identified two forms of harassment. *Quid pro quo*, which means "something for something" in Latin, indicates that a manager demands sexual favors or insists that an employee put up with sexually harassing behavior for a job benefit (Batty, 1993). The second form involves a *hostile work environment* that imposes Title VII liability on the employer if the workplace is rendered offensive by acts of sexual harassment that the employer knew about or should have known about

(Hamilton & Veglahn, 1992). Discrimination suits have been brought against restaurant owners who require waitstaff to wear sexually revealing uniforms; uniforms help define a restaurant's concept and create a certain atmosphere (Batty, 1993). Even though women are usually the victims of sexual harassment, some men are beginning to report sexual harassment by their female bosses.

Many restaurant owners have established a written policy that specifically describes what sexual harassment is, what will not be tolerated, and what disciplinary action will be taken against someone who harasses a colleague. Each employee is made aware of the policy, and the management team is trained to look for signs of sexual harassment. An investigation should be made immediately and the harasser, if still in the operation, should be closely monitored and the victim regularly checked to make sure she or he is comfortable with the work situation.

Immigration Reform and Control Act

In 1986, Congress passed the **Immigration Reform and Control Act (IRCA)** to control unauthorized immigration by making it unlawful for a person or organization to recruit or hire persons not legally eligible for employment in the United States. Employers must comply with the law by verifying and maintaining records on the legal rights of applicants to work in the United States.

IRCA requires employers to complete an I-9 form for each employee to verify the worker's identity and eligibility to work in the United States ("Employer Focus Groups," 1993). The law also prohibits employers of four or more persons from discriminating against employees or job applicants based on their national origin or citizenship status. Employers feel strongly that immigration paperwork is overwhelming, burdensome, and hard to cope with; they have little knowledge about a federal ban on job bias against workers who may look or sound foreign ("Employer Focus Groups," 1993). IRCA made headlines when journalists reported that Zoë Baird, a nominee for Attorney General, had hired an illegal immigrant as a nanny but neglected to comply with the Immigration Reform and Control Act and notify the Immigration and Naturalization Service about the nanny's status (Cheney, 1993a). Subsequently, she withdrew her nomination.

Americans with Disabilities Act

The **Americans with Disabilities Act (ADA)** is comprehensive legislation that creates new rights and extends existing rights for the 43 million Americans with disabilities; it also protects disabled Americans who are not covered under existing laws (Cross, 1993). The purpose of the act is to do the following:

- Provide a national mandate to eliminate discrimination against individuals with disabilities
- Provide consistent enforceable standards for those with disabilities
- Ensure that the federal government plays a central role in enforcing the standards
- Invoke congressional authority to address the major areas of discrimination faced by the disabled

The ADA's definition of *disability* is broad and provides civil rights protection for people in the following three categories ("ADA Employment," 1993):

- A physical or mental impairment that substantially limits one or more major life activities, such as walking, seeing, hearing, speaking, learning, or working
- A record of impairments (such as mental, emotional, or physical illness and recovering alcoholics or drug addicts) but have recovered or are recovering
- Regarded by others to have an impairment, for example, severe burns or rumored to have AIDS

The EEOC has provided examples of conditions that qualify as disabilities, among which are orthopedic, visual or speech impairments, HIV infection, acquired immune deficiency syndrome (AIDS), epilepsy, mental retardation, and former drug use (Cross, 1993). The ADA does not require hiring of disabled persons who are not qualified for a job in terms of skill, education, and experience, but it does require that disabled persons who meet the job qualifications should be given equal consideration for a job and equal treatment on the job (McLauchlin, 1992). The U.S. Justice Department reported that complaints filed against restaurants and bars account for just over 11% of the nearly 700 complaints filed in the first 10 months after the ADA's rules took effect ("Foodservice Accounts," 1993). The ADA is not to be feared, however, because it is an excellent tool to cement a long-term working relationship with a potential employee (Doyle, 1992). In fact, restaurants are the leading industry in the employment of people with disabilities. The restaurant industry has been making reasonable accommodations for years.

EMPLOYMENT PROCESS

Once an organization has an idea of its future human resources needs, the next step usually is recruiting new employees. The major phases in the employment process are recruitment, selection, and orientation.

Recruitment

Recruiting is the process of attempting to locate and encourage potential applicants to apply for existing or anticipated job openings (Sherman & Bohlander, 1992). Efforts are made to inform applicants about qualifications required for the job and career opportunities the organization can give them. A decision has to be made by management whether the job vacancy will be filled with someone inside or outside the organization. The decision depends on availability of employees, the organization's human resources policies, and requirements of the vacant job.

Inside the Organization

Most organizations try to fill vacancies above entry-level positions through promotions and transfers (Sherman & Bohlander, 1992). Promotion from within has several advantages. Employee morale and motivation are positively affected by internal promotions, assuming such promotions are perceived as being equitable. A transfer might not have the motivational value of a promotion, but it sometimes can protect employees from layoff and, at the same time, may broaden their experiences.

Organizations have a sizable investment in their employees; therefore, using the abilities of present employees to their fullest extent improves the return on this investment. Promotion from within or a transfer usually is less expensive than hiring from outside the organization. Individuals recruited from within also will be familiar with the organization and therefore may need less training and orientation.

Methods of Finding Candidates.   A number of methods are used for recruiting from within. Qualified candidates can be identified by computerized records, job posting and bidding, and recall of those who have been laid off (Sherman & Bohlander, 1992). Computers have made data banks that contain complete records and qualifications of each employee possible. This information can be retrieved in minutes.

Organizations may inform employees about job openings by using job posting and bidding. Vacancy notices can be posted on bulletin boards, in employee publications, in special handouts, and by direct mail and public-address systems. Employees who have been laid off may be recalled to their jobs if economic conditions improve and they left in good standing. Employee layoffs usually are based on seniority or ability. Unions generally recommend seniority because members should be entitled to certain rights proportionate to the number of years on the job. Using seniority may have an impact on women and minorities, who often have less seniority than other employees.

Limitations of Recruiting Within.   Sometimes requirements for a higher-level job may be very different from those of the current job. For example, a foodservice production supervisor may be promoted to unit manager because of excellent job performance in supervising production employees and operations. This same individual may fail, however, as unit manager of the foodservice operation because his or her scope of skills and knowledge may be inadequate for this higher-level position.

Another danger in promoting from within is that new ideas may not be heard. Even if an organizational policy is to promote from within, candidates from the outside occasionally should be considered to prevent inbreeding of ideas and attitudes.

Outside the Organization

Eventually, after promotions and transfers, a person outside the organization has to be recruited to fill a vacancy. When a promotion occurs, a chain reaction of promotions follows. The decision that a manager has to make is at which level a person should be recruited. The labor market varies with the type of job to be filled and the salary to be paid for the job.

One of the most widely used external sources for obtaining employees is help wanted advertisements in newspapers and trade journals. Radio, television, billboards, and posters also have been used. The statement *equal opportunity employer* generally is included in the advertisement.

Each state has an employment agency that is responsible for the unemployment insurance program; each agency is subject to regulations by the U.S. Employment Services (USES). Unemployed people must register at a local office and be available for employment to receive a weekly unemployment compensation check (Sherman & Bohlander, 1992). The local office has a computerized job bank for the area, and the USES has a nationwide one. These agencies also help employers with employment testing, job analysis, evaluation programs, and community wage surveys. Private employment agencies, which help job seekers find the right job, charge a fee that may be paid by either the employer or the job seeker or both. In contrast, executive search firms, often called *headhunters,* help employers find the right person for the right job. Fees, paid for by the employer, are usually a certain percentage of the annual salary for the position to be filled.

Educational institutions are good sources for candidates who have little work experience. High schools emphasize clerical and blue-collar jobs. Vocational schools and community colleges provide candidates for technical jobs and colleges and universities for technical and professional areas. Recruiters for various organizations often are sent to college campuses to seek candidates.

Other recruitment efforts can be helped by recommendations made by current employees or from unsolicited applications. Many professional organizations offer a placement center to their members. Labor unions can be an important source of applicants, especially for blue-collar jobs.

Temporary employment agencies are one of the fastest growing recruitment sources (Sherman & Bohlander, 1992). Temporary employees do not receive benefits, and they can be dismissed without filing unemployment insurance claims. Employee leasing is a process whereby an employer terminates a number of employees who are then hired by an employee leasing company that leases employees back to the original organization. The company takes care of all the human resources duties and in return is paid a fee, usually a percentage of payroll cost.

In all recruitment activities, organizations must be cognizant of the legal requirements. Organizational practices or policies that adversely affect equal employment opportunities for any group are prohibited unless the restriction is a justifiable job requirement.

Selection

The selection process begins after recruiting applicants for a job. The process includes a comparison of applicant skills, knowledge, and education with the requirements of the job and involves decision making by the organization and the applicant. Qualifications for each applicant must be compared to the job requirements that are identified in the job specification. After thorough screening, choosing an applicant who meets the criteria usually indicates that she or he will learn the job easily and adjust to it with a minimum of difficulty. Union agreements and civil service regulations that protect employees with job tenure provide management with an additional incentive for making a good selection. Their regulations on discharging protected employees can be very difficult.

The selection process for all workers is protected by equal employment legislation, court decisions, and the *Uniform Guidelines on Selection Procedures.* These guidelines recommend that an employer be able to demonstrate that selection procedures are valid in predicting or measuring performance in a particular job (Sherman & Bohlander, 1992). Factors that affect the process include a small pool of applicants, geographic immobility of career couples, and changing staffing needs because of promotion or turnover.

The steps in the selection process are depicted in Figure 20.1. In practice, the actual selection process varies among organizations and among levels in the same organization. For example, the selection interview for production and service employees may be perfunctory, whereas an extensive process may be necessary for choosing the unit manager for a foodservice operation, starting with submission of a résumé, then a formal application. The selection process should conform to accepted ethical standards, including privacy and confidentiality, as well as legal requirements. The intent of the process is to gather from applicants reliable and valid information that will predict their job success and then to hire the candidates likely to be most successful (Griffin, 1993).

Figure 20.1.   Steps in the selection process.

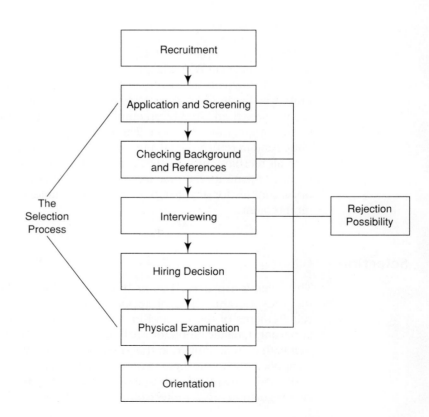

Application and Screening

The application form serves three purposes. First, it indicates that the applicant is interested in a position; second, it provides the interviewer with basic information to conduct an interview; and third, it becomes part of the file if the applicant is hired. In organizations with a human resources department, the application and initial screening steps will be conducted by that department.

The application form or blank is used to obtain information that will be helpful in reaching an employment decision. The application collects objective biographical information about an applicant, such as education, work experience, special skills, and references. Questions referring to an applicant's age, sex, race, religion, national origin, or family status may violate provisions of the Civil Rights Act of 1964. Information requested on an employment application should not lead to discrimination against applicants.

In some organizations, preliminary interviews may be used to screen out unqualified applicants. This is a quick way to identify those candidates worthy of more attention (Marvin, 1992). An inadequate educational level or experience record or obvious disinterest may be reasons for eliminating the individual from the applicant pool at this point in the process. The applicant may be asked questions on previous experiences, salary expectations, and other issues related to the job. In foodservice organizations, an important issue to discuss initially is the applicant's willingness to work weekends and holidays or early morning and late evening hours.

Background and Reference Checks

The practice in many organizations is to conduct background and reference checks after the initial screening interview. In others, however, this checking may be done after the employment interview. When the interviewer is satisfied that the applicant meets the qualifications for the job, information about previous employment and other information given by the applicant are investigated.

Most organizations use both the mail and telephone to check references. Telephone checks are used most often because they save time and responses are more candid. Supervisors who know the applicant's work habits and performance usually give the best information. Recently, some organizations have been charged with negligently hiring or retaining employees who later commit crimes (Sherman & Bohlander, 1992). These organizations are charged with failure to adequately check references, criminal records, or general background that would have shown the employee's behavior. Inadequate reference checking is one of the major causes of high turnover, employee theft, and white-collar crime.

Interviews

The employment interview has been the most widely used and probably the most important step in the selection process. The interview may be conducted by one person or a group of persons in the organization, as the guide to interviewing, Figure 20.2, shows. Some serious doubts have been raised about the validity of the interview as a selection method because of differences between interviewers.

Figure 20.2. A guide for interviewing.

INTRODUCTION

These guidelines have been prepared to help interviewers conduct fair and objective interviews. An interview should provide as much information as possible about an applicant's potential to perform the duties of a particular position. The most valuable interview is objective and permits the interviewer(s) to determine the knowledge, skills, and abilities of a prospective employee.

INTERVIEW DEVELOPMENT

Form the Interview Team

If feasible, use a team approach. The team approach is preferable because it saves time and allows for comparison of the applicant by the team members. The size of the interview team may vary, but generally two to three members are recommended.

Familiarize the Interviewer(s) With the Position

The interviewer(s) must be familiar with the major duties and responsibilities, and the essential knowledge, skills, and abilities of the position at entry level. Be sure that each interviewer reviews the position description carefully.

Establish Criteria for Selection

The selection criteria must be consistent with the complexity and level of the job. Focus on performance factors that can be demonstrated in the selection procedure. Understand the departmental and organizational goals as they relate to this position. Such criteria must be job-related and might include performance during the interview, relevant training, education and experience, affirmative action goals, etc. Example: To what extent is job success dependent upon effective oral communication skills, on-the-spot reasoning skills, and the ability to effectively present oneself to strangers?

Develop Job-Related Questions

"Nice to know" questions are not permitted! Lawsuits may result from applicants who are rejected on the basis of irrelevant questions asked by interviewers.

Develop Interviewing Strategies

There are may different interviewing strategies. Develop strategies that are appropriate for the position level and skill requirements.

Establish a System to Evaluate the Responses

It might be beneficial to set up a formula for rating or ranking the applicant's responses to the questions based on the selection criteria. Evaluating the responses in this manner will help make the selection process easier and more objective.

INTERVIEW SUGGESTIONS

Preparing Questions

When developing questions, always keep in mind that they must be job-related and appropriate for the complexity and level of the position. It is helpful to weigh the questions based on the importance

Figure 20.2.   *continued*

of each selection criterion. Below are six main categories of questions that are commonly used by interviewers. Different types of questions may be combined to obtain a certain response.

1. **Close-ended questions.** These questions may sometimes be helpful when an interviewer(s) wants to know certain information at the outset or needs to determine specific kinds of knowledge. Example: "Could you name the five specific applications involved in ...?"

2. **Probing questions.** These questions allow the interviewer(s) to delve deeper for needed information. Example: "Why?, "What caused that to happen?", or "Under what circumstances did that occur?"

3. **Hypothetical questions.** Hypothetical situations based on specific job-related facts are presented to the applicant for solutions. Example: "What would you do if ..."; "How would you handle ..."

4. **Loaded questions.** These questions force an applicant to choose between two undesirable alternatives. The most effective way to employ a loaded question is to recall a real-life situation where two divergent approaches were both carefully considered, then frame the situation as a question starting with, "What would be your approach to a situation where ..."

5. **Leading questions.** The interviewer(s) sets up the question so that the applicant provides the desired response. When leading questions are asked, the interviewer cannot hope to learn anything about the applicant.

6. **Open-ended questions.** These are the most effective questions, yield the greatest amount of information, and allow the applicant latitude in responding. Example: "What did you like about your last job?"

Determining Strategies

Although there are many different interviewing strategies, the following are examples of three different perceptive strategies.

1. **Situational interviewing.** This strategy is based on the assumption that the closer you can get to a real work situation, the better the evaluation will be. The situational interview could involve taking a tour of the workplace and asking the interviewee to actually perform some aspect of the job, or a closely related aspect of the job.

2. **Stress interviewing.** This strategy calls for the use of tough or negatively phrased questions. The interviewer(s) is trying to keep the candidate off balance while evaluating poise and quick thinking under pressure. This style would *not* be suitable if the employee will not face undue stress on the job.

3. **Behavioral interviewing.** The interviewer(s) is looking for a behavioral pattern. All questions are based on the past. The assumption is that a "leopard never changes its spots." The interviewer(s) may get an idea of what action the interviewee might take in the future based on what happened in the past.

Each of these strategies has its strengths and weaknesses. One strategy should not be used exclusively for all interviews. Different position levels might require different interview approaches. The sensible approach is to take the best aspects of each style and combine them to produce a comprehensive strategy.

Note: The interview process should not include the use of a testing device without prior approval from the Division of Personnel Services (K.A.R. 1–6–10).

Figure 20.2.   *continued*

Evaluating Responses

As part of evaluating the responses, the interviewer(s) should review the job description to ensure thorough familiarity with the requirements, duties, and responsibilities of the position. Furthermore, the interviewer(s) should review the work history and relevant educational credentials of each candidate and consider the intangible requirements of the job. Finally, the interviewer(s) should review the selection criteria, evaluate and rate the responses, and rank the applicants based on that criteria.

INTERVIEW PROCESS

Pre-Interview

1. Schedule interviews to allow sufficient time for post interview discussion, completion of notes, etc.
2. Secure an interview setting that is free from interruptions or distractions.
3. Review applications and resumes provided by the applicants.
4. Provide an accurate position description to each applicant and allow adequate time for reading before the interview begins.

Opening the Interview

1. Review the functions of the agency or unit in which the position is located.
2. Allow the applicant an opportunity to pose questions or seek clarification concerning the position.
3. Explain the interview process to the applicant.

Questioning

1. Question the applicant following the method established in the developing stage.
2. Be consistent with all applicants.
3. Allow the applicant sufficient time to respond to each question.
4. Record any relevant information elicited from the questions.

Closing the Interview

1. Inform the applicant when the decision will be made and how notification will occur.
2. Confirm the date of the applicant's availability to begin work.
3. Confirm the applicant's correct address and telephone number.
4. Give the applicant a final opportunity to raise any questions.
5. Obtain all necessary information from the applicant about references.

Post-Interview

1. Review the selection criteria.
2. Review and complete notes.
3. Avoid prejudgment and discussion of applicants between interviews.
4. Use the selection criteria established in the developing stages.
5. Rank the applicants based on the selection criteria.
6. When possible, decide upon a second and third choice in the event the first choice should decline the offer.
7. Document the basis for the final recommendation.
8. Notify all applicants interviewed of the results prior to announcing the selection.

Source: State of Kansas, Department of Administration, Division of Personnel Services.

For managerial and professional personnel, a series of interviews may be conducted. For example, a potential director of a department of food and nutrition services in a large medical center hospital would probably be interviewed by several different administrators, by the professional staff in the department, and perhaps by several of the other department heads. An applicant for a cook's position in this same hospital, however, might be interviewed by only the foodservice supervisor and the production dietitian.

Interviews can be either structured or unstructured. In the *structured interview,* the interviewer asks specific questions of all interviewees. The interviewer knows in advance the questions that are to be asked and merely proceeds down the list of questions while recording the responses. This interview technique gives a common body of data on all interviewees, allows for systematic coverage of all information deemed necessary for all applicants, and provides a means for minimizing the personal biases and prejudices of the interviewer. It is frequently used in interviews for lower-level jobs.

The *unstructured interview* allows the interviewer the freedom to ask questions he or she believes are important. Broad questions, such as "Tell me about your previous job," are asked. The unstructured interview may be useful in assessing such characteristics of an individual as ability to communicate and interpersonal skills. Comparison of answers across interviewees is difficult with unstructured interviews, however, because questions may be quite different or asked in a different context. Unstructured interviews are generally used with higher-level personnel in the organization because of the broad nature of these jobs.

More attention is being given to the structured interview because of EEOC requirements (Sherman & Bohlander, 1992). This type of interview is more likely to provide the type of information needed for making sound decisions. It helps to reduce the possibility of legal charges of discrimination. Employers need to realize that the interview is highly vulnerable to legal attack, and more litigation in this area can be expected in the future.

Regardless of the method used, the interview must be planned. First, interviewers must acquaint themselves with the job description and specification for the vacant job. Second, they must review the application file carefully, including the information from background and reference checks. Third, for structured interviews, the questions should be prepared in advance; for the unstructured interview, a general outline should be developed.

As with other aspects of the employment process, interviews are subject to numerous legal questions. The EEOC does not approve of questions related to race, color, age, religion, sex, or national origin. An interviewer can ask about physical handicaps, however, if the job involves manual labor, but not otherwise. Guidelines for employment inquiries for the State of Kansas are shown in Table 20.3. In conducting an interview, the interviewer should first attempt to establish rapport with the applicant. Sufficient time should be allowed for an uninterrupted interview, which should occur in a place where privacy is possible. In addition to avoiding questions that may be discriminatory, the interviewer should try not to ask either "yes/no" questions that do not require applicants to express themselves or leading questions to which the expected response is obvious.

Table 20.3.   Guidelines for employment inquiries in the State of Kansas

	Permissible Inquiries	Inquiries Which Must Be Avoided[a]
Name	Questions which will enable work and education records to be checked.	Inquiry about the name which would indicate lineage, ancestry, national origin, descent, or marital status.
Age	If age is a legal requirement, whether applicant meets the minimum or maximum age requirements; upon hire, proof of age can be required.	If age is not a legal requirement, any inquiry or requirement that proof of age be submitted must be avoided. **Note:** The Age Discrimination in Employment Act, as amended in 1986 prohibits discrimination against persons over age 40. The Kansas Act Against Discrimination prohibits discrimination against persons age 18 and over.
Race or Color	Race may be requested for affirmative action statistical recording purposes. Applicants must be informed that the provision of such information is voluntary.	Any inquiries which would indicate race or color.
Gender	Inquiry or restriction of employment is permissible only where a bona fide occupational qualification exists. (This BFOQ exception is interpreted very narrowly by the courts and EEOC.) The employer must prove that the BFOQ exists and that all members of the affected class are incapable of performing the job.	Any inquiry which would indicate gender.
Marital and Family Status	Whether applicant can meet specified work schedules and/or will be able to travel.	Any inquiry which would reveal marital status; information on applicant's children, child-care arrangements or pregnancy.
Disabilities	Under the provisions of the Kansas Act Against Discrimination, as amended, and the Americans with Disabilities Act of 1990, applicants may be asked if they are able to perform the essential duties of the position with or without reasonable accommodation.	Whether an applicant is disabled or inquiry about the nature or severity of the disability. Inquiries about any association with or relationship to a person with a disability. **Note:** Except in cases where undue hardship can be proven, employers must make reasonable accommodations for an employee's disability. Reasonable accommodation may include making facilities accessible, job restructuring, modified work schedules, modifying examinations, training materials or policies, acquiring or modifying equipment or devices, or providing qualified readers or interpreters.

Table 20.3.   *continued*

	Permissible Inquiries	Inquiries Which Must Be Avoided[a]
Religion	Employers may inform applicants of normal hours and days of work required by the job. **Note:** Except in cases where undue hardship can be proven, employers must make reasonable accommodations for an employee's religious practices. Reasonable accommodation may include voluntary substitutions, flexible scheduling, lateral transfer, change of job duties, or use of annual or vacation leave.	Any inquiry which would indicate applicant's religious practices and customs.
Address	Address may be requested so that the applicant can be contacted. Names of persons with whom applicant resides may be requested for compliance with the nepotism policy.	Any inquiry which may indicate ethnicity or national origin.
Ancestry or National Origin	Languages applicant reads, speaks or writes and the degree of fluency if a specific language is necessary to perform the job.	Inquiries into applicant's lineage, ancestry, national origin, descent, birthplace, or native language; how applicant learned a foreign language.
Arrest, Conviction and Court Records	Inquiry into arrest records and actual convictions which relate reasonably to fitness to perform a particular job. ARREST—The employer must consider whether the alleged conduct is job-related, the likelihood that the alleged conduct was actually committed and the time that has passed since the arrest. CONVICTION—The employer must consider the nature and gravity of the offense(s), the time that has passed since the conviction and/or completion of the sentence, and whether the conduct for which the applicant was convicted is job-related.	Ask or check into a person's arrest, court, or conviction record if not substantially related to functions and responsibilities of the particular job in question.
Birthplace and Citizenship	If United States citizenship is a legal requirement, inquiry about the citizenship of an applicant is permissible. The Employment Eligibility Verification (Form I–9) must be submitted by those who are hired to provide evidence of identity and employment eligibility.	Any inquiry which would indicate the birthplace of the applicant or any of the applicant's relatives.

Table 20.3. *continued*

	Permissible Inquiries	Inquiries Which Must Be Avoided[a]
Military Service	Type of education and experience gained as it relates to a particular job.	Type of discharge.
Photograph	Statement that a photo may be required after hire for purposes of identification.	Any requirement or suggestion that a photo be supplied before hiring.
Education	Applicant's academic, vocational or professional education; schools attended.	Any inquiry which would indicate the nationality, racial, or religious affiliation of a school; years of attendance and dates of graduation.
Experience	Applicant's work experience, including names and addresses of previous employers, dates of employment, reasons for leaving, and salary history.	Any inquiry regarding non-job-related work experience.
Financial Status	If required for business necessity, questions concerning financial stability. Examples of agencies that make inquiries into applicants' financial status are the Kansas Highway Patrol, Kansas Bureau of Investigation, and the Kansas Lottery.	If not required for business necessity, questions concerning financial stability.
Notice in Case of Emergency	Name and address of person(s) to be notified in case of accident or emergency may be requested after selection is made.	Name and address of relative(s) to be notified in case of accident or emergency.
Organizations	Inquiry into the organizations to which an applicant belongs and offices held relative to the applicant's ability to perform the job sought. NOTE—An applicant should not be required to provide the name of an organization which will reveal the religious, racial, or ethnic affiliation of the organization.	A list of all organizations to which the applicant belongs.
References	Names and addresses of persons who will provide professional and/or character references for applicant.	Requirement that a reference be supplied by a particular individual.
Relatives	Names of applicant's relatives already employed by the state agency in which employment is sought for compliance with the nepotism policy.	Name or address of applicant's relatives who are not employed by the state agency in which employment is sought.

[a] Any inquiry should be avoided which, although not specifically listed among the above, is designed to elicit information which is not needed to consider an applicant for employment.

Source: State of Kansas, Department of Administration, Division of Personnel Services.

Other common pitfalls that may be encountered in interviewing a job applicant are personal biases and the halo effect. *Personal biases* have an impact when they interfere with an interviewer's judgment about a candidate. For example, a qualified male applicant may be rejected because he has long hair. The *halo effect* occurs when a manager allows a single prominent trait to dominate judgment. For example, a person who is verbally fluent may impress an interviewer who might then overlook the applicant's poor employment record.

Medical Examination

A medical examination is the last step before hiring because of the expense. It is given to ensure that the prospective employee is healthy enough to perform the job. It can also be used as a baseline for future examinations. Physical requirements such as strength, agility, height, and weight tend to discriminate against women and have been questioned as being necessary for specific jobs. The preemployment physical is particularly valuable in the placement of handicapped persons; it also provides an opportunity through laboratory analysis to detect whether applicants use drugs.

Health regulations in some cities or counties specify that foodservice workers have food handler permits or cards prior to employment. Requirements for securing such a permit differ according to local regulations and may include successful completion of a sanitation training program, a test on food handling practices, and a limited physical exam. Because of the particular concern in foodservice organizations about preventing contamination of food and spreading communicable diseases, blood tests and stool cultures are sometimes required prior to employment and at periodic intervals thereafter. This practice is especially common in healthcare organizations.

Hiring Decision

The most critical step in the selection process is the decision to accept or reject applicants for employment. The final decision must be made carefully because of the cost of placing new employees on the payroll, the relatively short probationary period in most organizations, and affirmative action considerations.

While **equal employment opportunity law** is largely a policy of nondiscrimination, **affirmative action** requires employers to analyze their work force and develop a plan of action to correct areas of past discrimination (Sherman & Bohlander, 1992). Employers must make an affirmative effort to recruit, select, train, and promote minority candidates who qualify as well as those who could be helped to qualify for openings given training or physical accommodation. Organizations are beginning to adopt a "fair employment" philosophy toward protected classes that may increase opportunities for all those seeking work.

When all the information about an applicant has been assembled, a method must be developed for summarizing it. Checklists and summary forms are commonly used as a means of assuring that all the pertinent information is included in the evaluation of an applicant. Applicants are rank ordered according to total score, which provides an objective basis for evaluating them in the decision-making process. Although the decision is not based solely on scores, this technique will force the decision maker to analyze the relevant factors and the candidates in a systematic fashion.

The manager or supervisor to whom the new employee would report generally makes the hiring decision. In large organizations, approvals by higher level managers may be required. Before a final offer is made, the selection recommendation may also need to be reviewed by affirmative action personnel.

The offer will generally confirm the details of the job, working arrangements, and salary or wages, and specify a time limit in which the applicant must reach a decision. Individuals who are rejected should be notified immediately and given the reason for rejection.

The salary offered should be competitive with those of similar jobs in other organizations in the area and should be compatible with the existing salary structure in the organization. Too low an offer may cause the new employee to feel disgruntled; too high an offer may cause morale problems with current employees.

Orientation

The recruitment and selection of employees are important steps in the employment process, yet the careful planning and decision making will be negated if the orientation to the organization and the job is not carried out properly. *Orientation* is the formal process of familiarizing new employees with the organization, job, and work unit (Sherman & Bohlander, 1992). Orientation is designed to provide new employees with the information they need to function comfortably and effectively in the organization. Three types of information are typically included in orientation:

- Review of the organization and how the employee's job contributes to the organization's objectives
- Specific information on policies, work rules, and benefits
- General information about the daily work routine

In large organizations, the orientation tends to be more formal than in small organizations, but it is an important process for getting employees off to a good start regardless of the formality of the orientation or size of the organization. If an organization has a human resources department, staff in that department generally provides information about the overall organization policies and benefits. Orientation in the department then focuses on job-related issues.

If new employees are properly oriented, several objectives can be achieved. First, start-up costs can be minimized because new employees may make costly mistakes if they are not properly oriented. Second, anxieties can be reduced. Third, orientation can help create realistic job expectations.

Unfortunately, the importance of an orientation program is underestimated in many organizations. New employees often are given a policy and procedure manual and told to study it until given another assignment. Policies and procedures may have little meaning to new employees who are not familiar with the operation.

The first day of employment is crucial to the success or failure of a new employee. An actual incident that occurred in a hospital foodservice provides an example of how weak the orientation process may be in some situations. On the first day at work, after completing the formal 2-hour group orientation session conducted by staff in the human resources department, Bill, a new dishwasher, walked into the department of food and nutrition services, and the director told him, "There is the dishroom. Go find Joe; he'll show you the ropes." Needless to say, the anxiety of a new employee was probably not reduced by that experience!

Zemke (1989a) stated that creating a first-class orientation program and process is a job worth doing, but doing it well requires time, money, patience, some creativity, and an understanding of what has worked well for others. Puckett (1982) developed the checklist in Figure 20.3 for the initial orientation of new foodservice employees. She cautioned that new employees should not be overloaded with information on

Figure 20.3.   Orientation checklist.

_____ 1. Pre-arrange for the employee to be met at a given place and a given time. Meet the employee and *show* him or her the way to the department.

_____ 2. Welcome the new employee. Put the employee at ease. Show interest in him/her as a person. Start out slow and give the employee time to relax.

_____ 3. In a quiet area, explain the purpose of the orientation.

_____ 4. Go over any forms that the employee will need to sign. Explain what the forms are and the disposition of the forms once they are signed.

_____ 5. Using an organizational chart, show the employee where he/she fits into the organization. Supply the name of the employee's immediate supervisor. Tell the employee who the other managers are and give the employee a brief description of his or her responsibilities.

_____ 6. Explain the purpose and objectives of the department and present the employee with a typed copy.

_____ 7. Introduce the employee to coworkers.

_____ 8. Show the employee the work area and briefly define the job.

_____ 9. Give the employee a tour of the department. Be sure to include such areas as the breakroom, time clock area, cafeteria, restrooms, employee health nurse, and any other department this employee will deal with.

_____ 10. Interpret the main departmental policies, rules, and regulations that the employee will need to know in order to do the job (time clock, meal and break times, uniform policy, parking facilities, telephone usage-message area, payday, pay rate, overtime, schedules, safety procedures, fire procedures). Give the employee a typed copy of these policies.

_____ 11. Explain job duties to the employee, and give copies of the job description and work schedule to him/her.

_____ 12. *Demonstrate* major job duties to the employee.

_____ 13. Assign the new employee to another employee for planned observation and training. Check back periodically on the employee's progress.

Source: "Making or Breaking the New Employee" by R. P. Puckett, 1982, *Contemporary Administrator, 5*(10), p. 14. Used by permission.

Figure 20.4.   Checklist for follow-up orientation session with a new employee.

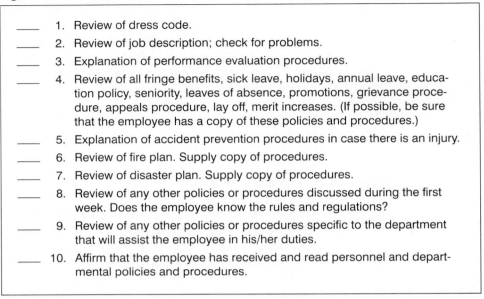

_____   1. Review of dress code.

_____   2. Review of job description; check for problems.

_____   3. Explanation of performance evaluation procedures.

_____   4. Review of all fringe benefits, sick leave, holidays, annual leave, education policy, seniority, leaves of absence, promotions, grievance procedure, appeals procedure, lay off, merit increases. (If possible, be sure that the employee has a copy of these policies and procedures.)

_____   5. Explanation of accident prevention procedures in case there is an injury.

_____   6. Review of fire plan. Supply copy of procedures.

_____   7. Review of disaster plan. Supply copy of procedures.

_____   8. Review of any other policies or procedures discussed during the first week. Does the employee know the rules and regulations?

_____   9. Review of any other policies or procedures specific to the department that will assist the employee in his/her duties.

_____   10. Affirm that the employee has received and read personnel and departmental policies and procedures.

Source: "Making or Breaking the New Employee" by R. P. Puckett, 1982, *Contemporary Administrator,* 5(10), p. 15. Used by permission.

the first day and stressed the need for follow-up sessions during the first week, with a review session during the first month on the job. Periodic monitoring and reinforcement during the early weeks on the job also are advisable. The checklist in Figure 20.4 outlines the follow-up orientation that Puckett suggested be held at the end of the second or third week on the job.

Proper orientation pays off in terms of decreased turnover and increased job performance. Hogan (1992) stated that there is no magic formula for reducing turnover of employees. One suggestion he had that could help reduce turnover is to establish an orientation program that is consistently followed and does much more than merely explain the mechanics of being an employee. For example, the organization's goals, philosophy, and culture should be explained to all new employees.

DEVELOPING AND MAINTAINING THE WORK FORCE

The employment process includes only the initial stages in building an efficient and stable work force. Employees need continual development if their potential is to be utilized effectively in an organization. This development process begins with orientation but should be continued throughout employment in the organization.

Employee development has become increasingly vital to the success of organizations. Changes in technology require that employees possess the knowledge and skills to cope with new processes and techniques.

Performance evaluation is another step in the development and maintenance of the work force. Other staffing functions include promotions, transfers, demotions, and separations as well as compensation management or the administration of salaries, wages, and fringe benefits.

Training and Development

Training is and must be the responsibility of all managers. The term *training* is frequently used to refer to the teaching of technical skills to employees, and **management development** refers to programs designed to improve the technical, human, and conceptual skills of managers. Many new employees have the knowledge, skills, and abilities for their job, but others might need extensive training before they can perform effectively. The primary purpose of a training program is to help the organization meet its goals. Equally important, however, is helping trainees meet their personal goals.

The need for training new employees and employees promoted to new jobs is obvious; however, training is needed for all employees to maintain standards in a foodservice operation. In healthcare facilities, training for all employees is mandated in federal regulations and accreditation standards. In other segments of the foodservice industry, the positive benefits of training in terms of increased productivity and morale are gradually being recognized.

In addition to training for separate jobs, many employers offer training programs that go beyond the immediate job requirements. Educational assistance for employees is given in some foodservice operations. One fast-food operation in a major city spent $10,000 in 1 year for tuition and other costs for employees attending a local community college; the operation saved much more in reduced turnover.

Another area of educational concern is illiteracy. Business is going to have to find its own solutions to workplace illiteracy (Sherer, 1990), just as it must respond to other social ills. Foodservice, because of its low-skill, low-paying jobs, attracts large numbers of functionally and marginally illiterate employees. Many foodservice managers have started literacy programs for employees and family members. Help is available from local school districts, adult and continuing education centers, state education departments, local colleges and universities, and the federal government's Job Training Partnership Act funds. Regardless of cost, basic skills and English as a second language training programs might become a way of life with restaurant operators. Most operators say that the cost is worth it; typical literacy programs foster increased morale, better communications, lower turnover, employee loyalty, and higher productivity.

In large organizations, a training staff may be available to assist with various aspects of a training program. Some training is conducted on a group basis, but individual training also is necessary. A needs assessment should reflect the kind of training needed and where it is needed.

After needs have been identified, managers should then initiate appropriate training efforts. **On-the-job training** methods include job rotation, internships, and apprenticeships. In *job rotation,* employees are assigned to work on a series of jobs over a period of time, thereby permitting them to learn a variety of skills. In an *internship,* job training is combined with classroom instruction. In an *apprenticeship,* employees are assigned to highly skilled co-workers responsible for their training.

Off-the-job training takes place outside the actual workplace. Methods used in off-the-job training range from laboratory experiences that simulate actual working conditions to other types of participative experiences, such as case studies or role playing, and to classroom activities such as seminars, lectures, and films. The programmed instruction method, in which subject matter is broken down into organized and logical sequences, requires responses by the trainee. Computer-assisted instruction also is used to deliver training material directly through a computer terminal in an interactive format. For training managers, management games can be used to make decisions affecting a hypothetical organization.

Employee work schedules and workloads need to be considered in planning inservice sessions because varied work shifts and alternating days off are common practices in foodservice operations. Training sessions may need to be held at the end of the work day or at the end of one shift and the beginning of another if the daily work routine does not allow time. Employees generally are given extra compensation for attending training sessions if they are scheduled outside the regular workday. In-service sessions conducted during the workday usually are fairly short, varying in length from 15 minutes to no more than 1 hour. Whenever possible, discussion should be supplemented with visual aids, demonstrations, or printed materials. Resource persons from other departments or from outside the organization will add interest to the program. Sessions should provide opportunity for employee involvement and should start and end promptly.

Regardless of the training method used, carefully formulated objectives and a well-planned outline are critical to the success of the training effort. An example of a simple form for developing a lesson plan is included in Figure 20.5.

Lippitt (1966) summarized several conditions for effective learning that also should be taken into consideration in employee training (Figure 20.6). Managers should be aware of several common pitfalls to avoid. Lack of reinforcement is a common training error; frequently, managers point out mistakes but do not give praise when an employee does the job correctly.

The phrase *practice makes perfect* applies to the training process. All too often, managers expect employees to perform a job perfectly the first time, forgetting that practice and repetition are important aspects of learning. Managers also forget that people learn at different rates—some learn rapidly and some more slowly. If a person is a slow learner, he or she is not necessarily a poor performer. Training must take into consideration the differences among individuals.

Development of managers and other professional staff may include continuing education opportunities provided by colleges and universities, professional and trade associations, or other organizations. These may include workshops, short courses, seminars, trade shows, or conventions. Some organizations have programs

Figure 20.5.   Form for planning inservice training session.

Training Session Title: Work Simplification		Date: August 10, 1994

Objective: After completing the training session, the employee will be able to apply work simplification principles to his or her job.

Content Outline	Learning Experience, Activities, Evaluation	References, Resource Material
1. What is work simplification? • Define work simplification. • Identify benefits of work simplification. 2. How do you apply work simplification? • Discuss questions for analyzing present methods—what, why, how, when, where, who. • Identify changes that can be made in present methods based on the analysis, e.g., materials used, finished appearance, steps in assembly. • Discuss types of changes that can be implemented and evaluation of effectiveness of change.	• Class discussion • Demonstration of several work simplification techniques in food production. • Written examination on principles of motion economy. • Observation of work performance.	Neil, C. A.: Working smarter not harder. *School Food Service Journal* 37(10):51, 1983.

to assist managers in obtaining degrees either while working or during an educational leave.

Even though some of these continuing education experiences may not be directly related to a manager's job, they may serve to renew enthusiasm, introduce the person to new ideas, or provide new personal contacts, which may later be a resource for information. The manager's participation may be fully or partially funded by the organization, and administrative leave is generally given. Funds for travel, however, are becoming limited in many organizations, especially in public institutions.

No training and development program is complete until evaluation is done. Employees may be asked to evaluate individual and group training in terms of satisfaction, benefit derived, and suggestions for future training. The true test of the effectiveness of any training effort, however, is improvement in employee performance.

Figure 20.6. Conditions for effective learning.

1. Acceptance that all people can learn.
2. The individual must be motivated to learn.
3. Learning is an active, not passive, process.
4. Normally, the learner must have guidance.
5. Approximate materials for sequential learning must be provided: hands-on experiences, cases, problems, discussion, reading.
6. Time must be provided to practice the learning, to internalize, to give confidence.
7. Learning methods, if possible, should be varied to avoid boredom.
8. The learner must secure satisfaction from the learning.
9. The learner must get reinforcement of the correct behavior.
10. Standards of performance should be set for the learner.
11. A recognition that there are different levels of learning and that these take different times and methods.

Source: "Learning Theories and Training" by L. This and G. Lippitt, 1966, *Training and Development Journal, 20*(4), p. 2. Copyright 1966, Training and Development, American Society for Training and Development, Reprinted with permission. All rights reserved.

Performance Appraisal

Performance refers to the degree of accomplishment of the tasks that make up an individual's job. Often confused with effort, performance is measured in terms of results. **Performance appraisals** take place in every organization, although they are not always formal. The success or failure of a performance appraisal program depends upon its philosophy and the attitudes and skills of those who manage it (Sherman & Bohlander, 1992). Gathering information is the first step in the process. The information about specific employees must be evaluated based on whether the employees are meeting organizational needs, and the information must be communicated to employees in hopes that it will result in their reaching high levels of performance.

Objectives

The primary objectives of a performance appraisal program are:

- To provide employees with the opportunity to discuss their performance with the supervisor or manager
- To identify strengths and weaknesses of the employee's performance
- To suggest ways the employee can meet performance standards, if they have not been met
- To provide a basis for future job assignments and salary recommendations

Research has shown that performance appraisals are used primarily for compensation decisions (Sherman & Bohlander, 1992). Employee placement decisions, such as transfers, promotions, or demotions, usually are based on performance appraisals. Many organizations are also using these appraisals to document employee performance as a protection against possible charges of wrongful termination or unfair employment practices.

Individuals want feedback about their performance, and performance appraisals can provide an opportunity to obtain such information. If performance compares favorably with that of other employees, an individual will obviously react favorably to performance review; however, feedback on poor performance is more difficult to accept.

In actual practice, appraisal programs have often yielded disappointing results. The focus too frequently is on past performance with little attention given to future performance.

Managers or supervisors usually are responsible for appraising employees in their units. They probably are in the best position to do this, but they often do not have time to observe the employees' performance adequately, which results in a less-than-objective appraisal. Often managers have to rely on others to complete an appraisal. Peer appraisals sometimes are used, but confidentiality can become a problem with this method. Managers often ask employees to rate themselves on the appraisal form before the performance interview, at which time the supervisor and the employee discuss job performance and agree on a final appraisal. At this meeting, future performance goals can be established for the employee, and these goals can be used as the basis for the next performance appraisal. The major difficulty with most performance appraisals is that the judgments frequently are subjective, relating primarily to personality traits or to observations that cannot be verified.

Subordinates often are asked to complete a performance appraisal form for their superiors. They can rate performance dimensions, such as leadership, oral communication, delegation of authority, coordination of team efforts, and interest in subordinates, but they should not rate performance of job tasks, such as budgeting and analytical ability (Sherman & Bohlander, 1992).

Criteria for measuring performance should be relevant to the job, understandable, and measurable. Job descriptions provide the basis for performance appraisal. The newer forms for job descriptions that include specifications and performance standards, such as that for a bartender presented in chapter 17, describe a base for appraising employee performance in terms of job responsibilities.

Methods

Commonly used performance appraisal methods include the following: checklist, rating scale, and critical incident. In the checklist method, the rater does not evaluate performance but merely records it. A list of statements or questions that are answered by yes/no responses is the basis for the evaluation. For example, a checklist for a chef might include the following items that can be answered with a yes or no: The chef

_____ is concerned about pleasing the customer
_____ checks with waitperson if order is not clear
_____ is concerned about presentation of the food
_____ adheres to food safety principles

A wide variety of **rating scales** have been developed. The traditional rating scale includes a number of dimensions on which an employee is rated on a three- to five-point scale for measuring quality of performance. These dimensions might include knowledge of work, initiative, quality of work, quantity of work, and interpersonal relationships. An excerpt from a performance appraisal for a chef in which a rating scale is used is shown in Figure 20.7. Raters also may be asked to provide comments or documentation on unsatisfactory performances and perhaps on outstanding performances.

The **critical incident** technique involves identifying incidents of employee behavior (Ingalsbe & Spears, 1979). An incident is considered critical when it illustrates that the employee has or has not done something that results in unusual success or failure on some part of the job. Therefore, critical incidents may include both positive and negative examples of employee performance. Although this method has the advantage of providing a sampling of behavior throughout the evaluation period, it is also a time-consuming technique.

Management by objectives (MBO) is a method for performance evaluation that is used primarily with managerial and professional personnel. Managers and their superiors agree on the objectives to be achieved, usually for a 1-year period. Periodic assessment of progress on the objectives is conducted at several intervals during the year, and objectives are revised if deemed appropriate. The degree to which these objectives are achieved, then, provides the basis for evaluation at the end of the period. MBO provides a great deal of objective feedback but requires an excessive amount of time and paperwork.

Foodservice managers and employees are beginning to look for new employee review methods that are in keeping with current philosophies of empowerment, teamwork, and open communication (Beasley, 1993). Traditional methods lump employees into performance categories, create excessive paperwork for managers, are geared to short-term goals, and are detrimental to teamwork. These methods concen-

Figure 20.7.   Performance appraisal for a chef using a five-point rating scale.

	Unsatisfactory	Needs Improvement	Satisfactory	Above Average	Superior
Quality of Work					
• Accuracy	☐	☐	☐	☐	☐
• Neatness	☐	☐	☐	☐	☐
• Organization	☐	☐	☐	☐	☐
• Thoroughness	☐	☐	☐	☐	☐

trate on old problems instead of future development. Some managers are eliminating forced rating scales and are evaluating job-based performance in terms of "satisfactory/unsatisfactory" or "meets standards/needs improvement." Space usually is available for managers and employees to write in goals or opportunities for improvement.

Another method is to have managers and employees discuss growth opportunities (Beasley, 1993). Managers are learning that spending only 10% of review time discussing past performance and 90% planning the employee's development is much more effective than the reverse. Managers and employees can identify the internal and external customers with whom the employee interacts and the barriers that prevent the employee from providing the best-quality products and services. They can then develop an action plan that will minimize the problems.

In foodservice operations, in which work teams have been established, some are beginning to conduct roundtable performance reviews of each member. To approach this level of openness, group members undergo training on how to give constructive feedback and how to handle such feedback themselves (Beasley, 1993).

Appraisal Interviews

The appraisal interview provides the manager with the opportunity to discuss an employee's performance and explore areas of improvement. Also, employees should have the opportunity to discuss their concerns and problems in the job situation. Because the purpose of the interview is to make plans for improvement, the manager should focus on the future rather than the past. In some instances, employees may be asked to complete a performance appraisal form prior to the interview, and this form is used during the interview session for comparison with the supervisor's assessment. Interviews should be scheduled several days in advance and conducted in a private setting because of the sensitivity involved.

The manager should observe the following points during the interview:

- Emphasize strengths on which the individual can build rather than stress weaknesses
- Avoid recommendations about changing personal traits; instead, suggest more acceptable ways of performing
- Concentrate on opportunities for growth within the employee's present job
- Limit plans for change or growth to a few objectives that can be accomplished within a reasonable period of time

The appraisal interview is probably the most important part of the performance appraisal. Trying to discuss the employee's past performance and plans for improvement in one session can be difficult. Dividing the interview into two sessions might be the answer. After all, the manager cannot be both an evaluator and a counselor in the same review period.

Personnel Actions

The performance appraisal process provides the basis for various types of personnel actions. If deficiencies in employee performance are noted, plans for improvement should be developed or appropriate disciplinary actions should be used. Changes in job placement may be suggested during evaluation and may include promotions, demotions, transfers, or even separations such as those by dismissal, resignation, or retirement. Leaves with or without pay are other personnel actions for which a manager must plan in order to maintain an adequate work force.

Promotion

A *promotion* is a change of assignment to a job at a higher level in the organization. The new job generally provides higher pay and status and requires more skill and responsibility. Advancement can serve as an incentive for improved performance for employees at all levels of the organization.

Promotions are tied closely to the performance appraisal and should be a way for recognizing good performance; however, individuals should not be promoted to positions in which their abilities are not appropriate. For example, an action-oriented unit or regional manager in a large multiunit foodservice organization who is a good decision maker might be misplaced in a staff position in the central office. Instead, this individual should probably be considered for a higher-level line position.

Two basic criteria for promotions are merit and seniority. Promotions should be fair, based on merit, and untainted by favoritism. Even when promotions are fair and appropriate, problems may occur, such as resentment by employees bypassed for promotion. If an organization is unionized, the union contract often requires that seniority be considered.

Demotion

A *demotion* consists of a change in job assignment to a lower organizational level in a job involving less skill, responsibility, status, and pay. Employees may be demoted because of reduction in positions of the type they are holding or because of reorganization. Demotion may also be used as a disciplinary action for unsatisfactory performance or for failure to comply with policies, rules, or standards in the organization. Acceptance and adjustment to loss of pay and status are difficult for most employees. Demoted employees may need special supervision or counseling as a result.

When the abilities or skills of a long-service employee decline, the solution may be to restructure a job rather than demote a loyal employee. In some instances the problem is solved by moving the person to a job with an impressive title but little authority or responsibility.

Transfer

Transfer involves moving an employee to another job at approximately the same level in the organization with basically the same pay, performance requirements, and status. A transfer may require an employee to change the work group, workplace, work shift, or organizational unit, or even move to another geographic area. A transfer can result from an organizational decision or an employee request. Transfers permit the placement of employees in jobs in which the need for their services is greatest. Also, transfers permit employees to be placed in jobs they prefer or to join a more compatible work group.

Separation

A *separation* involves either voluntary or involuntary termination of an employee. In instances of voluntary termination or employee resignations, many organizations attempt to determine why employees are leaving by asking them to complete a questionnaire or by conducting an exit interview. This interview can be a powerful tool for identifying a variety of personnel-related problems. According to human resources researchers, employees leaving a company can provide insight into problems encountered while in the organization's service. The exit interview provides employers an opportunity to learn why their employees are leaving and to devise plans for correcting turnover.

Involuntary separation, or firing of an employee, should be considered an action of last resort because of the investment the organization has in the employee. Training and counseling should be tried before termination is considered. Also, because of legal implications, thorough and complete documentation of poor performance is necessary when firing an employee.

Another type of separation may be the result of a decision to reduce the work force. In this instance, the layoff may be the result of elimination of jobs, new technology, or depressed economic conditions. Decisions on layoffs should be made carefully, and assistance in finding another job should be given to the laid-off employee, if at all possible.

Employee Discipline

Discipline is the action against an employee who fails to conform to the policies or rules established by the organization. It is used to aid in obtaining effective performance and to ensure adherence to work rules. It also serves to establish minimum standards of performance and behavior. Some of the more common problems with employees that necessitate disciplinary action are listed in Figure 20.8.

Foodservice operators agree that the key to effective discipline is keeping in touch with employees (Boyle, 1987b). Good communications begin with clearly defined expectations, which should be discussed during the job interview. Disciplining an employee is not easy, but it can be constructive. Once alerted to a problem, the employee may be able to solve it together with the manager, leading to substantial improvement in the employee's overall job performance.

Disciplinary Procedures.   Usually, the following several steps are involved in a disciplinary procedure:

- Unrecorded oral warning
- Oral warning with notation in an employee's personnel file
- Written reprimand
- Suspension
- Discharge

Figure 20.8.   Common disciplinary problems.

Attendance Problems	On-the-Job Behavior Problems
Unexcused absence	Intoxication at work
Chronic absenteeism	Insubordination
Unexcused/excessive tardiness	Horseplay
Leaving without permission	Smoking in unauthorized places
	Fighting
Dishonesty and Related Problems	Gambling
	Failure to use safe devices
Theft	Failure to report injuries
Falsifying employment application	Carelessness
Willful damage to company property	Sleeping on the job
Punching another employee's time card	Abusive or threatening language to
Falsifying work records	supervisors
	Possession of narcotics or alcohol
	Possession of firearms or other
	weapons

Source: Personnel Management. The Utilization of Human Resources, 6th Edition, by H. J. Chruden and A. W. Sherman, Copyright 1980, South-Western Publishing Co. Used by permission.

Exceptions do occur in the above steps, depending on the nature of the employee's offense. For example, a foodservice employee caught in the act of theft would probably be fired immediately, rather than be warned, orally or in writing, or suspended.

A key element in discipline is consistency. An employee should feel that any other employee would receive the same discipline under essentially the same circumstances. Disciplinary action is the consequence of an employee's behavior, and not of a personality conflict with the supervisor. Discipline must be administered in a straightforward, calm way without anger or apology. The manager should avoid argument and administer discipline in private to avoid embarrassing the employee. In a few instances, public reprimand may be necessary to enable a manager to regain control of a situation.

In most organizations, a grievance procedure has been established to ensure that employees have due process in disciplinary situations. In other words, grievance procedures allow an employee to have a fair hearing when he or she believes a personnel action has been administered unfairly. These procedures generally involve a multistep process whereby the employee first requests a review by the next highest level supervisor. If a satisfactory solution is not found at this level, then a higher-level review can be requested. Grievance committees are often appointed to review actions not settled through the chain of command. In unionized organizations, grievance procedures are specified in the union contract and include provisions for hearings by an outside arbitrator when problems cannot be solved within the organization.

Grievances and disciplinary action can be reduced in a number of ways. These methods include the following:

- Preparation of accurate job descriptions and specifications
- Selection of individuals with appropriate qualifications for job requirements
- Development of effective orientation, training, and performance evaluation systems
- Utilization of good human skills by supervisors

Identifying Causes.   An employee's immediate supervisor has the primary responsibility for preventing or correcting disciplinary problems. In attempting to uncover reasons for unsatisfactory behavior, the supervisor must first consider if the employee is aware of certain policies and work rules before initiating disciplinary action. If the employee is aware of these policies, the supervisor must then realize that understanding the causes underlying the problem is as important as dealing with the problematic behavior; any attempt to prevent recurrence requires an understanding of these causes.

Health problems, personal crises, emotional problems, or chemical dependency may be the source of unsatisfactory performance. Marital, family, financial, or legal problems are the predominant personal crisis situations that may affect employee performance.

Another factor affecting job performance may be stress (Heneman, Schwab, Fossum, & Dyer, 1989), an occupational health hazard that is now being increasingly recognized. Although most employees are able to cope with stress, some employees are not and become "troubled" employees. Impaired mental health, alcoholism, and drug abuse may be symptomatic of the troubled employee. Burnout, as discussed in chapter 15, is a stress-related condition experienced by many restaurant line managers, who typically are required to supervise employees, handle paperwork, and attend to customers' needs (Krone, Tabacchi, & Farber, 1989b).

Supervisors need to be alert to all these possible causes of problematic behavior. Because supervisors may not be prepared to assist or counsel an employee, many organizations provide them with some combination of training, policy guidance, or counseling. The counseling may be for supervisors or employees, and it may be with a specialized counselor or medical personnel both inside and outside the organization.

Some organizations are beginning to take a more positive approach in dealing with troubled employees. An **employee assistance program** typically provides diagnosis, counseling, and referral for advice or treatment, when necessary, for the problems related to alcohol or drug abuse, emotional difficulties, and marital or family difficulties (Sherman & Bohlander, 1992).

These programs typically include training of supervisors, internal staff counseling, referral services to community agencies, and involvement of the employee's dependents. Organizations are recognizing that discharging a troubled employee may be not only inhumane but also an ineffective solution because similar problems may occur with a new employee. Personal problems, especially alcoholism, are interfering with staff productivity and eroding the bottom line in the hospitality industry (Quick, 1989). Employee assistance programs, however, provide a cost-effective method for resolving these problems.

Compensation Management

Employee compensation represents a substantial part of the operating costs of an organization. Good working conditions, sound employment practices, and compensation appropriate for an individual's qualifications and the responsibilities of a job are essential for recruitment and retention of capable employees.

Compensation

Compensation is the financial remuneration given by the organization to its employees in exchange for their work and includes salaries or wages and benefits. *Salary* is the term used to refer to earnings of managerial and professional personnel; *wages* refer to hourly earnings of employees covered by the Fair Labor Standards Act, sometimes called the Minimum Wage or Wage-Hour Law. In establishing the rate of pay for employees, the employer has to look at many factors, including wage mix, before making a decision.

Wage Mix

A *wage mix*, shown in Figure 20.9, is a combination of both external and internal factors that can influence rates at which employees are paid (Sherman & Bohlander, 1992). Government legislation also influences wage mix.

External Factors.   The external factors that influence the wage mix include labor market conditions, geographic area, cost of living, collective bargaining, and government influence. Labor market conditions reflect the supply and demand of employees in the geographic area. A wage structure that is comparable with wages in the geographic area should be developed. If this information is not available in the area,

Figure 20.9.   External and internal factors affecting the wage mix.

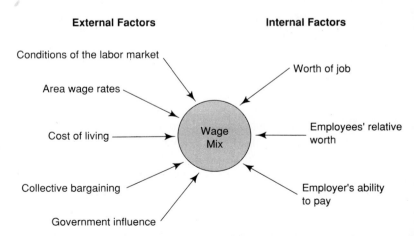

Source: Reproduced from *Managing Human Resources* by A. W. Sherman and G. W. Bohlander with the permission of South-Western Publishing Co. Copyright 1992 by South-Western Publishing Co. All rights reserved.

a national survey could be used. For example, the National Restaurant Association conducts a biannual survey on hourly wages for 18 foodservice occupations from buspersons and dishwashers to bakers and bartenders (Riehle, 1993). Results are available for the United States, regions, states, and substates/city areas.

Compensation rates during inflation have been increased periodically to permit employees to maintain their purchasing power (Sherman & Bohlander, 1992). Wage setters generally use the consumer price index, which is a measure of the average change in prices over time in a fixed so-called market basket of products and services. Labor unions use collective bargaining to increase compensation. The union's goal is to increase real wages by a percentage greater than the consumer price index to improve the purchasing power and standard of living of its members.

Internal Factors.   Often, if the organization does not have a compensation program, the worth of jobs is determined by people familiar with them. Pay rates can be influenced by the labor market or collective bargaining. Organizations with compensation programs use job evaluations to aid in setting pay rates. *Job evaluation* is the process of determining the relative worth of jobs in order to establish which jobs should be paid more than others. The first step of compensation is a job evaluation to identify the work content and requirements of each job: required responsibilities, education, experience, and the skill necessary for performing duties under the existing conditions.

Two methods used for evaluating jobs are job ranking and job classification. *Job ranking*, in which jobs are arranged on the basis of their relative worth from most to least complex, is the simplest method but is subject to the amount of skill and information that raters have. *Job classifications* classify and group jobs according to a series of predetermined wage classes or grades. This method is designed to place jobs of the same general value in the same pay grade, and it provides a smooth progression for advancement.

Employees' relative worth often is based on the premise that employees who possess the same qualifications should receive the same pay. Differences in performance are rewarded through promotion of various incentive systems. Merit raises should be based on an effective performance appraisal, but often they are given automatically. An employer's ability to pay is based on the willingness of the taxpayer to provide funds in the public sector or by the profits from products and services in the private sector.

Government Influence

Since the 1930s, a broad spectrum of federal laws has been enacted with significant impact on the compensation practices in organizations. Several of the laws that address compensation relating to nondiscrimination were described in the Human Resources Planning section of this chapter. Federal employment laws relate to either compensation or benefits. The Fair Labor Standards Act deals with compensation; Social Security, unemployment, and workers' compensation are insurance benefits. Private pension plans also are subject to federal regulations under the Employment Retirement Income Security Act (ERISA). In addition, in foodservice and other segments of the hospitality industry, legislation and regulations have been enacted that apply specifically to employees who receive tips, as described in chapter 15. Foodservice managers in operations in which employees frequently receive tips must stay up to date on current provisions of these laws and regulations as well as other legislation related to compensation and employment practices.

Compensation Regulations.   The **Fair Labor Standards Act** (FLSA) was passed in 1938 and has been amended many times since then. The major provisions of the FLSA are concerned with minimum wage rates and overtime payments, child labor, and equal rights. Through the years, amendments to the act have enlarged the number of work groups covered by the law and have steadily increased the minimum wage. Employees in the foodservice industry were not covered by the minimum wage legislation until the mid-1960s. Employers are required to pay 1½ times the regular rate for all hours worked in excess of 40 during a given week. If employees are given time off for overtime work, it must be 1½ times the number of overtime hours. The FLSA forbids the employment of minors under age 18 in hazardous occupations. Employment of minors under age 16 also is controlled under the act. The Equal Pay Act of 1963 and the federal Age Discrimination Act of 1967 prohibit discrimination of women and employees over the age of 40.

The act also defines specific occupations that are exempt from minimum wage and overtime requirements. Executive, administrative, professional, and outside salespersons are included in the exempt category. Employee exemption depends on the responsibilities and duties of a job and the salary paid. Definitions and requirements for job responsibilities and salary provisions are defined by the U.S. Department of Labor for these exempt categories.

Benefits Regulations.   The **Social Security Act** of 1935 established an insurance program to protect covered employees against loss of earnings resulting from retirement, unemployment, disability, or, in the case of dependents, from the death or disability of the person supporting them. The Social Security program is supported by means of a tax levied against an employee's earnings that must be matched by the employer.

In recent years, the tax rate and the maximum amount of earnings subject to the tax have periodically been adjusted upward to cover liberalization of benefits. In the mid-1960s, the Medicaid and Medicare programs, which provide health insurance benefits, were established by amendments to the Social Security Act.

Employees who have been working in employment covered by the Social Security Act and who are laid off may be eligible for unemployment insurance benefits for a period of up to 26 weeks. To receive these benefits, employees who become unemployed through no fault of their own must submit an application for unemployment compensation with the state employment agency, register for available work, and be willing to accept any suitable employment that may be offered to them. Each state has its own unemployment insurance law that defines the terms and benefits of its program. Funds for unemployment compensation come from a federal payroll tax based on wages paid to each employee. Most of the tax is refunded to the states that operate their unemployment compensation programs in accordance with minimum standards by the federal government.

The first U.S. legislation providing compensation to workers disabled in industrial accidents was enacted in 1911. Today, workers' compensation insurance covers most American workers. The theory is that the cost of work-related accidents and illnesses should be considered one of the costs of doing business and ultimately should be passed on to the customer. Disabilities may result from injuries or accidents, as well as from occupational diseases such as black lung, radiation illness, and asbestosis. Each state has its own laws and agencies to administer disabilities laws. In general, these laws provide income and medical benefits to victims of work-related accidents, reduce court delays arising out of personal injury litigation, encourage employer interest in safety and rehabilitation, eliminate payment of fees to lawyers and witnesses, and promote study of accident causes.

Most of the nonfarm work force now receives some form of retirement protection through either private pension plans, deferred profit-sharing plans, or savings plans. In recent years, these programs increasingly have been placed under federal regulation. To protect pension plans from failure, the Employment Retirement Income Security Act of 1974, commonly known as ERISA or the Pension Reform Law, was enacted. Although the act does not require employers to have a pension plan, it provides standards and controls for plans. This law establishes employer responsibilities, reporting and disclosure, employee participation and coverage, vested rights in accrued benefit, and other requirements.

Healthcare benefits are in a state of flux today. President Clinton on September 23, 1993, presented to Congress a plan with 12 major points to guarantee a generous, minimum package of health insurance to all Americans (Goodgame, 1993). Under this plan, all employers would contribute to the cost of their employees' healthcare. If this plan is accepted, states could begin setting up healthcare alliances as early as 1995.

Compensatory benefits include paid time off, such as vacation, sick leave, holidays, military and jury duty, and absences due to a death in the family or other personal leave. Leave time has shown a slow but steady increase over the years and accounts for the largest portion of payroll costs.

LABOR RELATIONS

Since the 1800s, labor unions have been an important force shaping organizational practices, legislation, and political thought (Sherman & Bohlander, 1992). Today, unions continue to be of interest because of their influence on organizational productivity, global competitiveness of the United States, the development of labor law, and human resources policies and practices. Like most organizations, unions are undergoing changes in both operation and philosophy. Labor has a new role in the United States that includes labor management programs, union company buyouts, and representation by the union on boards of directors.

Labor relations, therefore, are an important function of HRM in many types of organizations. **Labor relations** is a term referring to the interaction between management and a labor union. A closely related term is *collective bargaining,* which focuses on the negotiation for the settlement of the terms of a collective agreement between an employer and a union. Unions are challenging the ability of managers to direct and control the various functions of HRM. For example, union seniority provisions may influence who is selected for job promotions or training programs. Pay rates may be determined by union negotiations, or employee performance appraisal methods may have to be changed to meet union guidelines. Labor relations, therefore, is a specialized function of HRM to which employers must give consideration.

Historically, unions have had their greatest success in the manufacturing, mining, transportation, and construction segments of the economy. With the decline in these sectors and the relative growth in the service industries, unions have begun to focus organizational efforts on service employees. Retailing establishments and restaurants, once believed to be immune to unionization because of their largely unskilled labor force, have yielded large numbers of new union members. Legislation in the 1970s removed the previous exemption of nonprofit healthcare institutions from unionization of employees. These developments indicate that managers of foodservice operations must be knowledgeable about labor legislation, collective bargaining, and other aspects of labor-management relations.

Reasons for Joining Unions

Labor unions have usually developed as a reaction to management's decision-making power. The power of employees to bargain with their employer on an individual basis and to protect themselves from arbitrary and unfair treatment is limited. Research has indicated that employees join unions because of economic needs, general dissatisfaction with management policies, and as a way to fulfill social needs. As a result, many employees find that bargaining collectively with their employer through a union is advantageous. Moreover, this method of bargaining is protected by legislation. When employees are unionized, HRM is subject to negotiation with the union and to the terms of the agreement. Some employees join unions because they work in union shops, which require employees to join as a condition of employment in states where they are permitted.

Structure and Functions of Unions

Modern unionization has its roots in the guilds of artisans of the Middle Ages. For centuries, employees who have similar skills have formed organizations to govern entry to an occupation, define standards of occupational conduct, and regulate employment. In the United States, the first union activity dates from the late 1700s.

In 1955, the American Federation of Labor (AFL), composed primarily of craft unions of, for example, carpenters and plumbers, and the Congress of Industrial Organizations (CIO), made up of industrial unions, merged to form the **AFL-CIO**. Most unions belong to this federation; about 75% of all union members belong to the AFL-CIO. Most researchers divide labor organizations into three levels: AFL-CIO, national and international unions, and local unions belonging to a parent national union.

The AFL-CIO is a federation of 83 autonomous national and international unions. The AFL-CIO serves it members by (Sherman & Bohlander, 1992):

- Lobbying before legislative bodies on subjects of interest to labor
- Coordinating organizing efforts among its affiliated unions
- Publicizing concerns and benefits of unionization to the public
- Resolving disputes between different unions as they occur

The major difference between national and international unions is that the international union organizes employees and charters local unions in foreign countries. The national union provides technical assistance in negotiating and administering labor contracts, financial assistance during strikes, administration of union-sponsored pension plans and other fringe benefits, training programs for local union officers, and publications. In addition, national unions are generally active in lobbying efforts in the U.S. Congress and at the state level. Finally, the national union also establishes the rules and conditions under which the local unions may be chartered.

National unions are autonomous organizations that may opt voluntarily for affiliation with the AFL-CIO or remain independent. A primary benefit of affiliation in the AFL-CIO is some protection from membership raiding by other unions. Other advantages of affiliation include facilitation of coordinated bargaining, assistance in organizing efforts, and lobbying on behalf of organized labor.

Officers of local unions are responsible for negotiating local labor agreements and investigating member grievances. The **union steward** represents union members in their relations with an immediate supervisor or other managers. Stewards usually are elected by union members in their department and serve without pay. Because they are full-time employees of the organization, they often spend their own time on union-member problems. When stewards represent members at grievance meetings on organizational time, their lost earnings are paid by the union. The **business representative** is hired by the local union to manage the union and also to settle a member's grievance if the steward was not successful.

According to the Bureau of Labor Statistics, the following were the three largest national unions on a nationwide basis in the foodservice industries in 1992:

- Hotel and Restaurant Employees Union
- United Food and Commercial Workers Union
- Service Employees International Union

These unions belong to the AFL-CIO federation. Under the Retail Trade Union membership category, which includes all types of foodservices, 6.5% are union members. Employees in the foodservice industry may be affiliated with local unions and those in healthcare and public organizations are often members of unions specifically organized for those sectors.

Government Regulation

In the private sector, four major federal laws regulate employer and union conduct in labor-management relations: the Norris-LaGuardia, Wagner, Taft-Hartley, and Landrum-Griffin acts. In the public sector, state and local government employees are covered by state laws, and federal civil service workers are covered by the Civil Service Reform Act of 1978.

Norris-LaGuardia Act

The Norris-LaGuardia Act of 1932, also known as the Anti-Injunction Act, severely restricted the ability of employers to obtain a federal injunction forbidding a union from engaging in picketing or strike activities. The Norris-LaGuardia Act also nullified yellow-dog contracts, which were agreements that required workers to state they were not union members and promise not to join one.

Wagner Act

The 1935 Wagner Act, formally called the National Labor Relations Act, has had the most significant impact on labor-management relations of any legislation. The act placed the protective power of the federal government behind employee efforts to organize and bargain collectively through representatives of their choice.

The Wagner Act established the right of a union to be the exclusive bargaining agent for all workers in a bargaining unit. A *bargaining unit* is a group of jobs in a firm, plant, or industry with sufficient commonality to constitute an entity that can be represented in union negotiations by a particular agent.

The law declared the following to be unfair labor practices:

- Management support of a company union
- Discharge or discipline of workers for union activities
- Discrimination against workers making complaints to the NLRB
- Refusal to bargain with employee representatives
- Interference with the rights of employees to act together for mutual aid or protection

The primary duties identified for the NLRB were to hold secret ballot elections to determine representation and to interpret and apply the law concerning unfair labor practices. The courts may review decisions on unfair labor practices, but the NLRB's decisions on representation elections are final.

Taft-Hartley Act

The Taft-Hartley Act was passed in 1947 over President Truman's veto. The law amends the Wagner Act, although most of the major provisions of the earlier law were retained. The major thrust of the legislation was to balance the powers of labor and management. Before passage of the Wagner Act, employees had little power to organize and bargain, and therefore the early labor legislation restricted only employer activity. Because the bargaining power of unions increased significantly following the passage of the Wagner Act, restraints on union practices were considered necessary.

The following activities were defined as unfair union practices in the Taft-Hartley Act:

- Restraining or coercing employers in the selection of parties to bargain on their behalf
- Persuading employers to discriminate against any employees
- Refusing to bargain collectively
- Participating in secondary boycotts and jurisdictional disputes
- Attempting to force recognition from an employer when another union is already the representative
- Charging excessive initiation fees
- Practicing "featherbedding" or requiring payment of wages for services not performed

In 1974, Congress extended coverage of Taft-Hartley to private nonprofit hospitals and nursing homes. Provisions require unions representing hospital employees to give 90 days' notice before terminating a labor agreement—30 days more than in other industries. Also, a labor dispute in a healthcare facility is automatically subject to mediation efforts of the Federal Mediation and Conciliation Service (Sloane & Witney, 1988).

Landrum-Griffin Act

The Landrum-Griffin Act, also known as the Labor-Management Reporting and Disclosure Act, was passed in 1959 because the provisions in the Taft-Hartley Act did not cover labor racketeering. These investigations revealed that a few labor organizations and employers were denying employees' rights to representation and due process within their labor organizations.

The act requires that labor organizations hold periodic elections for officers, that members be entitled to due process both within and outside the union, that copies of labor agreements be made available to covered employees, and that financial dealings between union officials and companies be disclosed to the U.S. Department of Labor. The law tightens the Taft-Hartley Act restrictions against secondary boycotts. A secondary boycott occurs when a union asks firms or other unions to cease doing business with someone handling a product affected by a strike. The union can ask only that the product not be used.

Contract Negotiations

After a union wins recognition as the employees' bargaining agent, negotiations begin on a contract with management. Bargaining is a difficult and sensitive proceeding.

In local negotiations, the union side is represented by its local negotiating committee and may include a field representative of the national union with which it is affiliated. The management team may be made up of organizational staff, financial or operations managers, and perhaps an attorney. Negotiations at the national level are led by top-level officials from the national union and by top-level corporate managers and staff.

Major bargaining issues usually fall into five areas:

- *Economic issues* deal with such provisions as base pay, shift differentials, overtime pay, length of service increases, and cost of living allowances as well as with benefits such as pension plans, insurance, holidays, and vacations.
- *Job security* means an entitlement to work or, in lieu of work, to income protection. Procedures for handling layoffs, promotions, transfers, assignments to specific jobs, unemployment benefits, severance pay, and call-in pay are among the job security provisions generally included in union contracts.
- *Working conditions issues* include work rules, relief periods, work schedules, and health and safety.
- *Management rights issues* detail the rights of management to give direction and discipline employees.
- *Individual rights issues* concern establishment of grievance procedures for employees.

Careful preparation for negotiations is critical and should include planning the strategy and assembling data to support bargaining proposals or positions. The conditions under which negotiations take place vary widely among organizations and are dependent on the goals that each side seeks to achieve and the strength of their relative positions. First-time negotiations are usually more difficult than subsequent negotiating sessions after a union has become established. The negotiation of an agreement often takes on the characteristics of a game in which each side attempts to determine its opponent's position while not revealing its own.

Proposals or positions of each side are generally divided into those that it feels it must achieve, those on which it will compromise, and those submitted primarily for trading purposes. Throughout the negotiations, all proposals or positions submitted by both sides must be disposed of if agreement is to be reached; that is, each proposal or position must be accepted by the other side, either in its entirety or in some compromised form, or it must be withdrawn. To achieve its bargaining proposals or positions, each side presents arguments and evidence necessary to support it and exerts pressure on the other side. The outcome of this power struggle determines which side will make the greater concession to avoid a bargaining deadlock.

The power of the union may be exercised by striking, picketing, or boycotting the employer's product. An employer who is struck by a union may continue to operate or opt for a lockout, which is a shutting down of operations. If a strike or lockout occurs, both parties are affected. Therefore, both participants usually are anxious to achieve a settlement.

When the two parties are unable to resolve a deadlock, a third party, either a mediator or an arbitrator, may be called in to provide assistance. A mediator attempts to establish a channel of communication between the union and management but has no power to force a settlement. In mediation, compromise solutions are only offered by the third party. If the contract calls for binding arbitration, an arbitrator is called in to render a decision that is binding on both the union and the employer. After an agreement has been reached through collective bargaining, the written document must be signed by representatives of both parties.

A New Model of Unionism

Employees today are disgruntled with the workplace; relocations caused by takeovers, shutdowns, and rightsizing have increased mistrust of some businesses (Hoerr, 1991). Unhappy employees are filing record numbers of wrongful discharge suits and job discrimination and unfair labor practice complaints. In the workplace, employees are demanding more challenging work, a voice in decision making, and greater job security. In the past, these complaints would have led employees to join a union, but this is not happening today. Approximately 30% of private-sector employees belonged to unions 20 years ago, but today only 12.1% belong; by the year 2000, membership could drop to 5%.

Hoerr (1991) had the following three thoughts about unions:

- U.S. unions face the same crisis as U.S. management, which is dealing with global competition.
- Unions are not an obstacle to competitiveness but under the right conditions, they can be at the center of an organization's efforts to improve its competitiveness.
- U.S. unions must reinvent themselves as some organizations are trying to do.

Unions must develop a vision of how employees can help shape the technological and social revolution in the workplace by identifying new leverage points and improving their own human resources to place labor's new vision into practice. Meeting the needs of new diverse social groups including women, immigrants, and members of minority groups will be a challenge in the future.

Leverage points for redefining union influence include training, work redesign, employee ownership, and the new work force (Hoerr, 1991). Unions have begun to have a more active role in training employees for rapid technological changes. Training and work redesign are possible in organizations where unions exist. Training can bring together labor's concerns about social justice and the organization's concerns about competitiveness in the market. Unions also can begin to represent employee interests in redesigning projects. Employee ownership and the new diverse work force can be growth markets where unions do not exist. If unions can develop these skills while protecting their social vision, the economy could benefit. Challenges facing organized labor are not much different from challenges facing organizations. The major challenge is for both to reinvent organizations, thus making the economy competitive!

SUMMARY

Human resources planning is the process of anticipating and making provision for the movement of people into, within, and out of an organization. A strategic plan has to be developed, followed by forecasting future employee needs. A supply and demand analysis also is required. The Equal Employment Opportunity Commission (EEOC) is responsible for employment practices under which individuals are not excluded from any participation, advancement, or benefits due to race, color, religion, sex, national origin, or other nonmerit reason.

The employment process includes the recruitment, selection, and orientation of new employees. Recruiting is the process of locating and encouraging potential applicants to apply for existing or anticipated job openings. The selection process includes a comparison of applicant skills, knowledge, and education with the requirements of the job and involves decision making between the organization and the applicant. Orientation is the formal process of familiarizing new employees with the organization, job, and work unit.

Training is the responsibility of all managers. Training refers to teaching technical skills to employees, and management development refers to programs designed to improve the technical, human, and conceptual skills of managers. Training is needed for all employees to maintain standards in a foodservice operation. The true test of the effectiveness of any training program is improvement in employee performance. Performance refers to the degree of accomplishment of the tasks that make up an individual's job and is measured in results. Performance appraisal takes place in every organization. The appraisal interview provides the manager the opportunity to discuss an employee's performance and explore areas of improvement.

Personnel actions include promotions, demotions, transfers, and such separations as dismissals, resignations, and retirements. Discipline is the action against an employee who fails to conform to the policies or rules established by the organization. An employee's immediate supervisor has the primary responsibility for preventing or correcting disciplinary problems.

Employee compensation represents a substantial part of the budget. Compensation is the financial remuneration given by the organization to its employees in exchange for their work and includes salaries or wages and benefits. Several federal laws, most of which address issues relating to nondiscrimination, pertain to compensation or benefits. The Fair Labor Standards Act deals with compensation; Social Security, unemployment, and workers' compensation are insurance benefits.

Labor unions are undergoing changes in both operation and philosophy. Labor relations, the interaction between management and a labor union, are an important function of human resources management in many types of organizations. Unions must develop a vision of how employees can help shape the technological and social revolution in the workplace by identifying new leverage points and improving their own human resources. Challenges facing labor unions are not much different from those facing organizations.

BIBLIOGRAPHY

Aaron, T., & Dry, E. (1992). Sexual harassment in the hospitality industry. *Cornell Hotel and Restaurant Administration Quarterly, 33*(2), 92–95.

ADA employment rules to take effect July 26 for many employers. (1993). *Washington Weekly, 12*(20), 1, 4.

Batty, J. (1993). Preventing sexual harassment in the restaurant. *Restaurants USA, 13*(1), 30–34.

Beasley, M. A. (1993). Better performance appraisals. (1993). *Food Management, 28*(9), 38.

Blackburn, R., & Rosen, B. (1993). Total quality and human resources management: Lessons learned from Baldrige Award-winning companies. *Academy of Management Executive, 7*(3), 49–66.

Boyle, K. (1987a). Child care: The fringe benefit of the 1990s? *Restaurants USA, 7*(5), 20–22.

Boyle, K. (1987b). Effective employee discipline requires keeping in close touch. *Restaurants USA, 7*(10), 26–28.

Cheney, K. (1993a). Avoid the Zoë Baird trap. *Restaurants & Institutions, 103*(11), 153–154.

Cheney, K. (1993b). Sticky situations: Deal a blow to drug use. *Restaurants & Institutions, 103*(7), 121, 124, 126.

Conrad, P. J., & Maddux, R. B. (1988). *Guide to affirmative action: A primer for supervisors and managers*. Los Altos, CA: Crisp Publications.

Cross, E. W. (1992). AIDS: Legal implications for managers. *Journal of The American Dietetic Association, 92*, 74–77.

Cross, E. W. (1993). Implementing the American with Disabilities Act. *Journal of The American Dietetic Association, 93*, 272–275.

Dear National Restaurant Association member . . . (Issue on the Family and Medical Leave Act). (1993). *Washington Weekly, 13*(28), 1–12.

Deinhart, J. R., & Gregoire, M. B. (1993). Job satisfaction, job involvement, job security, and customer focus of quick-service restaurant employees. *Hospitality Research Journal, 16*(2), 29–43.

Doherty, R. F. (1989). *Industrial and labor relations terms: A glossary* (5th ed.). ILR Bulletin No. 44. Ithaca, NY: Cornell University Press.

Doyle, F. (1992). Hiring your way to a great staff. *Restaurants USA, 12*(6), 20–22.

Employer focus groups criticize immigration paperwork. (1993). *Washington Weekly, 13*(34), 4.

Ewing, D. W. (1989). *Justice on the job: Resolving grievances in the nonunion workplace*. Boston: Harvard Business School Press.

Foodservice accounts for about 11% of ADA charges at Justice Department. (1993). *Washington Weekly, 13*(4), 1.

Forrest, L. C. (1989). *Training for the hospitality industry* (2nd ed.). East Lansing, MI: Educational Institute of the American Hotel and Motel Association.

Goodgame, D. (1993). Ready to operate. *Time, 142*(12), 54–58.

Griffin, R. W. (1993). *Management* (4th ed.). Boston: Houghton Mifflin.

Hamilton, A. J., & Veglahn, P. A. (1992). Sexual harassment: The hostile workplace. *Cornell Hotel and Restaurant Administration Quarterly, 33*(2), 88–92.

Heneman, H. G., Schwab, D. P., Fossum, J. A., & Dyer, L. D. (1989). *Personnel/human resource management* (4th ed.). Homewood, IL: Richard D. Irwin.

Hoerr, J. (1991). What should unions do? *Harvard Business Review, 69*(3), 30–32, 36–39, 42, 44–45.

Hogan, J. J. (1992). Turnover and what to do about it. *Cornell Hotel and Restaurant Administration Quarterly, 33*(1), 40–45.

Ingalsbe, N., & Spears, M. C. (1979). Development of an instrument to evaluate critical incident performance. *Journal of The American Dietetic Association, 74*(2), 134–140.

Iwamuro, R. (1992). Disabled workers in foodservice. *Restaurants USA, 11*(6), 36–38.

Jenks, J. M., & Zevnik, B. L. P. (1989). ABCs of job interviewing. *Harvard Business Review, 67*(4), 38–39, 42.

Jesseph, S. A. (1989). Employee termination. II: Some dos and don'ts. *Personnel, 66*(2), 36–38.

Kavanaugh, R. R., & Andrews, M. (1988). Downsizing for the hospitality industry. *Cornell Hotel and Restaurant Administration Quarterly, 28*(4), 65–70.

Kohl, J. P., & Greenlaw, P. S. (1992). The ADA, part II: Implications for managers. *Labor Law Journal, 42*, 52–56.

Kohl, J. P., & Stephens, D. B. (1989). Wanted: Recruitment advertising that doesn't discriminate. *Personnel, 66*(2), 18–20, 22–26.

Kohl, J. P., & Stevens, D. B. (1989). Recruitment policies and practices of restaurants, hotels and clubs: A study of the personnel procedures of hospitality firms. *Hospitality Education and Research Journal, 13*(2), 45–50.

Koral, A. M. (1986). *Conducting the lawful employment interview: How to avoid charges of discrimination when interviewing job candidates* (rev. ed.). New York: Executive Enterprises Publications.

Krone, C., Tabacchi, M., & Farber, B. (1989a). A conceptual and empirical investigation of workplace burnout in foodservice management. *Hospitality Education and Research Journal, 13*(3), 83–91.

Krone, C., Tabacchi, M., & Farber, B. (1989b). Manager burnout. *Cornell Hotel and Restaurant Administration Quarterly, 30*(3), 58–63.

Lawler, E. E., III, & Mohrman, S. A. (1987). Unions and the new management. *Academy of Management Executive, 1*(3), 293–300.

Liberson, M. J. (1987). AIDS: A managerial perspective. *Cornell Hotel and Restaurant Administration Quarterly, 28*(3), 57–61.

Lorenzini, B. (1992). The accessible restaurant. Part II: Employee accommodation. *Restaurants & Institutions, 102*(12), 150–151, 154, 158, 162, 166, 168, 170.

Marvin, B. (1992). How to hire the right people. *Restaurants & Institutions, 102*(14), 60–61, 68, 73, 75, 77.

McCool, A. C., & Stevens, G. E. (1989). Older workers: Can they be motivated to seek employment in the hospitality industry? A preliminary report: Factors influencing older persons to retire or to work beyond the usual retirement age. *Hospitality Education and Research Journal, 13*(3), 569–572.

McLauchlin, A. (1992). Take note: ADA's employment rules take effect in July. *Restaurants USA, 12*(5), 9–10.

Muczyk, J. P., & Reimaon, B. C. (1989). MBO as a complement to effective leadership. *Academy of Management Executive, 3*(2), 131–138.

National Restaurant Association. (1987a). *A primer on how to recruit, hire, and retain employees.* Washington, DC: Author.

National Restaurant Association. (1987b). *A restaurateur's guide to the insurance crisis.* Current issues report. Washington, DC: Author.

National Restaurant Association. (1987c). *Substance abuse and employee assistance programs.* Current issues report. Washington, DC: Author.

National Restaurant Association. (1988). *A 1988 update: Foodservice and the labor shortage.* Current issues report. Washington, DC: Author.

National Restaurant Association. (1989a). *Foodservice employment 2000: Exemplary industry programs.* Current issues report. Washington, DC: Author.

National Restaurant Association. (1989b). *Model employment programs: Recruitment and retention of foodservice employees.* Current issues report. Washington, DC: Author.

National Restaurant Association. (1992). *Americans with Disabilities Act: Answers for foodservice operators.* Washington, DC: Author.

Puckett, R. P. (1982). Dietetics: Making or breaking the new employee. *Contemporary Administrator, 5*(10), 14–16.

Quick, R. C. (1989). Employee assistance programs: Beating alcoholism in the dish room and the board room. *Cornell Hotel and Restaurant Administration Quarterly, 29*(4), 63–69.

Riehle, H. (1993). Hourly wage survey for 1992. *Restaurants USA, 13*(1), 40–41.

Roseman, E. (1981). *Managing employee turnover: A positive approach*. New York: AMA-COM.

Rothman, M. (1989). *Employee termination. I: A four-step procedure*. Personnel, 66(2), 31–35.

Sherer, M. (1990). If you don't have a literacy program READ THIS NOW! *Restaurants & Institutions*, *100*(27), 75, 78, 82, 84, 88, 92.

Sherman, A. W., & Bohlander, G. W. (1992). *Managing human resources* (9th ed.). Cincinnati: South-Western Publishing.

Sloane, A. A., & Witney, F. (1988). *Labor relations* (6th ed.). Englewood Cliffs, NJ: Prentice-Hall.

Smart, B. D. (1989). *The smart interviewer*. New York: Wiley.

Smith, J. H., Jr. (1992). A study of handicapped employment in the hospitality industry. *Hospitality and Tourism Educator*, *4*(3), 16–25.

Spertzel, J. K. (1992). The Americans with Disabilities Act: Employment doors open a little wider for the disabled. *School Food Service Journal*, *46*(4), 36–37.

This, L., & Lippitt, G. (1966). Learning theories and training. *Training and Development Journal*, *20*(4), 2.

VanDyke, T., & Strick, S. (1988). New concepts to old topics: Employee recruitment, selection and retention. *Hospitality Education and Research Journal*, *12*(2), 347–360.

Weinstein, J. (1992a). Here's how smart restaurant companies turn training and development into . . . Personnel success. *Restaurants & Institutions*, *102*(29), 92–93, 96, 101, 104, 108, 113.

Weinstein, J. (1992b). Initial ADA complaints scarce. *Restaurants & Institutions*, *102*(25), 105–108.

What the new family and medical leave law will require. (1993). *Washington Weekly*, *13*(7), 2.

Wheelhouse, D. R. (1989). *Managing human resources in the hospitality industry*. East Lansing, MI: Educational Institute of the American Hotel and Motel Association.

Woods, R. H., & Kavanaugh, R. R. (1992). Here comes the ADA—are you ready? *Cornell Hotel and Restaurant Administration Quarterly*, *33*(1), 24–32.

Woods, R. H., & Macaulay, J. F. (1987). Exit interviews. How to turn a file filler into a management tool. *Cornell Hotel and Restaurant Administration Quarterly*, *28*(3), 38–46.

Zemke, R. (1989a). Employee orientation: A process, not a program. *Training*, *26*(8), 33–35, 37–38.

Zemke, R. (1989b). Workplace illiteracy. Shall we overcome? *Training*, 26(6), 33–39.

21

Management of Financial Resources

Effective management of financial resources is critical to the success of any foodservice operation. Control of costs is important in both profit and not-for-profit organizations. Foodservice managers must use accounting and financial management concepts to analyze financial performance. Preparation of financial statements and the interpretation of information in these statements are critical responsibilities of a foodservice manager. The computer can be a valuable tool to assist with the preparation and analysis of financial information.

USERS OF FINANCIAL STATEMENTS

Seven groups of users of financial statements have been identified in profit organizations: owners, boards of directors, managers, creditors, employees, governmental agencies, and financial analysts. Each user group has a different need for financial data. Owners have invested in the business, and their primary concern is the state of their investment. Boards of directors, elected to oversee operations and make business decisions for an organization, use financial statements to determine the effectiveness of managers supervising the daily operation. Managers are concerned with assessing the daily and long-term success of their decisions, and they use financial data to evaluate plans. Creditors, whose concern is the likelihood that payment obligations will be met, are those lending money or goods on credit to the operation. Employees have an interest in financial information to help assess the company's ability to meet wage and benefit demands. Governmental agencies are concerned with financial data as they relate to taxation and regulation. Financial analysts are persons outside the firm who desire information about a firm for their own or a client's purpose.

Not-for-profit organizations, which provide goods or services that fulfill a social need and are not operating for anyone's personal financial gain, have many of the same users of their financial information. In not-for-profit operations, such as hospitals and schools, excess revenues over expenditures are not distributed to those who contributed to support through taxes or voluntary donations but are used to further the purposes of the organization. In tax-supported organizations, the public is viewed as the owner, and its interest is the efficiency of the organization in using public monies. Boards of directors often oversee operations in not-for-profit organizations; they use financial statements to determine the effectiveness of operations. Suppliers are concerned with the ability of an organization to pay bills. Managers use financial statements to provide information for controlling expenditures in relation to available funds. Some not-for-profit organizations are established to provide services to a specific clientele for a fee that closely approximates the cost of providing the service. This clientele can be categorized as an additional user because of its concern about the organization's efficiency to assure that costs for services are maintained as low as possible.

BASIC FINANCIAL STATEMENTS

Primary financial statements used by foodservice managers are the balance sheet, income statement, and statement of change in financial position. Each provides unique information necessary for analyzing the effectiveness of operations.

The **balance sheet** is a statement of assets, liabilities or debts, and capital or owner's equity at a given time or at the end of the accounting period. The **income statement** is the financial report that presents the net income or profit of an organization for the accounting period. It provides information about the revenues and expenses that resulted in the net income or loss. The **statement of change in financial position** provides information about the source and use of funds. The balance sheet is considered a static statement because it presents the financial position at a specific date or time. The income and changes in financial position statements are considered *flow* or *dynamic statements* because operating results over time are presented.

Standard methods of accounting and presentation of financial statements, termed *Uniform Systems of Accounts,* have been established in a number of industries, including certain segments of the hospitality services industries. These uniform systems of accounts within a particular industry provide for the uniform classification, organization, and presentation of revenues, expenses, assets, liabilities, and equity. They include a standardized format for financial statements, which permits comparability of financial data within an industry. Financial statements can be prepared using computer spreadsheet or customized software programs.

Balance Sheet

The balance sheet, or statement of financial condition, is a list of assets, liabilities, and owner's equity of a business entity at a specific date, usually at the close of the last day of a month, quarter, or year. This statement is designated a *balance sheet* because it is based on a fundamental equation that shows that assets equal liabilities plus owner's equity. An example of a balance sheet for the Oakwood Country Club is shown in Figure 21.1.

Assets

The first section of the balance sheet is a list of assets, which are generally categorized as current or fixed. The example in Figure 21.1 shows three categories of assets for the Oakwood Country Club: current, fixed, and other assets. *Current assets* include cash and all assets that will be converted into cash in a short period of time, generally 1 year. The cash accounts are cash on hand and in checking accounts and cash in savings. Other current assets include accounts receivable, inventory, prepaid expenses, and entrance fees receivable. Any marketable securities held by the country club would have been included as a current asset.

Fixed, or *long-term, assets,* are those of a permanent nature, most of which are acquired to generate revenues for the business. Fixed assets are not intended for sale and include land, buildings, furniture, fixtures, and equipment, in addition to small equipment such as china, glassware, and silver. Because fixed assets generally lose value over their expected life, their value is reduced by a value termed *accumulated depreciation*.

Liabilities

Liabilities are categorized as current and long term. *Current liabilities* represent those that must be paid within a period of 1 year, including such items as accounts payable for merchandise, accrued expenses, and annual mortgage payment. Accrued expenses are due but not paid at the end of the accounting period, such as salaries, wages, or interest. *Fixed,* or *long-term, liabilities,* in contrast, are obligations that will not be paid within the current year. An example of a long-term liability is a mortgage for building and land; annual mortgage payments due during the current year

Figure 21.1. Example of a balance sheet.

Oakwood Country Club **BALANCE SHEET** December 31, 1994			
Assets		**Liabilities and Owner's Equity**	
CURRENT ASSETS		**CURRENT LIABILITIES**	
Cash		Accounts payable	$ 28,585
on hand and in checking	$ 14,271	Accrued expenses	8,590
in savings	5,521	Mortgage payable, current	24,000
Accounts receivable	67,278	Total Current Liabilities	$ 61,175
Inventory	5,330		
Prepaid expenses	5,688		
Entrance fees receivable	19,885		
Total Current Assets	$117,973		
FIXED ASSETS		**LONG-TERM LIABILITIES**	
Land	$375,000	Mortgage payable	$390,047
Clubhouse and structure	402,950	Less current portion	(24,000)
Furniture, fixtures, equipment	67,000	Total Liabilities	$366,047
China, glassware, silver	28,856		
Less depreciation	(117,000)		
Total Fixed Assets	$756,806		
OTHER ASSETS	$ 6,083	**OWNER'S EQUITY**	
		Capital stock	$ 47,000
		Additional paid in capital	365,503
		Retained earnings	41,137
		Total Owner's Equity	$453,640
TOTAL ASSETS	$880,862	**TOTAL LIABILITIES & OWNER'S EQUITY**	$880,862

are current liabilities and reduce the long-term mortgage liability, as shown in Figure 21.1.

Owner's Equity

The owner's equity or capital section of the balance sheet represents that portion of the business that is the ownership interest, as well as earnings retained in the business from operations. In profit-oriented enterprises, the ownership may be one of three kinds: a *proprietorship,* a business owned by a single individual; a *partnership*, a business owned by two or more people; or a *corporation,* a business incorporated under the laws of the state with ownership held by stockholders. In a not-for-profit corporation, the members may be the owners. As shown on the balance sheet from the

Oakwood Country Club (Figure 21.1), the owner's equity section includes the value of the members' capital stock, additional paid in capital, and the retained earnings.

Income Statement

The income statement (also known as the statement of income or the profit-and-loss statement) is a primary managerial tool reporting the revenues, expenses, and profit or loss as a result of operations for a period of time. In a not-for-profit organization, it is often called a statement of revenue and expense. An example of an income statement is shown in Figure 21.2.

Sales or revenues include the cash receipts or the funds allocated to the operation for the period. For example, in a college residence hall, revenues might include monies paid by residents for meal plans and cash sales to nonresidents who may eat in the foodservice facility.

In a foodservice establishment, the cost of sales section of the income statement reflects the cost of products sold that generated the revenue. Cost of sales often is calculated in the following manner:

Figure 21.2.   Example of an income statement.

The Corner Deli
INCOME STATEMENT
October 31, 1994

REVENUE		$100,000
COST OF SALES		37,000
GROSS PROFIT		63,000
CONTROLLABLE EXPENSES		
Salaries and wages	$27,000	
Employee benefits	5,500	
Direct operating expenses	3,200	
Marketing	1,950	
Energy and utility service	4,000	
Administrative and general	2,800	
Repairs and maintenance	1,800	
		46,250
INCOME BEFORE RENT, INTEREST, DEPRECIATION		16,750
RENT	7,200	
INTEREST	1,000	
DEPRECIATION	1,200	
		9,400
INCOME BEFORE TAXES		7,350
TAXES		1,000
NET INCOME		$6,350

Inventory at beginning of period	$ XXX
+ Purchases during the period	+ XXX
Total value of available food	$ XXX
– Inventory at end of period	– XXX
Cost of goods sold during period	$ XXX

Gross profit or income is determined by subtracting cost of goods sold from sales or revenue.

Net profit or loss is determined by subtracting expenses from gross profit. In a not-for-profit organization, net profit is often referred to as excess revenues over expenditures. Expense accounts include controllable expenses, rent, interest, depreciation, and taxes. Controllable expenses include such items as salaries and wages, employee benefits, marketing, and utilities in addition to the others shown in Figure 21.2.

Statement of Change in Financial Position

The basic objective of the statement of change in financial position (SCFP), or funds flow statement, is to explain changes in net working capital (funds) for a specified period of time. Net working capital is equal to current assets minus current liabilities. Common sources and uses of funds are identified in Table 21.1.

An example of the SCFP for the University Motel is shown in Figure 21.3. The SCFP shows funds provided internally from operations, externally from the sale of long-term assets such as equipment, and funds used for long-term investment activity like building additions, dividends, and mortgage payments.

TOOLS FOR COMPARISON AND ANALYSIS

The foodservice manager should use a variety of tools to analyze financial data, such as ratio analysis, trend analysis, common-size statements, and break-even analysis.

Table 21.1.   Sources and uses of funds

Sources	Uses
Assets	
• Decrease in fixed assets	• Increase in long-term assets
• Decrease in other assets	• Increase in other assets
Liabilities and Owner's Equity	
• Increase in long-term debt	• Decrease in long-term liabilities
• Increase in owner's equity	• Decrease in owner's equity
• Income from operations	

Figure 21.3.   Example of a statement of change in financial position.

University Motel
STATEMENT OF CHANGE IN FINANCIAL POSITION
Year ending December 31, 1994

Sources of Funds
Funds provided by operations

Net income	$100,000
Add expenses not requiring working capital	
Depreciation	150,000
Amortization	20,000
Proceeds from sale of equipment	10,000
Total Sources	280,000

Uses of Funds

Building addition	155,000
Dividends declared	20,000
Mortgage payable, reclassified	32,000
Total Uses	207,000
Net Increase in Working Capital	**$ 73,000**

The resulting operational indicators help managers understand financial information and compare performance to earlier periods. The manager should make several kinds of comparisons in reviewing and analyzing financial information. These comparisons can be categorized as internal and external. Managers can complete these comparisons more efficiently when using a computer to perform the calculations and comparisons.

Standards of Comparison

Internal standards of comparisons include a review of current performance in relation to budgeted performance, past performance, or preestablished department standards. Because the budget represents the plan for financial operations for a period of time, comparisons with the budget indicate whether or not operations are proceeding as planned. Comparison of current performance in relation to past performance provides information on changes occurring in the operation over time. Having preestablished department standards provides objective benchmarks for comparison of actual performance with expected performance.

External standards of comparison include a review of performance in relation to similar operations or comparisons with industry performance. For example, in multi-unit operations, managers may wish to compare the food cost of unit A with that of units B and C, or they may use industry data published annually by the National Restaurant Association.

Ratio Analysis

Most financial information is presented as a collection of totals or balances of accounts, the meaning or significance of which may not be readily apparent. A **ratio analysis**, or analysis of financial data in terms of relationships, facilitates interpretation and understanding. Computation of various types of financial ratios is an important tool in analysis.

A *ratio* is a mathematical expression of the relationship between two items that may be expressed in several ways:

- *As a common ratio.* The ratio between *x* and *y* may be stated as *x to y (x:y).* For example, an operation that has 12 full-time employees and 4 part-time employees would have a full-time to part-time employee ratio of 3 to 1 (3:1).
- *As a percentage.* The ratio may be expressed as a percentage, such as a percentage of sales.
- *As a turnover.* Some relationships are best expressed as a turnover or the number of times *x* must be "turned over" to yield the value of *y,* calculated by dividing *y* by *x.* The number of times inventory is turned over in a month is one example.
- *On a per unit basis.* The relationship may also be expressed in dollars per unit, such as sales dollars per stool at a restaurant counter.

Ratios frequently are categorized according to primary use. The major categories include the following:

- Liquidity
- Solvency
- Activity
- Profitability
- Operating

Examples of each of these major types are listed in Table 21.2.

Liquidity Ratios

Liquidity ratios indicate the organization's ability to meet current obligations—in other words, its ability to pay bills when due. An organization may be making a profit but have insufficient cash to pay current bills. Several ratios are used to analyze the liquidity of a business, two of the most common being the current ratio and the acid-test ratio. Data from the balance sheet are used in both calculations. The current ratio is the relationship between current assets to current liabilities and is computed as follows:

$$\text{Current ratio} = \frac{\text{Current assets}}{\text{Current liabilities}}$$

Creditors may consider a high current ratio as assurance they will receive payment for products and services, but managers may wish to maintain a lower current ratio by avoiding excess buildup of cash or inventories.

Table 21.2.   Ratios categorized by primary use

Types	Examples
Liquidity ratios	Current ratio Acid-test ratio
Solvency ratios	Solvency ratio Debt to equity ratio Debt to assets ratio
Activity ratios	Inventory turnover Occupancy percentage
Profitability ratios	Profit margin Return on equity Return on assets
Operating ratios	Analysis of the revenue mix Average customer check Food cost percentage Labor cost percentage Food cost per customer Meals per labor hour Meals per full-time equivalent Labor minutes per meal

The *acid-test ratio,* also called the *quick ratio,* is another comparison of current assets and current liabilities, but it yields a more accurate measure of bill-paying capability. Current liabilities are measured against cash and other assets readily convertible to cash, such as accounts receivable and marketable securities. The acid-test ratio is calculated as follows:

$$\text{Acid-test ratio} = \frac{\text{Cash} + \text{Accounts receivable} + \text{Marketable securities}}{\text{Current liabilities}}$$

Solvency Ratios

Solvency ratios are used to examine an establishment's ability to meet its long-term financial obligations and its financial leverage. The basic solvency ratio is the relationship between total assets and total liabilities, calculated as follows:

$$\text{Solvency ratio} = \frac{\text{Total assets}}{\text{Total liabilities}}$$

Other solvency ratios examine the relationship between liabilities and equity (debt to equity ratio) and between liabilities and assets (debt to asset ratio).

Activity Ratios

Activity ratios are designed to examine how effectively an organization is utilizing its assets. These ratios usually are expressed as either percentages or turnovers. The *inventory turnover ratio,* one of the most widely used activity ratios, shows the number of times the inventory is used up and replenished during a period. Inventory turnover is calculated as follows:

$$\text{Inventory turnover} = \frac{\text{Cost of goods sold}}{\text{Average inventory value}}$$

A high ratio indicates that a limited inventory is being maintained by a foodservice organization, and a low ratio indicates that larger amounts of money are tied up in inventories.

Percentage of occupancy is another activity ratio important in many segments of the industry. Hotels, hospitals, college or university residence halls, and nursing homes compute occupancy percentages regularly. These percentages indicate the relationship between the number of beds or rooms available and the number being used by the clientele of the operation:

$$\text{Percentage of occupancy} = \frac{\text{No. of beds/rooms occupied}}{\text{No. of beds/rooms available}}$$

Profitability Ratios

Profitability ratios measure the ability of an organization to generate profit in relation to sales or the investment in assets. Profit or net income is an absolute term expressed as a monetary amount of income remaining after all expenses have been deducted from income or revenue; profitability is a relative measure of the profit-making ability of an organization. Most foodservice operations, whether profit or not-for-profit, must generate some level of net income, as discussed earlier in this chapter; disposition of profit is a primary difference between these two types of operation. Also, the not-for-profit organization may have a lower expectation with regard to level of profit. Three major profitability ratios are utilized:

$$\text{Profit margin} = \frac{\text{Net profit}}{\text{Sales}}$$

$$\text{Return on equity} = \frac{\text{Net profit}}{\text{Equity}}$$

$$\text{Return on assets} = \frac{\text{Net profit}}{\text{Total assets}}$$

The *profit margin* is the most commonly used measure of operating profitability; it uses information from the income statement to assess overall financial efficiency. The *return on equity* measures the adequacy of profits in providing a return on owners' investments; the *return on assets* is a measure of management's ability to generate a return on the assets employed in generating revenue. Both calculations combine data from the income statement and balance sheet to assess financial efficiency.

Operating Ratios

Operating ratios are primarily concerned with analysis of the success of the operation in generating revenues and in controlling expenses. *Analysis of the revenue mix* of the operation is one type of sales analysis. For example, a restaurant may analyze its food and beverage sales to determine the relative proportion of revenues generated by these two aspects of the operation; percentage of revenues generated by food and beverage sales would be calculated by dividing the sales in each area by total sales. In a student union, the relative percentage of sales from catering, vending, and cafeteria operations may be of interest and similarly would be calculated by dividing the sales from each area by the total sales.

The *average customer check* is another measurement of generation of sales dollars; it is calculated by dividing total sales by the number of customer checks. Sales per seat in a restaurant, per meal, per waitress, and per menu item are other sales ratios that may be utilized in a foodservice establishment.

As stated initially in this chapter, control of costs is a primary responsibility of foodservice managers; their organizations' profitability is strongly related to their effectiveness in such control. In foodservice organizations, one of the main control areas is the level of various categories of expenses in relation to sales. For example, the *food cost percentage,* the cost of food divided by total sales, and the *labor cost percentage,* the cost of labor divided by total sales, are performance indicators commonly scrutinized by foodservice managers.

A percentage breakdown of the food dollar spent on various food categories also can be calculated by dividing the cost of food for each category by the total spent on food. Such analysis may reveal possible cost problems. If an analysis of expenditures indicates that dollars spent on meat have increased for two successive months, for example, the manager may need to examine menu offerings, purchase prices, inventory control procedures, production methods, portion controls, or selling prices.

Another performance indicator is *food cost per customer* (or per patient, or student, or the like), which often is used for analyzing food cost in healthcare or university foodservices. This indicator is determined by dividing total food cost by the number of customers (patients/students) served.

Also, various types of ratios are computed to analyze labor cost:

$$\text{Meals per labor hour} = \frac{\text{Total number of meals served}}{\text{Total labor hours to produce meals}}$$

$$\text{Meals per full-time equivalents} = \frac{\text{Total number of meals served}}{\text{Total full-time equivalents to produce meals}}$$

$$\text{Labor minutes per meal} = \frac{\text{Total labor minutes to produce meals}}{\text{Total number of meals served}}$$

Using Ratio Analysis

Ratio analysis constitutes an effective tool for evaluating financial stability and operating effectiveness, providing managers with the information they need to make decisions and control operations. Ratios selected for analysis of financial results must be appropriate to the operation, however, and the relationships measured must be understood by the foodservice manager.

Managers should use a variety of ratios in analyzing operations because a single ratio is insufficient for sound decision making. Ratios should be used in combination with trend analysis to compare changes over a period of time. Consistency in accounting methods, however, is required to yield comparable data for comparisons over a period of time.

Trend Analysis

Trend analysis is a comparison of results over several periods of time; changes may be noted in either absolute amounts or percentages. It also is used to forecast future revenues or levels of activity. Trend analysis may utilize several of the various types of ratios discussed in the preceding section.

Often a graphic analysis of the financial data over a period of time will help managers detect and understand changes. For example, many foodservice managers regularly plot on a graph the percentage of sales spent on food, labor, and other operating expenses. The percentage is depicted on the vertical axis and the period of time (days, months, years) on the horizontal axis.

Common-Size Statements

Comparison among financial statements for various periods or from different departments within an organization may differ because of varying levels of volume. If the financial data are expressed as percentages, however, meaningful comparisons can be made because data have a common base.

Financial statements in which data are expressed as percentages are called **common-size statements**. These are especially useful in comparing results of the income statement of an operation from one accounting period to another or for comparing results among units of a multiunit operation, such as a chain restaurant or foodservice centers in a residence hall. To prepare a common-size income statement, each figure on the statement is calculated as a percentage of total sales by dividing each figure by the total sales. A common-size balance sheet sets total assets equal to 100% and individual asset categories as percentages of the total 100%. Similarly, total liabilities and owner's equity are taken as 100% and individual categories become percentages of that 100%. Comparisons with industry performance are facilitated by expressing the financial statements in percentages.

Break-Even Analysis

Break-even analysis is another tool for analyzing financial data. The name of this technique is derived from the term *break-even point*, or the point at which an operation is just breaking even financially, making no profit but incurring no loss. In other words, total revenues equal total expenses.

Break-even analysis requires classification of costs into fixed and variable components. *Fixed costs* (FC) are those costs required for an operation to exist, even if it produces nothing. These costs do not vary with changes in the volume of sales but stay fixed, or constant, within a range of sales volume. The size of the existing physical plant and the equipment capacity define this volume. If either is expanded, however, fixed costs will increase. Insurance, rent, and property taxes are examples of fixed costs.

Variable costs (VC) are those costs that change in direct proportion to the volume of sales. As the volume of sales increases, a proportionately higher amount of these costs is incurred, as with direct materials or food cost.

A complication in defining fixed and variable costs is that some costs cannot be clearly classified as either entirely fixed or entirely variable. These costs are termed *semivariable* because a portion of the total cost will remain fixed regardless of changes in sales volume, and a portion will vary directly with changes in sales volume. Such costs often include labor, maintenance, and utilities. Managers need to divide these semivariable costs into their fixed and variable components before they can utilize break-even analysis for decision making. Once semivariable costs have been divided into their fixed and variable portions, the break-even point can be calculated by the following formula:

$$\text{Break-even point} = \frac{\text{Fixed cost}}{1 - \dfrac{\text{Variable cost}}{\text{Sales}}}$$

To illustrate, the following data are available for XYZ cafeteria:

$$
\begin{aligned}
\text{Fixed cost} &= \$ 28,000 \\
\text{Variable cost} &= \$ 60,000 \\
\text{Sales} &= \$100,000
\end{aligned}
$$

Substituting the figures in the formula, the break-even point for XYZ cafeteria is calculated as shown below:

$$\text{Break-even point} = \frac{\$28,000}{1 - \dfrac{\$60,000}{\$100,000}}$$

$$= \frac{\$28,000}{1 - .60}$$

$$= \frac{\$28,000}{.40}$$

$$= \$70,000$$

Figure 21.4. Break-even chart for XYZ cafeteria.

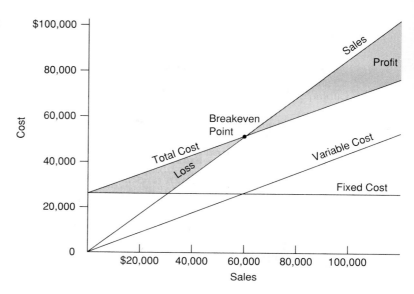

The denominator in this formula, or the ratio of variable cost to sales subtracted from 1, is referred to as the *contribution margin.* It represents the proportion of sales that can contribute to fixed costs and profit after variable costs have been covered. In the illustration, the contribution margin is 0.40 or 40%. The break-even point for XYZ cafeteria is shown graphically in Figure 21.4.

Break-even analysis is a tool for projecting income and expense as well as profit under several assumed conditions. It can assist the foodservice manager in understanding the interrelationships among volume, cost, and profits, but it requires that the operational costs of an organization be known and that they can be segmented into fixed and variable classifications. An expansion of this break-even concept is known as cost-volume-profit (CVP) analysis. Using the CVP formula, a manager can determine the volume required for a given level of profit. The CVP formula is as follows:

$$\text{Sales or revenue level} = \frac{\text{Fixed cost} + \text{Profit desired}}{1 - \dfrac{\text{Variable cost}}{\text{Sales}}}$$

BUDGETING

A **budget** is a plan for operating a business expressed in financial terms or a plan to control expenses and profit in relation to sales. Budgeting is the process of budget planning, preparation, control, reporting, utilization, and related procedures. Budget control involves the use of budgets and performance reports throughout the planned period to coordinate, evaluate, and control operations in accordance with the goals specified in the budget plan.

In a governmental operation, however, the budget represents something different: a schedule of authorizations for a given period or purpose and the proposed means of financing it. Funds are established, representing a sum of money or other resources segregated for carrying on specific activities or attaining certain objectives in accordance with specific regulations, restrictions, and limitations. These funds constitute an independent fiscal entity.

Budgeting provides an organized procedure for planning and for development of standards of performance in numerical terms. Planning, coordination, and control are the three primary objectives of budgeting. A well-constructed operating budget shows the overall structure of the operation and enables staff to visualize its place in it.

Budgets provide a basis for control, but they must be implemented by operational personnel within the organization. Budgets are a tool for periodic examination, restatement, and establishment of guidelines. A budget is a plan that should be well constructed and capable of attainment. To be effective, all staff members responsible for implementing the budget should have input.

Steps in Budget Planning

The first step in the budgeting process is the development of the *sales budget*. Various internal and external influences must be considered in constructing the sales budget, which includes an estimate of revenues expected during the budget period. Past experience should be examined, including both past performance and past budgets, and any unusual aspects clarified.

Changes in pricing also must be considered. For example, if board rates are increased for a college residence hall, the proportion allocated per student for the foodservice operation should be reviewed. National and local indicators also should be considered, such as new competitors, other new businesses in the area, industry trends, and changes in the national economy.

The profit objective of the organization should be considered. The projected profit from a given volume of sales should be estimated and the ability to attain the profit objective determined. Seasonal variations in sales also should be projected, and segmentation of sales in various units of the organization forecast. For example, in a hotel, the sales forecast for room service, catering, a coffee shop, and a dining room may be appropriate.

The second step in the budgeting process is development of the *expenditure budget*. Expenditures in relation to the projected level of revenues should be estimated for food, labor, and other operational expenses. In estimating food costs, anticipated changes in food prices should be considered.

Increments in salaries and wages and in payroll taxes are key elements in estimating the *labor budget*. Also, the foodservice manager must project the impact of menu changes or other operational changes that may affect the labor needed in the operations. Similar factors should be considered in projecting expenditures for other operating expenses.

The third step in the budget process involves development of the *cash budget*, which is a detailed estimate of anticipated cash receipts and disbursements throughout the budget period. The cash budget will assist management in coordinating cash inflow and outflow and in synchronizing cash resources with need. Seasonal effects on cash position thus become apparent and the availability of cash for taking advantage of discounts is identified. Also, the cash budget assists management in planning financial requirements for large payments, such as tax installments or insurance premiums, and indicates availability of funds for short-term investments.

The development of the *capital expenditure budget* is the fourth step in the budget planning process. Improvements, expansions, and replacements in building, equipment, and land are the major capital expenditures. These major capital investments may be for purposes of expanding or improving facilities.

Careful planning is particularly important in capital budgeting because of the magnitude of funds and the long-term commitments involved. Capital expenditures may be prorated over several budget periods. A number of sophisticated techniques are available for evaluating capital budgeting alternatives such as net present value (NPV) and internal rate of return (IRR). Discussion of these concepts is beyond the scope of this book.

The last or fifth step in the budget planning process is the compilation of a *forecast income statement*, often termed a *pro forma statement,* which is a composite of the sales and expenditure budgets and includes a projection of profit. The budget planner should compute various ratios to determine if the appropriate relationship of expenditures to sales or to revenues has been projected.

The basic steps of the budget process outlined in the foregoing paragraphs may vary in different organizations, and the foodservice manager may be involved in only certain aspects of budget planning. For example, in a very large organization, the cash budget and compilation of the total budget are often the responsibility of the controller and other top-management staff. Rather than compile the capital budget, the foodservice manager may develop only capital budget requests, which are rank ordered according to departmental priorities. In a small operation, however, the manager in consultation with the accountant is usually involved in the total process.

After the budget planning process has been completed, various reviews and approvals are needed before the budget document is finalized; these vary widely, depending on the size and type of the organization. The approved budget then becomes the standard against which operations are evaluated. Periodic reports are issued comparing operating results with budget estimates. The manager's job becomes one of using these comparative reports to bring about operational changes conforming to budget plans unless, however, data suggest that budget amendments are in order. An example of the budget for a not-for-profit club, approved by the board of directors, is shown in Figure 21.5. In this example, a net loss is anticipated because of some planned renovations and improvements to the club's facilities. By projecting this deficit, the board of directors can make financial decisions with it in mind.

Budgeting Concepts

Several budgeting concepts are used, the most common being the *fixed* or *forecast budget.* A budget is prepared at one level of sales or revenues. *Incremental budget-*

Figure 21.5.   Example of a budget.

City Club PROPOSED BUDGET 1994–1995		
Estimated Receipts		
Participating club dues	$ 23,000	
Annual member dues	244,000	
Profit from foodservice	35,000	
Donations	4,000	
Special events	5,000	
Interest income	3,000	
Miscellaneous income	7,000	
Total Estimated Receipts		$321,000
Estimated Expenditures		
Salaries	$ 58,300	
FICA taxes	3,000	
Note and interest payments	87,000	
Utilities	32,000	
Repairs and maintenance	2,700	
Office supplies and postage	3,500	
Insurance	21,000	
Lease on lights	1,080	
Dishes	4,000	
Newsletter	5,750	
Special events	7,900	
Accounting fee	1,750	
Christmas bonus	1,850	
Flowers and decorating	7,000	
Upholstering furniture	3,500	
Remodel kitchen	40,000	
Carpets	36,000	
Miscellaneous expenses	8,670	
Total Estimated Expenses		325,000
Estimated Net Loss		($ 4,000)

ing involves using the existing budget as a base and projecting changes for the ensuing year in relation to the current budget.

The *flexible budget* is adjusted to various levels of operation or sales useful for operations with varying sales or revenues throughout the year. A foodservice organization may develop budgets at two or three possible levels of sales to assist in adjusting expenditures to actual sales volume. For example, resort hotels and hospitals have wide variations in occupancy rates. A flexible budget will enable the management of these organizations to adjust expenditures in relation to occupancy levels.

SUMMARY

Effective management of financial resources is critical to the success of a foodservice organization. Foodservice managers must understand key accounting and financial management concepts to analyze financial performance. Accounting is an information processing activity designed to provide quantitative data on the operation of an organization.

The primary financial statements generated by the accounting system are the balance sheet, income statement, and statement of change in financial position. The balance sheet is a statement of assets, liabilities, and owner's equity at a given time, and the income statement is the financial report that presents the net income or profit for the accounting period. The statement of change in financial position is an explanation of the changes in net working capital for a specified time.

The foodservice manager should use a variety of tools to analyze financial data, including standards of comparison, ratio analysis, trend analysis, common-size statements, and break-even analysis. Standards of comparison can be either internal or external. In ratio analysis, financial data are presented in terms of relationships, which facilitate interpretation and understanding. Trend analysis is a comparison of results over several periods of time. Common-size financial statements are those in which data are expressed as percentages. Break-even analysis is another tool for analyzing financial data and refers to the point at which an operation is breaking even financially, making no profit but incurring no loss.

A budget is a plan expressed in financial terms for operating a business; budget control involves the use of budgets to coordinate, evaluate, and control operations in accordance with the goals specified in the budget plan. The steps in the budgeting process include development of the following budgets: sales, expenditure, cash, and capital expenditures. The last step is the compilation of a forecast income statement, which is a composite of the sales and expenditure budgets, and includes a projection of profit.

BIBLIOGRAPHY

Anthony, R. N., & Young, D. W. (1988). *Management control in nonprofit organizations* (4th ed.). Homewood, IL: Richard D. Irwin.

Baker, M. M. (1988). Meeting the challenge in foodservice management: Enhanced quality at less cost. *Journal of The American Dietetic Association, 88*, 441–442.

Barwise, P., Marsh, P. R., & Wensley, R. (1989). Must finance and strategy clash? *Harvard Business Review, 67*(5), 85–90.

Coltman, M. M. (1991). *Hospitality management accounting* (4th ed.). New York: Van Nostrand Reinhold.

DeMicco, F. J., Dempsey, S. J., Galer, F. F., & Baker, M. (1988). Participative budgeting and participant motivation: A review of the literature. *FIU Hospitality Review, 6*(1), 77–94.

DeYoung, R., & Gregoire, M. B. (1993). Use of capital budgeting techniques by foodservice directors in for-profit and not-for-profit hospitals. *Journal of The American Dietetic Association, 93*, 67–69.

Dittmer, P. R., & Griffin, G. G. (1989). *Principles of food, beverage, and labor cost controls for hotels and restaurants* (4th ed.). New York: Van Nostrand Reinhold.

Educational Institute of the American Hotel and Motel Association. (1987). *A uniform system of accounts and expenses dictionary for small hotels, motels, and motor hotels* (4th ed.). East Lansing, MI: Author.

Everett, M. D. (1989). Managerial accounting systems: A decision-making tool. *Cornell Hotel and Restaurant Administration Quarterly, 30*(1), 46–51.

Fromm, B., Moore, A. N., & Hoover, L. W. (1980). Computer-generated fiscal reports for food cost accounting. *Journal of The American Dietetic Association, 77*, 170–174.

Horngren, C. T., & Foster, G. (1991). *Cost accounting: A managerial emphasis* (7th ed.). Englewood Cliffs, NJ: Prentice-Hall.

Hotel Association of New York City. (1986). *A uniform system of accounts for hotels* (8th ed.). New York: Author.

Keiser, J., & DeMicco, F. J. (1993). *Controlling and analyzing costs in food service operations* (3rd ed.). New York: Macmillan.

Kwansa, F., & Evans, M. R. (1988). Financial management in the context of the organizational life cycle. *Hospitality Education and Research Journal, 12*(2), 197–214.

Moncarz, E. S. (1988a). Understanding annual reports of hospitality firms. *FIU Hospitality Review, 6*(1), 32–42.

Moncarz, E. S. (1988b). Understanding annual reports of hospitality firms: Part II. *FIU Hospitality Review, 6*(1), 32–42.

Moncarz, E. S., & O'Brien, W. G. (1990). The powerful and versatile spreadsheet. *The Bottom Line, 5*(4), 16–21.

National Restaurant Association. (1990). *Uniform system of accounts for restaurants* (3rd ed.). Washington, DC: Author in cooperation with Laventhol & Horwath.

Rose, J. (Ed.). (1984). *Handbook for healthcare foodservice management*. Rockville, MD: Aspen Publishers.

Schmidgall, R. S. (1990). *Hospitality industry managerial accounting* (2nd ed.). East Lansing, MI: Educational Institute of the American Hotel and Motel Association.

Schmidgall, R. S., Geller, A. N., & Ilvento, C. (1993). Financial analysis using the statement of cash flows. *Cornell Hotel and Restaurant Administration Quarterly, 34*(1), 47–53.

Schmidgall, R. S., & Ninemeier, J. D. (1986). Food-service budgeting: How the chains do it. *Cornell Hotel and Restaurant Administration Quarterly, 26*(4), 51–55.

Sneed, J., & Kresse, K. H. (1989). *Understanding foodservice financial management*. Rockville, MD: Aspen Publishers.

Stinson, J. P., & Guley, H. M. (1988). Use of the microcomputer to determine direct costs of menu items. *Journal of The American Dietetic Association, 88*, 586–590.

Implementing TQM Philosophy at The Fairfax

Organizations are in a constant state of change, and foodservice organizations are no different from others in that respect. Many large industries have hired total quality management (TQM) consultants to convert organizations from being traditional with many levels of management and a rigid chain of command to a new type in which middle management is being phased out and replaced by associates who work as teams in making decisions. In addition to understanding the structure and changes in the foodservice organization, knowledge of human and financial resources is necessary to solve the case problem.

Background Information

Marriott is one of the largest employers in America and has a work force of more than 200,000. It has been widely recognized for its successful human resources programs. Marriott regards its employees, which it calls associates, as internal customers and listens to them. The corporation is committed to the 7 Habits of Highly Effective People, a concept developed by the Covey Leadership Center, which has its basis in the Golden Rule, Do unto others as you would have them do unto you. Among recent honors, Marriott has been named one of the best places to work by *Working Mother, Black Enterprise* and *Hispanic* magazines. Approximately 40% of its employees are minorities and over half are female, with management made up of approximately 15% minority and 40% female.

The Marriott Corporation declared in its 1991 annual report that the TQM philosophy is being adopted in all its properties. A 4-hour session on TQM was presented by the executive vice president for the Marriott Service Division in the recent *Continuous Improvement—Continuous Learning & Change* conference for senior living services. (See the case overview following chapter 4.) Many foodservice operations are changing their traditional management structures by examining human and financial resources to determine if their organization charts reflect what is happening.

Human and Financial Resources

Human resources management is an extension rather than a rejection of personnel management and is important in a TQM organization. Because of the value Marriott places on human resources management, it changed its terminology from *employees* to *associates* at all its properties a few years ago. Team effort in producing menu items to satisfy customers was the beginning of human resources management in many foodservice operations including the F&B Department in The Fairfax Senior Living Services community. The F&B director is ultimately responsible for the menu, but he becomes a team member along with the chef, cooks, purchasing agent, financial advisor, consulting dietitian, dietary manager, waitstaff, and residents. Each member of the team has expertise and experience in his or her area and is empowered to make decisions that lead to customer satisfaction. The F&B director admits that empowering associates has been difficult for him to accept because he developed his management style in the 1980s, when top management made decisions and everyone in the organization abided by them. He admits, however, that he is becoming very comfortable with the TQM philosophy and that associates make more realistic decisions as a team than he could ever make as a boss. An outcome of empowerment is the daily staff meeting to discuss current problems for all associates in the F&B Department.

The F&B director has made disciplinary action a positive experience for foodservice associates. He has instigated what he calls coaching and counseling sessions for associates whose performance does not meet expectations. Instead of condemning associates for poor performance, he concentrates on helping them improve. Most associates appreciate the director's handling of the situation and, therefore, improve their performance.

Financial accountability is required if the organization is to survive. Most of the retirement communities are owned and managed by profit-making organizations. Satisfying the external customer can make the difference between staying in or going out of business. Good financial records, such as food and labor costs, are essential in a TQM organization.

Organization Charts

Figure 1 is the organization chart for top-level management positions in the Marriott Corporation. J. W. Marriott, Jr., the chairman, president, and chief executive officer, represents the board of directors. Three executive vice presidents—lodging, finance, and service divisions—and one vice chairman of government affairs report directly to him. Note that Senior Living Services, which has 18 communities, is in the Marriott service division. Five other vice presidents—general counsel, architecture and construction, lodging, human resources, and corporate relations—serve as counsels to Mr. Marriott.

Each community consists of a number of departments, which have their own functions. Figure 2 shows the nine departments in The Fairfax; each department has a director or manager, assistant managers, and associates. Figure 3 is the organization chart for The Fairfax F&B Department, in which the dining room manager, executive chef, and healthcare foodservice manager are at the second level and the F&B director at the top of the hierarchy. Even though the TQM philosophy is starting to be practiced, the concept is not apparent in the organization chart.

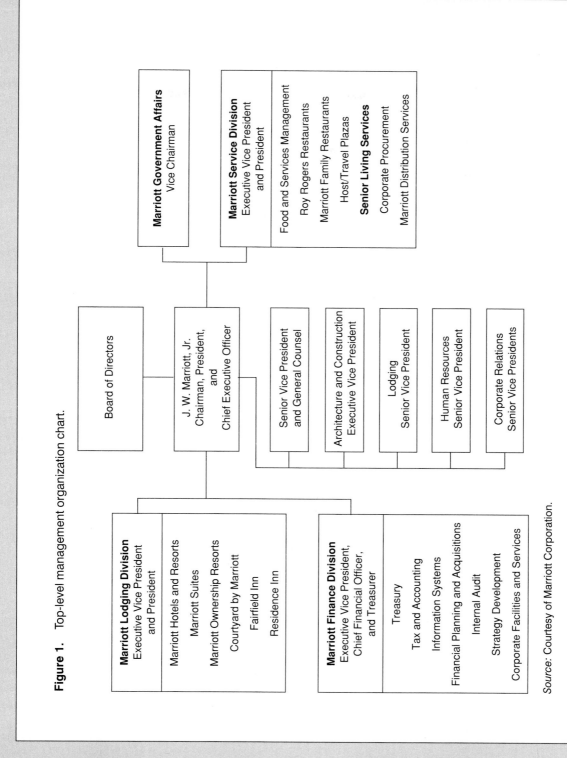

Figure 1. Top-level management organization chart.

Source: Courtesy of Marriott Corporation.

The chart contents (organized top-level management):

Marriott Government Affairs
Vice Chairman

Marriott Service Division
Executive Vice President and President
- Food and Services Management
- Roy Rogers Restaurants
- Marriott Family Restaurants
- Host/Travel Plazas

Senior Living Services
- Corporate Procurement
- Marriott Distribution Services

Board of Directors

J. W. Marriott, Jr.
Chairman, President, and Chief Executive Officer

Senior Vice President and General Counsel

Architecture and Construction
Executive Vice President

Lodging
Senior Vice President

Human Resources
Senior Vice President

Corporate Relations
Senior Vice Presidents

Marriott Lodging Division
Executive Vice President and President
- Marriott Hotels and Resorts
- Marriott Suites
- Marriott Ownership Resorts
- Courtyard by Marriott
- Fairfield Inn
- Residence Inn

Marriott Finance Division
Executive Vice President, Chief Financial Officer, and Treasurer
- Treasury
- Tax and Accounting
- Information Systems
- Financial Planning and Acquisitions
- Internal Audit
- Strategy Development
- Corporate Facilities and Services

Figure 2. Department organization chart for The Fairfax and Belvoir Woods Health Care Center.

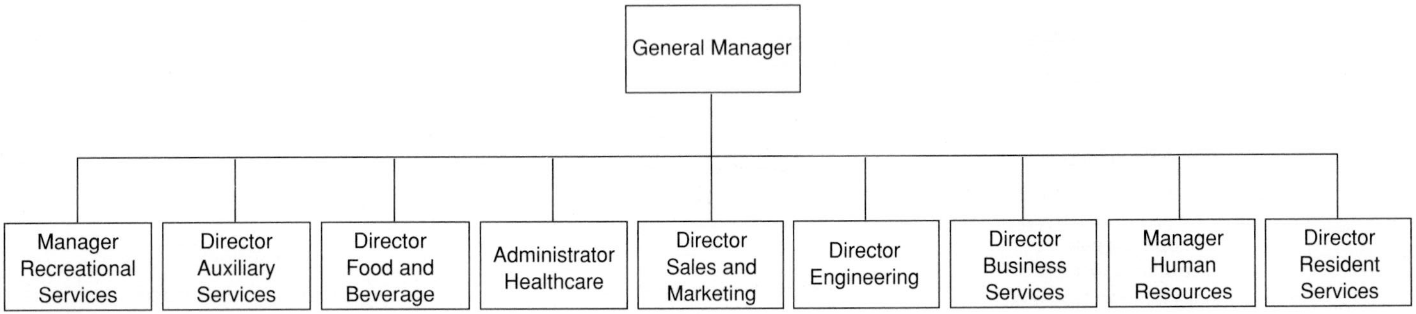

Source: Courtesy of Marriott Corporation.

Figure 3. Organization chart for the Food and Beverage Department at The Fairfax.

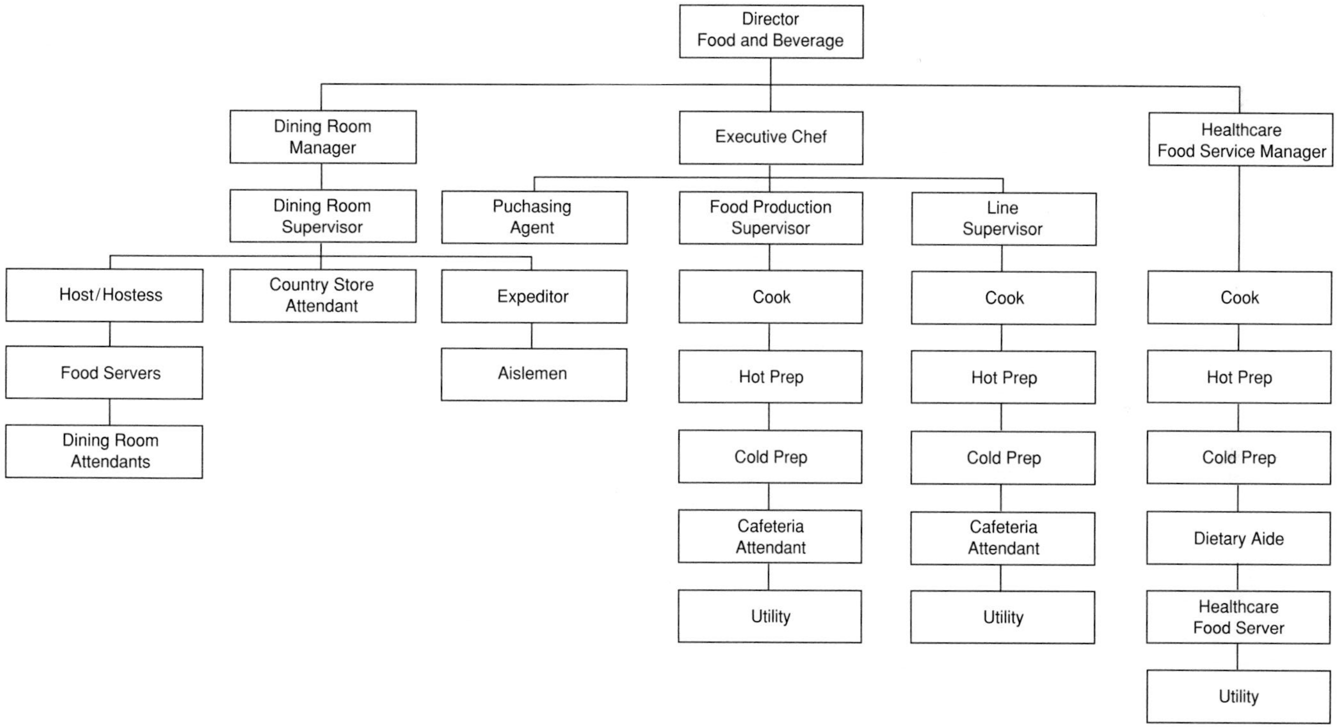

Source: Courtesy of Marriott Corporation.

The Problem

J. W. Marriott, Jr., at the Board of Directors' suggestion, has sent out requests to consulting companies to submit bids on changing the Marriott Corporation to a TQM operation. After studying the organization charts, he was concerned that TQM was given only lip service but was not being practiced because he saw no evidence of the concept in the charts. With the background information and an in-depth analysis of the management of the community, develop a strategic plan with time lines to convert the current Marriott organization into a TQM one. As one of the competing consultants, a significant part of your plan would be to revise the organization charts (Figures 1, 2, and 3) to make them reflect what actually will happen in the organization when management changes are made.

Points for Discussion

- If you were J. W. Marriott, Jr., how would you sell the idea of a TQM organization to top-level management?
- What are the criteria for a TQM organization?
- How will resistance to change affect the strategic plan?
- What plans will have to be made to protect people in positions no longer needed?
- How can you help associates realize they are a very important part of The Fairfax retirement community?

Appendix A: Resources for Writing Specifications

GOVERNMENT PUBLICATIONS

Institutional Meat Purchase Specifications (IMPS). Available from: Livestock and Meat Standardization Branch, USDA, AMS, LS Room 2603-S, P.O. Box 96456, Washington, DC 20090-6456, (202) 720-4486.

- General Requirements ($2.00)
- Quality Assurance Provisions ($5.00)
- Fresh Beef (Series 100) ($3.00)
- Fresh Lamb and Mutton (Series 200) ($2.00)
- Fresh Veal and Calf (Series 300) ($2.00)
- Fresh Pork (Series 400) ($2.00)
- Cured, Cured and Smoked, and Fully Cooked Pork Products (Series 500) ($2.00)
- Cured, Dried and Smoked Beef Products (Series 600) ($2.00)
- Variety Meats and Edible By-Products (Series 700) ($2.00)
- Sausage Products (Series 800) ($2.00)
- IMPS Set—Consists of Series 100–800, General Requirements, and Quality Assurance Provisions ($15.00)

Manuals for Food Service Supervisors (1992). These manuals can be purchased from: Food Information Service Center, 21050 S.W. 93rd Lane Road, Dunnellon, FL 32630, (904) 489-8919.

- Volume 1: *Catalog of Specifications*
- Volume 2: *Contract Purchasing*
- Volume 3: *Food Facts*
- Volume 4: *Directory of Information Sources*
- Volume 5: *Storage and Care of Food Products*
- Volume 6: *Purchasing French Fry Potatoes*
- Volume 7: *USDA Donated Foods Program*
- Volume 8: *Guidelines for Food Purchasing and Meal Cost Management* (designed for schools and institutions)
- Volume 9: *Guidelines for Food Purchasing and Meal Cost Management* (designed for use with nutrition programs for the elderly)
- Volume 10: *Food Identifications and Standards*

U.S. Department of Agriculture, Food and Nutrition Service. (October 1985). *Directory of Information Sources for Food Products and Markets*. Technical Assistance Manual Series. Volume IV.

BOOKS

The Almanac of the Canning, Freezing, Preserving Industries. (Annual). Westminister, MD: Edward E. Judge & Sons.

Dietary Managers Association. (1991). *Professional Procurement Practices: A Guide for Dietary Managers*. Lombard, IL: Author. (Developed by J. D. Ninemeier.)

Kotschevar, L. H., & Donnelly, R. (1994). *Quantity Food Purchasing* (4th ed.). New York: Macmillan.

National Association of Meat Purveyors. (January 1992). *The Meat Buyers Guide*. Reston, VA: Author.

Peddersen, R. B. (1981). *Foodservice and Hotel Purchasing*. New York: Van Nostrand Reinhold.

Produce Marketing Association. (1994). *Foodservice Guide to Fresh Produce, 1994*. (Available from the Produce Marketing Association. P.O. Box 6036, Newark, Delaware, 19714-6036. $5.95.)

Reed, L. S. (1993). *SPECS: The Comprehensive Foodservice Purchasing and Specification Manual*. New York: Van Nostrand Reinhold.

Stefanelli, J. M. (1992). *Purchasing: Selection and Procurement for the Hospitality Industry* (3rd ed.). New York: John Wiley.

Virts, W. B. (1987). *Purchasing for Hospitality Operations*. East Lansing, MI: The Educational Institute of the American Hotel and Motel Association.

Warfel, M. C., & Cremer, M. L. (1990). *Purchasing for Food Service Managers* (2nd ed.). Berkeley, CA: McCutchen Publishing.

Appendix B: Standards for Food Products

BEVERAGES

Cocoa   Cocoa or hot chocolate should be a rich brown color. Flavor should be delicately sweet with the rich, aromatic flavor of cocoa predominant. Cocoa powder should be well mixed in the milk with no visible flecks of undissolved cocoa. Cocoa should have a cooked starch flavor, be free of skin, and be slightly thicker than milk (Kotschevar, 1988).

Coffee   Brewed coffee, no more than 30 minutes old, should be evaluated for flavor including taste and aroma, which should be balanced and sufficiently strong to give a pleasing flavor. Taste should be properly balanced between bitterness (astringency), acidity, and sweetness. Aroma should be fragrant, mellow, heavy, and rich with coffee bouquet. Brewed coffee should be bright and clear and the color rich and deep brown. It should be full bodied and sensed in the mouth as denser than water. Coffee should be served at 160° to 200°F (Kotschevar, 1988). Iced coffee should have the same characteristics, but the original brew should be stronger to permit dilution by ice.

Tea   Tea is judged by flavor, strength, clarity, and color. Tea generally has a slightly bitter flavor caused by tannin, some sweetness, and acidity. Aroma should be fruity and fragrant. The term *brisk* often is used to describe the zestful, stimulating quality of tea. The type of tea determines the color of the brew. For example, green tea produces a delicate, greenish yellow pale liquor and rather fruity flavor; oolong is darker, less bitter, still fruity, but has a softer flavor than green tea; black tea is a copper-colored brew with a soft, mild flavor, but slightly acidic (Kotschevar & Donnelly, 1994). Iced tea should have the same characteristics, but the original brew should be stronger to permit dilution by ice.

BREADS, QUICK

Biscuits, Baking Powder   A baked biscuit should be round in shape with vertical sides and a fairly smooth level top. Size should be two to three times that of an unbaked biscuit. Crust should be light brown in color with some evidence of flour. Interior can be peeled in thin layers and should be creamy white with a fine even-grained cell structure and a moist, fluffy crumb. Flavor should be mild with a slight taste of baking powder.

Biscuits, Cinnamon Raisin   Addition of cinnamon, sugar, and plumped raisins to the baking powder biscuit dough yields a light brown biscuit with a spicy, sweet taste. Flecks of

735

brown cinnamon are evident in the biscuits. Raisins should be whole, moist, and evenly distributed. Cell structure is slightly more compact than in a plain biscuit. A thin powdered sugar glaze on the baked biscuit gives additional sweetness and a smooth top.

Bread, Corn   Crust is crisp, shiny, pebbly, and slightly brown on top with darker sides and bottom. Interior should have a coarse, moist, light, and tender crumb and be free of tunneling. When cut, slight crumbling may occur. Color should be characteristic of type of cornmeal used, that is, white, yellow, or blue. Flavor should have a well-balanced corn taste. Sweetness depends on regional preferences.

Bread, Loaf   The muffin batter bread should be loaf-shaped with an even, medium brown crust, and a crack down the middle. Interior grain, from coarse to very fine, depends on the formulation. The bread should be free from tunnels and have a moist crumb. Flavor should be slightly sweet, due to the sweetener, and mild with a characteristic flavor of any additions (e.g., bananas, cranberries, nuts, dates, spices). Color depends on added ingredients, which should be evenly dispersed and identifiable.

Cake Doughnuts   Exterior should have a rich, golden brown, crisp crust and an inner core that resembles a baked product more than a fried food. Interior should be pale yellow. The doughnut should be evenly rounded with a slightly wrinkled or star-shaped center hole. Crumb should be fine, tender, and moist with uniform-sized cells. Volume should be large in proportion to the weight of the doughnut. Flavor should be delicate and well-balanced. Glaze or coating, if used, should be light to enhance but not dominate the flavor.

Coffee Cake   Golden brown crust appears sugary or shiny and may be slightly crisp. Top may be uneven depending on topping. Cell size is slightly larger than in a butter cake, yet smaller than in a muffin. Flavor, aroma, and color should be characteristic of ingredients, especially additional ones, such as, extracts, fruit, nuts, and spices. Any addition should be evenly dispersed and identifiable. Coffee cake should be served warm because freshness is lost rather rapidly.

BREADS, YEAST

Bread, French   Exterior surface should be relatively thin, hard and crusty, and an even, golden brown color with three to five diagonal slashes across the top. The loaf shape should be a cylinder, a minimum of 12 inches long and 3 inches in diameter. Interior is snowy white with a moist, soft, elastic crumb. Cell walls are uniformly thin and of varying sizes. Air pockets are desirable. Flavor is bland with slight yeast emphasis.

Bread, Plain   The bread should have a symmetrical loaf shape with a golden brown top crust and uniformly brown sides, ends, and bottom. Crust should be tender and easily cut. Interior texture should be fine, even grained, free from large air bubbles, and have oval-shaped cells with thin walls. The creamy white crumb should be moist, silky, and slightly resistant to bite with no trace of being doughy. Flavor should be pleasant, nutlike, and have a hint of yeast. Density and color of bread will vary depending on type of flour used (such as cracked wheat, rye, white). Color also is dependent on added ingredients (for example, brown sugar, molasses, raisins, spices), which should be evenly dispersed.

Bread Sticks   Characteristics are identical to those of plain bread. Outer surface should be golden brown and crisp. Sticks should be approximately 5 inches in length and 1 inch in diameter.

Croissants Exterior surface is golden brown, flaky, and tender. Interior is very moist, light, and flaky and easily peels off in thin sheets. Croissants have a buttery flavor and color. Shapes vary from crescent to round to straight, depending on use.

Kolache Characteristics are identical to those of sweet rolls with added flavor of grated lemon peel. Kolaches are round with an indentation in the middle filled with prune or apricot, cottage or cream cheese, or poppy seed filling. Crust is golden brown, and filling is moist but solid. Top is covered with sifted powdered sugar.

Rolls, Cinnamon The rolls should have characteristics of sweet rolls and good volume in proportion to weight. Shape is distinctive owing to basic preparation from a sheet of dough sprinkled with cinnamon and sugar, rolled, and cut in slices of uniform thickness. Rolls should have a golden brown crust and are generally topped with a powdered sugar glaze. Flavor is sweet, wheaty, mild, and enhanced with cinnamon.

Rolls, Dinner Characteristics are identical to those of loaf bread. Size and shape may vary. Top crust should be golden brown, smooth, and round, and the bottom light brown.

Rolls, Pan Characteristics are similar to those of dinner rolls. Pan rolls should be brown on top and bottom and also on the sides of the rolls placed on the edge of the baking pan. Other sides resemble the interior of the rolls.

Rolls, Sweet Characteristics are similar to loaf bread except for the tinge of yellow egg color in the interior, moister crumb, and sweeter and more full-bodied flavor. Addition of other ingredients like fruit, nuts, and spices enhances the flavor. Glazes, icings, powdered sugar add to flavor and appearance of the rolls.

Rolls, Sticky Pecan Characteristics are similar to cinnamon rolls except for the baked-on brown sugar syrup topping with bite-sized pieces of pecans. The soft syrup topping should be shiny, clear, and smooth. Pecans should be evenly dispersed over the rolls and in the interior.

CEREAL, PASTA, RICE

Cereal, Cooked Color of the cereal depends on the color of the grain—tan for oatmeal, white for farina, and yellow for cornmeal. Cooked cereal should be tender with a cooked starch flavor. Consistency should be thick, and texture, smooth. Cooked cereal should be served hot.

Rice, Boiled, Steamed, Baked Rice should be pure white, and each grain should be intact and separate. Brown rice has a light brown color. Rice grains should be tender but firm and have a bland, cooked starch flavor.

Pasta, Cooked Pasta has an ivory color or a color characteristic of added ingredients— green from spinach, pink from tomatoes, black from squid, golden yellow from corn or saffron, medium yellow from egg yolks, light tan from whole wheat flour. Cooked pasta should be firm to the bite, al dente, and have a bland, cooked starch flavor that blends with various sauces.

Pasta Sauce, Creamy The white sauce base should be smooth and mild with a cooked starch flavor. If cream is one of the ingredients, the rich flavor is evident. Addition of chicken broth, garlic, herbs should blend with the white sauce. Ingredients, including cheese, poultry, seafood, and vegetables, should be identifiable and add characteristic flavor and color to the sauce.

Pasta Sauce, Light   Sauce is light in color and almost clear. Basil, garlic, olive oil, and parsley are the predominant flavors and enhance the mild-flavored pasta. Addition of herbs and freshly grated Parmesan cheese adds piquancy to the finished product. Minced clams add a characteristic flavor and chewy texture to the sauce.

Pasta Sauce, Red   The sauce should be an orange-red tomato color. Consistency should be sufficiently thick to preclude running when served over hot pasta. The tomato flavor should be predominant but enhanced with chopped garlic, green pepper, onions, and herbs. Browned ground meat or meatballs can be cooked in the sauce, giving a meaty flavor.

DESSERTS

Cake

Angel food   Rough, light brown crust is dry to the touch. Interior is snowy white. Crumb should be soft, moist, light, feathery, and slightly elastic. Cell size is somewhat uneven and cell walls are uniformly thin. Flavor should be mildly sweet and well-balanced and have a cooked egg white flavor.

Butter   Crust should be an even golden brown with a dull matte finish. Top should be slightly rounded, puffy, and free of cracks. Light yellow interior has a soft, moist, tender, velvety crumb and very small cells with uniformly thin walls. Flavor is sweet, mild, and buttery. Flavor and color depend on added ingredients (for example, brown sugar, chocolate, spices).

Carrot   Crust is soft, and browning is barely noticeable. Interior is a cinnamon brown color and is very moist and compact. Grated raw carrots are difficult to detect but add body and moisture to the product. Chopped nuts and cinnamon flavor are identifiable. The cake usually is topped with a mild-flavored cream cheese icing, which blends with the positive cinnamon flavor.

Chocolate   Characteristics except for color and crust are identical to those of a butter cake. Color may vary from reddish to deep brown. Browning of the crust is barely noticeable. Flavor intensity depends on type of chocolate, which may vary from mild sweet to strong bittersweet.

Chiffon   Rough, light brown crust is dry to the touch. Texture should be delicate, springy, and light. Small, fine, uniform cells with thin walls are desirable. Crumb should be moist and soft. Color is a pale yellow and flavor is mild, but both may be changed by additional ingredients and coloring and flavoring agents, such as chocolate, maraschino cherry juice, spices.

Icing   Cream icing should be smooth and easy to spread. It should be fluffy, yet maintain its shape, and adhere to the cake. Color and flavor depend on ingredients. Fat content for cake icings ranges from 10% to 25% to permit ease in cutting.

Pound   The loaf cake should have a light brown crust with a narrow split down the middle of the slightly rounded top. Interior is yellow with a smooth crumb. Texture is firm with fine, compact cells. Well-blended flavor is buttery and sweet.

Cookies

Bar   Bars should be approximately 3-by-2½ inches with a uniform height of 1/2 to 3/4 inches. Top and bottom crusts should be uniformly medium brown. Cut edges should be straight. Texture may vary from chewy to cakelike. Any additions to the bars (e.g., chocolate chips, coconut, dried fruits, nuts, spices) should be evenly dispersed and identifiable. Flavor should be rich, sweet, and characteristic of additions. A topping of glaze, icing, or powdered sugar may be added.

Brownie   Characteristics are similar to those of bar cookies. Exterior is dark brown and smooth. Interior is darker than the exterior. Brownie is tender and slightly moist with a moderately chewy texture free of crumbling. Nuts should be evenly dispersed and identifiable. A topping of icing may be added.

Drop   Shape is symmetrically round, approximately 3 inches in diameter. Height may vary from 1/4 to 1/2 inch depending on whether the cookie is chewy or cakelike. Top and bottom should be evenly browned with no evidence of sugar crystals unless added for specific varieties (e.g., sugar cookies). Flavor should be rich, sweet, and characteristic of any additions (e.g., butterscotch or chocolate chips, chocolate chunks, coconut, dried fruits, nuts, oatmeal, spices), which should be evenly dispersed and identifiable. Cookies may be topped with a glaze, icing, or powdered sugar.

Pie

Cream   The filling should be opaque, creamy, smooth, and free of lumps. Filling should retain its shape when cooled and sag only slightly at the cut edge. Quantity of filling should be adequate to fill the shell. Flavor and color are characteristic of the additional ingredients and coloring and flavoring agents, for example, banana, butterscotch, chocolate, coconut, vanilla. Cream pies must be refrigerated immediately after preparation.

Custard   Top surface film should be yellow with flecks of brown nutmeg. Pale yellow filling should be creamy, smooth, shiny, and tender. Gel should be uniformly coagulated and free of porosity. Egg and nutmeg flavors predominate. The pie should have a crisp crust with an adequate amount of filling without overflow. Custard pies must be served chilled.

Fruit   Thickened juice should be clear, bright, smooth, and have a slight cooked starch flavor. Each piece of bite-sized fruit should be distinct and identifiable. Filling should sag only slightly at the cut edge. Flavor is moderately sweet and characteristic of the specific fruit— apple, apricot, blueberry, cherry, gooseberry, peach, pineapple, and rhubarb.

Lemon   The filling should be a translucent gel with a color characteristic of lemon rind. Filling should be viscous and retain its shape when cut. Flavor is a pleasing blend of tart and sweet. Quantity of filling should be adequate to fill the crisp shell.

Meringue   Top is a pale golden color, and interior is snowy white. Meringue should be tender, fluffy, moist, and have a sweet flavor. It should adhere to and completely cover the pie filling, meeting the crust at the edge. Beads of liquid or weeping should not be present.

Pumpkin   Color should be orangish brown with noticeable flecks of spices. Filling should be viscous and retain its shape when cut. Texture should be slightly coarse. Flavor is a mild blend of pumpkin with pleasing notes of cinnamon, cloves, ginger, and nutmeg.

Shell, Graham Cracker   Shell should retain the shape of the pie pan. Shell should be crisp with no evidence of sugar crystals. Flavor is buttery, sweet, and characteristic of graham crackers.

Pastry Crust should be thin. Top surface should be rough and blistery with a soft luster. Center of the top crust should be light, golden brown with color deepening slightly toward the edges. Texture of both top and bottom crusts should be crisp, flaky, and fork tender, but not crumbly. Crust should cut easily. Flavor depends on the amount of salt and type of fat. Top crust may be brushed with egg wash and sprinkled with granulated sugar before baking.

Other

Cheese Cake The top is covered with a creamy white sour cream mixture, and sides and bottom are the brown color of a graham cracker shell. Interior has the smooth consistency of cream cheese and is uniformly coagulated. Cheese cake is pale yellow and has a very mild cheese, egg, and vanilla flavor. Variations in color and flavor depend on additional ingredients (e.g., chocolate, lemon, mint, pumpkin, spices). The baked cheese cake often is topped with a cherry, lemon, raspberry, or strawberry glaze.

Cream Pudding Pudding should be glossy, creamy smooth, and free of lumps and surface skin. It should mound slightly when dipped with a spoon. Color and flavor depend on added ingredients and coloring and flavoring agents (e.g., butterscotch, chocolate, vanilla).

Fruit Crisp A crunchy topping should be evenly distributed over a slightly thickened and sweetened fruit mixture (e.g., apples, blueberries, sour cherries). Topping should be delicate brown and have a coarse, crumbly consistency with a flavor characteristic of oatmeal, brown sugar, and butter. Fruit should be identifiable by color and flavor.

EGGS

Fried Appearance should be bright, glossy, and compact with rounded edges. A sunny-side-up egg should have a bright yellow or orange-yellow, well-rounded yolk. If the egg is turned over, the yolk is covered with a thin film of coagulated white. The egg should be shiny and soft underneath. Coagulated areas should be firm, yet tender (Kotschevar, 1988).

Omelet Bottom is a delicate light brown, and the top is pale yellow. The omelet should be moist and tender and have a delicate flavor. Omelets generally are filled with various ingredients (such as diced tomatoes, sautéed mushrooms, shredded cheese, small cubes of ham). The folded or rolled shape should be even and uniform (Kotschevar, 1988).

Poached The egg should have some shiny white adhering closely to the bright yellow yolk. It should be compact with smooth edges, rather than spread out. Coagulated areas should be slightly firm, yet tender (Kotschevar, 1988).

Scrambled Eggs should be bright and clear with a soft sheen and have no evidence of browning. The uniform pale yellow eggs should be in small pieces that are tender and moist. Flavor should be delicate and mild.

Soft or Hard Cooked Appearance should be bright and fresh with the colors natural and clear. Yolk of a soft-cooked egg should be warm throughout but liquid with one half to three fourths of the white coagulated, depending on degree of doneness. A hard-cooked egg has a solid, mealy yolk that is bright yellow or orange-yellow and is uniformly coagulated. White of a hard-cooked egg is firm and glossy (Kotschevar, 1988).

ENTRÉES, ACCOMPANIMENTS

Entrées

Chicken Fried Steak   Exterior should be golden brown with a crisply fried, well-seasoned, batter coating. The tenderized beef cutlet interior should be moist. Flavor should be characteristic of fresh beef. It is typically served with a milk gravy made from meat drippings.

Chicken Tetrazzini   Melted orange-yellow or grated Parmesan cheese tops spaghetti and chicken chunks in a pale yellow chicken sauce with identifiable ingredients—chopped green pepper, onion, parsley, pimiento, and sliced mushrooms. Spaghetti is tender but al dente with a cooked starch flavor. Sauce blends with the mild-flavored spaghetti. A portion flows slightly when spooned on a plate.

Chili con Carne   Color of coarsely ground meat, tomato, and bean mixture should be a reddish brown. Well-blended flavor should be characteristic of ingredients and seasonings, that is, chili powder, cumin, garlic, onion. Consistency when served should be thick.

Fajita   Color of the meat or poultry strips in the filling is light brown, contrasted with sautéed red and green pepper strips and sliced onions. Tender strips of meat or poultry are flavored with various marinades. Filling typically is rolled into a steamed flour tortilla and served with condiments, such as cheese, guacamole, salsa, sour cream.

Grilled Ground Beef Patty (Hamburger)   Exterior should be an even dark brown color and glossy. Interior should reflect desired degree of doneness ranging from slightly pink to light brown. Patty should be moist and tender with characteristic beef flavor. Fat content of cooked patty should reflect raw meat specification for lean/fat ratio. Patty should be served immediately.

Lasagna   Layers of ingredients should be identifiable: ribbon-edged lasagna noodles, pale yellow Parmesan and mozzarella cheese, white ricotta or cottage cheese, and tomato red sauce with flecks of browned meat and dark green herbs. Noodles are tender but al dente; the cheese has a soft, smooth consistency; and the sauce is thick. Lasagna should have a spicy tomato beef flavor blended with the mild mozzarella and white cheese and the piquancy of Parmesan. A lasagna serving has sharply cut edges and retains its shape.

Macaroni and Cheese   Macaroni should be al dente but tender and have a cooked starch flavor. Sauce should have a medium thick consistency with an American or cheddar cheese flavor. Each piece of the macaroni should be completely coated with the light yellow creamy sauce. A portion should spread slightly when spooned on a plate.

Meat Loaf   Ground meat mixture should have a dark brown, thin crust with an even brown interior. Meat loaf should be moist, yet firm enough to retain its shape when sliced. Flavor should be characteristic of ingredients (beef base, catsup, dried onion soup, garlic, onions).

Pizza   Color of baked pizza should be characteristic of ingredients. Yeast leavened, slightly blistered crust may be thin and crisp or thick and chewy, and should be well browned on the bottom and edges. Typically, sauce is thick and tomato based with well-blended seasonings of basil, fennel, garlic, and oregano, for example. Toppings may include any combination of cheese (usually mozzarella), fruit, meat, seafood, and vegetables. Portions can be individual round, wedge shaped, or square, depending on shape of pan.

Stew   Fork-tender, bite-sized meat or poultry cubes should be of uniform size and lightly browned. Bite-sized vegetables should be firm, yet tender, while maintaining characteristic flavor and shape. Meat or poultry and vegetables are covered with smooth, mild-flavored natural gravy of medium consistency. A variation is pot pie, in which stew is covered with a rich pastry or drop biscuits and baked until golden brown.

Tuna and Noodles   Lightly browned melted orange-yellow cheese tops ivory-colored noodles and tuna chunks in a chicken-based white sauce. Noodles should be tender but al dente and have a cooked starch flavor. White sauce should be smooth and mild enhanced by the characteristic flavor of tuna. Chopped celery and onion add crunchiness and also flavor. A portion should have an overall semi-solid consistency and flow slightly on a plate.

Accompaniments

Bread Dressing   Baked dressing should be light in weight for volume and tender. Moistness should complement the meat or poultry, dry with rich sauces and moist with dry meats. It should hold together when served. Chopped celery, onion, and seasoning should blend with the flavor of the meat or poultry. Day-old white bread is the major ingredient, but cornbread also is popular. Color of the dressing is characteristic of type of bread.

Fried Apples   Apple rings, approximately 1/2-inch thick, should be tender, glossy, and lightly browned. Tart apples should have caramelized or brown sugar and butter flavor. Apple flavor should predominate with only enough sweetness to complement the entrée.

Fried Rice   Color and flavor are characteristic of soy sauce, which blends well with the bland steamed rice. Chopped onions, frozen peas, shredded carrots, small pieces of tender but firm scrambled eggs are identifiable and give additional color and flavor to the rice.

Fritters   Crisp crust is golden brown with a yellow interior that resembles a baked product. Fritters usually are spherical in shape. Flavor is slightly sweet and mild and complements characteristic flavors of additional ingredients (such as apples, bananas, corn, peaches, or pineapple).

Rice Pilaf   Rice is pale yellow and tender with a cooked starch flavor. Grains are light, fluffy, and remain separate. Finely chopped onions and margarine or butter enhance the bland flavor of the rice in addition to the chicken broth and seasonings. Flavor also may be enhanced by additional ingredients (such as green peppers, mushrooms, pimiento, and water chestnuts).

FISH, MEAT, POULTRY

Fish, Cooked   Color of cooked fish should be reminiscent of the raw state (i.e., white for orange roughy, pinkish orange for salmon, pale tan for swordfish). Flavor and odor should be mild and characteristic of fresh fish. Flesh should be moist and should flake easily. Seasonings and sauces enhance the mild flavor of fish.

Meat, Cooked   Meat should have a pleasing appearance when cooked. Meat should be moist and drippings rich and of good color. Flavor and color should be natural to type of meat, style of cooking, and degree of doneness. Beef and lamb roasts should be well browned; veal should have a reddish brown surface, and pork should have a uniformly rich brown surface. Outer surfaces of lamb and pork roasts should be crisp. Sliced roasted meat should be firm, juicy, tender, and retain its shape. Broiled meats should have a rich brown color with well-developed flavor and aroma. Braised meat should be tender and juicy with a rich, glossy surface. Pieces of braised meat should be uniform, even, and symmetrical (Kotschevar, 1988).

Poultry, Cooked Poultry flesh should be moist, and drippings pale yellow to light brown. Flavor and color should be natural to type of poultry and style of cooking. Skin on roasted poultry should be crisp. Sliced roasted poultry should be firm, juicy, tender, and retain its shape. Flavor of simmered poultry is delicate and the color ivory. Flesh should be tender and moist and retain its shape. Color of the stock should be pale yellow.

Seafood, Cooked Color should be typical of seafood type (pink exterior and white interior for shrimp, lobster, and crab; gray oysters; white scallops). Shellfish should be free of sand, tender, and moist. Flavor should be mild, characteristic of type, and blend well with seasonings and sauces.

SALADS

Fruit Fresh fruits should be ripe but firm and selected for variety in flavor, texture, and color. Flavors should be zestful and well blended. Color should be clear, fresh, and natural. Fruit that oxidizes quickly should be dipped into lemon or other juices to prevent darkening. Fruit should be identifiable and neatly cut into bite-sized pieces. The leafy greens used as underliners should be chilled and crisp. Dressing for fruit salads is usually slightly sweet and flavored with fruit juices or honey. Canned fruit should be well drained before using in salads.

Molded Commercial dessert gelatins generally are used as the base. Gelatin should have bright sparkle, clear color, good setting ability, and a true, pleasing flavor. It should be firm enough to hold its shape, especially when cut, yet tender, quivery, and transparent. Ingredients should complement the slightly sweet, natural, fruit flavor of the gelatin. Ingredients should be cut into bite-sized pieces and evenly distributed through the gelatin, not exceeding 35% of total recipe weight. Underliner greens should be cold and crisp.

Tossed Greens should be distinct, bite-sized, neatly cut or torn, and of good proportion. More than one type of greens should be used to give variety in texture, color, and flavor (e.g., curly endive, escarole, iceberg or leaf lettuce, radicchio, romaine, spinach). Greens should be chilled, crisp, and dry. Other ingredients also may be added for interest (e.g., avocado, broccoli florets, carrot shreds, mushroom slices, tomato wedges, radish slices). Dressing and seasoning, added just before serving, should complement ingredients. Garnishes will give additional flavor and crispness to the salad (e.g., alfalfa sprouts, crisp bacon, seasoned croutons, sunflower seeds, toasted sesame seeds, toasted wheat germ).

Vegetable Raw and cooked vegetables should be selected for flavor, texture, and color. Flavors should be zestful, pleasing, and complement each other. Color should be clear, fresh, and natural. Raw vegetables should be crisp and cooked tender and firm. Vegetables should be neatly cut into bite-size pieces. Marinating in a piquant dressing adds flavor.

SANDWICHES

Cold

Cheese, Meat, Poultry   Bread should be fresh, firm, and moist with a close, smooth crumb, and good flavor. Bread should cut with a clean edge. Meat, poultry, or cheese should be thinly sliced or shaved. Lettuce, sliced pickles, and thin slices of tomato add crispness and flavor.

Soft Filling   Bread should be fresh, firm, and moist with a close, smooth crumb, and good flavor. Chicken, egg, ham, tuna, and other fillings should be identifiable, easy to spread, but firm enough to retain shape. Chopped celery, green pepper, and pickles add color and crispness to the filling. Salad dressing should coat all ingredients and serve as a binding agent. Various seasonings such as lemon juice for tuna and dry mustard for eggs complement the mild flavor of basic ingredients. Bread should cut with a clean edge. Sandwich should stay together when picked up, without losing filling.

Hot

Bacon, Lettuce, and Tomato   Bread should be fresh, firm, and moist with a close, smooth crumb, and good flavor. Golden brown toast should be hot and bacon crisp and hot. Lettuce should be cold and crisp, and the thinly sliced tomato cold. Toast and filling should cut with a clean edge. Mayonnaise usually is served separately.

Barbecued Meat   Bread or roll should be fresh, firm, and moist with a close, smooth crumb, and good flavor. Meat should be tender, thinly sliced, and covered with a reddish brown, spicy barbecue sauce. Flavor of meat should be characteristic of type, and flavor of sauce should complement it.

Cheeseburger   Round hamburger bun should be freshly baked and have a golden brown exterior and a soft white interior. Grilled hamburger should have a fresh beef flavor and be juicy and tender with a brown and glossy exterior. It should be completely covered with melted cheese.

Grilled Cheese   Bread should be fresh, firm, and moist with a close, smooth crumb, and good flavor. Grilled bread should be crisp and have a buttery flavor. Cheese should be kept within limits of the bread and completely melted.

Roast Beef   Sandwich may be either closed or open-face and consists of one or two slices of bread, roast beef, and gravy. Beef is juicy, thinly sliced, hot, and the flavor characteristic. Gravy, often served separately, is either au jus or thickened meat juice.

SOUPS

Hot

Bean   The slightly thick soup is light tan with small cubes of pink ham. Beans and ham should be very tender but identifiable. Ham flavor is predominant and blends well with the mild navy or Great Northern beans. Seasonings of black pepper, chopped celery, onion enhance the flavor.

Beef Rice Transparent brown beef stock has body and minute globules of fat on the surface. Each grain of rice should be separate. Chopped celery and onions add characteristic flavor to the broth. Chicken rice soup, pale yellow in color, is a variation.

Bisque Bisque is a variation of a cream soup and contains chopped shellfish, such as crab, lobster, or shrimp. The slightly thick soup is a pale orange color. Flavor is characteristic of type of seafood, cream, and various seasonings, quite often dry sherry wine.

Bouillon Bouillon is a clear, light soup, with a rich meat, poultry, or vegetable flavor. Color depends on the basic ingredient. Surface may have a few minute globules of fat.

Cheese Thin white-sauce-based soup is the orange-yellow color characteristic of cheddar cheese. Chopped onions and finely diced carrots and celery are tender but crisp and enhance the flavor of the chicken stock and cheese. Consistency of soup is similar to moderately heavy cream and lightly coats a spoon.

Clam Chowder, Manhattan Chowder should be a tomato red color. Chopped clams and small pieces of sautéed onion and golden brown salt pork cracklings or bacon should be identifiable and give a characteristic flavor. Chopped potatoes and celery and diced carrots should be tender but firm. Flavor should be spicy but not overpowering, using bay leaves, black pepper, thyme, Worcestershire sauce.

Clam Chowder, New England The chowder should be a light cream color with the consistency of heavy cream. Chopped clams and small pieces of lightly sautéed chopped onion and golden brown salt pork or bacon should be identifiable and give a characteristic flavor. One-half-inch cubes of potato should be tender but firm.

Consommé Consommé is a clear, concentrated broth, pale yellow or light brown with a subtle chicken, fish, game, meat, or vegetable flavor. Surface may have a few minute globules of fat.

Cream Color and flavor depend on type of mashed, strained, or finely chopped meat, poultry, or vegetables added to a white sauce base. Consistency should be similar to moderately heavy cream and thick enough to delicately coat a spoon.

French Onion Dark brown beef broth and thin slices of sautéed onions are topped with a slice of toasted French bread, completely covered with melted Swiss and Parmesan cheeses. Soup has a slightly thick consistency and is a blend of strong and spicy flavors, such as black pepper, sautéed onion, Worcestershire sauce.

Oriental Chicken Mildly flavored and transparent stock should be pale yellow and the color of vegetables distinct and bright. Tender chicken should be in one-half-inch cubes and green onions in one-eighth-inch slices. Other vegetables (e.g., celery, mushrooms, spinach) should be crisp and in bite-sized pieces. Thin noodles or chicken wontons should be tender.

Potato The creamy white soup is thickened primarily with pureed potatoes. Chopped celery and onions blend well with the mild flavor of the potatoes and milk. Butter and cream may be added to give a rich flavor. Seasonings, especially white pepper, enhance the flavor.

Split Pea Soup is a green pea color with small cubes of pink ham. Split peas should be very soft or pureed and the ham tender. Characteristic flavor of ham is predominant and blends with the mild-flavored peas. Chopped carrots, onions, and potatoes are identifiable and add flavor to the soup. Variations include black beans, lentils, yellow split peas.

Vegetable Beef   Soup should have a reddish brown color with chopped or one-half-inch cubes of beef and numerous identifiable vegetables. Soup traditionally has chopped celery, onions, and tomatoes and cubed carrots and potatoes, which are tender and firm with a characteristic flavor.

Cold

Beet Borscht   Chilled soup is a beet-red color topped with a dollop of sour cream. Ingredients are chopped, and the flavor is a combination of beets, celery, onions in a consommé stock. Sour flavor is derived from lemon juice or red wine vinegar and sour cream.

Fruit   Color and predominant flavor of the chilled soup depend on type of fruit or berry puree (i.e., blueberry, peach, raspberry, strawberry). Flavor is slightly sweet with a piquancy provided by lemon juice and a dollop of sour cream. Consistency is smooth and slightly thick.

Gazpacho   Chilled soup is a tomato-red color with chopped pieces of celery, cucumbers, green peppers, parsley, sautéed mushrooms, and onion. Characteristic flavor of tomato is predominant and blends well with the other ingredients. Sour flavor is derived from red wine vinegar and spicy flavor from Worcestershire and tabasco sauces.

Vichyssoise   Chilled soup is creamy white and topped with chopped chives or parsley. Flavor is a blend of chicken broth, white potatoes, and onions. Texture is creamy smooth.

VEGETABLES

General

Deep-fat Fried   Batter coating or breading should be golden brown and crisp and the vegetable soft but not saturated with oil. Vegetables most often deep-fat fried are eggplant, mushrooms, okra, onions, and zucchini squash. Cauliflower, carrots, parsnips, and sweet potatoes are steamed before frying. Fat or oil should be bland. Dipping sauces enhance the flavor of the vegetables.

Grilled   Identifiable pieces of vegetables, such as green onions, eggplant, mushrooms, peppers (green, orange, red, yellow), and zucchini squash, should be soft and slightly browned showing darker grill marks. Oil, most often olive, for basting should be mild flavored. Herbs may be added to the oil to enhance flavor.

Steamed   Color should be natural and clear and the form distinct and uniform. Flavor should be natural, sweet, pleasant, and free of any trace of rawness. Seasonings should mildly contrast but blend with the flavor of the vegetable. Texture varies with the type. High-moisture vegetables, like green beans and cabbage, should be slightly underdone. Legumes and potatoes should be soft (Kotschevar, 1988).

Stir-fried   Bite-sized pieces of vegetables are identifiable and retain natural color and some original crispness. Flavors should be characteristic of vegetable type (e.g., broccoli, carrots, cauliflower, celery, mushrooms, onions, peppers, snow peas, tomatoes, zucchini squash). Seasoning enhances the flavor of the vegetables (fresh ginger, garlic, sesame oil, soy sauce). Vegetables can be coated with a light, clear, cornstarch glaze.

Legumes

Beans, Baked   Cooked navy beans should be soft and tender and the color of the added ingredients (catsup, molasses, mustard, onion, salt pork). Flavor of the added ingredients should permeate the beans. Baked beans are moist and retain shape.

Beans, Refried   The product should be pinkish brown, the color of pinto beans. Seasonings should enhance, but not overpower, the natural flavor of the beans (chili powder, garlic, jalapeño peppers, onion, tabasco sauce). Flavor of cooking oil should be unobtrusive. Refried beans should have a mashed, slightly lumpy consistency and be slightly moist.

Cooked   The natural shape of legumes should be preserved in cooking. Cooked legumes should be soft and tender, but retain identity, and the color typical of the fresh product (such as black, pinto, red, white beans; black-eyed and split green or yellow peas; brown or red lentils). Flavor varies with the type of legume and should be distinctive. Various seasonings and ingredients complement the flavor.

Potatoes

Au Gratin   Surface should have a lightly browned soft yellow sheen characteristic of melted cheese. Cooked cubed or sliced potatoes should be uniform in size and tender but firm. Cheese sauce should be smooth and velvety with a light yellow color and thick enough to coat the potatoes. The product should retain its shape when served.

Boiled or Steamed   White, slightly opaque potatoes should have a mild flavor that blends well with seasonings, sauces, and other ingredients. Potato should be tender, but firm, and retain its shape when cubed, shredded, or sliced.

Baked   Skin should be the typical brown potato color and should be crisp, with doneness indicated by touch. Interior should be white with a soft and mealy texture. The potato should be split open immediately after removing from the oven to let steam escape and prevent sogginess. Flavor is mild and blends well with toppings, such as bacon, butter, cheese, chives, meat sauces, vegetables, sour cream.

Deep-fat Fried   Potatoes cut into varying lengths and shapes may be crinkle-cut or smooth. Exterior should be golden brown and crisp and interior white and soft. Cooking oil should be bland with no distinctive taste.

Mashed   Color should be creamy white and texture soft and smooth throughout. The blend of salt, butter, and milk should enhance the natural potato flavor without being predominant.

Salad, German   Cooked potato slices should be soft but firm. Chopped onion, crisp bacon pieces, diced celery, pimiento, and parsley are identifiable in color and flavor. The tart vinegar sauce should be adequately sweet and thick enough to coat all ingredients. Salad should be served hot.

BIBLIOGRAPHY

Kotschevar, L. H. (1988). *Standards, principles, and techniques in quantity food production* (4th ed.). New York: Van Nostrand Reinhold.

Kotschevar, L. H., & Donnelly, R. (1994). *Quantity food purchasing* (4th ed.). New York: Macmillan.

Glossary

ABC inventory method   Tool for classifying products as A, B, or C according to value.

Acceptance and preference sensory test   Consumer panel answers questions about whether or not people will like the menu item.

Accident   Event that is unexpected resulting in injury, loss, or damage.

Accountability   State of being responsible to one's self, to some organization, or even to the public.

Achievement-Power-Affiliation   McClelland's theory emphasizing needs that are learned and socially acquired as the individual interacts with the environment.

Activity analysis   Continuous observation for a chronological record of the nature of activities performed by individual workers.

Activity sampling   See Occurrence sampling.

Actual cost   Pricing method in which food, labor, variable, and fixed costs plus profit equal the price of the menu item.

Actual purchase price   Inventory valuation method involving pricing the inventory at the exact price of each product.

Administrative intensity   Degree to which managerial positions are concentrated in staff positions.

Affective sensory test   Preference, acceptance, and opinions of a product are evaluated by consumers who have no special sensory training.

Affirmative action   Employers are required to analyze their work force and develop a plan of action to correct areas of past discrimination.

AFL-CIO   Merger of the American Federation of Labor (AFL), composed primarily of craft unions, and the Congress of Industrial Organizations (CIO), made up of industrial unions.

Agency   Business relationship between the agent and principal.

Agent   Individual who has been authorized to act on behalf of another party, known as the principal.

À la carte   Method of pricing a menu.

All-or-nothing bid   Suppliers are required to quote the best price on a complete list of items to be awarded the order.

American Dietetic Association (ADA)   Professional organization serving the public through promotion of optimum nutrition, health, and well-being.

American Public Health Association (APHA)   Professional organization representing all disciplines and specialties in public health.

American School Food Service Association (ASFSA)   National association for federally sponsored child nutrition programs.

Americans with Disabilities Act (ADA)   Comprehensive legislation that creates new rights and extends existing rights for the 43 million Americans who are not covered under existing laws.

Analytical sensory test   Differences and similarities of quality and quantity of sensory characteristics are evaluated by a panel of specially trained persons.

Approved brand specifications   Quality indicated by designating a product or known desirable characteristics.

Artificial intelligence   Computer program that attempts to duplicate the thought processes of experienced decision makers.

As purchased (AP)   Amount of food purchased before processing to give the number of edible portions for serving customers.

Assembly/serve   Foodservice in which convenience, or minimal cooking, menu items require only storage, assembly, heating, and service.

Atmospherics   Physical elements in an operation's design that appeal to customers' emotions and encourage them to buy.

Authority   Delegation from top to lower levels of management and the right of managers to direct others and take action because of their position in the organization.

Baby boomers   Persons born between 1946 and 1964.

Baby busters   Persons born after the baby-boom years.

Balance   Linking process in the foodservice system that refers to management's ability to maintain organizational stability while adapting to changing economical, political, social, and technological conditions.

Balance sheet   Statement of assets, liabilities or debts, and capital or owner's equity at a given time.

Barbecue smoker   Compact-size oven with racks to smoke, using wood chips, up to 100 pounds of meat at one time.

Batch cooking   Variant of production scheduling in which the total estimated quantity of menu items is divided into smaller quantities and then cooked as needed.

Behavioristic variables   Market segmentation by customers' attitudes and needs for a product.

Benchmarking   Label for the concept of setting goals based on knowing what has been achieved by others.

Bid buying   Buyer decides which supplier will be chosen for the order based on price quotations on specifications.

Biochemical spoilage   Type of spoilage caused by natural food enzymes, which are complex catalysts that initiate reactions in foods.

Biological solution   Bacteria are used by a bioremediation company to break down animal fats and food products that clog drains.

Boundaries   Limits of a system that set the domain of organizational activity.

Brand   Trademark indicating quality of a product.

Branding evolution   Limited-menu restaurant chains contract with other foodservice operations to set up their own store, cart, or kiosk in the facility.

Break-even analysis   Technique for looking at financial data to determine the point at which profit is not being made and losses are not being incurred.

Broiling   Similar to grilling except the heat source is above the rack with the food item.

Brokers   Third parties who are in business for themselves and do not take title to the goods being sold.

Brunch   Hybrid of breakfast and lunch.

Budget   Plan for operating a business expressed in financial terms.

Burnout   Emotional exhaustion.

Business ethics   Self-generating principles of moral standards to which a substantial majority of business executives gives voluntary assent.

Business representative   Hired by the local union to manage the union and also to settle a member's grievance if the steward was not successful.

Buyer   A person who purchases goods or services.

Causal forecasting model   Based on the assumption that an identifiable relationship exists between the item being forecast and other factors, such as selling price, number of customers, and market availability.

Center-of-the-plate   Most prominent menu item, usually the entrée, on which the menu is based.

Centralized meal assembly   Trays in a hospital foodservice operation are assembled in a central location before being distributed and served to patients.

Centralized purchasing   Purchasing activity is done by one person or department.

Chain of command   Clear and distinct lines of authority that need to be established among all positions in the organization.

Charbroiler   Either gas or electric equipment with a bed of ceramic briquettes above the heat source and below the grid.

Chemical spoilage   Type of food spoilage of certain ingredients in a food or beverage resulting from interaction with oxygen or light.

Clamshell   Hinged or removable top with its own heat source has been added to a gas or electric grill that permits cooking both sides of the food at one time.

Clean   Free of physical soil and with an outwardly pleasing appearance.

Code of ethics   Set of rules for standards of professional practice or behavior established by a group.

Combination convection oven/steamer   (Referred to as the "combo" or "combi") Flow of both convected air and steam through the oven cavity produce a super-heated, moist internal atmosphere.

Combustion   Form of solid waste recycling in which the energy value of combustible waste materials is recovered.

Commissary foodservice   Centralized procurement and production facilities with distribution of prepared menu items to several remote areas for final preparation and service.

Common-size statement   Financial statement in which data are expressed as percentages for comparing results of the income statement from one accounting period to another.

Communication   Oral, written, or computer-generated information whereby decisions and other information are transmitted.

Compensation   Financial remuneration given by the organization to its employees in exchange for their work.

Conceptual framework   Loosely organized set of ideas, some simple and some complex, that provides the fundamental structure of an organization.

Conceptual skill   Ability to view the organization as a whole, recognizing how various parts depend on one another and how changes in one part affect other parts.

Conditions of certainty   Adequate information is available to assure results.

Conditions of risk   Probability techniques are necessary for estimating the likelihood of events occurring in the future.

Conditions of uncertainty   Occurrence of future events cannot be predicted.

Conduction   Transfer of heat through direct contact from one object or substance to another.

Continuous care retirement communities (CCRC)   Living centers for retired persons who want quality residential services with healthcare available.

Continuous quality improvement (CQI)   Focused management philosophy for providing leadership, structure, training, and an environment to continuously improve all organizational processes.

Controlling   Management function ensuring that plans are being followed and taking corrective action, if needed.

Convection   Distribution of heat by the movement of liquid or vapor and may be either natural or forced.

Convection oven   Fan on the back wall of an oven creates currents of hot air within the cooking chamber.

Convenience stores (C-stores)   Retail operations that sell products including food to customers 24 hours a day.

Conventional foodservice   Foods are purchased in various stages of preparation for an individual operation, and production, distribution, and service are completed on the same premises.

Cook-chill   Method in which menu items are maintained in the chilled state for as short a time as possible before heating for service.

Cook-freeze   Method in which menu items are stored in the frozen state for periods generally ranging from 2 weeks to 3 months before tempering and heating for service.

Cooking loss   Decrease in yield of many foods in production primarily because of moisture loss.

Coordination   Process of linking activities of various departments in the organization.

Corporate culture, or organizational culture   Shared philosophies, values, assumptions, beliefs, expectations, attitudes, and norms that knit an organization together.

Cost center   Department expected to manage expenses but not generate profits for the organization.

Cost-effectiveness   Technique that provides a comparison of alternative courses of action in terms of their cost and effectiveness in attaining a specific objective.

Critical control points   Locations in the food product flow where mishandling of food is likely to occur.

Critical incident   Technique for identifying when an employee has or has not done something that results in unusual success or failure.

Cross-training   Technique being used to involve employees in the total customer value concept by giving them experience in other areas, thus removing barriers between employees and creating a good climate for customers.

Customer   Anyone, external or internal, who is affected by a product or service.

Cycle menu   Series of menus offering different items daily on a weekly, biweekly, or some other basis, after which the menus are repeated.

Decentralized meal assembly   Menu items in a hospital foodservice facility are produced in one location and transported in bulk to various locations for assembly in areas near to the patients.

Decisional roles   According to Mintzberg, a manager can commit the organization to new courses of action and determine strategy; includes entrepreneur, disturbance handler, resource allocator, and negotiator roles.

Decision making   Selection by management of a course of action from a variety of alternatives.

Decision tree   Technique for assessing consequences of a sequence of decisions with reference to a particular problem.

Deck oven   Pans of food are placed on metal decks, usually with separate heat sources, that may be stacked on each other.

Deep fat fryer   Tank of oil or fat heated by gas or electricity into which foods are immersed.

Delegation Process of assigning job activities and authority to a specific employee within the organization.

Delphi group Decision-making technique for developing a consensus of expert opinion.

Delphi technique Panel of experts who individually complete questionnaires on a chosen topic until a consensus is reached.

Demographic variables Population characteristics that might influence product selection, such as age, sex, race, income, education, occupation, and religion.

Departmentalization Process of grouping jobs according to some logical arrangement.

Descriptive sensory test Information about certain sensory characteristics, such as taste, aroma, texture, tenderness, and consistency, that is needed for quality control and recipe development.

Dietary Guidelines for Americans Seven recommendations for good health revised in 1990 by the USDA and the U.S. Department of Health and Human Services.

Direct energy Expended within the foodservice operation to produce and serve menu items at safe temperatures.

Direct reading measurement tables Recipe adjustment method that uses printed tables for adjusting weight and volume ingredients for recipes that are divisible by 25.

Discrimination sensory test Detectable differences among food items are determined.

Distribution and service The third subsystem in the foodservice system.

Downsizing See Rightsizing.

Dry heat method Heat is conducted by dry air, hot metal, radiation, or a minimum amount of hot fat.

Dry storage Holding of food not requiring refrigeration or freezing but requiring protection from the elements, insects, and rodents as well as from theft.

Dynamic equilibrium Continuous response and adaptation of a system to its internal and external environment.

Echo-boom children Offspring of baby boomers ranging from newborns to people in their 20s.

Economic order quantity (EOQ) Inventory concept derived from a sensible balance of ordering cost and holding cost.

Edible portion (EP) Weight of a food item that is available for consumption by a person, that is, minus bones, skin, and cooking loss.

Effectiveness "Doing the right things." Degree to which quality of product and service in exchange for money satisfies the customer (qualitative).

Efficiency "Doing things right." Minimization of resources that an organization must spend to achieve customer satisfaction (quantitative).

Employee assistance program Provides diagnoses, counseling, and referral for advice or treatment for problems related to alcohol or drug abuse, emotional difficulties, and marital or family difficulties.

Empowerment Level or degree to which managers allow employees to act within their job descriptions.

Environment All internal and external conditions, circumstances, and influences affecting the system.

Equal employment opportunity law Policy of nondiscrimination.

Equifinality Same or similar output can be achieved by using different inputs or by varying the transformation process.

Ethics Principles of conduct governing an individual or a business.

Expectancy Theory based on the belief that people act in such a manner as to increase pleasure and decrease displeasure.

Expert systems   Artificially intelligent computer software that shows significant promise for improving the effectiveness and efficiency of management decision making.

Exponential smoothing forecasting model   Time series model in which an exponentially decreasing set of weights is used, giving recent values more weight than older ones.

External audit   Inspection program performed by governmental or nongovernmental agencies.

External customers   Recipients of food and services that do not belong to the organization that produces them.

Factor pricing method   Also known as the *mark-up method*, the difference between cost and selling price.

Factor recipe adjustment method   Ingredients are changed from measurements to weights and multiplied by a conversion factor determined by dividing the desired yield by the recipe yield.

Fair Labor Standards Act   Concerned with minimum wage rates and overtime payments, child labor, and equal rights.

Fast-food restaurant   Establishment primarily selling limited lines of refreshments and prepared food.

Feedback   Those processes by which a system continually receives information from its internal and external environment.

FIFO (first in, first out)   Inventory valuation method based on the assumption that pricing closely follows the physical flow of products through the operation.

5-A-Day program   Sponsored by the Produce for Better Health Foundation and the National Cancer Institute to increase consumption of fruits and vegetables to at least five servings a day.

Focus group   Qualitative marketing technique bringing a group of people together for a one-time meeting to discuss a specific aspect of an operation.

Food and Drug Administration (FDA)   Agency within the U.S. Department of Health and Human Services that enforces the Federal Food, Drug and Cosmetic Act, which ensures that foods other than meat, poultry, and fish are pure and wholesome.

Foodborne infection   Caused by activity of large numbers of bacterial cells carried by the food into the gastrointestinal tract.

Foodborne intoxication   Caused by toxins formed in food prior to consumption.

Food Guide Pyramid   Complex illustration with many different food and nutrition messages, developed by the USDA in 1992.

Food product flow   Alternate paths within foodservice operations that food and menu items may follow, initiating with receiving and ending with service to the customer.

Food production   Preparation of menu items in the needed quantity and the desired quality at a cost appropriate to the particular operation.

Food Safety and Inspection Service (FSIS)   Responsible for ensuring that meat and poultry products destined for interstate commerce and human consumption do not pose any health hazards.

Forecasting   Art and science of estimating events in the future, which provides a data base for decision making and planning.

Forecasting model   Mathematical formula to predict future needs, which is used as an aid in determining quantity.

Frying   Cooking in fat or oil; includes sautéing, pan frying, and deep frying.

Full-service restaurants   Establishments that sell food and service to customers who prefer to eat at a table on premises and includes casual, theme, family dining, and fine dining restaurants.

Full-time equivalent (FTE)   Minimum number of employees needed to staff the operation.

Functional manager   Responsible for only one area of organizational activity.

Game theory   Competition in decision making is accomplished by bringing into a simulated decision situation the actions of an opponent.

General manager   Responsible for all activities of a unit.

Genetically engineered food   Food can be altered to reduce spoilage and improve flavor by splicing in new genes or eliminating existing genes.

Geographic variables   Climate, terrain, natural resources, population density, and subcultural values that influence customer product needs.

Goal   Aim or purpose of the foodservice system.

Grade   Level of quality of agricultural products as defined by the USDA.

Grade standards   Requirements, which are voluntary, that must be met by a product, such as fruits, vegetables, eggs, dairy products, poultry, and meat, to obtain a particular USDA grade.

Grapevine   Informal communication that may facilitate organizational communication and meet the social needs of individuals within the organization.

Grappa   A pungent brandy distilled from the pulp that remains after grapes have been pressed and the juice run off.

Grazing   Eating small amounts of food throughout the day.

Griddle   Extra-thick steel plate with a ground and polished top surface with raised edges, either gas or electric, usually set on a counter top.

Grilling   Similar to broiling only the heat source is below the rack with the food item.

Gross Domestic Product (GDP)   Total national output of goods and services valued at market prices. (GDP is becoming the primary measure of production in the 90s in the System of National Accounts, the set of international guidelines for economic accounting.)

Gross National Product (GNP)   Measurement of all currently produced final goods and services; the term is being phased out in the 90s in favor of GDP.

Group decision making   More than one person is needed to solve complex problems for which specialized knowledge is required from several fields.

Group purchasing   Bringing together managers from different operations for joint purchasing.

Groupthink   Reaching agreement becomes more important to group members than arriving at a sound decision.

Handling loss   Decrease in the yield of a recipe because of the amount of food wasted in the preparation of the product.

Hazard analysis   Identification of specific foods that are at risk of contamination.

Hazard Analysis Critical Control Point (HACCP)   Systematic analysis of all process steps in the foodservice subsystems, starting with food products from suppliers and ending with consumption of menu items by customers.

Heterogeneity of service   Variation and lack of uniformity in people's performance.

Hierarchy   Characteristic of a system that is composed of subsystems of a lower order and a suprasystem of a higher order.

Historical records   Past purchasing and production records provide the base for forecasting quantities.

Horizontal division of labor   Emphasis on encouraging employees to share ideas across all levels and departments in the new organization.

Human resources   Skills, knowledge, and energies of people required for the system to function.

Human resources management   Extension, by having an understanding of personnel functions performed in accordance with organizational objectives, rather than rejection of traditional requirements for managing personnel effectively.

Human resources planning Process of making provision for the movement of people into, within, and out of an organization.

Human skill, or interpersonal skill Concerns working with people and understanding their behavior.

Immigration Reform and Control Act (IRCA) Control of unauthorized immigration by making it unlawful for a person or organization to hire persons not legally eligible for employment in the United States.

Income statement Financial report that presents the net income or profit of an organization for the accounting period.

Independent purchasing Unit or department of an organization that has been authorized to purchase.

Indirect energy Energy expended to facilitate functions that use energy directly.

Induction Use of electrical magnetic fields to excite the molecules of metal cooking surfaces.

Induction-heat grill Cooks magnetically and has no open flame or thermostat for control.

Informational roles According to Mintzberg, communication may be the most important aspect of a manager's job and includes monitor, disseminator, and spokesman roles.

Infrared waves Type of radiation used in food production that has a longer wavelength than visible light.

Ingredient room Ingredient assembly area designed for measuring ingredients to be transmitted to the various work centers.

Input Any human, physical, or operational resource required to accomplish objectives of the system.

Inseparability Situation in which the customer is involved in the production and delivery processes because services are produced at the same time they are consumed.

Intangibility of services Inability of services to be seen, touched, tasted, smelled, or possessed before buying.

Integration Parts of the system sharing objectives of the entire organization.

Interacting groups Decision-making technique in which members discuss, argue, and agree upon the best alternative.

Interdependency Reciprocal relationship of the parts of a system.

Internal audit Responsibility of management of a foodservice organization and may be part of a self-inspection program for sanitation or a component of a TQM program.

Internal customers Employees and suppliers of products who also belong to the organization that produces them.

Interpersonal roles According to Mintzberg, the focus is on relationships and includes figurehead, leader, and liaison roles.

Interpersonal skill See Human skill.

Inventory Record of material assets owned by an organization.

Inventory control Technique of maintaining assets at desired quantity levels.

Invoice Document prepared by the supplier that contains the same essential information as the purchase order, namely, quantities, description of items, and price.

Irradiation Exposure of foods to gamma rays classified as a food additive and regulated by the FDA.

Job analysis Process of obtaining information about jobs by determining what the duties and tasks are.

Job characteristics Job dimensions that could improve the efficiency of organizations and job satisfaction of employees.

Job description List of duties, working conditions, and tools and equipment needed to perform the job.

Job design   Concerned with structuring jobs to improve organization efficiency and employee job satisfaction.

Job duties   Statements that usually are arranged in order of importance that should indicate the weight or value of each job.

Job enlargement   Increase in the total number of tasks employees perform.

Job enrichment   Increase in the variety and number of tasks as well as the control the employee has over the job.

Job identification   Such information as location of job, to whom the employee reports, and often the number of employees in the department and the Directory of Occupations code number.

Job satisfaction   Pertains to individuals' general attitude toward their occupations.

Job specification   List of abilities, skills, and other credentials needed to do the job.

Job title   Indication of level in the organization and degree of authority the job possesses.

Joint Commission on Accreditation of Healthcare Organizations (JCAHO)   Regulatory agency to determine the degree to which the organization complies with established control standards.

Just-in-time (JIT)   Philosophy and strategy that has effects on inventory control, purchasing, and suppliers.

Just-in-time purchasing   Purchasing of products for immediate production and consumption by the customer without having to record it in inventory records.

Labor relations   Interaction between management and a labor union.

Latest purchase price   Inventory valuation method in which the last price paid for a product is used.

Law of Agency   Buyer's authority to act for the organization, as well as the obligation each owes the other and the extent to which each may be held liable for the other's actions.

Law of Contract   Agreement between two or more parties requiring the buyer to be certain that each agreement is legally sound.

Law of Warranty   Guarantee by the supplier that an item will perform in a specified way.

Leadership   Process of influencing activities of an individual or group in efforts toward goal achievement.

Leading   Management function particularly concerned with individual and group behavior.

Lead time   Interval between the time that a requisition is initiated and receipt of the product.

LIFO (last in, first out)   Inventory valuation method based on the assumption that current purchases are largely made for the purpose of meeting current demands of production.

Limited-menu restaurant   Classification based on the Standard Industrial Classification (SIC), Eating and Drinking Places, and used by the NRA for a fast-food restaurant.

Line and staff   Organizational relationship of line personnel, those with a linear responsibility, to staff personnel, those who serve in an advisory capacity to line managers.

Linear programming   Technique used in determining an optimal combination of resources to obtain a desired objective.

Line-item bidding   Each supplier quotes a price on each product on the buyer's list, and the one offering the lowest price receives the order for that product.

Line position   In the direct chain of command, a position that is responsible for the achievement of an organization's goals.

Linking process   Coordination of the characteristics of a system in the transformation of resources into goals.

Low-temp cooking and holding ovens   Cooking temperatures are from 100°F to 325°F and holding temperatures are from 60°F to 200°F.

Low-temperature storage   Holding perishable food in refrigerated or frozen storage for preservation of quality and nutritive value immediately after delivery and until use.

Maître d'hôtel   Master of the house, head steward, majordomo, or head waiter.

Make-or-buy decisions   Procedure for deciding whether to purchase from oneself (make) or purchase from suppliers (buy).

Malcolm Baldrige National Quality Award   Recognition given to U.S. organizations that excel in quality management by meeting criteria requiring major shifts in management philosophy, practices, and policies.

Management   Process whereby unrelated resources are integrated into a total system for accomplishment of objectives.

Management by objectives   Technique sometimes used for performance appraisal in which managers and their superiors agree on the objectives to be achieved usually for a 1-year period.

Management development   Programs designed to improve the technical, human, and conceptual skills of managers.

Management functions   Integral component of the transformation element, including planning, organizing, staffing, leading, and controlling, performed by managers to coordinate subsystems in accomplishing goals.

Manufacturers' representatives   Middlemen who do not take title, bill, or set prices and usually represent small equipment manufacturing companies.

Market   Medium through which a change in ownership moves commodities from producer to customer.

Market regulation   Protection for the consumer by the U.S. Department of Health and Human Services and the U.S. Department of Agriculture that does not stifle economic growth.

Market segmentation   Process of dividing a total market into groups of customers who have similar product needs.

Market share   Percentage of industry sales for a product.

Marketing   Activities that facilitate satisfying customers in a dynamic environment through the creation, distribution, promotion, and pricing of goods, services, and ideas.

Marketing channel   Exchange of ownership of a product from the producer, through the manufacturer or processor and the middleman to the customer.

Marketing concept   Management philosophy that affects all activities of an organization and emphasizes satisfying needs of customers.

Marketing cost analysis   Classifies costs to determine which are associated with specific marketing activities.

Marketing environment   Consists of all forces outside the organization that influence marketing activities and exchanges.

Marketing mix   Specific combination of marketing elements to achieve goals and satisfy the target market.

Marketing objective   Statement of what is to be accomplished through marketing activities.

Marketing plan   Written document governing an organization's marketing activities.

Marketing research   Systematic gathering, recording, and analyzing of data about problems relating to the marketing of goods and services.

Marketing strategy   Encompasses selecting and analyzing a target market and creating and maintaining an appropriate marketing mix that will satisfy that market.

Markup   Difference between cost and selling price.

Master schedule   Overall plan for employee scheduling, including days off.

Master standard data (MSD)   Seven basic elements of work are combined into larger, more condensed elements.

Materials management   Unifying force that gives interrelated functional subsystems a sense of common direction.

Maturity theory   Argyris's theory that a number of changes take place in the personality of individuals as they develop over the years into mature adults.

Meal equivalent   Number of snacks prepared and served to equal staffing for one meal. (A meal equivalent has the same effect as a meal in any situation involving meal count.)

Memory   All stored information that provides historical records of a system's operations.

Mentor   Teacher, tutor, or a coach, depending upon the relationship between the mentor and the protégé.

Menu   List of items available for selection by a customer and the most important internal control of the foodservice system.

Menu engineering   Computer-assisted management information tool that focuses on the contribution to profit of a menu item by comparing the popularity of the item with its profit.

Menu evaluation   Continuing process and should be conducted during production and service of each meal and after major menu planning sessions.

Menu pricing   Cost of a menu item to a customer and includes food and labor, operating, and profit.

Mesquite grill   Smoke from mesquite wood, the source of heat, gives a smoky and somewhat subtle tangy flavor to grilled meats.

Method   Relates to only one step of a procedure.

Microbiological spoilage   Type of food spoilage caused by microorganisms, such as bacteria, molds, yeast, viruses, rickettsiae, protozoa, and parasites.

Microbreweries   Place where beer is brewed, which may be sold on- or off-premises.

Microwaves   Waves have a very short length and are generated by an electromagnetic tube.

Mini-max (Minimum-maximum) method   Tool for controlling inventory by establishing lower and upper levels for each product in storage.

Mission statement   Summation of the purpose of the organization, competition, target market, product and service, and recipients of the service including customers, employees, owners, and the community.

Model   Conceptual simplification of a real situation in which extraneous information is excluded and analysis is simplified.

Moist heat method   Heat is conducted to the food product by water or steam.

Motivation   All the inner striving conditions, such as wishes, desires, and drives and an inner force that activates a person.

Moving average forecasting model   Time series model that uses a repetitive process for developing a trend line by averaging the number of servings for a specified number of times for the first point on the line and then dropping the oldest and adding the newest number of servings for subsequent points.

Mystery shoppers   Persons unknown to customers and employees who eat at a restaurant and evaluate their own experiences and those of other customers.

National Association of College and University Food Services (NACUFS)   Professional organization that defines standards and criteria to be used for self-monitoring or for a voluntary peer review program.

National Council of Chain Restaurants (NCCR)   Professional organization that has an active role in shaping food safety policy for the future.

National Marine Fisheries Service (NMFS)   A voluntary inspection program for seafood safety.

National Sanitation Foundation International (NSF International)   Nonprofit, noncommercial organization that develops minimum sanitation standards for foodservice equipment.

Need hierarchy   Maslow's theory that states people are motivated by their desire to satisfy specific needs in the following ascending order: physiological, safety, social, esteem, and self-actualization.

Negotiation   Process of working out a purchasing and sales agreement, mutually satisfactory to both buyer and supplier, and reaching a common understanding of the essential elements of a contract.

Network   Graphic representation of a project, depicting the flow as well as the sequence of defined activities and events.

New organization   Organizations in which employers are challenged to improve the quality of work life and to develop a corporate culture.

Nominal group   Structured technique for generating creative and innovative alternatives or ideas.

Nonprogrammed decision   Relatively unstructured and takes more time than a programmed decision because the focus is on qualitative problems that take a higher degree of judgment.

Nutrition labeling   Regulations issued by the FDA and the USDA's Food Safety and Inspection Service (FSIS) state that any nutritional health claims made about food have to be documented on labels and menus.

Observation   Method requiring trained observers to estimate visually the amount of plate waste.

Occurrence sampling   Method for measuring work and nonwork time of employees in direct and indirect activities and for measuring operating and downtime of equipment.

Offer-versus-serve   A meal plan in school foodservice programs in which students must choose three of five menu components for the lunch to be reimbursed by the federal government.

Off-the-job training   Occurs outside the actual workplace.

Omnibus Budget Reconciliation Act (OBRA) of 1987   Federal nursing home regulation designed to improve the quality of life and care for residents in nursing homes.

One-stop shopping   Customers can purchase almost anything from food products, service supplies and equipment, and office supplies to furniture from one distributor.

On-the-job training   Includes job rotation, internships, and apprenticeships.

Open systems   Organizations that are in continual interaction with the environment.

Operational resources   Money, time, utilities, and information.

Organization   Group of people working together in a structured and coordinated way to achieve goals.

Organization chart   Graphic portrayal of the organization structure that depicts basic relationships of positions and functions while specifying the formal authority and communication network.

Organizational change   Any substantive modification to some part of the organization.

Organizing   Management function of grouping activities, delegating authority, and providing for coordination of relationships, horizontally and vertically.

Orientation   Formal process of familiarizing new employees to the organization, job, and work unit.

Output   Result of transforming input into achievement of a system's goal.

Overproduction   Extra costs are generated because salvage of excess food items is not always feasible.

Paradigm   Mental frame of reference that dominates the way people think and act.

Paradigm shift   New set of rules going from quality to total customer value occurring in organizations.

Pareto analysis   80–20 rule—approximately 80% of the total sales volume comes from approximately 20% of the customers.

Participative management Involves empowering employees to share in decisions about their work and employment conditions.

Percentage method Recipe adjustment method in which measurements for ingredients are converted to weights and then the percentage of the total weight for each ingredient is computed.

Performance Degree of accomplishment of tasks that make up an individual's job.

Performance appraisal Comparison of an individual's performance with established standards for the job.

Performance specifications Quality is measured by the effective functioning of large or small equipment, disposable paper and plastic items, or detergents.

Performance standards Desired results at a definite level of quality for a specified job.

Perishability of services Services that cannot be stored for future sale.

Permeability of boundaries Characteristic of an open system that allows the system to be penetrated or affected by the changing external environment.

Perpetual inventory Purchases and issues continuously are recorded for each product in storage, making the balance in stock available at all times.

Personal ethics Person's religion or philosophy of life that is derived from definite moral standards.

Personnel management Performance of basic functions such as selection, training, and compensation of employees without much regard for how they relate to each other.

pH value Degree of a food's acidity or alkalinity that affects bacterial growth.

Physical inventory Periodic actual counting and recording of products in stock in all storage areas.

Physical resources Materials and facilities.

Physical spoilage Type of food spoilage caused by temperature changes, moisture, and dryness.

Pilferage Inventory shrinkage.

Planning Management function of determining in advance what should happen.

Planning for production Establishment of a program of action for transformation of resources into products and services.

Plate waste Amount of food left on a plate, which often is used as a measure of food acceptability.

Policy General guide to organizational behavior developed by top-level management.

Portion control Achievement of uniform serving sizes not only for control of cost but also for customer satisfaction.

Power Resource that enables a leader to induce compliance or influence others.

Pressureless convection steamer Fan directs the steam flow throughout the steamer cavity, encircling the food.

Preventive maintenance Keeping equipment and facilities in a good state of repair.

Pricing psychology Psychological aspects of pricing affect customer perceptions, which then influence the purchase decision.

Prime cost Pricing method consisting of raw food cost and direct labor cost and is very labor intensive.

Principal Person who needs an agent to work on his or her behalf.

Prix fixe Fixed price for a complete meal.

Procedure Chronological sequence of activities.

Procurement Managerial responsibility for acquiring material for production; the first subsystem in the transformation element of the foodservice system.

Production Second subsystem in the transformation element of the foodservice system and defined as the process by which products and services are created.

Production schedule   Worksheet that activates the menu, the major control of the production subsystem, and provides a test of forecasting accuracy.

Production scheduling   Time sequencing of events required by the production subsystem to produce a meal.

Productivity   Ratio of output to input.

Productivity level   Meals per labor hour in foodservice operations.

Profit center   Any department that is assigned both revenue and expense responsibilities.

Programmed decision   Concerned with concrete problems that require immediate solutions that are quantitative and reached in a short time.

Psychographic Variables   Motives and life-styles that can be used to segment the market.

Pulper   Water-filled tank in which solid waste is broken down into a slurry by a shredding device and then water is pressed out of it.

Purchase order   Document, based on information in the requisition, completed by the buyer, who gives it to the supplier.

Purchasing   Activity concerned with the acquisition of products.

Quality   Features and characteristics of a product or service that focuses on its ability to satisfy a customer's given needs.

Quality assurance (QA)   Procedure that defines and ensures maintenance of standards within prescribed tolerances for a product or service.

Quality attributes   Characteristics of a food, such as microbiological, nutritional, and sensory, that require control throughout the procurement/production/service cycle to maintain them.

Quality circles   Problem-solving process by small groups of employees who choose a project and focus on quality improvement.

Quality control   Continuous process of checking to determine if standards are being followed, and if not, taking corrective action.

Quality improvement process (QIP)   Structured problem-solving approach that focuses on operating processes and involves staff in analyzing current situations and developing recommendations for problem resolution.

Quality of work life   How work is organized by the manager and how jobs are designed.

Quantity   Part of the decision process when the ability to produce in the desired amount is required.

Queuing theory   Cost of waiting in lines is balanced against the cost of expanding facilities.

Radiation   Generation of heat energy by wave action within an object.

Range oven   Part of a stove, generally called a *range,* located beneath the cooking surface.

Rating scales   Number of dimensions on which an employee is rated on a three- to five-point scale for measuring quality of performance.

Ratio analysis   Mathematical expressions of the relationship between two items are categorized according to primary use, such as liquidity, solvency, activity, profitability, and operating, to facilitate understanding.

Ready prepared foodservice   Menu items are produced and held chilled or frozen until heated for serving.

Receiving   Activity for ensuring that products delivered by suppliers are those that were ordered.

Recipe   Formula by which weighed and measured ingredients are combined in a specific procedure to meet predetermined standards.

Recipe format   Definite pattern or style for recording a recipe.

Recipe standardization   Process of tailoring a recipe to suit a particular purpose in a specific foodservice operation.

Recruitment   Process of attempting to locate and encourage potential applicants to apply for existing or anticipated job openings.

Recycling   Act of removing materials from the solid waste stream for reprocessing into valuable new materials and useful products.

Regression analysis forecasting models   Causal models that are based on the assumption that the linear relationship between variables will continue for a reasonable time in the future.

Reinforcement   Skinner's theory that people behave in a certain way because they have learned that certain behaviors are associated with positive and others with negative outcomes.

Reorder point   Lowest stock level that safely can be maintained to avoid emergency purchasing.

Requisition   First document in the purchasing process that may have originated in any one of a number of units in a foodservice operation.

Resources   Human and physical inputs into the foodservice system that are transformed to produce outputs.

Responsibility   Obligation to perform an assigned activity or see that someone else performs it.

Restaurant-type menu   Static menu in which the same items are offered every day.

Retailers   Final middlemen in the marketing channel who sell products to the ultimate buyers, the customers.

Rightsizing   Eliminating unneeded managers and empowering employees can increase productivity in a time of economic slowdown.

Risky shift   Tendency of individuals to take more risk in groups than as individuals.

Rules   Specification of action, stating what must or must not be done.

Safety stock   Backup supply to ensure against sudden increases in product usage rate.

Sales analysis   Detailed study of sales data for the purpose of evaluating the appropriateness of a marketing strategy.

Sanitary   Free of disease-causing organisms and other contaminants.

Sanitation and maintenance   Fourth subsystem in the foodservice system.

Scheduling   Having the correct number of employees on duty, as determined by staffing needs.

Scientific management   Theory developed by Frederick W. Taylor in the early 1900s emphasizing the systematic approach for improving worker efficiency by using performance standards and time studies.

Selection   Process of comparing applicant skills, knowledge, and education with the requirements of the job, and involves decision making.

Self-reported consumption   Technique for measuring plate waste in which individuals are asked to estimate their plate waste using a scale similar to one used by trained observers.

Sensory analysis   A science that measures the texture, flavor, and appearance of food products through human senses.

Service   Wide variety of intangible factors influencing the satisfaction of the buyer.

Service management   Philosophy, thought process, set of values and attitudes, and a set of methods to transform an organization to a customer-driven one.

Service profit chain   Goal of profit rather than customer satisfaction because profit is most closely linked to customer retention that results from customer satisfaction.

SERVSAFE®   Crisis management program developed by the Educational Foundation of National Restaurant Association that concentrates on three areas of potential risk—food safety (based on the HACCP model), responsible alcohol service, and customer safety—and focuses on the manager's role in assessing risks, establishing policies, and training employees.

Sexual harassment   Form of sex discrimination, thus in violation of Title VII of the Civil Rights Act of 1964.

Shift schedule   Staffing pattern for each block of time, usually 8 hours, regardless of idle time in the operation.

Simulation   Use of a model for imitating a real-life occurrence and studying its properties, behavior, and operating characteristics.

Single-use menu   Menu that is planned for service on a particular day and not used in the exact form a second time.

Situation analysis   Identification of an organization's internal strengths and weaknesses as well as external opportunities and threats.

Skill   As defined by Katz, an ability that can be developed and is manifested in performance.

Skills inventory   Information on each employee's education, skills, experience, and career aspirations.

Social caterers   Establishments primarily engaged in serving prepared food and beverages at another facility rather than at a fixed business location, such as a restaurant.

Social Security Act   Insurance program to protect covered employees against loss of earnings resulting from retirement, unemployment, disability, or from death or disability of the person supporting them.

Sociocultural factors   Customs, mores, values, and demographic characteristics of the society in which the organization functions.

Solid waste   Products and materials discarded after use in homes, commercial establishments, and industrial facilities (U.S. Environmental Protection Agency).

Source reduction   Design and manufacture of products and packaging with minimum toxic content, minimum volume of material, and a longer useful life (U.S. Environmental Protection Agency).

Sous vide   A process of cooking menu items in a vacuum-sealed bag and then chilling or freezing the product before heating for service.

Specification   Statement understood by buyers and suppliers of the required quality of products, including allowable limits of tolerance.

Split shift scheduling   Assigning employees to work only for peak times of business.

Spoilage   Denotes unfitness for human consumption due to chemical or biological causes.

Staff position   Intended to provide expertise, advice, and support for line positions.

Staffing   Management function that determines the appropriate number of employees needed by the organization for the work that has to be accomplished.

Staffing tables   Pictorial representation of all jobs with the number of employees in those jobs and future employment requirements.

Staggered schedule   Employee work hours that begin at varying times to meet the needs of the foodservice operation.

Standard   Result of the managerial process of planning that is defined as the dimensions of what is expected to happen.

Standard Industrial Classification (SIC)   1987 manual used to promote the comparability of foodservice establishment data.

Standardized recipe   Recipe that consistently delivers the same quantity and quality of product when followed precisely.

Statement of changes in financial position   Information about the source and use of funds.

Static menu   Same menu items are offered every day, that is, a restaurant-type menu.

Steam-jacketed kettle   Space is created by the jacket surrounding the kettle through which steam, usually self-contained, is introduced to provide heat for cooking.

Storage   Holding of products under proper conditions to ensure quality at time of use.

Strategic business units (SBUs)   Division, product-line, or product department that sells specific products or services to a target market and competes against competitors.

Strategic plan   Deals with decisions, allocation of resources, and integration of the organization in the environment.

Subjective forecasting model   Based on the idea that little relationship exists between the past and long-term future and, therefore, forecasters must rely on opinions and quantitative information that might be related to the item being forecast.

Subsystem   Complete system in itself but not independent, an interdependent part of the whole system.

Superdistributor   Middleman who offers an extremely wide variety of products thus giving the customer the opportunity to engage in one-stop shopping.

Supplier   Term used by the National Association of Purchasing Management in the *International Journal of Purchasing and Materials Management* for a person who offers products for sale.

Synergy   Units or parts of a system sharing objectives of the entire organization.

System   Collection of interrelated parts or subsystems unified by design to obtain one or more objectives.

Systems analysis   Decision-making process aiding the manager in making the best choice among several alternatives.

Systems approach   Keeping the organization's objectives in mind throughout the performance of all activities.

Systems management   Application of systems theory to managing organizations.

Systems philosophy   Way of thinking about an event in terms of parts or subsystems with emphasis on interrelationships.

Table d'hôte   (French for "host's meal") A complete meal consisting of several courses at a fixed price (prix fixe).

Target market   Customers for whom an organization creates a marketing mix that specifically meets the needs of that group.

Technical skill   Understanding of, and proficiency in, a specific kind of activity, particularly one involving methods or techniques.

Technical specifications   Applicable to products for which quality may be measured objectively and impartially by testing instruments.

Theft   Premeditated burglary.

Theory X   McGregor's theory that motivation is primarily through fear and the supervisor is required to maintain close watch of employees if organizational goals are to be met.

Theory Y   McGregor's theory that emphasizes managerial leadership by permitting employees to experience personal satisfaction and to be self-directed.

Tilting fry pan   Floor-mounted rectangular pan, sometimes called a *braising pan,* with gas- or electric-heated flat bottom, pouring lip, and hinged cover.

Time and temperature   Critical elements in quantity food production that must be controlled to produce a high-quality product.

Time series forecasting model   Based on the assumption that actual occurrences follow an identifiable pattern over time.

Total customer value   Combination of the tangible and intangible experienced by customers that become their perception of doing business with an organization.

Total quality management (TQM)   Management philosophy directed at improving customer satisfaction while promoting positive change and an effective cultural environment for continuous improvement of all organizational aspects.

Total quality service (TQS)   Based on the assumption that all quality standards and measures should be customer referenced and help employees guide the organization to deliver outstanding value to customers.

Traditional organization   Organization in which lines of authority, which create order, are established.

Training   Teaching of technical skills to employees.

Transformation   Any action or activity utilized in changing inputs into outputs.

Trend analysis   Comparison of results over several periods of time to forecast future revenues or levels of activity.

Two-factor   Herzberg's theory of work motivation focusing on outcomes of performance that satisfy needs.

Underwriters Laboratory, Inc. (UL)   An organization responsible for the compliance of equipment with electrical safety standards.

Underproduction   Production of less than needed can increase labor and food costs because additional labor is needed to prepare a substitution often at a higher price.

Union steward   Representative of union members in their relations with an immediate supervisor or other managers.

United States Department of Agriculture (USDA)   Federal department with an important role in the food regulatory process by grading, inspecting, and certifying all agricultural products.

United States Department of Health and Human Services (USHHS)   Federal department that administers the Food, Drug and Cosmetic Act and the Public Health Service Act.

Usage rate   Determined by past experience and forecasts of a product.

Value   Perceived relationship between quality and price.

Value added   Increase in value caused by both processing or manufacturing and marketing exclusive of the cost of products or overhead.

Value Analysis   Methodical investigation of all components of an existing product or service with the goal of discovering and eliminating unnecessary costs without interfering with the effectiveness of the product or service.

Values and Life-Styles (VALS) program   Classification system for segmenting customers by demographic and life-style factors.

Vegan-approved menu items   Those with no animal meat or by-products for strict vegetarians.

Vertical division of labor   Based on lines of authority.

Warehouse club purchasing   No-frills approach to purchasing because the wholesale warehouse is a self-service, cash and carry operation.

Warewashing   Process of washing and sanitizing dishes, glassware, flatware, and pots and pans either manually or mechanically.

Weighted average   Inventory valuation method in which a weighted unit cost is used, and is based on both the unit purchase price and the number of units in each purchase.

Wholesaler   Middleman who purchases from various plants, provides storage, and sells products to retailers.

Work design   Industrial engineering term for productivity improvement to assist the worker to work more efficiently without expending more effort.

Work measurement   Method of establishing an equal relationship between the amount of work performed and the human input to do the work.

Yield   Amount of product resulting at the completion of various phases of the procurement/production/service cycle that is usually expressed as a weight, volume, or serving size.

Index